T0343293

Developments in Soil Science – Volume 33

GEOMORPHOMETRY

Concepts, Software, Applications

Developments in Soil Science

SERIES EDITORS: A.E. Hartemink and A.B. McBratney

On the Cover

The figure on the cover shows an unsupervised classification of topography from SRTM30 DEM data by an iterative nested-means algorithm and a three part geometric signature (Iwahashi and Pike, 2007 — available at http://gisstar.gsi.go.jp)

Developments in Soil Science – Volume 33

GEOMORPHOMETRY

Concepts, Software, Applications

Edited by

TOMISLAV HENGL
Institute for Biodiversity and Ecosystem Dynamics
University of Amsterdam
Amsterdam, The Netherlands

HANNES I. REUTER
Institute for Environment and Sustainability
DG Joint Research Centre
Land Management and
Natural Hazards Unit – European Commission
Ispra, Italy

ELSEVIER

Amsterdam • Boston • Heidelberg • London • New York • Oxford
Paris • San Diego • San Francisco • Singapore • Sydney • Tokyo

Elsevier
Radarweg 29, PO Box 211, 1000 AE Amsterdam, The Netherlands
The Boulevard, Langford Lane, Kidlington, Oxford OX5 1GB, UK

First edition 2009
Reprinted 2009

British Library Cataloguing in Publication Data
A catalogue record for this book is available from the British Library

Library of Congress Cataloging-in-Publication Data
A catalog record for this book is available from the Library of Congress

ISBN: 978-0-1237-4345-9

For information on all Elsevier publications
visit our website at www.elsevierdirect.com

Printed and bound by CPI Group (UK) Ltd, Croydon, CR0 4YY

Transferred to digital print 2013

09 10 11 12 10 9 8 7 6 5 4 3 2

This book is dedicated to all geographers and earth scientists, from which two must be singled out for special mention: Waldo R. Tobler — methodological revolutionary, conceptualiser of Analytical Cartography, and hero to numberless quantitative geographers; and Peter A. Burrough — one of the founders of geoinformation science and mentor to a generation of GIS scientists.

CONTENTS

AUTHORS

This book is the joint effort of a number of specialists and researchers. Information about the authors and their affiliations is given in the following section.

Jürgen Böhner is a Professor of Physical Geography and head of the SAGA-GIS developer team at the Department for Earth Sciences, University of Hamburg. He graduated from the University of Göttingen in geography, meteorology, bioclimatology and botany, where he gained his Ph.D. in 1993. Until 2004, he was a scientific assistant at the Department of Physical Geography at Göttingen University, coordinating and participating in national and international research projects on climatic variability, applied meteorology, remote sensing and process modelling. His Habilitation thesis in 2004 mirrors his major interests in modelling topoclimates. More recently, his major focus had been the creation of complex land-surface parameters for regional climate modelling purposes and the modelling of soil related processes (wind and water erosion, translocation and deposition processes).
Current employer: Institut für Geographie, University of Hamburg, Germany
Contact: jboehner@geowiss.uni-hamburg.de

Endre Dobos is an Associate professor at the University of Miskolc, Department of Physical Geography and Environmental Sciences. He did a Ph.D. in Soil Mapping with GIS and Remote Sensing at the Agronomy Department of Purdue University, Indiana, USA, in 1998 and an M.Sc. in GIS and Environmental Survey at the Faculty of Civil Engineering of the Technical University of Budapest in 1996. Endre has chaired the working group on Digital Soil Mapping of the European Commission and co-chaired the Digital Soil Mapping working group of the International Union of Soil Sciences.
Current employer: University of Miskolc, Miskolc, Hungary
Contact: ecodobos@uni-miskolc.hu

Stefan Emeis is a meteorologist with main emphasis on turbulent transport processes in the atmospheric boundary layer. Stefan did a post-doc at the Institute of Meteorology and Climate Research of the University of Karlsruhe/Forschungszentrum Karlsruhe, and a habilitation in Meteorology at the University of Karlsruhe in 1994. He has long-year experience in numerical modelling of atmospheric flows

and chemistry over orography which includes the use of digital terrain and land use data. The outcome of these modelling studies made contributions to the understanding of pressure drag exerted on the atmosphere by sub-gridscale orography and to the understanding of air pollution in mountainous areas. His present specialisation is surface-based remote sensing of the atmospheric boundary layer.
Current employer: Forschungszentrum Karlsruhe GmbH, Garmisch-Partenkirchen, Germany
Contact: stefan.emeis@imk.fzk.de

Ian S. Evans is credited with a number of innovations in geomorphometric concepts and techniques, starting with formalising a system of surface derivatives to replace many more arbitrary measures, and calculating them from DEMs. Ian has worked on a number of projects involving data analysis in Geography, including the specific geomorphometry of glacial cirques and drumlins, using both manually measured indices and DEM-based attributes. Ian worked in the NERC/RCA Experimental Cartography Unit from 1968 to 1970 and from then onwards in the Geography Department at Durham, from 1979, as Senior Lecturer, and from 1999, as Reader. He has held research grants from the US Army, from ESRC and from the Ministry of Defence and various offices in professional organisations, including Chairman (1996–1997) of the British Geomorphological Research Group.
Current employer: Durham University, Durham, UK
Contact: i.s.evans@durham.ac.uk

Paul Gessler is an Associate Professor of Remote Sensing and Spatial Ecology and Co-Director of the Geospatial Laboratory for Environmental Dynamics in the College of Natural Resources at the University of Idaho, USA. He completed a Ph.D. in Environmental Modeling at the Australian National University, Canberra, and remote sensing (M.Sc.) and soils (B.Sc.) degrees from the University of Wisconsin — Madison. Paul pursued research in soil–landscape modelling with the CSIRO Division of Land & Water in Australia for seven years before starting in academia. He's involved in a diversity of research and teaching involving the remote sensing, characterisation and monitoring of forest ecosystems along with wildland fire fuels and fire hazard mapping, airborne sensor development, and soil–landscape modelling and digital soil mapping. All activities involving the analysis of complex terrain and an integrated element of geomorphometry.
Current employer: University of Idaho, Moscow, Idaho, USA
Contact: paulg@uidaho.edu

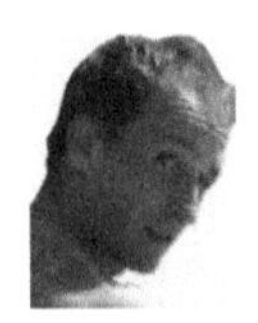

Stephan Gruber firmly believes that topography makes happy. He currently lives in Switzerland. His main research interests are the measurement and modelling of high-mountain cryosphere phenomena (permafrost, glaciers and snow). Geomorphometry is one of the techniques he uses — it is often very powerful due to the dominating influence that topography has on surface and near-surface processes.

He currently works at the University of Zürich, where he received his Ph.D. Previously, he has done research in other places: the University of Giessen, Germany; the Arctic Centre/University of Lapland, Finland; the ITC in the Netherlands; the National Snow and Ice Data Center in Boulder, Colorado, USA and the Université de Savoie, France. Stephan is still contemplating a suitable land-surface parameter that quantifies the happiness caused by topography. Like in many other cases it is probably somehow related to the first derivative.
Current employer: University of Zürich, Zürich, Switzerland
Contact: stgruber@geo.unizh.ch

Peter L. Guth is a Professor in the Department of Oceanography at the United States Naval Academy. Peter was trained as a field geologist at the Massachusetts Institute of Technology, and has worked for a number of summers in the Sheep Range of southern Nevada. He shifted his research focus to microcomputer land-surface analysis while teaching at the U.S. Military Academy, and has been developing the freeware MicroDEM program for over 20 years. He has used MicroDEM to investigate software algorithms such as slope, line of sight, and viewshed computations; for looking at anomalies in digital elevation models such as contour-line ghosts; for quantifying the degree of land-surface organisation and fabric; and for looking at geomorphometric land-surface characteristics computed for the United States and the entire world during the Shuttle Radar Topography Mission. He has also worked with Geodesy Base, a small company that locates fires, using web-based tools and GIS in lookout towers.
Current employer: U.S. Naval Academy, Annapolis, MD, USA
Contact: pguth@usna.edu

Tomislav Hengl is a GIS scientist with special interests in soil mapping, land-surface modelling and the use of statistical techniques for spatial analysis in general. He studied at the Faculty of Forestry in Zagreb, then received a scholarship for a post-graduate study abroad. He finished his M.Sc. in 2000 and Ph.D. degree in 2003, both of them in the Netherlands, at the International Institute for Geoinformation Science and Earth Observation and Wageningen University. He joined the Joint Research Centre of the European Commission, as a post-doctoral researcher, in June 2003. He has published several research papers that focus on the preparation of land-surface parameters, their improvement using different filtering techniques, and on the use of land-surface parameters in soil–landscape modelling, including lecture notes on extraction of DEM-parameters in ILWIS GIS. His recent interests are development of automated predictive mapping techniques and integration of geostatistics, geomorphometry and remote sensing.
Current employer: Faculty of Science, University of Amsterdam, Netherlands
Contact: hengl@science.uva.nl

Jaroslav Hofierka is an associate professor of Physical Geography and Geoecology and Head of the GIS Laboratory in the Department of Geography and Regional Development at the University of Presov, Slovakia. He received a Ph.D. degree in Cartography and Geoinformatics from Comenius University, Bratislava, Slovakia in 1998. His main research activities have been focused on digital terrain modelling and applications, spatial interpolation and the modelling of landscape processes (water erosion, solar radiation) using GIS. He has been participating in the development of Open Source GRASS GIS since 1992. His other research areas include renewable energies, spatial and temporal landscape changes and municipal information systems.

Current employer: Department of Geography and Regional Development, University of Presov, Presov, Slovakia
Contact: hofierka@geomodel.sk

Sven D. Jelaska is an assistant professor of Plant Ecology in the Department of Botany of Faculty of Science at the University of Zagreb, Croatia. His main interest is in scientific research dealing with flora and vegetation spatial distribution, including the issues of biological diversity. Using the GIS, statistical methods (CCA, CART, DA, logit, etc.) and other technologies (e.g. RS, GPS, hemispherical canopy photos) he integrates various types of data relevant for description and explanation of spatial distribution of biological entities. These were backbones of his M.Sc. and Ph.D. thesis, both in ecology, accepted at the Faculty of Science, University of Zagreb in 1999 and 2006, respectively. He managed the creation of preliminary ecological network of Croatia, and co-managed the late "*Ecological Network of Croatia as a part of PEEN and NATURA2000 network*". He was actively involved in project "*Mapping of habitats of the Republic of Croatia*". As a biodiversity expert he participated in "*National report on climate change 1996–2003*". He published over 20 peer-reviewed papers on various aspects dealing with vascular flora and vegetation.

Current employer: Faculty of Science, University of Zagreb, Zagreb, Croatia
Contact: sven@botanic.hr

John Lindsay is a lecturer in physical geography and geocomputation at the University of Manchester. He studied geography at the University of Western Ontario, Canada, where he completed an M.Sc. and a Ph.D. in the areas of fluvial geomorphology and digital land-surface analysis, respectively. John's research area has focused on DEM preprocessing, particularly in relation to topographic depressions, and the extraction of DEM-derived channel networks and network morphometrics. John also has considerable interest in the development of software and algorithms for digital land-surface analysis and is the author of TAS GIS.

Current employer: Department of Geography, University of Guelph, Guelph, Ontario, Canada
Contact: jlindsay@uoguelph.ca

Robert A. MacMillan is a private sector consultant who earns his living applying geomorphometric techniques to map and model natural landscapes. Bob has a B.Sc. in Geology from Carleton University, an M.Sc. in Soil Science from the University of Alberta and a Ph.D. in GIS and hydrological modelling from the University of Edinburgh. Bob graduated as a geologist but trained as a soil surveyor with both the national and Alberta soil survey units in Canada. Bob spent more than 10 years as an active field soil surveyor (1975–1985) with experience in Alberta, Ontario, East Africa, Nova Scotia and New Brunswick. From 1980 onwards, Bob was increasingly involved in developing and applying computer-based procedures for enhancing soil survey products, including statistics and geo-statistics, analysis of soil map variability and error, use of GIS to both create and apply soil map information and use of DEMs to assist in the creation of maps. Bob led the first project to use GIS for soil information in Alberta in 1985 and created his first DEM in 1985 in which soil attributes from a grid soil survey were related to terrain attributes computed from the DEM. Bob led the design effort that resulted in production of the seamless digital soils database for Alberta (AGRASID). Since 1994 Bob has operated a commercial consulting company (LandMapper Environmental Solutions Inc.) that has completed numerous projects that used automated analysis of digital elevation models and ancillary data sources to produce maps and models for government and private sector clients. Bob developed the LandMapR toolkit to provide a custom, in-house, capacity to analyse the land surface to describe and classify landforms, soils, ecological and hydrological spatial entities in an automated fashion. The LandMapR procedures have been used to produce ecological and landform maps for millions of ha in BC and Alberta and to classify hundreds of agricultural fields. The toolkit has been used by more that 50 individuals, private sector companies, universities and major government organisations in Canada and internationally.
Current employer: LandMapper Environmental Solutions Inc., Edmonton, AB, Canada
Contact: bobmacm@telusplanet.net

Andrew Nelson is a geographer with interests in the Multi Scale Modelling of environmental issues, Geographically Weighted Statistics, Biodiversity Mapping and Analysis, Accessibility Models, Neural Networks, Population and Poverty Modelling and Watershed Modelling. He has previously worked at the World Bank, UNEP, CGIAR and is currently a post-doctoral researcher at the EC Joint Research Centre in Italy. Andy has worked on hole-filling algorithms for the SRTM data, and multi-scale land-surface parameter extraction using geographically weighted statistics to identify appropriate scales for environmental modelling.
Current employer: European Commission, Directorate General JRC, Ispra, Italy
Contact: andrew.nelson@jrc.it

Victor Olaya Ferrero is a GIS developer with an interest in computational hydrology and land-surface analysis. He studied Forest Engineering at the Polytechnic University of Madrid and received an M.Sc. degree in 2002. After that, he created a small company dedicated to the development of software for forest management. Victor is currently employed as a Ph.D. student at the University of Extremadura, where he leads the development of the SEXTANTE project — a GIS specially developed for forest management purposes. Victor has developed several applications containing land-surface parametrisation algorithms. He is also the author of the *"A gentle introduction to* SAGA *GIS"*, the official manual of this GIS.
Current employer: University of Extremadura, Plasencia, Spain
Contact: volaya@unex.es

Scott D. Peckham is a research scientist at INSTAAR, which is a research institute at the University of Colorado in Boulder. Scott has been honoured to pursue research as a NASA Global Change Student Fellow (1990–1993) and a National Research Council Research Associate (1995–1998). His research interests include physically-based mathematical and numerical modelling, watershed-scale hydrologic systems, coastal zone circulation, source-to-sink sediment transport, scaling analysis, differential geometry, theoretical geomorphology, grid-based computational methods, efficient computer algorithms and fluvial landscape evolution models. Scott is also CEO and founder of Rivix LLC, which sells a software product for land-surface and watershed analysis called RiverTools, and is also the primary author of a next-generation, spatially-distributed hydrologic model called TopoFlow.
Current employer: University of Colorado at Boulder and Rivix LLC, Broomfield, CO, USA
Contact: peckhams@rivix.com

Richard J. Pike has dedicated his entire career to land-surface quantification. His earliest research in continuous-surface morphometry (in 1961 on mean valley depth) was as a student of Walter F. Wood, the pioneering terrain analyst of the *"quantitative revolution"* in American geography. Inspired by astrophysicist Ralph B. Baldwin, he subsequently became expert in the specific morphometry of planetary impact craters. Richard was educated both as a geologist (Tufts, B.Sc.; The University of Michigan, Ph.D.) and a geographer (Clark, M.A.). He has worked for USGS since 1968, when he organised creation of the Agency's first DEMs and morphometric software. Among his many contributions are lunar surface-roughness data for the Apollo Roving Vehicle Project, the concept of the geometric signature, co-authorship of the celebrated digital shaded-relief map of the United States, Supplementband 101 of the Zeitschrift für Geomorphologie, and a 7000-entry annotated bibliography.
Current employer: U.S. Geological Survey, Menlo Park, CA, USA
Contact: rpike@usgs.gov

Hannes I. Reuter is a geo-ecologist, who graduated from Potsdam University with majors in Soil Science and GIS/remote sensing. He obtained his degree in soil science from the University of Hannover, while working at the Leibniz-Centre for Agricultural Landscape Research (ZALF) on precision farming topics. He used land-surface parameters in his Ph.D. thesis to investigate relationships between relief, soil and plant growth, using a couple of ArcInfo Scripts. His interest is in improving the understanding of landscape processes at different scales. He is currently working on finding optimal methods for filling in data voids in the SRTM data model.
Current employer: European Commission, Directorate General JRC, Ispra, Italy
Contact: hannes.reuter@jrc.it

Arnaud Temme holds M.Sc. degrees in Soil Science and Geoinformation Science (*cum laude*), both obtained at Wageningen University. In 2003, he became a Ph.D. student there, under the supervision of the chairs of Soil Inventarisation and Land Evaluation. His main interest is the dynamic landscape, and the methods for studying it. His Ph.D. study area is in the foothills of the Drakensberg, South Africa, where he studies the evolution of a 100-ka landscape as a function of climatic change and endogenous feedbacks. In his first paper, he presented an algorithm for dealing with sinks in DEMs, within landscape evolution models, so that it would no longer be necessary to remove the sinks before running the model. Arnaud has a part-time Ph.D. job to enable him to pursue, simultaneously, a career in mountaineering.
Current employer: Wageningen University and Research Centre, Wageningen, The Netherlands
Contact: Arnaud.Temme@wur.nl

Jo Wood has been a Senior Lecturer in the Department of Information Science at City University London since 2000. Between 1991 and 2000, he was a lecturer in GIS at the University of Leicester, in the Department of Geography. He gained an M.Sc. in GIS at Leicester in 1990 and then studied there for his Ph.D. in geomorphometry, which he was awarded in 1996. His teaching and research interests include land-surface analysis, spatial programming with JAVA, geovisualisation and GI Science. Jo gained his Ph.D. on *"The Geomorphological Characterisation of Digital Elevation Models"* in 1996. This thesis, one of the first studies to incorporate the multi-scale land-surface parametrisation of DEMs, won the 1996 AGI Student-of-the-Year Award. The approach suggested by the thesis was later incorporated into some of the GIS GRASS modules and also led to the development of the land-surface analysis GIS, LandSerf, which Jo has been authoring for 9 years. He is currently supervising Ph.D. students in the areas of Ethno-physiography and in object-field representations of geographic information.
Current employer: City University, London, UK
Contact: jwo@soi.city.ac.uk

CO-AUTHORS

Oleg Antonić
Current employer: Rudjer Bošković Institute, Zagreb, Croatia
Contact: oantonic@irb.hr

Lieven Claessens
Current employer: Wageningen University and Research Centre, Wageningen, The Netherlands
Contact: lieven.claessens@wur.nl

Olaf Conrad
Current employer: University of Göttingen, Göttingen, Germany
Contact: oconrad@gwdg.de

Peter V. Gorsevski
Current employer: Bowling Green State University, Bowling Green, Ohio, USA
Contact: peterg@bgnet.bgsu.edu

Gerard B.M. Heuvelink
Current employer: Wageningen University and Research Centre, Wageningen, The Netherlands
Contact: gerard.heuvelink@wur.nl

Christian Huggel
Current employer: University of Zürich, Zürich, Switzerland
Contact: chuggel@geo.unizh.ch

Ben H.P. Maathuis
Current employer: International Institute for Geo-Information Science and Earth Observation (ITC), Enschede, The Netherlands
Contact: maathuis@itc.nl

Kurt C. Kersebaum
Current employer: Leibniz Centre for Agricultural Landscape and Land Use Research (ZALF), Müncheberg, Germany
Contact: ckersebaum@zalf.de

Hans R. Knoche
Current employer: Institut für Meteorologie und Klimaforschung, Garmisch-Partenkirchen, Germany
Contact: hans-richard.knoche@imk.fzk.de

Helena Mitášová
Current employer: North Carolina State University, Raleigh, NC, USA
Contact: helena_mitasova@ncsu.edu

Markus Neteler
Current employer: Fondazione Mach – Centre for Alpine Ecology, 38100 Viote del Monte Bondone, Trento, Italy
Contact: neteler@cealp.it

Jeroen M. Schoorl
Current employer: Wageningen University and Research Centre, Wageningen, The Netherlands
Contact: jeroen.schoorl@wur.nl

Peter A. Shary
Current employer: Institute of physical, chemical and biological problems of soil science, Moscow region, Russia
Contact: p_shary@mail.ru

Pierre Soille
Current employer: European Commission, Directorate General JRC, Ispra, Italy
Contact: pierre.soille@jrc.it

Alicia Torregrosa
Current employer: US Geological Survey, Menlo Park, CA, USA
Contact: atorregrosa@usgs.gov

Lichun Wang
Current employer: International Institute for Geo-Information Science and Earth Observation (ITC), Enschede, The Netherlands
Contact: lichun@itc.nl

FOREWORD

WHY GEOMORPHOMETRY?

We began to think about a geomorphometry book in the summer of 2005 following a request to suggest auxiliary data that would assist the automated mapping of soils. The first thing that came to mind, of course, was — Digital Elevation Models (DEMs). The longer we considered our response to the request, the more we realised that a substantial gap had opened between the formal discipline of land-surface quantification and a vast informal, and rapidly growing, community of DEM users.

The practical aspects of morphometric analysis seemed to us neglected in the literature. Apart from Wilson and Gallant's "*Terrain Analysis: Principles and Applications*" and Li, Zhu and Gold's "*Digital Terrain Modeling: Principles and Methodology*", few textbooks are suited both for training and for guiding an inexperienced DEM user through the various steps, from obtaining a DEM to carrying out analyses in packaged software. It was our experience that, although irreplaceable, Wilson and Gallant's book is not ideal for either purpose; not only it is primarily a compilation of research or review papers, but it relies heavily on Ian Moore's TAPES software, a comprehensive package to be sure but just one of many now available. Meanwhile, new parameters and algorithms for processing DEMs were circulating in the scientific literature; an update and summary of the field seemed increasingly appealing. Richard Pike later told us that he (and others) had pondered a geomorphometry text for many years. We also discovered that there is quite some disorder in the field. A major problem is the absence of standards for extracting descriptive measures ("*parameters*") and surface features ("*objects*") from DEMs. Many users are confused by the fact that values of even basic parameters such as *slope gradient* may vary — depending on the mathematical model by which they are calculated, size of the search window, the grid resolution... although the measures themselves might appear quite stable. Serious issues also exist over operational principles, for example, pre- and postprocessing of DEMs: should unwanted depressions (sinks, or pits) be filtered out, or not? which algorithms should be used to propagate DEM error through subsequent analyses? should DEMs be smoothed prior to their morphometric application or not, and if so, by how much? These and other questions got us thinking about many aspects of land-surface quantification.

In November 2005, we prepared the initial draft of a Table of Contents and immediately agreed on three things: the book should be (1) practical, (2) comprehensive, and (3) a fully integrated volume rather than an *ad hoc* compilation of

FIGURE 1 Participants in the first meeting of the authors, Plasencia, Spain, 18–22 May 2006.

papers. We also knew that our goals would be more likely achieved in collaboration with a number of co-authors. Initially, we invited ten colleagues to join us but the number slowly grew, along with interest in the book. Our third objective posed difficulties — how to synchronise the output of well over a dozen authors? To solve this problem, we launched an online editorial system that allowed us to exchange documents and data sets with all the authors, thereby encouraging transparent discussion among everyone in the group. It became clear that there would be many iterations before the chapters were finalised and authors sent in their last word.

Our action leader at JRC, Luca Montanarella, soon recognised the importance of this project and supported us in organising the first authors' meeting, which was kindly hosted by Victor Olaya and Juan Carlos Gimenez of the Universidad de Extremadura in Plasencia, Spain. At this meeting, we found ourselves convinced of the effectiveness of a group approach to the writing; enthusiasm for the book was overwhelming. In response to last-minute invitations, Paul Gessler and Ian Evans joined the group (Paul took less than 24 hours to decide to make the 12,000 kilometre trip from the western U.S., even though the meeting would convene in just 4 days) and immediately provided useful feedback.

It was Ian Evans who *rocked the boat* by opening a discussion on some of the field's terminology. First to be scrutinised, and heavily criticised, was "*terrain*". Gradually we began to see the problems arising from its use and elected to adopt less ambiguous language. We understand that whatever our arguments, the wider user community will not readily abandon *terrain* and *terrain analysis* in favour of our preferred *land surface* and *geomorphometry* (indeed, there is not 100% agreement among this book's authors), but we hope that the reader will at least agree to think along with us. The Plasencia meeting further revealed that most authors were in

FIGURE 2 Geomorphometrists are easily recognised by their obsession with shape — explaining a morphometric algorithm often requires much use of the hands.

favour of pricing the book at a non-commercial rate, thereby opening it up to the widest possible readership — yet without jeopardising its scientific and technical content.

The meeting also led us to suspect a *"gender gap"* in the field. Despite their many contributions over the years, women geomorphometrists were absent at Plasencia. We hasten to add that we invited several women colleagues to join us, but only four were able to participate in preparing this first edition. We look forward to an improved balance in the next, and succeeding, editions of this book and take encouragement from Peter Shary, who reported from the 2006 Nanjing Symposium on Terrain Analysis and Digital Terrain Modelling that the number of younger women now working with DEMs (at least in Asia) is clearly on the rise.

During final editing of the book's initial draft we decided to prepare a state-of-the-art gallery of land-surface parameters and objects, to assist less experienced readers in applying DEMs to their best advantage, and then to support an independent Web site to encourage further evolution of the Geomorphometry Project. You are now invited to visit this site, post comments on it, evaluate software scripts and packages, upload announcements of events or jobs, and eventually post your own articles. The floor is open to all.

WHAT CAN YOU FIND IN THIS BOOK?

The volume is organised in three sections: theoretical (concepts), technical (software), and discipline-specific (applications). Most of the latter are in the environmental and Earth sciences, so that the book might best be compared with that of Wilson and Gallant (2000). Our book differs, however, in that it offers technical details on a variety of software packages and more instruction on how to carry out similar data analyses yourself.

This book is more about the surface properties that can be extracted from a DEM than about creating the DEM itself. To appreciate our chosen operational focus, a basic acquaintance with geographical information systems (GIS) (Burrough, 1986) and (geo)statistics (Goovaerts, 1997) will be helpful. Readers who require added technical information on DEMs and how to generate them should consult the books by Li et al. (2005) *"Digital Terrain Modeling: Principles and Methodology"* and Maune (2001) *"Digital Elevation Model Technologies and Applications: The DEM Users Manual"*.

Each of the book's three sections consists of nine or ten chapters that follow a logical sequence from data processing to extraction of land-surface parameters and objects from DEMs. Many chapters overlap in both content and examples, illustrating not only the many *types* of land-surface parameters, but also their *variants* — differing parameter values calculated from an identical DEM by different software. Links to external sources and important literature can be found at the end of each chapter, and well over 100 text boxes flag (important) *remarks* throughout the book. All major types of land-surface parameters and objects, together with a quick reference to their significance and interpretation, are listed in the gallery of parameters and objects available on the Geomorphometry Web Site. A list of references and an index are provided at the end of the book.

Part I: Concepts

The book's opening Chapter 1 will first orient you to the field of geomorphometry, its basic concepts and principles, and major applications. This introduction is followed by a historical review of the discipline, from before the first contour lines to the computer programs by which early DEMs were processed. You will also find a detailed description of the Baranja Hill case study, which is used to demonstrate algorithms and applications throughout the book.

Chapter 2 in Part I is a mathematical introduction to modelling the land surface. Following a discussion of the most important model properties, including surface-specificity, is a list of mathematical models and data structures to represent topography and its intrinsic attributes, such as scale dependence, multi-fractality, and the fit of a model to the true land surface. Special attention is accorded formulas for calculating first- and second-order surface derivatives.

The most common sources of digital elevation data are reviewed in Chapter 3. Each DEM source is described in terms of the equipment or hardware used to collect elevation data, as well as the advantages and disadvantages of postprocessing in converting the raw data into a DEM. Also compared are such key characteristics of the different sources as cost per km^2, typical footprints, postprocessing requirements, and data accuracy and precision.

Chapter 4 is devoted to techniques for improving the quality of DEMs prior to geomorphometric analysis. Included are algorithms to: detect artefacts, systematic errors, and noise in DEMs; deal with missing values (voids), water bodies, and tree-canopy distortion (e.g. in SRTM data); and filter out spurious DEM depressions. The chapter closes with a discussion of simulation techniques to minimise DEM error.

A geostatistical technique to model uncertainty in DEMs and analyse its impact on the calculation of land-surface parameters (slope, wetness index, soil redistribution) is introduced in Chapter 5. The focus is on propagation of DEM error through subsequent analyses using the sequential Gaussian simulation.

Chapter 6 is an overview of *"basic"* morphometric parameters, measures derived directly from DEMs without added special input. The measures range from local land-surface parameters (slope, aspect, solar aspect, curvature) to regional parameters (catchment area, slope length, relative relief) and statistical parameters such as terrain roughness, complexity, and anisotropy. Each measure is illustrated by the Baranja Hill test site.

Following in Chapter 7 are hydrological land-surface parameters for quantifying water flow and allied surface processes. This overview will guide you through the key concepts behind DEM-based flow modelling, again, illustrated by our Baranja Hill case study. Methods for parameterising the physics involved in moving mass (water, sediment, ice) over an irregular surface (topography) are explained, as well as related parameters and objects derived from modelled flow.

Chapter 8 contains an extensive review of solar radiation models and approaches to quantifying exposure of the land surface to climatic influences. First discussed are algorithms by which incoming solar radiation may be estimated from DEMs. Topo-climatic modelling is then extended to the estimation of land-surface temperature, precipitation, snow-cover, and exposure to wind and the flow of cold air.

The final Chapter 9 in Part I introduces landform types and elements and their relation to continuous topography *versus* specific geomorphic features. Next described are techniques for extracting landform classes, either from a list of predefined geomorphic types or by automated extraction of generic surface facets from DEMs. An extensive comparison of approaches to landform classification highlights the value of geomorphometric standards and data-systems that could win wide (international) acceptance.

Part II: Software

Chapter 10 opens the middle third of the book with a general inventory and prospect of all packaged computer programs suited to geomorphometry (of which we are aware), including software not demonstrated in this book. The remaining chapters illustrate eight well-known packages currently available for land-surface analysis, ranging from commercial (ArcGIS) to medium-cost (RiverTools) and freely-available (including open-source) (SAGA, GRASS, ILWIS, LandSerf, TAS, MicroDEM) software. Five chapters are authored by the originators of the software, and three by later developers or expert users; each chapter follows a common structure:

- Description of the software, its origins and target users, and how to acquire the package and install it.
- Using the software package for the first time — what it can, or cannot do; where and how to get support.
- How to import and display DEMs, using our Baranja Hill case study.
- Which land-surface parameters and object-parameters can be derived from the package, and how they are calculated.

- How particular land-surface parameters and objects can be interpreted and applied.
- Summary of strong and weak points of the software, any known bugs, and how the package may be expected to evolve.

We intend that each chapter serve a dual purpose, as a user manual and as a review of scientific information. For readers requiring further support, links to original user guides, mailing lists, and technical documentation and where to download them are given in each chapter.

Part III: Applications

The final section of the book exemplifies the role of geomorphometry in geo- and environmental sciences ranging from soil and vegetation mapping, hydrological and climatic modelling, to geomorphology and precision agriculture. Chapter 19 introduces the role of digital land-surface analysis in creating maps and models across a broad spectrum of disciplines. It explains why DEM analysis has become so essential for quantifying and understanding the natural landscape. The chapter reviews basic concepts underlying the many uses of geomorphometry as well as how these applications incorporate automated mapping and modelling. It also describes some of the mathematical, statistical, and empirical methods by which predictive scenarios have been modelled using land-surface data.

Subsequent chapters of Part III describe specific cases of automated DEM analysis in various disciplines. These examples are not necessarily all-encompassing, but illustrate some of the many different approaches to using geomorphometry to generate and interpret spatial information. Each of the next eight chapters follows a common structure:

- Introduction to state-of-the-art applications, explaining the importance of geomorphometry in this field and reviewing recent research.
- Guided analysis of an example, usually the Baranja Hill case study, including an interpretation of the results.
- Summary of opportunities and limitations as well as suggestions for future research.

In considering the prospect for geomorphometry, the book's closing chapter *peers into a crystal ball* — what breakthroughs might emerge from future advances in technology? Which concepts, applications, and societal needs are likely to drive the discipline? How dramatic an increase in detail and accuracy can be expected of future DEMs? The chapter also includes a proposal for the design and operation of a geomorphometric atlas of the world that could provide a reference data-repository for most applications of DEM-derived information.

CLOSING THOUGHTS AND ACKNOWLEDGEMENTS

This book is intended primarily for (a) universities and research institutes where graduate or post-graduate courses are conducted in geography and other environmental and geo-sciences, and (b) GIS specialists and project teams involved in mapping, modelling, and managing natural resources at various spatial scales. We believe, moreover, that it will prove its worth as a tutorial and reference source to anyone involved in the analysis of DEMs.

It is not our intention that this volume deliver an exhaustive synthesis of geomorphometry. A reader with a background in civil engineering, for example, will quickly note applications and technical areas that are under-represented or absent. This does not mean that we did not think it worthwhile to include them, but rather that other books are better suited to the task. Nonetheless, we hope that a diverse readership will come to regard our book as a worthwhile source of information on the methods and applications of modern geomorphometry. We offer the book not so much as a stand-alone achievement, but rather as part of an initiative to promote development of the science so that not only researchers in geomorphometry, but also the wider community of DEM users, will apply it wisely. We offer our apologies if we have inadvertently and unintentionally omitted anyone's contributions to geomorphometry.

We wish to thank our science reviewers, *Bodo Bookhagen* (Stanford University, School of Earth Sciences, Stanford, CA, USA), *Peter Burrough* (University of Utrecht, The Netherlands), *Ian S. Evans* (Durham University, Durham, UK), *Peter Fisher* (City University, London, UK), *John Gallant* (CSIRO Land and Water, Canberra, Australia), *Gerard B.M. Heuvelink* (Wageningen University and Research Centre, Wageningen, The Netherlands), *Robert A. MacMillan* (LandMapper Environmental Solutions Inc., Edmonton, AB, Canada), *Richard Pike* (U.S. Geological Survey, Menlo Park, CA, USA), *David Tarboton* (Utah State University, Logan, UT, USA), *Stephen Wise* (University of Sheffield, Sheffield, UK), and *Ole Wendroth* (University of Kentucky, Kentucky, US). Their numerous comments and suggestions for improving and extending various chapters have been invaluable in bringing this project to a successful conclusion.

We are especially grateful to *Richard Pike* and *Ian S. Evans* (two fathers of modern geomorphometry) for providing the support and encouragement during the last phases of line-editing. We are also grateful to *Roko Mrša* (the Croatian State Geodetic Department) for organising a licence to use the Baranja Hill datasets. Last, but not least, we thank JRC colleagues *Nicola Lugeri* for cross-checking over 1000 references, *Nadine Bähr* for her tips'n'tricks of graphical editing, *Pierangello Principalli* and *Alessandro Piedepalumbo* for their professional-quality printing and binding of v1.0 and v2.0 of the book, our secretary *Grazia Faber* for providing continual remedy for the inevitable bureaucratic headaches, and many other colleagues within JRC and farther afield who have supported us in this endeavour.

Every effort has been made to trace copyright holders. We apologize for any unintentional omissions and would be pleased to add an acknowledgment in future editions.

Tomislav Hengl and Hannes I. Reuter
Ispra (VA), July 2007

Part I

Concepts

CHAPTER 1

Geomorphometry: A Brief Guide

R.J. Pike, I.S. Evans and **T. Hengl**

basic definitions · the land surface · land-surface parameters and objects · digital elevation models (DEMs) · basic principles of geomorphometry from a GIS perspective · inputs/outputs, data structures & algorithms · history of geomorphometry · geomorphometry today · data set used in this book

1. WHAT IS GEOMORPHOMETRY?

Geomorphometry is *the science of quantitative land-surface analysis* (Pike, 1995, 2000a; Rasemann et al., 2004). It is a modern, analytical-cartographic approach to representing bare-earth topography by the computer manipulation of terrain height (Tobler, 1976, 2000). Geomorphometry is an interdisciplinary field that has evolved from mathematics, the Earth sciences, and — most recently — computer science (Figure 1). Although geomorphometry[1] has been regarded as an activity within more established fields, ranging from geography and geomorphology to soil science and military engineering, it is no longer just a collection of numerical techniques but a discipline in its own right (Pike, 1995).

It is well to keep in mind the two overarching modes of geomorphometric analysis first distinguished by Evans (1972): *specific*, addressing discrete surface features (i.e. *landforms*), and *general*, treating the continuous land surface. The morphometry of landforms *per se*, by or without the use of digital data, is more correctly considered part of *quantitative geomorphology* (Thorn, 1988; Scheidegger, 1991; Leopold et al., 1995; Rhoads and Thorn, 1996). Geomorphometry in this book is primarily the computer characterisation and analysis of continuous topography. A fine-scale counterpart of geomorphometry in manufacturing is *industrial surface metrology* (Thomas, 1999; Pike, 2000b).

The ground beneath our feet is universally understood to be the interface between soil or bare rock and the atmosphere. Just what to call this surface and its science of measurement, however, is less obvious. Numerical representation of the

[1] The term, distinguished from morphometry in other sciences (e.g. biology), dates back at least to Neuenschwander (1944) and Tricart (1947).

Developments in Soil Science, Volume 33 © 2009 Elsevier B.V.
ISSN 0166-2481, DOI: 10.1016/S0166-2481(08)00001-9.

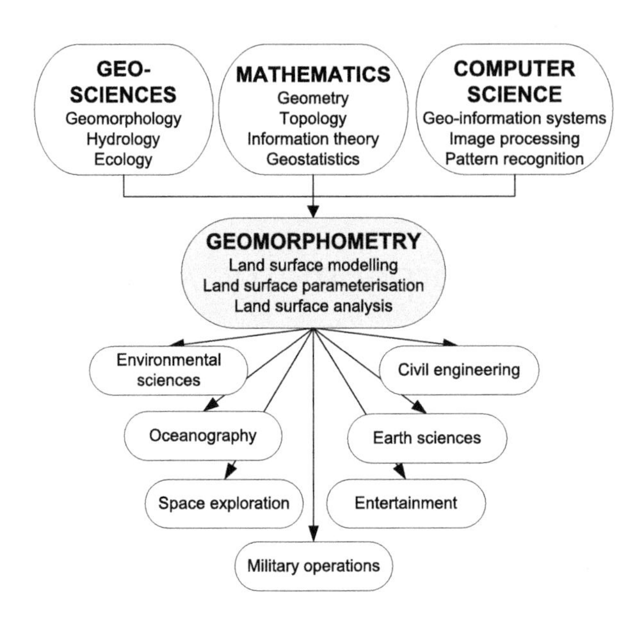

FIGURE 1 Geomorphometry and its relation to source and end-user disciplines. Modified after Pike (1995).

land surface is known variously as *terrain modelling* (Li et al., 2005), *terrain analysis* (Wilson and Gallant, 2000), or the *science of topography* (Mark and Smith, 2004).[2] Quantitative descriptors, or measures, of land-surface form have been referred to as *topographic attributes* or *properties* (Wilson and Gallant, 2000), *land-form parameters* (Speight, 1968), *morphometric variables* (Shary et al., 2002), *terrain information* (Martinoni, 2002), *terrain attributes* (Pennock, 2003), and *geomorphometric attributes* (Schmidt and Dikau, 1999).

REMARK 1. *Geomorphometry is the science of topographic quantification; its operational focus is the extraction of land-surface parameters and objects from digital elevation models (DEMs).*

Despite widespread usage, as a technical term *terrain* is imprecise. *Terrain* means different things to different specialists; it is associated not only with land form, hydrographic features, soil, vegetation, and geology but also (like *topography*) with the socio-economic aspects of an area (Li et al., 2005). *Terrain*[3] also can signify an area of ground, a region... unrelated to *shape* of the land surface. The much used *terrain analysis* (Moore et al., 1991a; Wilson and Gallant, 2000) is confusing (unless preceded by *quantitative*), because it has long denoted qualitative (manual) stereoscopic photo- or image-interpretation (Way, 1973). Nor does the more precise *digital terrain modelling* (Weibel and Heller, 1991) escape ambiguity, as *terrain modelling* can infer measurement or display of surface heights, unspecified quantification of topography, or any digital processing of Earth-surface features.

[2] The most frequent equivalents of *geomorphometry* in Google's online database appear to be *surface* or *terrain modelling*, *terrain analysis* and *digital terrain modelling* (Pike, 2002).

[3] *Terrain* is from the Latin *terrenum*, which might be translated as *"of the earth"*.

Additionally, in many countries (e.g. France, Spain, Russia, Slovakia) *relief*[4] is synonymous with morphology of the land surface (King et al., 1999). This usage is less evident in Anglophone regions (e.g. Great Britain, North America), where *relief*, usually prefixed by *relative* or *local*, has come to denote the difference between maximal and minimal elevation within an area (Partsch, 1911; Smith, 1953; Evans, 1979), "*low*" and "*high*" relief indicating small and large elevation contrasts respectively.[5]

To minimise confusion, the authors of this book have agreed to consistently use *geomorphometry* to denote the scientific discipline and *land surface*[6] to indicate the principal object of study. Digital representation of the land surface thus will be referred to as a *digital land surface model* (DLSM), a specific type of *digital surface model* (DSM) that is more or less equivalent to the widely-accepted term *digital elevation model*[7] (DEM).

An area of interest may have several DSMs, for example, surface models showing slope gradient or other height derivative, the tree canopy, buildings, or a geological substrate. DSMs from laser altimetry (LiDAR, light detection and ranging) data can show more than one *return surface* depending on how deep the rays penetrate. Multiple DLSMs are usually less common but can include DEMs from different sources or gridded at different resolutions, as well as elevation arrays structured differently from square-grid DEMs (Wilson and Gallant, 2000). Objects of the built environment are of course not part of the land surface and must be removed to create a true bare-earth DLSM.

Digital elevation model (DEM) has become the favoured term for the data most commonly input to geomorphometry, ever since the U.S. Geological Survey (USGS) first began distribution of 3-arc-second DEMs in 1974 (Allder et al., 1982). Even *elevation* is not unique as it can also mean surface uplift (e.g. the Himalayas have an *elevation* of 5 mm/year). However, the alternative terms are less satisfactory: *height* is relative to a nearby low point, and *altitude* commonly refers to vertical distance between sea level and an aircraft, satellite, or spacecraft. Thus *digital height model* and *altitude matrix* (Evans, 1972) are avoided here.

REMARK 2. *The usual input to geomorphometric analysis is a square-grid representation of the land surface: a digital elevation (or land surface) model (DEM or DLSM).*

In this book, DEM refers to a gridded set of points in Cartesian space attributed with elevation values that approximate Earth's ground surface (e.g. Figure 5, below). Thus, contour data or other types of sampled elevations, such as a triangular array, are not DEMs as the term is used here. "DEM" implies that elevation is available continuously at each grid location, at a given resolution. See Chapter 2 for a detailed treatment of topography and elevation models.

[4] fren. *Topographie*, germ. *Relief*, russ. рельеф, span. *Relieve*.

[5] This quantity is also known as *reliefenergie* (Gutersohn, 1932), particularly in Germany and Japan.

[6] fren. *Surface terrestre*, germ. *Gelände*, russ. земная поверхность, span. *Topografía*. A term that became widely known through the morphometric work of Hammond (1964).

[7] fren. *Modèle numèrique de terrain*, germ. *Digitales Gelände Model*, russ. цифровая модель рельефа, span. *Modelo de elevación digital*.

Finally, we define *parameter* and *object*, the two DEM-derived entities fundamental to modern geomorphometry (see, e.g., Mark and Smith, 2004). A *land-surface parameter*[8] is a descriptive measure of surface form (e.g. slope, aspect, wetness index); it is arrayed in a continuous field of values, usually as a raster image or map, for the same referent area as its source DEM. A *land-surface object*[9] is a discrete spatial feature (e.g. watershed line, cirque, alluvial fan, drainage network), best represented on a vector map consisting of points, lines, and/or polygons extracted from the square-grid DEM.

It is also important to distinguish parameters *per se*, which describe the land surface at a point or local sample area, from quantitative *attributes* that describe objects. For example, slope gradient at a given point refers only to its x, y location, whereas the volume of, say, a doline (limestone sink) applies to the entire area occupied by that surface depression; slope is a land-surface parameter, while depression volume over an area is an attribute of a land-surface object. Each of these quantities can be obtained from a DEM by a series of mathematical operations, or *morphometric algorithms*.

2. THE BASIC PRINCIPLES OF GEOMORPHOMETRY

2.1 Inputs and outputs

The fundamental operation in geomorphometry is *extraction of parameters and objects from DEMs* (Figure 2). DEMs, i.e. digital land-surface models, are the primary input to morphometric analysis. In GIS (geographic information system) terms, a DEM is simply a raster or a vector map showing the height of the land surface above mean sea level or some other referent horizon (see further Section 2 in Chapter 2).

Geomorphometry commonly is implemented in five steps (Figure 2):

1. *Sampling the land surface* (height measurements).
2. *Generating a surface model from the sampled heights.*
3. *Correcting errors and artefacts in the surface model.*
4. *Deriving land-surface parameters and objects.*
5. *Applications of the resulting parameters and objects.*

Land-surface parameters and objects can be grouped according to various criteria. Parameters commonly are distinguished as primary or secondary, depending on whether they derive directly from a DEM or additional processing steps/inputs are required (Wilson and Gallant, 2000). In this book, we will follow a somewhat different classification that reflects the purpose and type of analysis. Three main groups of land-surface parameters and objects are identified:

- *Basic morphometric parameters and objects* (see Chapter 6);
- *Parameters and objects specific to hydrology* (see Chapter 7);
- *Parameters and objects specific to climate and meteorology* (see Chapter 8);

[8] fren. *Paramètre de la surface terrestre*, germ. *Reliefparameter*, russ. характеристика рельефа, span. *Variable del terreno*.
[9] fren. *Object de la surface terrestre*, germ. *Reliefobjeckt*, russ. объект земной поверхности, span. *Elemento del terreno*.

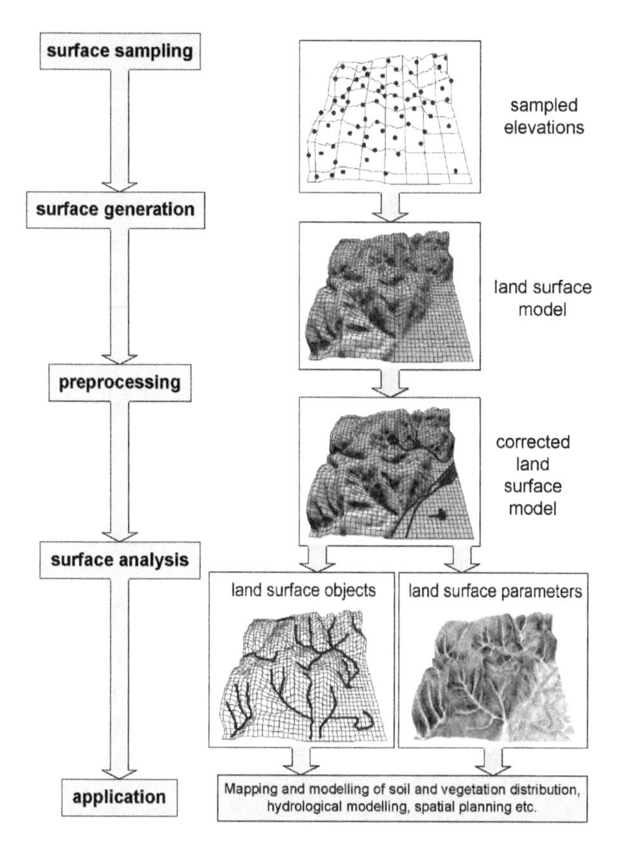

FIGURE 2 The operational focus of geomorphometry is extraction of land-surface parameters and objects from DEMs.

Basic parameters and objects describe local morphology of the land surface (e.g. slope gradient, aspect and curvature). Hydrological or flow-accumulation parameters and objects reflect potential movement of material over the land surface (e.g. indices of erosion or mass movement). The third group of parameters and objects is often calculated by adjusting climatic or meteorological quantities to the influence of surface relief.

A special group of land-surface objects — geomorphological units, *land elements* and *landforms* — receives its own chapter (Chapter 9). A landform is a discrete morphologic feature — such as a watershed, sand dune, or drumlin — that is a functionally interrelated part of the land surface formed by a specific geomorphological process or group of processes. Each landform may be composed of several landform elements, smaller divisions of the land surface that have relatively constant morphometric properties.

> REMARK 3. *A landform element is a division of the land surface, at a given scale or spatial resolution, bounded by topographic discontinuities and having (relatively) uniform morphometry.*

Recognition of landforms and less exactly defined tracts, commonly referred to as *land-surface types*, from the analysis of DEMs is increasingly important. Many areas of the Earth's surface are homogeneous overall or structured in a distinctive way at a particular scale (e.g. a dune field) and need to be so delineated (Iwahashi and Pike, 2007). In the special case of landforms extracted as *"memberships"* by a fuzzy classification algorithm, such forms can be considered to *"partake"* of a particular land-surface object — instead of directly mapping, say, a stream channel, we can obtain a *"membership value"*[10] to that landform.

2.2 The raster data structure

Many land-surface representations, such as the background topography seen in video games and animated films, are modelled by mass-produced surface heights arrayed in some variant of the surface-specific *triangulated irregular network* (TIN) model (Blow, 2000; Hormann, 1969; see Chapter 2, Section 2.1). Most geomorphometric applications, however, use the square-grid DEM model. To be able to apply the techniques of geomorphometry effectively, it is essential to be familiar with the concept of a raster GIS and its unique properties.

Although the raster structure has a number of disadvantages, including a rectangular data array regardless of the morphology of the study area, large data-storage requirements, and under- and over-sampling of different parts of a diverse study area, it will remain the most popular format for spatial modelling in the foreseeable future. This structure is especially advantageous to geomorphometry because most of its technical properties are controlled automatically by a single measure: spatial resolution, *grid size* or *cell size*,[11] expressed as a constant x, y spacing (usually in metres) (Hengl, 2006).

In addition to grid resolution, we also need to know the coordinates of at least one grid intersection (usually marking the lower left-hand corner of the entire DEM array) and the number of rows and columns, whereupon we should be able to define the entire map (Figure 3). This of course assumes that the map is projected into an *orthogonal system* where all grid nodes are of exactly equal size and oriented toward cartographic North.

Accordingly, the small 6×6-pixel DEM in Figure 5 (see below) can also be coded in an ASCII file as an array of heights:

```
ncols 6
nrows 6
xllcorner 0
yllcorner 0
cellsize 10.00
nodata_value -32767
10 16 23 16 9 6
14 11 18 11 18 19
19 15 13 21 23 25
20 20 19 14 38 45
24 20 20 28 18 49
23 24 34 38 45 51
```

[10] Such a value has been designated by the rather clumsy term *channelness*.

[11] *Cell size* is a more appropriate term than *grid size* because *grid size* can also imply size of the whole grid.

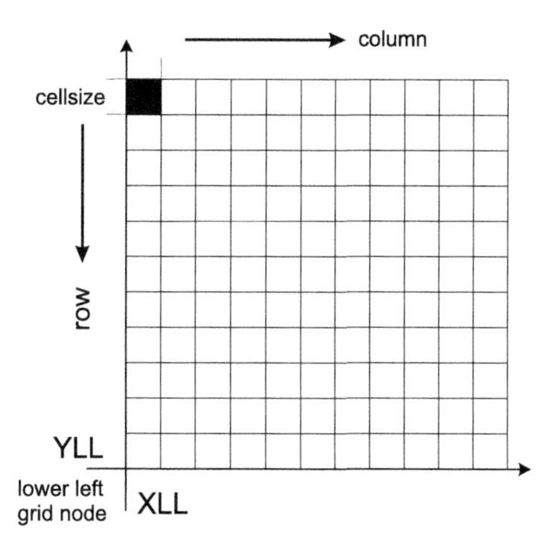

FIGURE 3 An orthogonal raster map can be defined by just five parameters: (a & b) number of rows and columns; (c & d) coordinates of the lower left corner and (e) cell size.

where `ncols` is number of columns, `nrows` is number of rows, `xllcorner` is the western edge of the map, `yllcorner` is the southern edge of the map, `cellsize` is grid resolution in metres, `nodata_value` is the arbitrary value used to mask out locations outside the area of interest and `10, 16, 23, 16, 9, 6` are the elevation values in the (first) row. This is the standard format for ASCII grid files used by ESRI Inc. for its ArcInfo and ArcGIS software. It is necessary to define the initial point of the grid system correctly: there is a difference in x, y location of half the `cellsize`, depending on whether the first coordinate is at the lower left-hand corner of the lower left-hand grid cell (`llcorner`) or at the centre of that cell (`llcenter`).

REMARK 4. *The principal advantage of a raster GIS over other spatial data structures is that a single measure — the cell or pixel size — automatically controls most technical properties.*

2.3 Geomorphometric algorithms

Performing morphometric operations within a raster GIS usually involves calculating intermediate quantities (over the same grid of interest) which are then used to compute the final output. Most morphometric algorithms work through the *neighbourhood operation* — a procedure that moves a small regular matrix of cells (variously termed a *sub-grid* or *filter window*) over the entire map from the upper left to the lower right corner and repeats a mathematical formula at each placement of this sampling grid.

Neighbouring pixels in a sampling window are commonly defined in relation to a central pixel, i.e. the location for which a parameter or an object membership is derived. In principle, there are several ways to designate neighbouring pixels, most commonly either by an identifier or by their position relative to the central

(*a*)

Z_{NB1}	Z_{NB2}	Z_{NB3}
Z_{NB4}	Z_{NB5}	Z_{NB6}
Z_{NB7}	Z_{NB8}	Z_{NB9}

Z_{NB1}	Z_{NB2}	Z_{NB3}	Z_{NB4}	Z_{NB5}
Z_{NB6}	Z_{NB7}	Z_{NB8}	Z_{NB9}	Z_{NB10}
Z_{NB11}	Z_{NB12}	Z_{NB13}	Z_{NB14}	Z_{NB15}
Z_{NB16}	Z_{NB17}	Z_{NB18}	Z_{NB19}	Z_{NB20}
Z_{NB21}	Z_{NB22}	Z_{NB23}	Z_{NB24}	Z_{NB25}

(*b*)

-1, 1	0, 1	1, 1
-1, 0	0, 0	1, 0
-1, -1	0, -1	1, -1

-2, 2	-1, 2	0, 2	1, 2	2, 2
-2, 1	-1, 1	0, 1	1, 1	2, 1
-2, 0	-1, 0	0, 0	1, 0	2, 0
-2, -1	-1, -1	0, -1	1, -1	2, -1
-2, -2	-1, -2	0, -2	1, -2	2, -2

FIGURE 4 The common designation of neighbours in 3×3 and 5×5 window environments: (a) by unique identifiers (as implemented in ILWIS GIS), (b) by row and column separation (in pixels) from the central pixel (as implemented in the ArcInfo GIS).

pixel (Figure 4). The latter (e.g. implemented by the DOCELL command in ArcInfo) is the more widely used because it can readily pinpoint almost any of the neighbouring cells anywhere on the map [Figure 4(b)].

Computing a DEM derivative can be simple repetition of a given formula over the area of interest. Consider a very small DEM of just 6×6 pixels. You could zoom into these values (elevations) and derive the desired parameter on a pocket calculator (Figure 5). For example, using a 3×3 sampling window, slope gradient at the central pixel can be derived as the average change in elevation. Three steps are required; first, the difference in relative elevation is calculated in x and y directions, whereupon slope gradient is obtained as the average of the two quadratics (Figure 5). By the Evans–Young method[12] (Pennock et al., 1987), slope gradient is calculated (see further Chapter 6):

$$G = \frac{z_{NB3} + z_{NB6} + z_{NB9} - z_{NB1} - z_{NB4} - z_{NB7}}{6 \cdot \Delta s}$$

$$H = \frac{z_{NB1} + z_{NB2} + z_{NB3} - z_{NB7} - z_{NB8} - z_{NB9}}{6 \cdot \Delta s}$$

[12] Often, one land-surface parameter can be calculated by several different formulas or approaches; we caution that the results can differ substantially!

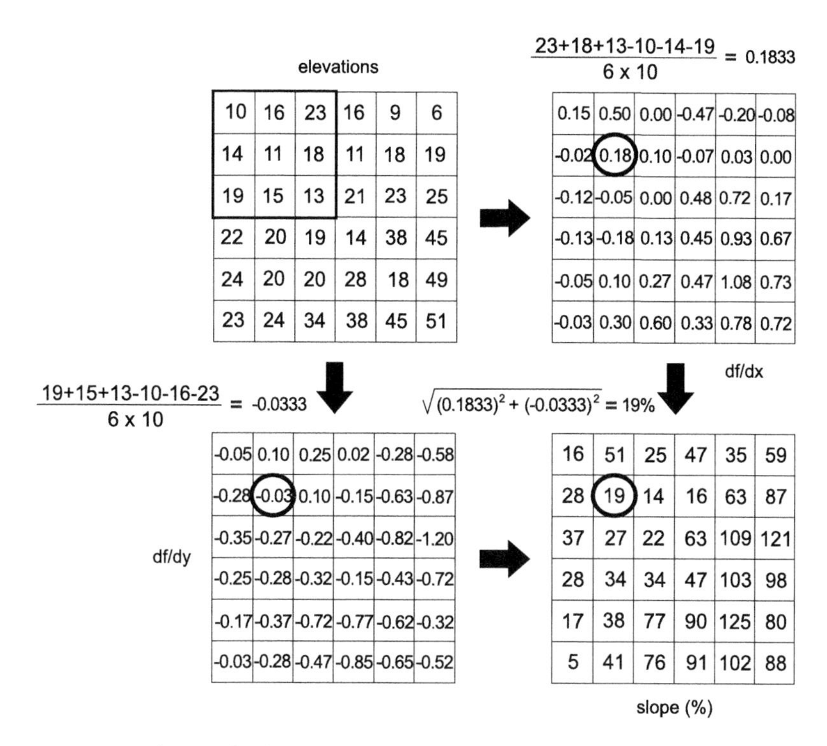

10	16	23	16	9	6
14	11	18	11	18	19
19	15	13	21	23	25
22	20	19	14	38	45
24	20	20	28	18	49
23	24	34	38	45	51

0.15	0.50	0.00	-0.47	-0.20	-0.08
-0.02	0.18	0.10	-0.07	0.03	0.00
-0.12	-0.05	0.00	0.48	0.72	0.17
-0.13	-0.18	0.13	0.45	0.93	0.67
-0.05	0.10	0.27	0.47	1.08	0.73
-0.03	0.30	0.60	0.33	0.78	0.72

-0.05	0.10	0.25	0.02	-0.28	-0.58
-0.28	-0.03	0.10	-0.15	-0.63	-0.87
-0.35	-0.27	-0.22	-0.40	-0.82	-1.20
-0.25	-0.28	-0.32	-0.15	-0.43	-0.72
-0.17	-0.37	-0.72	-0.77	-0.62	-0.32
-0.03	-0.28	-0.47	-0.85	-0.65	-0.52

16	51	25	47	35	59
28	19	14	16	63	87
37	27	22	63	109	121
28	34	34	47	103	98
17	38	77	90	125	80
5	41	76	91	102	88

FIGURE 5 Numerical example showing slope tangent (in %) extracted from a DEM using a 3×3 window.

where G is the first derivative in the x direction (df/dx), H is the first derivative in the y direction (df/dy), z_{NB5} is the (central) cell for which the final value of slope is desired, $z_{\text{NB1,2,3,4,6,7,8,9}}$ are the eight neighbouring cells, and Δs is pixel size in metres (Figure 5). The slope gradient as a tangent is finally computed as:

$$\text{SLOPE} = \sqrt{H^2 + G^2}$$

Note that the example in Figure 5 shows values of slope gradient for rows and columns at the edge of the map, although we did not actually have the necessary elevation values outside the map area. Normally, a neighbourhood operation is possible only at a grid location surrounded by its eight immediate neighbours. Because keeping to this practice loses the outermost rows and columns, the expedient solution illustrated in this example is to estimate missing neighbours by duplicating cells at the edges of the DEM and tolerating the (usually) modest error in the final result. By so doing, the output map retains exactly the same size as the input map.

REMARK 5. *Because most land-surface parameters vary with spatial scale, or can be calculated by different algorithms and sampling grids, no map computed from a DEM is definitive.*

Adjustments such as these differ among software packages, so that almost always some small differences will be found in outputs from exactly the same mathematical formulas. To avoid confusion, in referring to various types of general land-surface parameters and objects we will consistently specify (1) the algorithm (reference), (2) size of the sampling window and (3) the cell size. The example above would be slope (*land-surface parameter type*) calculated by the Evans–Young method (Pennock et al., 1987) (*variant*) in a 3×3 window environment (*sub-variant*) using a 10 m DEM (*cell size*). The rounding factor also can be important because some intermediate quantities require high precision (many decimal places), while others must never equal zero or take a negative value.

Finally, in Figure 5 we can see that the pixel with highest slope, 125%, is at location `row` = 5, `column` = 5 and the lowest slope, 5%, is at location `row` = 6, `column` = 1. Of course, in a GIS map the heights are rarely represented as numbers but rather by colour or greyscale legends.

3. THE HISTORY OF GEOMORPHOMETRY

Before exploring data, algorithms and applications in detail, it is well to step back and consider the evolution of geomorphometry, from the pioneering work of German geographers and French and English mathematicians to results from recent Space Shuttle and planetary-exploration missions. While its ultimate origins may be lost in antiquity, geomorphometry as we know it today began to evolve as a scientific field with the discoveries of Barnabé Brisson (1777–1828), Carl Gauss (1777–1855), Alexander von Humboldt (1769–1859), and others, reaching maturity only after development of the digital computer in the mid- to late-20th century.

> REMARK 6. *Geomorphometry evolved from a mix of mathematics, computer processing, civil and military engineering, and the Earth sciences — especially geomorphology.*

The earliest geomorphometry was a minor sub-activity of exploration, natural philosophy, and physical geography — especially geomorphology; today it is inextricably linked with geoinformatics, various branches of engineering, and most of the Earth and environmental sciences (Figure 1). In the following sections we will briefly describe the approaches and concepts of pre-DEM morphometry as well as analytical methods applied to contemporary data. Additional background is available in Gutersohn (1932), Neuenschwander (1944), Zakrzewska (1963), Kugler (1964), Hormann (1969), Zavoianu (1985), Krcho (2001), and Pike (1995, 2002).

3.1 Hypsometry and planimetric form

Geomorphometry began with the systematic measurement of elevation above sea level, i.e. land surveying — almost certainly in ancient Egypt.[13] Height measurement by cast shadows is ascribed to the Greek philosopher Thales of Miletus

[13] Land surveying that focuses on measurement of terrain height is often referred to as *hypsometry*, from the Greek χυπςος — height.

(ca. 624–546 B.C.). The concept of the elevation contour to describe topography dates to 1584 when the Dutch surveyor Pieter Bruinz drew lines of equal depth in the River Spaarne; but this was an unpublished manuscript (Imhof, 1982). In 1725 Marsigli published a map of depth contours in the Golfe du Lion, i.e. the open sea. In 1737 (published in 1752) Buache mapped the depth of the Canal de la Manche (English Channel), and in 1791 Dupain-Triel published a rather crude contour map of France (Robinson, 1982, pp. 87–101/210–215).

In 1774, British mathematician Charles Hutton was asked to summarise the height measurements made by Charles Mason,[14] an astronomer who wanted to estimate the mass of Earth. Hutton used a pen to connect points of the same height on the Scottish mountain Schiehallion, developing the *isohypse* (or isoline) concept. This has proved very effective in representing topography and is one of the most important innovations in the history of mapping by virtue of its convenience, exactness, and ease of perception (Robinson, 1982). DeLuc, Maskelyne, Roy, Wollaston, and von Humboldt were among many early investigators who used the barometer invented by Evangelista Torricelli (1608–1647) and developed by Blaise Pascal (1623–1662) to measure elevation; see also Cajori (1929) and de Dainville (1970).

With the spread of precise surveying in late 18th- and early 19th-century Europe, illustrations ranking mountain-top elevations and the lengths of rivers began to appear in atlases.[15] Mountain heights and groupings were studied qualitatively, often by military engineers (von Sonklar, 1873), as *orography*, their heights and derived parameters as *orometry* (Figure 6). Early 19th-century German geographers such as von Humboldt (recently cited in Pike, 2002, and Rasemann et al., 2004) compared summit heights in different ranges. Von Sonklar (1873), and earlier regional monographs, went further and considered the elevations of summits, ridges, passes and valleys as well as relative heights, gradients and volumes. *Orometry* — with emphasis on mean slope, mean elevation and volume, planimetric form, relative relief, and drainage density — became a favoured dissertation topic for scores of European geographers (Neuenschwander, 1944). The overarching charter of geomorphometry was nicely captured many years ago by the German geographer Alfred Hettner (1859–1941), when he wrote in a brief consideration and critique of 19th-century orometry: "*But it is more important to enquire whether we cannot express the entire character of a landscape numerically*" (Hettner, 1928, p. 160; republished in 1972).

Before the wider availability of contour maps in the mid-19th century,[16] most quantitative analyses of topography were of broad-scale linear features: rivers and coasts. The concavity of longitudinal river profiles, adequately determined from spot heights, came to be represented by exponential and parabolic equations (Chorley et al., 1964, §23). Carl Ritter (1779–1859) introduced indices of *Küstenentwicklungen* (*Coastal Development*) to distinguish intricate coastlines such as fjords from simpler ones such as long beaches. Some indices were more descriptive than

[14] This is the same Charles Mason who, with Jeremiah Dixon, surveyed the Mason–Dixon Line in the USA between 1763 and 1767.

[15] Tufte (1990, p. 77) reproduces just such a detailed 1864 diagram from J.H. Colton.

[16] Because early topographic maps represented relief by hachures, not contours, analysis of slope required detailed field survey and thus was rare.

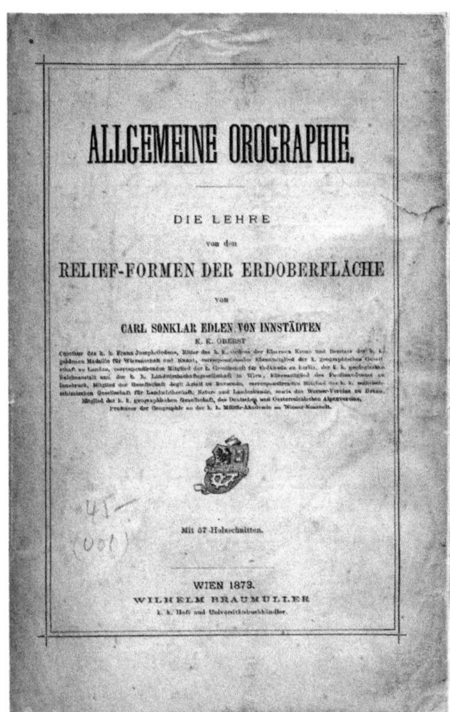

ALLGEMEINE OROGRAPHIE.

DIE LEHRE

von den

RELIEF-FORMEN DER ERDOBERFLÄCHE

von

CARL SONKLAR EDLEN VON INNSTÄDTEN

K. K. OBERST

Mit 57 Holzschnitten.

WIEN 1873.

WILHELM BRAUMÜLLER

k. k. Hof- und Universitätsbuchhändler.

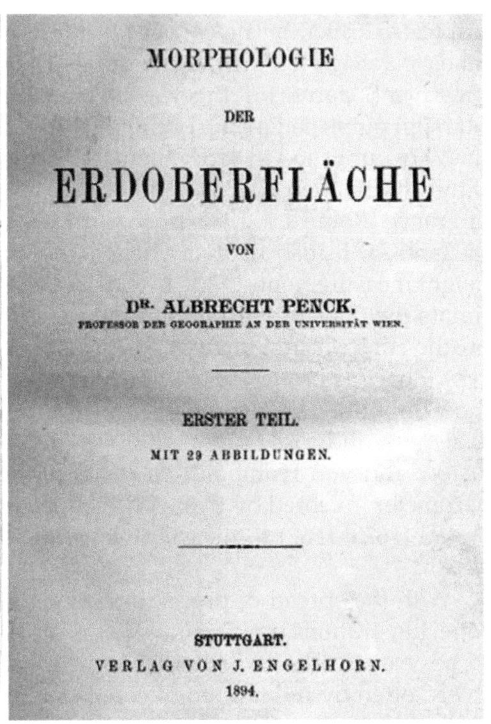

MORPHOLOGIE

DER

ERDOBERFLÄCHE

VON

DR. ALBRECHT PENCK,

PROFESSOR DER GEOGRAPHIE AN DER UNIVERSITÄT WIEN.

ERSTER TEIL.

MIT 29 ABBILDUNGEN.

STUTTGART.

VERLAG VON J. ENGELHORN.

1894.

FIGURE 6 Two landmarks of early geomorphometry from Germany and Austria, arguably the cradle of geomorphometry. The brief 19-page chapter on *orometrie* in von Sonklar's 1873 textbook (left) presented twelve quantitative measures of mountain morphology, which stimulated much publication on land-surface characterisation. One of the best summary treatments of early geomorphometry (including criticism of Sonklar!) was a much longer and wider-ranging chapter in Penck's 1894 textbook (right). Photos by R. Pike.

others; the ratio of an island's area to the square of its perimeter, for example, combined coastal sinuosity with compactness, whereas the ratio of its area to area of the smallest circumscribed circle was only an inverse measure of elongation, not *circularity* as claimed.

The impossibility of agreeing on a definitive length for a section of coastline eventually led to Richardson's (1961) establishment of a scaling relation between step length (i.e. measurement resolution) and estimated line length, and later the *fractal concepts* (Mandelbrot, 1967, 1977) of self-similarity and non-Euclidean form. As Mandelbrot's (1967) title implies, these widely applied scaling concepts were firmly rooted in coastal geomorphometry.[17]

Once contour maps were more available, relief analysis flourished. Measurement of highest and lowest points within a sample area (commonly a square or circle) quantified the vertical dimension as *relief* (*Reliefenergie* in German), which developed from the need to express relative height (Gutersohn, 1932). Partsch

[17] Much further evidence could have been found in Volkov (1950), not cited by Mandelbrot (see also Maling, 1989, pages 277–303, and pages 66–83 citing the 1894 measurements of A. Penck on the Istrian coast).

(1911) used elevation range per 5×5 km square to produce what probably is the first quantitative map of local (relative) relief. Other definitions expressed relief for a hillslope (ridge crest to valley floor) or for a fluvial drainage basin: "catchment" or "watershed" relief (Sherman, 1932). Attempts to define relief as the separation between an upper *relief envelope* or summit surface and a valley or *streamline surface* (reviewed in Rasemann et al., 2004) were less successful because of scale variations. Working for the U.S. Army, W.F. Wood (1914–1971) quantified the dependency of relief upon area by statistical analysis of 213 samples measured on U.S. contour maps (Wood and Snell, 1957).

Geographers and later geomorphologists planimetered the areas enclosed by contours to generate plots of elevation versus area. Estimates for the entire globe by Murray (1888) were rough but sufficient to establish the bimodality of Earth's elevations, peaking near 0 and −4600 m, which posed numerous questions for geologists and geophysicists. This *hypsographic curve* could be cumulated and integrated for comparative studies of regions (de Martonne, 1941). The histograms of de Martonne (1941) are misleading, however, because he used two class intervals with the same linear vertical scale.

The dimensionless hypsometric integral, first applied to *landforms* (cirques) by Imamura (1937) and to regions by Péguy (1942), approaches zero where a few high points rise above a plain, and 1.0 where most surface heights cluster near the maximum. Although this device is useful morphologically and in geomorphology, hydrologic and other applications often require retention of landform dimensions. Strahler (1952) popularised an integral of the hypsometric curve, which later was proven identical to a simpler measure as well as the approximate reciprocal of elevation skewness[18] (Pike and Wilson, 1971). Péguy (1948) called further for a more conventional statistical approach and proposed the standard deviation of elevation as a measure of relief because of instability of the maximum. He asserted: "*Like all adult science, the geography of the second half of this century will be called to make more and more continuous appeal to mathematical methods*" (Péguy, 1948, p. 5).

Clarke (1966) critically reviewed hypsometry, clinometry and altimetric analysis, which had often been used in the search for old erosion (planation) surfaces over the prior 40 years. He showed that several types of clinographic curves, going back to the earliest examples by Sebastian Finsterwalder and Carl Peucker in 1890, can be misleading in their attempts to plot average slope gradient against elevation.

3.2 Drainage topology and slope frequency

In 1859, Alfred Cayley published "*On contour lines and slope lines*", which laid out the mathematical foundation of geomorphometry.[19] In this extraordinary paper, the land surface is considered in the gravitational field, and thus certain lines and points are more significant than others. Cayley defined *slope lines* as being always at right angles to contours. On a smooth, single-valued surface, all slope lines run from summits to pits (ultimately the ocean), except those joining summits (*ridge*

[18] See further Figure 4 in Chapter 28.

[19] He was preceded by even earlier French mathematicians and geometers (Pike, 2002).

lines) and those joining pits (*course lines*). *Passes* are the lowest points on the former, and *pales* are the highest points on the latter. Each pass and pale is located at the intersection of a ridge line and a course line.

James Clerk Maxwell (1870) further noted that each territory defined by these special lines was part of both a *hill* whose lines of slope run down from the same summit, and a *dale* whose slope lines run down to the same pit. Hills are bounded by course lines, and dales by ridge lines. These pioneering *semantics* remained neglected until their rediscovery by Warntz (1966, 1975) and Mark (1979). They have since been again rediscovered by the engineering-metrology community (Scott, 2004).

Fluvial geomorphometry evolved from concepts of stream frequency (and its reciprocal, drainage density) and stream order, notably in the pioneering work of Ludwig Neumann and Heinrich Gravelius (Neuenschwander, 1944). The quantitative study of rivers and river networks initially was dominated by hydraulic engineers rather than geographers or geomorphologists, the work of Horton (1932, 1945) on network topology and related geometric attributes of drainage basins being especially influential. His revolutionary 1945 synthesis of hydrology and geomorphology rapidly evolved into the sub-field of drainage network analysis in the 1950s and 1960s (Shreve, 1974), which grew to such an extent that elaboration of stream-order topology overshadowed geometric analysis of the land surface.

Many geomorphological studies from the 1960s through the 1980s sought to relate hillslopes to streams (see later section) and in so doing exhaustively parameterised the shape and relief of individual drainage basins (Zavoianu, 1985; Gardiner, 1990). The *drainage basin* is Earth's dominant land-surface object and its analysis is, strictly speaking, a branch of specific geomorphometry. However, fluvial networks occupy so high a fraction of Earth's surface that the analysis of distributed drainage systems has come to dominate the more process-oriented implementations of general geomorphometry (Rodríguez-Iturbe and Rinaldo, 1997).

Statistical analysis of large samples of slopes began with Strahler's (1950) work in southern California, leading to the Columbia School of quantitative and dynamic fluvial geomorphology (Morisawa, 1985). Strahler measured maximum slope down a hillside profile (flow-line) and mean (overall) gradient, and related both to the gradient and topological order of the stream below. Tricart and Muslin (1951) advocated measuring large samples of 100 to 200 slope gradients from crest to foot on maps, in degrees rather than percentage; histograms for a homogeneous sample area tended to be symmetric and conspicuously peaked. Adapting a technique from structural geology, Chapman (1952) added a third dimension to slope analysis by treating planar surfaces as '*poles to the plane*'. He constructed radial plots of slope gradient against aspect (calculated from a gridded sample of points) to visually interpret asymmetry and lineation, an approach subsequently incorporated in the MicroDEM package (Guth et al., 1987).

The adoption of frequency distributions and statistical tests represented considerable progress and was promoted by Chorley (1957, 1966) for both drainage basins and individual slope segments. Tricart (1965) critically reviewed slope and fluvial morphometry, asserting that scale cannot be ignored if river profiles and channel incision are to be related to slope processes (Schumm, 1956). Yet despite such advances, the more dominant view among geologists and geographers in the early- to mid-1950s remained: *"mathematical analyses of topographic maps... are tedious, time-consuming, and do not always yield results commensurate with the amount of time required for their preparation"* (Thornbury, 1954, p. 529).[20]

Hormann (1969) brought a more distributed context to topographic analysis by devising a Triangulated Irregular Network (TIN), linking selected points on divides, drainage lines and breaks in slope to interrelate height, slope gradient, and aspect. Rather than individual data points, Hormann plotted averages over intervals, but also was able to consider valley length, depth, gradient, and direction. Criticised by one German colleague as excessively coarse and mechanistic, Hormann's TIN model was successfully developed in North America (Peucker and Douglas, 1975). Its surface-specific vector structure, complementary to the raster square-grid model, has since become a staple of both geomorphometry and GIS packages (Jones et al., 1990; Weibel and Brandli, 1995; Tucker et al., 2001).

Slopes had been profiled in the field (down lines of maximum gradient) in the 19th century (Tylor, 1875), but early geomorphometricians calculated slope from the contour spacing on maps[21] (as illustrated in Figure 7). As geomorphologists grew dissatisfied with the inadequacies of contour maps, field measurement of gradients and profiles became widespread in the 1950s. Slope profiling developed especially in Britain where many contours were interpolated yet photogrammetry was regarded as inadequate by the official mapping agency. Slope profiles were surveyed either in variable-length segments or with a fixed 1.52 m frame (Young, 1964, 1972; Pitty, 1969)[22]; still, a truly random sample of sinuous lines from a rough surface proved elusive. One motive for plotting frequency distributions of slope gradient was to discover characteristic slope angles, and upper and lower limiting angles relevant to slope processes (Young, 1972, pp. 163–167). Parsons (1988) reviewed further developments in slope profiling and slope evolution.

Local shape of the land surface is largely a function of curvature, or change of slope, a second derivative of elevation (Minár and Evans, 2008). Its importance in both profile and plan for hydrology and soils has long been recognised (Figure 7) and it forms the basis of a generic nine-fold (3×3) classification into elementary forms that are *convex*, *straight* or *concave* in plan, and in profile (Richter, 1962). This appealing taxonomy is useful, but precisely what constitutes a *straight* (i.e. planar) slope must be defined operationally; e.g. Dikau (1989) used a 600 m radius of curvature as the threshold of convexity and concavity (see further Figure 7 in Chapter 9).

The breaks and inflections of slope that delimit elementary forms or facets of the land-surface form the basis of morphographic mapping, a subset of geomorphological mapping which we shall not review in detail here (Kugler, 1964; Young, 1972; Barsch, 1990). Morphography is based on field mapping and air-

[20] Even more severe was the criticism of Wooldridge (1958), who wrote disparagingly: *"At its worst this is hardly more than a ponderous sort of cant... If any best is to result from the movement, we have yet to see it'..."*

[21] Average slope could be estimated from the density of contour intersections with a grid (Wentworth, 1930).

[22] Equal spacing of profiles along a mid-slope line provided better coverage than starting from the slope crest or foot (Young, 1972, p. 145).

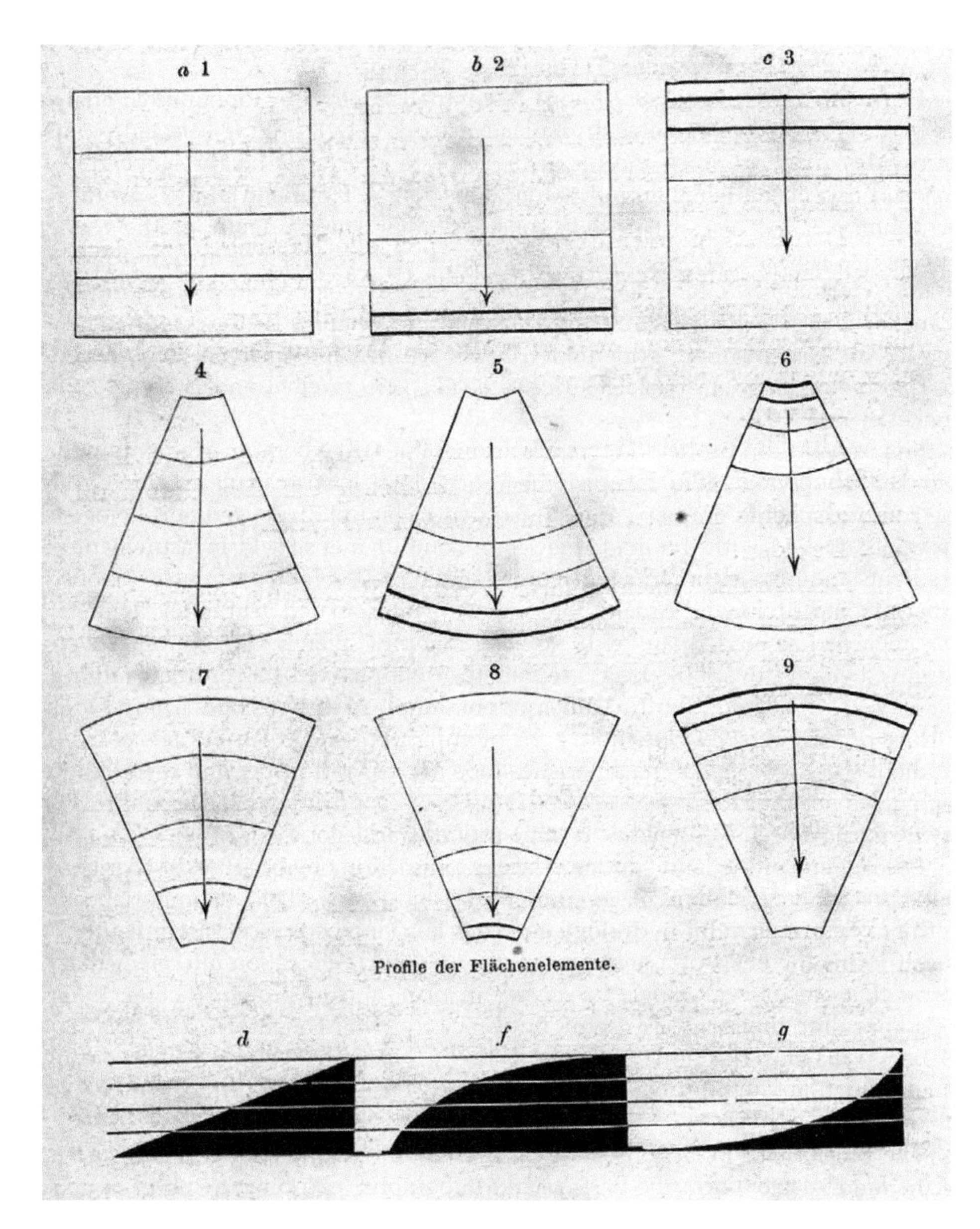

FIGURE 7 Illustration of the nine basic elements of surface form in the 1862 textbook on military geography by an Austrian army officer, long pre-dating 20th-century morphometry (see further Chapter 9). Photo by R. Pike.

photo interpretation, but a number of recent papers have attempted to automate the practice from DEMs, with varying success (see further Chapter 22).

3.3 Early DEMs and software tools

World War II innovation in technology set the stage for postwar advances in geomorphometry, many of which were inaccessible or poorly circulated due to defence-related sponsorship. Pike (1995) asserted that the field is unlikely to have developed as it did without the Cold War (1946–1991) and its space exploration offshoots[23] (Cloud, 2002). Some of the limited-distribution American reports from the 1950s and 1960s that stimulated general geomorphometry are listed by Zakrzewska (1963) and Pike (2002). Wood and Snell (1960), for example, manually measured six factors (in order of importance: average slope, grain, average elevation, slope direction changes, relative relief, and the elevation–relief ratio) from contour maps for 413 sample areas in central Europe, to delimit 25 land surface regions — a model for subsequent multivariate regionalisation by computer. Before the end of the decade W.F. Wood, M.A. Melton (1958), and others were beginning to tabulate topographic data on punched cards.

With emergence of the digital computer in the early- to mid-1950s, the progress of geomorphometry accelerated rapidly. The first input data were not DEMs but point elevations and topographic profiles. *Trend-surface analysis*, for example, numerically separates scattered map observations into two components, regional and local. The technique assumes that a spatial distribution can be modelled numerically as a continuous surface, usually by a polynomial expression, and that any observed spatial pattern is the sum of such a surface plus a local, *random*, term. Much used on subsurface data in petroleum exploration, by the 1960s it had attracted the attention of geomorphologists, notably to confirm planation surfaces or enhance local surface features (Krumbein, 1959; King, 1969). Trend-surface analysis commonly yields results as a square-grid array, but the polynomial fits to elevation data frequently oversimplified real-world variations in the topography.

The early numerical descriptions of topographic profiles were carried out by *spectral analysis*, a mathematical technique from signal processing and engineering that displayed the observations by spatial frequency (Bekker, 1969). First used to quantify the roughness of aircraft runways from surveyed micro-relief elevation profiles (Walls et al., 1954), elevation spectra were calculated from lunar surface measurements to support design of the Moon-landing program's Roving Vehicle (Jaeger and Schuring, 1966). To target lunar imaging missions, J.F. McCauley and colleagues at USGS had earlier (1963–64) computed slope gradient from topographic profiles generated through Earth-based photoclinometry ("*shape from shading*") of the Moon's surface (Bonner and Schmall, 1973). These data were also used to quantify the scale-dependency of slope gradient. Although single linear profiles capture apparent rather than true (maximum gradient) slopes[24] and do not deliver

[23] For example, the U.S. Navy funded Strahler and E.H. Hammond, and later T.K. Peucker and David Mark (in Canada). Ian Evans' early work was supported by the U.S. Army and that of Pike by the Army and the National Aeronautics and Space Administration; the library of small DEMs (Tobler, 1968) that inspired both of us was funded by the Army.

[24] *Mean* apparent slope is correctable to its true value by multiplying by 1.5708 (i.e. $\pi/2$).

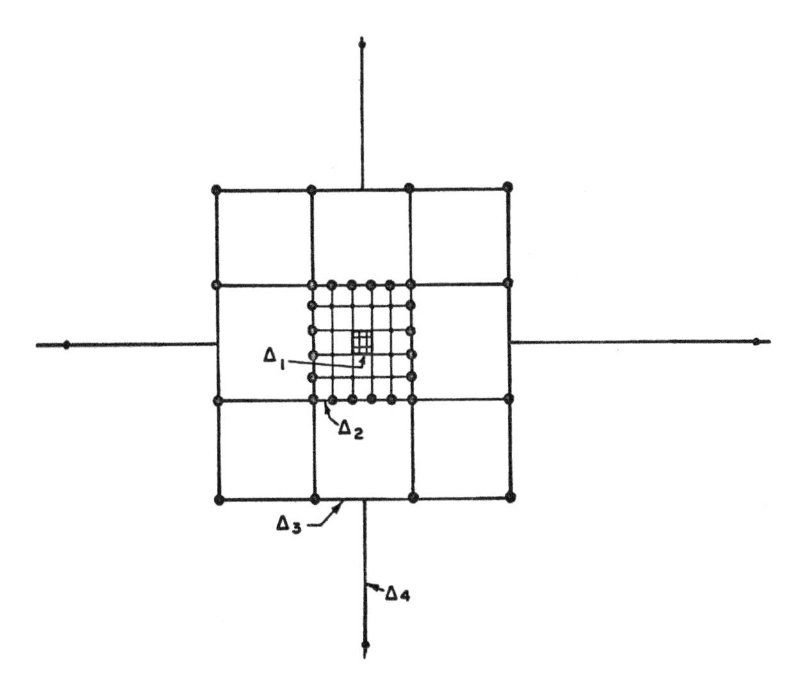

FIGURE 8 The earliest representation of a gridded x, y DEM designed to quantify variation in line-of-sight visibility with spatial scale. Grid spacings $\Delta_1 - \Delta_3$ of nested arrays (each 34×34 elevations) were 180, 800, and 9650 m. From unclassified 1959 American Association for the Advancement of Science symposium presentation by Arthur Stein.

the full 3-D character of a surface, spectral analysis continued to support morphometric objectives, such as delimiting morphologic regions of the seafloor (Fox and Hayes, 1985).

By the mid- to late-1950s, arrays of gridded elevations were being prepared by geophysicists for gravity correction, by civil engineers for highway location, and by the military in classified research on tactical combat doctrine. The DEM concept was first described openly by Miller and Laflamme (1958) at the Massachusetts Institute of Technology[25] but did not come into general use until the 1960s. Its potential and importance were clouded by limitations of the mainframe computers of the day. Although some DEMs were prepared from direct photogrammetry or field survey, most of them were laboriously interpolated by hand from existing contour maps[26] (e.g. Tobler, 1968). Semi-automated digitising of the entire United States at a grid resolution of about 63 m from 1:250,000-scale contour maps over 1963–1972 (Noma and Misulia, 1959; U.S. Army Map Service, 1963), later distributed by the USGS, marked a breakthrough in DEM availability. First and second surface derivatives (of gravity data) had aided in petroleum exploration; their calculation for

[25] Cloud (2002) writes: *"Much of the primary development work was done by staff at the MIT Photogrammetric Laboratory, under contract to the Army/Air Force nexus"*; see also Figure 8.

[26] By 1964, W.F. Wood at Cornell Aeronautical Laboratory was creating DEMs to model line-of-sight calculations.

the land surface by Tobler (1969) from manually-digitised DEMs marked another milestone, for it provided the basis for systematising general geomorphometry.

Evans (1972, 1980) criticised the pre-DEM fragmentation of the field (Neuenschwander, 1944), especially its many diverse and unrelated indices calculated or measured by hand from contour maps. Using a manually interpolated DEM and building upon the work of Tobler (1969), Evans (1972) showed that a point (or small x, y neighbourhood) could be characterised by elevation and its surface derivatives slope gradient and curvature, the latter in both plan and profile. Krcho (1973, 2001) independently provided a full mathematical basis for a system of surface derivatives in terms of random-field theory. These parameters could then be summarised for an area by standard statistical measures: mean, standard deviation, skewness, and kurtosis. Following the lead of W.F. Wood, in 1968 Pike and Wilson (1971) began to create USGS' first (manual) DEMs and computer software to calculate an extensive suite of parameters, including the hypsometric integral (Schaber et al., 1979) and values of (apparent) slope and curvature at multiple profile and grid resolutions.

About the same time, Carson and Kirkby (1972) demonstrated the relevance of elevation derivatives to geomorphological (mainly slope) processes, laying the basis for a more mathematical, modelling, approach to geomorphology that was intrinsically quantitative. Measures of surface position and catchment area already had been estimated manually by Speight (1968) to characterise *landform elements*. Pike (1988) subsequently proposed automating the multivariate approach to surface characterisation from DEMs and introduced the concept of the geometric signature of landform types.

Early maps and diagrams of geomorphometric results were limited to low-resolution displays by cathode-ray tube and then to 128 typed characters per line on computer printout-paper 38 cm in width — convenient for tables but clumsy for maps (Chrisman, 2006). With replacement of these crude output devices by pen-driven vector plotters and then high-resolution raster plotters, first in black and finally in colour, computer mapping came of age (Clarke, 1995). Among the most effective displays for topography is the shaded-relief (also *reflectance*) map, which shows the shape of the land surface by variations in brightness. Relief shading originated in the *chiaroscuro* of Renaissance artists. It was highly refined by Imhof (1982) and then automated by his Israeli student Pinhas Yoeli (1967). Comparable techniques[27] are now standard on virtually all GIS and geomorphometric packages. For comprehensive summaries of manual and automated relief shading see http://www.reliefshading.com and Horn (1981).

Computer programs suited to the statistical analysis of topographic data became increasingly available in the 1960s. Particularly useful to the geomorphometrist for sorting out descriptive parameters were techniques of multiple-correlation and factor and principal-components analysis (Lewis, 1968). With the rise of numerical taxonomy in the biological sciences (Sokal and Sneath, 1963) came the complementary multivariate technique *cluster analysis*, wherein observations were

[27] The first detailed large-format shaded-relief image published as a paper map (Thelin and Pike, 1991) portrayed the conterminous United States from a 12,000,000-point DEM (0.8-km resolution).

automatically aggregated into groups of maximum internal and minimum external homogeneity (Parks, 1966). Cluster analysis proved adept at automating the identification of topographic types and delimiting land-surface regions from samples of land-surface parameters (Mather, 1972).

> REMARK 7. *Development of the digital elevation model (DEM), first publicly described in 1958 by American photogrammetrists at MIT, has paralleled that of the electronic computer.*

Although geomorphometry was taking advantage of the computing revolution[28] in the 1970s and 1980s, limited computer power still held back more ambitious calculations. The constraints on morphometric analysis by 1980s computers are nicely illustrated by Burrough (1986) for a land evaluation project in Kisii, Kenya, where several land-surface parameters were derived from a DEM by the "*Map Analysis Package*" (MAP). Computing capabilities of this pioneering software, developed in FORTRAN by Dana Tomlin at Harvard, were restricted to 60×60 grid cells (see also Figure 9).

A major goal was accurate capture of surface-specific lines from DEMs, the most essential being stream networks. Early efforts at drainage tracing were rather crude: the widely implemented D8 approach routed flow only in eight directions (Figure 7 in Chapter 7), often creating bogus parallel flow lines oblique to the natural ground slope (Jenson, 1985; Jenson and Domingue, 1988). This problem equally reflects inferior DEMs and low-relief topography. Improved methods soon were devised (Fairfield and Leymarie, 1991) to split the flow into adjacent grid cells, yielding more realistic networks, whereupon the *DEM-to-watershed transformation* (Pike, 1995) rapidly grew into an active sub-field that still shows lively development.

By the end of the 1980s, it was possible to process DEMs over fairly large areas. The executable DOS package MicroDEM (Guth et al., 1987), for example, could extract over ten land-surface parameters and visualise DEMs together with remote sensing images. Martz and de Jong (1988), Hutchinson (1989) and Moore et al. (1991a) further advanced hydrological modelling and practical applications in morphometry. Since the early 1990s and the personal computer revolution, algorithms have been implemented in many raster-based GIS packages (see Chapter 10 for a review) and point-and-click geomorphometry on desktop and laptop machines is now the everyday reality.

3.4 The quantification of landforms

Recognition and delimitation of such discrete features as *drainage basins* (Horton, 1932, 1945), *cirques* (Evans, 2006), *drumlins* (Piotrowski, 1989), and *sand dunes* (Al-Harthi, 2002) on a continuous surface is more difficult than that of elementary forms and thus *Specific Geomorphometry* remains the more subjective practice

[28] Mark (1975a, 1975b), Grender (1976), and Young (1978) were among the pioneers who developed operational programs to calculate slope, aspect, and curvatures from gridded DEMs. See also Schaber et al. (1979), Horn (1981), and Pennock et al. (1987).

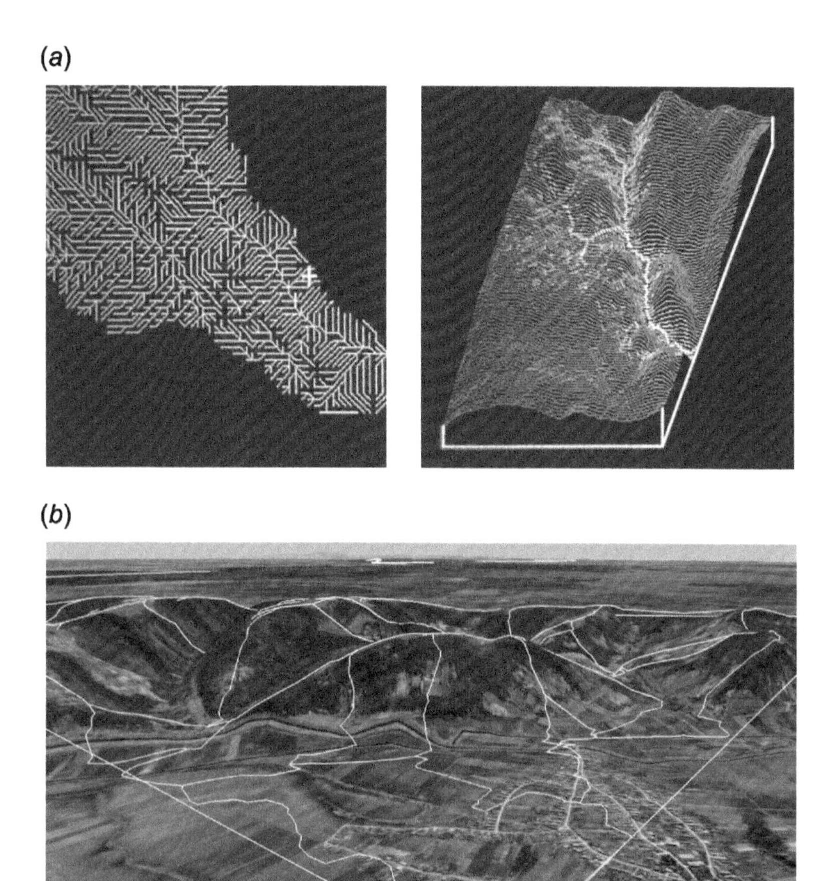

FIGURE 9 Geomorphometry then and now: (a) output from late-1980s DOS programme written to display land-surface properties: (left) map of local drainage direction, (right) cumulative upstream drainage elements draped over a DEM rendered in 3-D by parallel profiles. Courtesy of P.A. Burrough; (b) watershed boundaries for the Baranja Hill study area overlaid in Google Earth, an online geographical browser accessible to everyone. (See page 708 in Colour Plate Section at the back of the book.)

(Evans and Cox, 1974). While this book does not delve deeply into this area (Evans, 1972; Jarvis and Clifford, 1990), it warrants brief mention here.

Astronomy was the first science to quantify, so it is no surprise that the earliest scientific measurement of a landform involved not Earth but the craters on its Moon (Pike, 2001b). An impact crater is rather easy to distinguish from the surrounding land surface and its axial symmetry enables its shape to be captured completely by only a few simple parameters. Not all landforms are so favoured; alluvial fans, landslides, dolines, and other features all require good operational definitions to ensure their proper characterisation. The introduction of DEMs has not eased this requirement, and the added precision (not necessarily accuracy!)

comes at the cost of measurement complexity (Mouginis-Mark et al., 2004). While the automated definition of, say, *valleys* and *valley heads* from DEMs can be tested against their visual recognition (Tribe, 1991, 1992b), the low accuracy of many DEMs can spoil such an exercise (Mark, 1983). Regardless, more Earth scientists are now using DEMs as their primary source of data for landform measurement (e.g. Walcott and Summerfield, 2008).

4. GEOMORPHOMETRY TODAY

DEM-based geomorphometry continues to evolve from a number of the themes described above. Geostatistical analysis has established spatial autocorrelation, quantification of the 'First Law of Geography' — "*Everything is related to everything else, but near things are more related than distant things*" (Tobler, 1970) — as a routine technique (Bishop et al., 1998; Iwahashi and Pike, 2007). Fractional dimensionality (Mandelbrot, 1967) and self-similarity (Peckham and Gupta, 1999) still appear to be useful for representing drainage networks and other spatial phenomena, although their extension to land-surface relief z thus far has been modest (Klinkenberg, 1992; Outcalt et al., 1994). Multi-resolution modelling of the land surface is a vital topic of study (Sulebak and Hjelle, 2003), and recent analysis of fluvial networks on Mars continues to extend the utility of DEMs (Smith et al., 1999). Further examples of contemporary geomorphometry will be found in the following chapters of this book, especially by way of software development in Part II and their applications in Part III.

The maturing of GIS and remote-sensing technology has enabled geomorphometry to emerge as a technical field possessing a powerful analytical toolbox (Burrough and McDonnell, 1998). At the outset of the 21st century, geomorphometry is not only a specialised adaptation of surface quantification (mainly geometry and topology) to Earth's topography, but an independent field comparable to many other disciplines (Pike, 1995, 2000a).

With today's rapid growth in sources of mass-produced DEMs, such as the Shuttle Radar Topographic Mission (SRTM) and laser ranging (LiDAR) surveys (see also Chapter 3), land-surface parameters are finding ever-increasing use in a number of areas. These range from precision agriculture, soil–landscape modelling, and climatic and hydrological applications to urban planning, general education, and exploration of the ocean floor and planetary surfaces. Earth's topography has been sufficiently well sampled and scanned that global DEM coverage now is available at resolutions of 100 m or better. Good DEM coverage is available beyond Earth. In fact, among Solar System planets, Mars has the most accurate and consistent DEM, with vertical accuracy up to ± 1 m (Smith et al., 1997; Pike, 2002).

Geomorphometry has become essential to the modelling and mapping of natural landscapes, at both regional and local scales (see further Chapter 19). Applications in the restricted sense of parameter and object extraction are distinguished from the use of DEMs for landscape visualisation or change detection. All varieties of spatial modelling are available, stochastic (e.g. spatial prediction) as well as process-based (e.g. erosion modelling). Because land-surface parameters and objects are now relatively inexpensive to compute over broad areas of interest, they can be used — with due caution — to replace *some* of the boots-on-the-ground field sampling that is so expensive and time-consuming.

REMARK 8. *Geomorphometry supports Earth and environmental science (including oceanography and planetary exploration), civil engineering, military operations, and video entertainment.*

The many uses of geomorphometry today can be grouped into perhaps five broad categories:

Environmental and Earth science applications Land-surface parameters and objects have been used successfully to predict the distribution of soil properties (Bishop and Minasny, 2005), model depositional/erosional processes (Mitášová et al., 1995), improve vegetation mapping (Bolstad and Lillesand, 1992; Antonić et al., 2003), assess the likelihood of slope hazards (Guzzetti et al., 2005), analyse wildfire propagation (Hernández Encinas et al., 2007), and support the management of watersheds (Moore et al., 1991a). Geomorphometric analyses further aid in deriving soil–landscape elements and in providing a more objective basis for delimiting ecological regions. Recent developments include automated methods to detect landform facets by unsupervised fuzzy-set classification (Burrough et al., 2000; Schmidt and Hewitt, 2004). Land-surface parameters even play a role in automatically detecting geological structures and planning mineral exploration (Chorowicz et al., 1995; Jordan et al., 2005).

Civil engineering and military applications Both fields were early users of DEMs (Miller and Laflamme, 1958). Today, engineers frequently employ DEM calculations to plan highways, airports, bridges, and other infrastructure, as well as to situate wind-energy turbines, select optimal sites for canals and dams, and locate microwave relay towers to maximise cell-phone coverage (Petrie and Kennie, 1987). Li et al. (2005, §14) review recent applications. Land-surface quantification is crucial to any number of military activities (Griffin, 1990); DEMs are used to simulate combat scenarios, actively guide ground forces as well as terrain-following missiles, and to automate line-of-sight and mask-angle calculations for concealment and observation (Guth, 2004; http://terrainsummit.com). Viewshed algorithms operating on DEMs have been found superior to simplistic sightline analysis for siting air-defence missile batteries (Franklin and Ray, 1994). As in the past (see above), much defence-related geomorphometry is classified and thus unavailable to the wider scientific community.

Applications in oceanography Measurement of seafloor topography is the province of *bathymetry*. DEMs — or rather DDMs (*digital depth models*[29]) — of the seafloor figure prominently in coastal geomorphology, geophysical analysis of global tectonics, the study of ocean currents, design of measures to protect shorelines from erosion, mineral exploration, and fisheries management (Burrows et al., 2003;

[29] See http://dusk2.geo.orst.edu/djl/samoa/ for an example of an archive of GIS data from multibeam bathymetry and submersible dives supporting a marine sanctuary in Samoa.

Giannoulaki et al., 2006). Surface parameters and objects computed for the seabed from DDMs have been used to optimise fish farming and to improve the mapping of marine benthic habitats (Bakran-Petricioli et al., 2006; Lundblad et al., 2006). Finally, seafloor morphometry plays a critical role in the navigation and concealment of nuclear submarines.

Applications in planetary science and space exploration A scientific understanding of Earth's Moon and the solid planets increasingly depends upon DEMs. LiDAR data from the 1994 Clementine[30] mission to the Moon produced two broad-scale global DEMs (Smith et al., 1997); their modest spatial resolutions of 1 and 5 km revealed previously unknown giant impact scars (Williams and Zuber, 1998; Cook et al., 2000). Grid resolution of the global DEM resulting from the spectacularly successful 1998–2001 Mars Orbital Laser Altimeter (MOLA[31]) mission exceeds that of Earth[32] (Smith et al., 1999)! Geomorphometry is well suited to take advantage of these results, as demonstrated by Dorninger et al. (2004) and by Bue and Stepinski (2006), who used the MOLA DEM to test algorithms for the automated recognition of landforms.

Applications in the entertainment business Mass-produced DEMs are essential to video game and motion picture animation, where geomorphometry is referred to as *terrain rendering*[33] (Blow, 2000). Usually structured in TIN arrays, these DEM applications range from creating background scenery to simulating landscape evolution and modelling sunlight intensity (often using Autodesk's 3ds Max package). Pseudo-realistic rendering is sufficient to create a visually convincing product, so exact reproduction of real-world landscapes is rarely necessary. Because the industry is highly competitive, design teams do not always publish their methods, making it difficult to follow the latest innovations.

Not all applications of geomorphometry are well developed or supported. Terrain rendering for computer games, for example, commands more financial resources than all environmental land-surface modelling combined (Pike, 2002)! Other generously-funded areas in the past have included military operations and space exploration. Any soil- or vegetation-mapping team would be grateful for the access to technology and data available to game developers or military surveillance agencies.

5. THE "BARANJA HILL" CASE STUDY

To enhance understanding of the algorithms demonstrated in Part II of this book, we will use a small case study consistently[34] throughout. In this way, you will be able to compare land-surface parameters and objects derived from different

[30] http://pds-geosciences.wustl.edu/missions/clementine/.

[31] http://wwwpds.wustl.edu/missions/mgs/megdr.html.

[32] The current global Mars DEM is at resolution of 1/128 of a degree, which at the equator is about 460 m. Locally, resolution is much better than that.

[33] See also the http://vterrain.org project.

[34] We were inspired mainly by statistics books that demonstrated several processing techniques on the same dataset, such as Isaaks and Srivastava (1989).

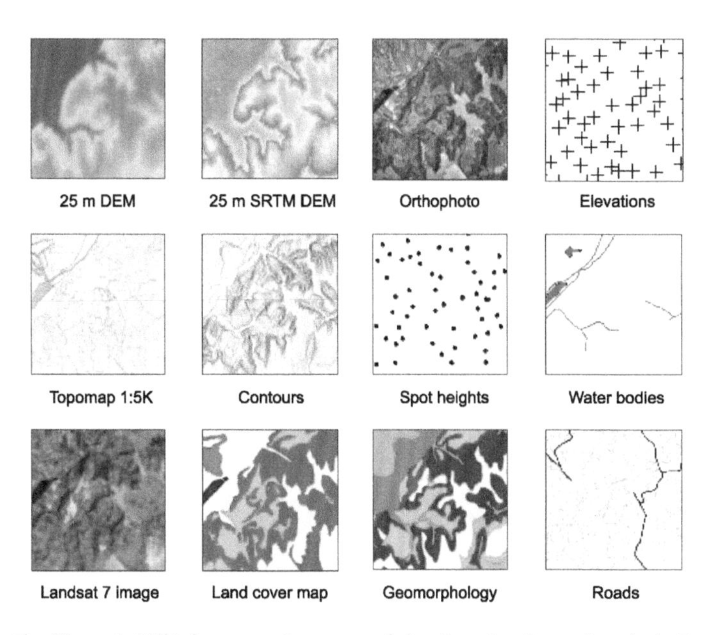

FIGURE 10 The "Baranja Hill" datasets. Courtesy of the Croatian State Geodetic Department (http://www.dgu.hr). (See page 709 in Colour Plate Section at the back of the book.)

algorithms and software packages and thus more easily find the software best suited to your needs.

The *"Baranja Hill"* study area, located in eastern Croatia, has been mapped extensively over the years and several GIS layers are available at various scales (Figure 10). The study area is centered on 45°47′40″N, 18°41′27″E and corresponds approximately to the size of a single 1:20,000 aerial photo. Its main geomorphic features include hill summits and shoulders, eroded slopes of small valleys, valley bottoms, a large abandoned river channel, and river terraces (Figure 11).

The Croatian State Geodetic Department provided 50k- and 5k-scale topographic maps and aerial photos (from August 1997). An orthorectified photo-map (5-m resolution) was prepared from these source materials by the method explained in detail by Rossiter and Hengl (2002). From the orthophoto, a land cover polygon map was digitised using the following classes: agricultural fields, fish ponds, natural forest, pasture and grassland, and urban areas. Nine landform elements were recognised: summit, hill shoulder, escarpment, colluvium, hillslope, valley bottom, glacis (sloping), high terrace (tread) and low terrace (tread).

Contours, water bodies, and roads were digitised from the 1:50,000 and 1:5000 topographic maps. Contour intervals on the 1:50,000 topographic map are 20 m in hill land and 5 m on plains, and on the 1:5000 map they are 5 and 1 m respectively. From the 1:5000 contours and land-survey point measurements, a 5 m DEM was derived by the ANUDEM (TOPOGRID) procedure in ArcInfo (Hutchinson, 1989), and then resampled to a 25 m grid. For comparison, the 30 m SRTM DEM (15′×15′ block) obtained from the German Aerospace Agency (http://eoweb.dlr.de) was resampled to 25 m (Figure 6 in Chapter 3). The total area of the case study is

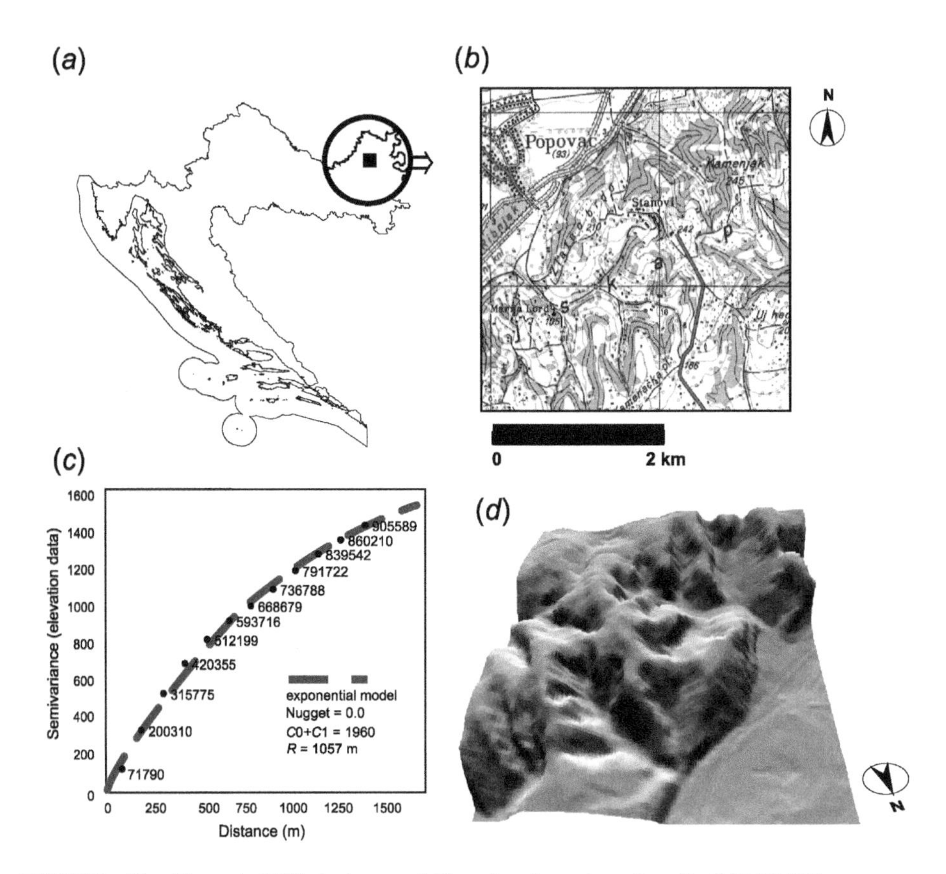

FIGURE 11 The "*Baranja Hill*" study area: (a) location in eastern Croatia; (b) 1:50,000 topographic map (reduced) showing main features; (c) omnidirectional variogram from the elevation point data; and (d) perspective view of the area. Courtesy of State Geodetic Administration of Republic of Croatia.

13.69 km^2 or 3.6×3.7 km. Elevation of the area ranges from 80 to 240 m with an average of 157.6 m and a standard deviation of 44.3 m. Both 25-m DEMs have been brought to the same grid definition with the following parameters: `ncols` = 147, `nrows` = 149, `xllcorner` = 6,551,884, `yllcorner` = 5,070,562, `cell-size` = 25 m. We used the local geodetic grid (Croatian coordinate system, zone 6) in the Transverse Mercator projection on a Bessel 1841 ellipsoid ($a = 6{,}377{,}397.155$, $f^{-1} = 299.1528128$). The false easting is 6,500,000, central meridian is at 18° east, and the scale factor is 0.9999. Note also that, to have proper geographic coordinates, you will need to specify a user-defined datum of $\Delta X = 682$ m, $\Delta Y = -199$ m and $\Delta Z = 480$ m (Molodensky transformation). The projection files in various formats are available on this book's website. The complete "Baranja Hill" dataset[35] consists of (Figure 10):

DEM25m 25-m DEM derived from contour lines on the 1:5000 contour map;

[35] You can access the complete "Baranja Hill" dataset via the geomorphometry.org website.

`DEM25srtm` 25-m DEM from the Shuttle Radar Topographic Mission;

`DEM5m` 5-m DEM derived from stereoscopic images;

`contours5K` Map of contours digitised from the 1:5000 topo-map;

`elevations` Point map ($n = 853$); very precise measurements of elevation from the land survey;

`wbodies` Layer showing water bodies and streams;

`orthophoto` Aerial (orthorectified) photo of the study area (pixsize = 5 m);

`satimage` Landsat 7 satellite image with 7 bands from September 1999;

`landcover` Land-cover map digitised from the orthophoto;

`landform` Polygon map of the principal landform elements (facets);

`fieldata` Field observations at 59 locations are available in report form.

6. SUMMARY POINTS

Geomorphometry is the science of quantitative land-surface analysis. A mix of Earth and computer science, engineering, and mathematics, it is a new field paralleling analytical cartography and GIS. It evolved directly from geomorphology and quantitative terrain analysis, two disciplines that originated in 19th-century geometry, physical geography, and the measurement of mountains.

Classical morphometry (orometry) was directed toward hypsometry and plan form, and calculating average elevation and slope, volume, relative relief, and drainage density from contour maps. Later work emphasised drainage topology, slope-frequency distribution, and land-surface classification. Techniques have ranged from trend-surface and spectral analysis of surveyed elevations and profiles to geostatistical and fractal analysis of 3-D elevation arrays.

Modern geomorphometry addresses the refinement and processing of elevation data, description and visualisation of topography, and a wide variety of numerical analyses. It focuses on the continuous *land surface*, although it also includes the analysis of *landforms*, discrete features such as watersheds. The operational goal of geomorphometry is extraction of measures (*land-surface parameters*) and spatial features (*land-surface objects*) from digital topography.

Input to geomorphometric analysis is commonly a *digital elevation model* (DEM), a rectangular array of surface heights. First described in 1958, DEMs developed along with the electronic computer. Many DEMs are prepared from existing contour maps; because all DEMs have flaws and even advanced technologies such as LiDAR introduce errors, DEMs must be corrected before use. The growth in sources of mass-produced DEMs has increased the spread of geomorphometric methods.

Geomorphometry supports countless applications in the Earth sciences, civil engineering, military operations, and entertainment: precision agriculture, soil–landscape relations, solar radiation on hillslopes, mapping landslide likelihood, stream flow in ungauged watersheds, battlefield scenarios, sustainable land use, landscape visualisation, video-game scenery, seafloor terrain types, and surface processes on Mars.

Geomorphometric analysis commonly entails five steps: sampling a surface, generating and correcting a surface model, calculating land-surface parameters or objects, and applying the results. The three classes of parameters and objects (basic, hydrologic, and climatic/meteorological) include both landforms and point-measures such as slope and curvature. Landform *elements* are fundamental spatial units having uniform properties. Complex analyses may combine several parameter maps and incorporate non-topographic data.

The procedure that extracts most land-surface parameters and objects from a DEM is the *neighbourhood operation*: the same calculation is applied to a small sampling window of gridded elevations around each DEM point, to create a complete thematic map. Processing is simplified by the *raster* (grid-cell) structure of the DEM, which matches the file structure of the computer. Because parameters can be generated by different algorithms or sampling strategies, and vary with spatial scale, no DEM-derived map is *definitive*. To encourage readers to compare maps created by the different software packages demonstrated in this book, several digital datasets for a small test area (Baranja Hill) are available via the book's website.

IMPORTANT SOURCES

Burrough, P.A., McDonnell, R.A., 1998. Principles of Geographical Information Systems. Oxford University Press Inc., New York, 333 pp.

Mark, D.M., Smith, B., 2004. A science of topography: from qualitative ontology to digital representations. In: Bishop, M.P., Shroder, J.F. (Eds.), Geographic Information Science and Mountain Geomorphology. Springer–Praxis, Chichester, England, pp. 75–97.

Pike, R.J., 2002. A bibliography of terrain modeling (geomorphometry), the quantitative representation of topography — supplement 4.0. Open-File Report 02-465. U.S. Geological Survey, Denver, 116 pp. http://geopubs.wr.usgs.gov.

Pike, R.J., 2000. Geomorphometry — diversity in quantitative surface analysis. Progress in Physical Geography 24 (1), 1–20.

Zhou, Q., Lees, B., Tang, G. (Eds.), 2008. Advances in Digital Terrain Analysis. Lecture Notes in Geoinformation and Cartography Series. Springer, 462 pp.

http://geomorphometry.org — the geomorphometry research group.

CHAPTER 2

Mathematical and Digital Models of the Land Surface

T. Hengl and **I.S. Evans**

conceptual models of the land surface · land surface from a geodetic perspective · land-surface properties and mathematical models · vector and grid models of the land surface · cell size and its meaning · how to determine a suitable grid resolution for DEMs · how to sample and interpolate heights · land surface and geomorphometric algorithms

1. CONCEPTUAL MODELS OF THE LAND SURFACE

1.1 Orography, topography, land surface

The objects of study of *orography*[1] are the undulations on the surface of the Earth. Although orography literally means the study of the Earth's relief, this term is only used by geographers and is often connected with mountainous areas. *Topography*[2] is also commonly related to the morphometric characteristics of land in terms of elevation, slope, and orientation. However, in land survey, topographic or topo-maps also contain information on land cover, infrastructure, etc., so that the term *topography*, strictly speaking, refers to all that is shown on topographic maps (Peuquet, 1984). In this book, we will also refer to topography as the description of the shape of the land surface (commonly presented as contours and hill-shading on topo-maps). In principle, the key interest of geomorphometry is in the *land surface* and its shape, and not in the elevation measurements nor topographic features, as such.

For a non-specialist, the most important aspect of the land surface that needs to be clarified first is its scale-dependency. Traditionally, geomorphologists have focused on the surface, smoothed at a scale of a few metres (*human scale*). In theory, algorithms and concepts of geomorphometry are applicable to all scales,

[1] From the Greek words ορος (mountain) and γραφειν (to draw).

[2] Topography is the study of Earth's surface features, including not only relief, but also vegetative and human-made features, and even local history and culture. From the Greek words τοπος (place) and γραφειν (to draw).

Developments in Soil Science, Volume 33
ISSN 0166-2481, DOI: 10.1016/S0166-2481(08)00002-0.

including microscopic scales, where the size of a study area is in millimetres. The latter, important in analysing frictional wear, is the province of surface engineering (Pike, 2000b). In earth sciences, the lower limit of the (real) land surface scale is one or two metres, and it relates to continuous bodies or aggregates of material rather than individual particles. Geomorphologists are, of course, concerned with smaller features, but they usually find other means of analysis appropriate for what they call *micro-relief*. Some examples of micro-relief are *overhangs* in weathering pits, *tafoni* and *gilgai* terrains, and *patterned ground* due to the action of frost or salt.

In surveys of slope profiles, Young (1972, p. 146) recommended that "*no measured length shall be more than 20 m or less than 2 m... for topography of normal scale*". Gerrard and Robinson (1971) analysed gradients for fixed measured lengths from 1.5 to 10 m, and discussed the effects of small protrusions and depressions (micro-relief) that can give variations of a few degrees for measuring lengths of a few metres. Pitty (1969) advocated fixed lengths and used a frame giving a constant slope length of 5 feet (1.52 m) for gradient measurement. The *size of the yardstick* determines many other properties of a DEM, including its spatial accuracy, vertical precision and applicability. Debate concerning fixed versus variable measuring lengths along profiles continued (Evans and Cox, 1999) and had some parallels with the debate about fixed grids, adaptive grids and irregular triangulations as bases for DEMs. Geomorphometrists traditionally exclude individual particles (stones), and repeated *micro-relief* such as earth hummocks from DEMs. This might change in the coming years as the more finely-detailed DEMs become widespread.

"*Every thing has a surface*" (Rana, 2004), but is the *land surface* a clear concept? There are clear difficulties in defining the land surface precisely. This was less of a problem when most DEMs were coarse, with measurement errors in metres. With more accurate DEMs, e.g. from LiDAR, algorithms are necessary to filter out vegetation and some human-made features, and what exactly constitutes the land surface then becomes more problematic. The top of a building is not the land surface, but neither is the floor of a cellar. The bottom of a river channel is part of the land surface, but, because it is difficult to survey, it is often omitted. The way we conceptualise the surface is becoming more and more important. Schneider (1998, 2001) shows how any surface model is an abstraction, and that uncertainty of shape (local form) is unavoidable as we interpolate between data points, or fit a smoother surface to compensate for data error.

In this chapter, we will introduce the land surface concept from both the geodetic and statistical perspectives, and review ways to represent it. We will also discuss ways of producing models of the land surface, from sampling procedures to DEM gridding techniques. At the end of the chapter, you can find an extensive comparison of the methods used to derive first and second order derivatives from DEMs.

1.2 The land surface from a geodetic perspective

As mentioned previously, in Section 2.1 of Chapter 1, the basis of geomorphometry is the quantitative analysis of the shape of a land surface, starting from its

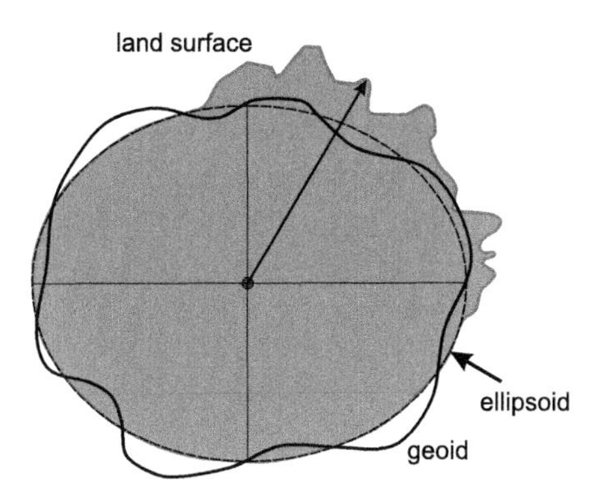

FIGURE 1 The difference between height above sea level (geoid) and height above the surface of the ellipsoid. Over the entire surface of the globe, this difference varies from −100 to 70 m. After De By (2001, p. 106).

heights (elevations, altitudes). By *height* we mean vertical distance from the *reference level-surface* with a height of 0. The heights on topo-maps typically relate to a local reference surface also known as a *vertical geodetic datum* (the mean sea level at a reference location). These reference surfaces differ from country to country.

Cartographers and geodesists attempt to estimate the reference level-surface of the globe by fitting to the Earth's ocean surface. This results in a complex, but smooth, 3D model called the *geoid* (De By, 2001). Any variation from the ellipsoid model is due to tidal forces and gravity differences between locations. From the recent satellite gravity missions, we will soon have a global definition of the geoid at a level of accuracy measured in centimetres. Meanwhile, the only truly global reference level-surface is the surface of the World Geodetic System[3] (WGS84) ellipsoid.

Note that there is a difference between the *height above sea level* (geoid) and the *height above the ellipsoid*[4] (Figure 1). For example, with your GPS, you could read a WGS84 height of 0 metres, but still be tens of metres above sea level. The difference between the height above the geoid and the height above WGS84 over the globe is in the range of −100 to 70 m.

Another important cartographic concept in land-surface analysis is the definition of horizontal space, i.e. *projection system*. In a standard GIS, a DEM should always be in Euclidean space, i.e. presented in a Cartesian coordinate system (in which the x, y axes are orthogonal to each other). In the case of gridded DEMs, this means that the size of grids is absolutely equal for each part of the study area. Because the formulas for extraction of land-surface parameters are derived using Euclidean mathematics, the input DEM should also be in such a system, which means that the DEM needs first to be projected to some coordinate system.

In practice, derivation of DEM parameters is possible also in *geographical coordinates*. In fact, many geomorphometrists suggest (see further Section 1.1 in

[3] http://earth-info.nga.mil/GandG/wgs84/.

[4] An ellipsoid is a mathematical model of the Earth used in cartography to project points from geographic coordinates to a Cartesian system. Ellipsoids are mathematically simple models of the Earth, while the geoid cannot be defined with just a few parameters.

Chapter 15) that DEM parameters and objects should always be derived from the native, unprojected DEMs because resampling of grids to some projection system can lead to systematic differences. For example, original GTOPO DEMs are distributed in geographical coordinates, with a fixed resolution of 30 arcsec. Before extracting geomorphometric parameters, z-values in DEMs with geographic coordinates (grid spacing in degrees) need to be scaled to the same degree-system. Bolstad (2006) suggests that the elevation values can be scaled to degrees using a simple formula:

```
degree_dem = [metric_dem] * 0.0000090
```

where `degree_dem` is a new grid with elevations in decimal degrees and `[metric_dem]` is the same map with original heights in metres. The grid spacing in geographic system is in fact inconstant (different grid spacing for different latitudes). In the case of the GTOPO DEM, the ground distance at equator in East/West direction is 928 m, at 60° is 465 m, and at 82° is 130 m. The ground distance in North/South direction is more or less constant: 921 m at equator, 929 m at 60° and 931 m at 82°. For datasets in geographical coordinates, a cell size adjustment can be estimated for each grid cell and then factored into calculation (Guth, 1995, p. 32). For example, the horizontal grid spacing (Δx) can be roughly estimated[5] as a function of the latitude and spacing at the equator:

$$\Delta x_{\text{metric}} = F \cdot \cos(\varphi) \cdot \Delta x^{0}_{\text{degree}} \tag{1.1}$$

where Δx_{metric} is the East/West grid spacing estimated for a given latitude (φ), $\Delta x^{0}_{\text{degree}}$ is the grid spacing in degrees at equator, and F is the empirical constant used to convert from degrees to metres. For example, for 1 arcsec DEM (0.000278°), to convert from degrees to metres, one needs to use $F = 111{,}319$ m. In the case of fine grid resolutions (<100 m) and for local neighbourhood analysis, it really does not matter much if we are using DEMs in metric (projected) or degree (geographic) systems.

An important issue concerning land surface conceptualisation is the water-surface problem. Ideally, in geomorphometric applications, we would like to use *complete* DEMs, i.e. maps that show the land surface both above and below sea level. In the past, because it was expensive to map the seafloor, very few samples were available, so water bodies were simply masked out. Today, with techniques such as LiDAR, radar, and sonar, providing digital bathymetry and satellite altimetry, the seafloor and sea surface is being mapped accurately and this information has already been used for various studies (Smith and Sandwell, 1997; Maune et al., 2001). Nevertheless, most of the DEMs in current use have constant values assigned to sea and lake surfaces, and no heights for the beds of rivers and lakes. A complete global DEM of the world is the 1-minute General Bathymetric Chart of the Oceans (www.gebco.net). This DEM is available at a resolution of 1-minute, which corresponds to about 1.8 km at 45° latitude.

A misconception about the land surface is that it needs to be surveyed only once. Although often considered as being a static entity, the land surface is not

[5] For more exact conversion functions see Vincenty (1975).

completely so. In fact, the land surface of the Earth is constantly changing — both gradually and due to catastrophic events such as volcanic eruptions, floods and landslides. The complexity and variety of the surface of planet Earth is due to the variety of materials of which it is composed, to the range of processes that have fashioned them, and to how these processes have changed over time:

- *Tectonic and volcanic processes* — including the movement of tectonic plates. These cause the folding and faulting of geological materials.
- *Erosional processes* — including erosion by catchment water, sea water, wind and gravitational stresses.
- *Processes controlled by living organisms* — including the building of coral reefs and termite mounds, biological weathering and many microbiological soil processes, and the direct or indirect effects of human actions.
- *Extra-terrestrial processes* — including the impacts of meteorites.

For example, due to the movement of tectonic plates, the distance between North America and Europe is constantly growing (by about 2 metres in 70 years). Likewise, each continent is slowly shifting, some mountain ranges, e.g. the Himalayas are rising due to tectonic activity, others are being eroded away. However, tectonic change of the land surface is fairly slow, so one detailed topographic survey, e.g. at a scale of 1:25,000, is good enough for a period of at least 50 years. The main exceptions are volcanic and landslide-prone areas, glaciers, dune systems, river floodplains and coasts composed of weak materials. This is in addition, of course, to areas of quarrying, dumping or construction. Areas of rapid change are sometimes mapped annually or even more frequently.

The term *land surface* also implies certain *topological restrictions* or simplifications. Although nowadays we have the technology to produce 3D shapes on the Earth's surface (see further Section 2.2 in Chapter 3), in applied geomorphometry DEMs usually do not show points with multiple heights. This means that we never use DEMs to represent features such as caves, overhangs or boulders (Shary, 1995). Boulders can be several metres long, for example in rockfall or glacial deposits. If they sit on the surface, they can be excluded from the concept of land surface in the same way as trees or buildings. However, if the boulders are close together or partly buried (and they can be in all degrees of burial), the tendency is to pass a surface through the boulders, either averaged, if the boulders are close together, or interpolated from the surrounding surface, if this is relatively smooth.

REMARK 1. *In geomorphometry, the land surface is commonly modelled as a single-sided surface. This means that no point in the projected space (x, y) can have multiple heights.*

1.3 Land-surface elements and properties

In geomorphometry, the land surface is often presented using universal elements or features (in GIS terms: polygons, lines and points), which can be recognised, regardless of scale or type of terrain. Imagine a land-surface model in a one-dimensional space [Figure 2(a)] — this surface will have several local minima

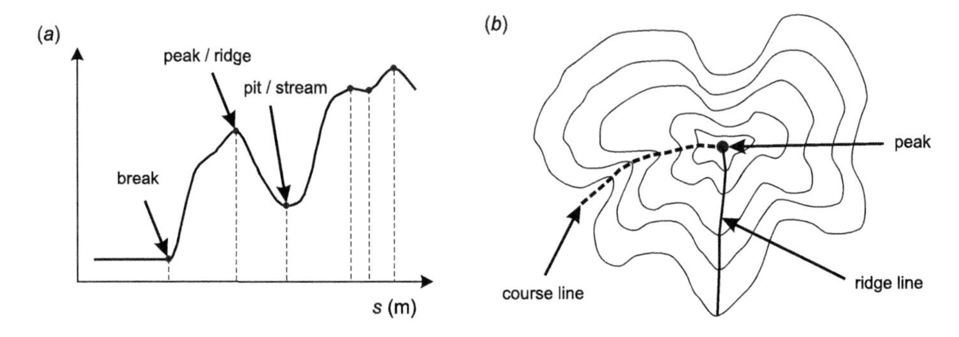

FIGURE 2 Common elements of a land surface: (a) surface-specific points and (b) surface-specific lines. Reprinted from Li et al. (2005, p. 23). With permission from Taylor & Francis Group.

and maxima and other similar *surface-specific elements*, also known as *morphologically important points* (Rana, 2004). The most important surface-specific elements are (Cayley, 1859; Maxwell, 1870; Li et al., 2005):

- *pits* — local minima (bottoms of depressions, holes);
- *peaks* — local maxima (tops of hills and mountains, summits);
- *ridge lines* — lines connecting points that are local maxima in transverse section;
- *course lines* — lines connecting points that are local minima in transverse section (river valleys, flowlines, ravines);
- *passes* — crossing points of ridge and course lines;
- *break lines* — where the slope change is sudden.

More general surface-specific lines and areas are (Evans and Cox, 1999; Hutchinson and Gallant, 2000):

- *contours* — lines of equal height above the geoid;
- *slope lines* — lines of downslope gravitational flow, at right angles to the contours;
- *plains* — areas of low relief where all altitude values are equal; but note that their frequency increases with the degree of rounding of height data (e.g. nearest metre).

The surface-specific elements are sufficient to approximate the land surface over an area, because all other intervening points[6] can be estimated by interpolation. This means that if we map the position of surface-specific elements, which is what land surveyors often do in the field, we can reconstruct the land surface to a first approximation (for further information see Section 3.1). Obviously, the greater the surface roughness, the more surface-specific elements there will be in an area. In reality, it is almost impossible to record all surface-specific elements because their density depends on how close we look. We can only try to determine the majority of elements on a given scale.

[6] These points are referred to as *random points* and can be approximated using a smooth fitting function, such as splines. See further Section 3.2.

REMARK 2. *A land surface consists of a finite number of surface-specific elements: pits, peaks, ridge lines, course lines, passes and break lines.*

Note also that, if the land surface is known, each of these surface-specific elements can be detected automatically[7] by analysing the change of elevation (sign, direction and the magnitude of slope change) in a local neighbourhood.

Another interesting property of topography is its possible self-similarity, i.e. *fractal structure*. Mandelbrot (1967) was among the first to recognise the fractal property of topography and suggest measures that can be used to describe it. The most important measure is the *fractal dimension*. The dimension of a smooth (Euclidean) line is 1, and that of a plane area is 2. A non-smooth line may look more or less like a band, and occupy a larger proportion of an area as it becomes more convoluted. Mandelbrot (1967) suggested expressing the complexity (*'noisiness'*) of a line by using a single dimension between 1 and 2 for a line (or between 2 and 3 for a rough surface).

Self-similarity implies that a line or surface is equally rough at all scales (horizontal wavelengths) and is statistically similar, however much it is enlarged or reduced. Evans and McClean (1995) evaluated the use of fractal measures to describe various DEMs, and concluded that simple fractal models can only be used to approximate a land surface over a limited range of scales. They can also be used to simulate a rough surface, as a starting point for modelling. *Multi-fractal*[8] models, on the other hand, are more complex, but may be more suited for analysing topography.

Unlike many other terrain variables, such as soil, vegetation or hydrographic features, the land surface has a special property in that, generally, it exhibits smoothness. It is almost always positively autocorrelated, and although its fractal dimension is above 2.0, it is well below 3.0. This is mainly because gravitational forces and erosional agents normally tend to level down differences in relief. This assumption of a degree of smoothness is important for the pre-processing of DEMs and the derivation of parameters and objects. Typically, geomorphometrists permit the smoothing of measured heights, the use of smooth interpolators, smooth functions for the land surface, and variograms with zero nugget and long range, because they assume that the smoothed model is always closer to reality. Evans and McClean (1995) compared variograms of heights for several study areas and concluded that most are non-linear, continuously curved even when both distance and variance are plotted on logarithmic[9] scales.

1.4 Mathematical models of the land surface

In the most simple terms, a land surface can be described as $z = f(x, y)$ or $z = f(s)$, which means that elevation z depends solely on planar coordinates x, y (Cayley,

[7] DEMs are frequently used to extract automatically contour lines, medial axes, stream lines, watershed boundaries and similar objects.

[8] Multi-fractals are non-uniform, interwined fractals, with continuously varying scaling exponents. They permit local fluctuation of the fractal dimension.

[9] Typically, variograms of heights show no upper bound, so it is more practical to display the distance ordinate using a logarithmic scale. Log–log variograms are standard in fractal studies (Evans and McClean, 1995).

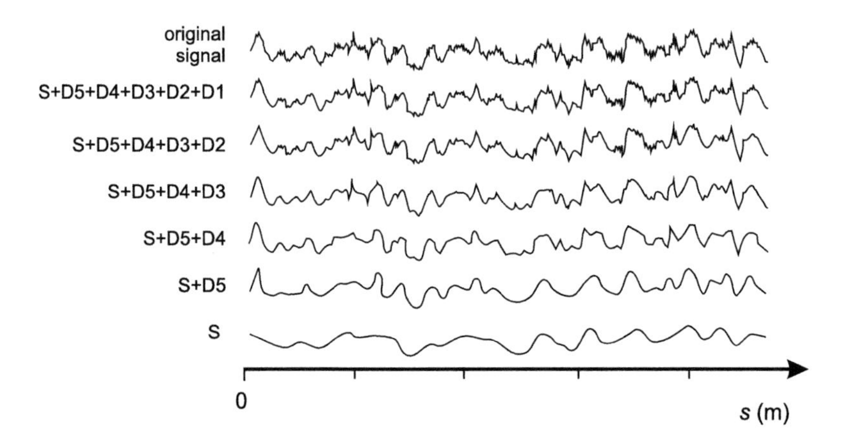

FIGURE 3 The filtering of an original signal to give six levels of generalisation. These can be compared to DEMs prepared for various scales. Reprinted from McBratney (1998). With permission from Springer Science and Business Media.

1859). From an information theory perspective, topography can viewed as a 2D, non-stationary[10] signal consisting of *multi-resolution components*. This can be expressed mathematically as (McBratney, 1998):

$$z^*(s) \cong S(s) + D_J(s) + D_{J-1}(s) + \cdots + D_1(s) \tag{1.2}$$

where $S(s)$ is the smooth(est) component of the signal and $D_J(s), \ldots, D_{J-k}(x)$ are progressively more detailed components. $S(s)$ is the topography at the coarsest possible scale. To this deterministic signal, increasingly detailed components can be added to represent variation at more and more detailed scales. $D_1(s)$ is the most detailed component corresponding to the finest scale resolvable. This procedure can also be run in the opposite direction: a DEM at a very fine/detailed scale can be partitioned into smoothed components by sequentially removing the finer-scale components (Figure 3). This is the basic principle of *Fourier analysis*, which has several practical applications in geomorphometry (Jordan and Schott, 2005). First, a Fourier analysis of a finer-scale signal can be used to filter out high-frequency noise and identify the optimal number of components needed to code that signal. After we have separated signal from noise, we can determine which landscape processes are influenced by which components (see also further Section 1.2 in Chapter 14).

From a statistical perspective, the land surface can be viewed as a combination of deterministic and random components (Tomer and Anderson, 1995; Oksanen, 2006a):

$$z(s) = z^*(s) + \varepsilon'(s) + \varepsilon''(s) \tag{1.3}$$

where $z^*(s)$ is the *deterministic component*, $\varepsilon'(s)$ is the random but *spatially correlated error* and $\varepsilon''(s)$ is the *pure noise* (measurement error). This model can be made

[10] Statistical term meaning: not having constant inherent properties such as local variance, variogram model, etc., throughout the area of interest.

more sophisticated for elevation measurements from various sources. For example, a land-surface model for SRTM DEMs can be formulated as (Gallant and Hutchinson, 2006):

$$z_{\mathrm{SRTM}}(s) = z(s) + m(s) \cdot g(s) + h(s) + \varepsilon_{\mathrm{SRTM}}(s) \tag{1.4}$$

where $z(s)$ is the actual land surface, $g(s)$ is the height of the canopy (trees, buildings, etc.), $m(s)$ is the mask value multiplier (0 or 1), $h(s)$ is the systematic offset due to the difference in geodetic datums used and $\varepsilon_{\mathrm{SRTM}}$ is the random observation error (assumed to be normally distributed). Such models can be useful for planning the filtering of DEMs (see Chapter 4) or for modelling the influence of propagated uncertainty on DEM derivatives (see also Chapter 5).

In practice, it is not easy to distinguish the random components (ε' and ε'') of a DEM from the deterministic components of the signal [Equation (1.2)]. Consider for example the following situation: a peak or a pit in it could be an error feature but also a real feature. In order to analyse deterministic and error components, we need to separate them, which leads to a sort of *chicken–egg* problem (Hengl et al., 2008). This also explains why there are still hardly any practical applications of wavelet analysis on DEMs. A further explanation is that, so far, wavelet analysis has been applied mainly to 1D sampled signals. The processing of 2D surface models is much more complex and computationally more demanding. However, in the near future, we can anticipate the development of more powerful tools for the Fourier analysis of DEMs.

To understand the complexity of a land surface, the question arises of whether it can be easily simulated. Clark et al. (1997), for example, provide an algorithm to generate virtual DEMs in three main steps: (1) first a number of ideal surfaces (e.g. ellipsoid, cone, ridge) are allocated randomly over the area; (2) these geometric objects are then disturbed to simulate their dissection and (3) surface roughness is introduced to represent small-scale variation. Another software where virtual terrains can be generated (even with continental shelves) is the Wilbur[11] package. In Wilbur, the inputs include abstract variables such as type of fractal surface, lacunarity, spherical area, spherical centre and radii (see also Figure 19 in Chapter 17). Although the outputs from such models may be very similar, visually, to real terrains, it is obvious that many elements and land-surface processes (see Section 1.2) will be oversimplified or ignored.

2. DIGITAL MODELS OF THE LAND SURFACE

In computer science, land surfaces are commonly presented as *Digital Elevation Models*,[12] which are complete representations of the continuous surface. Recall from Section 1 of Chapter 1: a DEM implies that heights need to be calculable for any point in the area of interest. Otherwise, we are merely dealing with sam-

[11] See e.g. http://profantasy.com for the Fractal terrains generator.

[12] See also Section 1 in Chapter 1 for a detailed discussion on the use of terminology.

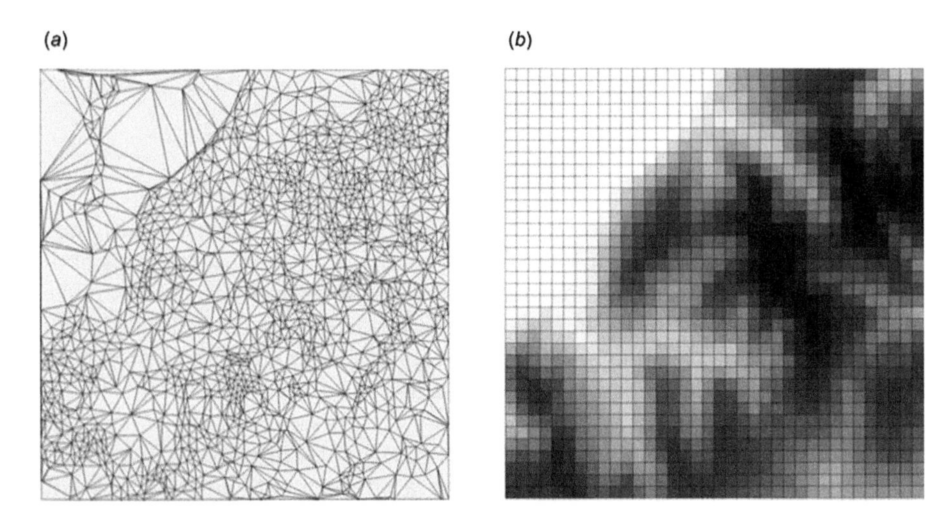

FIGURE 4 DEM models: (a) TIN DEM (1:50,000 scale), and (b) gridded DEM (100 m) for Baranja Hill. In this case, the TIN DEM is much more accurate in representing the shapes, although the number of spatial elements (TIN node, grid node) in both maps is about the same.

ples of heights at discrete locations (points, or lines such as contours), and not with models of a land surface.

> REMARK 3. *A land surface is commonly represented as a Digital Elevation Model (DEM). A DEM is a complete representation of a land surface, which means that heights are available at each point in the area of interest.*

Considering their design, all DEMs can be categorised into two groups: *raster-based or regular* and *vector-based or irregular* DEMs (Weibel and Heller, 1991; Rana, 2004; Li et al., 2005). Because heights need to be available for the whole area of interest, any type of DEM can be transformed into a point dataset (x, y, z).

2.1 Vector (irregular) models

For areas with high relief or rougher surfaces, irregular DEMs can use smaller spacing between points, and larger spacing where relief is lower or where the surface is smoother. In this way, they can more accurately describe geological faults and other sharp elevation changes, using the same number of points as grids (Figure 4). The original data of ground geodetic surveys, including LiDAR surveys are commonly represented using irregular point maps.

Transforming irregular point samples into complete DEMs can be achieved by triangulation (Akima, 1978). This involves calculating triangles with apices at the given points. Following Delaunay's criterion, triangles are positioned so that they are closest to those with angles of similar values (Peuquet, 1984; Deren and Xiao-Yong, 1991; Okabe et al., 2001). The resulting *Triangulated Irregular Networks* or TINs [Figure 4(a)] are based on point DEMs, and added on to this are data describing triangle parameters. TINs may be drawn in plan, or they may be

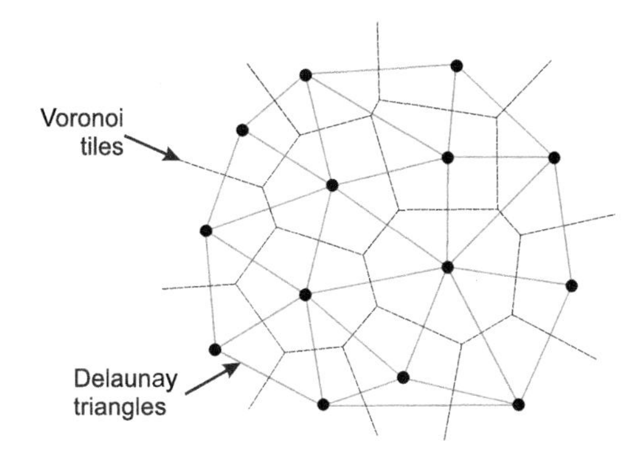

FIGURE 5 The Delaunay triangulation as compared with the Voronoi tesselation.

generalized into 3D structures (*curved TINs*) by adding elevation to each triangle apex. TINs require a smaller number of points to describe sharp elevation changes (Weibel and Heller, 1991; Li et al., 2005). Figure 5 shows how the Delaunay triangulation and Voronoi's (1907) tessellation can be derived from a point-sampled elevation data set.

Mark (1975a) was among the first to compare TIN versus grid DEMs. He concluded that, if grid models are used, twice as much memory is required to produce equally good estimates of surface parameters. Note that, by definition, a TIN-based DEM with planar triangular facets is discontinuous: the values between triangles change abruptly, which will often be an artefact (an artificial break line). An overview of alternative ways of representing topographic surfaces using lattices, facets or polygonal facets is given by Rana (2004).

TINs are more accurate in representing discrete changes of topography, i.e. land-surface objects, but this will be misleading where the topography is rather smooth. To solve this problem, many software packages now allow the user to define the *hardness* of the points and lines that are used to generate a DEM. In the 3D analyst of ArcGIS (for further information, see Section 2.2 in Chapter 11), the user can select which inputs are hard features (e.g. roads, streams, and shorelines) and which are soft features (e.g. contours, or the ridge lines on rolling hills), so that the true smoothness of surface can be adjusted locally. TINs are widely used in perspective representations of surfaces, especially in dynamic fly-through displays where the foreground is represented in full detail, but the background can be simplified as larger triangles.

Some TINs are *surface-specific* in that points are deliberately sampled along break, ridge and course lines. Surface-specificity is taken further in the contour and slope-line model developed at CSIRO in Australia and implemented as the TAPES-C program (Hutchinson and Gallant, 2000). The surface is represented mainly by a set of quadrilateral elements formed by segments of two adjacent contours, and the two slope lines running approximately orthogonal to these contours. In areas of convergence (pits and valley-heads) and divergence (peaks and fans), one contour segment may be replaced by a point, which produces a triangle.

The advantage of such a model is its relevance to gravitational movements, and especially to hydrological applications (Moore and Grayson, 1991). It does, however, require special calculation and optimisation, with some data reformatting and user interventions; the inputs include high points and saddle points as well as contours digitised in a consistent direction and ordered from the lowest to the highest. Intervention is required where contours are widely spaced. Flow is spread across downslope elements, which may be problematic if these are broad. This makes the explicit inclusion of channels desirable where the contributing area (drained) is greater than a user-specified threshold (this is done in TAPES-C).

2.2 Raster (regular) models

In raster[13] DEMs, elevations are stored using a regular structure that is absolutely consistent in each part of the study area. A regular DEM is essentially a rectangular matrix of heights for which plan coordinates (x, y) can be calculated *on-the-fly* due to the regular spacing of the grid points [Figure 4(b)].

> REMARK 4. *The key advantage of using gridded DEMs is that they have a simple structure, which makes them more suited for implementing geomorphometric algorithms and map displays.*

Although the debate on whether to use vector or raster models of DEMs in geomorphometry is still unresolved, in most applications, gridded DEMs are used as a standard. In comparison to alternative models, gridded DEMs have the following advantages:

- Grids have a simple structure and can easily be reconstructed.
- It is considerably easier to design land-surface parameters and objects using grids, because simpler algorithms can be used.
- Grids have a uniform spatial structure. Almost all properties of gridded DEMs are defined by a single characteristic — cell size.
- A grid model is more suited to the computer models used in image processing and for printing.

However, gridded DEMs also have the following disadvantages:

- Grids under-sample topography in areas where the topography is complex, and they over-sample smooth topography.
- Re-projection of a grid is slow and leads to a loss of accuracy. This is because the initial grid loses its regular structure in a new projection, and so it has to be re-calculated.[14]
- The different distances between grid centres in cardinal and diagonal directions have a negative impact on the precision of much hydrological modelling.

[13] *Grid* is probably a more suitable term than *raster* because *raster* is commonly related to technology, while *grid* is a mathematical concept. However, 'raster map' seems to be more popular among the GIS community.

[14] This procedure in GIS is often referred to as *re-sampling*. It can be quite time-consuming, because a new value has to be estimated at each new grid node.

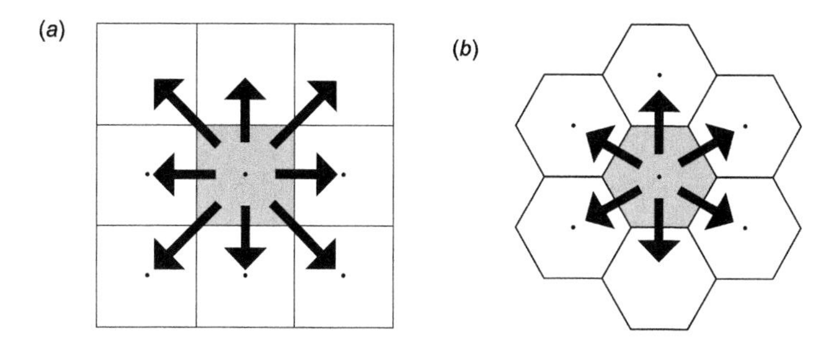

FIGURE 6 Square-tiled grids (a) and hexagonal-tiled grids (b). Note that hexagonal grid points only have six neighbours, while squared have eight. After de Sousa et al. (2006).

Although gridded DEMs appear to be smooth and continuous, they are also a discrete representation of topography. Once we define the size of the grid, we do not know any more whether the spatial variation between the grid cells is smooth or not. Similarly, a grid does not necessarily have to be regular. In computer games, irregular grids and lattices are frequently used to increase the speed of calculations. For those points closer to the observer, small cell sizes are used, while those far away are handled using larger cell sizes,[15] in order to keep the same appearance at a lower processing cost.

When grids are regular, they are stored as arrays, but when they are irregular they are stored using data structures such as *QuadTrees*. Vector models can also be regular. A regular lattice of points stored as a vector layer is regular. This distinction between regular and irregular DEMs can be quite confusing, so perhaps it is better just to think of raster and vector DEMs.

There have recently been some attempts to replace the regular square-tiled grid models with *hexagonal models* (de Sousa et al., 2006), which have a higher capacity for maintaining the original flow directions and have a slightly higher compaction capacity (Figure 6). Indeed, hexagonal grids even maintain more visual detail, and reflect the organisation of most biological optical organs. Still, we expect that, in the coming years, square-tiled grid models will probably remain standard for DEMs, due to their simple and widely accepted format.

2.3 Cell size, support size, pixel size

As noted previously, in Chapter 1, *cell size* — the distance between two grid nodes expressed in ground metres — defines most of the technical characteristics of a gridded DEM. Cell size can be closely related to the level of detail or spatial precision of a map, which, in cartography, is often related to the concept of *scale*. As a rule of thumb, cell size should equal 0.5 mm on a paper map, which means that a 25 m resolution grid would correspond to a 1:50,000 scale map (Hengl, 2006). Enlarging the cell size leads to *aggregation* or upscaling; decreasing the cell size leads to *disaggregation* or downscaling. As the grid becomes coarser, the overall infor-

[15] This is known as the *Constant Level of Detail* (CLOD) in the game development jargon.

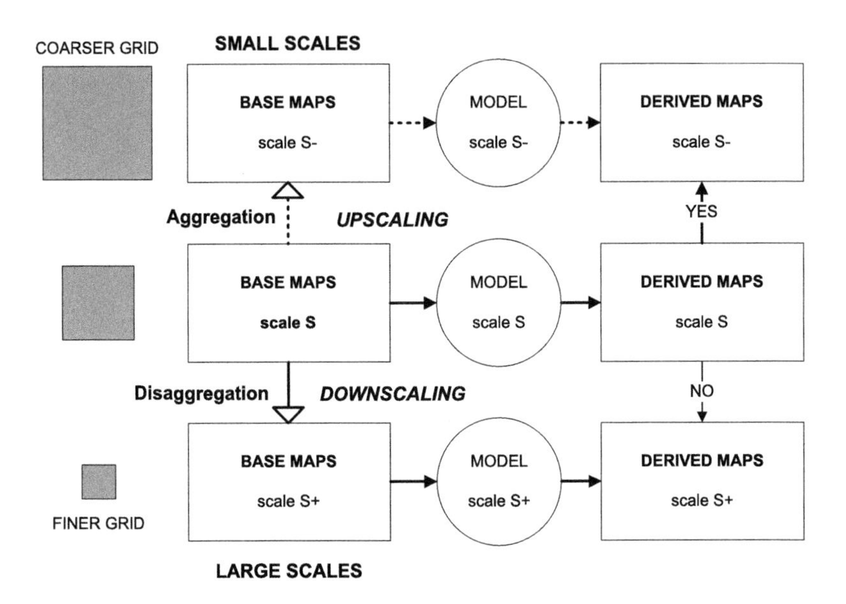

FIGURE 7 Upscaling and downscaling in a grid-based GIS. S indicates scale: S^- are smaller scales and S^+ are larger scales. Reprinted from Hengl (2006). With permission from Elsevier.

mation content in the map will progressively decrease, and vice versa (McBratney, 1998; Kuo et al., 1999; Stein et al., 2001).

In cartography, coarser cell sizes go together with smaller scales and larger study areas, and finer cell sizes with larger scales and smaller study areas. The former convention often confuses non-cartographers because *bigger* pixels mean *smaller* scale, which usually means a *larger* study area (Figure 7). The term *scale* is probably more ambiguous that the term resolution, because *scale* can be related to at least three concepts: (a) generalisation level, (b) temporal and spatial resolution and (c) spatial extent (size of the area). Also note that, although often used as synonyms, there is a difference between cell size and *grid resolution* — by increasing the grid resolution, the cell size will become smaller and the other way around.

Note in Figure 7 that both aggregation and disaggregation can be carried out before or after geomorphometric analysis. If the model is linear, the two routes should yield the same results (Heuvelink and Pebesma, 1999); if not, there can be serious differences. Many land-surface parameters will show different features when derived from DEMs of different resolutions (Kienzle, 2004).

A textbook example of the influence of cell size on land-surface parameters is the derivation of slope. If the cell size is fine enough (e.g. a few centimetres), we will be able to detect even very steep slopes in an area that might seem absolutely[16] flat, such as a plain or river terrace. As the grid resolution becomes coarser, average values of slope gradient decrease. Even if you are mapping slopes in the Himalayas, if the grid resolution is coarse enough, the slope will become much

[16] Unlike absolute relief, *apparent* or *virtual relief* is just an impression of a landscape. It can easily be exaggerated by stretching the z-coordinate.

gentler than you might experience in the field. Once the grid resolution becomes so coarse that the whole study area can be fitted into one pixel, then the slope map will equal zero. Of course, there is no need to derive a land-surface parameter out of a map consisting of a single pixel, but we hope that this example has illustrated the scale-relativity of land-surface parameters.

Another important issue to clarify before going further with geomorphometric algorithms is the distinction between cell size and support size. The *support size* is typically a fixed area or volume of the land that is being sampled. Support size can be increased by using composite samples, or by averaging point-sampled heights belonging to the same blocks of land. *Grid point* or *grid node* is the more appropriate term when we are definitely dealing with a regular point sample — each value applies to a point, not an average over a cell (or pixel).

In geostatistics, one can control the support size of the outputs by averaging multiple predictions over regular blocks of land. This is known as *block kriging* (Heuvelink and Pebesma, 1999). This means that we can sample elevation at point locations, and then interpolate them for blocks of e.g. 10×10 m. The latter often confuses GIS users because as well as using point elevation measurements to interpolate values at regular point locations (point kriging), and then display them using a raster map, we can also make interpolations for blocks of land (block kriging) and display them using the same raster model (Bishop et al., 2001).

DEMs derived from topographic (airborne or satellite) images have a support size that equals the original scanning resolution [Figure 8(a)]. The values shown at pixel nodes represent the average value of all elevations in those pixels. On the contrary, LiDAR measurements have a relatively small support size (they are almost point measurements), so that it is more accurate to represent them first as a point map [Figure 8(b)]. Such densely measured point values can easily be aggregated to some suitable support size, by averaging values that fall in the same grid block. The distinction between the support and cell size is less important for the visualisation of DEMs, but it can be very important for the validation of simulation models, or for the assessment of uncertainty in elevation measurements (see also Chapter 5). For practical reasons, one should always try to select a cell size that matches the support size of the sampling, and avoid using a cell size that is finer than the support size.

The projected grid (cell size on the map) is referred to as *grid mesh*. Grid mesh and pixel size only coincide when 2D map-image magnification corresponds with a situation in which each grid mesh is represented by one pixel. The concept of *pixel size* is completely arbitrary for grids: pixel size may have any value for a given grid resolution, depending on the map-image magnification. However, many people often refer to the cell size as the pixel size, which is not such a big mistake, as long as the ground metres are mentioned (e.g. pixel size of 25 m).

2.4 Suitable cell size

The key problem when selecting a cell size for geomorphometry is that there can be significant differences between the interpolated surface elevation and the actual land surface, meaning that some peaks and channels might disappear (or be

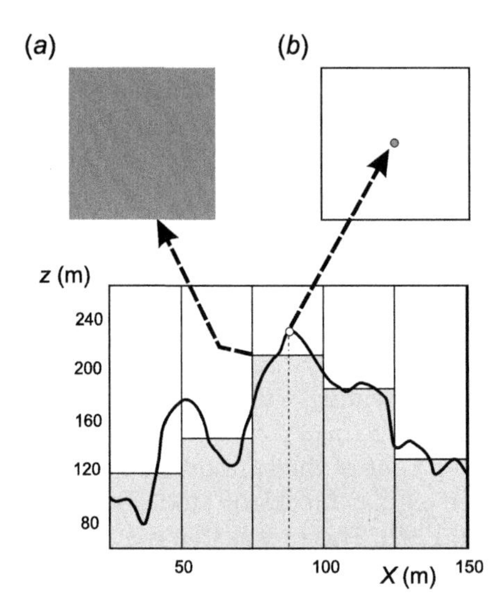

FIGURE 8 The distinction between pixel size and support size in relation to the original distribution (the *signal*): (a) support size equals the resolution of an image (the averaged measurements); (b) support size is infinitely small (the point measurement at the centre of a pixel).

displaced) when represented in a coarse raster DEM. In general, a finer DEM resolution will also mean more accurate land-surface parameters and a higher information content (Kuo et al., 1999). The amount of increase, however, depends on the general variability of the landscape. A smooth, generally regular, landscape does not need a fine resolution DEM. If the DEMs are coarser than the scale of landscape processes, any results or indices derived from DEMs must be treated with caution (Pain, 2005). Even more so, if the cell size is too fine in relation to the vertical accuracy, it might introduce local artefacts and slow down the computation of land-surface parameters. In most cases, the computational cost of preparing, storing and using the map increases exponentially, the more the cell size decreases (Barr and Mansager, 1996). Obviously, we need a grid resolution that optimally reflects the complexity of a terrain; one that can represent the majority of the geomorphic features (Borkowski and Meier, 1994; Kienzle, 2004; Pain, 2005; M.P. Smith et al., 2006).

A suitable cell size can be derived for a given set of sampled elevations (e.g. contours or points) based on the complexity of terrain. Imagine a one-dimensional topography with a number of inflection points (Figure 9). The problem of selecting a suitable grid resolution can be related to the Nyquist–Shannon sampling theorem (Hengl, 2006). In our one-dimensional example above, the land surface is the signal and its frequency is determined by the *density of inflection points*. Hence, the cell size should be at least half the average spacing between inflection points or finer:

$$\Delta s \leqslant \frac{l}{2 \cdot n(\delta z)} \tag{2.1}$$

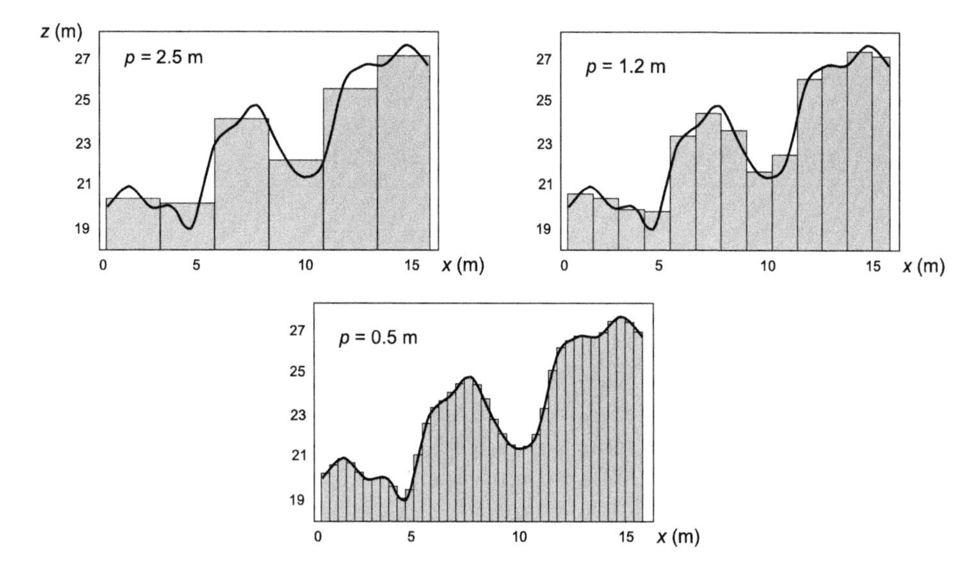

FIGURE 9 A schematic example showing the effect of grid resolution on the representation of topography: a cell size that is too coarse ($\Delta s = 2.5$ m) will misrepresent the topography; while a finer cell size ($\Delta s = 0.5$ m) will be more effective in representing all the peaks and channels present. Reprinted from Hengl (2006). With permission from Elsevier.

where l is the length of a transect and $n(\delta z)$ is the number of inflection points observed. We can also be more precise and look for the (low) 5% quantile of spacing distribution between the inflection points. In the example in Figure 9, there are 20 inflection points and the average spacing between them is 0.8 m, suggesting a cell size of at least 0.4 m or finer. The cumulative distribution of distances between the inflection points shows that the 5% threshold for the smallest spacing between the inflection points is 0.2 m. Hence, a cell size of between 0.4 and 0.2 m is recommended.

In the 2D case, if the DEM is based on digitised contours, a suitable grid resolution can be estimated from the total length of the contours. Here, we do not actually have a map of inflection points, but we can approximate them using the contour map. Assuming that the contours were selected to present changes of surface function, a suitable cell size is:

$$\Delta s = \frac{A}{2 \cdot \sum l} \tag{2.2}$$

where A is the total size of the study area and $\sum l$ is the total cumulative length of all digitised contours. A more precise approach than in Equation (2.2) is to evaluate the density of contour lines in an area and derive the 5% probability of the smallest spacing between the contours (Hengl, 2006). Note that the density of contours on a topo-map is controlled by scale — if we work with more detailed scales, the vertical spacing between the contours will be smaller, so the density will increase and we will be able to use finer cell sizes. This means that there is no universally suitable cell size, but only a cell size that is *'good enough'* for a certain working scale.

> REMARK 5. *A suitable cell size for a DEM can be selected to fit the complexity of terrain and/or the scale of work for the targeted application.*

3. THE SAMPLING, GENERATION AND ANALYSIS OF LAND SURFACES

3.1 Sampling of heights

Given the properties of a defined land surface, we can plan a sampling strategy to produce an accurate surface model. A successful sampling plan will guarantee a quality DEM, which, at the same time, will also improve the quality of any application of that DEM. Li et al. (2005) recognise three main groups of height-sampling methods, corresponding to the three types of DEM recognised above:

(1) statistical (random) sampling;
(2) regular sampling;
(3) feature-based sampling.

According to Katzil and Doztsher (2000), five sampling approaches have been used to generate DEMs. They all differ in whether the points are placed:

- systematically (e.g. a regular grid);
- along parallel profile lines;
- along contours;
- on all minima and maxima (all tops and pits);
- along surface-specific lines, such as topographical breaks, ridge and course lines.

We learned in Section 1.3 that a land surface viewed at a given scale consists of a finite number of surface-specific elements. This means that the most economic sampling technique would be to focus on sampling these elements only. This is not an easy task, because we first need to detect such features. A more serious problem with feature-based sampling is that the number of features is scale-dependent, so, in addition, we first need to define a threshold value and specify which objects are large enough to be recognised as ridges, peaks and similar. To avoid these problems, most new topographic surveys are, in fact, based on dense, regular, sampling techniques, and the surface features are then determined *a posteriori*.

When heights are sampled sparsely over the area of interest (at spot heights, or along contour lines), the success of the land-surface modelling will depend upon which DEM interpolation method is chosen.

3.2 The generation of DEMs from sampled heights

The conversion of sampled heights to raster DEMs is also often referred to as *gridding*. There are many possibilities for gridding: the techniques range from nearest point, triangulation, inverse distance, minimum curvature and splines to various kriging algorithms. For example, the software package **Surfer** (www.ssg-surfer.com) offers dozens of interpolation techniques: Inverse Distance, Kriging, Minimum Curvature, Polynomial Regression, Triangulation, Nearest Neighbour, Shepard's Method, Radial Basis Functions, Natural Neighbour, Moving Average, Local Polynomial, etc. Similarly, **TNTmips** (http://microimages.com) offers surface fitting, contouring, triangulation and profiling techniques, including surface fitting by Quintic method, contouring by iterative thresholding, and triangulation with breaklines. An inexperienced user will often be confused by which technique to select in order to generate a DEM that is suitable for their needs. In the following paragraphs, we will try to clarify the differences between the various gridding techniques and elaborate procedures to select a suitable DEM generation technique.

The least complicated gridding technique is *simple linear interpolation*. In this case, height at some grid node (x_0,y_0) is estimated using:

$$\hat{z}(x_0, y_0) = a_0 + a_1 \cdot x_0 + a_2 \cdot y_0 = \mathbf{a}^\mathrm{T} \cdot \mathbf{s}_0 \tag{3.1}$$

where $\hat{z}(x_0, y_0)$ is the interpolated value at the new grid node and $\mathbf{a} = [a_0, a_1, a_2]$ are the unknown coefficients that define the plane. So if we have (at least) three points with known heights (z_1, z_2, z_3) and coordinates, the coefficients can be solved using (Li et al., 2005, p. 117):

$$\begin{bmatrix} a_0 \\ a_1 \\ a_2 \end{bmatrix} = \begin{bmatrix} 1 & x_1 & y_1 \\ 1 & x_2 & y_2 \\ 1 & x_3 & y_3 \end{bmatrix}^{-1} \cdot \begin{bmatrix} z_1 \\ z_2 \\ z_3 \end{bmatrix} \tag{3.2}$$

or using matrix algebra:

$$\mathbf{a} = \mathbf{s}^{-1} \times \mathbf{z} \tag{3.3}$$

where $\mathbf{s}$ is the matrix with coordinates and their transforms. Note that the model coefficients $\mathbf{a}$ need to be solved for each local polygon, which makes this method fully local. In the case of contour data, we can estimate the height at a new grid node by using only two values from the two neighbouring contours (Gorte and Koolhoven, 1990). The height is then estimated simply as a weighted average:

$$\hat{z}(x_0, y_0) = \frac{w_{\mathrm{N1}} \cdot z_{\mathrm{N1}} + w_{\mathrm{N2}} \cdot z_{\mathrm{N2}}}{w_{\mathrm{N1}} + w_{\mathrm{N2}}} \tag{3.4}$$

where z_{N1} and z_{N2} are the closest points on the two neighbouring contours and the weights are estimated based on the inverse shortest distance from the new grid node to the contour (Gorte and Koolhoven, 1990):

$$w_{\mathrm{N}i} = \frac{1}{d_{0,\mathrm{N}i}} = \frac{1}{\sqrt{(x_0 - x_{\mathrm{N}i})^2 + (y_0 - y_{\mathrm{N}i})^2}} \tag{3.5}$$

Equation (3.1) can be extended to four unknowns using the bilinear model:

$$\hat{z}(x_0, y_0) = a_0 + a_1 \cdot x_0 + a_2 \cdot y_0 + a_3 \cdot x_0 \cdot y_0 \tag{3.6}$$

and further to bicubic:

$$
\begin{aligned}
&= a_{01} + a_{02} \cdot x_0 + a_{03} \cdot x_0^2 + a_{04} \cdot x_0^3 \\
&+ a_{05} \cdot y_0 + a_{06} \cdot y_0^2 + a_{07} \cdot y_0^3 + a_{08} \cdot x_0 \cdot y_0 \\
&+ a_{09} \cdot x_0^2 \cdot y_0^2 + a_{10} \cdot x_0^2 \cdot y_0 + a_{11} \cdot x_0 \cdot y_0^2 \\
&+ a_{12} \cdot x_0^3 \cdot y_0 + a_{13} \cdot x_0 \cdot y_0^3 + a_{14} \cdot x_0^3 \cdot y_0^2 \\
&+ a_{15} \cdot x_0^2 \cdot y_0^3 + a_{16} \cdot x_0^3 \cdot y_0^3
\end{aligned}
\tag{3.7}
$$

which now requires 16 known points.

Interpolations from digitised contours are known to generate poor DEMs (Wise, 1998). When the contour is on one side only — as in a 'contour isolation' around a peak or pit — a flat area is generated at that contour height. On sloping ridges ('*noses*'), interpolation along two cardinal axes will often miss the contour above, giving an artificially stepped surface. Likewise, in valleys, the contour below may be missed. These artefacts will show up in perspective views and in the 'spikiness' of a detailed altitude histogram.

Point measurements of heights are also suitable for interpolation by averaging. In this case, we need to know the distances from the new point to the sampled points:

$$\hat{z}(x_0, y_0) = \sum_{i=0}^{n} \lambda_i \cdot z_i \tag{3.8}$$

where λ_i is the weight for neighbour i. The sum of weights needs to equal one to ensure an unbiased interpolator. We can take into account all sampled points (global interpolation) or set a threshold distance (search radius) to speed up the calculation. Equation (3.8) in matrix form is:

$$\hat{z}(x_0, y_0) = \boldsymbol{\lambda}_0^{\mathbf{T}} \times \mathbf{z} \tag{3.9}$$

The simplest approach for determining the weights is to use the *inverse distance*:

$$\lambda_i = \frac{\frac{1}{d_i^{\beta}}}{\sum_{i=0}^{n} \frac{1}{d_i^{\beta}}}, \quad \beta > 1 \tag{3.10}$$

where d_i is the distance from the new point to a known sampled point and β is a coefficient that is used to adjust the weights. The higher the β, the less importance will be put on distant points. Note also that Equation (3.8) is just a general case of simple linear interpolation with two nearest neighbours [Equation (3.4)]. The problem is how to estimate β objectively so that it reflects the inherent properties of a dataset.

One solution for estimating the weights in Equation (3.8) objectively is to analyse the autocorrelation structure in the height measurements and then fit a variogram that can reflect the autocorrelation structure more objectively. In this case, the weights can be determined using (Isaaks and Srivastava, 1989):

$$\boldsymbol{\lambda}_0 = \mathbf{C}^{-1} \times \mathbf{c}_0, \quad \sum_{i=1}^{n} \lambda_i(s_0) = 1 \tag{3.11}$$

where $\mathbf{C}^{-1}$ is the inverse matrix of covariances between all points and $\mathbf{c}_0$ is the vector of covariances for the new point; both estimated using the fitted variogram model. This technique, known as *kriging*, is one of the most widely used stochastic interpolation techniques for point-sampled data. It provides an objective measure of the interpolation error and can be used for both interpolation and simulation. However, kriging is not really appropriate for interpolating heights, mainly due to three problems: (a) it will often oversmooth the values; (b) it ignores the hydrological connectivity of a terrain; (c) it is sensitive to *hot-spots*,[17] which can cause many artefacts.

Another stochastic approach to interpolation is by fitting a local mathematical surface, such as a higher order polynomial, to a larger group of points (the number of points need to be larger than the number of parameters). This group of methods is referred to as *moving surface* interpolation methods. The algorithm will determine coefficients by maximising the local fit:

$$\sum_{i=1}^{n} (\hat{z}_i - z_i)^2 \rightarrow \min \tag{3.12}$$

This can be achieved by the least squares solution:

$$\mathbf{a} = \left(\mathbf{s}^{\mathrm{T}} \times \mathbf{s}\right)^{-1} \times \left(\mathbf{s}^{\mathrm{T}} \times \mathbf{z}\right) \tag{3.13}$$

In practice, for each output grid node, a polynomial surface is calculated by a moving least squares fit within the specified limiting distance. Most algorithms will also include a weight function to ensure that points close to an output pixel will obtain greater weight than points which are farther away.

A special group of interpolation techniques is based on *splines*. Splines[18] are preferable to simple polynomial interpolation because the interpolation error can be minimised, even when using low degree polynomials. There are many versions and modifications of spline interpolators. The most widely used for gridding are *thin-plate splines* (Hutchinson, 1989) and *smoothing splines with tension* (Mitášová and Hofierka, 1993). In the case of regularized spline with tension and smoothing (implemented in GRASS GIS), the predictions are obtained by (Mitášová et al., 2005):

$$\hat{z}(x_0, y_0) = a_1 + \sum_{i=1}^{n} w_i \cdot R(v_i) \tag{3.14}$$

where a_1 is a constant and $R(v_i)$ is the radial basis function determined using (Mitášová and Hofierka, 1993):

[17] Locations where extremely high or low values are observed that are statistically different from the population.

[18] A spline is a special type of piecewise polynomial. Splines are immune to the *Runge* and *Gibbs* phenomena — severe artefacts that commonly occur when polynomial interpolation is used.

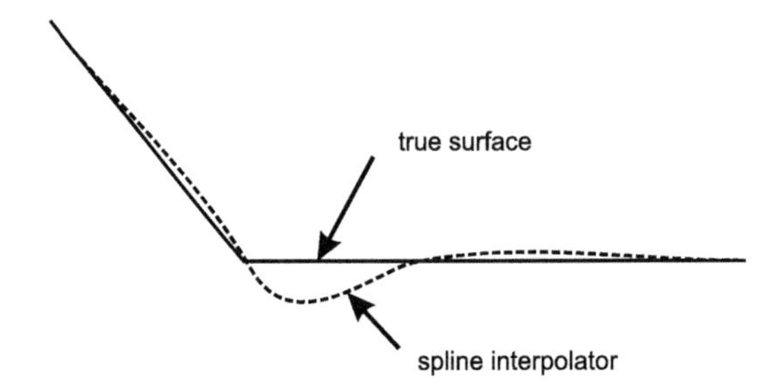

FIGURE 10 Overshooting the true surface line — a problem with spline interpolation.

$$R(\upsilon_i) = -\left[E_1(\upsilon_i) + \ln(\upsilon_i) + C_E\right]$$
$$\upsilon_i = \left[\varphi \cdot \frac{\mathbf{h}_0}{2}\right]^2 \tag{3.15}$$

where $E_1(\upsilon_i)$ is the exponential integral function, $C_E = 0.577215$ is the Euler constant, φ is the generalized tension parameter and $\mathbf{h}_0$ is the distance between the new and interpolation point. The coefficients a_1 and w_i are obtained by solving the system:

$$\sum_{i=1}^{n} w_i = 0$$
$$a_1 + \sum_{i=1}^{n} w_i \cdot \left[R(\upsilon_i) + \delta_{ij} \cdot \frac{\varpi_0}{\varpi_i}\right] = z(\mathbf{s}_i) \tag{3.16}$$

where $j = 1, \ldots, n$, ϖ_0/ϖ_i are positive weighting factors representing a smoothing parameter at each given point $\mathbf{s}_i$. The tension parameter φ controls the distance over which the given points influence the resulting surface, while the smoothing parameter controls the vertical deviation of the surface from the points (see further Section 1.3 in Chapter 17). By using an appropriate combination of tension and smoothing, one can produce a surface which accurately fits the empirical knowledge about the expected variation (Mitášová et al., 2005). Splines have problems in representing discrete transitions — they often 'overshoot' at the edges of flood plains or other breaks in slopes, even generating elevations which are outside the range of the input data (Figure 10).

All interpolation methods can be grouped according to three aspects: (a) *the smoothing effect*, (b) *the proximity effect* and (c) *stochastic assumptions*. With respect to the smoothing effect, an interpolator can be either exact or approximate; and the proximity effect can be either global (all sampled points are used to estimate the value at each grid node) or local (only a subset of sampled locations is used to estimate the value at each grid node). An exact interpolator, such as linear interpolation, preserves the values at the sampled data points and is usually based on local values, within a neighbourhood. Interpolators such as kriging adjust completely to observed spatial auto-correlation structure and allow objective

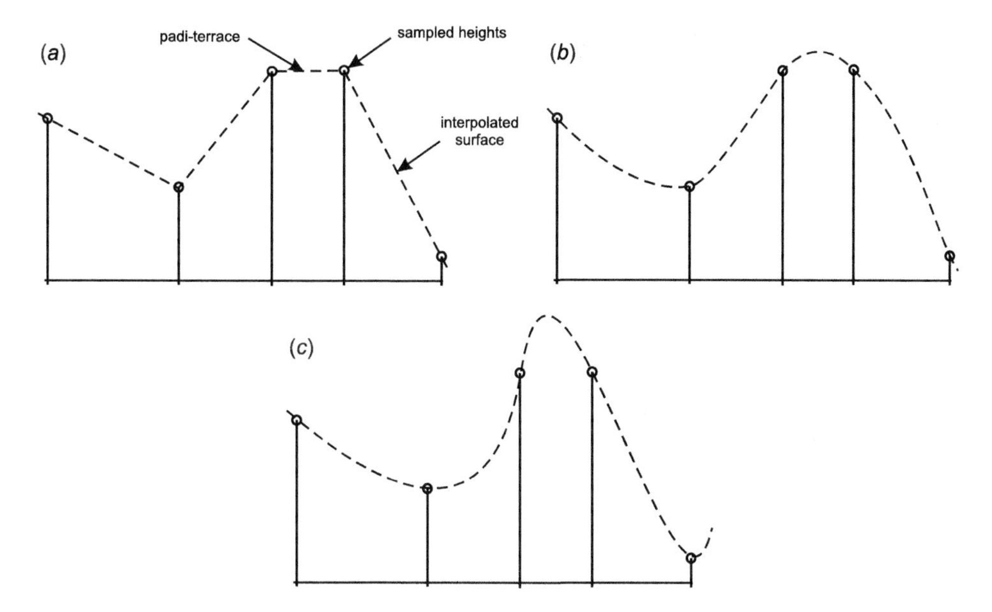

FIGURE 11 A schematic example, showing the effect of choice of interpolation algorithm on the quality of output: (a) pure linear interpolation can cause obvious artefacts such as multiple terraces and flat summits; (b) splines are often very successful interpolators, because they can represent the shapes correctly; (c) if unrealistic parameters are used with splines, the final result can be even poorer than if a simple linear interpolator had been used.

incorporation of stochastic assumptions, while purely mechanical gridding techniques require human intervention on selecting of the smoothing effect, etc.

For a summary comparison of gridding techniques, see Table 1. Artefacts may be formed whichever interpolation technique is used, but each technique differs considerably in how sensitive it is to the spatial distribution of both the samples and the errors associated with them.

The quality of DEMs can be significantly improved by using an appropriate interpolator. The diversity of input data is a further aspect that distinguishes those interpolators that are only able to consider point-elevation measurements from those that can distinguish between soft and hard break lines, positions of streams and human-built objects (such as the previously mentioned gridding technique in ArcGIS; Section 2.2 in Chapter 11).

> REMARK 6. *ANUDEM (based on the discretised thin-plate spline technique) is an iterative DEM generation algorithm that produces hydrologically-correct DEMs.*

There has been interest in finding the *optimal DEM interpolation method* for quite some time. The quality of any interpolator can be estimated using cross validation or jackknifing (Smith et al., 2005). Extensive comparisons of techniques suitable for interpolating heights can be found in Wood and Fisher (1993), Mitášová et al. (1996), Carrara et al. (1997) and Wise (2000). Many will agree that

TABLE 1 Summary comparison of the methods used for interpolating the land surface from sampled heights

Method	Smoothing effect	Local/ Global	Deterministic/ Stochastic	Requirements/ Inputs	Possible problems
Linear interpolation	Low	Local	Deterministic	None	No error assessment; cut-offs and similar artefacts
Inverse distance interpolation	Low	Local/ Global	Deterministic	Weighting function, search radius	No error assessment; over-smoothing
Ordinary kriging	Medium	Local/ Global	Stochastic	Variogram model, search radius	Over-smoothing; statistical assumptions
Moving surface	High	Global	Deterministic/ Stochastic	Polynomial order, search radius	Possible over-fitting; over-smoothing
Splines	High	Local	Deterministic	Smoothness factor, search radius	Overshooting; over-smoothing
ANUDEM	High	Local/ Global	Deterministic	Smoothness factor, search radius, streams	Over-smoothing

there is no universal gridding technique that is clearly superior, and appropriate for all sampling techniques and DEM applications (Weibel and Heller, 1991; Li et al., 2005). Mitas and Mitášová (1999) evaluated various interpolation approaches to elevation data, and concluded that the most important aspects are how well smoothness and tension are described, and how well streams and ridges are incorporated. They ultimately suggested that regularized splines in conjunction with a tension algorithm would be the most suitable DEM interpolator (Mitášová and Hofierka, 1993). Indeed, splines (implemented in GRASS GIS and ANUSPLIN[19]) commonly produce smooth surfaces, which often fit reality (Mitas and Mitášová, 1999). Another flexible solution for interpolating contour data is the minimum curvature method (Fogg, 1984), which, for example, is implemented in Surfer.

Although there is no absolutely ideal DEM interpolator, it is important to implement algorithms that can incorporate secondary information (such as layers representing pits, streams, ridges, scarps or break lines) where available. One

[19] A programme developed at the Australian National University (ANU) for thin-plate spline smoothing (Hutchinson, 1995).

such widely advocated and applied hybrid technique is the ANUDEM algorithm by Hutchinson (1988, 1989, 1996), implemented as the TOPOGRID function in the grid module of ArcInfo. ANUDEM uses an iterative finite difference interpolation technique — starting with a coarse grid, drainage conditions are enforced, then the spatial resolution is increased, then drainage enforcement is performed again, and so on, until the desired resolution is reached. It is essentially a discretised thin-plate spline technique (Wahba, 1990), in which the roughness penalty has been modified to allow the fitted DEM to follow abrupt changes in the land surface, such as streams and ridges. Another possibility for generating DEMs using secondary information is regression-kriging (Hengl et al., 2008). This has an advantage over ANUDEM because the model parameters can be determined based on the statistical properties of the point data.

The success of a DEM interpolator depends very much on how it is applied: if the application is for water or mass-movement modelling, it is important to prepare a DEM that is *hydrologically correct*. Yet if the DEM is used to produce ortho-photos, the absolute accuracy of elevation values will be more important — even if some drainage paths are incorrect.

Many geomorphometrists believe that Hutchinson's (1989) modification of thin-plate splines, that adjusts for the correct pathway of water across the surface, should be the preferred gridding method for producing DEMs for geomorphometric analysis (Table 1). One advantage of ANUDEM over any other interpolation algorithm is that a hydrologically-correct land-surface model is enforced. However, even ANUDEM can produce poorer results than a simple linear interpolator, if unrepresentative input parameters are selected (Wise, 2000).

A suitable interpolator can best be selected by analysing the properties of input data and the characteristics of an application (Pain, 2005). For example, if the heights were measured with a very precise device, then we need to employ an exact interpolator. If the samples were located accurately and the heights were measured with high precision, then we should employ an interpolator that preserves all these features. If the measurements were noisy, then we should consider employing interpolators that can smooth this noise (e.g. smoothing splines, kriging or fitting moving surfaces). Many properties of the target surface, such as the short-range variation, anisotropy, etc., can be estimated objectively from the sampled heights. For example, Figure 12 compares variograms for Baranja Hill fitted from field-sampled heights with those from heights measured by a scanning device (SRTM). In this case, the SRTM DEM is much less precise than the field-sampled heights. This means that its heights should also be filtered for noise prior to any geomorphometric analysis.

The geostatistical models of land surface are often highly *non-stationary* (see also Section 2 in Chapter 5). A land surface can exhibit both abrupt and gradual changes, and both perfect smoothness and considerable dissection. All this can happen even within a small study area. Therefore, hybrid and local gridding methods should generally be preferred to purely mechanical or geostatistical and/or global techniques (Smith et al., 2005). The problem is that there are still very few[20]

[20] With the exception of Surfer and TNTmips.

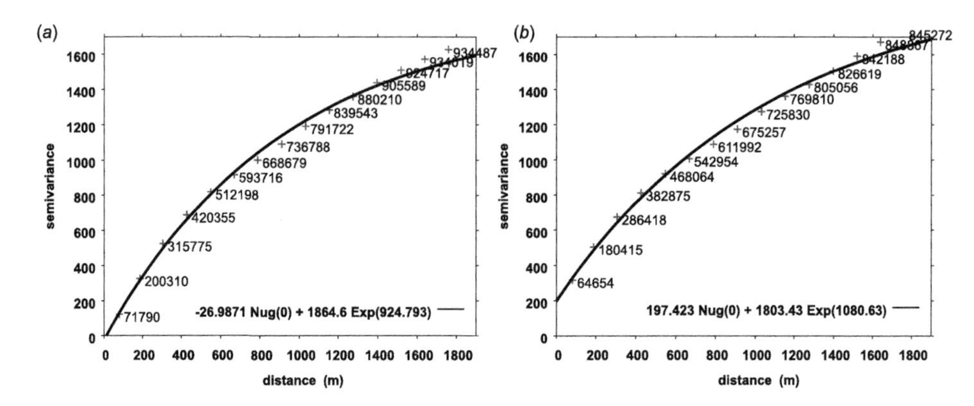

FIGURE 12 A comparison of variograms for sampled heights at Baranja Hill, derived from (a) field-sampled heights and (b) SRTM DEM; both fitted automatically using the gstat package. The SRTM DEM shows much higher nugget (197), i.e. unrealistic surface roughness. Compare also the output DEMs in Figure 5 of Chapter 3.

software packages that generate realistic land surfaces by interactively specifying a variety of inputs for incorporating our knowledge about the surface, and helping to minimise artefacts.

3.3 Land-surface analysis algorithms

Understanding mathematical models of the land surface helps us to design geomorphometric algorithms that avoid artefacts and inaccuracies. This will now be illustrated by deriving first and second derivatives for calculating slope, aspect and curvatures. Consider a small portion of a DEM — a 3×3 neighbourhood (see also Figure 4 in Chapter 1):

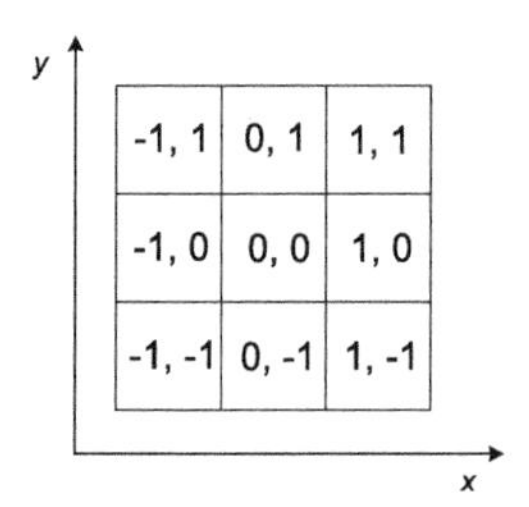

We can fit these 9 points exactly using any polynomial function with 9 fitted coefficients, for example (Unit Geo Software Development, 2001):

$$\begin{aligned} f(x,y) = {} & a_0 + a_1 \cdot x + a_2 \cdot x^2 + a_3 \cdot y \\ & + a_4 \cdot x \cdot y + a_5 \cdot x^2 \cdot y + a_6 \cdot y^2 \\ & + a_7 \cdot x \cdot y^2 + a_8 \cdot x^2 \cdot y^2 \end{aligned} \tag{3.17}$$

For the derivatives in the x direction we can substitute y with 0 and obtain:

$$f(x,y) = a_0 + a_1 \cdot x + a_2 \cdot x^2 \tag{3.18}$$

and the first and second derivatives then equal:

$$\frac{df}{dx} = a_1 + 2 \cdot a_2 \cdot x$$
$$\frac{d^2f}{dx^2} = 2 \cdot a_2 \tag{3.19}$$

Because we are interested in derivatives at the central pixel where $x = 0$, the equations modify to:

$$\frac{df}{dx} = a_1$$
$$\frac{d^2f}{dx^2} = 2 \cdot a_2 \tag{3.20}$$

The elevations at locations $-1, 0$, $0, 0$ and $1, 0$ equal:

$$z_{(-1,0)} = a_0 - a_1 + a_2$$
$$z_{(0,0)} = a_0 \tag{3.21}$$
$$z_{(1,0)} = a_0 + a_1 + a_2$$

and the coefficients a_1 and a_2 can then be solved using:

$$a_1 = \frac{z_{(1,0)} - z_{(-1,0)}}{2}$$
$$a_2 = \frac{z_{(-1,0)} - 2 \cdot z_{(0,0)} + z_{(1,0)}}{2} \tag{3.22}$$

so the filters for first and second derivative in x direction will look like this:

$\frac{df}{dx}$

0	0	0
-1	0	1
0	0	0

$\frac{d^2f}{dx^2}$

0	0	0
1	-2	1
0	0	0

Calculating the matrix coefficients for the 5×5 filter (see Figure 4 in Chapter 1) follows the same method as for the 3×3 filters. However, there are now more unknown coefficients, so it is a bit more complicated to derive the formulae. The polynomial function in the x direction is now:

$$f(x, y) = a_0 + a_1 \cdot x + a_2 \cdot x^2 + a_3 \cdot x^3 + a_4 \cdot x^4 \tag{3.23}$$

and the first and second derivatives are then:

$$\frac{df}{dx} = a_1 + 2 \cdot a_2 \cdot x + 3 \cdot a_3 \cdot x^2 + 4 \cdot a_4 \cdot x^3$$
$$\frac{d^2f}{dx^2} = 2 \cdot a_2 + 6 \cdot a_3 \cdot x + 12 \cdot a_4 \cdot x^2$$

By substituting x in the equations with the values $-2, -1, 0, 1$ and 2, and then restructuring and simplifying them, we get:

$$a_1 = \frac{z_{(-2,0)} - 8 \cdot z_{(-1,0)} + 8 \cdot z_{(1,0)} - z_{(2,0)}}{12}$$

$$a_2 = \frac{-z_{(-2,0)} + 16 \cdot z_{(-1,0)} - 30 \cdot z_{(0,0)} + 16 \cdot z_{(1,0)} - z_{(2,0)}}{24}$$

so the filters for first and second derivatives in x direction for a 5×5 moving window will look like this:

$\frac{df}{dx}$

0	0	0	0	0
0	0	0	0	0
1	-8	0	8	-1
0	0	0	0	0
0	0	0	0	0

$\frac{d^2f}{dx^2}$

0	0	0	0	0
0	0	0	0	0
-1	16	-30	16	-1
0	0	0	0	0
0	0	0	0	0

Note that the 5×5 filters[21] are a little bit more complicated, but for noisy data, they will produce more stable results, because they are less sensitive to local outliers. Assuming full accuracy of the elevation data, one does need to fit a surface as in Equation (3.17). Each grid square can be split into two triangles and the slope aspect and gradient can be calculated from the plane surface that exactly fits each trio of points. As there are two diagonals, there are two versions of these detailed maps. Of course, this approach cannot provide corresponding curvatures.

The most popular algorithms for deriving first and second derivatives are those of Evans (1972), Shary (1995), Zevenbergen and Thorne (1987), and the modified Evans–Young (Shary et al., 2002) method. The Evans–Young algorithm (Evans, 1972; Young, 1978; Pennock et al., 1987) fits a second order polynomial to the 3×3 neighbourhood filters:

$$z = \frac{r \cdot x^2}{2} + s \cdot x \cdot y + \frac{t \cdot y^2}{2} + p \cdot x + q \cdot y + z_0 \tag{3.24}$$

where p, q, r, s, t, z_0 are coefficients determined using:

$$\begin{aligned}
p &= \frac{z_3 + z_6 + z_9 - z_1 - z_4 - z_7}{6 \cdot \Delta s} \\
q &= \frac{z_1 + z_2 + z_3 - z_7 - z_8 - z_9}{6 \cdot \Delta s} \\
r &= \frac{z_1 + z_3 + z_4 + z_6 + z_7 + z_9 - 2 \cdot (z_2 + z_5 + z_8)}{3 \cdot \Delta s^2} \\
s &= \frac{-z_1 + z_3 + z_7 - z_9}{4 \cdot \Delta s^2} \\
t &= \frac{z_1 + z_2 + z_3 + z_7 + z_8 + z_9 - 2 \cdot (z_4 + z_5 + z_6)}{3 \cdot \Delta s^2} \\
z_0 &= \frac{5 \cdot z_5 + 2 \cdot (z_2 + z_4 + z_6 + z_8) - (z_1 + z_3 + z_7 + z_9)}{9}
\end{aligned} \tag{3.25}$$

[21] Alternatively, the results of applying a 5×5 filter can be achieved by applying a 3×3 filter twice over.

In accordance with the polynomial formula, here the coefficients p, q, r, s, t approximate the following partial derivatives:

$$p = \frac{\partial z}{\partial x} \qquad q = \frac{\partial z}{\partial y} \qquad r = \frac{\partial^2 z}{\partial x^2}$$
$$s = \frac{\partial^2 z}{\partial x \partial y} \qquad t = \frac{\partial^2 z}{\partial y^2} \tag{3.26}$$

Horn (1981) proposed using a third-order finite difference estimator, so that the east–west and south–north derivatives equal:

$$p = \frac{(z_3 + 2 \cdot z_6 + z_9) - (z_1 + 2 \cdot z_4 + z_7)}{8 \cdot \Delta s}$$
$$q = \frac{(z_1 + 2 \cdot z_2 + z_3) - (z_7 + 2 \cdot z_8 + z_9)}{8 \cdot \Delta s} \tag{3.27}$$

Having only 6 coefficients, an Evans–Young polynomial does not necessarily pass through any of the 9 original elevations, but normally, it will be close to them. Its elevation at the central point is given by z_0. In the algorithm of Shary (1995), the following polynomial is used:

$$z = \frac{r \cdot x^2}{2} + s \cdot x \cdot y + \frac{t \cdot y^2}{2} + p \cdot x + q \cdot y + z_5 \tag{3.28}$$

where p, q, r, s, t are the coefficients that need to be defined and z_5 is the observed height at the central point. Fitting this equation to the sub-grid 3×3 by least squares, one obtains:

$$p = \frac{z_3 + z_6 + z_9 - z_1 - z_4 - z_7}{6 \cdot \Delta s}$$
$$q = \frac{z_1 + z_2 + z_3 - z_7 - z_8 - z_9}{6 \cdot \Delta s}$$
$$r = \frac{z_1 + z_3 + z_7 + z_9 - 2 \cdot (z_2 + z_8) + 3 \cdot (z_4 + z_6) - 6 \cdot z_5}{5 \cdot \Delta s^2}$$
$$s = \frac{-z_1 + z_3 + z_7 - z_9}{4 \cdot \Delta s^2}$$
$$t = \frac{z_1 + z_3 + z_7 + z_9 - 2 \cdot (z_4 + z_6) + 3 \cdot (z_2 + z_8) - 6 \cdot z_5}{5 \cdot \Delta s^2} \tag{3.29}$$

Shary's polynomial differs from that of Evans–Young in that it has to pass through the central point. Apart from this adjustment, the algorithms are the same, except for the r and t coefficients (Schmidt et al., 2003).

In the Zevenbergen and Thorne (1987) algorithm, the following partial quartic polynomial is used:

$$z = A \cdot x^2 \cdot y^2 + B \cdot x^2 \cdot y + C \cdot x \cdot y^2 + \frac{r \cdot x^2}{2}$$
$$+ s \cdot x \cdot y + \frac{t \cdot y^2}{2} + p \cdot x + q \cdot y + D \tag{3.30}$$

where $A, B, C, D, p, q, r, s, t$ are the coefficients that need to be defined. Here we have 9 coefficients and 9 elevations, so the polynomial passes exactly through all data points and its coefficients are:

$$\begin{aligned}
p &= \frac{z_6 - z_4}{2 \cdot \Delta s} \\
q &= \frac{z_2 - z_8}{2 \cdot \Delta s} \\
r &= \frac{z_4 + z_6 - 2 \cdot z_5}{\Delta s^2} \\
s &= \frac{-z_1 + z_3 + z_7 - z_9}{4 \cdot \Delta s^2} \\
t &= \frac{z_2 + z_8 - 2 \cdot z_5}{\Delta s^2} \\
A &= \frac{(z_1 + z_3 + z_7 + z_9) - 2 \cdot (z_2 + z_4 + z_6 + z_8) + 4 \cdot z_5}{4 \cdot \Delta s^4} \\
B &= \frac{(z_1 + z_3 - z_7 - z_9) - 2 \cdot (z_2 - z_8)}{4 \cdot \Delta s^3} \\
C &= \frac{(-z_1 + z_3 - z_7 + z_9) - 2 \cdot (z_6 - z_4)}{4 \cdot \Delta s^3} \\
D &= z_5
\end{aligned} \tag{3.31}$$

where p, q, r, s, t approximate the same partial derivatives [Equation (3.26)].

Unlike the Zevenbergen–Thorne algorithm, the Evans–Young and Shary algorithms provide a modest smoothing of the input data. Using the Zevenbergen–Thorne algorithm, the first derivative is derived as:

$$p = \frac{z_6 - z_4}{2 \cdot \Delta s} \tag{3.32}$$

In both the Evans–Young and Shary algorithms, this is replaced by the average of the three finite differences along axis x:

$$\begin{aligned}
p &= \frac{1}{3}\left(\frac{z_3 - z_1}{2 \cdot \Delta s} + \frac{z_6 - z_4}{2 \cdot \Delta s} + \frac{z_9 - z_7}{2 \cdot \Delta s}\right) \\
&= \frac{z_3 + z_6 + z_9 - z_1 - z_4 - z_7}{6 \cdot \Delta s}
\end{aligned} \tag{3.33}$$

In the presence of any error or rounding in the data, this is clearly more broadly-based, and thus more stable.

Shary et al. (2002) suggested that, before calculating the DEM derivatives, a parametric isotropic smoothing should be performed to reduce the local errors:

$$z_5^* = h \cdot \frac{z_2 + z_4 + z_6 + z_8}{9} + \left[1 - h \cdot \frac{4}{9}\right] \cdot z_5 \tag{3.34}$$

where $h \in (0, 1 - 2^{-0.5})$ is the smoothing factor. This filter will replace the elevation z_5 at the central point of the 3×3 neighbourhood portion with the new value z_5^*. Values of $h < 0.293$ are sufficient for *weak* smoothing, while a stronger smoothing

($h > 0.293$) can be achieved by:

$$z_5^* = k \cdot \frac{z_1 + z_3 + z_7 + z_9}{9} + h \cdot \frac{z_2 + z_4 + z_6 + z_8}{9} + \left[1 - (k + h) \cdot \frac{4}{9}\right] \cdot z_5 \tag{3.35}$$

where $k = 1 - 2^{-0.5} \cdot (1 - h)$ is the smoothing factor in diagonal directions. In practice, Shary et al. (2002) suggest using $h = 0.5$, which gives good-enough results for maps of curvatures for practically any type of terrain. Equation (3.35) then simplifies to:

$$z_5^* = \frac{z_2 + z_4 + 41 \cdot z_5 + z_6 + z_8}{45} \tag{3.36}$$

When all nine elevations of the 3×3 sub-grid have been replaced by their smoothed values, the original Evans–Young algorithm is applied to calculate the derivatives p, q, r, s, t. This modified Evans–Young algorithm is based on the 5×5 rather than 3×3 sub-grid. According to Peter A. Shary (http://www.giseco.info), the averaging (smoothing) in these algorithms increases in the following order:

(1) Zevenbergen–Thorne;
(2) Evans–Young and Shary;
(3) modified Evans–Young.

Skidmore (1989a) compared various early approaches for deriving derivatives and showed that the quadratic algorithm (Evans–Young) was the best for gradient (because it has the lowest standard error and mean error). For aspect, Horn's (1981) third-order finite difference method gave a somewhat lower standard error, but a much higher mean error than Evans–Young's quadratic algorithm. Guth's (1995) results showed the superiority of an *eight neighbours, unweighted* algorithm, for slope gradient and aspect. Its output suffers less from quantization, compared with Horn's eight neighbours, weighted and simpler techniques. Burrough and McDonnell (1998, after Jones) gave preference to the Zevenbergen–Thorne algorithm.

Florinsky (1998) compared four algorithms theoretically, and assumed that deviations in LSPs can be represented by the first term of the polynomial series. He used a Root Mean Square Error (RMSE) criterion to compare the algorithms, and found that the Evans algorithm was the most precise for calculating partial derivatives and the coefficients p, q, r, s and t, compared with the Zevenbergen–Thorne and Shary methods. Recently, Schmidt et al. (2003) compared the Zevenbergen–Thorne, Evans–Young and Shary algorithms experimentally, and concluded that the Evans–Young and Shary algorithms provide more precise results for curvatures (i.e. the smallest deviations from straight lines on their plots), in contrast to Zevenbergen–Thorne's.

The derivation of the formulae to extract first and second-order derivatives from DEMs illustrates the importance of making proper assumptions about the nature of the land surface. In practice, smooth models of topography and a small amount of smoothing of DEMs prior to geomorphometric analysis have proved more popular among geomorphometricians (as the heights carry measurement error anyway), although nobody can really claim that any of the above listed approaches is absolutely superior for all data sets and study areas. Recall from Section 2.3 of Chapter 1: *there can be many valid slope maps of the same study area* (Gerrard and Robinson, 1971).

4. SUMMARY POINTS

Land surface is a unique natural feature that cannot be simulated in any simple way. It needs to be measured, systematised and represented. This is mainly because landscape-forming processes alternate and interact, resulting in unpredictable features at local, regional and global scales. Even if our view of the surface is simplified to a single-valued function of latitude and longitude (with no caves, overhangs or vertical slopes), and human modifications are excluded, the land surface cannot be represented accurately by any mathematical model with a small number of parameters (Evans and McClean, 1995). Mathematical models (e.g. fractal or spectral; also Fourier series and other polynomials) of the land surface have their uses, but it can be dangerous to regard them as being universally applicable, or even as capturing the essence of a real land surface.

Understanding the concept of the land surface and its specific properties is a first step towards successful geomorphometric analysis. Many people underestimate the complexity of the land surface, and generate DEMs or run analyses blindly, without cross-checking their assumptions. There are many choices that need to be made before the actual extraction of land-surface parameters and objects commences, such as: How can heights be sampled and then interpolated? (or which data source should one choose?); Which gridding technique should one use?; Should heights be smoothed or not?; Which search radius should one use to run geomorphometric analysis? All these decisions need to be adjusted to specific datasets and case studies.

Ignoring aspects such as the correct definition of a reference vertical datum, the density and distribution of the initial height observations, and the accuracy of measurement, can lead to serious artefacts and inaccuracies in the outputs of geomorphometric analysis. For this reason, the design of DEM production and the steps in the analysis, from sampling and gridding to geomorphometric analysis, need to be adjusted to the properties of a specific terrain. For example, a preliminary variogram of heights in a study area can provide considerable information about surface smoothness and/or measurement error that can help us to filter out missing values or assumed artefacts (see further Chapter 4). The distances between contours, and field estimates of the spatial scale of processes and the density of hydrological features, can be used for making decisions about a suitable grid resolution or to perform an additional sampling of heights.

> REMARK 7. *In the near future, DEMs will consist of multiple layers that will carry information about sub-grid properties and the local uncertainty of heights.*

Digital models of the land surface might also be improved, but we do not anticipate that the currently dominant gridded DEM model will be replaced by better models in the near future. There are two obvious reasons for this: (1) increasingly, heights are being sampled by scanning devices, and those on airborne or satellite platforms produce regular sets of heights (images); and (2) gridded models fit the current design of computer programs too well to change them. With the exception of drainage routing, it is much easier to program algorithms for geomorphometric analysis using grid models, i.e. rasters or matrices of heights, than to use TINs or surface-specific models.

Laser scanning can already provide at least two versions of the surface — the vegetation canopy (first returns) and the ground surface (last returns). We anticipate that, in the near future, DEMs will provide not only layers of a single variable, but will consist of multiple layers. One additional layer that is likely to be added is the estimated height-error, but we could also attach layers that define a surface's sub-grid properties: for example, polynomial coefficients that can be used to rebuild the land surface on a finer grid; or local measures such as surface roughness or grain-size statistics. Information should also be added about the beds of water bodies (channels and lakes), as well as their surfaces, and perhaps also about how they have changed over time. In turn, this will require more sophisticated geomorphometric algorithms — ones with the capacity to include this type of information in the derivation of land-surface parameters and objects.

IMPORTANT SOURCES

Li, Z., Zhu, Q., Gold, C., 2005. Digital Terrain Modeling: Principles and Methodology. CRC Press, Boca Raton, FL, 319 pp.

Rana, S. (Ed.), 2004. Topological Data Structures for Surfaces: An Introduction for Geographical Information Science. Wiley, New York, 214 pp.

Hutchinson, M.F., Gallant, J.C., 2000. Digital elevation models and representation of terrain shape. In: Wilson, J.P., Gallant, J.C. (Eds.), Terrain Analysis: Principles and Applications. Wiley, pp. 29–50.

Mitas, L., Mitášová, H., 1999. Spatial interpolation. In: Longley, P., Goodchild, M.F., Maguire, D.J., Rhind, D.W. (Eds.), Geographical Information Systems: Principles, Techniques, Management and Applications, vol. 1. Wiley, pp. 481–492.

Weibel, R., Heller, M., 1991. Digital terrain modeling. In: Maguire, D.J., Goodchild, M.F., Rhind, D.W. (Eds.), Geographical Information Systems, vol. 1. Longman, London, pp. 269–297.

CHAPTER 3

DEM Production Methods and Sources

A. Nelson, H.I. Reuter and **P. Gessler**

the most common data sources for DEM data · comparison between ground and remote sensing-based techniques · the most frequently used and contemporary DEM production methods · production and digitising of topographic maps · LiDAR and SRTM DEMs · Geoscience Laser Altimeter System (GLAS) · ASTER and SPOT DEMs · the strengths and weaknesses of DEM data derived from different sources and methods

In general, there are three sources of DEM data:

- *Ground survey techniques* — the accurate surveying of point locations (latitude, longitude and elevation or x, y, z values). We will look at traditional and high-tech ground survey techniques.
- *Existing topographic maps* — the derivation of contours, streams, lakes and elevation points from hardcopy topographic maps. We will focus on digitising (using a digitising table, or on-screen) and semi-automatic scanning to convert raster images of topographic maps into vectors.
- *Remote sensing* — the interpretation of image data acquired from airborne or satellite platforms. We will pay particular attention to: (i) Photogrammetric/stereo methods (encompassing both airborne and satellite); (ii) Laser (mostly airborne at present, but will be from satellites in the future); and (iii) Radar (both airborne and satellite — using interferometry).

1. GROUND SURVEY TECHNIQUES

The horizontal and vertical location of points on the Earth's surface can be geolocated with an accuracy of a few millimetres. Such surveys are carried out using *theodolites* (instruments for measuring angles in horizontal and vertical planes), notebooks and triangulation methods (calculating distances and angles between

Developments in Soil Science, Volume 33 © 2009 Elsevier B.V.
ISSN 0166-2481, DOI: 10.1016/S0166-2481(08)00003-2.

points) to create a dense mesh of triangles with observation points at each apex. Plotting these observations of location and elevation on paper or digitally, results in an accurate, scaled representation of the features of the terrain. This is a time-consuming process requiring highly skilled and meticulous surveyors. Though no less skilled, the advent of electronic theodolites, total stations, and Electronic Distance Measuring (EDM) which measures the characteristics of a LASER (*L*ight *A*mplification by the *S*timulated *E*mission of *R*adiation) fired between an observation point and a reflecting target point has speeded up the collection of these observations. When these observations are then combined with computer modelling, surveys with accuracies of a few millimetres over many kilometres can be created.

A drawback is that the complexity and cost of surveying with such equipment requires dedicated surveying teams, and these are often beyond the means of small mapping projects. An alternative source of survey data can be derived from Global Positioning System (GPS) units, although they are less accurate. To increase the accuracy of the GPS signal, differential GPS (DGPS) can be used to transmit (by radio or satellite) the error of the GPS signal measured at a stationary location. Manufacturers of such systems quote vertical and horizontal accuracies as being within the ranges of 4–20 and 8–40 m for GPS and 1–3 and 2–6 m for DGPS readings, respectively. In good conditions, i.e. with five or more satellites, these ranges are very conservative, as a horizontal accuracy of less than a metre, and vertical accuracy of 1–2 m, can be easily achieved.

To locate observation points, whereas triangulation surveys require some planning, GPS surveys can be carried out quickly, by simply traversing the study area and taking a reading at regular intervals. The tabular data from both types of survey consists of at least point identifiers and Easting, Northing and Height measurements. To view them, the x (Easting), y (Northing) and z (Height) coordinates for each point can be converted into common GIS formats (Figure 1). These point data are used to create a Triangular Irregular Network (TIN) (see Section 2 in Chapter 2), or they can be interpolated either into contours or into a continuous gridded representation of the terrain.

Some of the advantages of ground survey information are high accuracy (elevations can be determined to an accuracy of around, or even less than, 1 cm); flexibility (the measurement density can be varied, depending on the terrain); and very little processing is required after the measurements have been taken. The major drawbacks are that the equipment is expensive, and intensive effort, and a lot of time, is required. For this reason, large surveys are almost exclusively performed as part of detailed construction or monitoring projects (for dams, roads, and bridges, etc.). In the past, National Mapping Agencies relied on these surveying techniques for creating topographic maps, but nowadays these have been largely superseded by remote sensing methods (Smith, 2005).

1.1 Topographic maps

There will be situations where individuals, agencies or jurisdictions will not have access to DEM data generated from expensive (and more accurate) methods.

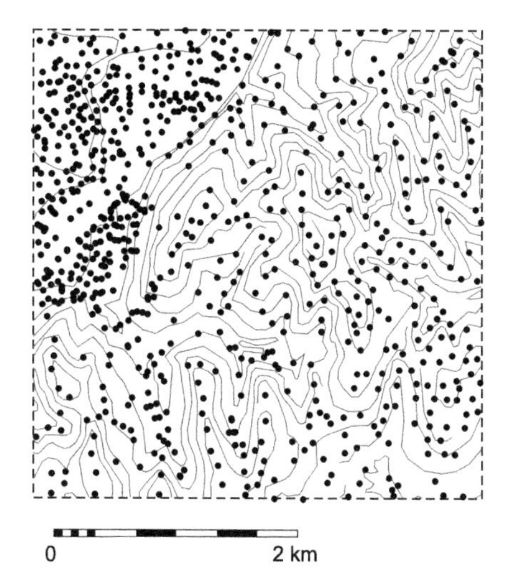

FIGURE 1 Spot heights from survey data of the Baranja Hill study area.

In these cases adequate DEMs can be extracted from contours as presented in existing topographic maps. The following sections are meant to provide guidance and instructions on methods that can be followed to create DEM surfaces starting from a paper topographic map. These methods represent a worst-case scenario and are no longer the main source of DEM data.

1.2 Manual digitising of topographic maps

The conversion of data presented in analogue form, such as maps and aerial photographs, into a digital form, is normally done manually by a human operator using a digitiser, although there are also automated and semi-automated digitising methods (Burrough and McDonnell, 1998). A longstanding technique for creating DEMs is to digitise heights from topographic maps, although it is becoming less common. A typical topographic map (Figure 2) will contain several thematic layers of information, such as contours, spot heights, intersecting features (such as cliffs, road and railway dissections, and dikes) and different types of water bodies such as rivers, coastlines and lakes. These thematic layers can be digitised into separate vector datasets and used as inputs to interpolation algorithms for creating a DEM.

Carrara et al. (1997) provide a comprehensive comparison of techniques for generating DEM data from contour lines. Various interpolation methods are described in Chapter 4, while the quantification of uncertainty, and errors in the resulting DEMs, is discussed further in Chapter 5. Details about interpolation procedures for specific software packages can be found in the various software chapters.

As mentioned above, digitising can be performed using a large dedicated digitising table and a cross-hair mouse linked to a computer installed with the appropriate software. The topographic map is placed on the table, georeferenced using

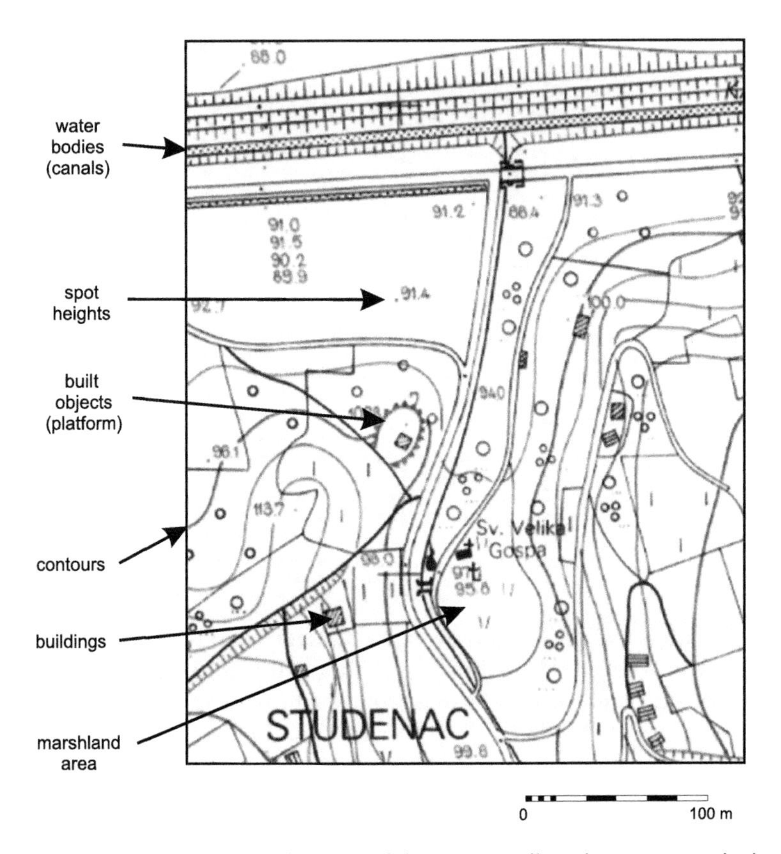

FIGURE 2 The 1:5000 scale topographic map of the Baranja Hill study area. Note the large number of features that emerge by showing a combination of contours, spot heights, water bodies and the type of land cover. Courtesy of State Geodetic Administration of Republic of Croatia.

corner coordinates, and the mouse is used to trace each map feature in turn. A single click will digitise a point feature, whereas linear or polygonal features require multiple clicks along their length or perimeter. The number of clicks needed will depend on the size and complexity of these features. The attribute information for each feature is added later, during the editing and labelling process. In theory, the denser the mouse clicks, the more detailed will be the information captured during the digitising process. Alternatively, the topographic map can be scanned using a large-format scanner to create a high-resolution geo-referenced image. The features shown on the map can then be digitised manually from this image *on-screen*.

Using a digitising table has the advantage that it can easily accommodate large sheets of topographic maps. It is also a better option when digitising old or worn-out sheets of maps, since features that may have discoloured will be easier to detect with the naked eye on the original map sheet than they would be on a scanned image on a monitor screen.

An advantage of on-screen digitising is that the digitiser can zoom into particular regions of the map, thus producing a more accurate digital map. Another advantage is that the data can be simultaneously digitised and edited, unlike the two-stage process required when using a digitising table. There is also a trade-off between the physical discomfort experienced by some operators at a digitising table and the potential eye strain that may result from on-screen digitising. In addition, we could also mention the mental strain for the many people who find digitising a tedious task!

1.3 The automated digitising of cartographic maps

Another approach is a semi-automated process for extracting features, using dedicated feature-recognition software to perform raster-to-vector conversion. This software automatically identifies the different thematic layers (often using colour classification algorithms) in the scanned image of the topographic map and splits them into separate raster layers.

These layers are then edited manually to clean up the features and correct them, before they are converted into vectors, for further editing and labelling. There are several software products for semi-automated digitising. One of these is the **R2V** software for converting rasters into vectors (http://ablesw.com/r2v/). Typically, this kind of software can input different scanned-image formats (such as `GeoTIFF`, `JPG`, `BMP`, etc.), process automatic images to extract the different thematic layers from the scanned image, trace lines automatically and export them into common GIS vector formats (including ESRI Shapefiles `SHP`, MapInfo `MIF`, AutoCAD `DFX` and Scaleable Vector Graphic or `SVG`), edit and label vectors, by georeferencing and rubber-sheeting them.

Raster-to-vector conversion algorithms typically use two approaches, depending on the map element being scanned: *optical character recognition* (OCR) and *skeletonising*. OCR is used to convert the map text (place names, labels, contour intervals, map information and metadata, etc.) into text that can be machine edited for later use when labelling and tagging map elements semi-automatically. The algorithms are trained to recognise most print fonts, so the OCR for roman printed or typed text (as opposed to italics or hand-written text) is generally very accurate and reliable. Skeletonising is used on line elements (rivers, roads, contours, power-lines, etc.) for converting scanned lines into vectors. Scanned lines may have a varying pixel width along the length of the line. In addition, in some map-symbol systems, single rivers and roads are represented by two or more lines. In these cases, skeletonising reduces the width of such elements to one pixel (typically positioned along the centre of the scanned line) so that they can be converted into vectors.

It is essential to have access to a high quality, large-format scanner. For large projects, scanning facilities are often available in-house. For smaller projects, scanning-service providers are more cost-effective. If the map is in colour, to be able to classify the colour accurately so that it can be separated into thematic layers, the scanned images should be in 24-bit colour. For black-and-white maps,

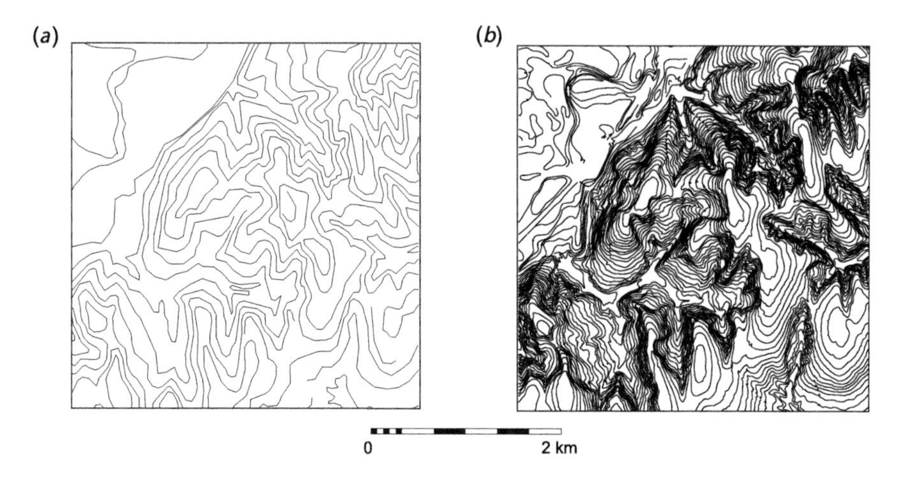

FIGURE 3 Contours extracted from a 1:50,000 scaled topographic map (left) and from a 1:5000 scaled map (right).

an 8-bit grey scale may be sufficient for separating ranges of grey scale into distinct layers.

The resolution should be between 200 DPI (for on-screen digitising) and 800 DPI (dots per inch[1]) depending on the quality and level of detail in the source map, but some experimentation will be required to ensure that the scanned image is of the highest quality possible. For semi-automated processing, the lines should be at least 3–4 pixels thick (about 600 DPI). The quality of the input image (in terms of clarity, sharpness, colour separation and contrast) can often be improved by using built-in image processing algorithms in the software, or by using third-party image processing software.

The advantages of digitising and scanning are that they can be carried out on any hard-copy topographic map of any scale and, assuming the availability of a suitably scaled topographic map, for more or less any size of study area. Figure 3 demonstrates the different levels of detail in contour information derived from (a) a 1:50,000 scaled topographic map, and (b) a 1:5000 scaled map. The disadvantages are that digitising and scanning are both expensive and time consuming (major expenses in projects are often incurred for data retrieval and processing) and the accuracy and skill of the operator determine their quality. For example, a map is usually geo-registered using the tick marks from the reference grid. This, in itself, can be a challenging task, which may result in significant positional errors in the horizontal plane. Again there is a trade off between manual and semi-automated methods. Though very laborious, data retrieval in digitising is a very accurate process, so it is much quicker and easier to edit. In contrast, retrieving data by scanning, though fast and semi-automated, often produces image layers that require a lot of data cleaning and editing before they can be converted into vector formats. These, then, also have to be edited. In Figure 3, some of the contour

[1] 200 DPI means that one pixel is about 0.17 millimetre, which is about the required cartographic standard for Maximum Location Accuracy (Rossiter and Hengl, 2002).

lines in both the 1:50,000 and 1:5000 scaled maps are discontinuous. This is partly due to lack of data in the stereographic process. This occurs where the ground elevation could not be determined, due to buildings, shadows, and clouds, etc.

Hard-copy topographic maps in a range of scales can be obtained from: the cartographic/surveying departments of local authorities, National mapping agencies,[2] university map libraries, and many good map and book stores. Whether they will be useful for a particular application will depend on their scale, timeliness, level of consistency, and the physical condition of the map (because paper maps can suffer from many distortions, due to humidity, handling, etc.).

Original map sheets in Mylar give the best results. Mylar is an extremely robust and stable polyester film which can be used to make high quality, flexible, transparent map sheets. Unfortunately, however, it is often very difficult to acquire maps in this format. However, no matter which format is used, in digitising, it is essential to honour the Copyright Law by adhering to the restrictions for reproducing mapped information in a digital format.

2. REMOTE SENSING SOURCES

In contrast to the methods shown above, remote sensing methods can rapidly cover large areas. The platform for remote sensing can be either airborne or situated in space (in a satellite), and the resulting imagery can be derived from three types of sources: *aerial photography*, *LiDAR* and *RADAR*. We will discuss each of these sources in turn and the most common DEM products derived from them.

2.1 Photogrammetric land-surface models

Aerial photographs are essentially high-resolution, high-quality photographs taken from airborne platforms. The photographs are usually natural colour, black-and-white or occasionally infra-red images. By using survey data and *Ground Control Points* (GCP), these photographs can be geo-referenced, digitally. A single photo might give an excellent visual overview of the terrain, but it does not provide information about the elevation. If several flight lines, or blocks of images for a geographic region with sufficient overlap — typically 60% (Figure 4) — can be acquired, then stereo photos, and the stereo models associated with them, can be derived. To do this, ground control points and photogrammetric principles are used to extract the necessary elevation information (Wolf and Dewitt, 2000). This same information can also be used for *orthorectification*. The resulting image is known as an orthophoto — an accurate representation of the location of objects in the photo. An overview of photogrammetry is provided by Smith (2005), and a thorough review of the process of generating DEMs from digital stereo imagery is provided by Lane et al. (2000).

[2] See also Smith (2005, pp. 158–159) for a list of agencies.

FIGURE 4 A series of three aerial photographs used to derive a DEM for Baranja Hill. The arrows indicate the line of flight. Courtesy of State Geodetic Administration of Republic of Croatia.

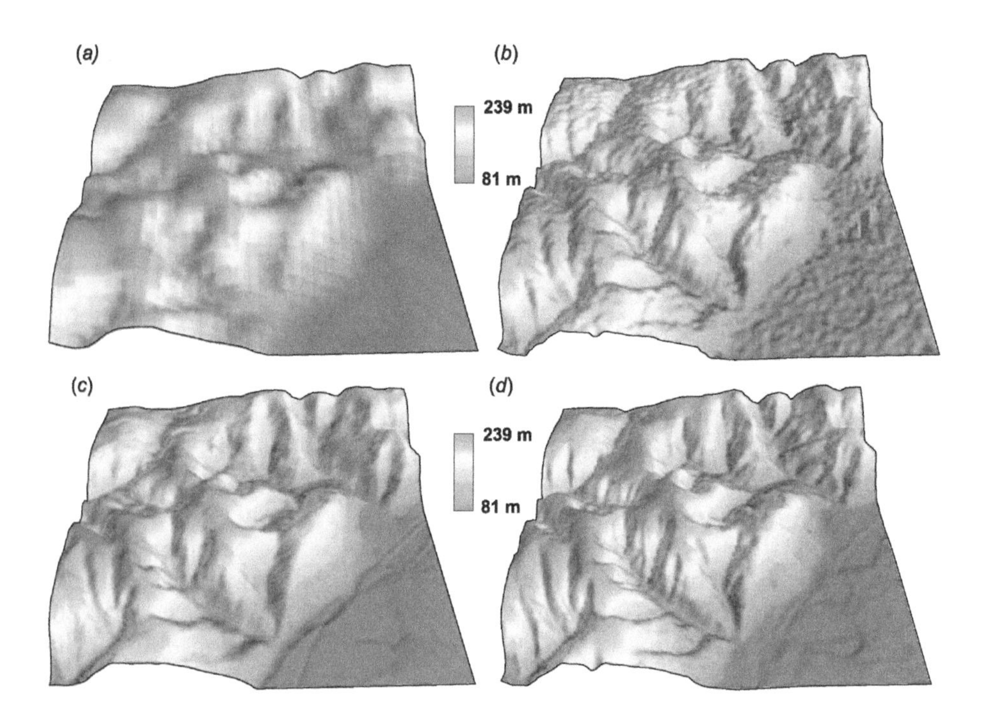

FIGURE 5 Comparison of DEMs from main sources for Baranja Hill: (a) 90 m resolution SRTM DEM, (b) 30 m resolution SRTM DEM, (c) DEM from 1:50,000 topo-map, and (d) DEM from 1:5000 topo-map. (See page 710 in Colour Plate Section at the back of the book.)

REMARK 1. *Digital photogrammetry requires huge hardware and software resources because huge data volumes have to be processed and stored (one standard 23×23 cm colour aerial photo in digital format requires approximately 1 GB).*

An advantage of photogrammetric DEM is that it is a standard approach. It is one that has been in use for several decades, and is still improving. For example, the amount of manually identified GCPs has been reduced by using in-flight *Real-Time Kinematic GPS* (RTKGPS) systems, whereby real-time corrections, accurate to within a few centimetres, can be made from just one reference station. Another advantage is that creating LSM is usually a self-contained process. It creates a visual record of the surface, and although fewer points are collected, the result is more focused than in a LiDAR approach (Molander et al., 2006).

There are several drawbacks to using stereo-photos to create DEM data. It is possible that there will be a systematic over-estimation of elevation due to camera distortion. The resulting DEM will often have spikes or pits in places where the DEM generating algorithm incorrectly matches two points from the stereo-pair. Aerial photography captures Earth's surface cover rather than the Earth's surface itself, so the final results will include tree-top canopies and buildings. This gives higher elevation values, rough surfaces and high slope values. Finally, the method requires GCPs which may not be available, or which may not be very accurate (Zukowskyj and Teeuw, 2000). Typically, aerial photography missions are undertaken when there is no snow-cover, and preferably when there is little or no leaf-cover. Hence they are limited by both seasonal and weather conditions.

2.2 LiDAR

The first commercial topographic mapping systems to use airborne laser scanning or LiDAR (*Li*ght *D*etection *a*nd *R*anging) appeared in the early 1990s. LiDAR is a type of active sensor, whereby the sensor transmits a signal (in near infra-red, or sometimes visible green part of the spectra) towards the ground and then records the reflection returning from that signal. The time delay between the transmission and reflection of the signal determines the distance between the sensor and the ground. Typically, for each second, between 5000 and 100,000 x, y, z data points are collected. In general, LiDAR data have been estimated to have measurement errors of around 15 cm in the vertical plane (Huising and Gomes-Pereira, 1998) and 50 to 100 cm in the horizontal plane (Smith, 2005). The majority of LiDAR systems use a near infra-red laser which is unable to penetrate fog, smoke or rain (Fowler, 2001; Norheim et al., 2002). LiDAR also has a relatively small footprint (90 km^2 per hour), so it is costly to create LiDAR-based DEMs for areas much larger than 20,000 km^2 (X. Li et al., 2001; Smith, 2005).

REMARK 2. *The biggest advantages of using LiDAR are high density of sampling, high vertical accuracy and a possibility to derive a set of surface models.*

The most common sensors in LiDAR systems are the discrete return sensors. These are able to receive multiple return signals, in the form of a sub-randomly distributed 3-D point cloud, from one single transmission. For the basic relations and formulae, please refer to Baltsavias (1999b). For example, the first object that the signal 'hits' could be forest canopy, so it would be this surface that would reflect the first return signal to the sensor. However, if the canopy was sparse, then some of the signal would continue down towards the ground surface, hitting any other objects in its path, which could also return signals. So the data received could be in the form of $x, y, z_1, z_2, z_3, \ldots, z_n$, where z_1 is the first (highest elevation) return signal and z_n is the last (lowest elevation) return signal.

The last return signal is often from ground elevation, but if sufficiently dense, it could be from the vegetation cover instead, as, in that case, no signal would be able to penetrate down to ground level. The importance of multiple signals, therefore, is that they usually record the character of both the ground surface and the vegetation or any other structures above the ground. For developing DEMs, and for other applications too, such as forestry, where estimates of timber volume or biomass have to be made using points above ground level, methods and algorithms that separate ground returns from those recorded above ground are critical (Axelsson, 1999).

One of the most recent datasets that has been provided to the international community is the elevation information from the *Geoscience Laser Altimeter System* (GLAS) instrument on the NASA ICESat satellite. This spaceborn LiDAR instrument emits 40 pulses per second and can generate elevation measurements with a vertical accuracy of 15 cm for a 60 m footprint, with measurements spaced 170 m apart (Zwally et al., 2002).

The production time for LiDAR DEMs is typically shorter than that for photogrammetrically generated DEMs (Baltsavias, 1999a). Despite reports in the literature (e.g. Kraus and Pfeifer, 1998) that LiDAR-derived DEMs tend to be smooth because of the filtering algorithms typically applied to them, from our own work on an agricultural area, we have not found this to be the case (Reuter, 2004). We would therefore prefer a model created in this way to a smooth, conventionally created DEM.

The disadvantages of LiDAR data are that they produce a very dense and detailed land-surface model, which could be difficult to handle during the production process. Also, the accuracy of the readings varies according to the characteristics of the terrain. For example, it may be impossible to measure very steep slopes accurately (Smith, 2005).

Nevertheless, LiDAR is definitely the method of the future. Several countries (e.g. Belgium, the Netherlands, etc.) have already produced national LiDAR DSM/LSM at resolutions of 2–5 m. In the terms of both spatial and vertical accuracy, this type of data is far superior to comparable DEMs derived from topographic maps or remote sensing imagery.

2.3 Radar

Radar systems can be either airborne or satellite-based. Platforms in space are particularly attractive, because, from them, large areas can be mapped within a short span of time, irrespective of whether there is access to that airspace or not. We will give a brief overview of radar systems and the issues that are common to all radar sources, and then we will discuss several systems in detail — one airborne system and the most common space-based sources.

For space-based, radar scanning systems, we distinguish between repeat-pass interferometry (e.g. ERS-1/2 tandem data), where the same scene is acquired with a short time-frame (typically one day) or single-pass interferometry (e.g. the SRTM mission), where a second, passive, antenna is deployed to synthesise a second image. In terms of DEM processing, interferometry is the creation of DEMs based on the phase difference between two recorded radar images, together with the flying height of the antennas.

As well as covering large areas rapidly, the longer radar wave-lengths are able to penetrate smoke, fog and rain, and, as a result, are almost independent of weather conditions. The downside is that radar data often contain errors and omissions. Compared with topographic DEMs, radar-based DEMs contain a lot of speckling (noise) and certain features, such as towers or mountains, can be mislocated due to a foreshortening effect whereby features that are tilted towards the direction of the radar signal are compressed. Finally, where there is no return signal because the ground target is obscured by a nearby tall object, shadowing occurs. This leaves a hole in the data, since no elevation values can be computed for the locations in the shadow. In the course of the interferometric process, height-map errors can be generated. These include measurement errors (inaccuracies in point determination) and geometrical errors in imaging (inaccuracies in orientating exteriors). The German Space Agency (X-SAR), for example, provides a product for height errors, whereas the NASA-SRTM output does not provide any such data, due to security restrictions.

REMARK 3. *Compared with topographic DEMs, radar-based DEMs contain a lot of speckling (noise) and towers or mountains will be mislocated due to a foreshortening effect.*

Based on techniques described in Graham (1974), the first commercial airborne RADAR scanning system — *InSAR/IfSAR* (Interferometric Synthetic Aperture RADAR) appeared in 1996. As InSAR/IfSAR systems can be flown at greater altitudes and at faster speeds resulting in larger footprints than LiDAR, these systems do not suffer the same problems. However, the longer wave-lengths also mean a loss in resolution and less accuracy than LiDAR (Hensley et al., 2001; Norheim et al., 2002). By using interferometry,[3] the steps in processing the radar data are much more sophisticated.

3. FREQUENTLY-USED REMOTE-SENSING BASED DEM PRODUCTS

3.1 SRTM DEM

One of the biggest and most complete missions in terms of coverage was the *Shuttle Radar Topography Mission* (SRTM), which was carried out between the 11th and

[3] For an introduction to radar interferometry, see Li et al. (2005, pp. 39–50).

20th of February 2000, onboard the space shuttle, '*Endeavour*' (Rabus et al., 2003). The area covered — between 60° North and 58° South — was recorded by X-Band Radar (NASA and MIL, covering 100% of the total global area) and C-Band Radar (DLR and ASI, covering 40%). The publicly available NASA global dataset has a resolution of approximately 90 m (3 arcsec). The non-public DLR–ASI data is available with a resolution of approximately 30 m (1 arcsec). Unlike the C-band system, the X-band could not steer its beam, so it could not operate in ScanSAR mode and therefore could not obtain full coverage of the Earth (Figure 7). Its 50 km swath offered nearly complete coverage at high latitudes, though. It also has a little better vertical accuracy — around 5 m.

The SRTM data[4] is projected geographically, with elevation reported in metres. It is referenced to the WGS84 EGM96 geoid and is georeferenced in the horizontal plane to the WGS84 ellipsoid. Original data from the SRTM mission are provided as binary `.HGT` files, but without any header information. In addition, worldwide, there are several other datasets, each incorporating different types of improvement (e.g., water body and coastline identification, void filling and mosaicing).

Because the X-Band SRTM data is freely available at a resolution 100 times greater than was previously the case (e.g., the 30″ resolution global GTOPO30 and GLOBE DEMs), and the coverage is almost global, it has attracted a lot of interest from third parties who are also distributing variants of the global SRTM data (Rabus et al., 2003). These include:

1. *USGS* (United States Geological Survey), which provides 1°×1° un-projected (Plate Caree) tiles of the unfinished (version 1) data and finished (version 2) data in `HGT` binary format (but, with no header!) as well as the Small Water Body Shape files. USGS also supplies SRTM in 4 different formats for user-selected regions via http://seamless.usgs.gov.
2. *CGIAR-CSI* (Consultative Group on International Agricultural Research-Consortium for Spatial Information), which provides 5°×5° un-projected tiles of topographically correct, void filled and coastline clipped version 2 SRTM data in `GeoTIFF` and `ASCII` format as well as the voids (for reference purposes) and Small Water Body Shape files.
3. *GLCF*, which provides the version 1 and version 2 data as `GeoTIFF`s on 1°×1° un-projected tiles and on much larger WRS-2 tiles in UTM projection (to match the WRS-2 Path/Row specification).
4. *WWF/USGS*, which is developing a worldwide set of SRTM derivates (HydroSHEDS), including river networks, watershed boundaries, drainage directions, and flow accumulations — which can be seen as an improvement on the GTOPO30 derivates (HYDRO1K).

`HGT`, `GeoTIFF` and `ASCII` formats can be read by most geomorphometric analysis packages, or can be converted into other formats using the GDAL (Geospatial Data Abstraction Library) conversion tools.

[4] A complete technical description of the SRTM data is available at: http://www2.jpl.nasa.gov/srtm/srtmBibliography.html.

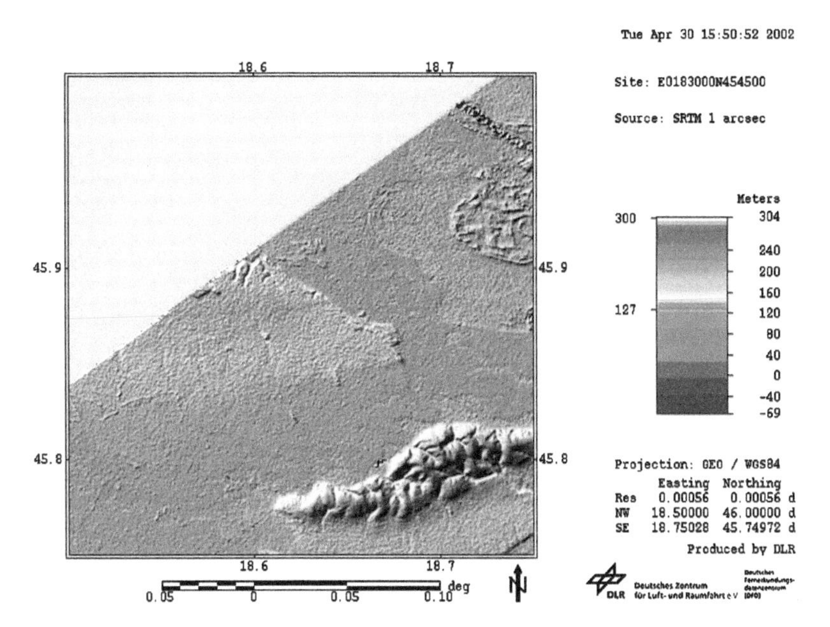

FIGURE 6 Example of a $15' \times 15'$ block of 1 arcsec SRTM DEM ordered for Baranja Hill. Courtesy of German Space Agency (DLR). (See page 710 in Colour Plate Section at the back of the book.)

REMARK 4. *The 3 arcsec SRTM DEM is one of the most consistent, most complete and most used environmental datasets in the world.*

What is often forgotten about SRTM is that the elevation data represent a DSM (see also Chapter 2), not a bare-earth model. This means that dense canopy forests and built-up areas are included. The presence of such features can be quite problematic, for example, in hydrological modelling. Other problems arise because of the nature of the interferometric process used to generate the DEMs. For example, at the land-water interface, there may be areas, known as voids, where there is no data, and in desert and mountain areas, problems can occur due to foreshortening and shadowing (Rodriguez et al., 2005). See Figure 12 for an example of surface detail in SPOT DEMs.

Figure 5(b) shows a DEM of the Baranja Hill study area, derived from the C-Band[5] product. Even when printed at this scale, when compared to the TOPO DEM in Figure 5(d), the speckled appearance of the SRTM DEM is obvious. A more obvious comparison can be made by simply subtracting the SRTM DEM from the TOPO DEM on a pixel by pixel basis (Figure 9). In this case, we are convinced that the SRTM DEM shows heights of canopy and not of land surface.

The range of elevation in the TOPO DEM is between 85 and 243 m, whereas as the SRTM DEM has a range of between 35 and 250 m. The differences, varying

[5] The $15' \times 15'$ block (Figure 6) was ordered from the German Aerospace Agency (http://eoweb.dlr.de).

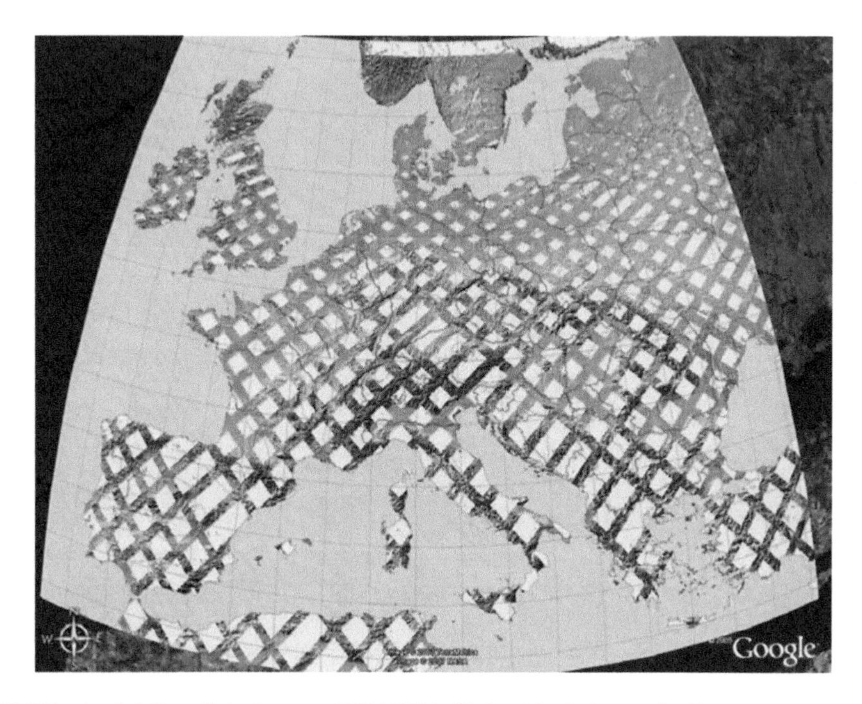

FIGURE 7 Availability of the 1 arcsec SRTM DEMs (C-Band Radar) over the European continent. Missing areas are were not acquired due to an energy shortage at the end of the mission (Rabus et al., 2003). To load the Google Earth map, visit geomorphometry.org. (See page 711 in Colour Plate Section at the back of the book.)

FIGURE 8 Availability of the 30 m ASTER DEMs over the European continent (before January 2006). (See page 711 in Colour Plate Section at the back of the book.)

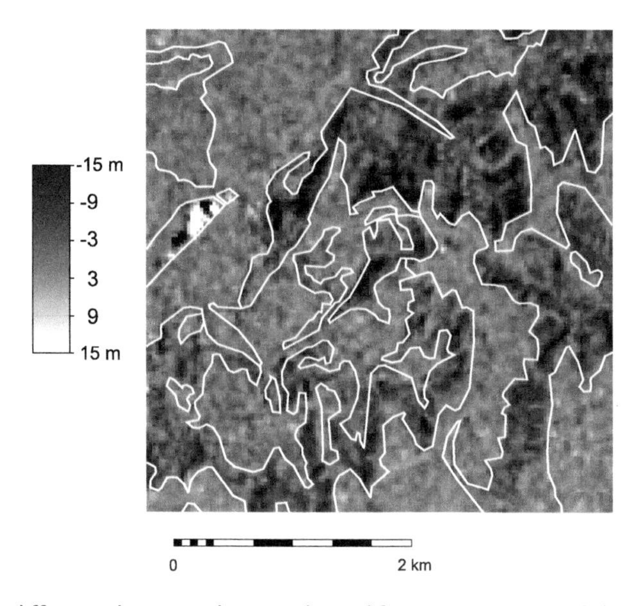

FIGURE 9 The difference between the DEM derived from a topo-map and the SRTM DEM usually reflects the natural vegetation. Forest borders and land-cover units such as water bodies are overlaid.

TABLE 1 The differences in elevation by land-cover classes between the topo- and the SRTM DEM

Land cover	Area in ha and as a %		Mean and std. of the difference in m	
Urban	46	(3%)	−0.59	(1.71)
Agriculture	455	(33%)	−0.16	(2.41)
Water	17	(1%)	4.71	(22.73)
Grassland pastures	235	(17%)	−0.66	(2.73)
Natural forest	610	(45%)	−4.66	(4.36)

from −54 to +140%, are normally distributed, but are not randomly distributed across the study area. This can be seen in the map and in Table 1 and Figure 10. The differences are clearly concentrated in two land-cover classes: (i) the forested areas where, on average, the SRTM DEM elevations are 4.66 m (2.49%) higher than the TOPO DEM elevations due to the forest canopy (Figure 10), and (ii) the water-body areas where the difference is −4.71 m (5.41%). The huge standard deviation — 22.73 m and 25.88% — over these areas reflects the difficulty of generating reliable results from radar-derived DEM data over water. The average differences are minimal for other land-cover classes. The mean differences there are <1 m (<1%) and the standard deviations are between 1.71 and 2.73 m (1.48 to 1.67%).

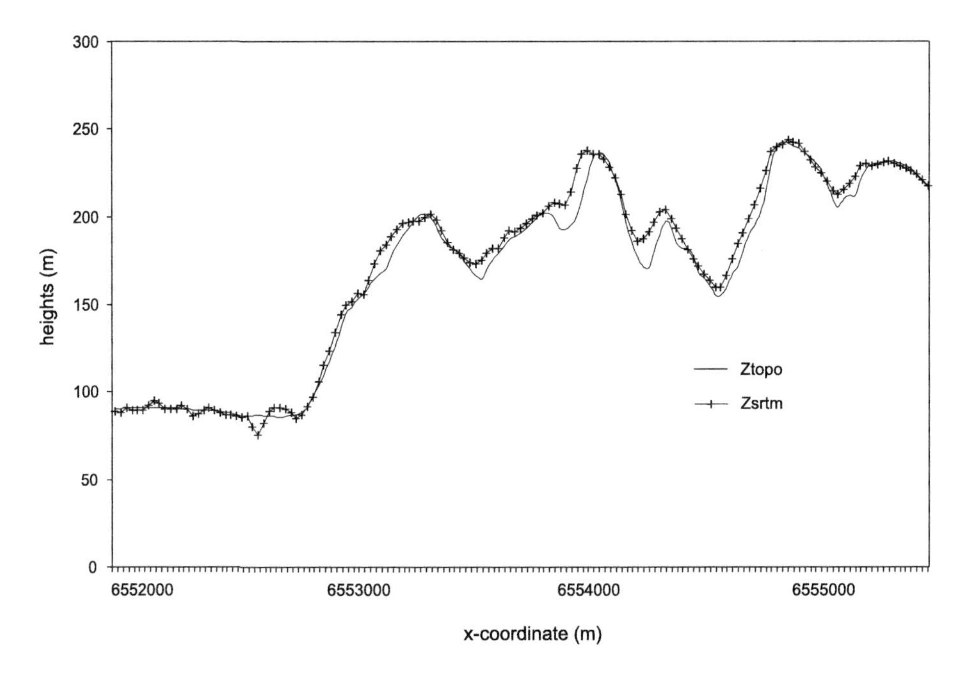

FIGURE 10 Transect of the SRTM and TOPO DEMs for Baranja Hill from West to East at $y = 5{,}073{,}012$. Note that the difference between the two DEMs can be used to map the height of canopy. In this case, we can also notice a difference between the two DEMs which is due to the foreshortening effect of the radar technology (systematic error).

3.1.1 ASTER DEM

The *Advanced Spaceborne Thermal Emission and Reflectance Radiometer* (ASTER, http://asterweb.jpl.nasa.gov) instrument is situated onboard the TERRA Satellite, which was launched in December 1999 as part of NASA's Earth Observing System (EOS). See Fujisada et al. (1999) for further information about this platform. ASTER DEMs are generated using 3N (nadir-viewing) and 3B (backward-viewing) bands of an ASTER Level-1A image acquired by the Visible Near Infra-Red (VNIR) sensor. The resulting DEM has a resolution of 30 m at the Equator. The vertical and horizontal accuracy is 30 m, and, without using any GCPs, the level of confidence is 95% (this is known as the *relative DEM product*).

If user-supplied GCPs are employed (the *absolute DEM product*), the vertical accuracy can increase to 7 m (Fujisada et al., 2005). Each image covers an area of 60 km^2 (Figure 8) and in January 2006, cost $80 USD per scene if downloaded (or it can be purchased by media order[6] for about $90 USD).

Since all the ASTER bands can be downloaded, it is possible to create your own DEM, if you have suitable software for processing the 3N and 3B bands in conjunction with available GCPs. One such software package, PCI Geomatica **OrthoEngine**, can extract a DEM from the two ASTER bands automatically. Using **OrthoEngine**, it is a straightforward process. It sets the geo-referencing and pixel

[6] http://edcimswww.cr.usgs.gov/pub/imswelcome/, http://glovis.usgs.gov.

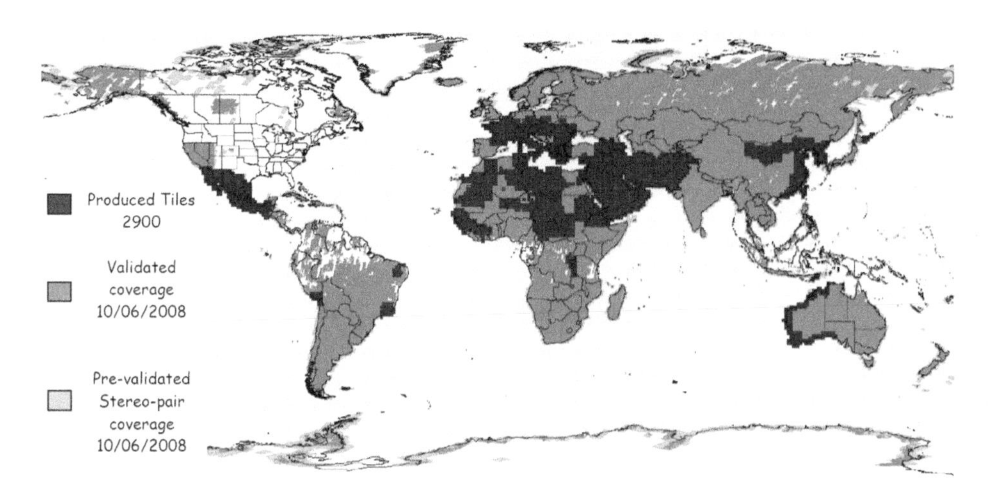

FIGURE 11 SPOT's validated HRS map showing the global availability of SPOT DEMs on 10th of June 2008. For un update, go to www.spotimage.fr (section SPOT 3D).

spacing information (at UTM and 15 m, respectively); reads the two ASTER channels from the CD or downloaded file; geo-references each band separately; selects the GCPs; and then creates a stereo image from which the 30 m DEM can be extracted.

An exciting development is the upcoming ASTER Global Digital Elevation Model (ASTER G-DEM), a joint collaboration between The Ministry of Economy, Trade and Industry of Japan (METI) and NASA. METI and NASA are undertaking the massive task of generating a seamless DEM from thousands of ASTER scenes to create the ASTER G-DEM product. ASTER G-DEM promises to be an improvement over the SRTM DEM in terms of resolution (one arc second instead of three arc seconds) and spatial coverage (83°N to 83°S instead of 60°N to 58°S), and will have comparable accuracy (20 m vertically and 30 m horizontally compared to 16 m and 20 m respectively for SRTM). The initial ASTER G-DEM will likely suffer from missing data — as does the SRTM DEM — because of cloud cover, but since there is an open ended time period for data acquisition (from year 2000 onwards), these missing data areas will be much easier to fill when sufficient numbers of new cloud free ASTER scenes become available. The ASTER G-DEM is scheduled for release in 2009, http://www.ersdac.or.jp/GDEM/E.

3.1.2 SPOT DEM

SPOT5 (http://spotimage.fr) was launched in 2002. It includes a *High Resolution Stereoscopic* (HRS) instrument, with a resolution of 1 arcsec (approximately 30 m at the Equator). SPOT DEMs are derived by automatically correlating stereopair images from forward and aft facing sensors in the HRS instrument. They have a vertical accuracy of at least 10 m and a horizontal accuracy of at least 15 m at a confidence level of 90%, with no GCPs. Each image swath is 600 km by 120 km and the price (in 2005) was €2.30 per km^2 for a minimum order of 3000 km^2

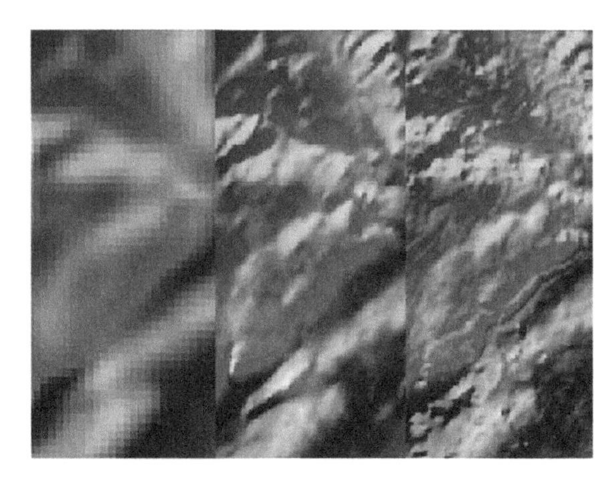

FIGURE 12 A comparison of three DEMs of the same area: (a) a DEM produced from a 1:100,000 topo-map; (b) a 30×30 m DEM interpolated from 1:25,000 topo-maps and (c) a 10×10 m highly accurate topography derived by using the new HRS SPOT scanner (courtesy of SPOT Image^{TM}).

(cf. Belgium, which has an area of about 30,000 km^2, and one SPOT swath, which covers 72,000 km^2).

The coverage available (in 2007) was over 110 million km^2, which is about 2/3rds of the world's land surface[7] (Figure 11). Note that it is usually not possible to obtain the stereo-pairs used for producing the SPOT DEM product.

4. SUMMARY POINTS

In this chapter, the aim has been to give an overview of DEM data sources, their processing requirements, and their advantages, disadvantages and appropriate usage (i.e. applications), without getting bogged down in detail or recommending one source or another. Links to important books and online sources of DEM data and other relevant information are provided below. As with all online sources, we have done our best to ensure that these links are correct. However, we cannot guarantee their stability or persistence.[8]

Table 2 summarises the key features of the data sources described in this chapter. The following caveats apply:

- All costs have been estimated in Euros, but some of these have been converted from other currencies using conversion rates from June 2006. The absolute prices of each source per km^2, and the relative differences in prices between sources, are bound to change. Table 2 will give you an idea of the relative costs in 2006.
- The accuracies (horizontal and vertical) are approximate figures, aimed at describing the relative differences in accuracy between the various sources,

[7] The global coverage map of SPOT DEM can be accessed from http://www.spotimage.fr, under section SPOT 3D.
[8] The Internet archive http://www.archive.org is a good place to start when tracking down old online documents.

TABLE 2 Summary of the key characteristics of data sources

Source	Resolution (pixel size in metres)	Accuracy	Footprint (km^2)	Cost (in €/km^2)	Post-processing requirements	Elevation/Surface
Ground survey	Variable but usually <5 m	Very high vertical and horizontal	Variable, but usually small	Very high	Low	Elevation
GPS	Variable but usually <5 m	Medium vertical and horizontal	Variable, but usually small	Low	Low	Elevation
Table digitising	Depends on map scale and contour interval	Medium vertical and low horizontal	Depends on map footprint	–	Medium	Elevation
On-screen digitising	Depends on map scale and contour interval	Medium vertical and low horizontal	Depends on map footprint	–	Medium	Elevation
Scanned topo-map	Depends on map scale and contour interval	Medium vertical and low horizontal	Depends on map footprint	–	Considerable	Elevation
Ortho-photography	<1	Very high vertical and horizontal	–	100 to 200 (depends on required accuracy)	Considerable	Surface
LiDAR	1–3	0.15–1 m vertical, 1 m horizontal	30–50/hour	25–50 (depends on required accuracy)	Considerable	Surface
InSAR/IfSAR	see below	see below	see below	see below	see below	see below

(*continued on next page*)

TABLE 2 *(continued)*

Source	Resolution (pixel size in metres)	Accuracy	Footprint (km^2)	Cost (in €/km^2)	Post-processing requirements	Elevation/ Surface
SRTM C BAND	90 (30)	16 m vertical, 20 m horizontal	Almost global 60N to 58S	Free via FTP	Potentially considerable	Surface
SRTM X BAND	30	16 m vertical, 6 m horizontal	Similar to C-Band, but only every 2nd path is available	€400 per tile 15″×15″	Potentially considerable	Surface
ASTER	30	7–50 m vertical, 7–50 m horizontal	3600	0.02 (min. of 3600 km^2/€64)	Medium	Surface
SPOT	30	10 m vertical, 15 m horizontal	72,000 per swath	2.30 (min. of 3000 km^2/€7000)	Medium	Surface

rather than guaranteeing the accuracy of any particular source. For a number of reasons (as described throughout this chapter), accuracy will vary from dataset to dataset. However, with the development of better post-processing methods, the source accuracy is likely to increase in the future.

Data sources and processing methods for generating DEMs have developed rapidly over the last few decades — from ground surveying to passive methods of remote sensing, and more recently to the emergence (and some may argue the dominance) of active sensors, such as LiDAR and RADAR (Molander et al., 2006). Higher levels of accuracy, more detail, and a larger, and more rapid, areal coverage, but at lower cost, has been a natural consequence of these new technologies. We are also starting to see automated and semi-automated methods of DEM generation from these sources, which will lead to further cost reductions and reductions in production times.

IMPORTANT SOURCES

Longley, P.A., Goodchild, M.F., Maguire, D.J., Rhind, D.W., 2005. Geographical Information Systems: Principles, Techniques, Management and Applications, 2nd edition abridged. Especially Part 1(a)-9: Representation of terrain (M.F. Hutchinson and J.C. Gallant).

Smith, S.E., 2005. Topographic mapping. In: Grunwald, S. (Ed.), Environmental Soil–Landscape Modeling: Geographic Information Technologies and Pedometrics, vol. 1. CRC Press, New York, pp. 155–182.

Rabus, B., Eineder, M., Roth, A., Bamler, R., 2003. The shuttle radar topography mission — a new class of digital elevation models acquired by spaceborne radar. Photogrammetric Engineering and Remote Sensing 57 (4), 241–262.

Wolf, P.R., Dewitt, B.A., 2000. Elements of Photogrammetry: With Applications in GIS, 3rd edition. McGraw–Hill, Boston, 624 pp.

Maune, D.F. (Ed.), 2001. Digital Elevation Model Technologies and Applications: The DEM Users Manual. American Society for Photogrammetry and Remote Sensing, Bethesda, MD, 539 pp.

http://users.erols.com/dlwilson/gps.htm — David L. Wilson's GPS Accuracy Web Page.

http://lidar.cr.usgs.gov — A guide to LiDAR data and other LiDAR websites.

http://www.intermap.com/customer/papers.cfm — Many articles on LiDAR, InSAR and SAR data and applications.

http://srtm.csi.cgiar.org — Hole-filled DEM derived from final SRTM product in GeoTIFF and ASCII formats in 5°×5° tiles.

http://hydrosheds.cr.usgs.gov — HydroSheds: hydrological data and maps based on shuttle elevation derivatives at multiple scales.

http://edcimswww.cr.usgs.gov/pub/imswelcome/ — ASTER DEM data are available from the Earth Observing System Data Gateway.

CHAPTER 4

Preparation of DEMs for Geomorphometric Analysis

H.I. Reuter, T. Hengl, P. Gessler and **P. Soille**

DEM quality issues · the detection, quantification and reduction of errors in DEMs · the preparation of DEMs for geomorphometric analysis · the automatic correction of SRTM DEMs by using land-cover information · filtering out local outliers and undefined pixels in output maps · filtering of noise in SRTM DEMs · dealing with water bodies in DEMs · removing spurious sinks · using different interpolation methods to filter missing data (voids) · implementing simulations to improve the reliability of land-surface parameters and objects

1. INTRODUCTION

In this chapter, guidance will be given on how to prepare elevation data for geomorphometric analysis. We first outline some common errors in raw DEM data sources and then suggest approaches for systematically improving the quality of DEMs. We start with height samples and end with the final DEMs used for geomorphometric analysis. More precisely, we will describe both simple and more advanced algorithms that can be used to reduce systematic and random errors, and enrich the quality of DEMs by incorporating auxiliary information on land cover and the hydrological properties of an area. These algorithms are implemented and described further in the software packages.

The nature of DEM-preprocessing algorithms very much depends on the type[1] of input data. For this reason, not all algorithms are applicable to raw DEMs. There is no superior DEM preparation procedure, however, you will certainly be better off if you (1) employ detailed auxiliary information (such as physiographic breaks, canals, land cover, and stream network, etc.), (2) choose preparation methods that fit your application and (3) clean data errors, but leave the real features in your data sets.

[1] Especially on whether the height data originates from field measurements or from platform based scanning devices.

Developments in Soil Science, Volume 33 © 2009 Elsevier B.V.
ISSN 0166-2481, DOI: 10.1016/S0166-2481(08)00004-4.

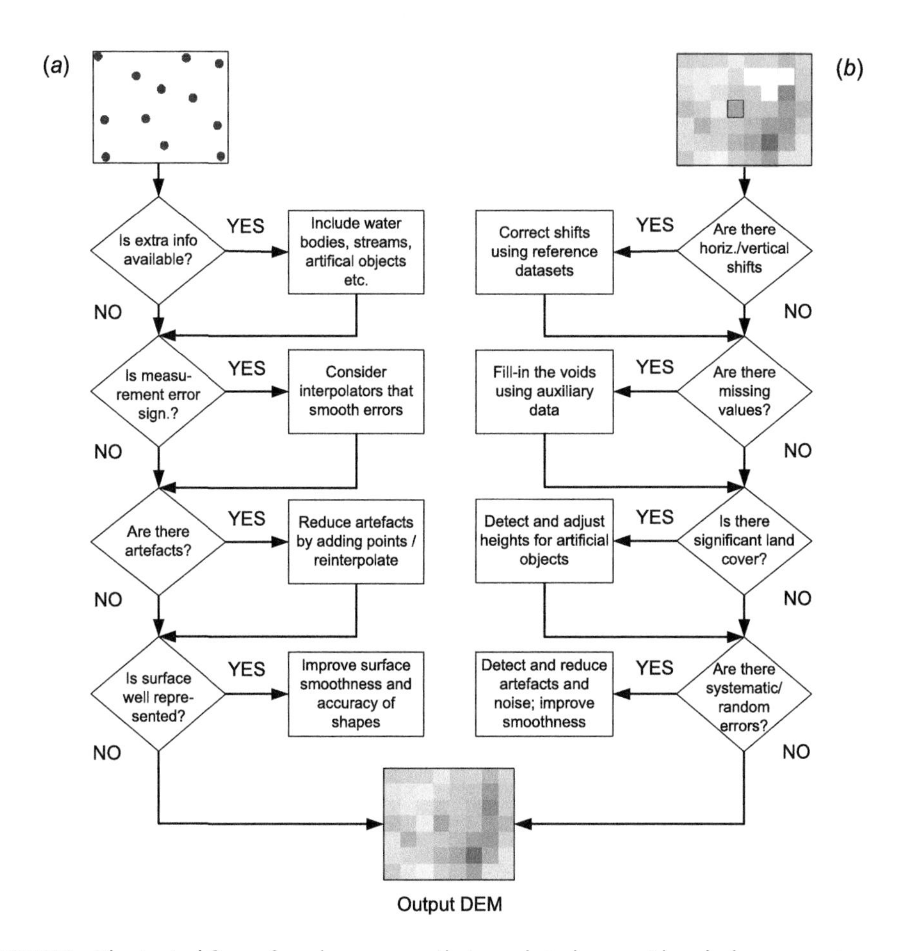

FIGURE 1 The typical flow of work processes that needs to be considered when preparing DEMs for geomorphometric analysis: (a) heights from point-measuring devices and (b) heights from airborne scanning devices.

Figure 1 shows a typical framework for pre-processing raw height measurements and producing DEMs with reduced errors. You may not need to apply all these steps, but you should be aware of the possible problems that may arise in the DEM outputs, if some of the pre-processing steps are omitted. DEM pre-processing can be quite time consuming and creating the 'perfect' DEM is sometimes rather difficult. Nevertheless, such efforts usually significantly improve the quality of outputs (Carlton and Tennant, 2001).

1.1 The quality of DEMs

The quality of the DEMs determines the quality of the geomorphometric analysis. Even the most sophisticated geomorphometric algorithm will be unable to rectify severe artefacts and errors in the input DEMs. The quality of land-surface para-

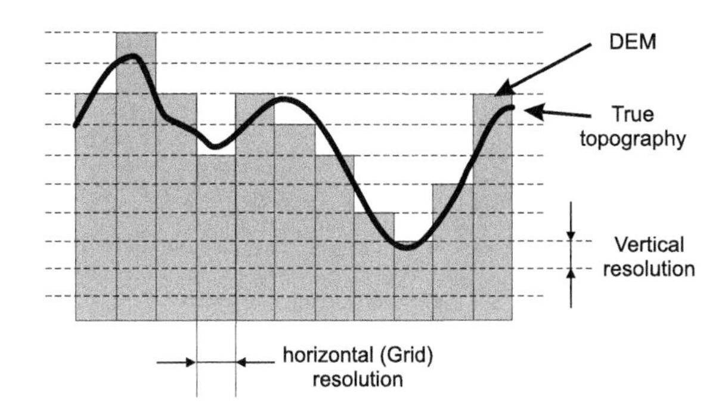

FIGURE 2 The horizontal and vertical resolution of DEMs.

meters and objects and geomorphometric applications depends on several factors (Florinsky, 1998):

- *the roughness of the land surface,*
- *the sampling density* (i.e. which method was used to collect elevation data),
- *the grid resolution* or sampling support,
- *the DEM gridding algorithm,*
- *the vertical resolution* (Figure 2),
- *the type of geomorphometric analysis.*

The land-surface roughness factor can not be controlled, all other factors can be fine-tuned, or even statistically optimised, to improve the quality of the outputs. The first five above-listed factors are objects of land-surface generation and are easily forgotten by analysts. Instead, they focus on improving the algorithms for geomorphometric analysis.

In geomorphometry, it is known that some algorithms are more precise in estimating the values of parameters and objects in an area. For example, functions of higher order DEM derivatives may encounter serious problems, due to their excessive sensitivity to DEM noise (Zhou and Liu, 2004). The sensitivity of local parameters to DEM noise (as well as their dependence on grid resolution) increases in the following order (Shary et al., 2002):

(1) *functions of first derivatives;*
(2) *linear functions of second derivatives* (simple curvatures);
(3) *square functions of second derivatives* (total curvatures).

A careful comparison of various geomorphometric analysis algorithms using controlled conditions allows us to see how sensitive they are to the input properties of DEMs, and help us select the most precise, most robust technique (Florinsky, 1998; Zhou and Liu, 2004).

Even the most accurate, most robust geomorphometric algorithms will result in poor outputs if the input DEMs are of low quality or inadequate for the targeted application. Prior to the actual extraction of land-surface parameters and objects, various procedures can be followed to improve DEMs. These procedures are available in many geomorphometric packages under the name of *DEM pre-*

processing or *DEM preparation*. In principle, any pre-processing of DEMs has three main objectives:

- *to remove gross errors and artefacts,*
- *to make a better approximation of the land surface,*
- *to make a better approximation of the hydrological/ecological processes* (such as flow, radiation, deposition, etc.).

Thus, the true applicability of DEMs for geomorphometric analysis can be assessed by providing an answer to the following questions:

- how accurately is the surface roughness presented (micro- and meso-relief)?
- how accurately is the hydrological shape of the land surface presented (concave and convex shapes, erosion or deposition, water divergence or convergence)?
- how accurately can the real ridges and streams be detected?
- how consistently are elevations measured over the whole area of interest?

Each of these aspects can have an impact on the final results of the DEM application. For example, if DEMs are constructed from contour lines, the surface roughness will be under-represented, so it is probably not very wise to extract geostatistical land-surface parameters (see Section 2.2 in Chapter 6). Similarly, if a radar-based DEM of an area with dense vegetation has not been filtered for forest areas, it might not be a good idea to try to derive a map of ridges and streams, because the forest clear-cuts and roads may show up as artificial channels.

REMARK 1. *The main objective of DEM-preprocessing is to remove artefacts, improve representation of shapes and hydrological/ecological processes.*

The four aspects of quality in DEMs, listed above, can not be assessed easily, because there are no reference maps of land-surface parameters and objects. Many researchers (Bolstad and Stowe, 1994; Giles and Franklin, 1996) have tried measuring simple land-surface parameters and objects, such as slope and aspect, in the field, to validate the accuracy of the land-surface parameters and objects. Florinsky (1998) believes that *"there is no reason to suppose that these reference-data measurements and computations are correct"*. Indeed, it will never be easy to measure *'true'* slope gradient in the field, because slope is related to the scale of the research (see Chapter 2). In some cases, however, it is possible to evaluate the spatial accuracy of objects on the land surface — e.g. to assess the spatial accuracy of predictable streams and watersheds (Wise, 2000) or even to cross-check solar radiation models (Reuter et al., 2005) — but the general impression is that there are still no consistent guidelines for checking the accuracy of land-surface parameters in the field. Most evaluations of DEMs and their land-surface parameters are still made visually.

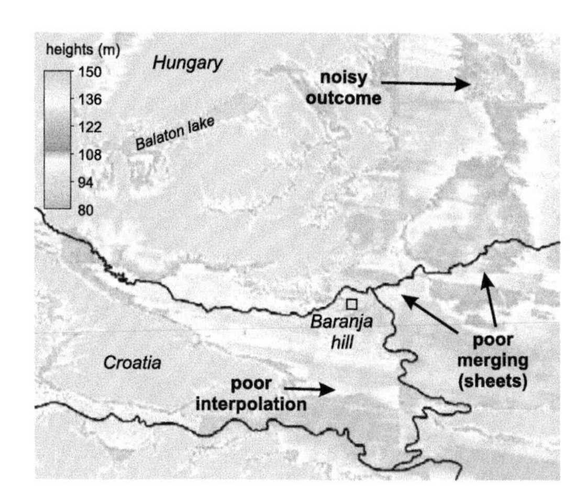

FIGURE 3 An example of local artefacts in part of the GTOPO DEM (1 km resolution). Such artefacts are only visible after careful inspection. (See page 712 in Colour Plate Section at the back of the book.)

1.2 The types of errors in DEMs and DEM derivatives

The errors in DEM and DEM-derived products can be grouped roughly into (Wise, 2000):

- *artefacts, blunders or gross errors,*
- *systematic errors,*
- *random errors* or noise.

Artefacts are especially important for land-surface parameters derived from second order derivatives (curvatures), aspect map and/or hydrological parameters. They are hard to detect in the elevation of the input DEMs alone, but will certainly be visible, either as missing or unrealistic values, in DEM-derived products. For example, zero slopes (terraces) will not be visible in the input DEM, but if you try to derive aspect or wetness index in these areas, the algorithm will either fail due to division by zero, or result in completely unrealistic features.

In other cases, you will be able to derive a physical value, but this would be completely erroneous (seen as ghost lines, stripes, or something similar). Because artefacts are commonly distinct erratic features, most of them can be detected visually in 3D views, by using sun shading or simple GIS operations (Figure 3). Examples of common artefacts are holes in a DEM due to wrongly coded elevation contours, or a mountain appearing on top of a small hill due to a positioning error (Figures 3 and 4, see also Brown and Bara, 1994; Garbrecht and Martz, 1997).

REMARK 2. *Artefacts and gross-errors in DEMs can be easily detected even visually. On contrary, the systematic errors need to be detected using specific statistical procedures.*

Systematic errors, on the other hand, reflect either a bias inherent in the data collection method, limitations in the methods used to derive a DEM, or sensitivity

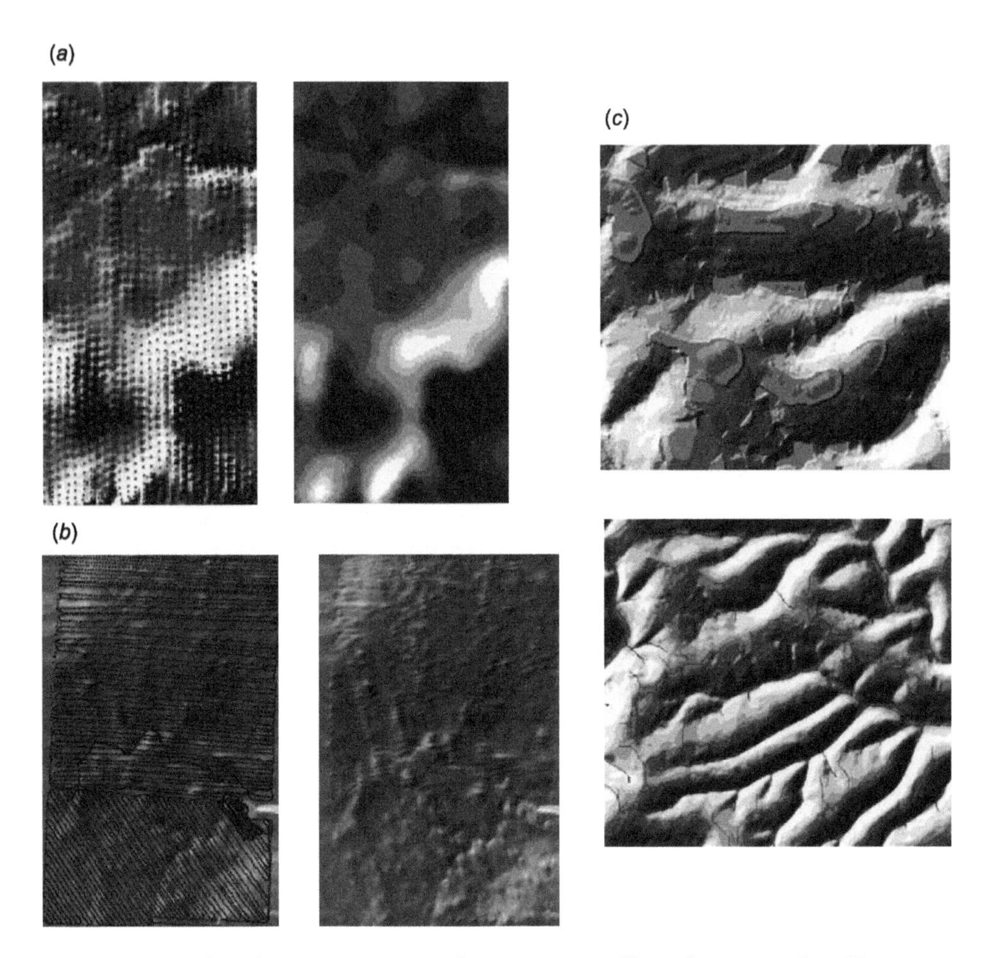

FIGURE 4 Examples of some common artefacts in DEMs and how these are reduced by filtering: (a) local noise arising from inverse weighted interpolation; (b) artificial patterns in DEMs from GPS input sources; and (c) systematic errors (terraces) in a DEM that has been interpolated from scanned contour data. After MacMillan (2004).

of the geomorphometric algorithm to computation parameters (e.g. a too narrow search window). Unlike artefacts, systematic errors are not clearly erroneous and you cannot rely just on a visual inspection of the end results. For example, a DEM derived from an aerial photo-pair will show much higher levels of inaccuracy in the shaded areas (e.g. North–East expositions) and near the edge of the photographs (Leica Geosystems GIS & Mapping, 2003), but this type of error is invisible to the naked eye. To detect them, more sophisticated statistical techniques are required.

The systematic errors can often be estimated by reconstructing the physical model of data acquisition. For example, a 30 m resolution SRTM X-band from the German Space Agency is distributed with a dataset showing an estimated height error for the same area of interest (see also the Baranja Hill dataset). This error is calculated according to formulae that takes into account (Bamler, 1997): the measurement of the accuracy of the phase, the imaging geometry resulting from the orbit and its accuracy, and the atmospheric distortions that influence wave propagation. Such information can be very useful for DEM preparation because it will allow you to depict the areas that (i) may need to be either re-measured or (ii) processed for specific filtering to eliminate errors (Hofer et al., 2006).

Systematic errors can be reduced by preparing the surveys carefully and by using processing techniques, such as increasing the contrast in the shaded areas or replacing the unreliable elevations with elevations from other, more reliable, data sources. One costly approach to reducing systematic errors is to plan several independent topographic surveys (e.g. various flying directions, on different dates) using the same technique and then derive an average elevation value for each pixel (see further Section 2.11).

Raaflaub and Collins (2006) evaluated the effects of DEM errors on the most common land-surface parameters (slope, aspect, upslope area, topographic index). They concluded that the systematic errors can be reduced if a higher number of neighbours is used to derive local land-surface parameters (e.g. slope). In the case of derivation of aspect, the average error sensitivity to DEM error is not dependent on the size of the neighbourhood.

Random errors are often associated with the measurement error, i.e. signal noise. They can be significant for the DEMs produced using remote sensing-based imaging instruments, but are inherent in any measurement process. Typically, we want to quantify such errors first, before we consider any pre-processing of DEMs. Otherwise, we might filter out true surface roughness and finish with over-smooth DEMs. Many instruments have constant measurement error (precision), which makes it easier to distinguish between the noise and local variability. For example, a SRTM DEM will show a noisy pattern that is constant in the whole area of interest. This is obviously not the true surface roughness as we would expect that the DEM is non-uniform — it should show different surface roughness in different parts of the study area (see also Section 1.4 in Chapter 2).

1.3 Specific error types in different data sources

The number of errors, and their impact on land-surface parameters and objects, largely depends on how the elevation measurements have been derived and how the DEMs have been produced. Although some of these issues have already been discussed in Chapter 3, we will address some common situations:

DEMs derived from digitised contours Spatial under-sampling and a generalisation of topography are the biggest problems in DEMs that are derived from digitised contours (Thompson et al., 2001; Reuter et al., 2006). Contours often show a different density in different parts of the study area, and such data is very sensitive to an interpolation algorithm. If very primitive gridding technique is used, the DEMs can show many erratic features (the most common are *terraces*, *ghost lines* and *tiger stripes*), and these propagate onto the other derivatives (Burrough and McDonnell, 1998). The second serious problem in DEMs derived by this method is that the elevation values become somewhat generalised, so that the final DEM does not reflect the actual surface roughness (the micro- and meso-relief). This problem can only be avoided either by supplementing the contours with additional point measurements of elevations, or by degrading the effective scale of a DEM. For example, you can use contours from a 1:5000 topo-map to produce a DEM with a coarser grid resolution (e.g. 30 m).

DEMs derived from ground-level devices The DEMs derived from measuring devices at ground level, such as DGPS measurements in precision agriculture (Bishop and Minasny, 2005), also have problematic features. For example, if a GPS device is placed in a vehicle, the density of observations will be heavily biased towards the relatively flat and open areas, e.g. the agricultural areas. Inaccessible areas, such as relatively steep slopes (>15%) and areas with dense vegetation cover or where there are large expanses of stagnant water, will be systematically under-sampled, and this will influence the results of the geomorphometric analysis. In precision agriculture, the relative accuracy of one recording run is usually sufficient. However, if there are pauses in the readings (e.g. because the machinery breaks down, or for lunch breaks), slight shifts can occur between the elevation and location of measurement points, which, if not processed and filtered correctly, can lead to artefacts such as a 45-degree slope in a completely flat field.

DEMs from airborne scanning devices DEMs derived from airborne scanners and airborne imagery have the big advantage over alternative techniques that their sampling density is not only very high but usually very regular. They also have a constant support size.[2] Scanning devices should produce complete images, without the need for interpolating the values sampled. In practice, appropriate methods need to be found to fill in areas of shadows or clouds (*voids*), and the spaces between adjacent strips of imagery (see Section 2.7). Because these densely sampled elevations usually represent a surface topography rather than the land surface, they may nevertheless be less accurate and less usable than sparsely sampled elevations. A property of the radar DEMs, for example, is that the natural vegetation and man-made objects are seen as small hills, escarpments or islands. In general, topographic imaging (especially on satellite platforms) requires a number of filtering steps, either to remove noise, interpolate missing values, or to filter out the impact of land cover (Figure 1).

1.4 A quantitative description of DEM errors

There are a number of ways to quantify errors in a DEM data set. One of the standard methods is to specify the height errors in the vertical plane. This involves comparing the height of a location, or even a patch, against the height, or a patch, at a reference location. To calculate the vertical error, Root Mean Square Error (RMSE) at a number of spots (n) is defined as:

$$\mathrm{RMSE} = \sqrt{\frac{\sum_{i=1}^{n}[z(s_i) - z_{\mathrm{REF}}(s_i)]^2}{n}} \tag{1.1}$$

[2] See Section 2.3 in Chapter 2 for the difference between the grid resolution and support size.

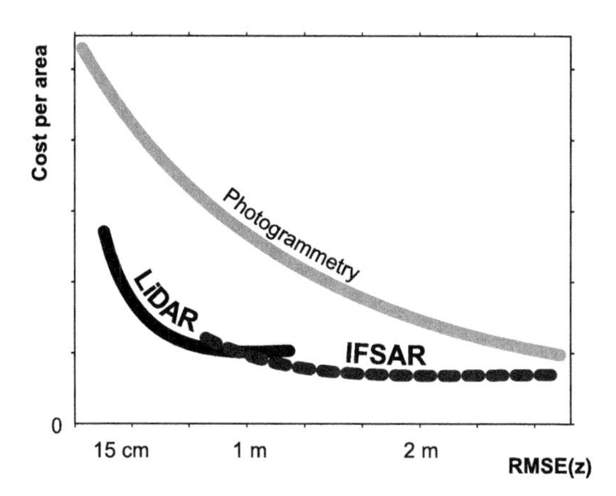

FIGURE 5 The general relationship between the absolute accuracy of height measurements and the cost of topographic surveys. Adapted from Maune (2001, p. 443).

where $z(s_i)$ is the elevation at the spot location, and $z_{\mathrm{REF}}(s_i)$ is the reference elevation at same location. Such reference points should be: (a) distributed evenly across the area of interest, (b) representative of the landscape and (c) measured to a much higher precision than the DEM being tested.

RMSE can usually be closely linked with the technique used for producing the data, i.e. price of height measurements per area unit (Figure 5). This is usually only globally reported (from this point onwards we will use RMSE), so that no indication is given of the spatial distribution of the error (Li, 1988, 1992). For Britain, for example, the error is assumed to be constant over the entire land area (Ordnance Survey, 1992).

To achieve a more complete statistical description of how the errors are distributed, Fisher and Tate (2006) suggested the Mean Error (ME) and the Error Standard Deviation (SD):

$$\mathrm{ME} = \frac{\sum_{i=1}^{n}[z(s_i) - z_{\mathrm{REF}}(s_i)]}{n} \tag{1.2}$$

$$\mathrm{SD} = \sqrt{\frac{\sum_{i=1}^{n}[z(s_i) - z_{\mathrm{REF}}(s_i) - \mathrm{ME}]^2}{n-1}} \tag{1.3}$$

Although ME and SD provide a more detailed evaluation of the DEM error, they give little insight into their spatial distribution, e.g. they give no indication of whether the error is correlated locally or not (see further Section 2 in Chapter 5).

> REMARK 3. *For geomorphometry, the main concern is the accuracy of the outputs (parameters and objects) and not the absolute accuracy of measured elevations.*

DEMs can also contain uncorrelated errors from remote sensing if measurement errors are independent of each other, and these can lead to small-scale distortions (e.g. slopes reflect measurement noise, if SRTM is not filtered). Correlated errors preserve the general shape of the landscape better, but, on larger scales, distortions become evident. For this reason, an interpolated surface might have a higher ME and error SD than a LiDAR DEM, whereas the opposite is the case for land-surface parameters.

Another quantitative parameter is the *Bias* between two data sets. It is the mean offset between the observed value of a DEM and another '*true*' reference surface:

$$\text{BIAS} = \text{avg}[\text{DEM} - \text{DEM}_{\text{REF}}] \tag{1.4}$$

This offset can be quite significant, if reported for different types of DEMs, e.g. a Surface DEM (SRTM) compared with a topographic DEM (Guth, 2006; Hofton et al., 2006).

The *horizontal accuracy* of DEMs (e.g. if the top of a mountain is at the correct location or offset by 250 m to the northeast) can be computed by generating a *displacement vector* ($\overrightarrow{DV}(s_i, s_j)$):

$$\overrightarrow{DV}(s_i, s_j) = \max_{\text{corr}}\left[s(z_i), s(z_j)\right] \tag{1.5}$$

whereby, the cross correlation of a reference surface patch (e.g. 100×100 cells) is compared with a DEM created for different offsets (s_i, s_j). The offsets with the highest correlation are recorded and used to assess the accuracy of the DEM (Figure 6).

Apart from the standard methods of evaluating the absolute accuracy of DEMs, in geomorphometry, we are more interested in the accuracy of land-surface parameters and objects and the resulting maps that can be derived from them. This has created a general misconception about the quality of DEMs for geomorphometric analysis: the absolute accuracy of the elevation values in a sample is not the most important indicator of high quality DEMs. For example, even where the elevation values are sampled very accurately (e.g. LiDAR can achieve an accuracy of ± 0.15 m, Figure 5), the results of the geomorphometric analysis may still be poor (e.g. because the DEM is too noisy; or the canopy is unfiltered). For high-quality geomorphometric analysis, it is more important that a DEM accurately resembles the actual shapes and flow/deposition processes of the land surface. This resemblance is often referred to as the *relative accuracy* or *geomorphological accuracy* of DEMs (Schneider, 1998; Wise, 2000). A SRTM DEM of the same area, at a coarser resolution (90 m), can often reflect the actual shapes much better than a fine-resolution DEM derived from a topo-map by using a poor gridding technique.

The true geomorphological accuracy can only be assessed by measuring land-surface parameters and objects such as drainage lines, landforms or view-sheds in the field, and then comparing their shapes, distribution and location with the values obtained from a geomorphometric analysis (Fisher, 1998).

2. REDUCING ERRORS IN DEMS

2.1 Orthorectifying DEMs

The first step to prepare a DEM for geomorphometric analysis is to improve its horizontal accuracy by assessing the shifts. Depending on the size of the DEM and

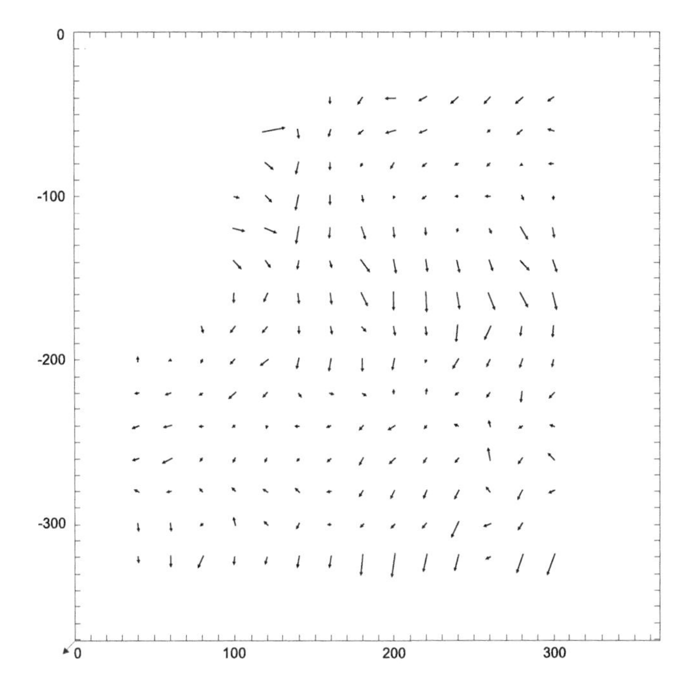

FIGURE 6 Horizontal displacements for X-SAR SRTM DEM versus TOPO DEM for Baranja Hill. Size of the vector has been exaggerated for better display. A library of IDL/ENVI routines for this application has been developed by W. Mehl.

how much computing power is available, the displacement vector ($\overrightarrow{DV}$) can be computed for every 50, 100 or 200th cell. This gives an insight into the horizontal and vertical accuracy of the DEM.

Figure 6 shows the horizontal $\overrightarrow{DV}$ for a comparison between the XSAR-SRTM dataset and a TOPO DEM. The vertical offset between both is reported as the difference (in meters) between both DEMs. If the observed cross correlations for each offset are fitted further using a minimum curvature interpolation (Hill and Mehl, 2003), the sub-scale[3] accuracy can also be determined. If a sufficient number of $\overrightarrow{DV}$s are computed, the DEM can be warped into the correct positions using the $\overrightarrow{DV}$s as a ground-control point for the warping process.

2.2 Reducing local outliers and noise

Local outliers can be defined as small, very improbable features, which could just be a noise in the data collection method (this is very common for remote sensing-based instruments). They can be detected and quantified by using the statistical approach suggested by Felícisimo (1994b) and elaborated further by Hengl et al. (2004a). In this case, the original elevation is compared with the values estimated from neighbouring elevations to derive a probability of finding a certain elevation

[3] This means that this method can also be used to determine the correct spatial location of heights, even at a smaller scale than that of the input data.

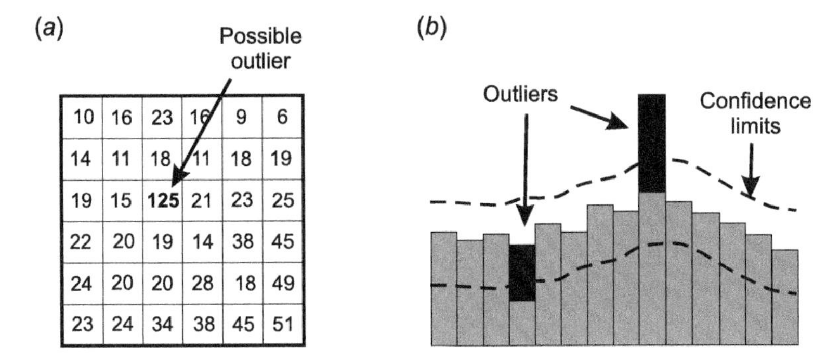

FIGURE 7 Schematic example showing the expected local outliers and the result of filtering. The black colour in (b) indicates change in the original values derived using the formula in Equation (2.6).

in the given neighbourhood:

$$\delta(s_i) = \hat{z}^{\mathrm{NB}}(s_i) - z(s_i), \quad i = 1, \ldots, n \tag{2.1}$$

where $\delta(s_i)$ is the difference between the original and estimated value and $\hat{z}^{\mathrm{NB}}(s_i)$ is the elevation estimated from the neighbouring values. Assuming a Gaussian distribution ($\bar{\delta}$ and s_δ), we can derive a probability of observing this difference for each pixel:

$$p(t_i) = \frac{1}{\sqrt{2 \cdot \pi}} \int_{t}^{+\infty} e^{-\frac{1}{2} \cdot t_i^2} \cdot dt, \quad t_i = \left| \frac{\delta_i - \bar{\delta}}{s_\delta} \right| \tag{2.2}$$

If the t goes >3 (three times the standard error), the probability of finding this value in the DEM will already be $<0.27\%$. It is probably wiser to use the original RMSE instead of the estimated s_δ to get an unbiased estimate of the measurement error (Hengl et al., 2004a).

Kriging is a statistically sound method for estimating the central value of neighbouring pixels. Only two types of distances are present in a 3×3 window environment (assuming the isotropic variation): those in the cardinal (2, 4, 6, 8) directions and those in the diagonal directions (1, 3, 7, 9). The predictions are made by:

$$\begin{aligned} \hat{z}^{\mathrm{NB}} &= w_{\mathrm{B}} \cdot [z_{\mathrm{NB1}} + z_{\mathrm{NB3}} + z_{\mathrm{NB7}} + z_{\mathrm{NB9}}] \\ &\quad + w_{\mathrm{A}} \cdot [z_{\mathrm{NB2}} + z_{\mathrm{NB4}} + z_{\mathrm{NB6}} + z_{\mathrm{NB8}}] \end{aligned} \tag{2.3}$$

where w_{A} is the weight in the cardinal direction and w_{B} the weight in the diagonal direction.

In a general case (a $k \times k$ window), the predictions are made by:

$$\hat{z}^{\mathrm{NB}} = \sum_{c=1}^{k^2} w_c \cdot z_{\mathrm{NB}c}, \quad c = 1, \ldots, k \tag{2.4}$$

gain=0.001

	-2	-1	0	1	2
2	-6	-15	-25	-15	-6
1	-15	50	260	50	-15
0	-25	260	0	260	-25
-1	-15	50	260	50	-15
-2	-6	-15	-25	-15	-6

FIGURE 8 The filter used to predict the central value following a given spatial auto-correlation structure (we used the variogram from Chapter 1). To ensure an unbiased estimator, the sum of weights, when multiplied by the gain, should equal 1.

where w_c is the weight at the cth neighbour and $w_\times$ the weight at the central pixel, so that:

$$w_\times = 0, \qquad \sum_{c=1}^{k^2} w_c = 1 \tag{2.5}$$

In the case of anisotropy, different weights can be used for different directions. The (kriging) weights are solved using the covariance function and the relative distances between all pixels. Finally, the corrected DEM is derived from the original DEM and the estimated elevations as a weighted average (Hengl et al., 2004a):

$$z^+(s_i) = p(t_i) \cdot z(s_i) + \left[1 - p(t_i)\right] \cdot \hat{z}^{\text{NB}}(s_i) \tag{2.6}$$

where $z^+(s_i)$ is the filtered elevation map and $p(t) \in [0, 1]$ is the normal probability of observing this difference. This means that we will take the average value between the observed and fitted values by using the probability of observing a value as the weights (Figure 7).

We used a 3×3 window to demonstrate the main principles. In practice, we recommend using a 5×5 window environment with 24 neighbours and five types of weights (see Figure 8). This is good enough to detect all the local outliers.

However, the 5×5 filter will only detect errors that have a maximum width of 2–3 pixels (lines or single pixels). If we deal with a group of outliers (a patch) covering a significant proportion, this algorithm will not be able to filter out the problematic area. In that case, it is better first to detect such problematic areas using some other filtering procedure and then to mask them out.

There are two different approaches for identifying such areas. First, as demonstrated by Axelsson (1999), problematic areas or high canopy areas can be detected automatically, providing they contain a specific structure and there is a clear definition of the land surface (e.g. isolated houses in a suburb, forests). Otherwise, more advanced algorithms employing data from different sensors (e.g. DEM + multispectral + other auxiliary information) need to be used (El-Hakim et al., 1998;

Zhang, 1999; Sequeira et al., 1999). This type of DEM + RS analysis gives a much higher confidence in the masks that are derived.

A faster approach to detect possible problematic areas is simply to average elevations for a 3×3 window and derive the difference between the elevation at the core cell. If the difference (in this case, a global threshold value) was larger than a threshold (e.g. 100 m for SRTM data), single pixels probably appear as spikes and pits that can be masked out. However, by applying a static threshold, we do not take into account that the errors (and the elevation) in any DEM also have properties that vary spatially. Therefore, we suggest modifying the threshold spatially, in order to identify the spikes and pits, e.g. a spike of 50 m in a flat desert area will probably be a data error, whereas, in a mountainous area, verifying the error will not be as straightforward (Lopez, 2000). A threshold of ±2 or 3 times the standard deviation in a given window (s_δ) would be appropriate for detecting local outliers (Albani and Klinkenberg, 2003):

$$e_L \leftarrow \begin{bmatrix} \text{if } z(s_i) < \hat{z}^{\text{NB}} - 2 \cdot s_\delta \\ \text{if } z(s_i) > \hat{z}^{\text{NB}} + 2 \cdot s_\delta \end{bmatrix} \tag{2.7}$$

2.3 Filtering water surfaces

Ideally, we would like to work with DEMs representing land surface both above and below the *water surface*. However, due to the higher costs of documenting DSMs of surfaces below water, this is often not possible. Instead, water bodies are usually masked, using a constant value for lake surfaces, or, for large river surfaces, values that are stepped down, monotonically. The constant value used to record the elevation of these lake surfaces is the local minimum elevation at their shorelines. However, these values must be used with care when hydrological networks are modelled connected to ocean or brackish water, or in boundary conditions like those that occur, for example, in the Netherlands.

2.4 Filtering of pure noise

The advantage of working with DEMs produced using airborne or satellite-based scanning devices (radar- or LiDAR-based) is that they have a very high and regular sampling density (no gridding is needed), so that the model of a land surface will be more accurate in depicting meso- and micro-relief. On the other hand, radar and LiDAR-based devices are subject to measurement errors defined by the physical limitations of the instrument. They also reflect only the topography of the surface objects and not of the bare earth, so that many pre-processing steps are needed before they can be used for geomorphometric analysis.

The raw DSM obtained from SAR or LiDAR data contains a severe amount of *noise*. For this reason, it cannot be used directly for geomorhometric analysis. In the case of the SRTM DEM, thermal noise is eminent in the DEMs (Selige et al., 2006), so that the absolute accuracy of measured heights is relatively poor (±15 m). If you just compare the 3 arcsec SRTM DEM with the DEM derived from a topo-map (Figure 5 in Chapter 3), you will see that the noise in the SRTM DEM is obvious.

The amount of (pure, uncorrelated) noise can also be estimated by fitting the variogram for the elevations (as shown in Figure 12 of Chapter 2). In the case of the 1 arcsec SRTM DEM, the nugget variance is about 200, which means that the short-range variation (measurement error) is about ± 14 m, which complies with the standards (Rabus et al., 2003; Miliaresis and Paraschou, 2005).

> REMARK 4. *The raw DEMs obtained from scanning devices typically contain a severe amount of noise and systematic errors. Many pre-processing steps are needed before they can be used for geomorphometric analysis.*

A solution to filter the noise in the SRTM DEMs is to use the phase noise filter designed by Lee et al. (1998) and further elaborated by Selige et al. (2006). This is a 16-directional anisotropic filter that will smooth the values in the SRTM DEM while preserving small and subtle features such as dikes and ditches. The only input needed for this filter is an estimate of the variance of the noise, which controls the amount of smoothing. The weighting function used is:

$$f(z_{d1}^{+}, \ldots, z_{d16}^{+}) = \beta \cdot z + (1 - \beta) \cdot \bar{z} \tag{2.8}$$

where z_{d1}^{+} is the filtered DEM value in one of the sixteen directions, β is the weight and $\bar{z}$ is the smoothed value derived in each of the sixteen directions. The weights are derived using:

$$\beta = \frac{\sigma_z^2 - \varepsilon^2}{\sigma_z^2} \tag{2.9}$$

where σ_z^2 is the variance derived in each of the 16 directions and ε^2 is the estimated phase noise. In the case of the 1 arcsec SRTM DEM for the Baranja Hill, we can use $\varepsilon = 4$ m, which is the smallest (background) value in the error map provided together with the 1 arcsec SRTM DEM.

Other approaches to overcome noise in the DEM imagery range from simple *median* filter (can be run iteratively, as well as with different window sizes, Albani and Klinkenberg, 2003), *tension splines* (Mitášová and Hofierka, 1993), *power spectrum* (Russell et al., 1997), *Fast Fourier Transformation* (Harrison and Chor-Pang, 1996), and *Wavelet Analysis* (Yu et al., 2004). Gallant and Hutchinson (2006) also tested the use of the *Kalman filter* for producing smoothed DEMs.

2.5 Filtering forests in SRTM DEMs

For geomorphometric applications, natural vegetation and man-made objects in radar/LiDAR DEMs can be quite problematic. In radar imagery, forest areas pose one of the biggest problems, because they appear as small elevated plateaus (Miliaresis and Paraschou, 2005). If the vegetation is short and with small leaves and branches, and if radar is applied when it is dry, a SRTM C-band (wave length $\sim$5 cm) will be able to penetrate it. Radar beams also reach the ground, providing the canopy cover of a forest is not dense. In dense, leafy forest, with medium to large branches, however, the radar waves will scatter and deflect upwards again, from near the top of the canopy[4] — from 0.5 to 0.75 of the height of the trees.

[4] See also Figure 10 in Chapter 3 and Equation (1.4) in Chapter 2.

LiDAR data usually contains information about the different returns of the Laser-Beam. This can be used to identify the surface of the ground in vegetated regions.

Generally, in heavily vegetated areas, a DEM derived from topographic maps based on a field survey will be much more reliable than those derived from radar. An SRTM DEM, on the other hand, always shows much more local detail (such as micro- and meso-relief; Figure 10 in Chapter 3). The optimal DEM is obviously the one that combines the best of the radar/LiDAR and TOPO DEMs. We can first detect problematic areas, such as a canopy, automatically, by deriving the difference between the two DEMs after they have been co-registered or *warped*:

$$\delta_i = z_i^{\text{TOPO}} - z_i^{\text{SRTM}} \tag{2.10}$$

then we can again use Equation (2.2) to estimate the probability of observing these values. Note that the reference TOPO DEM should be at least twice as precise as the SRTM DEM (e.g. DEMs from 1:25,000 topo-maps). Although a DEM derived from stereo-photogrammetric digitising is also influenced by (forest) vegetation, the surveyors will still try to locate the measurement points on the ground. In areas where the probability becomes rather small (i.e. where the difference is not accidental), we will certainly be dealing with natural or man-made objects (Figure 10 in Chapter 3).

The TOPO DEM is more accurate in areas with dense vegetation, but the SRTM DEM shows more detail. The average elevation can be computed as a weighted Average between the DEM derived from the contours of 1:25,000 topographic maps and the SRTM DEM [see also Equation (1.4) in Chapter 2]. We recommend the following procedure:

$$z_i^{+} = \left\{ \begin{matrix} z_i^{\text{TOPO}} & \text{if } p(t) \leqslant \nu \\ \frac{w_i^{\text{TOPO}} \cdot z_i^{\text{TOPO}} + w_i^{\text{SRTM}} \cdot z_i^{\text{SRTM}}}{w_i^{\text{TOPO}} + w_i^{\text{SRTM}}} & \text{if } p(t) > \nu \end{matrix} \right\}$$

where w_i^{TOPO} is the estimated uncertainty of the TOPO DEM, w_i^{SRTM} is the estimated uncertainty of the SRTM DEM and ν is the threshold[5] probability value for the difference between the two DEMs. This means that, in the areas assumed to have little vegetation, we will take the average value between the TOPO DEM and the SRTM DEM. In the areas where the presence of forests or similar objects [see Figure 9(a)] can be assumed, we will use the values from the topographic DEM.

A more sophisticated approach, discussed by Selige et al. (2006), is to use the land cover information to estimate the height of objects/canopy. Assuming that the different land cover types have a standard, fixed height, the land cover map can be used to filter the SRTM DEM accordingly.

2.6 Reducing padi terraces

As mentioned previously, if the elevations (e.g. contour lines) are only sparsely available, then the DEMs interpolated from them might show numerous artefacts, large and prominent enough to propagate onto geomorphometric applications.

[5] We suggest using a 0.1% threshold.

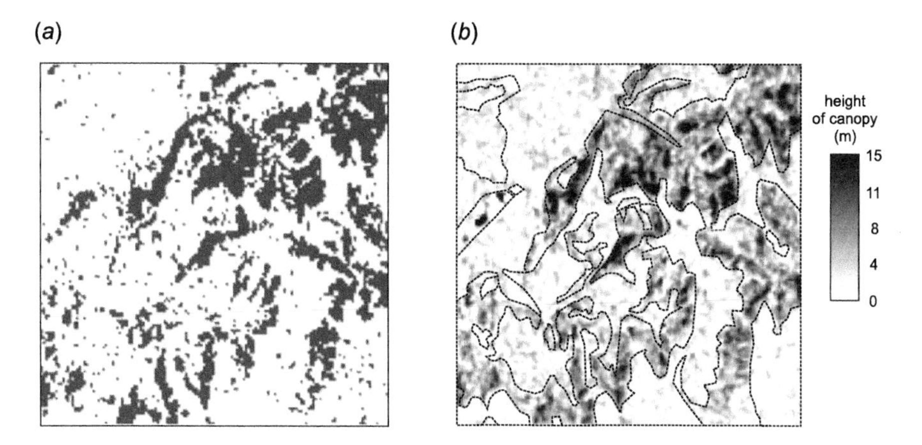

FIGURE 9 DEM filtering of the Baranja Hill study: (a) low probability values ($<0.1\%$) indicate forest areas that have been detected automatically; (b) the difference between the SRTM DEM and the TOPO DEM gives a direct estimate of the height of the canopy.

The most common artefacts in DEMs created from such data are so-called *padi terraces* or cut-offs. These are areas typical of closed contours (Figure 10) where all the surrounding pixels show the same value. All values inside the closed contours will be assigned the same elevation, because either the hill tops, small ridges and valley bottoms are not recorded in the topographic map, or no elevation value is attached to them. Terraces usually occur because a linear interpolator has been used, but they can also occur when smoother interpolators, such as splines, have been used.

Terraces can be reduced by accounting for features, such as break-lines indicating ridges or valley bottoms, that are not shown on the contours. This reduction can be achieved by digitising supplementary contour lines, spot and bottom heights representing small channels, small depressions (e.g. sinks), hilltops and ridges that can be inferred, but are not indicated on the original topographic maps. The proportion of artefacts may be fairly high, especially over flat terrains, which makes manual digitisation a time consuming process.

An alternative to digitising hill tops and valley bottoms manually is to detect the *medial axes* between the closed contour lines automatically (see a further imple-

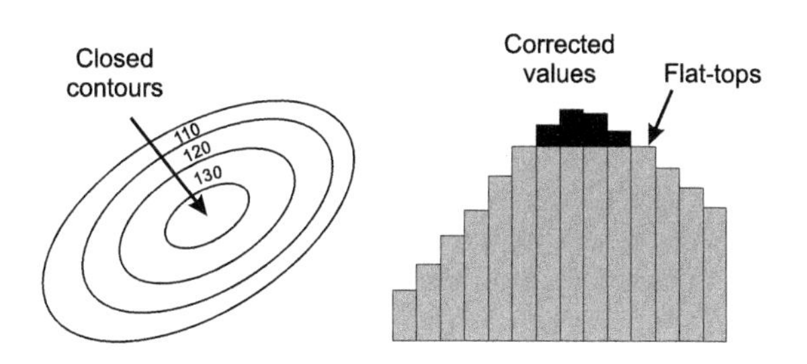

FIGURE 10 A schematic example of the terrace problem in DEMs.

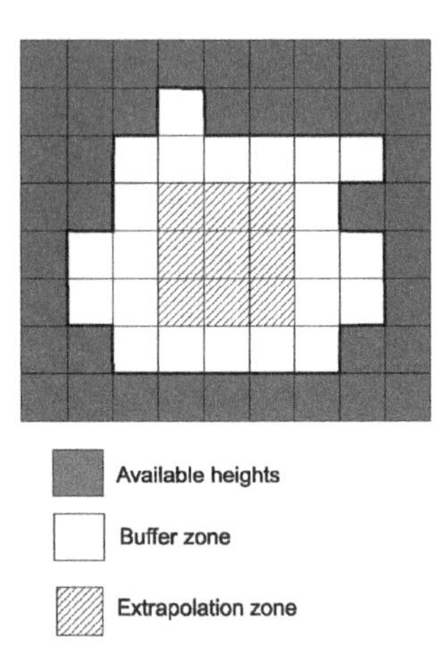

FIGURE 11 The general principle of a void-filling algorithm. The void is divided into a buffer zone and an extrapolation zone. The extrapolation zone can be estimated using auxiliary data.

mentation of this algorithm in Section 2.1 of Chapter 13). Elevation can be assigned to the medial axes by adding (convex) or subtracting (concave) a threshold elevation value (Hengl et al., 2004a):

$$z^{+}(s_i) = \begin{Bmatrix} z(s_i) + \Delta z & \text{if } \tau = \text{convex} \\ z(s_i) - \Delta z & \text{if } \tau = \text{concave} \\ z(s_i) & \text{otherwise} \end{Bmatrix} \tag{2.11}$$

where Δz is the estimated difference between the original DEM and the expected elevation.

2.7 Processing voids

Voids or NODATA areas occur for various reasons: where data recordings that do not overlap, in the case of cloud cover and/or in the case of the failure of a recording device. To filter such data, one approach would be to plan an additional campaign and then process it. However, new land surveys are rather expensive, especially because of the geographical distribution of the voids in the study area.

REMARK 5. *Voids are patches of missing values in the raw DEMs. They are most commonly removed by void filling operations.*

If a void is reasonably small, it can be filled using nearby elevations (see Figure 11). To generate a seamless DEM, different interpolation methods (see Section 3.1 in Chapter 2) are applied, depending on the area of the void. When auxiliary information is used, it should only be applied to void areas that are large enough to be on the same support. For example, it makes no sense to apply the 1 km GTOPO DEM as auxiliary information for filling in a void of 180 m in the ASTER DEM (Section 3 in Chapter 3).

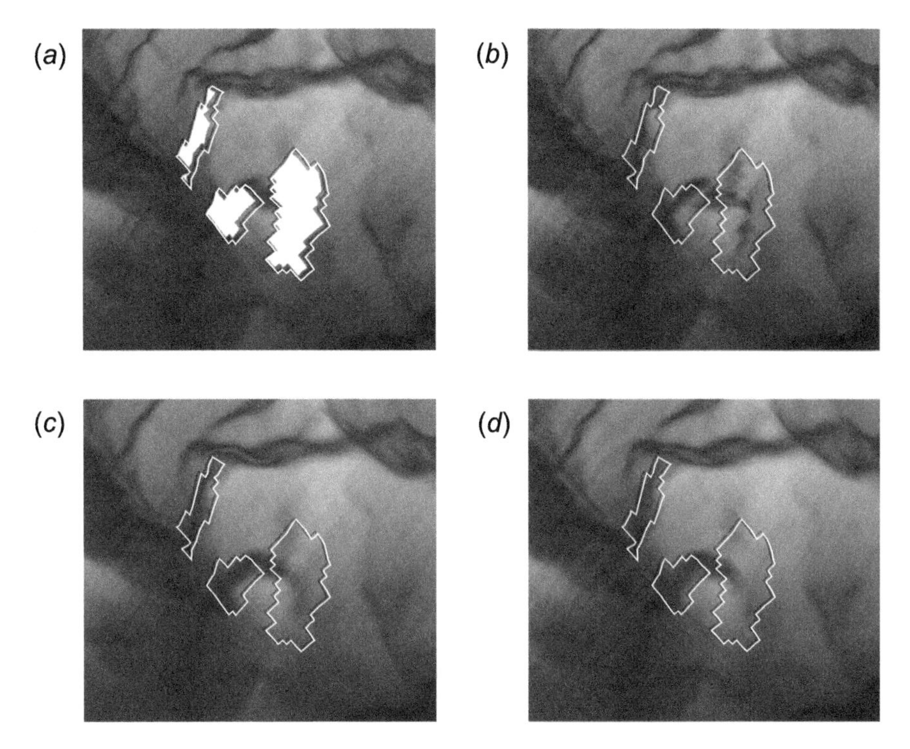

FIGURE 12 The results of three void-filling algorithms: (a) void, (b) moving window, (c) kriging, (d) ANUDEM.

Generally, if data from different DEMs are merged to create a seamless mosaic, it is necessary to adjust the DEMs to each other, keeping in mind several aspects:

- the difference in spatial resolution;
- horizontal and vertical shifts (e.g. due to different vertical datum);
- first or second-order trends;
- the differences in the models and production techniques used for the DEMs;
- the spatial distribution of errors (see, for example, Hutchinson, 1989; Fisher and Tate, 2006);
- the difference in absolute and relative accuracy.

Several interpolation methods (see also Section 3.1 in Chapter 2) can be applied to support *void filling* and any additional auxiliary information. Jet Propulsion Laboratory (http://www.jpl.nasa.gov), in the processing of the SRTM data, used a threshold of 16 posts for applying a moving window interpolation from the surrounding elevation, and ANUDEM has been applied for other larger void areas (see http://srtm.csi.cgiar.org). The methods, as shown in Figure 12, can be divided into:

- geostatistical void filling algorithms (e.g. kriging or spline) — see Figure 12(c);

- an algorithm where the void is simply filled in using auxiliary information (fill and feather void filling);
- an algorithm where the area is filled in using a moving average of the surrounding data — see Figure 12(b);
- a number of points and connections to describe a landscape that are built onto a topology (TIN);
- multi-scaled hydrological procedures, such as the one used by Hutchinson (1989) — see Figure 12(d).

From the algorithms listed above, the **ANUDEM** approach (Hutchinson and Gallant, 2000) seems to be the most suitable approach for filling of voids. It creates a hydrologically correct DEM — which is often a necessity for many land-surface parameters, because it ensures that ridges are retained, streams enforced and spurious sinks removed.

2.8 Filling in sinks

Sinks[6] are features that may have been introduced when generating the DEM. Removing erroneous (spurious) sinks is a pre-processing step driven mainly by hydrological applications in which a hydrologically correct network is required that simulates the flow of water on the surface of the ground (Tarboton et al., 1991; Wise, 2000). In this section, we therefore assume that the elevation data that are analysed are DEMs representing terrain elevations. Consequently, as explained in Section 2.5, DSMs should be filtered beforehand to remove all vegetation heights. This should also simultaneously suppress sinks that correspond to closed depressions of the forest canopy, or clearings within a forest.

> REMARK 6. *A spurious sink is often an erratic feature that does not correspond with actual features of the terrain. It needs to be removed to allow accurate hydrological modelling.*

Ideally, all man-made structures above the ground revealed on DEMs, that do not actually impede the flow of water, should also be removed. In highly urbanised areas, it may prove very difficult to create a DEM, those flow paths of which agree with the actual flow paths. For example, a large bridge should be set to the ground elevation below that bridge.

DEM sinks can be classified as genuine or spurious depending on whether they represent actual terrain features or not. On planet Earth, large genuine sinks are filled with expanses of water, the largest of which are the oceans. Other large genuine sinks correspond with the outlets of major basins such as the Caspian Sea and Lake Chad. Medium-sized *genuine sinks* are common in arid, glacial, and Karst landscapes. For example, in the Baranja region, rivers sometimes disappear underground and are connected to an underground network of streams. In such cases, it can even be desirable to introduce sinks. Genuine small to medium-sized sinks are found in regions protected by dikes (e.g. polders and flood plains), and

[6] Also called a pit or a closed depression, or a *spurious sink*, if a sink is very unprobable at a certain location.

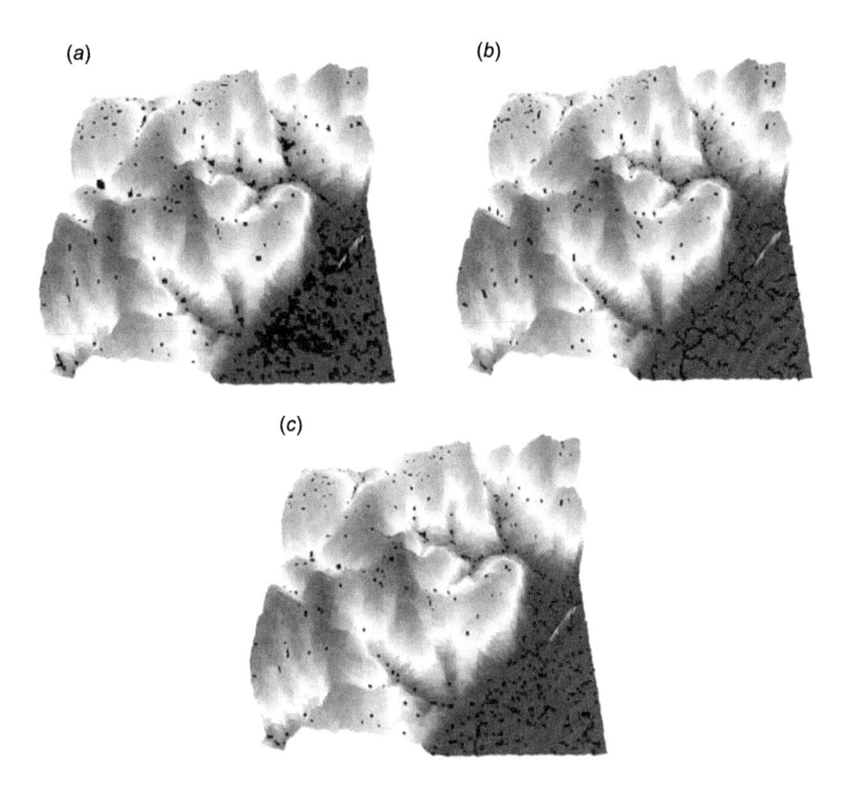

FIGURE 13 Three approaches to removing spurious sinks: (a) sink filling, (b) carving, and (c) the optimal combination of filling and carving. The detected sinks are indicated black. (See page 712 in Colour Plate Section at the back of the book.)

other anthropogenic terrain features, such as open-mine pits and dams. Smaller genuine sinks are ubiquitous, since every single closed depression on the terrain, however small, corresponds to a genuine sink. Nevertheless, these small natural sinks can only be resolved by DEMs which have a very high horizontal and vertical resolution.

Contrary to genuine sinks, spurious sinks found in DEMs are those that do not correspond with actual features of the terrain. They are mainly caused by measurement errors and the lack of density of sampling points on the continuous surface of the terrain, e.g. where sampling points fall on the sides of a narrow valley and miss the valley bottom. In temperate climates, and where there is sufficient energy in the relief, most sinks with vertical resolutions, within the range of a meter, that appear on DEMs, can be considered spurious (Rieger, 1998). Therefore, it is necessary first to remove all these spurious sinks before modelling water run-off on these DEMs, so that a fully connected river network can be generated, using flow simulation and drainage-area calculations (see Chapter 7).

The need for this pre-processing step was recognised already in 1984 by Mark and Aronson who advocated smoothing by using a moving average to eliminate many small depressions. However, this approach not only alters all elevation values, but could also generate new sinks. More radical approaches have been proposed since then. Essentially, three approaches can be considered, as illustrated in Figure 13:

Sink filling Whereby sinks are progressively filled (by increasing their elevation values) until the elevation of their lowest outflow point is reached [see Figure 13(a)]. Sink filling is the oldest and still the most frequently used method for removing spurious sinks (Jenson and Domingue, 1988; Martz and de Jong, 1988; Soille and Gratin, 1994), but although successful in doing this, artificially flat regions are created, should the sink have been caused by an unresolved valley bottom. For example, pit filling applied to raw SRTM data fills a large part the Panonian Plain because, where the Danube leaves this plain, the valley bottom is unresolved.

Carving Whereby a descending path is created from the bottom of the sink, by carving the terrain along this path until the nearest point is reached which has an elevation lower than that of the bottom of the sink [see Figure 13(b)]. Carving decreases the elevation of pixels occurring along a path (Soille et al., 2003; Soille, 2004b). Sink filling, by contrast, increases the elevation of the entire region below the sink's pouring point. This effect can be observed in Figure 13 by comparing the masks of the pixels modified by these two procedures. *Sink unblocking* (Morris and Heerdegen, 1988), *phenomenon-based approach* (Rieger, 1992), and *breaching* (Martz and de Jong, 1988; Martz and Garbrecht, 1999) procedures for removing spurious sinks are related to the carving procedure, in the sense that they too operate by decreasing the elevation values of specific pixels. However, in contrast to carving, sink unblocking and breaching procedures fail to suppress complex sink configurations such as nested sinks. In Rieger (1992), it is suggested that the phenomenon-based approach may also be applied to nested depressions, by handling them in decreasing order.

The combined approach The optimal approach combines sink filling and carving, thereby minimising the sum of the differences in elevation between the input DEMs and the output DEMs that do not have sinks [Figure 13(c)]. In the combined approach, sinks are filled up to a certain level, and carving proceeds from this level (Soille, 2004b). The method is optimal, in the sense that the level at which sink filling stops and carving takes over is defined, which minimises the sum of the differences in elevation between the input DEMs and the output DEMs that do not have sinks. Indeed, it can be shown that the sum of the elevation differences between carving and sink filling always displays a unique minimum. However, other minimisation criteria can be used instead, for example, the number of modified pixels may be considered (Soille, 2004a). The impact reduction approach proposed in Lindsay (2005) selects either sink filling or breaching for each sink, depending on which method distorts the DEM the least (TAS allows selective filling of depressions, Figure 3 in Chapter 16). Therefore, in contrast to the optimal solution, the impact reduction approach does not advocate combining filling and carving at the level of a single sink. Table 1 summarises the sum of elevation differences and the number of modified pixels for each method applied to the Baranja test DEM. As expected, the hybrid method gives the minimum sum of elevation

TABLE 1 The sum of elevation differences (which is proportional to a volume) and the number of modified pixels (which is proportional to an area) for pit filling, carving, and the optimal hybrid, where the sum of elevation differences between input DEMs and DEMs which have no sinks is minimised

	Sum of differences (m)	Number of modified pixels
Pit filling	4119	1475
Carving	2009	703
Optimal hybrid (volume-based)	1370	667
filling part	635	233
carving part	735	434

Measurements were performed on the Baranja Hill DEMs processed in Figure 13.

differences: i.e. 1370 m compared with 4119 m for sink filling and 2009 m for carving.

2.9 Mosaicing of adjacent DEMs

One point which applies to all processes that generate DEMs is the overlap between adjacent DEMs. No algorithm can create reliable information, if only half of the information is available. For example, the profile curvature of DEMs created from different topographic map sheets simply can not be merged. It is therefore advisable to use a DEM that is actually around 10–20 cells wider than the actual study area (see also Figure 11 in Chapter 7).

One of the simple methods is to take an average between different DEMs, however this method will almost certainly create artefacts. An advanced method creates cut-lines along certain features,[7] which are then used to make single non-overlapping patches. Adjacent grid tiles are mosaicked so that the geometric discontinuities can be minimised by automatically selecting the most *salient seam line*[8] in the overlapping domain between both DEMs.

2.10 Filtering LiDAR DEMs

The post-processing required to filter out vegetation cover, buildings, power poles and other structures has been a major challenge when preparing a DEM from LiDAR data. Commercial vendors do not usually divulge the methods they use for filtering. This is an area of active research and a variety of post-processing algorithms have been under development, e.g. Sequeira et al. (1999), Forstner (1999) for urban areas, Gomes Pereira and Wicherson (1999) for fluvial zones, Petzold et

[7] Such as mountain ridges, roads, and waterways (Soille, 2006).

[8] A cut-line which is used in mosaicking of images or DEMs. One looks for profound objects in the landscape (rivers, ridges, or roads in the case of remote sensing images) along which the cut-lines can be defined.

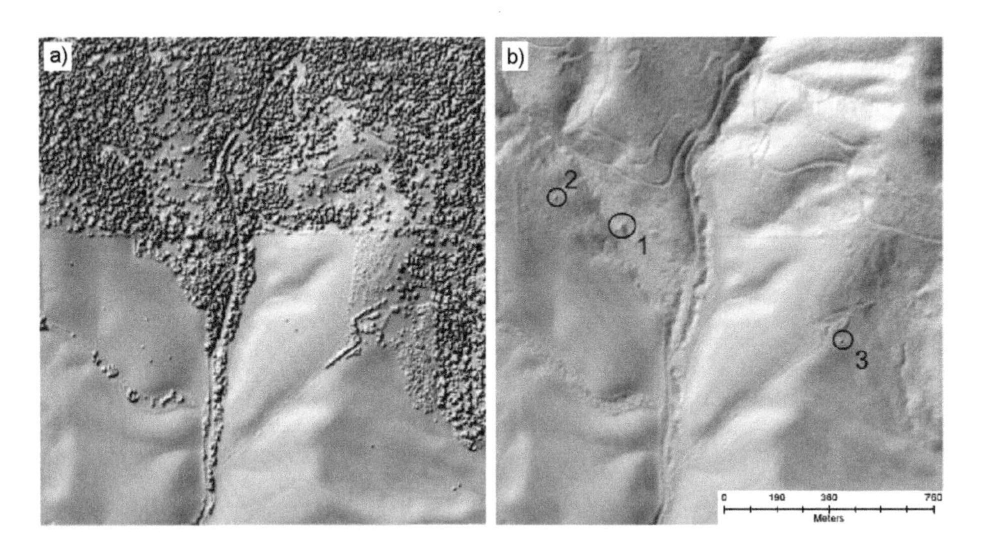

FIGURE 14 Hillshade of a LiDAR-derived elevation dataset (a) with a multiscale curvature filter technique applied; (b) to identify ground returns from discrete return LiDAR in a forested environment. The remaining problem areas are numbered. Reprinted from Evans and Hudak (2007). With permission from IEEE Transactions on Geoscience and Remote Sensing.

al. (1999) for state-agency mapping purposes and Haugerud and Harding (2001) for forest applications.

Evans and Hudak (2007) have recently published the details of a *Multiscale Curvature Filter* algorithm used in forested environments (Figure 14). This technique has great potential as it builds on previous research (Hutchinson, 1989; Haugerud and Harding, 2001; Mitášová et al., 2005), is mostly automated, and the software is freely available as an ArcGIS AML (Evans and Hudak, 2007).

2.11 Reducing errors by averaging DEMs from various sources

Probably the most robust (and most expensive) approach for improving the quality of DEMs is to average elevation values from different sources and campaigns, and weight them using the spatially varying error in each single DEM. One requirement for this sort of application is that all DEMs should be orthorectified, and therefore do not contain any offset. The following equation can then be applied:

$$\hat{z}^{+}(s_i) = \frac{\sum_{u=1}^{p} w_u(s_i) \cdot z_u(s_i)}{\sum_{u=1}^{p} w_u(s_i)}, \quad i = 1, \ldots, n \tag{2.12}$$

where $\hat{z}^{+}$ is the improved elevation, $z_u(s_i)$ is the elevation value from the uth DEM at the s_i location, p is the total number of input DEMs, and w_u are the weights (maps or global values) of the input DEM determined by RSME [see

Equation (1.1)]:

$$w_u(s_i) = \left(\frac{1}{\mathrm{RMSE}_u(s_i)}\right)^2 \tag{2.13}$$

The advantage of using this approach is that it can be applied to elevation data based on grids, as well as on point or transect measurements (Knöpfle et al., 1998).

> REMARK 7. *The most robust approach to improving the quality of DEMs is to average elevation values from different sources or plan multiple topographic mapping campaigns for the same area.*

3. REDUCTION OF ERRORS IN PARAMETERS AND OBJECTS

3.1 Filtering missing or zero values

The absolutely flat areas present a problem when a parameter is derived by division with slope (e.g. TWI). There are two different approaches to deal with the 0 degree slope gradients. The simple approach is to assign a value of half the pixel size to any catchment area that is equivalent to zero, and which has a value of 0.01 for any zero slope value. The results of the TWI then provide a more complete picture. We can also approximate TWI by averaging the slope and catchment-area maps iteratively, from surrounding pixels, until all zero values are replaced by low values. This creates realistic pools of high TWI in the plains (Hengl and Rossiter, 2003).

An alternative is to avoid using TWI in flat areas completely, because the conceptual model of water distribution does not apply to flat terrain. In such areas, water movement is driven by sub-surface gradients of water heights, which do not correspond to topographic gradients. For example, the multi-resolution index approach (*MrVBF*) might identify areas where TWI is of no help (Gallant and Dowling, 2003).

In hydrological applications, the quality of land-surface parameters is usually improved by adjusting the interpolation to the existing network of streams and ridges, or by removing sinks. If contour lines of the area are available, errors in land-surface parameters can be filtered by overlying contour lines on the land-surface parameters that have been extracted, and then digitising any additional features such as elevations, streams,and ridges, etc. For example, an extraction of a drainage-network map, or a TWI, can be corrected by carefully inspecting how well features match reality. As this (mental) process can also be very time-consuming, the expectation is that many will make use of more automated procedures.

3.2 Reducing errors by averaging the simulations

Given the uncertainty in elevation values, the soundest approach for reducing the errors in land-surface parameters is to average a set of equi-probable realisations[9] (Burrough et al., 2000; Hengl et al., 2004a; Raaflaub and Collins, 2006).

The method works as follows: for each of the m simulated DEMs, land-surface parameters are derived m times and are then averaged per pixel:

$$\bar{\text{LSP}}(s_i) = \frac{\sum_{j=1}^{m} \text{LSP}[z^{*j}(s_i)]}{m} \tag{3.1}$$

where $\bar{\text{LSP}}(s_i)$ is the averaged map of a land-surface parameter and $\text{LSP}[z^{*j}(s_i)]$ is the jth realisation of the land-surface parameter, calculated from the simulated elevation map z^{*j}. An estimate of the propagated uncertainty is derived from the RMSE error of several simulations:

$$\text{RMSE(LSP)} = \sqrt{\frac{\sum_{j=1}^{m} [\text{LSP}(z^{*j}) - \bar{\text{LSP}}]^2}{m}} \tag{3.2}$$

By averaging land-surface parameters derived from multiple realisations of the same DEM we can decrease the uncertainties caused by the limitations of the geomorphometric algorithm. This will eventually lead to a *stable image* of a land-surface parameter (see e.g. Figure 16). In addition, the map of propagated uncertainty (RMSE(LSP)) can be used to depict problematic areas and to digitise additional contours.

In practice, there are three main techniques that can be used to generate simulations of DEMs (Figure 15):

1. *Pure Monte-Carlo simulations*: simulate the errors in sampled heights and then re-interpolate DEMs.
2. *Direct geostatistical simulations of heights*: run conditional geostatistical simulations using the default variogram and directly generate DEMs.
3. *Geostatistical simulations of errors*: assess the errors in the heights and then simulate the error surfaces using the point dataset; then add them back to the original DEM (see further Section 3 in Chapter 5).

3.2.1 Monte-Carlo simulations of 3D points

In the case the DEM will be generated from sparsely located point measurements of heights, the equiprobable DEMs can be produced by simulating multiple 3D point[10] datasets and then re-interpolating them in a GIS package (see also a script for Monte-Carlo simulations in TAS in Chapter 16). For each sampled point we simulate independently the value for x, y and z by using the inverse normal probability function (Banks, 1998):

$$z_i^{\text{SIM}} = z_i + \text{RMSE}(z) \cdot \sqrt{-2 \cdot \ln(1 - A)} \cdot \cos(2 \cdot \pi \cdot B) \tag{3.3}$$

[9] The resulting realisation of a DEM should have the same internal properties (histogram, variogram) as the original land surface.

[10] The same method can be applied also to contour lines.

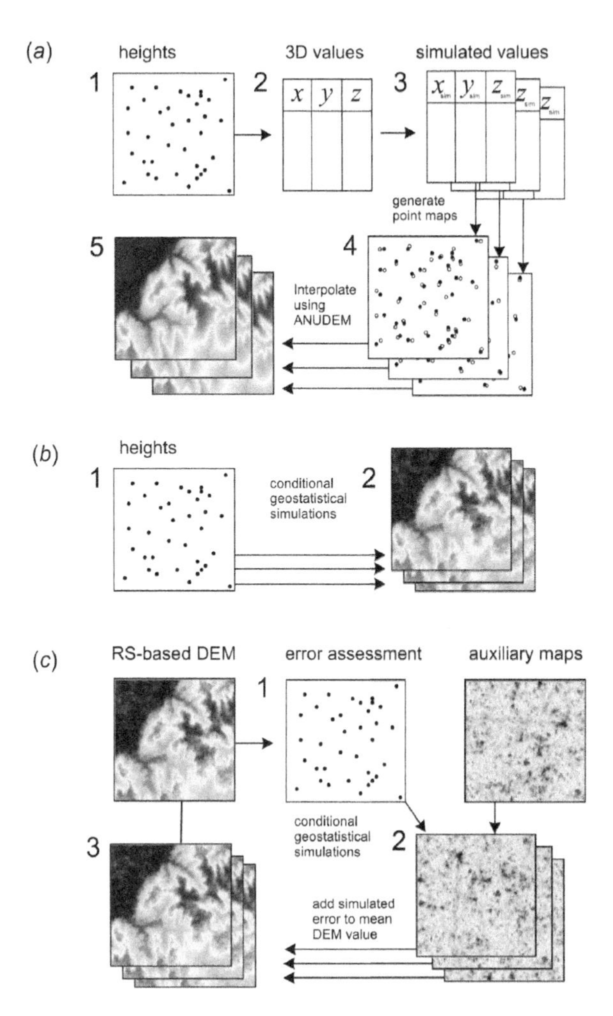

FIGURE 15 Approaches to simulate equiprobable realisations of DEMs: (a) simulation of 3D errors in input data and re-interpolation; (b) direct conditional geostatistical simulations using a point map; (c) geostatistical simulations of the errors using accurate field measurements.

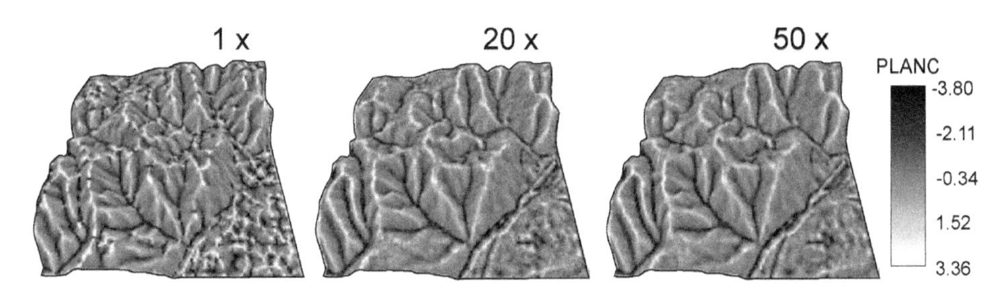

FIGURE 16 A comparison of PLANC, calculated by using a single, 20 and 50 realisations. See also http://geomorhometry.org for an animated display of error reduction.

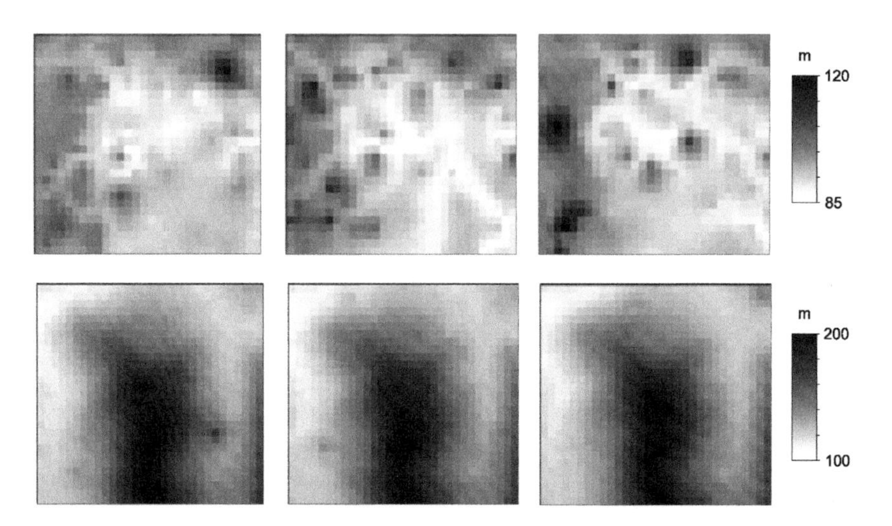

FIGURE 17 Comparison of three equiprobable realisations of DEMs produced using the ANUDEM procedure with default settings in ArcInfo GRID module: in the area of low relief (above) and in the area of high relief (below). Each block is about 750×750 m.

where z_i^{SIM} is the simulated elevation with induced error, A and B are the independent random numbers within the 0–0.99... range, z_i is the original value at ith location, and RMSE(z) is the standard error of measured heights. The same way we can simulate the values for x and y and produce m point maps with simulated values x_i^{SIM}, y_i^{SIM}, z_i^{SIM} [see the flowchart in Figure 15(a)].

Because we will use in this example height measurements from the 1:5000 topo-maps, the standard errors in horizontal (RMSE(x), RMSE(y)) and vertical spaces (RMSE(z)) can be estimated by using the cartographic standards. For example, the planimetric error on topo-maps is typically:

$$\mathrm{RMSE}(x, y) = 0.2\ \mathrm{mm} \cdot S \tag{3.4}$$

where S is the scale number. So, for the 1:5000 scale, the planimetric error is about ± 1 m. RMSE(z) can be also estimated using cartographic standards — contour interval h and local slope (Li, 1994):

$$\mathrm{RMSE}(z) = B \cdot h + \mathrm{RMSE}(xy) \cdot \tan\beta \tag{3.5}$$

where B is empirical number (commonly used values are within the 0.16–0.33 range), RMSE(xy) is the (empirical) planimetric error and β is the local slope (Pilouk and Tempfli, 1992). If the contour interval is 5 m and $B = 0.25$ then we can estimate the RMSE(z) to be ± 1.5 m in flat terrains. This also means that we can take global values for the planimetric errors (x, y), but locally adjust the vertical errors using a preliminary slope map. Also note that these simulations are purely statistical simulations since we do not really have to know the spatial dependence structure of heights.

The example in Figure 17 shows the differences for DEMs produced using the procedure shown in Figure 15(a). In this case, we used the default settings in ArcInfo's ANUDEM procedure to generate 100 DEMs of Baranja Hill at 25 m grid: the tolerances used to adjust the smoothing of input data and the removing of sinks in the drainage enforcement process have been set at 2.5 m, horizontal standard error has been set at 1 m and number of iterations has been set to 30. Figure 17 shows how striking are the difference between the areas of low and high relief. In the case of low relief, although we see very distinct stream channels and ridges, they are unrealistic.

3.2.2 Geostatistical simulations of heights

Assuming that height is a regionalised stationary random field, the sampled heights can be used to generate equiprobable realisations of DEMs by conditional *geostatistical simulations*. This approach has been extensively discussed by Kyriakidis et al. (1999) and Hengl et al. (2008). First we need to estimate the variogram model for point-sampled heights [Figure 12(a) in Chapter 2]. A realisation of the DEM can then be derived by directly running conditional geostatistical simulations on the sampled heights. This can be done efficiently in the R[11] statistical computing environment using:

```
> heights.sim <- krige(heights~1, data=points, newdata=dem25m,
model=heights.vgm, nmax=50, nsim=1)
```

where `heights.sim` is the output simulated DEM map, `krige` is the **gstat** function, `heights` are the point measurements of heights in table `points` interpolated over the grid map `dem25m` using the `heights.vgm` variogram and neighbourhood of 50 closest points. Both geostatistical simulations and geomorphometric analysis can be further combined in R using the RSAGA package (see www.geomorphometry.org for sample scripts).

The problems of this techniques are as follows. First, we will generate a land surface using a gridding technique which is not considered suitable for generation of land surfaces (see discussion in Section 3.2 of Chapter 2), second, geostatistical simulations commonly result in DEMs which show artificial surface roughness (see further discussion). As can be seen in Figure 18(b), the conditional geostatistical simulations will produce noisy outputs (even if the nugget is set to zero!), unlike the ANUDEM procedure [Figure 18(a)] that results in (more realistic?) smooth realisations of a land surface.

3.2.3 Conditional geostatistical simulations of errors

If a point map with accurate measurements of heights is available (*error assessment dataset*), it can be used to estimate the spatial structure of the error (variogram) and produce equiprobable realisations of a DEM (this technique is extensively discussed further in Chapter 5). After we estimate the error values using the error assessment dataset, we can again run conditional geostatistical simulations to produce equiprobable error surfaces. An example script to run conditional simulations in R is given down-below. First, we need to import the point map (heights) and the SRTM DEM, then overlay the point map and the SRTM DEM, and de-

[11] You will first need to load the sp and gstat packages in R (Pebesma, 2004).

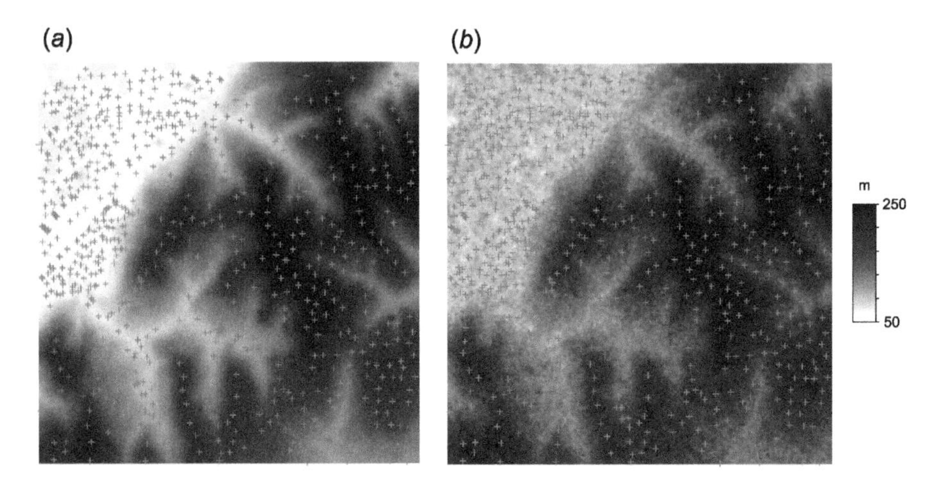

FIGURE 18 Comparison of two approaches to simulation of DEMs: (a) derived using interpolation of simulated 3D points and ANUDEM procedure, and (b) derived directly using geostatistical simulations on points. The input spot heights are overlaid.

rive deltas. We can fit a variogram automatically for deltas and run conditional simulations in **gstat** package using:

```
> delta.smv <- variogram(deltas~1, data=points) delta.vgm <-
> fit.variogram(delta.smv, vgm(1, "Exp", 1000, 1))
```

where `delta.vgm` is the fitted variogram model, `fit.variogram` is the **gstat** function for automated fitting of variograms, `vgm` is definition of the initial variogram and `Exp` is the type of the variogram model (Exponential).

Note that we can also use some auxiliary map to improve[12] the simulations of the errors. For example, we can assume that the errors are positively correlated with the slope gradient (`slope`) or solar insolation (`solin`). If these maps are available, they can be added to the conditional simulations as covariates or predictors. In the case of the 1 arcsec SRTM DEM, the command in **R** then modifies to (Hengl et al., 2008):

```
> deltas.sim <- krige(deltas~slope+solin, data=points,
newdata=dem25m, model=deltas.vgm, nmax=50, nsim=1)
```

so the final output of the simulated error surface will be adjusted to local variability of relief, which is more realistic then if we use a global estimate of the errors [Figure 19(d)].

Compare the results of mapping TWI using a single and an average from 100 simulations for DEMs derived from simulated points and simulated SRTM DEMs in Figure 20. Obviously, a land-surface parameter derived from the SRTM DEM will be much less precise than the one derived from the precise field measurements. This happens due to a high proportion of noise in the SRTM DEM that

[12] This assumes that the errors are significantly correlated with the auxiliary map, which is often the case.

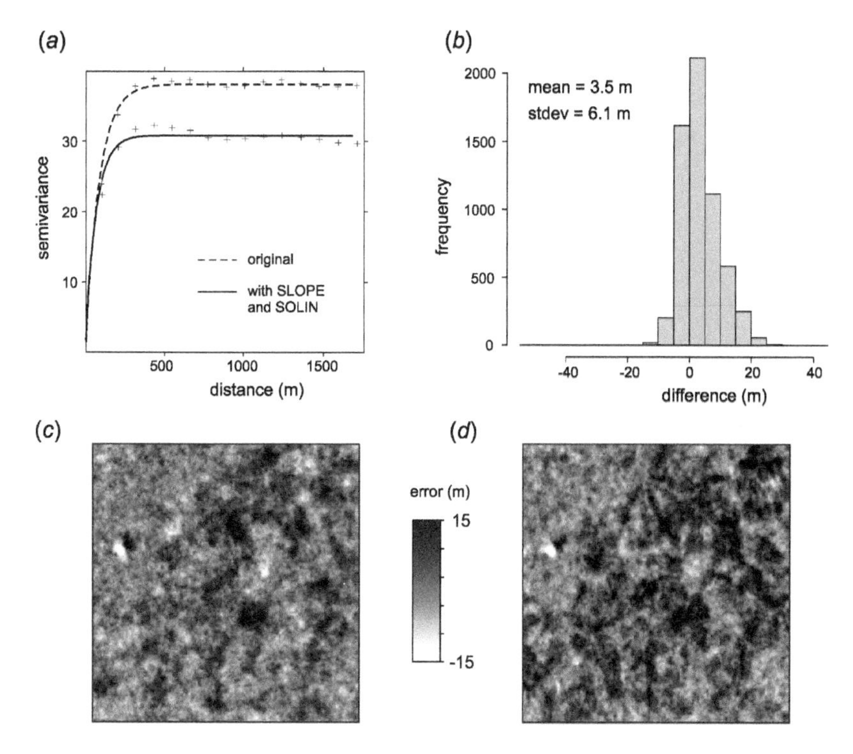

FIGURE 19 Modelling of errors in the SRTM DEM: (a) variograms of errors before and after addition of SLOPE and SOLIN; (b) histogram of errors; (c) simulated errors without any auxiliary information; and (d) simulated errors using SLOPE and SOLIN as auxiliary predictors.

probably needs to be filtered out prior to the geomorphometric analysis. Note also that the simulated error surface is added to the mean value [Figure 15(c)], i.e. the input SRTM DEM, but it should be added to the deterministic DEM (which is unknown). By adding the simulated error surface to a DEM that already consists of error components, we create a realisation of the SRTM DEM which is noisier than the original DEM (Hengl et al., 2008).

4. SUMMARY

This chapter has aimed to give an overview of DEM errors and of the algorithms that can remove/reduce them. We focused on the most-frequently used algorithms, rather than attempting to give an complete overview. From all listed approaches to pre-processing DEMs to reduce errors, three different groups can be distinguished:

(a) *The empirical methods* — here improving the accuracy of a DEM is primarily based on the knowledge of features. The limitation of the empirical approach is that it is time-consuming and almost impractical for large areas and detailed scales.

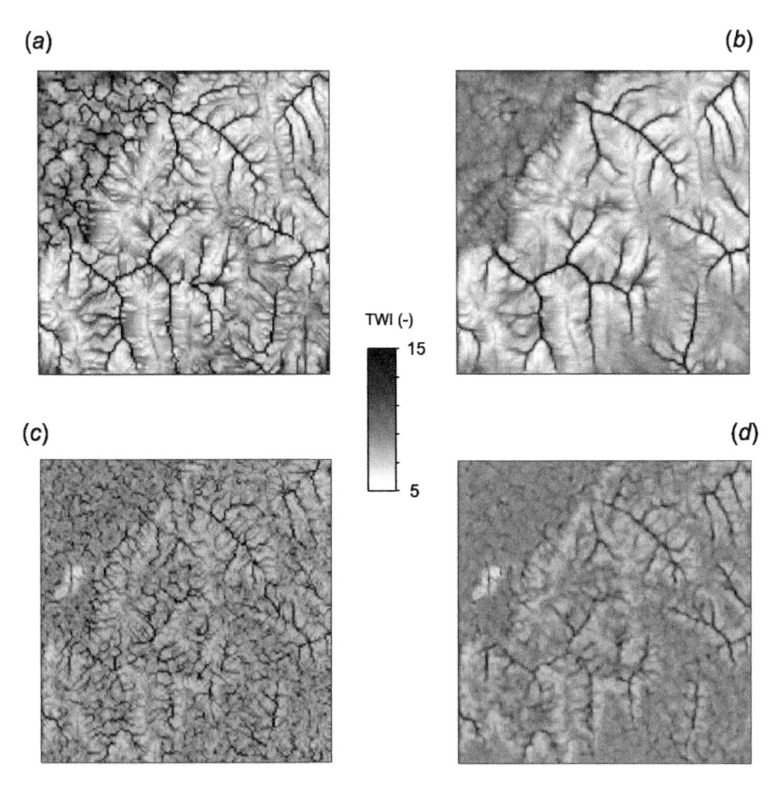

FIGURE 20 Different realisations of TWI: (a) a single realisation derived from simulated point map and (b) an average from 100 simulations; (c) a single realisation derived from the simulated SRTM DEM and (d) an average from 100 simulations.

(b) *The filtering methods* — here erratic features are filtered using various filter-based algorithms. Filtering the outliers, can also be adjusted, using geostatistical analysis and the correct variogram models for land-surface parameters and threshold values.
(c) *The simulation methods* — this approach is based on geostatistical simulations and is fully data-driven. The errors are reduced by calculating the average value of the land-surface parameter, calculated from multiple equiprobable realisation of the DEM. In general, this creates a more natural and more contiguous picture of the geomorphology. The advantage of averaging is that there is no need to calculate the filtering weights or to select the window size.

From the approaches listed above, the simulation approach seems to be especially interesting, as it can be fully automated. It is also attractive because it offers a (propagated) measure of the uncertainty of deriving a parameter. We have reviewed three main methods to produce DEM simulations in Section 3.2. An advantage of using the pure Monte-Carlo simulations is that all the simulated output DEMs will be as smooth as the original DEM. The advantage of the geostatistical simulations, on the other hand, is that they take into account spatial autocorrelation structure and are statistically more sophisticated. Where extra information is available, we should also try to localise the errors — either by using the slope or exposition maps to adjust the errors instead of using the global values that might overestimate the errors in the areas of low relief and underestimate them in the areas of high relief.

The precision of algorithms used to derive the same land-surface parameters and objects will vary because their sensitivity to various[13] features in DEMs is also different. Raaflaub and Collins (2006), for example, demonstrated that land-surface parameters derived with larger window sizes and by using more realisations will be more reliable and will show fewer erratic features, but such procedures will also be more time-consuming. For example, the 100 realisations of TWI using the point data and **ANUDEM** procedure [Figure 15(a)] for a small area of Baranja Hill took about 2–3 hours on a standard PC.

The computational complexity and lack of simulation wizards will remain the main drawback of wider use of error propagation techniques. The benefits are, on the other hand, obvious — by averaging multiple realisations we are able to produce a more generalised image of land-surface parameters and objects and filter out improbable features. We are also able to estimate the propagated uncertainty of deriving parameters and object out of DEMs coming for various sources, and then further use this information in spatial analysis.

DEM preprocessing methods have changed quite strongly over the last few years, and will continue to do so. Finally, we can anticipate the following trends with regard to DEM preparation techniques:

- *Integration of topographic and auxiliary information* — We believe that the use of multi-spectral information from different sensors and data sources will influence DEM preparation quite strongly (e.g. the location of lakes, streams, ridges and/or breaks will be identified from satellites and included in the processing chain). Additional to that, we are curious to see how water subsurface elevation models will be created and analysed in the next couple of years.
- *Standardisation of the error assessment procedures* — Future DEMs will need to provide error estimates of their recorded elevation in a standard protocol, otherwise their usefulness and reliability will be limited.
- *Assessment of the temporal component* — As LiDAR data from space becomes available for the research community (GLAS mission), we may need to develop algorithms to incorporate a time factor into our processing algorithm (e.g. to account for whether the water level of the lake is high or low, because of the season).
- *Data processing automation* — One of the next highlights will be to see how to develop algorithms that can process DEMs efficiently and in an automated way.

[13] Such as the distinctness of relief, surface roughness, abrupt changes in value, etc.

IMPORTANT SOURCES

Oksanen, J., 2006a. Digital elevation model error in terrain analysis. Ph.D. Thesis. Faculty of Science, University of Helsinki.
Carlton, D., Tennant, K., 2001. DEM quality assessment. In: Maune, D.F. (Ed.), Digital Elevation Model Technologies and Applications: The DEM Users Manual. American Society for Photogrammetry and Remote Sensing, Bethesda, MD, pp. 395–440.
Wise, S., 2000. Assessing the quality for hydrological applications of digital elevation models derived from contours. Hydrological Processes 14 (11–12), 1909–1929.
Garbrecht, J., Martz, L.W., 1997. The assignment of drainage direction over flat surfaces in raster digital elevation models. Journal of Hydrology 193 (1–4), 204–213.
Hengl, T., Bajat, B., Reuter, H.I., Blagojevic, D., 2008. Geostatistical modelling of topography using auxiliary maps. Computers & Geosciences. http://dx.doi.org/10.1016/j.cageo.2008.01.005.
Hutchinson, M.F., 1989. A new procedure for gridding elevation and stream line data with automatic removal of spurious pits. Journal of Hydrology 106, 211–232.

CHAPTER 5

Geostatistical Simulation and Error Propagation in Geomorphometry

A.J.A.M. Temme, G.B.M. Heuvelink, J.M. Schoorl and **L. Claessens**

use of geostatistics to model errors in DEMs · the second order stationarity assumption · stochastic spatial simulation techniques · stochastic simulation of error maps · Monte-Carlo methods · propagation of errors from a DEM to land-surface parameters · interpretation of error propagation analysis for DEMs

1. UNCERTAINTY IN DEMS

DEMs can be produced in many ways, as was discussed in Chapter 3. When DEMs are generated from measured point data or from digitised contour lines, interpolation is required. Measurement error and digitisation error will affect the accuracy of the DEM, and so will interpolation error. Several approaches exist to assess the accuracy of DEM elevation values, derived from point data or contours, calculated from aerial photographs or measured with airborne devices like LiDAR. Most researchers use a simple comparison between a set of heights from the DEM against *'real'* elevation values, mostly taken from a more accurate source of topographic data. The root mean square error of elevation (RMSE) is then calculated as a first indication of the difference between estimated and true values (Wise, 2000). Alternatively, Carrara et al. (1997) suggest five simple criteria to evaluate DEM quality and to compare elevation histograms:

- The DEM should have (almost) the same values as contours when close to contour lines.
- DEM values must be in the range given by the bounding contour lines.
- DEM values should vary almost linearly between the values of the bounding contour lines.
- DEM patterns must reflect realistic morphology in (almost) flat areas.
- Artefacts should be limited to a small proportion of the data set.

Developments in Soil Science, Volume 33 © 2009 Elsevier B.V.
ISSN 0166-2481, DOI: 10.1016/S0166-2481(08)00005-6.

Although certain interpolation methods tend to perform better for specific data sources and specific topographically based applications (Weibel and Brandli, 1995; Desmet, 1997), Wilson et al. (2000) conclude that attempts to make generalisations of 'best' methods to generate DEMs from contour lines are very difficult. Simple interpolation methods will give satisfactory results as long as the input data are densely sampled. Sophisticated algorithms may produce unsatisfactory results when applied to sparse data (see also Section 3.1 in Chapter 2). DEM quality may be improved by combination with higher-quality point data (Kyriakidis et al., 1999).

Errors in DEMs will propagate to derived land-surface parameters and modelling results in a way that is not easily predicted. Relatively small errors can have a large impact in some cases while large errors may cancel out in other cases. Land-surface parameters such as slope, aspect or curvature may be more useful measures of the quality of a DEM, because they are important derived properties of DEMs and sensitive to artefacts (Wise, 1998). Bolstad and Stowe (1994) compared slope and aspect values calculated with a third-order finite difference method from different DEM interpolations with field measurements. They found substantial and statistically significant errors for both attributes and a positive correlation between slope error and slope, meaning larger errors on steeper slopes.

REMARK 1. *Errors in DEMs will propagate to derived land-surface parameters and modelling results in a way that is not easily predicted.*

Van Niel et al. (2004) demonstrated that solar radiation, a variable derived directly from a DEM, is less affected by DEM error than aspect and slope. Also, topographic position was less affected by DEM errors than topographic wetness index. Desmet (1997) tested the outcome of a transport-limited erosion model applied to DEMs created with nine different interpolation techniques. He found extreme sensitivity of erosion predictions (up to two orders of magnitude) to topographic differences due to curvature, this being a determining factor in the model. Some other analyses of DEM error assessment and how DEM errors propagate to derived products are given by Hunter and Goodchild (1997), Holmes et al. (2000), Endreny and Wood (2001), Aerts et al. (2003), Fisher and Tate (2006).

It is intuitively appealing to think that DEMs with high spatial resolution, i.e. small cell size, have low uncertainty. After all, they are better able to describe small scale landscape variation than DEMs with lower resolution (see also Chapter 2). However, it is too simplistic to equate high resolution to high accuracy and coarse resolution to poor accuracy. A high resolution DEM may still have a greater uncertainty than a low resolution DEM if we are less certain of its attribute values. High resolution DEMs have a higher potential to describe the landscape than low resolution DEMs, but whether or not this potential is actually achieved depends mainly on the accuracy of the attribute value.

REMARK 2. *A high resolution DEM may still have a greater uncertainty than a low resolution DEM if we are less certain of its attribute values.*

The purpose of this chapter is to demonstrate how uncertainty in DEM attributes can be quantified using geostatistical methods and to show how the propagation of errors to DEM derived products may be computed. Next to attribute errors (wrong elevation values), DEMs may have positional errors like a shift along one or both coordinate axes, rotational errors, scaling errors, projection errors or a combination of these. In this chapter, we will only consider attribute errors. In order of increasing complexity, we will consider the propagation of error from DEMs to three derivatives, namely slope (a local land-surface parameter), topographic wetness index (a regional land-surface parameter) and soil redistribution resulting from water erosion (a complex model). We describe the uncertainty propagation analysis in detail and outline how interested readers may implement the procedure in their own work.

2. GEOSTATISTICAL MODELLING OF DEM ERRORS

Consider an elevation map (DEM) that is defined on a spatial domain of interest D:

$$z^* = \langle z^*(s) \mid s \in D \rangle \tag{2.1}$$

We are uncertain about the true elevation Z, which we represent as the sum of our representation z^* and an unknown error ε:

$$Z(s) = z^*(s) + \varepsilon(s) \tag{2.2}$$

Here, $Z(s)$ and $\varepsilon(s)$ are random variables and $z^*(s)$ is a deterministic variable. The unknown error $\varepsilon(s)$ is composed of an autocorrelated part and a pure noise part [Equation (1.3) in Chapter 2], but for now, we will consider their sum. We take a probabilistic approach because we do not know the true error ε and can only quantify it in terms of a probability distribution. In order to characterise the statistical properties of ε, we need to determine the shape of its probability distribution and associated parameters. A sensible choice, supported by the *Central Limit Theorem* from statistics (Heuvelink, 1998), is to assume that ε is normally distributed. The normal distribution has two parameters, the mean μ_ε and standard deviation σ_ε. In many cases it may be sensible to assume that μ_ε is zero, because $z^*(s)$ is ideally constructed such that it is free of systematic error or bias.

Alternatively, we may assume that the systematic error is constant over the spatial domain and can thus be estimated from a sample of control points, simply by taking the average of the observed errors at control points (see further Section 4). The standard deviation σ_ε signifies the random error in $z^*(s)$ and will be greater than zero, unless the DEM is completely error-free. If we assume that σ_ε is spatially invariant (a constant) then it too can be estimated from a sample of observed errors at control points. This assumption can be relaxed by dividing the spatial domain in subregions with different values for σ_ε. For example, mountainous parts of the area may have larger DEM errors than flat parts and this can be accommodated by assigning larger standard deviations to the DEM error in the mountainous parts.

DEM errors are usually *spatially autocorrelated*. This means that when the true elevation at some location is overestimated in the DEM, then it is likely that the elevation at a neighbouring location is also overestimated. This is a consequence of the ways in which DEMs are constructed. Contour drawing, digitisation and interpolation all cause spatial correlation in the associated errors.

If we assume that the degree of spatial correlation between $\varepsilon(s)$ and $\varepsilon(s + \mathbf{h})$ only depends on the distance of vector $\mathbf{h}$ between the locations, then it can be characterised by the correlogram $\rho_\varepsilon(|\mathbf{h}|)$ or the (semi)variogram $\gamma_\varepsilon(|\mathbf{h}|)$. Note that this means that correlation is assumed independent of the direction of vector $\mathbf{h}$ between the locations, i.e. we assume isotropy. The correlogram can be estimated from observed DEM errors at control points, provided the number of observations is sufficiently large (roughly about 60 observation points or more). The theory of geostatistics has developed a large body of methods for estimation of correlograms or related functions such as the variogram (Goovaerts, 1997).

REMARK 3. *DEM errors are usually spatially autocorrelated — when the true elevation at some location is overestimated in the DEM, then the elevation at a neighbouring location will also be overestimated.*

The assumptions of constant mean μ_ε, constant standard deviation σ_ε and spatial autocorrelation $\rho_\varepsilon(|\mathbf{h}|)$ which only depends on the distance between points are jointly termed the *second-order stationarity* assumption. This assumption is often invoked in geostatistics, simply because without it, it would be difficult to define an estimable probability model. Under the second-order stationarity assumption, the statistical parameters of the DEM error ε can be reliably estimated using observed DEM errors at a sufficiently large number of control points. We will demonstrate this in the case study.

Alternatively, the parameters may also be derived from expert knowledge by taking an 'educated guess', but it is important to realise that this is no substitute for an objective assessment on measured data. When observation data are abundant, one could consider relaxing the second-order stationarity assumption, for example by allowing the standard deviation $\sigma_\varepsilon(s)$ to vary in space or by using a less stringent model for the spatial autocorrelation ρ_ε.

SRTM DEMs are an important group of DEMs that require some extra attention. First, there is a conceptual problem when we compare control points of terrain altitude to a SRTM DEM of surface altitude (see Section 1 in Chapter 1). Preferably, error propagation analysis with SRTM DEMs is done after correction for the difference between surface and terrain altitude. Second, for SRTM DEMs, error information is available in the form of a map of RMSE. When RMSE varies in space, we can no longer assume that standard deviation σ_ε is constant in space. Taking RMSE as our estimate of σ_ε and estimating the mean μ_ε and spatial autocorrelation ρ_ε from control points as described above, we can make a sensible error model for SRTM DEMs. By comparing σ_ε, measured from control points, with the RMSE values in the map we can check for consistency. If we do not have control points but only the RMSE map, we will have to guess μ_ε and ρ_ε. Then, we cannot check for consistency. More elaborate error models are provided by Goovaerts

(1997). Fisher (1998) presents an alternative way to deal with scarce error information.

3. METHODS FOR ERROR PROPAGATION ANALYSIS

When spatial data are used in GIS operations, the errors in the input maps will propagate to the output of the operation. Although users may be aware that errors propagate through their analyses, in practice they rarely pay attention to this problem. Experienced users will know that the quality of their data is not reflected by the quality of the graphical output in a GIS, but they cannot truly benefit from this knowledge unless the uncertainties are formally defined and explored through an uncertainty propagation analysis.

Here we will describe how propagation of attribute errors in spatial modelling can be computed using the *Monte-Carlo method*. This method is the most often used error propagation method because it is generic, flexible and intuitively appealing. The uncertainty propagation problem can be formulated mathematically as follows (Heuvelink, 1998). Let U be the output of a GIS operation g on m uncertain inputs A_i:

$$U = g(A_1, \ldots, A_m) \tag{3.1}$$

The operation g may take various forms, but here we focus on the derivation of land-surface parameters, in which case g could be a moving-window operation to compute slope and aspect from a gridded DEM, a more complex function to remove sinks, or an erosion prediction model based on transport capacity equations. In DEM analyses the number of inputs m to the operation g will usually be just one (namely the DEM), but here we present the more general situation. Indeed, an erosion prediction model may have additional uncertain inputs such as soil erodibility or rainfall.

> REMARK 4. *When spatial data are used in GIS operations, the errors in the input maps will propagate to the output of the operation.*

The objective of the uncertainty propagation analysis is to determine the uncertainty in the output U, given the operation g and the uncertainties in the input attributes A_i. The Monte-Carlo method computes the result of $g(a_1, \ldots, a_m)$ repeatedly, with input values a_i that are randomly sampled from their joint probability distribution. The results form a random sample from U, such that parameters of the probability distribution of U can be estimated from the sample. Thus, the Monte-Carlo method comprises:

(1) repeat N times:
   (a) generate a set of realisations a_i from the joint probability distribution of $A_i, i = 1, \ldots, m$;
   (b) for this set of realisations a_i, compute and store the output $u = g(a_1, \ldots, a_m)$;

(2) compute and store sample statistics from the N outputs u.

A random sample from the m inputs A_i can be obtained using an appropriate pseudo random number generator (e.g. Van Niel and Laffan, 2003). The number of realisations N must be sufficiently large to obtain stable results, but exactly how large N should be depends on how accurate the results of the uncertainty analysis should be. The accuracy of the Monte-Carlo method is proportional to the square root of the number of runs N. Therefore, to double the accuracy one must quadruple the number of runs. This means that although many runs may be needed to reach stable and accurate results, any degree of accuracy can be reached by taking a large enough sample N.

> REMARK 5. *The objective of the uncertainty propagation analysis is to determine the uncertainty in the output U, given the operation g and the uncertainties in the input attributes A_i.*

Lindsay and Creed (2006) found that with their method for distinguishing between actual and artefact depressions in a DEM, about 300 realisations yielded a stable result. Heuvelink (1998) finds that in many practical cases, N should be at least one hundred. This implies that the Monte-Carlo method is computationally demanding, particularly when the operation g takes much computing time.

Using a minimal N can help to reduce computing time. If the parameters of the error model are known, we can calculate the required N for a given accuracy of the DEM. If we want to find a minimal N for a given accuracy of a land-surface parameter, we can perform a large number of random simulations of a DEM (e.g. $N = 400$) and compute the output standard error of the land-surface parameter at every cell. Decreasing the number of runs with one at a time, we repeat this process until the mean difference in standard error between subsequent simulations becomes larger than the chosen accuracy. Monte-Carlo analysis can then be performed with the N from this last simulation. Note that this method itself is computationally demanding and that it may be faster instead to take a safe guess for N.

Application of the Monte-Carlo method to uncertainty propagation with operations that involve spatial interactions, such as sediment transport or slope calculation, requires that the spatially distributed uncertain inputs are generated in a way that takes their spatial correlation into account. Various techniques can be used for stochastic spatial simulation, perhaps the most attractive one being the *sequential Gaussian simulation* algorithm (Goovaerts, 1997). Briefly, this method works as follows.

Each location (grid cell) of the spatial domain is visited in a random sequence. At each location, the conditional probability distribution of the variable (e.g. the DEM error) is computed. For the first location, this is simply the (normal) distribution with pre-defined mean μ_ε and standard deviation σ_ε. A value from this probability distribution is drawn using an appropriate random number generator and assigned to the location. At the second location, the conditional probability distribution is computed by conditioning the variable at the location to the value that was sampled at the first location. This is done using simple kriging.

Simple *kriging*, like all kriging methods, not only produces an estimate of the attribute value but also quantifies the uncertainty attached to it, by means of a kriging standard deviation. The conditional probability distribution at the second location has the kriging estimate as its mean μ_ε and the kriging standard deviation as its σ_ε. From this distribution, a value is drawn. At the third location, the conditional probability distribution is calculated again, now using the two previous locations as conditioning data in simple kriging. This process is repeated until values for all locations have been drawn.

Note that in an early stage, when values have been drawn at only few locations, it is possible that the next random location is outside the range of influence of the other locations. In that case, the simple kriging estimate and standard deviation equal the pre-defined mean and standard deviation. At a later stage, previously drawn locations are more likely to be within the range and influence mean and standard deviation.

The sequential simulation method can also be used in cases where the variable is a priori known at some locations (such as when there are observed DEM errors at control points), simply by visiting these locations first, and assigning the observed value to the simulated values instead of simulating from a probability distribution, and adding these values to the conditioning data set.

REMARK 6. *Various techniques can be used for stochastic spatial simulation — the most attractive one being the sequential Gaussian simulation.*

4. ERROR PROPAGATION: BARANJA HILL

We now apply the methodology described above to the Baranja Hill area. The analysis is separated into two parts. First, we build and fit a geostatistical model that characterizes the uncertainty in the Baranja Hill 25 m DEM. For this we use a set of ground control points. Second, we analyse how the errors in the Baranja Hill DEM propagate to three land-surface parameters, namely slope, topographic wetness index and soil redistribution resulting from water erosion. For this, we use the Monte-Carlo method on 100 simulated DEMs, both unfilled (with sinks) and filled (without sinks). Figure 1 summarizes the procedure.

4.1 Geostatistical modelling

The Baranja Hill DEM for which we want to quantify the uncertainty is the 25 m resolution DEM derived from the 1:5000 contour map. To assess its uncertainty we use a data set of 5867 precise elevation measurements (see the description of Baranja Hill in Chapter 1). From this data set, we considered only one measurement per DEM cell, resulting in 3633 cells that had an elevation measurement. At these locations, the differences between the DEM elevation and the measured elevation were computed. A histogram of the differences is given in Figure 2.

The mean and standard deviation of the observed errors are 0.75 and 7.45 m, respectively. These values were assigned to the parameters μ_ε and σ_ε of the error

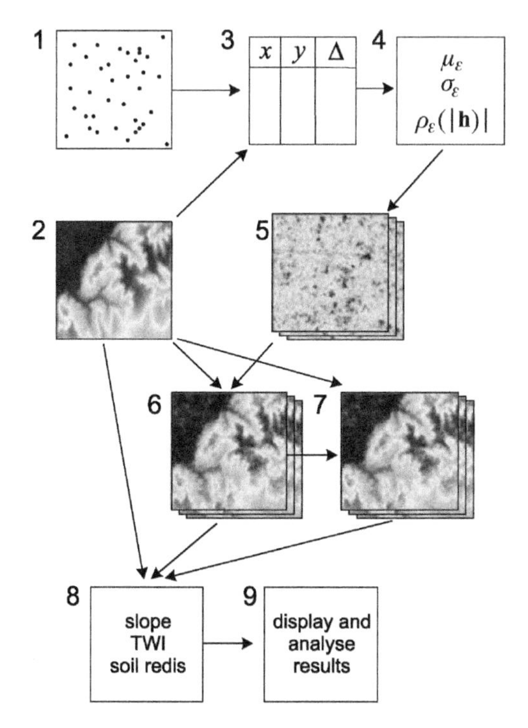

FIGURE 1 Flowchart of the procedure followed for the Baranja Hill case study. 1: Control points. 2: Original DEM. 3: Differences between control points and DEM. 4: Error model of DEM, calculated from (3). 5: 100 simulations of possible error using model from (4). 6: 100 simulations of possible DEM by adding original DEM (2) to simulated errors (5). 7: Filled versions of original DEM and 100 simulated DEMs. 8: Calculation of land-surface parameters from original and simulated unfilled and filled DEMs. 9: Report on results.

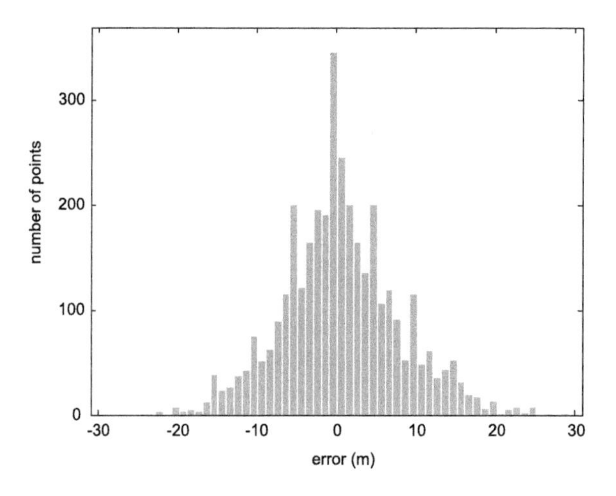

FIGURE 2 Histogram of observed errors at 3363 control points.

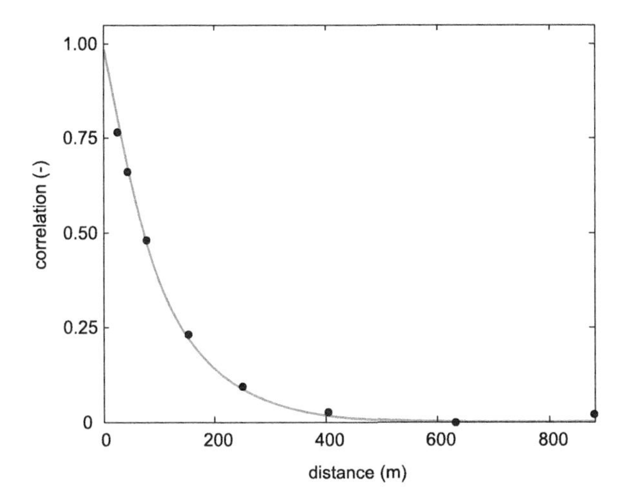

FIGURE 3 Correlogram of DEM errors at control points. Dots are observed correlations, solid line represents the fitted negative exponential correlogram.

model introduced in Section 2. The correlogram of the DEM error is given in Figure 3 as a fitted line through observed correlations as a function of distance. The fitted[1] model is a negative exponential function satisfying:

$$\rho(\mathbf{h}) = e^{-\frac{\mathrm{h}}{103}} \tag{4.1}$$

This implies that the DEM error is spatially autocorrelated up to distances of about 300 m [$\rho(300) \approx 0.05$]. The variogram corresponding to Figure 3 could be calculated with:

$$\gamma(\mathbf{h}) = \left[1 - \rho(\mathbf{h})\right] \cdot \sigma_\varepsilon^2 \tag{4.2}$$

Note that the correlogram is a continuous line starting at $\rho(\mathbf{0}) \equiv 1$ that has no sudden downward jump at the origin (no *'nugget'* variance), meaning that errors are strongly correlated at short distances. Simulated maps of the error will therefore have little noise. However, having a non-zero grid resolution implies that the correlation between adjacent grid cells is not equal to 1, so that grids may still appear to have noise. Also note that in building the statistical model of DEM error, we have assumed that it satisfies the second order-stationarity assumption, although the number of control points would have allowed us to relax this assumption and build a more elaborate model (see e.g. Oksanen, 2006b, for a discussion).

The simulated error map was added to the deterministic DEM, thus creating a single possible reality of the uncertain DEM. For the subsequent Monte-Carlo uncertainty propagation analysis, 100 of these simulated DEMs were generated. Assuming that the statistical error model and second-order stationary assumption are valid, the differences between the 100 DEMs generated in this way reflect our

[1] We initially fitted the variogram with gstat, using weighted least squares with the number of pairs in each class as weights. However, the correlogram is preferred here because it is a more general expression of autocorrelation.

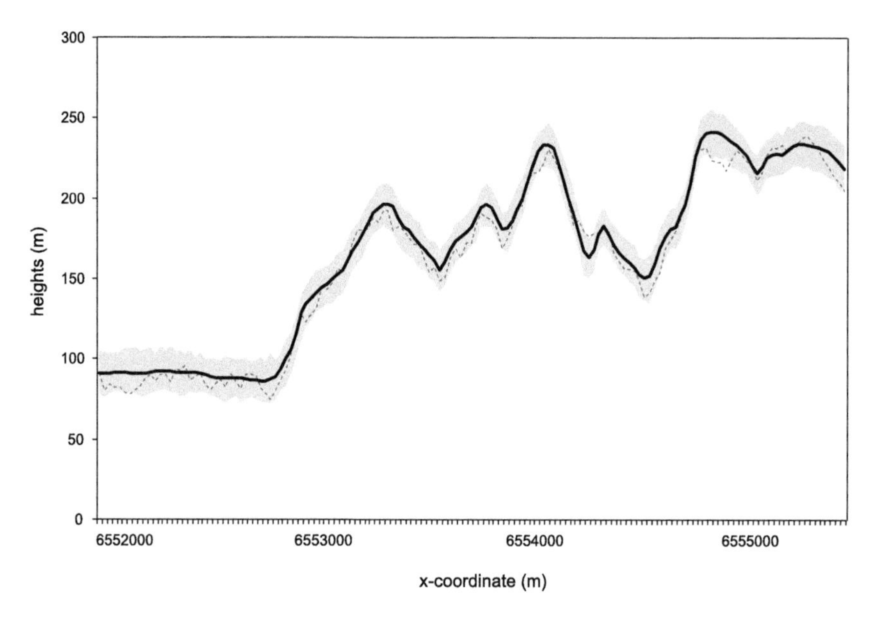

FIGURE 4 Transect of the deterministic and simulated DEMs from West to East at $y = 5{,}073{,}012$. Solid line represents the deterministic DEM, dashed line is one the 100 simulated DEMs. Gray area is the 90% confidence interval of the 100 simulated DEMs.

uncertainty about the true elevation. The transect in Figure 4 presents this uncertainty. If our assumptions are valid, any of the 100 DEMs could be the true DEM, it is just that we do not know which one it is. In fact, it is more likely that none of them corresponds with the true DEM because the 100 maps are just a small sample taken from the infinite population of possible realities to which the true DEM belongs.

However, Figure 4 hints that our assumptions are not valid. The simulated DEM is varying in a way that seems unrealistic. Two reasons for this come to mind:

- The second order stationarity assumption assumes an error model that is constant over the area of the DEM. In reality, the landscape itself is not at all constant over the area of the DEM, and the same is probably true for the error model.
- The precise elevation measurements that we have used to calculate the error model, may not have been taken at random locations but at meaningful locations like summits, minima, breaks in slope, etc. That may lead to an overestimation of error in the original DEM and an overly variable error map.

On the other hand, it would be wrong to expect a simulated DEM that is exactly as smooth as the deterministic DEM. In DEM creation, interpolation methods are used that minimise roughness (like ANUDEM) and that may result in a DEM that is smoother than reality. Therefore we should expect simulated DEMs to vary more than deterministic DEMs.

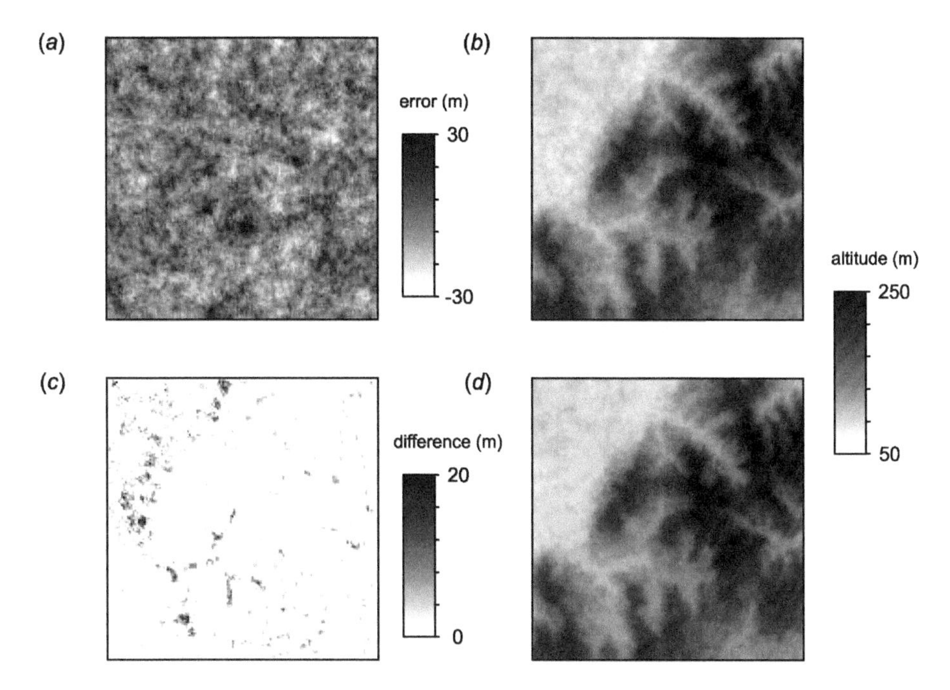

FIGURE 5 Simulations with elevation data: (a) example simulated error map and (b) corresponding example simulated DEM; (c) additions to this DEM to remove sinks and (d) resulting filled DEM.

The mean of the 100 simulated DEMs is almost equal to the sum of the deterministic DEM plus the systematic error $\mu_{\varepsilon} = 0.75$ m and their standard deviation is almost equal to $\sigma_{\varepsilon} = 7.45$ m.

Figure 5(a) shows a simulated error map that was generated using sequential Gaussian simulation. A problem with the resulting DEM shown in Figure 5(b) is that it is not geomorphologically realistic. By adding the error map to the deterministic DEM we have created many small sinks that in reality do not exist. Note that this problem would have been even larger if we would have chosen a non-zero nugget value. The simulated DEMs have more local variation in altitude than the (perhaps overly) smoothed deterministic DEM and thus steeper slopes. To obtain simulated DEMs that make more sense, we have also modified each of the 100 simulated DEMs using a sink removal algorithm. Many procedures exist to remove sinks from DEMs (see also Section 2.8 in Chapter 4). We used the method of Planchon and Darboux (2001) and imposed an arbitrary slope of 0.01% on the flat surfaces resulting from the filling of the sinks. The result and the difference for one example realisation are shown in Figure 5(d) and (c).

Removing sinks from deterministic DEMs is common when preparing for geomorphometrical analysis (see also Chapter 4), although we have to keep in mind that sinks may also be real features of a landscape, like filled puddles, ponds and lakes, or dry craters, karst and hummocky terrain (Temme et al., 2006). In the Monte-Carlo analysis below, we will perform the analysis both for the 100 unfilled

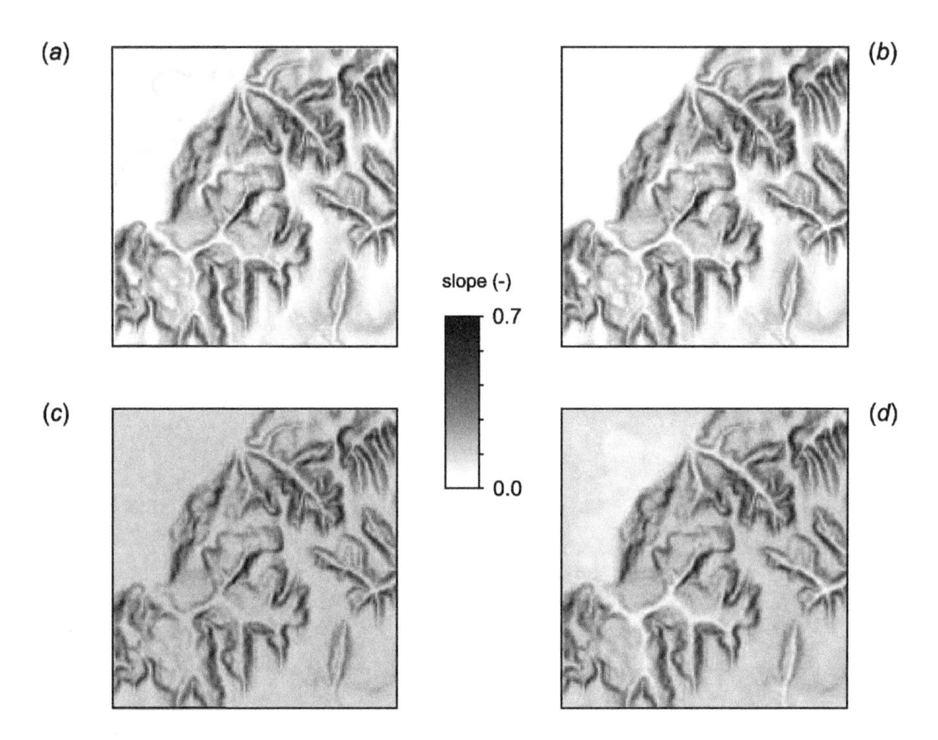

FIGURE 6 Steepest descent slope map (tan Δ) based on (a) deterministic unfilled DEM and (b) filled DEM. Mean slope map (tan Δ) based on 100 simulated slope maps of (c) unfilled DEMs and (d) filled DEMs.

DEMs and the 100 filled DEMs. Thus, we will also be able to draw conclusions on the influence of sink removal on uncertainty propagation.

4.2 Monte-Carlo uncertainty propagation analysis

As a first order derivative of DEM, we calculated the *tangent of slope* for the unfilled and filled DEMs using the function:

$$\tan\Delta = \sqrt{\left(\frac{\partial z}{\partial x}\right)^2 + \left(\frac{\partial z}{\partial y}\right)^2} \tag{4.3}$$

where $\partial z/\partial x$ and $\partial z/\partial y$ are estimated using a 3×3 window (see also Chapter 6). This function is available in most GIS software, but for automation purposes we wrote it in a C++ shell. The deterministic results and the means for the 100 unfilled and the 100 filled DEMs are given in Figure 6. The coefficients of variation (i.e. standard deviation divided by the mean) of slope for the 100 unfilled and the 100 filled DEMs can be seen in Figure 7.

The mean slope maps of the 100 simulations on unfilled and filled DEMs closely resemble the deterministic slope maps, although the mean slope of the simulated DEMs is noticeably greater than the slope of the deterministic DEMs

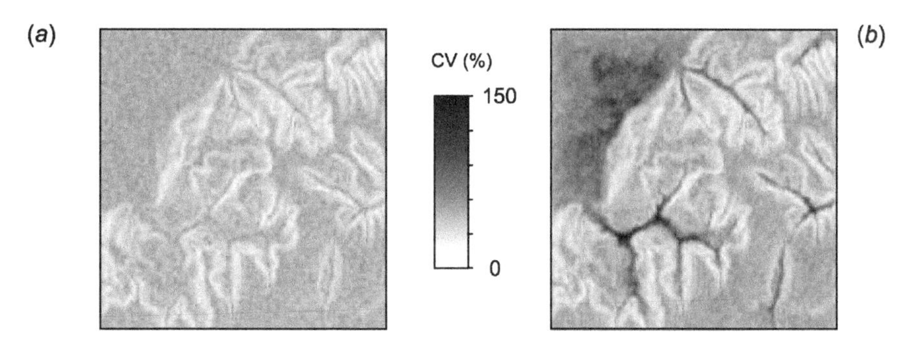

FIGURE 7 Coefficient of variation of slope (tan Δ) based on (a) 100 simulated unfilled DEMs and (b) filled DEMs.

in flat areas. The larger slopes in simulated DEMs result from adding different error values to different cells. On slopes, this can lead to a decrease or an increase in steepness. On flat areas, this can only lead to an increase in steepness. The simulated DEMs have more local spatial variation than the (overly) smoothed deterministic DEM and thus steeper slopes.

Differences in mean slope between the unfilled and filled versions are visible in the bottoms of the valleys and in the flat areas in the northwest. Slopes of unfilled DEMs are larger in these areas than slopes of filled DEMs. This is a direct result of the filling of sinks to almost flat areas in filled DEMs. It is also visible in the deterministic DEMs. The mean coefficient of variation of slope is 42% for unfilled and 49% for filled DEMs. These large coefficients indicate that uncertainty in slope is large. Apparently, DEM errors propagate strongly to this land-surface parameter. Not surprisingly, the coefficient of variation of slope is largest in the flat terrain and in valley bottoms for filled DEMs. In these areas, mean slope is small and standard deviation is large. In general, however, we find a rather uniform coefficient of variation, meaning more uncertainty in slope in steeper areas. Slightly smaller coefficients of variation are found for the steepest slopes.

As a second-order derivative of DEM, we calculated the Topographic Wetness Index (TWI) for the filled and unfilled DEMs:

$$\mathrm{TWI} = \ln\left(\frac{A}{\tan\Delta}\right) \tag{4.4}$$

with A (m^2) being the contributing area, and $\tan\Delta$ being the slope as discussed before. We calculated the contributing area using the multiple flow direction principle (Holmgren, 1994). The deterministic results and the means for the 100 unfilled and the 100 filled DEMs are given in Figure 8. The coefficients of variation of TWI for the 100 unfilled and the 100 filled DEMs are given in Figure 9.

Both the unfilled and filled deterministic maps display large TWI values in the flat areas and valley bottoms and small values on the steepest slopes where the valleys end. Intermediate TWI values are found on the plateaux above the valleys. This is a logical result: the steepest slopes combine a large tangent of slope with a small contributing area, resulting in a small TWI. In contrast, slopes are small

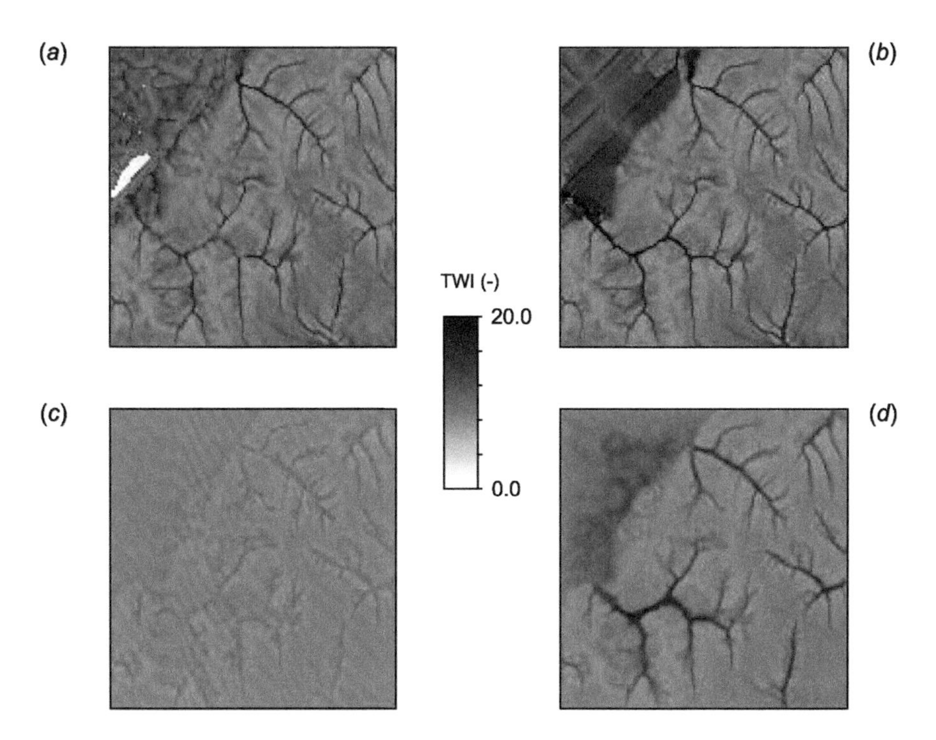

FIGURE 8 TWI map based on (a) deterministic unfilled DEM and (b) filled DEM. Mean TWI map based on (c) 100 simulated unfilled DEMs and (d) filled DEMs.

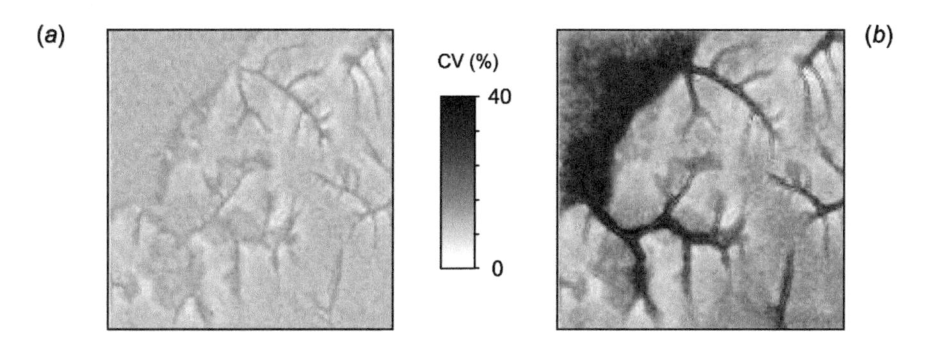

FIGURE 9 Coefficient of variation of TWI based on (a) 100 simulated unfilled DEMs and (b) filled DEMs.

and contributing area is large in the valleys and flat areas, resulting in a large TWI. Differences between the deterministic filled and unfilled results are visible in the valleys and the flat areas.

In the valleys, the TWI of the unfilled DEM shows the lack of connectivity in the drainage pattern through TWI values that gradually increase downstream and then suddenly decrease. In flat areas, the TWI of the filled DEM shows an unrealistic diagonal pattern resulting from the filling of depressions. Also, in the TWI of the unfilled DEM, we find some NODATA values. These result from a completely flat area and therefore a division by zero in the calculation of the TWI.

Small TWI values are found for the simulated unfilled DEMs. Because of the very limited connectivity of the drainage system, the contributing area is small and as a result, TWI values are small. The connectivity is restored but roughness remains in the simulated filled DEMs, leading to TWI values similar to but smaller than those of the filled deterministic DEM. In the valleys and flat areas, this is also the result of changing drainage patterns between simulations.

For each of the 100 simulated unfilled DEMs, NODATA values for TWI occurred at different locations, resulting from a zero-slope in the DEM. In calculating the mean and standard deviation, we have ignored these values. The mean coefficient of variation of TWI is 10% for unfilled and 16% for filled DEMs. These are smaller than the corresponding coefficients of variation of slope. Thus, when judged on the coefficient of variation, the uncertainty in TWI as a result of DEM uncertainty is smaller than the uncertainty in slope. Apparently, TWI is less sensitive to errors in the input DEM than slope.

The spatial pattern of the coefficient of variation is comparable to that of the mean TWI values for both unfilled and filled simulations. This means that an increase in mean TWI leads to a relatively larger change in standard deviation of TWI. Thus, even relatively speaking, we are less certain of high TWI values. The main reason for variation between simulations is the change in drainage pattern and this leads to more variation where TWI values are larger. Coefficients of variation for TWI (between 0 and 0.4) are smaller than they were for slope (between 0 and 1.5), but they display the same pattern: higher uncertainties in the flat areas and valley bottoms.

As a complex derivative of DEM, we simulated the *erosion and sedimentation* in the Baranja Hill case study for 10 years. For this simulation, we used the water-erosion module of multi-process landscape evolution model LAPSUS (*LA*ndscape *P*roces*S* modelling at m*U*lti dimensions and *S*cales, described in detail by Schoorl et al., 2000). The water erosion module first simulates overland waterflow using the multiple flow direction algorithm that was also used for the calculation of TWI (Holmgren, 1994). Using waterflow and slope, LAPSUS calculates a sediment transport capacity. By comparing this transport capacity to the actual amount of sediment in transport, the amount of erosion or deposition is calculated. We used a default set of parameters because we are not interested in the actual modelling of the redistribution, but in the influence that DEM uncertainty has on it.

The module deals with flows of water and sediment into sinks using the algorithm of Temme et al. (2006). Therefore, it can deal with both the unfilled and the filled DEMs. Schoorl et al. (2000) have shown that LAPSUS is sensitive to DEM resolution, but did not address its sensitivity to errors in DEMs. The deterministic results and the means for the 100 unfilled and the 100 filled DEMs of soil redistribution are given in Figure 10. The coefficients of variation for the 100 unfilled and the 100 filled DEMs can be seen in Figure 11.

The deterministic redistribution for the unfilled DEM shows that erosion occurs mainly in the upper valleys, where surplus of transport capacity is apparently

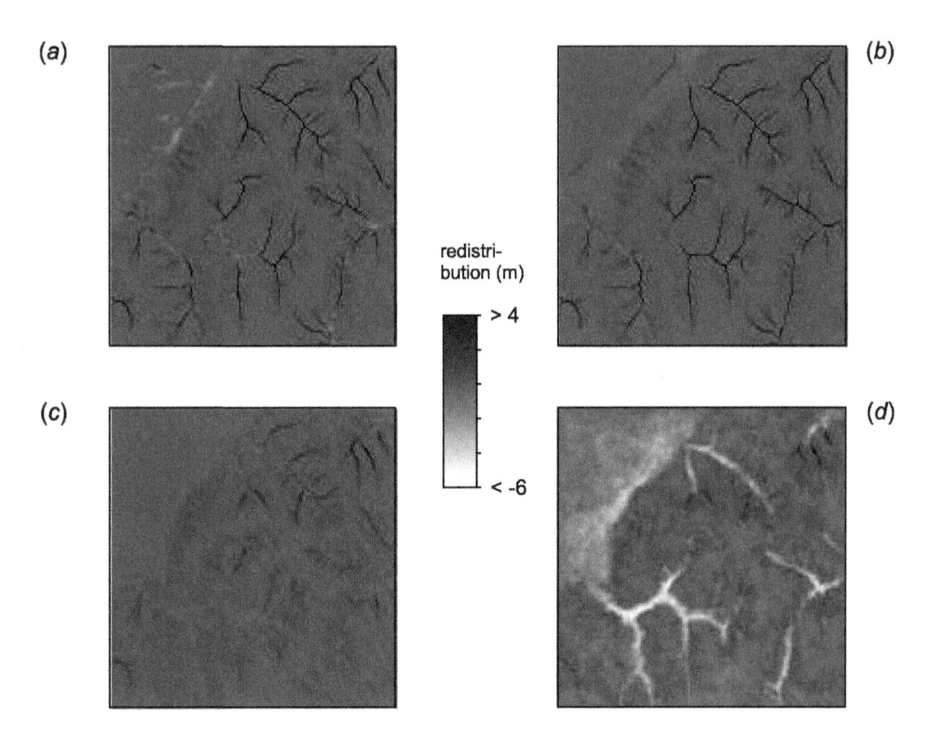

FIGURE 10 Soil redistribution map after 10 years based on (a) deterministic unfilled DEM and (b) filled DEM. Mean soil redistribution map based on 100 simulated soil redistribution maps on (c) unfilled DEMs and (d) filled DEMs. Soil redistribution is in meters. Positive values indicate erosion, negative values indicate deposition.

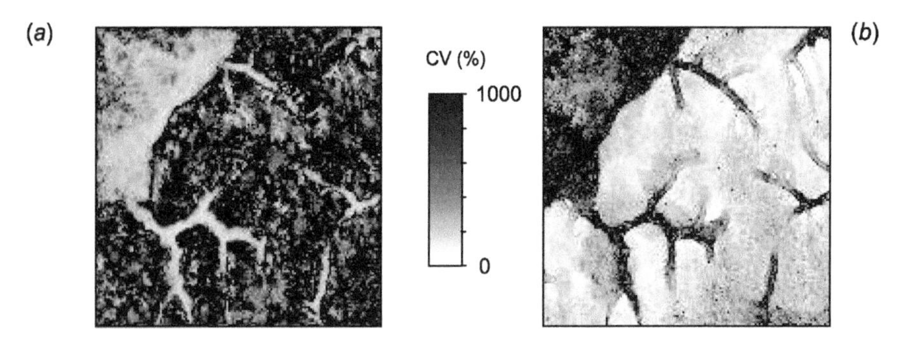

FIGURE 11 Coefficient of variation based on 100 simulated soil redistribution maps on (a) unfilled DEMs and (b) filled DEMs.

large. In depressions and flat areas in the valleys, and when exiting the valleys, deposition occurs because of a deficit in transport capacity. The redistribution for the filled DEM shows a similar picture, albeit with less deposition in the valleys and flat areas. This results from removing the depressions from the DEM. Also, two unrealistic straight lines of erosion are visible in the flat area in the northwest for the filled DEM. This is caused by the unrealistic drainage pattern imposed by filling.

The mean soil redistribution maps of the 100 simulations on unfilled DEMs show considerably more deposition and less erosion, but the overall pattern resembles the redistribution of the deterministic unfilled DEM. Erosion occurs in the upper valleys, and deposition occurs in flat areas. The different amount of deposition and erosion results from adding the error map. Both an increase in the overall amount and size of depressions, and an increase in general roughness of the landscape create more space for redeposition while decreasing erosion. Also, the mean soil redistribution of filled DEMs shows less erosion and slightly more deposition than the soil redistribution of the filled deterministic DEM. Filling does not change the increased roughness of slopes and the almost flat areas that result from filling still function as places for sediment deposition.

Differences between the mean redistribution of unfilled and filled DEMs are clearly visible in the valley bottoms and flat area in the northwest, where mean deposition is large in depressions of the unfilled DEMs. In the filled DEMs, these sinks were filled before the simulation, resulting in less deposition. On parts of the plateaux above the valleys, deposition occurs in unfilled DEMs, whereas a small amount of erosion occurs in filled DEMs. The average deposition in unfilled DEMs masks the fact that values on the plateaux flip between erosion and deposition, depending on the topography of the individual simulated DEM. In filled DEMs, this is much less pronounced because all sinks have been filled.

We have taken the absolute value of soil redistribution to prevent negative values in the calculation of coefficients of variation. So instead of erosion and sedimentation as before, we are looking at the coefficient of variation of the amount of soil redistribution. The mean coefficient of variation of soil redistribution is 4600% for unfilled and 1000% for filled DEMs. This is two degrees of magnitude larger than the coefficients of variation of slope and TWI. Thus, when judged on coefficient of variation, the uncertainty in soil redistribution as a result of uncertainty in the DEM is much larger than for slope or TWI. Soil redistribution is extremely sensitive to errors in input DEMs. However, here it must be noted that when the estimated soil redistribution is close to zero (neither erosion nor sedimentation), then even a small uncertainty in soil redistribution may lead to an extremely large coefficient of variation. This is a disadvantage of the coefficient of variation.

For soil redistribution and perhaps also for slope and TWI, it may alternatively be worthwhile to quantify the propagation of errors in terms of the standard deviations associated with the predicted values (i.e. consider absolute errors instead of relative errors, see Heuvelink, 1998). However, this has disadvantages when trying to compare the sensitivities of outputs with different units, as we do in this chapter.

The maps of coefficients of variation of soil redistribution seem to mirror each other. For the plateaux above the valleys, the very high coefficients of variation (>1000%) in unfilled DEMs result from the large differences between individual predictions of sedimentation and erosion relative to the mean redistribution of almost zero.

Conversely, in filled DEMs this part of the landscape has the lowest coefficients of variation (<300%). That is because in these DEMs, this part of the landscape consistently experiences a bit of erosion.

For the valleys and low-lying flat areas, coefficient of variation is small for unfilled DEMs. Unlike on the plateaux, where redistribution values flip between erosion and sedimentation in these DEMs, sedimentation is the normal process in these areas as visible from the small values in the mean map (<300%). Standard deviation is greater than it was on the slopes, but its importance relative to the mean has decreased: small coefficients of variation. For filled DEMs, these places display large coefficients of variation. This is result of the changes in drainage pattern between the individual DEMs. Redistribution (erosion or sedimentation) is large in the drainage channels, and close to zero next to them. A very small mean with large standard deviation results in very large coefficients of variation (>1000%).

Coefficients of variation are larger and more spatially variable for soil redistribution than they are for TWI and slope. Next to the sensitivity of coefficient of variation for values with a mean around zero as we observed earlier, this results from the fact that, besides DEM, also TWI (contributing area) and slope are non-linear factors in the model. These inputs are themselves sensitive to errors in the DEM as we have demonstrated before. The output of the model is thus sensitive to errors in the input DEM via three pathways. Given the importance of these three inputs in the model, we stack uncertainty on uncertainty, resulting in the observed high and variable coefficients of variation.

5. SUMMARY POINTS

Error propagation analysis in geomorphometry can be done and delivers informative results. However, it takes considerable effort because we must statistically model the error in a DEM (not routinely known or only partially known, just a RMSE is not enough) and we must run a Monte-Carlo analysis.

In our case study, we demonstrated how errors propagate from the DEM to three land-surface parameters: slope, TWI and soil redistribution. For these derivatives, we have shown the differences between deterministic and simulated outcomes in terms of pattern and value. We did that for simulated DEMs containing sinks and for simulated DEMs without sinks.

Analysing the results of the increasingly complex land-surface parameters, we see that the mean result of the simulations for slope resembles the deterministic result more closely than those for TWI and soil redistribution. This can be understood when we realise that contributing area, which is an important variable in the calculation of TWI and soil redistribution, is strongly dependent on continuity of the drainage pattern. This pattern is perturbed by adding error and is not restored in the same way for all simulations by filling sinks.

This leads to a difference between mean TWI and soil redistribution, and deterministic TWI and soil redistribution. Moreover, slope is an important variable in TWI and soil redistribution. Differences in the individual simulations of slope lead to differences in TWI and soil redistribution in a non-linear way, because in calculating TWI we take a logarithm of the inverse of slope, and in calculation of soil redistribution we repeatedly use a power of slope. This also leads to a difference between the mean values of TWI and soil redistribution and their deterministic counterparts.

We observed that the maximum coefficient of variation of the 100 simulations is larger and less uniformly distributed for soil redistribution than for slope and TWI. The former, more complex derivative is therefore more sensitive to errors in DEMs. Slope is less sensitive, and TWI is least sensitive, although the coefficient of variation for TWI is spatially more variable than that of slope. Considering the influence of DEM filling on the mean results, we saw that this is largest for TWI and soil redistribution. Here, the effect of restoring the drainage pattern is large. The influence of DEM filling on the coefficients of variation is largest for soil redistribution.

Whereas mean slope results closely resemble deterministic results, the mean results of TWI and soil redistribution are more realistic than deterministic results. For these land-surface parameters, a Monte-Carlo analysis is not only recommended for analysing how errors propagate, but also to correct for possible bias due to using overly smoothed DEMs. Also, artefacts in the deterministic DEM may lead to a wrong result. For TWI, soil redistribution and other land-surface parameters, we recommend a standard derivation as a mean from multiple realisations of a DEM (see further Section 3.2 in Chapter 4). Software tools that semi-automatically do this Monte-Carlo derivation (by, for instance, taking the mean value of a parameter from 50 realisations of an input) are not yet available however, although PCRaster (Karssenberg and De Jong, 2005) offers considerable functionality

It is important that sufficient validation data are available because they are needed to compute the error in DEMs in a statistically valid way. However, with sufficient validation data one still needs assumptions to arrive at a statistical error model. The assumption of second-order stationarity in DEM errors does not seem very realistic because it creates DEMs that do not exist in the real world. This problem is not completely solved by filling depressions, as Temme et al. (2006) have demonstrated. On the other hand, deterministic DEMs themselves are smoothed representations that do not perfectly correspond to reality either (see also Chapter 2). We advocate that alternative error models and stochastic simulation methods are developed to simulate uncertain DEMs that better represent the real world situation

IMPORTANT SOURCES

Fischer, P.F., Tate, N.J., 2006. Causes and consequences of error in digital elevation models. Progress in Physical Geography 30 (4), 467–489.

Goovaerts, P., 1997. Geostatistics for Natural Resources Evaluation. Applied Geostatistics. Oxford University Press, New York, 496 pp.

Heuvelink, G.B.M., 1998. Error Propagation in Environmental Modelling with GIS. Taylor & Francis, London, UK, 144 pp.
Schoorl, J.M., Sonneveld, M.P.W., Veldkamp, A., 2000. Three dimensional landscape process modelling: the effect of DEM resolution. Earth Surface Processes and Landforms 25 (9), 1025–1034.
http://www.sil.wur.nl/UK/LAPSUS/ — LAPSUS model.

CHAPTER 6

Basic Land-Surface Parameters

V. Olaya

morphometric parameters that can be extracted from the DEM without further adjustment · local and regional morphometric land-surface parameters · computation of slope and aspect from the DEM · computation of curvatures · computation of viewshed · computation of flow accumulation, catchment height and slope, flow length · computation of surface roughness and anisotropy · computation of shape complexity index

1. INTRODUCTION

Although all geomorphometric parameters relate to the morphology of the land surface, a number of them can be derived directly from a DEM without further knowledge of the area represented. We refer to those measures as *basic land-surface parameters*. Although many of them overlap with more application-specific measures, basic parameters differ in that they represent the raw shape of the land surface, regardless of how that surface relates to formational processes.

Once we have the DEM and it has been pre-processed, the first and most immediate analysis that can be performed on it is the one related to its geometrical and topological properties. This is not only the most immediate, but also one of the most useful ones, since parameters such as slope or insolation, to name a few, are related to many different practical fields, from hydrology to forest science, from geology to biology.

Some basic land-surface parameters are analysed locally,[1] while others — regional ones — also need to consider other (or all!) parts of the DEM apart from the exact point where they are to be calculated. The regions needed for regional analysis are defined by the movement of flows downhill and the relation it establishes between cells. It is when we add the effect of the gravitational field to the purely geometrical ideas of local analysis that differential geometry really turns into geomorphometry. Some of these parameters are introduced in Table 1.

[1] Typically, these measures were meant to be computed within a small vicinity — e.g. 3×3 window.

Developments in Soil Science, Volume 33 © 2009 Elsevier B.V.
ISSN 0166-2481, DOI: 10.1016/S0166-2481(08)00006-8. All rights reserved.

TABLE 1 Some basic land-surface parameters

LSP	Type	What does it describe?
Slope	Local	Flow rate
Aspect	Local	Flow-line direction
Tangential curvature	Local	1st accumulation mechanism
Profile curvature	Local	2nd accumulation mechanism
Catchment area	Regional	Flow magnitude
Hypsometry	Regional	Distribution of height values
Catchment height/slope	Regional	Flow characteristics
Insolation	Regional/local	Intensity of direct solar irradiation
Visual exposure	Regional	Extent of visible area
Roughness	Local	Terrain complexity

In this chapter, we will present different formulations (each one of them with a different degree of mathematical complexity) to extract basic morphometric parameters from the DEM itself and then discuss importance and interpretation of each parameter. For each type of land-surface parameter, a detailed description will be given in the following sections, including further information on the physical phenomena described by each one. We will first introduce some well-known parameters such as slope or aspect, and then extend that to some less common ones such as the Shape Complexity Index, the Anisotropic Coefficient of Variation, geostatistical land-surface parameters or fractal-based ones, which represent the most actual trend in this field.

REMARK 1. *Basic land-surface parameters are measures that can be derived directly from a DEM without further knowledge of the area represented. There can be local (geometric and statistical) and regional parameters.*

2. LOCAL LAND-SURFACE PARAMETERS

The study of morphometric land-surface parameters should start with those related to the local morphometry of each point. We can divide them into two main groups: (1) geometric and (2) statistical measures.

2.1 Geometric parameters

Geometric parameters are those based on an analysis of the geometrical properties of the land surface. This includes slope and aspect, as well as curvatures and other values derived from them. Among all the parameters that are used to characterise a land surface, the ones that are described in this part of the chapter are probably those with the strongest mathematical foundation. Do not forget that a land surface itself is a surface in a mathematical sense and, as such, can be analysed using

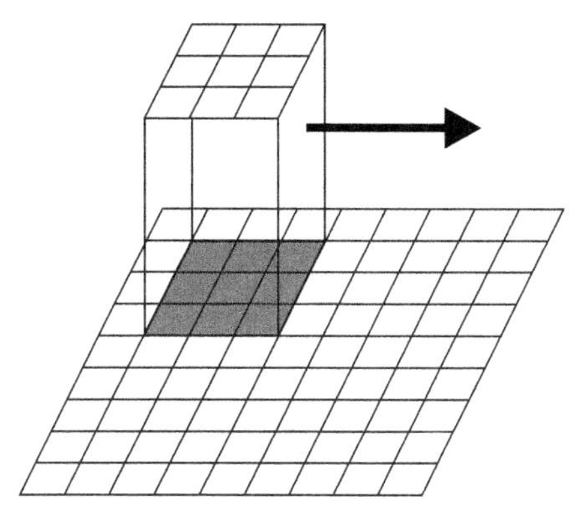

FIGURE 1 A complete analysis of a DEM is normally obtained by moving it across the DEM using a 3×3 rowing window. See also Section 2.2 in Chapter 1.

all the concepts of differential geometry. Parameters like slope or aspect are easy to calculate with just a few basic geometrical ideas.

To perform this mathematical analysis, first we need a mathematical function describing the land surface, and from this, we can extract new parameters. As was previously explained, we can describe the land surface locally by using a mathematical function, and this function is, therefore, a basic tool for extracting local morphometric land-surface parameters and objects.

In Section 3.3 of Chapter 2, several land-surface models were presented and the advantages and disadvantages of each one were explained. We also learnt that these models have a significant influence on the precision of the land-surface parameters and objects extracted from them. However, they are all treated in a unique way, simply by applying mathematical ideas; and the same mathematical principle is implemented, regardless of the land-surface model behind the derivation of the formulae.

Examples will be given for the Evans method, but the same reasoning can be applied to the equations formulated by Zevenbergen–Thorn, Pennock, or for the method devised by Shary, or any of the others methods introduced in Chapter 2. In all cases, to calculate a new function in each location, the local analysis window (usually 3×3) has to be moved across the DEM so that it corresponds with the value of first one function and then another, until all the functions have been measured. This process can be seen schematically in Figure 1.

Land-surface parameters for the target cell (the central cell of the core) can be computed for each function. At the edges, special formulations are needed because, due to the lack of data outside the limits of the grid, the 3×3 window cannot be defined.

This results in a new grid, which has the same dimensions as the DEM, but with a different parameter. It is worth mentioning at this point that some of the local land-surface parameters derived from this mathematical function of the DEM (or rather, a small part of the DEM), can also be calculated using a different approach, that avoids the need for a mathematical function. However, the methods based on a formal mathematical approach have been proved to give better results (see e.g. Jones, 1998) and many more parameters can be obtained, in a rigorous way. We will provide a concise description of these other methods when we discuss each of the geomorphometric parameters separately. These alternative methods have a notable importance historically and, moreover, they are rather easy to interpret, especially in relation to some hydrological parameters that are presented in later chapters of this book.

2.1.1 Functions of first derivatives

The first morphometric properties of a terrain than can be studied are those derived from the first partial derivatives of its surface. A basic concept in vector calculus implying those first partial derivatives is the *gradient*. Given a scalar field — such as the elevations contained in the DEM — the gradient is a vector field pointing in the direction in which maximal variation in the values of that scalar field occur. The expression of the gradient in two dimensions is:

$$\nabla\overline{\mathbf{Z}} = \left(\frac{\partial z}{\partial x}, \frac{\partial z}{\partial y}\right) \tag{2.1}$$

The two main geometrical properties that can be derived from the gradient are its length (modulus) and its direction. When these concepts are applied to geomorphometry, they constitute two of the most important land-surface parameters: slope and aspect.

Slope gradient reflects the maximal rate of change of elevation values and is defined as:

$$\mathrm{SLOPE} = \arctan\left(|\nabla\overline{\mathbf{Z}}|\right) \tag{2.2}$$

which indicates the angle between the horizontal plane and the one tangential to the surface. Note that the tangential plane is defined by the gradient vector itself and is, therefore, normal to the surface. It is also normal to the contour line passing through the point. Slope gradient can be expressed in radians or degrees, but it is usual (and more practical in many fields) to reflect its values in %, using the expression:

$$\mathrm{SLOPE}(\%) = \tan(\mathrm{SLOPE}) \cdot 100 \tag{2.3}$$

The value of a slope gradient at each cell can be obtained by calculating the partial derivatives of the approximated function. For example, values resulting from the original Evans method are:

$$\begin{aligned}\frac{\partial z}{\partial x} &\approx \frac{\partial(a \cdot x^2 + b \cdot y^2 + c \cdot xy + d \cdot x + e \cdot y + f)}{\partial x}\\ &= 2 \cdot a \cdot x + c \cdot y + d\end{aligned} \tag{2.4}$$

$$\begin{aligned}\frac{\partial z}{\partial y} &\approx \frac{\partial(a \cdot x^2 + b \cdot y^2 + c \cdot xy + d \cdot x + e \cdot y + f)}{\partial y}\\ &= 2 \cdot b \cdot y + c \cdot x + e\end{aligned} \tag{2.5}$$

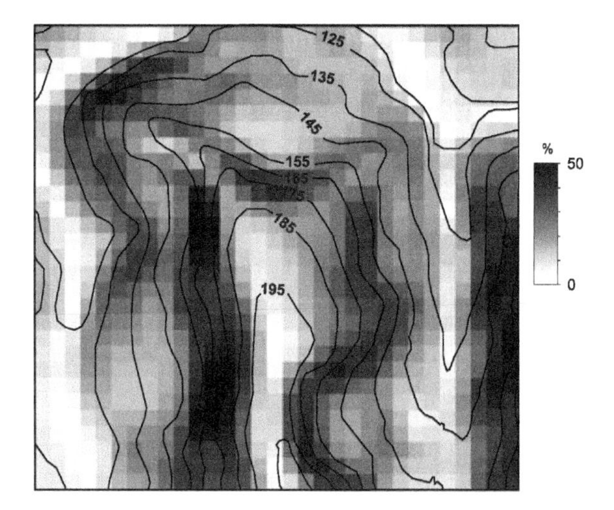

FIGURE 2 Slope gradient map for the Baranja Hill area, overlaid with contours. Grid resolution = 25 m.

and since we want to calculate the value at the origin of the coordinates (i.e. in the central cell of the analysis window), we should make $x = y = 0$.

$$d = \frac{\partial z}{\partial x} = \frac{z_3 + z_6 + z_9 - z_1 - z_4 - z_7}{6 \cdot \Delta s} \tag{2.6}$$

$$e = \frac{\partial z}{\partial y} = \frac{z_1 + z_2 + z_3 - z_7 - z_8 - z_9}{6 \cdot \Delta s} \tag{2.7}$$

Then, by applying Equation (2.2) to these gradient values, we get the unique value of the slope at the cell (Figure 2):

$$\text{SLOPE} = \arctan\left(\sqrt{d^2 + e^2}\right) \tag{2.8}$$

The structure of a DEM as a grid is equivalent to a raster image. Thus geomorphometry is similar, in some ways, to image analysis. In fact, many operations, such as computing gradient values, have their counterparts in image analysis (see also Section 2.2 in Chapter 1). The reader familiar with image analysis will have no difficulties in recognising that the values obtained in Equations (2.6) and (2.7) are the result of applying, respectively, the convolution kernels:

$\frac{1}{-6 \cdot \Delta s}$	0	$\frac{1}{6 \cdot \Delta s}$
$\frac{1}{-6 \cdot \Delta s}$	0	$\frac{1}{6 \cdot \Delta s}$
$\frac{1}{-6 \cdot \Delta s}$	0	$\frac{1}{6 \cdot \Delta s}$

$\frac{1}{-6 \cdot \Delta s}$	$\frac{1}{-6 \cdot \Delta s}$	$\frac{1}{-6 \cdot \Delta s}$
0	0	0
$\frac{1}{6 \cdot \Delta s}$	$\frac{1}{6 \cdot \Delta s}$	$\frac{1}{6 \cdot \Delta s}$

which are simple *Prewitt kernels*; ones that are frequently used in image analysis, for edge detection (Seul et al., 2000). Combining ideas from image analysis and geomorphometry can be a fruitful way of developing new parameters and gaining a better understanding of the information contained in the DEM. More examples of the close relationship between these two disciplines will be shown in other parts of this chapter.

Returning to the concept of slope gradient, because many different disciplines can use slope values for their particular concerns, it is extremely important. Slope gradient can also be used to derive many new parameters, such as those for soil erosion and deposition, soil wetness, flow speed, and many others. These are explained in detail further in Chapter 7.

> REMARK 2. *Combining ideas from image analysis and geomorphometry can be a fruitful way of developing new parameters and gaining a better understanding of the information contained in the DEM.*

An important parameter related to slope gradient is *surface area*. If we consider a cell with grid resolution Δs, its area can be calculated as $A_1 = (\Delta s)^2$. However, this is not the real area (it is a *planimetric* area), since we are neglecting the influence of slope on the cell. The *real* area, the one, for example, that is available to the animals living on that cell, or the one that has to be used to calculate the cost of reforesting the cell, should consider the influence of slope. The value of this surface area is useful, among others, for landscape analysis and wildlife habitat studies. A simple way of calculating it, is to consider the slope of the centre cell only. The area of a cell with SLOPE would be (Berry, 1996):

$$A = \frac{A}{\cos(\text{SLOPE})} \tag{2.9}$$

Hobson (1972) considered this method to over-estimate the surface area (just the opposite of what happens when slope is not considered), so to achieve a more precise estimation, a new method was needed. Jenness (2004) proposed an alternative formulation. They used eight 3-dimensional (3D) triangles to connect the central point of each cell to the central points of the eight surrounding cells. They then calculated and summed the areas of the portions of each triangle that lay within the cell boundary.

Regarding *aspect*, its value depends on which direction is taken as the origin (i.e. the direction in which aspect equals 0). Usually, it is measured in degrees, read clockwise from north, ranging from 0 to 360°. Aspect is defined as (Gallant and Wilson, 1996):

$$\text{ASPECT} = 180 - \arctan\left(\frac{q}{p}\right) + 90 \cdot \frac{p}{|p|} \tag{2.10}$$

where arctan() is in degrees and p and q are first-order derivatives:

$$p = \frac{\partial z}{\partial x}, \qquad q = \frac{\partial z}{\partial y} \tag{2.11}$$

Gradient values obtained for the Evans method can be used here as well. According to this formula, aspect is not defined in points where $\partial z / \partial y = 0$, but there

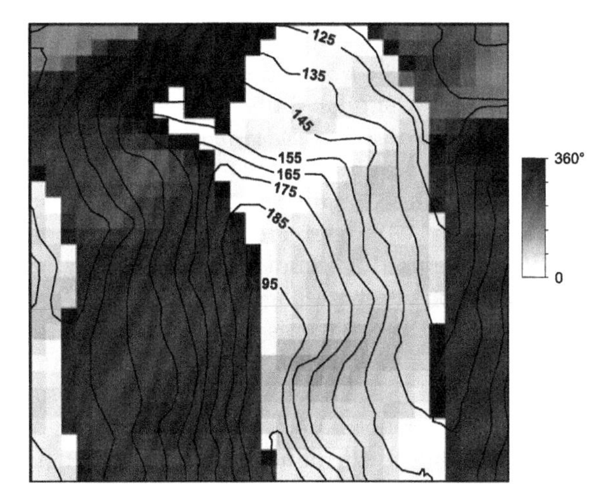

FIGURE 3 Aspect map for the Baranja Hill area overlaid with contours. Grid resolution = 25 m.

is no physical substantiation of this. For this reason, another formula of aspect can be used instead (Shary et al., 2002):

$$\begin{aligned} \text{ASPECT} = & -90 \cdot [1 - \text{sign}(q)] \cdot [1 - |\text{sign}(p)|] + 180 \cdot [1 + \text{sign}(p)] \\ & - \frac{180}{\pi} \cdot \text{sign}(p) \cdot \arccos \frac{-q}{\sqrt{p^2 + q^2}} \end{aligned} \tag{2.12}$$

where the sign(x) equals either:

$$\text{sign}(x) = \begin{cases} 1 & \text{if } x > 0 \\ 0 & \text{if } x = 0 \\ -1 & \text{if } x < 0 \end{cases} \tag{2.13}$$

Clearly, aspect cannot be defined in special points where gradient equals zero. The aspect map for the Baranja Hill area is shown in Figure 3.

Aspect indicates the flow-line direction. When water (or any other flowing material) moves downhill under the influence of gravitational force, it will follow the direction specified by aspect. This is the basis, for the most part, of flow direction algorithms. Aspect is a *circular land-surface parameter*, i.e. m and $m + 360°$ describe the same slope direction. For this reason, some authors use $\cos(a)$ as slope *northerness*, and $\sin(a)$ as slope *easterness*, especially in statistical comparisons (e.g. King et al., 1999).

REMARK 3. *Slope gradient and aspect are the most standard local geomorphometric measures. They can be derived using various algorithms.*

Aspect can also be used to determine whether a cell should be considered *sunny* or not. For example, in the northern hemisphere, slopes facing south (i.e. with an aspect of 180°) receive more insolation than those facing north (i.e. with an aspect of 0°), so a relation between these two parameters could be established. This is, however, a very simple approach. For a more precise characterisation of the insolation see Section 3.1 in Chapter 8.

Here we will just see how a similar (but simpler) combination of slope and aspect can be used to generate so-called *shaded-relief* maps. These maps try to create images that reflect the topography of the DEM, so they are clearer and more intuitive to the human eye. These have been used routinely since the 1980s for improving the visual appearance of relief (Horn, 1981).

Shaded relief maps can be used as stand-alone images, or, better, they can be used to alter the representation of other images containing different parameters of different kinds of information. This effect is achieved using a shaded-relief map and modifying the brightness of the pixels according to the values in it. This gives the user an impression of 3D space.

The values for each cell are computed by applying simplified insolation models, neglecting the influence of the surrounding relief and considering insolation as a local parameter.[2] Considering the sun at a position defined by its azimuth (ϕ) and elevation (θ) over the horizon (Figure 2 in Chapter 8), the *insolation* of a cell with aspect a and slope s is given by (Shary et al., 2005):

$$F = \frac{100 \cdot \tan(s)}{\sqrt{1 + \tan^2(s)}} \left[\frac{\sin(\theta)}{\tan(s)} - \cos(\theta) \cdot \sin(\phi - a) \right] \tag{2.14}$$

which gives a value in the range 0–100. Apart from constituting useful (and eye-catching) representations, such DEM-derivatives can be used to correct the effect of relief in remote-sensing images (Riaño et al., 2003). To obtain better results and enhance the content of such images, for these tasks, more elaborate models, which include light reflectance, can be applied (Felícisimo, 1994a).

As mentioned in the previous section, some of the land-surface parameters already described can be calculated, without needing to represent a cell and its neighbourhood mathematically. One of the most productive fields of geomorphometry is its application to hydrology. This has resulted in specific formulations for calculating slope and aspect, based on hydrological concepts.

The ideas behind the most basic (and earliest) algorithm to calculate flow directions, the so-called Deterministic 8 or D8 (O'Callaghan and Mark, 1984), are the same as those behind the maximum-slope method for calculating slope (Travis et al., 1975), and also an early formulation for extracting basic morphometric variables. The slope from the central (core) cell of any given cell to each one of 8 surrounding cells can be calculated using basic geometrical ideas:

$$s = \frac{\delta z}{d} \tag{2.15}$$

where d is the distance between the centre of the problem cell and the centre of the surrounding one. For cells z_2, z_4, z_6 and z_8, d equals the cell size Δs, while for the remaining ones $d = \sqrt{2} \cdot \Delta s$. From the eight values obtained, the largest one can be taken as the slope of the cell. The direction in which this slope is measured can be used as aspect. Note that, while slope measurement using this method is not very

[2] Although, in fact, it is not a local land-surface parameter, because it is affected by the relief of any other cells around that can cast shades or reflect light.

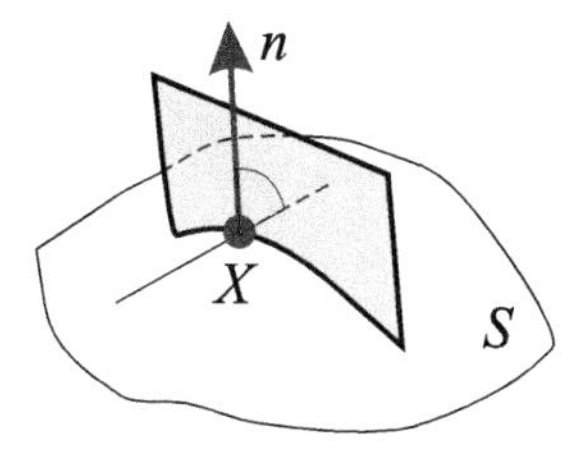

FIGURE 4 A normal section of a surface S at a point X is a curve that results from the intersection of the surface and a plane passing through vector n orthogonal to S. The bold curve in the above scheme is one possible normal section; others arise as a result of this plane rotating around n.

accurate, conceptually, and its value is usually over-estimated (Srinivasan and Engel, 1991), slope gradient values can cover the same range as those obtained from the other, more elaborate, approaches introduced earlier in the book. On the other hand, aspect values can only be multiples of 45°, thus restricting their range and discretising the slope measurement. This is the major drawback of this conception, especially when it is applied to the calculation of flow directions.

An improvement of the above method is the one proposed by Tarboton (1997), who suggested the so-called D-infinity method, to account for discretisation. The method proposed has been named D∞, to indicate clearly that it constitutes a revision of D8. The D∞ method takes the cell that defines the steepest downward direction, and also the two adjacent ones, whose values are used to calculate a more precise aspect value. The range is again from 0 to 360°, without any restriction (see also the detailed discussion on flow algorithms in Section 3.2 of Chapter 7).

2.1.2 Functions of second derivatives

The next step in the analysis of local morphometric variables is the one involving second derivatives. These are related to the concavity and convexity of the surface. The parameter describing this is called *curvature*. An important concept for understanding the geometrical meaning of curvatures is the so-called normal section of a smooth surface (Figure 4).

A normal section is a plane curve, the curvature of which is defined as $1/R$, where R is the radius of a circle best fitted to this curve at a given point (Figure 5). The curvature k of a plane curve $z(x)$ is given by the formula:

$$k = \frac{\frac{\mathrm{d}^2 z}{\mathrm{d}z^2}}{\left[1 + \left(\frac{\mathrm{d}y}{\mathrm{d}x}\right)^2\right]^{1.5}} \qquad (2.16)$$

A general way to introduce (simple) curvature onto a smooth surface is to define a plane curve on it and then use the above formula to represent this curvature as a function $f(p, q, r, s, t)$ of the first and second partial derivatives of elevation. Consequently, the dimensions of curvature are $[\mathrm{L}^{-1}]$.

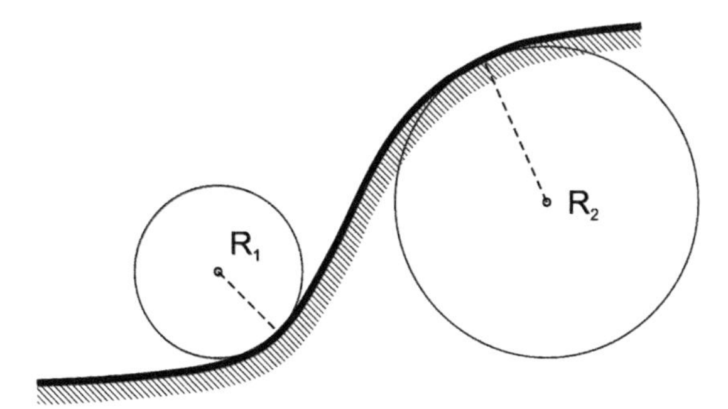

FIGURE 5 Curvature $1/R$ of a plane curve is the inverse of the radius R of a circle that is best fitted to this curve at a given point. It is agreed in Earth sciences that the sign of curvature is positive for a convex surface shape ($R_2 > 0$), and negative for a concave one ($R_1 < 0$).

> REMARK 4. *The profile (or vertical) curvature and tangential (or horizontal) curvature can be used to distinguish (locally) convex and concave shapes. Concave tangential curvature indicates convergence and convex divergence of flow lines. Convex profile curvature indicates acceleration of flow.*

An effective way to introduce simple curvatures onto a slope is to consider some chosen (or naturally marked) directions on it (Shary et al., 2002). There are four such directions on a smooth surface (Figure 6).

Two of them (aa' and bb') are physically distinguished by the gravitational field of the Earth, while the other two are marked by the surface itself: cc' by the maximal value of the normal section curvature and dd' by the minimal value of it. Curvatures of the corresponding four normal sections are known as:

(1) *the profile (or vertical) curvature*, PROFC, — that of normal section aa';
(2) *the tangential (or horizontal) curvature*, TANGC, — that of the normal section bb';

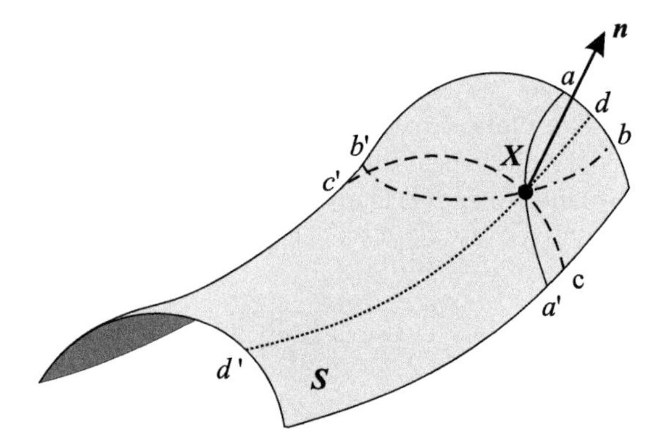

FIGURE 6 The four directions naturally marked on the surface S. n – the normal vector to S at point X; aa' – the gradient line, bb' – the contour line, dd', cc' – the main normal sections.

(3) *the maximal curvature*, MAXC, — that of normal section cc';
(4) *the minimal curvature*, MINC, — that of the normal section dd'.

From this point, we will use the following notation to simplify the expressions, along with the equivalences introduced in Equation (2.11):

$$r = \frac{\partial^2 z}{\partial x^2}, \qquad s = \frac{\partial^2 z}{\partial x \partial y}, \qquad t = \frac{\partial^2 z}{\partial y^2} \tag{2.17}$$

The new values r, s and t take the following values for the Evans method:

$$\begin{aligned} r &= \frac{z_1 + z_3 + z_4 + z_6 + z_7 + z_9 - 2 \cdot (z_2 + z_5 + z_8)}{3 \cdot \Delta s^2} \\ s &= \frac{z_3 + z_7 - z_1 - z_9}{4 \cdot \Delta s^2} \\ t &= \frac{z_1 + z_2 + z_3 + z_7 + z_8 + z_9 - 2 \cdot (z_4 + z_5 + z_6)}{3 \cdot \Delta s^2} \end{aligned} \tag{2.18}$$

With the above designations, the formula of profile curvature is (Krcho, 1973; Young, 1978):

$$\text{PROFC} = -\frac{p^2 \cdot r + 2 \cdot p \cdot q \cdot r \cdot s + q^2 \cdot t}{(p^2 + q^2) \cdot \sqrt{(1 + p^2 + q^2)^3}} \tag{2.19}$$

It is not defined for special points (where slope equals zero), and it is field-specific. The tangential curvature can be derived using (Krcho, 1983; Shary, 1991; Mitášová and Hofierka, 1993):

$$\text{TANGC} = -\frac{q^2 \cdot r - 2 \cdot p \cdot q \cdot s + p^2 \cdot t}{(p^2 + q^2) \cdot \sqrt{1 + p^2 + q^2}} \tag{2.20}$$

This is not defined for special points either, and it is also field-specific. The curvature of the contour line is known as the plan curvature (Evans, 1972; Krcho, 1973). Its formula is:

$$\text{PLANC} = -\frac{q^2 \cdot r - 2p \cdot q \cdot s + p^2 \cdot t}{\sqrt{(1 + p^2 + q^2)^3}} \tag{2.21}$$

As with the previous curvatures, it is not defined for special points, and it is field-specific. It has the same sign as the tangential curvature, but differs from it by a positive factor that depends solely on the slope.

For understanding the physical meaning of these curvatures, we will use the two accumulation mechanisms: the first describes the convergence/divergence of flow-lines, while the second one describes the relative deceleration/acceleration of material flows.

In cells with concave plan curvature, the flow-lines converge, while they diverge in those with convex plan curvature. In much the same way, material flows experience relative deceleration in cells which have a concave profile curvature, and relative acceleration in those with a convex one. Corresponding theorems were proven later, with the same results (Shary, 1995; Shary et al., 2002), and this provided a substantiation of Aandahl's (1948) hypothesis.

The average of the curvatures of any mutually orthogonal normal sections is *mean curvature* (MEANC). It can be obtained using PROFC and TANGC (Young, 1805):

$$\begin{aligned} \text{MEANC} &= \frac{\text{PROFC} + \text{TANGC}}{2} \\ &= \frac{q^2 \cdot r - 2 \cdot p \cdot q \cdot s + p^2 \cdot t}{(p^2 + q^2) \cdot \sqrt{1 + p^2 + q^2}} \\ &- \frac{(1 + q^2) \cdot r - 2 \cdot p \cdot q \cdot s + (1 + p^2) \cdot t}{2 \cdot (1 + p^2 + q^2)^{3/2}} \end{aligned} \tag{2.22}$$

The mean curvature describes mean-concave and mean-convex terrains, which makes it especially interesting for geomorphologic studies. Positive values of mean curvature are associated with areas which have relative accumulation, while negative ones will appear at zones with relative deflection.

Shary (1995) further introduced a system of 12 curvatures, where the three basic curvatures: mean, unsphericity and difference curvature, are used to derive the remaining nine curvatures. The *unsphericity curvature* (UNSPHC) can be derived using (Shary, 1995):

$$\text{UNSPHC} = \frac{\sqrt{\begin{aligned} &\left(r \cdot \sqrt{\tfrac{1+q^2}{1+p^2}} - t \cdot \sqrt{\tfrac{1+p^2}{1+q^2}}\right)^2 \cdot (1 + p^2 + q^2) \\ &+ \left(p \cdot q \cdot r \cdot \sqrt{\tfrac{1+q^2}{1+p^2}} - 2 \cdot s \cdot \sqrt{(1 + q^2) \cdot (1 + p^2)}\right. \\ &\left. + p \cdot q \cdot t \cdot \sqrt{\tfrac{1+p^2}{1+q^2}}\right)^2 \end{aligned}}}{2 \cdot (1 + p^2 + q^2)^{\frac{3}{2}}} \tag{2.23}$$

This parameter describes how the surface differs from a sphere, and, consequently, has a value of 0, only if the surface is a sphere itself. The second independent curvature is the *difference curvature* (DIFFC), defined as a half of a difference between vertical and horizontal curvature Shary (1995):

$$\begin{aligned} \text{DIFFC} &= \frac{q^2 \cdot r - 2 \cdot p \cdot q \cdot s + p^2 \cdot t}{(p^2 + q^2) \cdot (1 + p^2 + q^2)^{1/2}} \\ &- \frac{(1 + q^2) \cdot r - 2 \cdot p \cdot q \cdot s + (1 + p^2) \cdot t}{2 \cdot (1 + p^2 + q^2)^{3/2}} \end{aligned} \tag{2.24}$$

DIFFC indicates which of the two curvatures is formally stronger. From the three independent curvatures (MEANC, DIFFC and UNSPHC), the following remaining curvatures can be derived: the minimal and maximal curvature (MINC, MAXC), *horizontal* and *vertical excess curvatures* (HEXC, VEXC), the *total Gaussian curvature* (TOTGC), *total accumulation curvature* (TOTAC) and *total ring curvature*

(TOTRC) (Shary, 1995; Shary et al., 2002):

$$\text{MINC} = \text{MEANC} - \text{UNSPHC} \tag{2.25}$$

$$\text{MAXC} = \text{MEANC} + \text{UNSPHC} \tag{2.26}$$

$$\text{HEXC} = \text{UNSPHC} - \text{DIFFC} \tag{2.27}$$

$$\text{VEXC} = \text{UNSPHC} + \text{DIFFC} \tag{2.28}$$

$$\text{TOTGC} = \text{MEANC}^2 - \text{UNSPHC}^2 \tag{2.29}$$

$$\text{TOTAC} = \text{MEANC}^2 - \text{DIFFC}^2 \tag{2.30}$$

$$\text{TOTRC} = \text{UNSPHC}^2 - \text{DIFFC}^2 \tag{2.31}$$

The total Gaussian curvature is rather simple and can be also derived directly using (Gauss, 1828):

$$\text{TOTGC} = \frac{r \cdot t - s^2}{(1 + p^2 + q^2)^2} \tag{2.32}$$

A detail explanation of each of the above-listed curvatures and their interpretation using sample case studies can be obtained from Peter A. Shary's website (http://www.giseco.info).

Florinsky (1998) uses also ROTOR — a curvature of flow lines which are perpendicular to contours:

$$\text{ROTOR} = \frac{(p^2 - q^2) \cdot s - p \cdot q \cdot (r - t)}{\sqrt{(p^2 + q^2)^3}} \tag{2.33}$$

which is a measure of twisting of flow lines. Flow lines turn clockwise when ROTOR > 0, while flow lines turn anti-clockwise when ROTOR < 0.

Although the meaning of curvatures is related mostly to the behaviour of the flows that go through the cell, this indirectly alters other parameters such as, for example, those related to soil properties. Therefore, curvatures can be of great value for understanding other different characteristics of the terrain that we are analysing. PROFC has a significant relation with soil moisture. It indicates the tendency of water to increase its speed (when the curvature reflects a convex form) or decrease it (when it reflects a concave one), thus also indicating whether a cell is prone to accumulating water or not (Shary et al., 2002). This tendency is a function not only of local morphology, but also of the morphology and area of the cells upslope. As we will see in Chapter 7, considering all the factors involved, there are certainly more accurate measures that can be used to describe water accumulation.

Regarding PLANC, it has been said that flows over cells with concave curvature tend to concentrate (converge), while those over cells with convex curvature tend to diverge. This also gives interesting information about the potential erosion that can be generated by that flow, and when combined with vertical curvature, this information can even be extended. For example, net erosion is more likely to occur in cells with concave plan curvature and convex profile curvature than

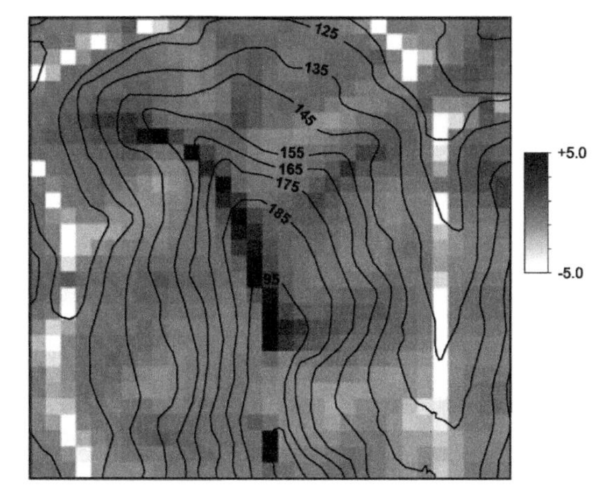

FIGURE 7 A mapped image of tangential curvature (in 100 rad/m) for the Baranja Hill area, overlaid with contours. Grid resolution = 25 m.

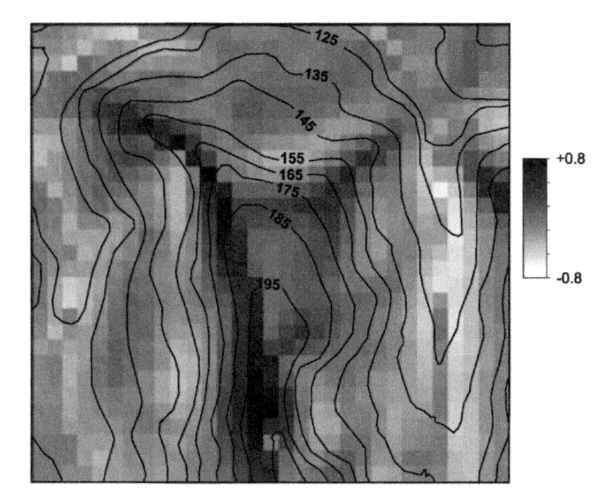

FIGURE 8 A mapped image of profile curvature (in 100 rad/m) for the Baranja Hill area, overlaid with contours. Grid resolution = 25 m.

in cells with different configurations. Curvatures are, however, a first approximation at a local level. As previously stated, to give an accurate estimation of these processes, other non-local parameters should also be considered.

Curvatures are not only interesting for hydrologists. For example, wildlife researchers can use curvature values to find out whether some parts of the terrain are protected (concave forms) or exposed (convex ones), as this obviously has an influence on the development of life forms. Mapped images of tangential and profile curvatures are shown in Figures 7 and 8, respectively.

Although their practical application still remains rather undefined, using third derivatives might also give some interesting morphometric information. However, it must be noted that additional orders of derivatives are extremely sensitive to errors (noise) in the DEM, and this sensitivity is then propagated onto the outputs. Schmidt et al. (2003) provides a more detailed discussion on this topic.

2.1.3 Visibility and visual exposure

In this section, we will show some other parameters that serve to analyse the morphometric characteristics of the DEM and its implication, and will also broaden the reader's view regarding the analysis mechanisms that can be used to extract useful information from a DEM.

Up to this point, we have been using a fixed size for our analysis window. This size was chosen according to the characteristics of the mathematical function used to define the DEM at each point. A discussion on the convenience of using one or other window size can be found in Chapter 14, along with some examples (Figure 5 in Chapter 14). However, if we do not need a mathematical definition of the DEM, and so do not need to adjust strictly to the minimum 3×3 window, then the size of the analysis window should depend on the particular characteristics and physical meaning of the land-surface parameters that we are calculating. Next, we will introduce some ideas related to visibility that will make this distinction a bit clearer. The concept of visibility is simple: from which other points can a single point be seen? Of course, the relation *A sees B* is reciprocal, so the above definition can also be rewritten as: given a single point, which other points can be seen from it? The set of points associated with that single point is called the *viewshed*. Calculating this visibility involves studying all the directions from which light rays reach (or leave) the analysed point, and defining a *line of sight* (LOS) for each one.

Following this line of sight from the analysed point, we can see if other points on it are visible, by simply checking if there are relief forms between them that block visibility. To do this, we calculate the angle α formed by the horizontal plane and a line connecting the two cells A and B to be analysed, using Equation (2.34):

$$\alpha = \arctan\left(\frac{z_B - z_A}{d_{AB}}\right) \tag{2.34}$$

where d_{AB} is the distance between cells A and B. If the angle formed by any other cell, B', situated closer to A is greater than the one formed by B, then B is not visible from A. A very simple numerical example can also be used to illustrate this point. Consider the 6×6 DEM shown in Figure 5 of Chapter 1, in which we define a line of sight from the upper left cell (with an elevation of 10) to the lower left one (with an elevation of 23).

Assuming a cell size of 1, the values of the angle α for the cells along the line of sight and whether they can be seen or not from the first cell (the upper left one) are shown in Table 2.

Although the concept of a viewshed is somehow similar to the concept of a watershed, which will be analysed in detail in the following section on regional land-surface parameters, it is worth noting that a viewshed is not necessarily a continuous polygon. It may be made up of as many different and disjointed locations as happen to be in view. Also, there is no direct relation between the cells

TABLE 2 A visibility analysis for a defined line of sight

Row, col	H	ΔH	$\Delta H/D$	Seen/not seen
1, 2	14	4	4	Seen
1, 3	19	9	4.5	Seen
1, 4	22	12	4	Not seen
1, 5	24	14	3.5	Not seen
1, 6	23	13	2.6	Not seen

that comprise it, but just a relation with the initial point from which lines of sight irradiate (which is the equivalent of the outlet point in a watershed). As previously pointed out, the extent of each line of sight (and, therefore, the analysis window that they implicitly define) is selected according to the parameter itself, and, in particular, to the physical meaning of visibility. Two points, 100 km apart, and with no obstacles between them, may be visible one from the other, but it is not reasonable to consider them as such, since the limitations of human sight should also be taken into account. However, the concept of visibility is not only applicable to light and to human sight, but to any emitter–receptor system and any form of radiation, such as, for instance, radio waves. In this case, a separation of 100 km might be within reach of the emitter–receptor system, and it can be assumed that each point can be *seen* from the other. Regardless of this, visibility maps are usually calculated by analysing all the cells in the DEM. The assumption, therefore, is that the emitter–receptor system works for any given distance.

The most basic conception of visibility just considers two possible values: *A can be seen from B* and *A cannot be seen from B* (in other words, B either belongs, or does not belong, to the viewshed defined by A). However, it is easy to extend this classification and turn visibility from a discrete parameter into a continuous one. The values that are frequently used are the distance between points, the angle of vision and the relative height. However, this last one can only be applied if we consider an object at point A with a defined height h (such as, for example, a building), and it tells us not only if that object can be seen from B, but also *how big* it looks from B. To calculate the relative size, the height and distance of the object are used:

$$\text{RELSIZE} = \arctan\left(\frac{h}{d_{AB}}\right) \tag{2.35}$$

where d_{AB} is the distance that separates A from B. The height of the object can also be used just to calculate visibility as a boolean parameter (i.e. to calculate a simple viewshed). Figure 9 shows the viewshed associated with a 20-metre object situated within the extent of the Baranja Hill area.

Note that adding the height of the object to the elevation of a cell might somehow alter the reciprocity of the *A sees B* relation, since seeing the object implies not only seeing the ground, but also the whole height of it. Also, note that the observer has a certain height as well.

FIGURE 9 A viewshed of a 20 m high object situated within the Baranja Hill area (an arrow indicates its approximate location). The black cells represent those from which the object cannot be seen.

REMARK 5. *Visibility, visual exposure and visibility index are relative measures that are based on simple geometric principles applied over the whole area or only for specific locations of interest.*

Using the concepts of *'visibility'* and *'line of sight'*, the local definition of insolation introduced earlier in this same chapter can be extended and improved. See further Chapter 8 that provides a complete description of solar radiation modelling.

If we calculate visibility not just for a single point, but for the whole DEM, we can obtain new land-surface parameters, such as the number of cells that can be seen from each cell, i.e. the number of different viewsheds to which a cell belongs. This is usually referred to as the *visual exposure* or *the visibility index.*

Calculating visual exposure is a very computationally intensive task. More efficient approaches have been developed, such as the one described in Franklin (1998). It is beyond the scope of this chapter to deal with parameters that are not exclusively related to the DEM itself and its morphometry, but it is interesting to note that visual exposure can be extended in many ways. This adds other variables related to the visual characteristics of each cell. By doing this, we can obtain, among others, parameters that are of great interest for visual impact assessment. Also, instead of considering all cells as possible locations from which a cell can be seen, a reduced set can be used (such as those in a road). Calculating the number of cells from the selected set that see each other in the DEM leads to the definition of a cumulative or additive viewshed. A simple introduction to this can be found in Berry (1996).

2.2 Statistical parameters

Before introducing the statistical land-surface parameters that have been created specifically for the analysis of DEMs, it is interesting to mention some basic ideas about the local analysis of raster layers. These are ideas that can be applied to any kind of grid, including, of course, DEM grids. That will lead us to the definition of some basic statistical parameters, some of which are clearly related to others that we have already seen, and might serve as indicators or first approximations of them.

Local (also know as neighbourhood) analysis, involves performing operations with the values of a given cell and the cells that surround it up to a certain distance. While, as we have seen, geometric morphometric parameters usually make use of a square 3×3 square window, generic local analysis may also use circular or angular neighbourhoods of different sizes. Whatever the size and shape of the neighbourhood considered, calculating derived parameters is nevertheless carried out in the same way, simply by using the values contained within its limits.

The first parameters to consider are the first four moments in the distribution of a value, namely, the *average value*, the *standard deviation*, the *skewness coefficient* and the *kurtosis coefficient* (Figure 6 in Chapter 28). Calculating the mean value of elevation serves as a filter to reduce noise and remove, among other things, spurious single-cell pits. However, there are more sophisticated methods for this, that give much better results, since they do not touch those cells that constitute the pits (see Section 2.8 in Chapter 4). Other statistical parameters, such as the median, can be used to obtain a similar result. Note that many statistical measures overlap with geometric measures. For example, standard deviation is strongly correlated with slope.

One should not forget that local analysis can be applied to grids other than the DEM, such as, for example, the slope or curvature grids introduced earlier in this section. The results that this analysis yields can also contain significant information about the configuration of the DEM.

Other interesting statistical measures are the *range of values*[3] or the *ruggedness* (Melton, 1965), which was originally developed in order to characterise basins:

$$\mathrm{RUGN} = \frac{\mathrm{RANGE}}{\sqrt{a}} \tag{2.36}$$

where a is the area covered by the local region being analysed.

Moving to some not-so-basic statistical parameters, another interesting one than can be included in this group is *terrain roughness*. Unlike the other land-surface parameters such as slope or aspect, there is no clear agreement in the way roughness should be measured (Felícisimo, 1994a), and methods differ significantly in their underlying concepts. The concept of terrain roughness is, however, simple and easy to understand: it indicates how *undulating* the terrain is, i.e. how complex it is.

The simplest way to compute terrain roughness is to use the standard deviation of the height in the cells of a given analysis window. High values of standard deviation indicate that the terrain is rather irregular around the cell being analysed, while low ones reflect a smooth terrain. It is interesting to note that, in the case of terrain roughness, the size of the analysis window plays a key role, as does the grid resolution. Depending on the kind of analysis to be performed (whether on

[3] The difference between the highest and lowest values.

a macro or micro scale), the spatial extent (not in cells, but in ground units) of the analysis window has to be chosen with care.

Using standard deviation is, however, not a very precise method, and it can produce incorrect results, such as assigning high values (rough terrain) to cells constituting a flat terrain within a slope. To avoid this, one solution is to fit a plane to the cells in the analysis window (much in the style of what was explained for the best-fit plane methods used to get a mathematical description of the DEM, locally), and then to calculate the standard deviation of the fitting instead (Sakude et al., 1998).

As previously stated, differences between methods are important. Completely different to the ones already described, Hobson (1972) proposed a vector approach to define the following *Surface Roughness Factor*:

$$\mathrm{SRF} = \frac{\sqrt{\left(\sum_{i=1}^{n} \mathrm{X}_i\right)^2 + \left(\sum_{i=1}^{n} \mathrm{Y}_i\right)^2 + \left(\sum_{i=1}^{n} \mathrm{Z}_i\right)^2}}{n} \tag{2.37}$$

where n is the sample size (i.e. the number of cells in the analysis window) and X_i, Y_i and Z_i are the components of the unit vector normal to the land surface at each one of the cells in the analysis window. These can be calculated from slope and aspect, using the following expressions:

$$\mathrm{X}_i = \sin(s) \cdot \cos(a)$$

$$\mathrm{Y}_i = \sin(s) \cdot \sin(a)$$

$$\mathrm{Z}_i = \cos(s)$$

Also related to slope gradient is the concept of surface area (see Section 2.1). The ratio between the surface area of a cell and its planimetric area can be used, as well as a measure of terrain roughness.

The surface roughness factor gives information about the DEM itself, since rough terrains constitute more complex entities that are harder to describe accurately and, therefore, the quality of a DEM created using interpolation techniques depends on it (Florinsky, 1998; Thompson et al., 2001). On the other hand, terrain roughness can be used just like any other parameter, by incorporating it into different models or by using it to derive new land-surface parameters. Studies related to wind analysis, for example, make frequent use of this parameter.

REMARK 6. *The most common statistical geomorphometric parameters are range, standard deviation, kurtosis, terrain roughness, anisotropy and fractal dimension.*

In the case of geostatistical analysis of densely measured elevations (e.g. LiDAR data), the *Surface Roughness Index* can be derived as the ratio between the fitted nugget variation and the local variance:

$$\mathrm{SRI} = \frac{\hat{C}_0}{\sigma^2_{\mathrm{NB}}} \quad \% \tag{2.38}$$

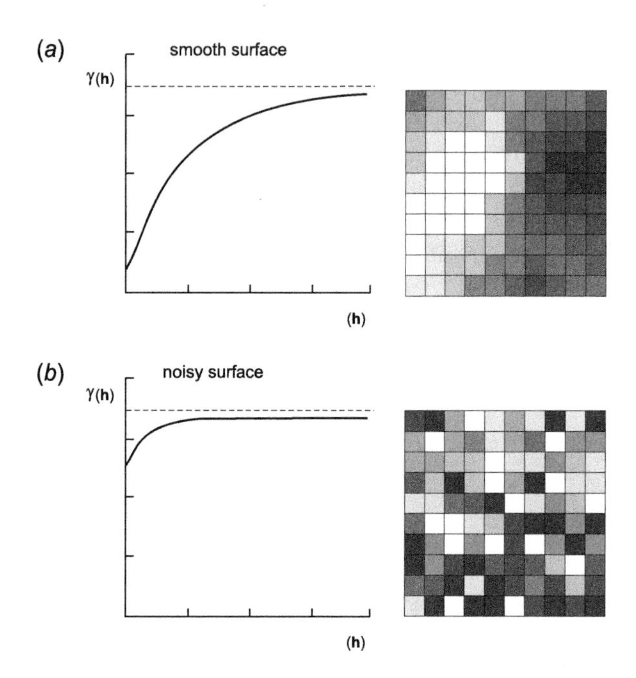

FIGURE 10 Two examples of variograms and the resulting images with (a) low and (b) high SRI.

where $\hat{C}_0$ is the locally fitted nugget parameter and σ^2_{NB} is the local variance in the predefined neighbourhood. The nugget parameter needs to be estimated using an automated (robust) variogram fitting method such as the one described in Walter et al. (2001). Note also, that in order to get a reliable estimate of the variogram parameters, one needs to use at least 100 point pairs (e.g. 10×10 window environment if the input data is in grid format). If SRI is 0%, this means that the topography is absolutely smooth, i.e. that the values are completely spatially correlated (Figure 10). Values for SRI above 100% and below 0% would also be possible, but it is conceptually better to assign them an undefined value.

Variogram modelling can also be repeated in various directions to see how isotropic the land surface is, locally. This type of analysis can be used to derive the *Anisotropy Index*, which can be defined as the ratio between the minimum and maximum range parameter of spatial dependence, fitted for various directions (Figure 11):

$$\mathrm{ANI} = \frac{\hat{R}_{\mathrm{min}}}{\hat{R}_{\mathrm{max}}} \tag{2.39}$$

where $\hat{R}_{\mathrm{min}}$ is smallest estimated range parameter and the $\hat{R}_{\mathrm{max}}$ is the highest estimated range parameter in various directions. Currently, such analysis can only be run in DiGEM (Bishop and Minasny, 2005). Operational packages to run geostatistical analysis on densely sampled raw elevation data are still missing.

A simplified version of the ANI is the *Anisotropic Coefficient of Variation* (ACV), which is defined as the difference of the first derivative in 4 directions (Figure 11):

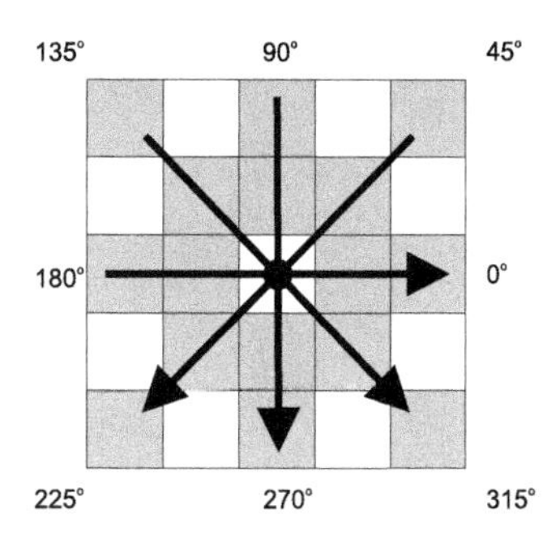

FIGURE 11 Directions in a 5×5 window environment.

$$\mathrm{ACV} = \log\left[1 + \frac{\sqrt{\frac{\sum_{i=1}^{8}(\partial z_{\mathrm{NB}i} - \partial z_{\mathrm{AVG}})^2}{8}}}{\partial z_{\mathrm{AVG}}}\right] \tag{2.40}$$

where $\partial z_{\mathrm{AVG}}$ is the average value of first derivative in 4 directions: east/west, north/south, north-east/south-west and north-west/south-east. The difference between the average derivative is then calculated for 8 neighbours (2× in each direction). The ACV (Figure 12) describes the general geometry of the local surface and can be used to distinguish elongated from oval landforms (see `G_landforms` in Chapter 13).

Fractals are another way of estimating terrain roughness, that not only serves this purpose, but opens up a whole new world of possibilities for characterising the morphometry of a terrain. Since surfaces with higher values of fractal dimension are more complex than those with lower ones, *Fractal Dimension* is clearly an

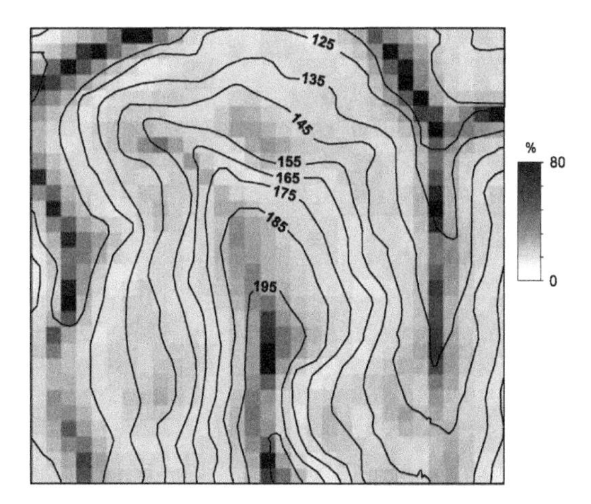

FIGURE 12 A mapped image of the Anisotropic Coefficient of Variation for the Baranja Hill area, overlaid with contours. Grid resolution = 25 m.

indicator of terrain complexity and, therefore, is closely related to terrain roughness.

The concept of the fractal dimension of a surface can be applied to the DEM itself as a whole, but a local fractal analysis is also of interest for obtaining a new layer of fractal dimension values, instead of a single one. Fractal dimension D can be measured using several different techniques, and here we found another concomitance with image analysis, since some of them are derived from the texture analysis algorithm of grey-scale images. The most widespread ones are probably box-counting (Falconer, 2003) and the fractional Brownian model (Mark and Aronson, 1984).

For the box-counting method, the usual $n \times n$ rowing window is substituted by a $n \times n \times n$ box centred on each cell. For each central cell, the analysis box is divided using several new cell sizes n' (the only valid values are those that make $q = n/n'$ an integer number). Voxels (the 3D equivalent of pixels) are then filled for each different value of n', and the number of filled *voxels* is counted. The fractal dimension is then estimated as the inverse of slope $P = \log(q)/\log(N)$, where N is the total number of filled voxels. Fractal ideas can be also applied to the analysis of polygons that result from *slicing* the DEM. Instead of performing a 3D fractal analysis, the 2D contour lines (polygon borders are nothing more than contour lines) extracted can be analysed to calculate the fractal dimension of the DEM surface indirectly. This clearly shows the close relation between these two approaches (Falconer, 2003). Statistical morphometric parameters are increasingly used in geomorphometry to quantify complexity of terrain. Gloaguen et al. (2007), for example, use fractal dimension derived for DEMs to automatically detect fault lines and similar geomorphological features.

2.2.1 Discrete analysis of the land surface

Another important trend in the development of new land-surface parameters is the application of pre-existing indices and parameters originally introduced for other types of data. Landscape metrics were originally developed for measuring spatial properties of landscape patterns, but their underlying ideas can also be applied to DEMs and used to generate new descriptors of land surfaces.

Since landscape metrics mostly deal with discrete information and the DEM contains a continuous variable (i.e. elevation), some kind of discretisation has to be performed before applying any calculation. *Slicing* the DEM into elevation classes is the most logical way of doing it. Once the DEM has been converted into a discrete layer, the shape and structure of each class (which defines at least one polygon) can be analysed and the results used to derive new indices related to terrain morphology. Hengl et al. (2003) suggested using the *Shape Complexity Index*, commonly used to describe polygons, on DEM slices. SCI indicates how compact (or oval) a feature is. It is derived using the perimeter-to-boundary ratio:

$$\mathrm{SCI} = \frac{P}{2 \cdot r \cdot \pi}, \quad r = \sqrt{\frac{A}{\pi}} \tag{2.41}$$

where P is the perimeter of the polygon, and A its area. r represents the radius of the circle with the same area. There is a direct relation between SCI and other ge-

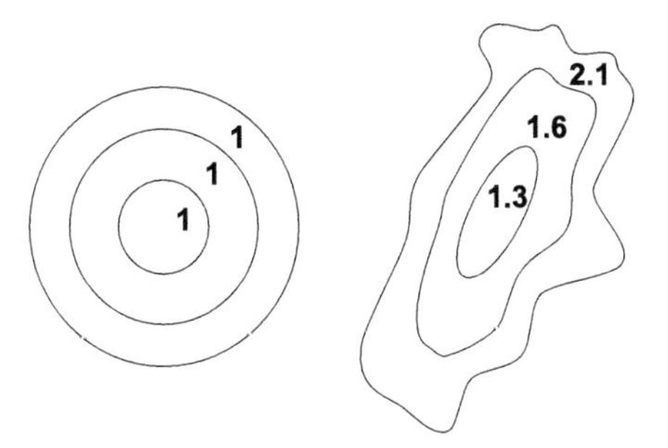

FIGURE 13 A comparison of the Shape Complexity Index values for a perfectly oval shape (left) and for different levels of complexity (right).

omorphometric ideas, such as the classification of landforms, since pits and peaks are more oval, while valleys and ridges are more longitudinal or dissected (Figure 13, and Figure 9 in Chapter 13).

3. REGIONAL LAND-SURFACE PARAMETERS

In the DEM, cells are not isolated one from each other. Gravity makes flows move downhill across the cells, going across them and establishing a topological relation between them. Analysing the morphometric properties of the whole set of cells and their relations leads to the definition of new land-surface parameters, some of which are described in this section.

In hydrology, a watershed is the region of land where water flows to a defined point, known as the outlet of the watershed. In other words, all the run-off generated within the watershed will eventually drain to the outlet. If we consider the land surface defined by a DEM, the above definition can be rewritten as the set of cells where water flows to a defined cell. Using flow direction algorithms, it is simple to calculate the watershed associated with a cell (which will constitute the outlet point), by just going upslope and adding all the cells *connected* with the outlet cell. From this point, it is assumed that the DEM is pre-processed and ready for its hydrological analysis (see also Section 2.8 in Chapter 4), so that *connectivity* is complete.

> REMARK 7. *Regional morphometric measures are mainly connected with hydrological properties of terrain. The most common parameters are: catchment area, flow-path length, slope length and proximity to local streams and ridges.*

This watershed has its own properties, and these can be used to describe the characteristics of the cell, thus constituting land-surface parameters themselves. The most important property is the area A of the watershed (i.e. the area of all the cells situated upslope of the outlet). This parameter is known as the *catchment area*, but it can also be found in the literature as *flow accumulation* or *upslope area*. A map image of the catchment area is shown in Figure 14.

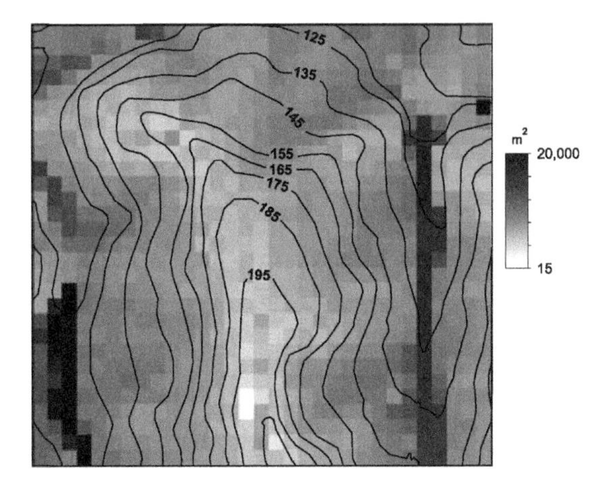

FIGURE 14 A catchment-area map image for the Baranja Hill area, overlaid with contours. Grid resolution = 25 m. A logarithmic scale has been used to improve representation.

The catchment area is very important. It can be used to extract channel networks and define several relevant indices. For a further explanation, see Section 6 in Chapter 7.

To estimate the catchment area of the outlet cell, we count the number of cells in the watershed and multiply the result by the area of a single cell. Of course, in that case, the area of all those cells is the same. However, we can also consider other parameters, where the values of each cell are different. If there are n cells upslope, and we denote this other parameter as α, then the generic expression shown in Equation (3.1) makes it possible to define a whole range of similar land-surface parameters, of which the catchment area is just one:

$$\text{LSP} = \sum \alpha_i \tag{3.1}$$

Clearly, catchment area is particularly significant in Equation (3.1) in which $\alpha_i = A, \forall i = 1, \ldots, n$:

$$\text{CA} = \sum_{i=1}^{n} A_i \tag{3.2}$$

If, for example, α is the runoff generated in each cell, then the total run-off that passes through the outlet cell can be calculated. To perform this calculation, a new grid with all those run-off values is needed. Calculating these parameters can be done using a recursive scheme, considering that, for any of the parameters P described in Equation (3.1), the value at a cell i equals the value of the parameter α at the cell, plus the values in all the surrounding ones that flow to i. Instead of just adding the values of the parameter for all the upslope cells, we can calculate an average:

$$\text{LSP} = \frac{\sum_{i=1}^{n} \alpha_i}{n} \tag{3.3}$$

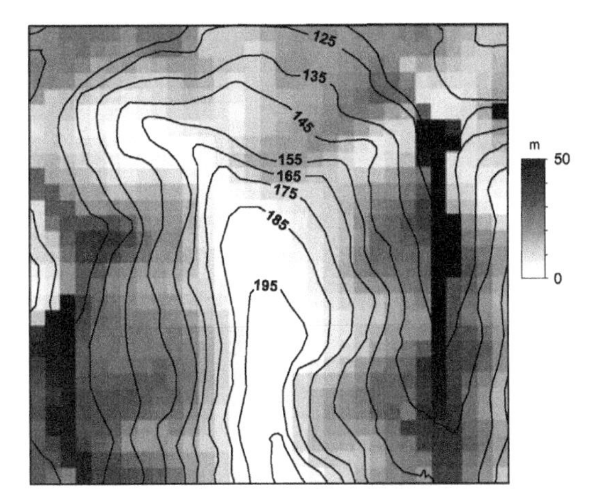

FIGURE 15 A mapped image of the catchment height for the Baranja Hill area. Grid resolution = 25 m.

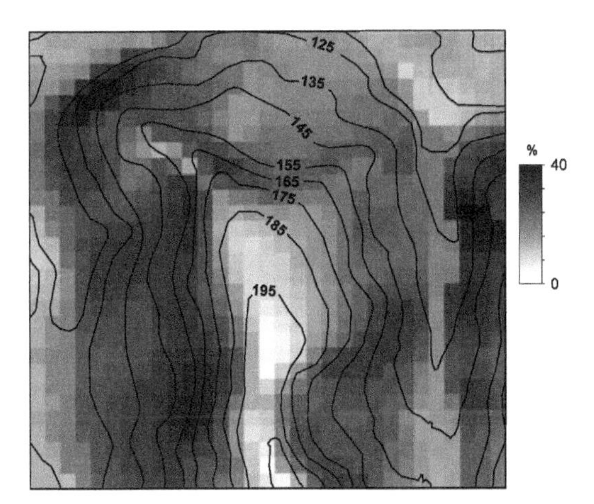

FIGURE 16 A mapped image of the catchment slope for the Baranja Hill Area, overlaid with contours. Grid resolution = 25 m.

And, to derive new, meaningful parameters, we can connect this equation with some of the local land-surface parameters that have already been introduced, or even with the DEM itself. Slope and height are the usual parameters for this, and, from them, two land-surface parameters emerge, namely *catchment height* and *catchment slope*. Figures 15 and 16 shows the map images of both of them.

Catchment height reflects the mean elevation (not absolute, but relative over the target cell) of all the cells upslope, thus constituting an indicator of the potential energy of all the flows that will eventually pass through the cell. A similar meaning can be associated with catchment slope, since it is directly related to the speed and power of those flows.

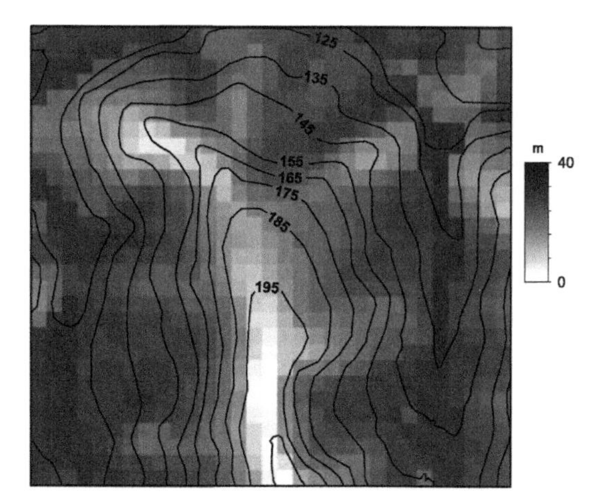

FIGURE 17 A mapped image of the flow-path length for the Baranja Hill Area, overlaid with contours. Grid resolution = 25 m.

By analysing the distribution of height values at all the cells upslope, the hypsometry of the catchment area can also be defined. An hypsometric curve gives information about the internal configuration of the catchment. Much in the same way, the *elevation–relief ratio* (Pike and Wilson, 1971) can be computed with the height values of all the cells upslope:

$$\mathrm{ERR} = \frac{Z_{\mathrm{avg}} - Z_{\mathrm{min}}}{Z_{\mathrm{max}} - Z_{\mathrm{min}}} \tag{3.4}$$

Apart from the regional land-surface parameters, based on surface measures, that consider the entire extent of the catchment, other ones based on linear measures can be defined based upon the same hydrological relations between catchment cells. The *flow-path length* is the most important of these variables (Figure 17). The flow-path length represents the total length of flow of all the flows upslope of a given cell, and it can be calculated using Equation (3.1).

A parameter that is similar, conceptually, to the length of the flow path is *slope length* (i.e. the maximum length of flow up to an *interruption* cell where the slope is considered to end). While the former is of significant interest for hydrological analyses, the latter is more frequently used in formulations related to erosion, such as the Universal Soil-Loss Equation (USLE) (Wischmeier and Smith, 1978).

Computing slope length for a cell is done by measuring flow lines in the direction opposite to the gradient ($a + 180°$), up to the closest interruption cell (Mitášová et al., 1996). It can also be carried out by considering not just the opposite direction, but all the cells upslope, by taking the maximum flow-path length from all of them, and adding to it the distance between that cell and the central cell (Griffin et al., 1988). Once again, both methods favour the usage of recursive algorithms for their computation.

The definition of those interruption cells can be done using a fixed-slope threshold (Mitášová et al., 1996), or a ratio between the slopes of a cell and the

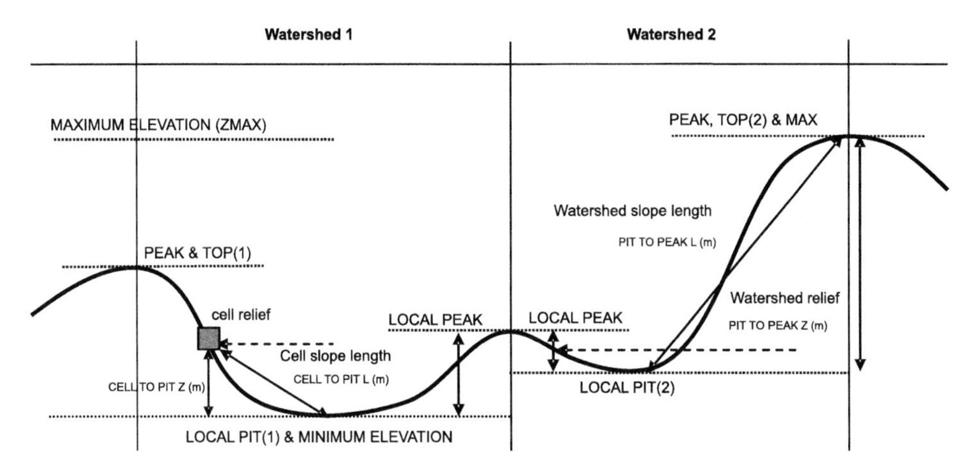

FIGURE 18 A schematic definition of horizontal (L) and vertical (Z) distances to local pits/streams and peaks/ridges. Courtesy of Robert A. MacMillan.

one situated upslope (Hickey et al., 1994), or their average or maximum uphill slope angle[4] (Griffin et al., 1988; Wilson, 1986).

The extraction of land-surface objects specific to hydrology is described in Chapter 7. Using those hydrological objects, new parameters can be defined that describe the spatial configuration of the DEM to which they are related. Among these, we can cite the following:

- the horizontal or vertical distance to the closest channel cell;
- the Euclidean distance to the closest channel cell;
- the flow distance (the distance following the flow path) to the closest channel cell.

Due to their proximity to streams, these distances can be related, for instance, to the wetness of the cells. They can also be used as a measure of local relative landform position. In this last case, values can be used to predict ecological soil types, as is explained in Chapter 23. Similar distances can be estimated using defined landform elements such as ridges and peaks, or pits (see Chapter 9). A summary of these measurements is shown in Figure 18, while mapped images of two of them (the percentages of vertical distances to streams and pits) are shown in Figures 19 and 20.

A last note on the accuracy and veracity of the land-surface parameters described in this section: for all these parameters that analyse the cells situated upslope, it is necessary to check that we are not ignoring cells that might be in the watershed above a cell, but not included in the DEM. This situation arises when a DEM does not extend far enough to cover all the cells. Cells in this situation are said to be affected by *edge contamination*, and their catchment-area values or other similar land-surface parameters should not be considered valid. Notice that edge contamination is also a land-surface parameter in the form expressed by Equa-

[4] This is a direct application of another land-surface parameter introduced previously.

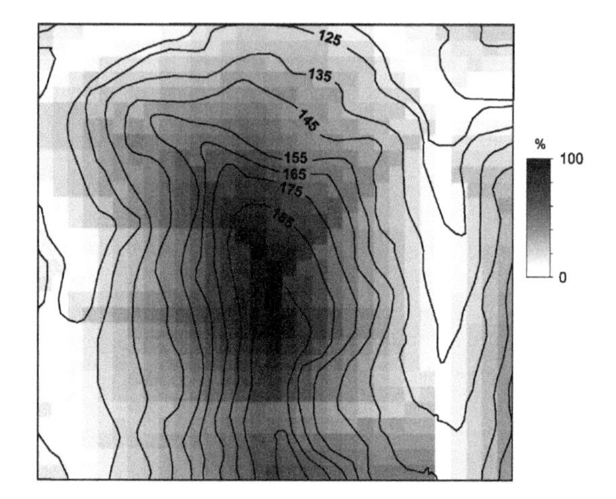

FIGURE 19 A measure of the regional context — percentage of vertical distance to a stream. Derived in LandMapR.

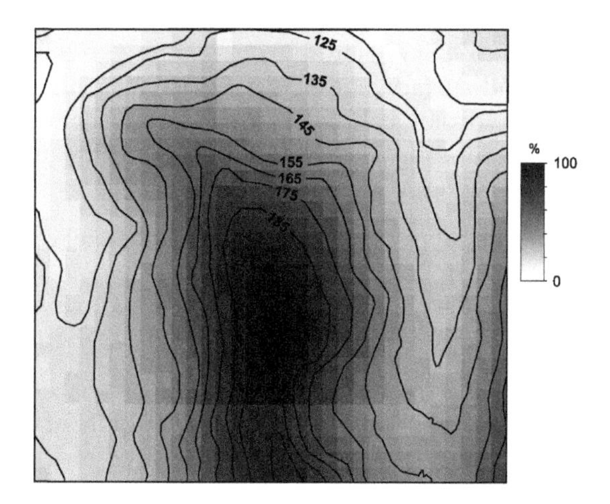

FIGURE 20 A measure of regional context — percentage of vertical distance to a pit. Derived in LandMapR.

tion (3.1), in this case α taking a value of 1 in border cells and 0 in the remaining ones.

4. SUMMARY POINTS

In this chapter we have looked at some of the most basic land-surface parameters, divided into two main groups: local and regional parameters. The local ones are calculated using a fixed size window around each cell, while the regional ones consider the relation between cells and study a non-fixed surrounding area for each cell.

Local land-surface parameters make use of geometrical or statistical concepts. For the former, we rely on a mathematical model of land surface and then employ general measures from differential geometry or (geo)statistics. First and second derivatives can be calculated, and their related parameters, such as slope or curvature, have proved to be useful for many different fields of application. For this analysis, the choice of the land-surface model significantly influences the parameters derived. In the case of statistical parameters, the set of values inside the local analysis window is used to extract statistical descriptors. These range from basic ones, such as the mean value or standard deviation, to complex, fractal-based ones, or the so-called *anisotropic coefficient of variation*.

Regarding regional parameters, they are linked with the hydrological configuration of the terrain. The most important of these is the catchment area. The areas implicitly defined by these parameters can be used to extract new parameters, such as the mean or extreme upslope values of an additional parameter, or related ones, such as the hypsometric curve or the elevation–relief ratio.

The measures of slope, aspect and curvatures that we have traditionally thought of as local measures are, in fact, and have increasingly become, focal measures computed within windows of many different sizes and shapes, and not just square 3×3 windows. Today, even the basic land-surface parameters have increasingly become multi-scale measures (see further Chapter 14) — they are often computed within windows of various dimensions (3×3, 5×5, . . . , 21×21) and shapes (circular, square, etc.). This makes the distinction between the local and regional parameters even more difficult.

IMPORTANT SOURCES

Shary, P.A., Sharaya, L.S., Mitusov, A.V., 2002. Fundamental quantitative methods of land surface analysis. Geoderma 107 (1–2), 1–32.

Schmidt, J., Dikau, R., 1999. Extracting geomorphometric attributes and objects from digital elevation models — semantics, methods, future needs. In: Dikau, R., Saurer, H. (Eds.), GIS for Earth Surface Systems — Analysis and Modelling of the Natural Environment. Schweizbart'sche Verlagsbuchhandlung, pp. 153–173.

Mitášová, H., Hofierka, J., Zlocha, M., Iverson, L.R., 1996. Modeling topographic potential for erosion and deposition using GIS. International Journal of Geographical Information Systems 10 (5), 629–641.

Evans, I.S., 1972. General geomorphometry, derivatives of altitude, and descriptive statistics. In: Chorley, R.J. (Ed.), Spatial Analysis in Geomorphology. Harper & Row, pp. 17–90.

CHAPTER 7

Land-Surface Parameters and Objects in Hydrology

S. Gruber and **S. Peckham**

phenomena related to the flow of water or other materials that can be parameterised using a DEM · basic principles and approaches to modelling of flow · differences between the diverse flow-modelling techniques available · advantages, disadvantages and limitations of the different approaches · why is parameterisation of surface flow a powerful technique?

1. HYDROLOGICAL MODELLING

Hydrology is the study of the movement, distribution, and quality of water throughout the Earth. The movement of water is primarily driven by gravity and to some degree modified by the properties of the material it flows through or flows over. The effect of gravity can mostly be approximated well and easily with a DEM. By contrast, surface and subsurface properties and conditions are rather cumbersome to gather and to treat. From this simple reasoning it is also evident, that in steep topography such parametrisation performs better than in very gentle topography where the relative importance of gravity decreases. Parametrisation means that we represent certain phenomena related to the flow of water with quantities (parameters) that are easy to calculate and/or for which data are readily available. In many cases we can deduce much information from the DEM, alone. However, one needs to be careful not to stretch these methods to applications that suffer from the inherent simplifications.

Land-surface parameters specific to hydrology have been applied to a multitude of different areas including:

- hydrological applications (Chapter 25);
- mapping of landforms and soils (Chapter 20);
- modelling landslides and associated hazard (Claessens et al., 2005);
- hazard mapping (ice/rock avalanches, debris flows) in steep terrain (Chapter 23);

Developments in Soil Science, Volume 33 © 2009 Elsevier B.V.
ISSN 0166-2481, DOI: 10.1016/S0166-2481(08)00007-X.

- erosion and deposition modelling (Mitášová et al., 1996);
- mass balance modelling on mountain glaciers (Machguth et al., 2006).

Most of these applications focus on steep terrain (hill slopes and headwaters), where topography clearly dominates the flow of water. Many hydrologic applications, however, also involve nearly horizontal terrain (channels and flood plains of large rivers) and require specific techniques to produce consistent results in areas where the flow of water is governed by features that are smaller than the resolution or uncertainty of the DEM.

The development and use of flow-based land-surface parameters gained importance in the late 1980s after the introduction of the D8 algorithm (O'Callaghan and Mark, 1984) and the 1990s have seen a number of multiple flow directions algorithms published and employed (Freeman, 1991; Quinn et al., 1991; Holmgren, 1994). Similarly, corresponding techniques for the treatment of ambiguous flow directions (Garbrecht and Martz, 1997) or the derivation of hydrologically-sound DEMs (Hutchinson, 1989) as well as sensitivity studies using existing algorithms (Wolock and Mccabe, 1995) were published. Methods based on original contour data (O'Loughlin, 1986) and TINs (Jones et al., 1990; Tucker et al., 2001) have some advantages over using gridded DEMs but have continued to play a subordinate role due to the wide availability and intuitive processing of raster data as well as the introduction of more advanced techniques for extracting information from raster DEMs.

While the development and refinement of methods is still ongoing, the near future will likely see much research dedicated to the optimal use of high-resolution and high-quality LiDAR elevation data sets that are currently becoming widely available.

2. FLOW DIRECTION AND ASPECT

2.1 Understanding the idea of flow directions

Flow direction is the most basic hydrology-related parameter and it forms the basis for all other parameters discussed in this chapter. Imagine you are standing somewhere in a hilly landscape that has a smooth ground surface. If you release one drop of ink on the ground, you intuitively expect it to flow down the steepest path at each place and to leave a trace on the ground that represents what is called a *flow line*. The physics of purely gravity-driven flow dictates that water will always take the steepest downhill path, such that flow lines cross contour lines at a right angle.

However, when we imagine a grid cell centred on a peak or *ridge line*, the flow direction is ambiguous, no matter how small we make the cell. In fact, flow direction for peaks and ridges is ambiguous even for mathematical surfaces with infinite resolution. Consistent flow distribution demands flow into opposite directions and thus violates the notion of having only one flow line or direction for each grid cell. In sloping terrain, such ambiguous flow directions are always *sub-grid* effects that cannot be represented at the present resolution. If, however, the surface is discretised (e.g. into a regular grid), then we are faced with the problem of how best to represent a continuous flow field with a regular grid. Then, the number of neighbouring directions that a drop can move to is limited and the best compromise needs to be found. This is the first problem of assigning one *single flow direction* to each grid cell in a regular grid that only has eight possible directions in multiples of 45° (Figure 1).

The second problem relates to the *divergence* (going-apart) — the opposite being *convergence* (coming-together) — of flow. If you release two drops on an inclined plane, they will keep flowing down slope, parallel to each other with constant spacing between their traces (flow lines). On the surface of a cone (plan-convex, see Section 2.1 in Chapter 6), the drops increase their spacing as they flow down slope — their flow lines are divergent. This means that there is the same mass (e.g. number of drops or volume of water) spread over a larger area. Similarly, on an inverted cone (plan-concave), two drops that are released nearby decrease their spacing — their tracks are convergent.

This entire section mainly deals with the formulation of how to move how much water into which neighbouring cells in order to have a representation of reality that is suitable for a given task. This can be pictured as many drops flowing from one cell to one or more adjacent cells, depending on their relative elevations. The partitioning of mass (or number of drops) contained in one cell to several lower neighbours may be justified by actual divergence or by the attempt to overcome the limits of having only 8 adjacent cells. If the local direction of steepest decent is not a multiple of 45°, then the flow may be partitioned between two neighbours to account for this. As a consequence, the water of one cell may be propagated into *multiple neighbour cells.*

However, the initial mass is then contained in two or more cells instead of one and thus dispersed over a larger area and a larger width along the contours. For some applications this may be inappropriate and is then termed over-dispersal. Now, we have assembled all four criteria by which to judge or select a flow direction algorithm:

(1) *handling of the discretization* into only eight possible adjacent flow directions (artifacts are sometimes called *grid bias*);
(2) *handling of divergence;*
(3) *handling of dispersal;* and
(4) *handling of sub-grid effects.*

At the same time it is evident that all four criteria are interconnected and that each algorithm will be a compromise between them. Often two more criteria are mentioned that we will not discuss in detail here but that can be very important for certain applications. One is the suitability for efficient computational evaluation and the other is the robustness of the method (i.e. its ability to describe all terrain shapes without exceptions). The basic types of *single- and multiple-neighbour flow algorithms* are fundamentally different: single-neighbour algorithms cannot represent divergent flow but for the same reason have no problem of over-dispersal. Multiple-neighbour algorithms can represent divergent flow but usually also suffer from some over-dispersal. Flow direction is ambiguous on peaks and ridges, which occur throughout fluvial landscapes and which are essentially singularities in the flow field.

2.2 Handling undefined flow directions

The assignment of flow directions relies on elevation difference between cells to drive the flow. This principle fails for local elevation minima (*pits*) that have no lower neighbours and for horizontal areas. Thus, an undefined drainage direction is often assigned to pits (no drainage direction) and horizontal areas (ambiguous or no drainage direction) resulting in the termination of flow accumulation in such cells. This effect may be:

- *real and wanted* (e.g. sinkholes in Karst);
- *real but unwanted* (e.g. if flow accumulation is desired to propagate though a lake);
- *artificial and unwanted* (e.g. pit artifacts in a DEM or falsely horizontal areas in large river plains).

If these effects are unwanted (in most cases they are), alternative methods[1] have to be employed for the designation of a flow direction in order to keep the physical quantities of derived land-surface parameters consistent. Horizontal areas are rare in real landscapes but can exist in DEMs where a cell is usually considered horizontal if it has the same elevation as its lowermost neighbour. Horizontal areas can originate from lakes, from interpolation artifacts or be the result of preprocessing during which depression have been filled. *Large rivers* also have very low slopes that usually are smaller than the DEM resolution and thus locally appear to be horizontal.

> REMARK 1. *In large river basins, special techniques can be required to calculate channel slope that is often lower than can be represented by the DEM.*

One approach to resolve ambiguous flow direction in flat areas is an iteration procedure during which flat cells are assigned a single flow direction to a draining neighbour cell (Jenson and Domingue, 1988) and the actual elevation values remain unchanged. In the first iteration this will only make cells next to outlets drain. In the second iteration, flat calls adjacent to the ones altered during the first step will receive a flow direction and so on. This approach has been extended to avoid unrealistic *parallel drainage lines* (Tribe, 1992a). The second approach is to make minute alterations to the elevation (Garbrecht and Martz, 1997) of the flat cell in order to impose a small artificial gradient (thus often called imposed gradient method). These artificial changes are made in an iterative way and result in topography that is also suitable for flow direction resolution by multiple-neighbour flow methods. However, in many cases this requires an increased numerical resolution of the DEM in computer memory and is often impractical for large river basins.

[1] A number of methods for the treatment of pits is discussed in Section 2.8 of Chapter 4.

2.3 Stream burning

Poor quality or simply the inherent generalisation of a DEM may cause drainage lines derived by digital delineation from gridded data to substantially differ from reality. Where vector hydrography information exists it can be integrated into the DEM prior to the actual analysis. This process is referred to as *stream burning* and can be effective in the digital reproduction of a known and generally accepted stream network. However, it has the disadvantage of locally altering topography in order to provide consistency between existing vector hydrography and the DEM. Several methods exist[2] (Hutchinson, 1989; Saunders and Maidment, 1996) but greatly differ in their success of improving, e.g. watershed delineation (Saunders, 1999). The pre-processing of the vector information required often represents an intensive effort.

2.4 Vertical resolution of DEMs and computation of slope

The above paragraph has discussed the assignment of drainage direction for areas that are horizontal in the DEM. Many times, these areas are not horizontal in reality. This section deals with the problem of assigning a slope to them because it is a key variable in many types of process-based hydrologic models. In the context of flow routing, for example, slope, water depth and roughness height are the main variables that determine the flow velocity. Here we will define slope as a dimensionless ratio of lengths (rise over run) or as the tangent of the slope angle ($\tan\beta$). When working with raster DEMs and computing slopes between grid cells, the ratio of the vertical and horizontal resolutions determines the minimum non-zero slope that can be resolved.

For example, a DEM with a vertical resolution of 1 m and a grid spacing of 30 m has a minimum resolvable slope of $1/30 = 0.0333$, while a DEM with a vertical resolution of 1 cm and a grid spacing of 10 m has a minimum resolvable slope of $1/1000 = 0.001$. This lower bound means that slopes on hillsides can usually be computed with a relatively small error, using any of several different local methods (as discussed in Section 3.3 of Chapter 2). However, slopes in channels are often much smaller than the numbers in these examples, and can even be smaller than 0.00001 for larger rivers. This is several orders of magnitude smaller than can typically be resolved and, as a consequence, these areas will appear horizontal in the DEM and require techniques for flow routing in horizontal areas.

One way to get better estimates of channel slope is to use the flow directions assigned to the horizontal DEM cells (see previous section) to identify a streamline or reach that spans a number of grid cells. The slope can then be computed as the elevation drop between the ends of the reach divided by its along-channel length. Depending on the size of the grid cells, this may yield an estimate of the valley slope instead of the channel slope. Channel sinuosity within the valley bottom will result in an even smaller slope.

[2] See also the AGREE — DEM surface reconditioning system (http://www.ce.utexas.edu/prof/maidment/gishydro/ferdi/); courtesy of Ferdi Hellweger.

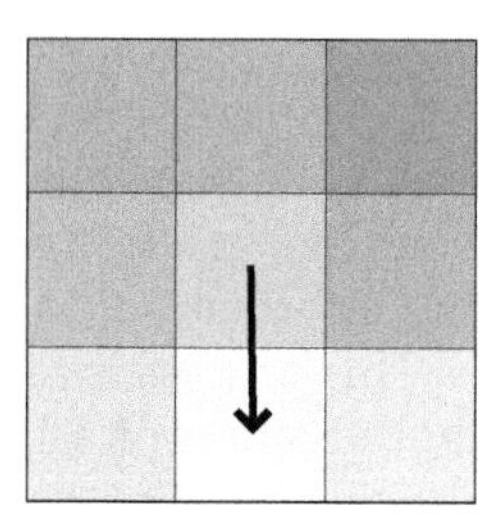

FIGURE 1 Single flow direction assigned to the central pixel in a 3×3 neighbourhood. Grey values represent elevation increasing with darkness of the cell.

3. FLOW ALGORITHMS

3.1 Single-neighbour flow algorithms

The most basic flow algorithm is the so-called *"D8"*, sometimes referred to as *method of the steepest descent* (O'Callaghan and Mark, 1984). From each cell, all flow is passed to the neighbour with the steepest downslope gradient (Figure 1) resulting in 8 possible drainage directions — hence the name D8. It can model convergence (several cells draining into one), but not divergence (one cell draining into several cells). Ambiguous flow directions (the same minimum downslope gradient is found in two cells) are usually resolved by an arbitrary assignment.

This method actually provides a very good estimate of the catchment area for grid cells that are far enough downstream to be in the fully convergent, channelised portion of the landscape. However, for grid cells on hillslopes or near peaks and divides where the flow is divergent, values obtained by this method can be off by orders of magnitude. The D8 method is widely used and implemented in many GIS software packages. Despite its limitations, it is useful for a number of applications such as extracting river network maps, longitudinal profiles and basin boundaries.

A number of other single-neighbour algorithms have been published. *Rho8* (Fairfield and Leymarie, 1991) is a stochastic extension of D8 in which a degree of randomness is introduced into the assignment of flow directions in order to reduce the grid bias. The drawback of this method is that — especially for small catchments — it produces different results if applied several times. The aspect-driven *kinematic routing algorithm*[3] (Lea, 1992) specifies flow direction continuously and assigns flow to cardinal cells in a way that traces longer flow lines with less grid bias than D8.

3.2 Multiple-neighbour flow algorithms

Only multiple-neighbour flow methods can accommodate the effects of divergent flow (spreading from one cell to several downhill cells, Figure 2) that are especially important on hill slopes. Four important multiple-neighbour flow algorithms as

[3] Also referred to as *"Lea's method or kinematic routing"*.

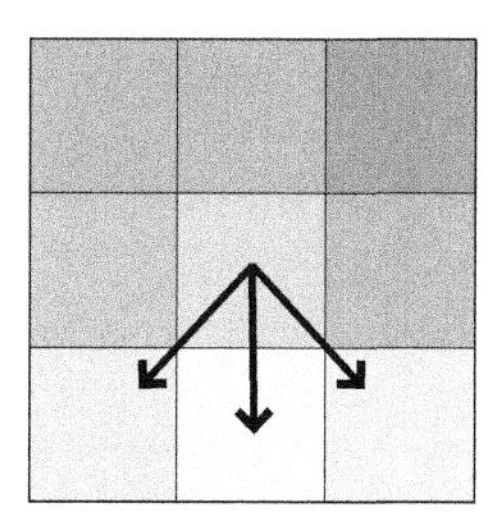

FIGURE 2 Multiple flow directions assigned to the central pixel in a 3×3 neighbourhood using MFD. Grey values represent elevation increasing with darkness of the cell. Multiple flow directions are assigned and a fraction of the mass of the central cell is distributed to each of the three lower cells that the arrows point to. All mass fractions together must sum to one in order to conserve mass.

well as the basic principles of their calculation are described here. This description is intended to highlight the important differences that exist between these approaches and thus help to judge their suitability for a given task.

> REMARK 2. *All flow-routing methods discussed in this chapter can represent convergent flow but only multiple-neighbour methods can accommodate divergent flow.*

3.2.1 Multiple Flow Direction (MFD) Method

A number of algorithms exist that handle divergent flow and partition the flow out of one cell to *all* lower neighbours (Freeman, 1991; Quinn et al., 1991, 1995; Holmgren, 1994). These algorithms do not have firmly-established names and are often simply referred to as MFD (multiple flow direction) methods, as the *TOPMODEL approach* (Quinn et al., 1991) or as *FD8* (Freeman, 1991). In a general formulation, the draining fraction d into neighbouring cell NBi is given by:

$$d_{\mathrm{NB}i} = \frac{\tan(\beta_{\mathrm{NB}i})^{v} \cdot L_{\mathrm{NB}i}}{\sum_{j=1}^{8}(\tan(\beta_{\mathrm{NB}j})^{v} \cdot L_{\mathrm{NB}j})} \tag{3.1}$$

The draining fraction d depends on the slope β (positive into lower cells and 0 for higher cells) into the neighbours, on different draining contour lengths L as well as an exponent v controlling dispersion. The drainage potentials into each neighbour are normalised to unity over the 3×3 kernel in order to preserve mass. In this way, different weights can be assigned to downstream pixels between which the flow is partitioned.

High values of v concentrate flow more toward the steepest descent and low values result in stronger dispersion (v must be $\geqslant 0$). Holmgren (1994) suggests values of $v = 4$–6 and equal L for cardinal and diagonal directions to produce best[4] results. In the widely used original TOPMODEL approach (Quinn et al., 1991), no exponent is used to control dispersion ($v = 1$), but differing contour lengths L are

[4] Freeman (1991) suggests $v = 1.1$, but it is unclear if he refers to slope in degrees or as the tangent so this has to be treated with care.

assumed somewhat arbitrarily for cardinal (0.50 × cell size) and diagonal neighbouring pixels (0.35 × cell size). The use of the exponent v makes this method very flexible, but, at the same time it is difficult to determine optimal values for it.

> REMARK 3. *The exponent in multiple flow direction methods only controls the amount but not the area of dispersion.*

Furthermore, it needs to be kept in mind that the exponent v only controls the amount of dispersion (how much volume is passed to each cell) but not the degree of dispersion (to which cells flow is propagated). Minute amounts (only limited by numerical precision) of flow will always be passed to each lower neighbour. A technique to restrict the lateral spreading in MFD methods is presented in Chapter 23.

MFD methods are powerful in handling sub-grid effects: a horizontal ridge pixel for instance will drain towards opposite sides. However, a well-known problem with this method, as pointed out by Costa-Cabral and Burges (1994), Tarboton (1997) and others, is that it produces over-dispersion. That is, this method causes flow to spread too much, with some fraction nearly flowing along the contours. For example, in the case of an inverted cone, some of the flow from a grid cell will eventually make its way to the opposite side of the cone.

3.2.2 D∞

In this approach proposed by Tarboton (1997) one draining flow direction is assigned to each cell. It is continuous between 0 and 2π radians and the *infinite* number of directions that can be assigned is reflected in the name D-Infinity or D∞. (In practice it is beneficial to handle drainage direction in degrees instead of radians to avoid truncation errors in the numerical representation of π leading to small errors in flow routing.) Based on this direction, the draining proportion d is then apportioned (applied to the discrete DEM grid, Figure 3) to the two pixels on either side of the theoretical drainage direction vector by:

$$d_1 = \frac{4 \cdot \alpha_2}{\pi}, \qquad d_2 = \frac{4 \cdot \alpha_1}{\pi} \tag{3.2}$$

The angles α are measured on a horizontal planar surface between the drainage direction vector and the vectors to the two pixels on either side of it ($\alpha_1 + \alpha_2 = 45°$).

The flow is thus partitioned between only two cells and the grid bias inherent in D8 as well as the over-dispersion to all lower neighbours inherent in MFD are avoided. The angle-weighted partitioning however is somewhat arbitrary. The derivation of the flow direction is based on planes defined by the eight point-triplets given by the centre pixels and two adjacent neighbour pixels (for details see Tarboton, 1997). The use of point triplets also avoids the problems associated to the local fitting of planes through four points as employed in the kinematic routing algorithm (Lea, 1992) and DEMON (Costa-Cabral and Burges, 1994). In situations of ambiguous drainage direction this approach assigns one direction arbitrarily. Drainage towards two sides (horizontal ridge) is therefore impossible.

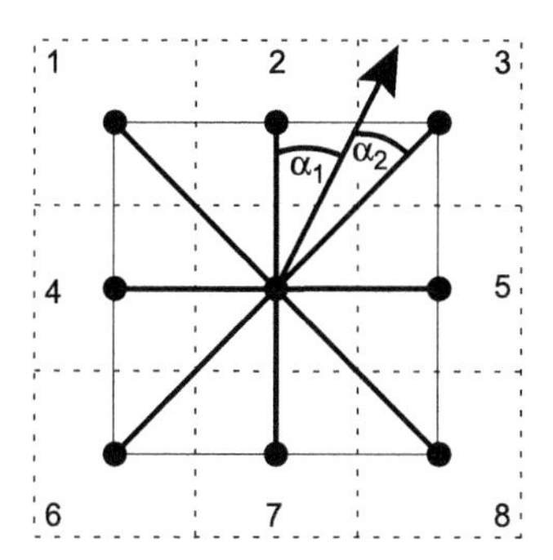

FIGURE 3 Concept of flow apportioning in D∞ (following Tarboton, 1997). A 3×3 pixel neighbourhood is given by the dashed lines and margin pixels are numbered 1 to 8. Pixel centres are represented by black points. The thick lines connecting the centres form eight triangles over which the drainage direction vector (arrow) is determined. Using this drainage direction vector, the flow is apportioned to the two pixels that bound the facet that the vector lies on. In this case, flow is distributed between pixels 2 and 3 [see Equation (3.2) where the subscripts 1 and 2 refer to pixels 2 and 3 in this example].

3.2.3 DEMON

This method relies on the construction of flow tubes based on best-fit planes through the four corners of a pixel and generally produces very realistic results in both convergent and divergent flow regimes (Costa-Cabral and Burges, 1994). However, the method that is used to determine aspect angle can lead to inconsistent flow geometry and does not address the ambiguity of flow direction on peaks and ridges. This method is implemented in only few software packages.

3.2.4 Mass-Flux Method (MFM)

The second author (S. Peckham) has developed another method called the Mass-Flux Method which is available in RiverTools (see Chapter 18). This method has so far not been published and evaluated in the scientific literature but both the promising results of its application as well as its basic concept warrant a brief description, here. The key idea of this method is to divide each grid cell into four *quarter pixels* and to define a continuous flow direction angle for each, using a grid that has twice the dimensions of the DEM. For each quarter pixel, the elevations of the *whole pixel* and two of its cardinal neighbours uniquely determine a plane and a corresponding slope and aspect (Figure 4).

While this removes the ambiguity of plane fitting and the associated problems, it also removes the ambiguity of flow direction for grid cells that correspond to peaks or ridges, since it allows flow from these grid cells to be routed in different directions. At the quarter-pixel scale, however, flow from each quarter pixel is only permitted to flow into one or two of its cardinal neighbours. The fraction that flows into these neighbours is determined by treating each grid cell as a control volume. Flow out of a control volume can only be through an edge. There can be no flow directly to a diagonal neighbour. The fraction of flow that passes through a given edge is computed as the dot product of the unit normal vector for that edge

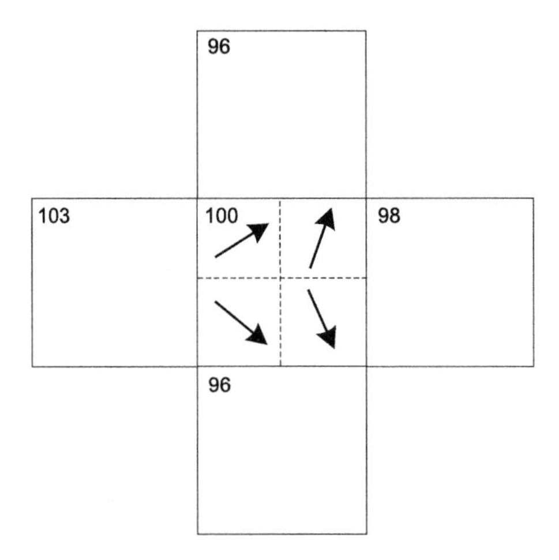

FIGURE 4 Flow directions assigned to quarter pixels using the Mass-Flux Method. Numbers refer to pixel elevations in this example. © 2005 Rivix LLC, used with permission.

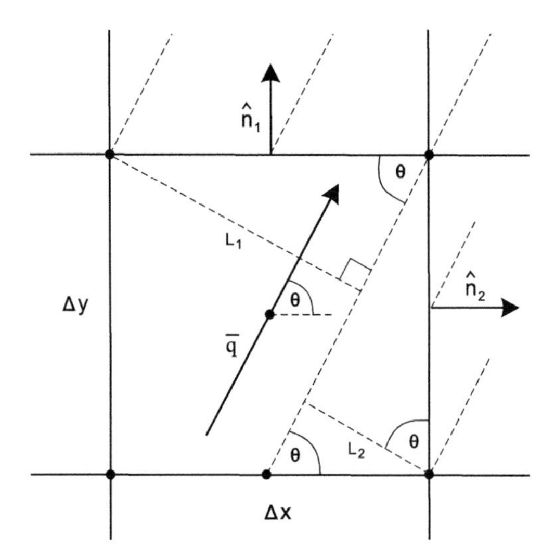

FIGURE 5 Flow apportioning between two cardinal neighbours in the Mass-Flux Method. L_1 and L_2 denote the projected flow widths into the upper and left neighbour and together equal the projected flow width w, $\hat{n}_1$ and $\hat{n}_2$ are vectors normal to the cell boundaries, $\bar{q}$ is the flow vector and θ is the flow direction. © 2005 Rivix LLC, used with permission.

and the continuous-angle flow vector, as shown in Figure 5. This is equivalent to decomposing the flow vector into two vector components along the grid axes.

Where flow is convergent, it is possible for two quarter-pixels to have a component of flow toward each other. This occurs because streamlines in the actual flow field are closer together than the grid spacing. While we know that streamlines cannot cross, the additional turning required for the streamlines to become paral-

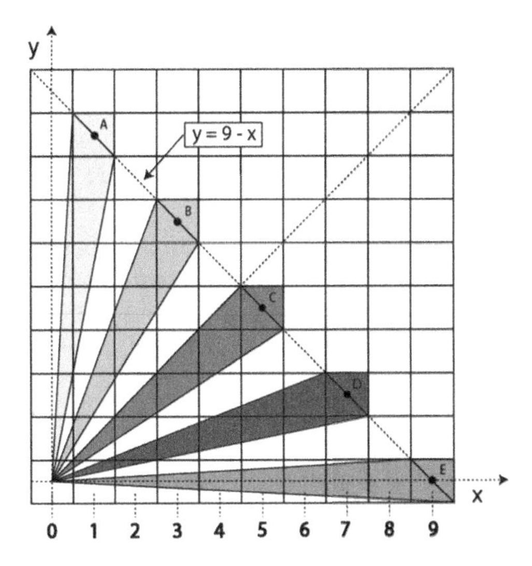

FIGURE 6 For the special case of a radially-symmetric surface such as a cone or a Gaussian hill, the TCA for pixels can be computed analytically. Each "*necktie*" region can be broken into two triangles of which the area can be computed. This shows that pixels A–E each have the same TCA. In this regular case ($\Delta x = \Delta y$), the flow width can thus vary between 1 and $\sqrt{2}$ multiplied by the grid resolution. © 2005 Rivix LLC, used with permission.

lel cannot be resolved. To address this streamline resolution problem, the grid of quarter-pixel aspect angles is scanned for these cases prior to computing the total contributing area (see below) and the angles are adjusted by the smallest amount that is necessary to produce a consistent vector field.

A grid of total contributing area values with the same dimensions as the DEM is found by integrating the contributions of the eight quarter-pixels that surround each whole pixel. Similarly, a whole-pixel grid of aspect angles is found using the vector sum of the quarter-pixel flow vectors.

3.3 Flow width

The flow width or effective contour length orthogonal to the outflow (w) is another important concept in hydrology and for flow-based parameters. For the D8 and MFD methods, flow widths to each of the eight neighbours must be defined in some manner, and a variety of different rules have been proposed. In the TOPMODEL approach (Quinn et al., 1991), different contour length factors (cardinal: $0.50\times\Delta x$, diagonal: $0.35\times\Delta x$) are accumulated over all draining directions. For multiple-neighbour methods that use a single, continuous flow angle such as Lea (1992) method, D-Infinity, DEMON and the Mass-Flux Method, the projected pixel width (Figures 5 and 6) can be computed as:

$$w = \left|\sin(\theta)\right| \cdot \Delta x + \left|\cos(\theta)\right| \cdot \Delta y \tag{3.3}$$

where θ is the aspect angle, and Δx and Δy are the grid cell sizes[5] along the two coordinate axes.

4. CONTRIBUTING AREA/FLOW ACCUMULATION

The concept of *contributing area* is very important for hydrologic applications since it determines the size of the region over which water from rainfall, snowfall, etc. can be aggregated. It is well known that the contributing area of a watershed is highly correlated with both its mean-annual and peak discharge. The *dendritic* nature of river networks results in water collected over a large area being focused to flow in a relatively narrow channel. Contributing area, also known as *basin area, upslope area* or *flow accumulation* is a planar area and not a surface area. It describes the spatial extent of a collecting area as seen from the sky.

When we speak of *Total Contributing Area* (TCA), we have an element of finite width in mind such as a grid cell or contour line segment and we are integrating the flow over this width. *Specific Contributing Area* (SCA) refers to area per unit contour length ($\text{SCA} = \text{TCA}/w$), and is the more fundamental quantity that must be integrated over some width to get the TCA. This distinction is analogous to how the terms discharge and specific discharge are used. In fact, in the idealised case of constant, spatially uniform rainfall rate, the TCA and SCA are directly proportional to the discharge and specific discharge. This correspondence makes it possible to recast the problem of computing contributing area as a steady-state flow problem.

Flow accumulation cannot only be used to accumulate contributing area but also other quantities such as the amount of contributing pixels, accumulated precipitation (spatially-varying input) or *accumulated terrain attributes* (e.g. elevation) that, if divided by the amount of contributing cells yield *catchment averages* of these properties. Flow accumulation is initiated with a starting grid that contains the input values to be propagated until they meet the DEM boundaries or end in sinks. A *starting grid* that has the value of 1 everywhere will yield the amount of cells in the catchment or when multiplied with the cell size squared the TCA draining through each cell as the final value.

The starting grid may also consist of individual areas or starting zones from which values are propagated that may correspond to contaminants or mass movements and have a value of zero elsewhere. From this, the amount of contaminant or mass passed though a cell can be determined. The *downslope area* of a single starting zone is made up of all cells that have a nonzero value in the flow accumulation grid. The *upslope area* of a certain zone can be determined using upward flow directions.

The principle of flow accumulation is simple: when the draining proportions d out of one cell into its neighbours (must sum to 1) are known, also the receiving proportions r draining into one cell are known. The receiving proportions determine, which fractions of each neighbouring cell are received. The amount of mass

[5] For most applications $\Delta x = \Delta y$.

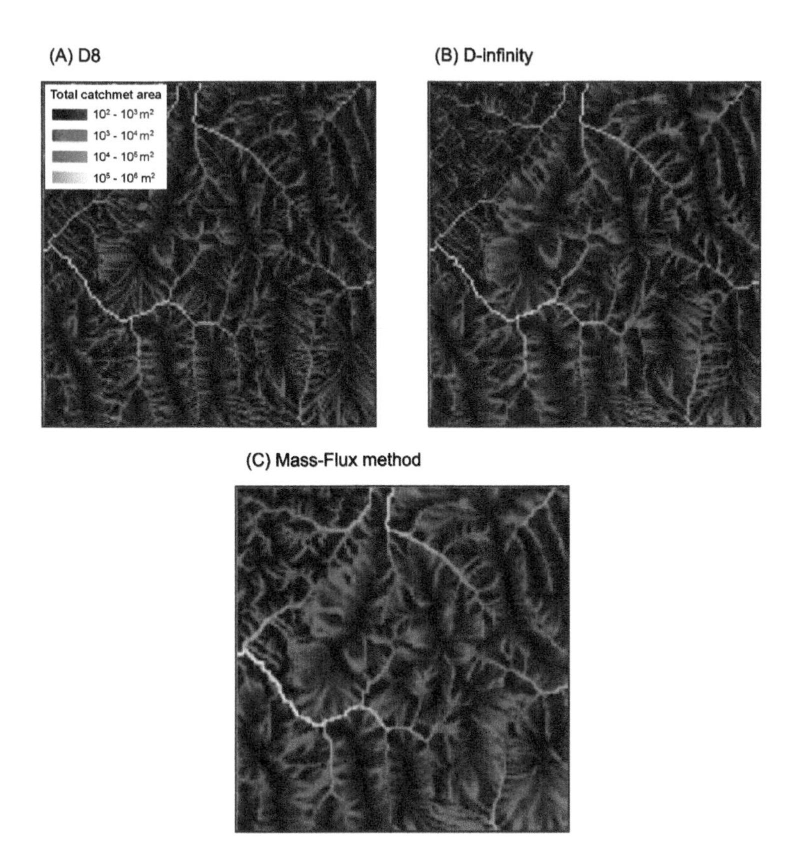

FIGURE 7 Total catchment area calculated for the Baranja Hill area using three different methods. (See page 713 in Colour Plate Section at the back of the book.)

(or volume, area or any other property) A that is accumulated in cell i is given by the sum of A in each neighbouring cell multiplied by the respective receiving fraction r plus the mass (or other quantity) input I in cell i itself:

$$A_i = \sum_{j=0}^{8}(A_{\mathrm{NB}j} \cdot r_{\mathrm{NB}j}) + I_i \tag{4.1}$$

Figures 7 and 8 show the spatial patterns resulting from the use of different flow direction methods for the calculation of TCA. D8 actually provides a very good estimate of the TCA for grid cells that are far enough downstream to be in the fully convergent, channelised portion of the landscape. However, for grid cells on hillslopes or near peaks and divides where the flow is divergent, values obtained by this method can be off by orders of magnitude. Especially here, in the hill slopes, the differences between the different approaches and between the values used for the dispersion coefficient in MFD are evident.

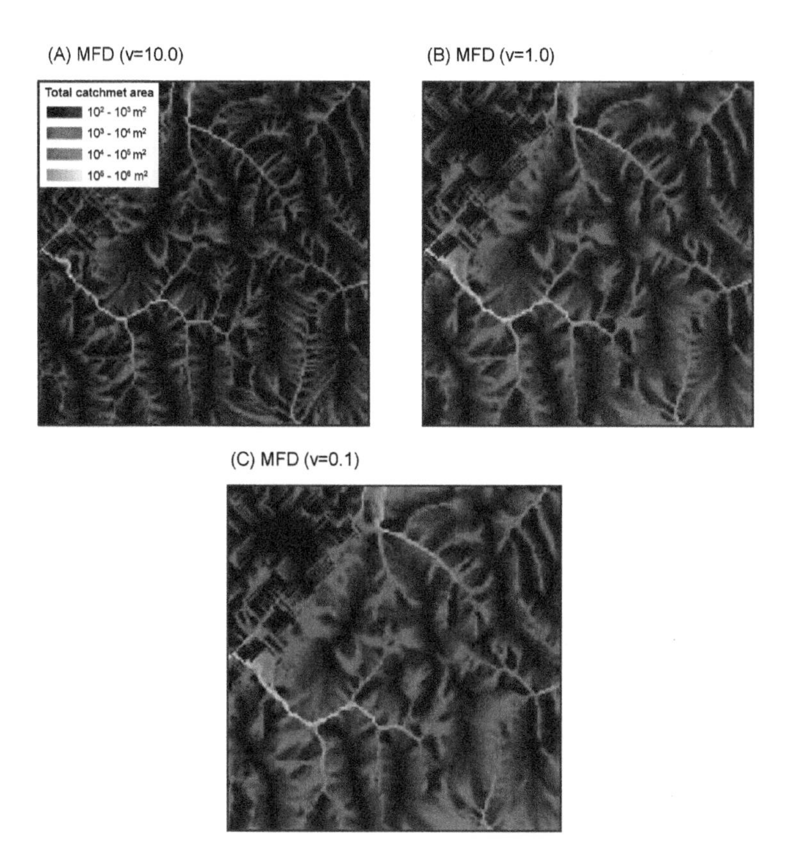

FIGURE 8 Total catchment area calculated for the Baranja Hill area using MFD and three different dispersion exponents. (See page 714 in Colour Plate Section at the back of the book.)

Figure 9 shows the result of applying D8, D-Infinity and MFM to the DEM of a cone. In parts (C) and (D), the MFM SCA grid shows a diamond pattern while the SCA grid is circular. Direct computation shows that a diamond pattern is the correct result — the area of each necktie-shaped polygon in Figure 6 is exactly the same.

In Figure 10 the propagation of one single mass input is displayed using different algorithms and different synthetic DEMs. The DEMs used are a sloping plane to show the handling of flow into a direction that is not a multiple of 45° and a sphere to demonstrate divergent flow.

> REMARK 4. *Calculation of catchment area or of accumulated terrain attributes based on catchment must be performed on DEMs that include the entire upslope area for all relevant pixels.*

Flow accumulation must be performed on the complete catchment of interest. The boundaries of the catchment should at least be one pixel away from the margin of the DEM to be sure of this. Otherwise, a contribution of unknown proportions is missing from the calculated results in the studied catchment. This edge contam-

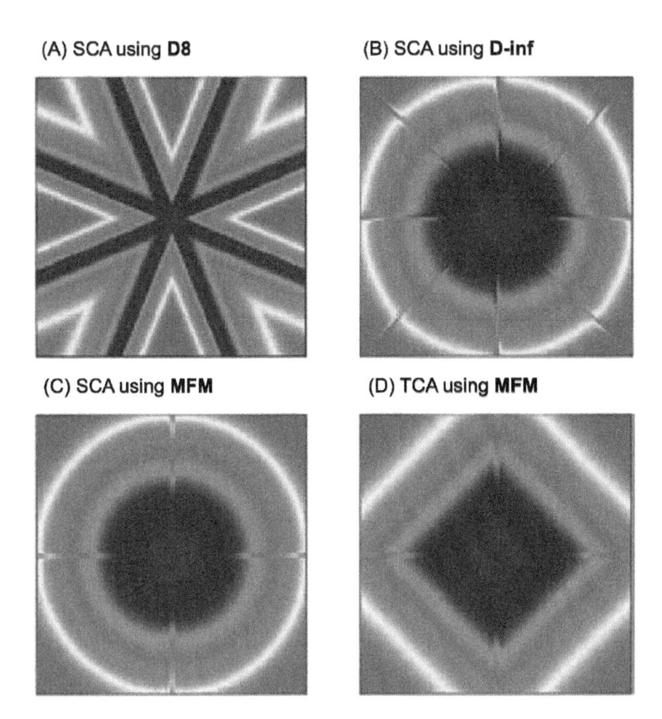

FIGURE 9 Parts (A)–(C) show the specific contributing area (SCA) calculated for the DEM of a cone sing D8, D-Infinity and MFM. The strong grid bias inherent in D8 is readily visible from the star pattern (A). Part (D) of this figure shows the total contributing area (TCA) calculated using MFM. This counter-intuitive result is correct because of the different flow widths of pixels (see Figure 6). When divided by the flow width, the SCA (C) shows the right circular pattern. (See page 715 in Colour Plate Section at the back of the book.)

ination effect can be assessed by propagating flow using a starting grid that only has a non-zero value in marginal pixels. All resulting pixels that have a value other than zero are affected by edge contamination and could thus contain an unknown error in their value of flow accumulation (Figure 11).

5. LAND-SURFACE PARAMETERS BASED ON CATCHMENT AREA

Catchment area is a powerful parameter of the amount of water draining though a cell that can be combined with other attributes to form compound indices. In the following we briefly describe the two most powerful and most frequently used indices: wetness and stream power.

The *Topographic Wetness Index*, also called Topographic Index or Compound Topographic Index (Quinn et al., 1991, 1995) is a parameter describing the tendency of a cell to accumulate water (Figure 12). The wetness index TWI is defined as:

$$\mathrm{TWI} = \ln\left[\frac{A}{\tan(\beta)}\right] \tag{5.1}$$

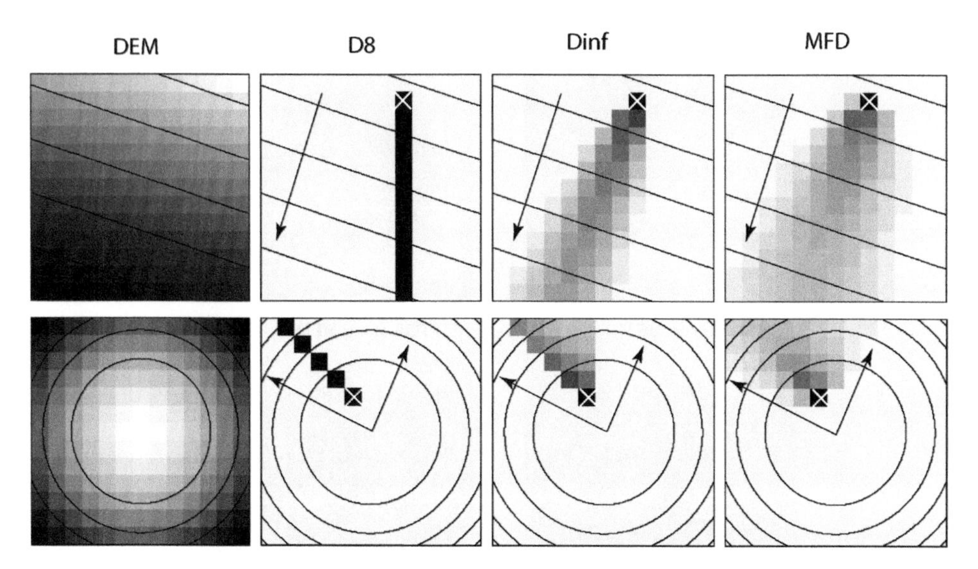

FIGURE 10 Graphic display of flow-propagation results using a synthetic DEMs (top: sloping plane, bottom: sphere). The first column (DEM) shows elevation values (dark: low, light: high) and isohypses. The remaining columns show topography by isohypses and arrows indicating the direction of drainage as well as grey values that correspond to the mass draining through one cell. In cells identified with a cross (starting zone), mass was inserted and propagated downwards. For D8, all downstream cells are black indicating that always the entire upstream mass was contained in the downstream cell. For D∞ and MFD, dispersion occurs and is indicated by grey cells where the upstream mass is divided into several downstream cells.

FIGURE 11 Edge-contaminated areas (white) have been removed from the calculated total contributing area. Both, the flow accumulation as well as the edge-contamination were computed using MFD. Other, less dispersive methods result in a smaller area of edge contamination. (See page 715 in Colour Plate Section at the back of the book.)

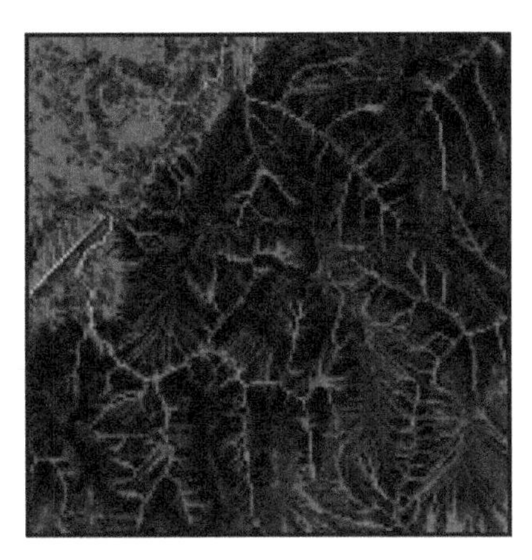

FIGURE 12 Wetness index calculated for the Baranja Hill. Values range from 3 (dark) to 20 (yellow); the data is linearly stretched. (See page 716 in Colour Plate Section at the back of the book.)

where A is the specific catchment area (SCA) and β is the local slope angle. It is based on a mass-balance consideration where the total catchment area is a parameter of the tendency to receive water and the local slope as well as the draining contour length (implicit in the specific catchment area) are parameters of the tendency to evacuate water. The TWI assumes steady-state conditions and spatially invariant conditions for infiltration and transmissivity. The natural logarithm scales this index to a more condensed and linear range. The original formulation also contained the lateral transmissivity of the soil profile that is usually omitted. This index is very powerful for a number of applications concerning vegetation, soil properties, landslide initiation and hydrology in hill slopes.

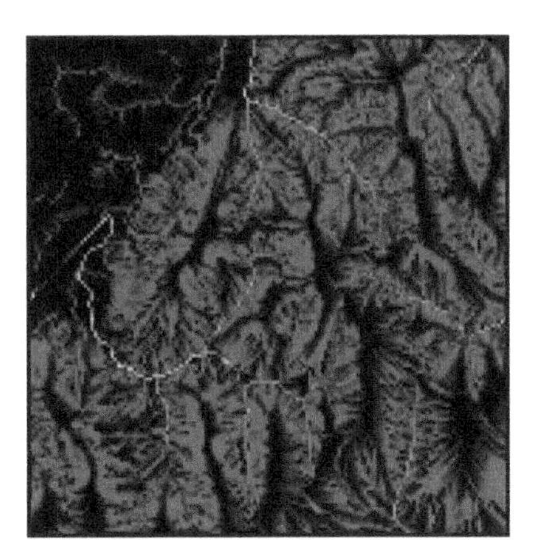

FIGURE 13 Stream power index calculated for the Baranja Hill. Values range from 1 (dark) to 12,000 (yellow); the data is stretched using logarithmic display. (See page 716 in Colour Plate Section at the back of the book.)

The *Stream Power Index* (Moore et al., 1988) can be used to describe potential flow erosion and related landscape processes (Figure 13). As specific catchment area and slope steepness increase, the amount of water contributed by upslope areas and the velocity of water flow increase, hence stream power and potential erosion increase. The stream power index SPI is defined as:

$$\mathrm{SPI} = A \cdot \tan(\beta) \tag{5.2}$$

A large number of other indices are proposed and discussed in the literature that use accumulated flow and relate to soil erosion (Moore and Burch, 1986) and landslide initiation (Montgomery and Dietrich, 1994). An overview and further discussion is provided by Moore et al. (1991a) and Wilson and Gallant (2000).

6. LAND-SURFACE OBJECTS BASED ON FLOW-VARIABLES

6.1 Drainage networks and channel attributes

One of the primary uses of the D8 method is the automated extraction of river network maps from raster DEMs. In addition to the map itself, a variety of attributes for each channel segment in a river network can be measured automatically. Figure 14 shows the *space-filling* drainage pattern that results from drawing a line segment between the centre of each grid cell and the neighbour grid cell that it flows towards, as determined by the D8 method. The drainage pattern is overlaid on an image which shows the locations of hills and valleys as resolved by the DEM.

Some grid cells are on hillslopes and some are in valleys. In order to create a map of the river network that drains this landscape, we need some method for *pruning* the dense drainage tree so that flow vectors on hillslopes are excluded. Many different pruning methods have been proposed, but no single method is best for all situations. A good pruning method should correctly identify the locations

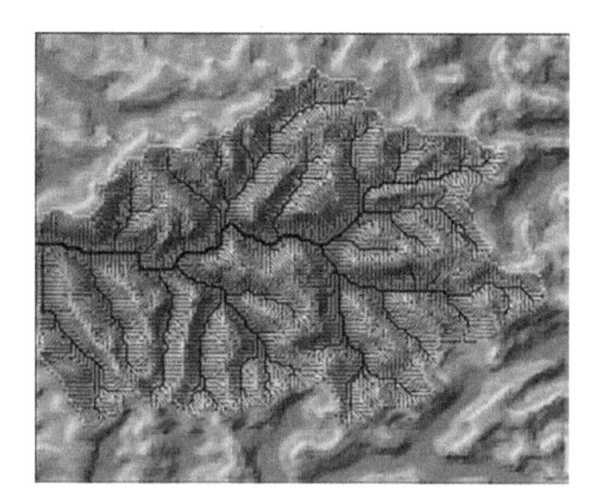

FIGURE 14 Complete drainage lines for one catchment. In the background, elevation is represented by colour. (See page 716 in Colour Plate Section at the back of the book.) © 2004 Rivix LLC, used with permission.

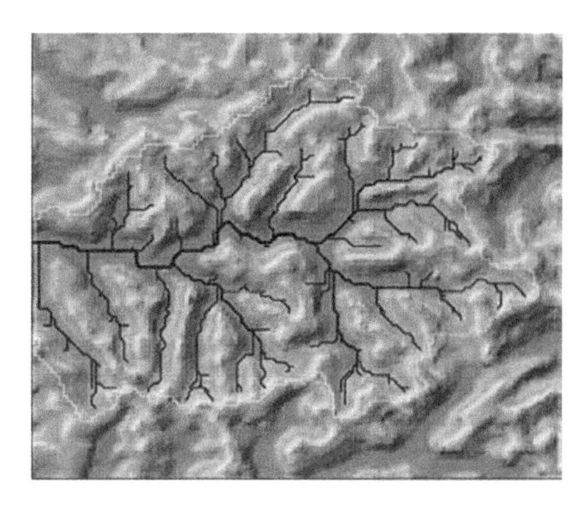

FIGURE 15 Drainage lines pruned by Horton–Strahler order. (See page 716 in Colour Plate Section at the back of the book.) © 2004 Rivix LLC, used with permission.

of channel sources as verified against a field survey. The most commonly-used pruning method is to first compute a grid of contributing areas (TCA) as explained in the previous section, and then remove the flow vector for any grid cell that has a TCA less than some specified threshold. A break in slope can often be identified in a scatter plot of slope versus area as explained by Tarboton et al. (1991) to identify this threshold. Sometimes, however, such a threshold is not apparent from the scatter plot.

Experience shows, however, that this simple method does not capture the natural variability that is present in real fluvial landscapes. The drainage density or degree of dissection is not spatially constant but varies with geology, elevation and other factors. A sometimes more robust method is to first create a grid of *Horton–Strahler order* for the dense drainage tree, and then remove flow vectors of grid cells that have orders less than some threshold value (Peckham, 1998), such as 3 (Figure 15).

REMARK 5. *Land-surface objects most commonly extracted from DEMs are: river networks, ridge lines, slope breaks and watershed boundaries. These can be further analysed for numerous attributes and properties including: relative position, distances, attached areas/volumes, or density.*

Unlike the TCA method, this method automatically *adapts* to the variability of the landscape. Horton–Strahler order cannot increase from order 1 to order 2 until a streamline intersects another streamline, which means that it provides a simple measure of flow convergence. So whether a hillslope happens to be long or short, this method more accurately identifies the toe of the slope. In general, any grid of values can be used together with a threshold to differentiate hillslopes from channels. However, the grid values must increase (or decrease) downstream along every streamline or a disconnected network will result. This is what happens when we attempt to use a TCA (or SCA) grid from the D-Infinity or Mass-Flux Methods. Grids computed as a function of both contributing area and slope have been proposed by Montgomery and Dietrich (1989, 1992) and others and appear to provide a process-based foundation for *source identification*.

Thresholds for network initiation work well in rugged terrain but produce spurious channels in flat areas (Tribe, 1992a). Once a pruning method has been applied to make a river network map, it is then possible to store the river network as an array of channel segments or links or Horton–Strahler streams, along with the network topology or connectedness and numerous attributes (Peckham, 1998).

Attributes can be computed for the channel segment itself, or for the basin that drains to its downstream end. Examples of attributes that can be computed and saved are: *upstream end pixel ID*, *downstream end pixel ID*, *stream order* (an integer-valued measure of stream hierarchy, Peckham and Gupta, 1999; Horton, 1932; Strahler, 1957), *contributing area* (above downstream end), *straight-line length*, *along-channel length*, *elevation drop*, *straight-line slope*, *along-channel slope*, *total length* (of all channels upstream), *Shreve magnitude* (total number of sources upstream of the pixel), *length of longest channel*, *relief*, *network diameter* (the maximum number of links between the pixel and any upstream source), *absolute sinuosity* (the ratio of the along-channel length and the straight-line length), *drainage density* (the ratio of the total length of drainage lines and the area drained by them, Horton, 1932; Tarboton et al., 1992; Dobos et al., 2005), *source density* (number of sources above the pixel divided by TCA), or *valley bottom flatness*.[6] Attributes for ensembles of sub-basins with the same Horton–Strahler order exhibit topological and statistical self-similarity. This property allows measurements at one scale to be extrapolated to other scales (Peckham, 1995a, 1995b; Peckham and Gupta, 1999).

6.2 Basin boundaries and attributes

D8 flow grids are also useful for extracting basin boundaries as polygons with associated attributes. Together, all of the grid cells that lie in the catchment of a given grid cell define a polygon. Numerous attributes, including its area, perimeter, diameter (the maximum distance between any two points on the boundary), mean elevation, mean slope and centroid coordinates can be computed. Many additional, flow-related attributes such as the maximum flow distance of any grid cell in the polygon to the outlet, or the total length of all channels within the polygon can also be computed.

The D8 method can also be used to partition a watershed into hydrologic subunits. Each subunit polygon represents the set of grid cells that contribute flow to a particular channel segment or reach. The set of subunit polygons fit together like puzzle pieces to completely cover the watershed. For exterior channel segments that terminate at sources, the polygons correspond to low-order sub-basins. For an interior channel segment, the polygon consists of two *wings*, one on each side of the segment, which often have a roughly triangular shape. Lumped hydrologic models can use these watershed subunits and their attributes to route flow through a watershed and compute *hydrographs* in response to storms. While

[6] An index computed as a multi-scale measure of flatness and lowness to identify depositional areas and valley bottoms (Gallant and Dowling, 2003).

lumped models are still in widespread use, spatially-distributed hydrologic models based on the D8 method (e.g. TopoFlow, Gridded Surface Subsurface Hydrologic Analysis — GSSHA) are starting to replace lumped models for many applications, and treat every grid cell as a control volume which conserves mass and momentum (see Chapter 25).

6.3 Flow distance, relief and longest channel length grids

D8 flow grids can be used to compute many other grid layers of hydrologic interest. One example is the along-channel *flow distance* from each grid cell to the edge of the DEM or to some other set of grid cells. A *relief grid* can also be defined, such that each grid cell is assigned a value as the difference between its own elevation and the highest elevation in the catchment that drains to it. Note that the relief of grid cells on drainage divides (peaks and ridges) is then simply zero. *Longest channel length* can also be computed as a grid layer, such that each grid cell is assigned a value as the length of the longest channel in the catchment that drains to it.

7. DEPOSITION FUNCTION

The concept of flow propagation is expanded by a deposition function to create a self-depleting flow that conserves mass between input and deposition in the *Mass Transport and Deposition* (MTD) algorithm (Gruber, 2007). This approach can be useful to model the re-distribution of eroded soil (Mitášová et al., 1996), the redistribution of snow by avalanches (Machguth et al., 2006) as well as other mass movements in steep topography (Chapter 23).

Similar concepts have also been applied to the delineation of lahar inundation zones (Iverson et al., 1998) and in the geomorphological model LAPSUS (Claessens et al., 2006; Schoorl et al., 2002). The key idea of the approach described here is that for each cell, a maximum deposition is pre-defined based on its slope (and possibly also other characteristics). During flow propagation, the flow though each cell is defined in a way similar to ordinary multiple flow direction methods and the local deposition is subtracted:

$$A_i = \sum_{j=0}^{8} (A_{\mathrm{NB}j} \cdot r_{\mathrm{NB}j}) + I_i - D_i \tag{7.1}$$

This means, that the flow passed though each cell A_i is equal to the sum of the flow received from its neighbours plus its own source term I_i, minus deposition D_i in this cell. Deposition D_i is limited by the amount of mass available $V_{\max}$ and the maximum deposition $D_{\max}$:

$$D_i = \min(D_{\max i}, V_{\max i}) \tag{7.2}$$

$$V_{\max i} = \sum_{j=0}^{8} (A_{\mathrm{NB}j} \cdot r_{\mathrm{NB}j}) + I_i \tag{7.3}$$

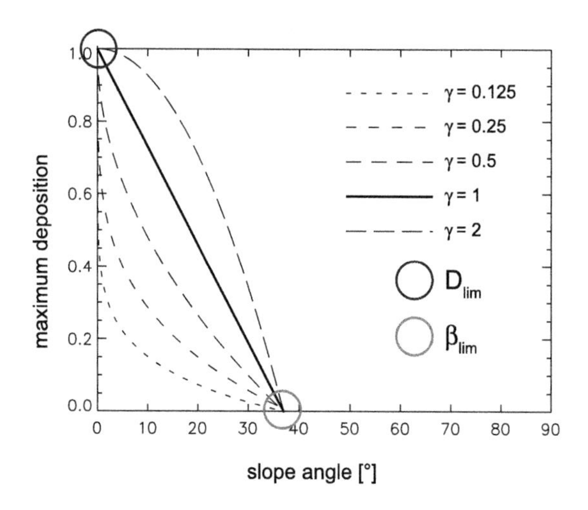

FIGURE 16 Maximum deposition as a function of slope.

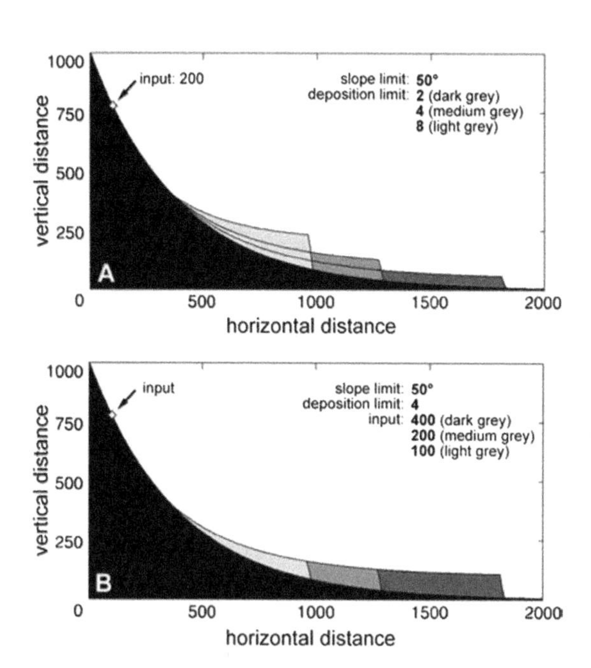

FIGURE 17 One-dimensional example of the influence that different deposition limits (A) and different amounts of mass input (B) have on the downslope deposition. Synthetic topography is black. Different deposits are shown in shades of grey. Reproduced from Gruber (2007) (see http://www.agu.org/pubs/copyright.html).

A generalised form of D_{max} can be described as a function of slope, e.g.:

$$D_{max} = \left[\left(1 - \left[\frac{\beta}{\beta_{lim}}\right]^{\gamma}\right) \geqslant 0\right] \cdot D_{lim} \tag{7.4}$$

where β_{lim} is the slope limit below which deposition can take place, γ is an exponent controlling the relative emphasis of steep and gentle slopes and D_{lim} is the deposition limit that describes the maximum possible deposition in horizontal areas (Figure 16).

The maximum deposition can also be made dependent on curvature, surface cover or altered manually — a reservoir or other safety structures for instance may be large sinks for debris flows. Important in this concept is the pre-definition of D_{max} for each cell. Figure 17 illustrates the influence of D_{max} and different amounts of mass input on the deposition pattern in a one-dimensional example. Both influence the runout distance of the mass movement. Chapter 23 provides further illustration of the use of this approach.

8. FLOW MODELLING USING TIN-BASED ELEVATION MODELS

The use of gridded DEMs dominates most applications in environmental science due to the relative ease of their processing and their widespread availability. However, the use of TIN data has several distinct advantages over gridded data for applications such as landscape evolution modelling, hydrologic modelling or the derivation of flow related-variables. The main advantages of TINs over gridded DEMs are: variable spatial resolution and thus dramatic reduction of the number of elements in most cases; suitability for adaptive resampling of dense topographic fields according to point selection criteria (Lee, 1991; Kumler, 1994; Vivoni et al., 2004) that optimise the topographic or hydrologic significance and the size of the data set; the suitability for dynamic re-discretisation (e.g. in response to landscape evolution and the lateral displacement of landforms); the effective drainage direction is not restricted to multiples 45° and grid-bias in the statistics of derived variables is absent or less pronounced; suitability to re-projection without data loss; and the possibility to constrain data sets by streams or basin boundaries precisely as needed.

These advantages come at the price of an increased complexity of data structures and algorithms that needs to be handled in the development of methods in a TIN framework. A number of hydrology-related algorithms (e.g., for flow routing, network extraction, handling of sinks) exist for TINs (Preusser, 1984; Palacios-Velez and Cuevas-Renaud, 1986; Gandoy-Bernasconi and Palacios-Velez, 1990; Jones et al., 1990; Nelson et al., 1994; Tachikawa et al., 1994; Tucker et al., 2001; Vivoni et al., 2005) and contour lines (Moore et al., 1988). While many of them route flow along the edges of triangles, Tucker et al. (2001) propose a method that uses Voronoi polygons to approximate effective contour width between two neighbouring nodes and this permits the solution of diffusion-like equations.

9. SUMMARY POINTS

Elevation dominates the movement of water and a multitude of associated phenomena at or close to the land surface. Because of the wide availability of DEMs, geomorphometric techniques are outstandingly powerful in the quantification, analysis, forecasting or parametrisation of phenomena related to the flow of water on the Earth's surface. However, the choice of methods depends on the task at hand (e.g., stream hydrology in large basins or geomorphology in steep headwaters) and on the data available. In this chapter we have given an introduction to the most important concepts in geomorphometry that relate to the flow of water.

The methods explained represent a selection of methods originating from a large and active research community. Most parameters described in this chapter can be computed using software packages such as SAGA GIS (Chapter 12), RiverTools (Chapter 18), TAS (Chapter 16), GRASS (Chapter 17) or ArcGIS (Chapter 11).

IMPORTANT SOURCES

Wilson, J.P., Gallant, J.C. (Eds.), 2000. Terrain Analysis: Principles and Applications. Wiley, New York, 303 pp.

Peckham, S.D., 1998. Efficient extraction of river networks and hydrologic measurements from digital elevation data. In: Barndorff-Nielsen, O.E., et al. (Eds.), Stochastic Methods in Hydrology: Rain, Landforms and Floods. World Scientific, Singapore, pp. 173–203.

Tarboton, D.G., 1997. A new method for the determination of flow directions and upslope areas in grid digital elevation models. Water Resources Research 33 (2), 309–319.

Quinn, P., Beven, K., Chevallier, P., Planchon, O., 1991. The prediction of hillslope paths for distributed hydrological modeling using digital terrain models. Hydrological Processes 5, 59–79.

Moore, I.D., Grayson, R.B., Ladson, A.R., 1991a. Digital terrain modeling: a review of hydrological, geomorphological, and biological applications. Hydrological Processes 5 (1), 3–30.

O'Callaghan, J.F., Mark, D.M., 1984. The extraction of drainage networks from digital elevation data. Computer Vision, Graphics, and Image Processing 28, 323–344.

CHAPTER 8

Land-Surface Parameters Specific to Topo-Climatology

J. Böhner and **O. Antonić**

how land surface influences climate and how we can use DEM to quantify this effect · land-surface parameters that affect direct, diffuse and reflected shortwave solar radiation · relation between land surface and longwave radiation patterns · integration of topographic effects on solar radiation · parameterising the thermal belt at slopes, thermal asymmetry of eastern and western slopes, and windward and leeward land-surface positions · modelling snow cover patterns using DEMs · estimating topographic exposure to wind

1. LAND SURFACE AND CLIMATE

Climate is usually defined as weather conditions averaged over a period of time, or, more precisely, the statistical description of relevant variables over periods from months to thousands or millions of years. *Climatology* is the study of climate. In contrast to *meteorology* (see Chapter 26), which studies short-term weather systems lasting up to a few weeks, climatology studies the frequency with which these weather systems occurred in the past. *Topo-climatology* is the part of climatology which deals with impacts of land surface (i.e. topography) on climate.

Land surface is widely recognised as a major control of the spatial differentiation of near-ground atmospheric processes and associated climatic variations. Advancements in all fields of climatic endeavour reveal a wide range of topographically induced or determined effects on atmospheric processes and climate, varying widely in terms of spatio-temporal scales and complexity. Particularly in weather forecasting, meteorologists commonly distinguish between different scales, referring to the characteristic horizontal extension of the phenomena to be observed and forecasted. Mid- to upper- troposphere planetary waves for example, the so-called Rossby waves, are assumed to be triggered by huge high mountain complexes such as the Rocky Mountains or the Tibetan Plateau and its bordering mountain ranges (Weischet, 1995; Böhner, 2006). With a typical wave

Developments in Soil Science, Volume 33
ISSN 0166-2481, DOI: 10.1016/S0166-2481(08)00008-1.

length of up to 10^4, Rossby waves are an example of orographic effects on the meteorological *macro α scale*.

The meteorological analysis of large- or *macro-scale* (>10^3 km) atmospheric motion systems such as planetary waves, high-pressure systems or trajectories of cyclones has been commonly referred to as synoptic meteorology, and that committed to *meso-scale* (10^1 to 10^3 km) processes and weather systems such as thunderstorms is referred to as *meso-meteorology*.[1] On the *macro β scale* (>10^3 km) to *meso β scale* (>10^2 km), the mass elevation effect with its typical uplift of vegetation belts due to enhanced heat surplus is a well-known effect in broad high mountain environments (Richards, 1981; Grubb, 1971).

The most marked variation of climatic pattern, however, is due to *boundary layer*[2] processes in topo-climatic scales with characteristic dimensions of not more than 10^1 km (*meso γ scale*) to minor 10^{-3} km (*micro γ scale*). Prominent examples are the influences of mountains and hills on the distribution pattern of precipitation, on the flow path of cold air and particularly the differential solar radiation income of sloping surfaces owing to varying aspects, slopes and horizon screening. These inter-relations between land-surface and topo-climatic variations are the main issues of this chapter.

REMARK 1. *Topo-climatology is the part of climatology which deals with impacts of land surface on climate. Land surface dominantly controls spatial differentiation of near-ground atmospheric processes and associated climatic variations.*

DEM-based land-surface parameters applicable as topo-climatic estimators (i.e. variables which can estimate spatial topo-climatic variability) can be divided into two logical groups. The first group comprises direct topo-climatic estimators which estimate real values of the particular topo-climatic variable (with exact units e.g. in °C, mm, $\mathrm{J\,cm^{-2}\,d^{-1}}$, etc.). Alternatively, the use of indirect topo-climatic estimators from the second group implies (on the basis of experience or logical consideration) that the particular estimator correlates with the examined topo-climatic variable. Testing this hypothesis, however, requires measurement data and a sufficient correlation between the estimator and the topo-climatic variable, in order to build an empirical model which converts an indirect estimator into a direct one. Land-surface parameters presented in this chapter are primarily grouped according to the main climatic variables, but with additional notation about belonging to the direct or indirect estimators.

Before discussing land-surface parameters in detail, in the following section, we present a brief overview of climate regionalisation approaches. We then introduce land-surface parameters relevant to assessing the short- and longwave radiation flux of the surface. The subsequent section deals with land-surface parameters, suitable to assessing the orographic effects on thermal conditions and cold air flow. Finally, the influences of the land surface on near-ground thermodynamics, on wind velocities and the closely related precipitation distribution are

[1] For further definitions of meteorological scales see Orlanski (1975) and Bendix (2004).

[2] The planetary boundary layer is the near surface layer of the atmosphere. It reaches up to about 2 km of height, depending on the orography.

discussed, emphasising the passive effects of terrain in particular. The more active terrain effects, such as orographically-triggered establishments of local circulation systems like slope breezes, valley and mountain wind systems, are more a subject of complex climate modelling approaches rather than a matter of geomorphometric analyses and are therefore discussed later in Chapter 26.

2. CLIMATE REGIONALISATION APPROACHES

Methods for spatially extensive, continuous estimations of *climatic variables* may generally be differentiated into: (1) interpolation techniques, (2) statistical regression analysis and (3) dynamic climate model-based approaches. Their order corresponds to their input data requirements, methodical complexity and computational demands.

For delineating spatial high-resolution climatic information from local observations, different interpolation techniques such as linear or inverse distance interpolations and geo-statistical kriging approaches form common and widely applicable GIS-routines (Hormann, 1981; Streit, 1981; Tveito et al., 2001). Currently, geostatistical *kriging interpolation* is favoured in climatologic applications as it includes additional statistical parameters such as the standard error of an estimated value for assessing the statistical precision of spatial estimates. Examples are discussed in Lloyd (2005) and Jarvis and Stuart (2001). As interpolation techniques only consider the coordinate variables of local observations, their application is limited to topographically simple regions with a more or less regular distribution of point source data. A major exception is the universal kriging approach (Goovaerts, 1997; Hengl et al., 2007a) which allows the integration of controlling land-surface parameters (indirect estimators) such as elevation, slope or aspect. Whilst it provides a powerful and suitable regionalisation strategy in high terrain, satisfactory results still require a more or less regular distribution of input data and a proper representation of topo-climatic settings.

> REMARK 2. *Climatic variables measured at climatic stations are most commonly mapped using kriging, universal kriging or splines.*

Regression analyses place fewer demands on input data distribution but have similar requirements in the representation of topo-climatic settings. The use of correlation (e.g. product-moment or canonical) and regression analyses aims to identify and quantify dependencies of spatial climatic variability from topographic variability (represented by indirect estimators). Commonly described as a statistical model, the regression equation serves as a transfer function (from indirect to direct estimator) for estimating a continuous climate surface dependent on topography. *Splining* is another deterministic spatial regression technique that locally fits a smooth mathematical function to point source data. For example, Fleming et al. (2000) used thin plate smoothing splines to estimate a baseline climatology for Alaska from sparse network observations.

Neural networks can also be used as the tool for development of empirical models (see e.g. Antonić et al., 2001b). In order to obtain a proper estimate of a continuous surface, regression approaches are often combined with interpolation techniques. In this case, local regression residuals are interpolated separately to obtain a correction layer that is added to the regression layer (Hormann, 1981; Antonić et al., 2001b).

Interpolation techniques and regression analyses are capable of delivering reliable continuous climate estimates in cases where proper point source data are available, however, both approaches limit the opportunities for constructing climatic scenarios to purely empirical temporal analogues (e.g. Rosenberg et al., 1993) only suitable for initial sensitivity studies (Carter et al., 1994; Von Storch, 1995; Gyalistras et al., 1997). Given the increasing need for case studies assessing possible future climate changes and their environmental and socio-economic implications, more advanced approaches integrate circulation variables from *General Circulation Model* (GCM) output, in order to enable an estimation of local to regional climate settings under climate change conditions.

Powerful, and frequently used approaches, in this context are the so-called *statistical downscaling*. The basic idea of statistical downscaling is to exploit the observed relationship between large-scale circulation modes (represented by GCM outputs) and local weather variations (observed at one or a set of meteorological stations). Using multivariate statistical analyses (e.g. product-moment or canonical correlation analyses) a set of suitable (optimally correlated), large-scale GCM variables is identified to obtain an empirical functions (e.g. a regression equations) which can predict the local weather variations of interest, depending on the controlling large-scale variations (Von Storch, 1995).

Although statistical downscaling is capable of connecting the simulation of regional weather variations directly with the physically consistent output of GCMs, a rather general criticism of this bottom–up modelling approach is due to its empirical character. More sophisticated dynamical downscaling approaches, instead, are commonly considered to be superior to pure statistical downscaling, in terms of physical consistency. These top–down modelling approaches are based on *Limited Area Models* (LAM), a physically based regional model type, nested in a coarse resolution GCM. Examples of these modelling approaches are discussed in Chapter 26.

3. TOPOGRAPHIC RADIATION

The *surface net radiation* and its components, the net shortwave radiation and the net longwave radiation are key factors in the climatology of the Earth. The fluxes of shortwave and longwave radiation predominately control the surface energy and water balance and thus affect the whole range of atmospheric dynamics in the boundary layer as well as most biophysical and hydrological processes at or near the Earth's surface. In its simplest form, the net radiation at the surface R_n is given by:

$$R_n = S_n + L_n \tag{3.1}$$

where S_n is the net shortwave radiation and L_n is the net longwave radiation.

There are three major causes of spatial variability of radiation at the land surface: (1) orientation of the Earth relative to the sun, (2) clouds and other atmospheric inhomogeneities and (3) topography. The first cause influences latitudinal gradient and seasons. The second cause is associated with local weather and climate. The third cause — such as spatial variability in elevation, slope, aspect and shadowing — can create very strong local gradients in solar radiation.

REMARK 3. *Calculation of net shortwave topographic solar radiation includes: (1) estimation of direct and diffuse component of total net shortwave solar radiation incoming at the unobstructed horizontal surface and (2) calculation of all effects caused by topography of this surface, specific for particular component.*

Although the importance of topographic effects on solar radiation has long been recognised, incorporation of these effects in the irradiance models was either neglected or simplified (e.g. Brock, 1981; Vardavas, 1987; Nikolov and Zeller, 1992), due to the complexity of formulation and the lack of suitable modelling tools. A decade ago, advances in DEM-based modelling together with analysis of remotely sensed data made it possible to include topographic effects in the *solar radiation models* at fine spatial scales over arbitrary periods of time (Dubayah and Rich, 1995).

In this section, we first describe topographic effects on solar radiation. Related land-surface parameters are only relative estimators which have to be weighted by the real solar radiation flux depending on local and seasonal climate peculiarities, which is discussed in a subsequent section.

3.1 Topographic exposure to radiation flux

The most important and probably the most relevant component for environmental applications in Equation (3.1), the *net shortwave radiation* S_n covers wavelengths from approximately 0.3 to 3.0 μm (shortwave to near infrared), and it can be expressed as:

$$S_n = S_s + S_h + S_t - S_r = (S_s + S_d + S_t) \cdot (1 - r) \tag{3.2}$$

Equation (3.2) comprises two alternative expressions for the total shortwave radiation. The first expression (left term) means that S_n at the given point is the sum of direct solar radiation received from sun disk (S_s), diffuse solar radiation received from the sky's hemisphere (S_h) and radiation received by reflection of surrounding land surface (S_t), decreased for radiation reflected off from the surface (S_r). An alternative and more frequently used expression [right term in Equation (3.2)] simply reduces the total shortwave radiation to the absorbed (not reflected) fraction, where r denotes surface reflectance factor (or the *surface albedo*). Reasonable reflectance r factors are widely available for numerous natural surfaces as tabulated standard values (Oke, 1988), or may be directly obtained from spatial extended remotely sensed datasets (for instance Landsat, SPOT, IRS).

Topographical effects on direct, diffuse and reflected radiation are not the same (see Figure 1), and therefore these effects have to be modelled separately for each

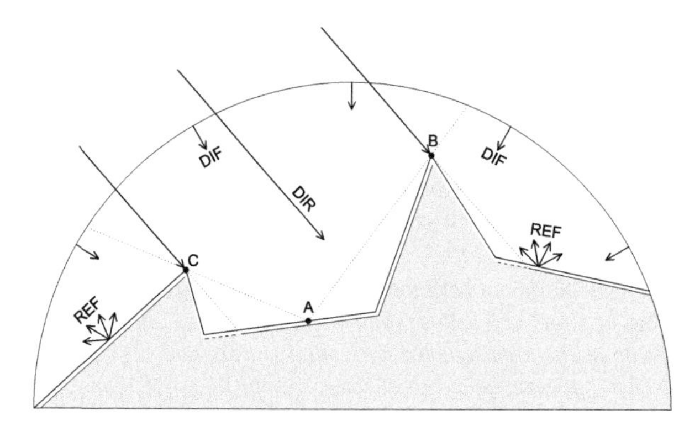

FIGURE 1 Schematic presentation of components of solar radiation: direct radiation from sun disk (DIR), diffuse radiation from sky hemisphere (DIF) and reflected radiation (REF). Bold line represents land surface, which is underlined by solid line where land surface is directly illuminated (i.e. receives direct solar radiation), and by hatched line where land surface is in a cast shadow. Absence of underline indicates self-shadowing. For the point A, the part of the visible sky hemisphere is controlled by points C and B. For the point B, the entire sky hemisphere is visible.

component. In other words, if we assume that S_n in Equation (3.2) relates to the ideal horizontal surface unobstructed by surrounding land surface (in which case S_t is obviously equal to zero), then net shortwave solar radiation S_n^* on the real land surface (which is not plain) can be expressed as:

$$S_n^* = \left(S_s^* + S_h^* + S_t\right) \cdot (1 - r) \tag{3.3}$$

where S_s^* and S_h^* are direct and diffuse solar radiation modified by topography, respectively.

For modelling of topographic effects on direct radiation over a year, sun elevation and azimuth (Figure 2) have to be calculated for each grid node in a DEM (usually hourly) using the following algorithms (Klein, 1977; Keith and Kreider, 1978):

$$\sin\theta = \cos\lambda \cdot \cos\delta \cdot \cos\varpi + \sin\lambda \cdot \sin\delta \tag{3.4}$$

$$\cos\phi = \frac{\cos\delta \cdot \cos\varpi - \sin\theta \cdot \cos\lambda}{\sin\lambda \cdot \cos\theta} \tag{3.5}$$

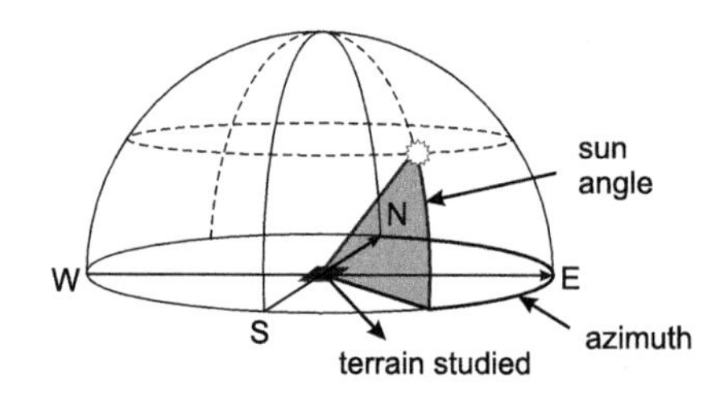

FIGURE 2 Direct solar radiation geometries.

$$\delta = 23.45 \cdot \sin\left(\frac{360^\circ \cdot [284 + J]}{365}\right) \tag{3.6}$$

$$\varpi = 15^\circ \cdot (12 - t) \tag{3.7}$$

where θ is the *sun elevation angle*,[3] ϕ is *sun azimuth*, λ is the *latitude*, δ is the *solar declination angle*, J is *Julian day number*, ϖ is the hour angle in degrees and the value $12 - t$ is equal to the distance of the given mid-hour from the true solar noon (0.5, 1.5, 2.5 h, etc.).

The angle between a plane orthogonal to sun's rays and terrain (*solar illumination angle*) has to be determined for each particular hour from:

$$\cos\gamma = \cos\beta \cdot \sin\theta + \sin\beta \cdot \cos\theta \cdot \cos(\phi - \alpha) \tag{3.8}$$

where β and α are surface slope and aspect calculated from DEM, respectively, γ is solar illumination angle for given surface (defined by β and α) and for a given sun position on the sky (defined by θ and ϕ). As long as $\sin\theta$ is >0, the point (i.e. cell in DEM) is directly illuminated, otherwise self-shadowing of land surface takes a place (see Figure 1). In addition to *self-shadowing*, the point can be also shadowed (i.e. without direct solar radiation) by shadow cast by neighbouring land surface (Figure 1). Determination of *cast-shadowing* is based on comparison of solar elevation angle and horizon angle in the solar azimuth, resulting in a binary mask (shadow/non-shadow) for each point and for each unit of daily time integration.

> REMARK 4. *Cosine of the solar illumination angle is the hourly topographic correction for direct radiation and can be used as estimator of direct radiation received at the surface at the given moment.*

If the horizon angle is greater than the solar elevation angle, the point is in shadow and its $\cos\gamma$ value has to be set to zero, regardless of β and α at this point. Horizon angle φ for any given point in DEM (with the elevation z) is defined as the maximum angle toward any other point in a given azimuth, within a selected search distance (see Figures 3 and 5), determined by:

$$\varphi = \arctan\left(\frac{\Delta z}{d}\right)_{\max} \tag{3.9}$$

where d is the distance to the point with higher elevation $z + \Delta z$ ($d \leqslant$ search distance). Figure 4 illustrates the effects of cast-shadowing.

Cosine of the solar illumination angle expressed by Equation (3.8) (after settings to zero for $\sin\theta \leqslant 0$ as well as for cast-shadowing) determines the distribution of unknown incoming direct radiation flux over a given surface at a given moment (i.e. unit of daily integration, usually hourly), and varies between 0 (shadow, i.e. without direct radiation) and 1 (land surface orthogonal to sun's rays). It can be also understood as *hourly topographic correction for direct radiation*, and can be used as an indirect estimator of direct radiation received at the surface (e.g. for some characteristic and interpretable moment such as winter/summer solstice at noon, or for specific purpose of satellite data topographic correction). In general, it can

[3] The elevation angle of the sun over the horizon; *solar inclination angle* is also widely used synonym.

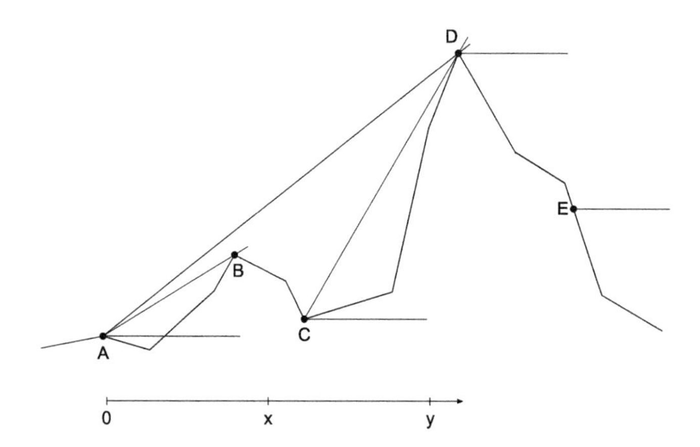

FIGURE 3 Some possible relationships between search distance and horizon angle. Bold line represents land surface, axis represents search direction and distance. For the point A under search distance x the critical point for horizon angle determination is B, and under distance y the critical point is D. For point C the critical point is D under both distances. For points D and E the horizon angle is set to zero.

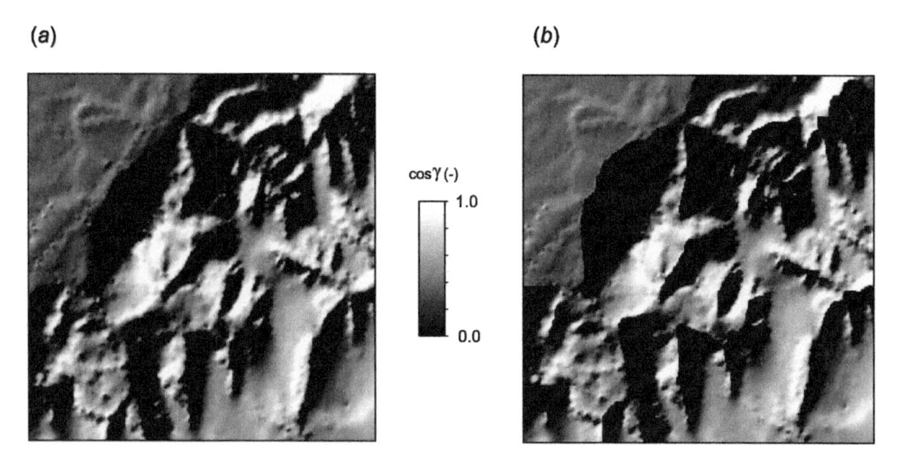

FIGURE 4 Cosine of the solar illumination angle for Baranja Hill area, under sun elevation and azimuth of 9 and 135° (SE), respectively: (a) — cast-shadowing ignored; (b) — cast-shadowing included.

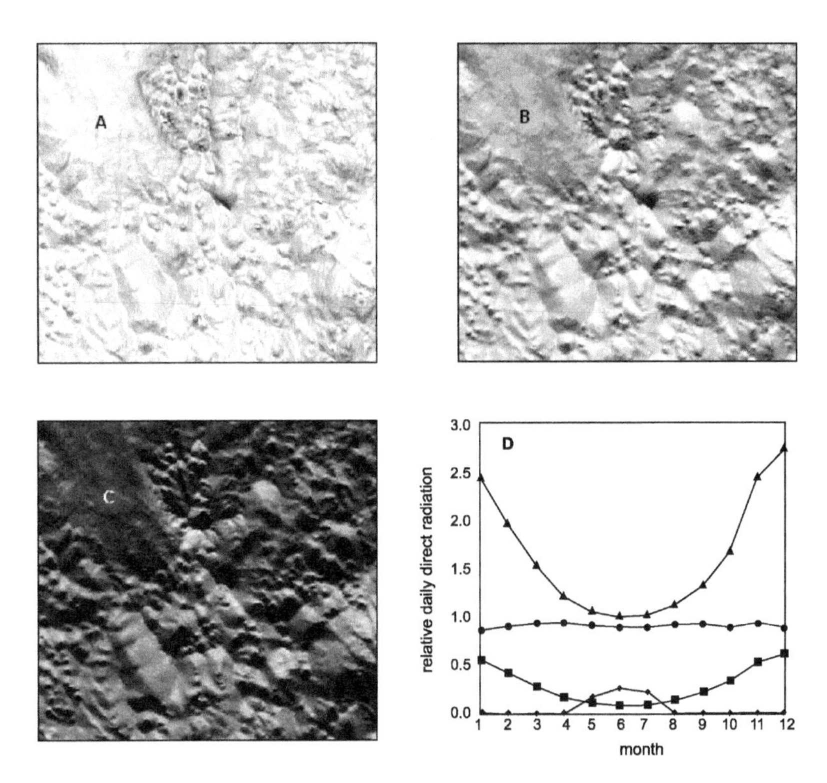

FIGURE 5 Spatial and annual distribution of monthly averaged topographic daily direct radiation relative to daily direct radiation on the unobstructed horizontal surface for the part of National park Risnjak, Croatia (~20 km^2, spatial resolution of 10×10 m, based on topography in a scale of 1:5000). A — June, B — September, C — December (values are stretched in the gray scale from minimum — black to maximum — white; see minimum and maximum values on D). D — basic statistics for all months (• — mean, ■ — standard deviation, ♦ — minimum, ▲ — maximum). Reprinted from Antonić (1998). With permission from Elsevier.

be stated that:

$$S^*_{S(h)} = \varsigma \cdot \frac{S_{S(h)}}{\sin\theta} \cdot \cos\gamma \tag{3.10}$$

where $S^*_{S(h)}$ represents hourly topographic direct radiation to the real land surface, ς denotes binary mask (shadow = 0, non-shadow = 1), and $S_{S(h)}$ denotes hourly direct radiation to the unobstructed horizontal surface. Division by $\sin\theta$ represents a recalculation from a horizontal surface to a surface orthogonal to the sun ray's.

Values of $\cos\gamma$ change during the day for each point in a DEM with exception of points in permanent shadow where they are equal to zero, according to the movement of the sun over the sky. Consequently, estimation of daily topographic direct radiation is an iterative procedure: the self-shadowing and shadows cast by surrounding land surface needs to be calculated from DEM for each unit of daily integration, following sun position on the sky (see also e.g. Dubayah and

Rich, 1995, or Antonić, 1998). Due to the fact that topographic effect on the direct component ($\cos\gamma$) is different for each daily integration unit, it has to be weighted (i.e. multiplied) during the daily integration by the amounts of direct radiation flux for respective daily integration unit:

$$S^*_{S(d)} = \sum_{i=1}^{n} S^*_{S(h)i} = \sum_{i=1}^{n} \varsigma_i \cdot \frac{S_{S(h)i}}{\sin\theta_i} \cdot \cos\gamma_i \tag{3.11}$$

where $S^*_{S(d)}$ represents daily topographic direct radiation to the real land surface defined by β and α, i denotes a particular hour and n denotes the number of hours during the day. Equation (3.11) clearly shows that $S^*_{S(d)}$ can be calculated as a direct estimator only if values of $S_{S(h)}$ for each particular hour during the day are known. However, Antonić (1998) showed how to produce a monthly averaged daily integration without requiring $S_{S(h)}$ data. For this purpose Equation (3.11) has to be expressed as:

$$S^*_{S(d)} = S_{S(d)} \cdot K_{S(d)} \tag{3.12}$$

$$K_{s(d)} = \sum_{i=1}^{n} \varsigma_i \cdot k_i \frac{\cos\gamma_i}{\sin\theta_i} \tag{3.13}$$

$$k_i = \frac{S_{S(h)_i}}{S_{S(d)}} \tag{3.14}$$

where $S_{S(d)}$ represents daily direct solar radiation to the ideal horizontal surface unobstructed by surrounding land surface, $K_{S(d)}$ can be understood as *daily topographic correction for direct radiation* (i.e. as cumulative topographic effect during the day), and k is the portion of $S_{S(h)}$ in $S_{S(d)}$ for each particular hour during the day. Antonić et al. (2000) presented a highly accurate empirical model, which estimates k (defined as the ratio between monthly mean hourly and monthly mean daily radiation) as a function of latitude, actual sun elevation angle (θ) and maximum sun elevation angle (θ_{max}) for the 15th day of the given month (at solar noon, which means that $t = 0$):

$$\begin{aligned} k = b_0 &+ \frac{b_1 \cdot \theta^3}{\theta_{max}} + \frac{(b_2 \cdot \theta + b_3 \cdot \theta^2 + b_4 \cdot \theta^3)}{\theta^2_{max}} \\ &+ \lambda \cdot \left[\frac{(b_5 \cdot \theta + b_6 \cdot \theta^2)}{\theta_{max}} + \frac{(b_7 \cdot \theta^2 + b_8 \cdot \theta^3)}{\theta^2_{max}} \right] \end{aligned} \tag{3.15}$$

where b_k are empirical parameters, derived using data measured at a number of pyranometric stations from the northern hemisphere (situated at $0° < \lambda < 70°$):

$$\begin{aligned} &b_0 = 0.321419, \quad b_1 = 0.005221, \quad b_2 = 53.902664, \\ &b_3 = 45.420267, \quad b_4 = -8.817633, \quad b_5 = -0.077503, \\ &b_6 = 0.001064, \quad b_7 = -0.252135, \quad b_8 = 0.002904 \end{aligned} \tag{3.16}$$

Testing of this model on independent data (including one station from the southern hemisphere) suggests its applicability worldwide (with the possible exception of polar zones).

It is clear that Equation (3.15) is meaningful only in the domain where actual θ is less than or equal to the respective θ_{max}. It has to be also noted that Equation (3.15) is applicable not only to specific mid-hour and average day of a given month, but also to any hour angle and any Julian day in the sense of a moving average of the empirically obtained values.

In the approach presented here, k is integrated over the day instead of $S_{S(h)}$ following the sun over the sky (under different topographic conditions), yielding the spatial distribution of $K_{s(d)}$, as radiation values relative to (i.e. multiplicators of) the unknown daily total of direct radiation. In cases when $S_{s(d)}$ is unknown, $K_{s(d)}$ can be used as an indirect estimator of the spatial distribution of monthly topographic direct solar radiation (for the area of interest).

Figure 5 illustrates the spatial and annual distribution of $K_{s(d)}$ on a part of the Croatian Karst (area of $\approx$20 km^2), showing that spatially averaged $K_{s(d)}$ is nearly constant over the whole year, maximum values increase towards the winter as well as total spatial variability, while minimum values are zero (some points are in permanent shadow over the entire day), except for the summer, when all points receive radiation. This shows that cumulative daily topographic effect on direct radiation can vary strongly during the year.

REMARK 5. *Sky view factor is an adjustment factor that is used to account for obstruction of overlying sky hemisphere by surrounding land surface.*

For modelling of topographic effects on *diffuse radiation*, the *sky view factor* (Ψ_s) has to be calculated for every point, in order to estimate an obstruction of overlying sky hemisphere by surrounding land surface (by a slope itself or by adjacent topography). This calculation is based on horizon angles (φ) in different azimuth directions (Φ) of the full circle, around each point in a DEM, following the expression [based on Dozier and Frew, 1990, but adapted according to the definition of φ given in Equation (3.9)]:

$$\Psi_S = \frac{1}{2\cdot\pi}\int_0^{2\pi}\big[\cos\beta\cdot\cos^2\varphi + \sin\beta\cdot\cos(\Phi-\alpha)\cdot(90-\varphi-\sin\varphi\cdot\cos\varphi)\big]\,\mathrm{d}\Phi \tag{3.17}$$

In practice, some azimuthal step (i.e. each 30°) is usually used:

$$\Psi_S = \frac{1}{N}\cdot\sum_{i=1}^{N}\big[\cos\beta\cdot\cos^2\varphi_i + \sin\beta\cdot\cos(\Phi_i-\alpha)\cdot(90-\varphi_i-\sin\varphi_i\cdot\cos\varphi_i)\big] \tag{3.18}$$

where N is the number of directions used to represent the full unit circle and φ_i is horizon angle in ith direction. Sky view factor varies from 1 for completely unobstructed land surface (horizontal surface or peaks and ridges) to 0 for completely obstructed land surface (only theoretical case). It is clear that the precision of the sky view factor calculation depends mostly on the number of directions

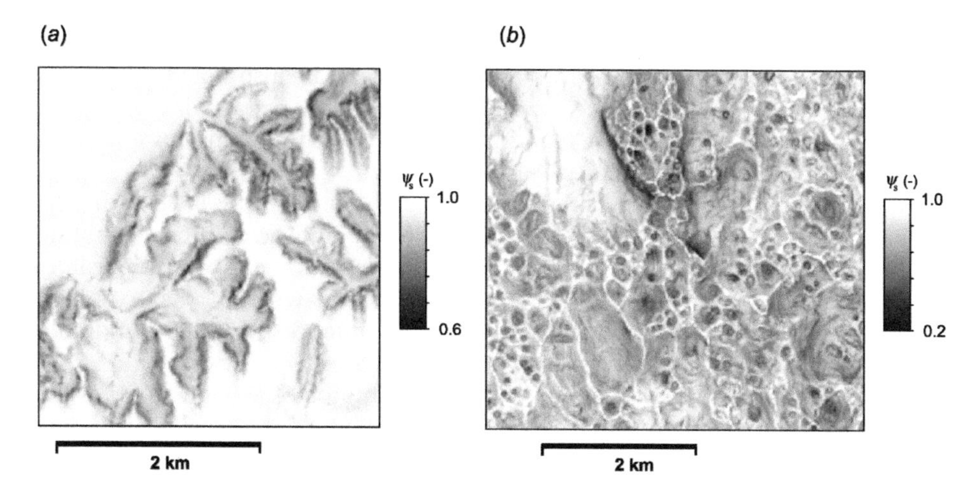

FIGURE 6 Spatial distribution of sky view factor for two distinct areas: (a) Baranja Hill area; (b) the part of National park Risnjak, Croatia (from Figure 5). In the Karst areas, due to a very dissected and irregular topography, a sky view factor of less than 0.3 (less then 30% of sky hemisphere is visible from the given point) can be observed.

used, but, conversely increasing the number of directions (and/or search distance) rapidly increases computational time. A general recommendation could be that more rugged land surface requires a denser sample of directions, but a smaller search distance (see also Figure 3). In undulating orography, a suitable simplification for calculation of Ψ_S is (Oke, 1988):

$$\Psi_S \approx \frac{1 + \cos\beta}{2} \tag{3.19}$$

Figure 6 shows the spatial distribution of Ψ_S [calculated by Equation (3.18)] for two areas with significantly different topography. Estimation of topographic effects on diffuse radiation usually assumes an isotropic sky, which means that each part of the sky has a hypothetically the same contribution to the total diffuse radiation. Under this assumption, the influence of topography on diffuse radiation can be expressed (for any chosen time unit) as:

$$S_h^* = S_h \cdot \Psi_S \tag{3.20}$$

In cases when S_h is unknown, Ψ_S can be used as an indirect estimator for the spatial distribution of diffuse solar radiation (for the area of interest). However, it has to be emphasised that the sky is not isotropic in general (for instance, the sky is often brighter near the horizon and near the sun). The consequence of an anisotropic sky is that accounting for topographic effects can not neglect which part of the sky is obstructed by land surface, and which is not (for a possible solution in this case see e.g. approach of Rich et al., 1994).

The surface radiation received by reflection from surrounding land surface is primarily influenced by the portion of the overlying hemisphere obstructed by surrounding land surface. Under an assumption of isotropy of surrounding terrain (which can rarely be expected to be realistic), the respective *terrain view factor* Ψ_t can be approximatively described by (Dozier and Frew, 1990):

$$\Psi_t \approx \frac{1+\cos\beta}{2} - \Psi_S \tag{3.21}$$

Anisotropy can be theoretically accounted for determining the geometric relationships between each particular point and all related points of surrounding land surface, but this is complex, and may not be worth the extra computation, due to the usually minor contribution of S_t in S_n^* (in comparison to contributions of S_s^* and S_h^*). Consequently, daily radiation received by reflection from surrounding land surface (S_t) can be adequately estimated for a chosen time unit by:

$$S_t \approx \Psi_t \cdot \left(S_{s(\text{avg})}^* + S_{h(\text{avg})}^*\right) \cdot r_0 \tag{3.22}$$

where $S_{s(\text{avg})}^*$ and $S_{h(\text{avg})}^*$ are direct and diffuse radiation for the same time unit, respectively, spatially averaged over the surrounding land surface visible from a given point, and r_0 is the spatially averaged reflectance (albedo factor) of the surrounding land surface. This calculation of S_t thus required identification of the surrounding visible land surface of each grid cell. However, an areal average of S_s and S_h for terrain with an elevation $>z$ (averaged for each grid cell with elevation z) may represent a sufficient and computationally efficient alternative.

Calculation of *net longwave radiation* L_n^* on the real surface of complex land surface takes into account previously introduced land-surface parameters:

$$L_n^* = L_n \cdot \Psi_S + L_{(\text{avg})} \cdot \Psi_t \tag{3.23}$$

The first term of Equation (3.23) integrates the sky view factor Ψ_S, in order to reduce net longwave radiation L_n (related to the surface completely unobstructed by topography) to the fraction unobstructed by real land surface. The second term estimates the longwave radiation emitted from the surrounding land surface towards the surface under consideration (L_t), as a function of terrain view factor Ψ_t and spatially averaged longwave radiation $L_{(\text{avg})}$ from the neighbouring visible surface.

3.2 Radiation at the unobstructed horizontal surface

The shortwave radiation components S_s and S_h are typically point source observations, mostly available from the regular meteorological station network, and thus require either physically based or empirical regionalisation strategies to obtain spatially extensive estimations of the *total incoming shortwave radiation*. Given the significant impact of the shortwave irradiance on the distribution pattern and growth characteristics of the vegetation in natural and managed ecosystems, the design and development of methods for the spatial prediction of shortwave irradiation has been subject to considerable modelling effort. Despite remarkable advancements in model development, however, deterministic radiation models have very diverse needs for necessary data input. Even under clear sky conditions, a proper estimation of direct insolation requires information on the vertical structure of the atmosphere and its chemical composition in different layers (Kyle, 1991).

If we simply assume the atmosphere to be homogeneous in terms of its vertical chemical composition, the direct shortwave solar radiation S_S on a horizontal surface at elevation z is given by:

$$S_S = \sin\theta \cdot S_C \cdot \tau \tag{3.24}$$

$$\tau = e^{-\frac{\tau_z}{\sin\theta}} \tag{3.25}$$

$$\tau_z = b \cdot \int_z^{\infty} \ell \cdot \Delta z \tag{3.26}$$

where S_C is the (exo-)atmospheric radiation (normally the solar constant), ℓ is the air density integrated over distance Δz from top of atmosphere to the elevation z. This model uses an atmosphere mass parametrisation approach according to Bouguer–Lamberts law (Malberg, 1994) to approximate the transmittance of atmosphere τ [Equation (3.25)] by an empirical estimation of its optical depth τ_z [Equation (3.26)]. The strength of atmospheric extinction is represented by the coefficient b, which, if not approximated by a radiative transfer model (Medor and Weaver, 1980; Kneizys et al., 1988; Dubayah, 1991), may be estimated by an empirical function of water vapour or precipitable water and calibrated using reference radiation data (Böhner and Pörtge, 1997; Böhner, 2006).

The direct calculation of τ in Equation (3.24) on the base of available pyranometer data is a frequently used option. However, the integration of Equation (3.26) in Equation (3.25) ensures the correct physical calculation of the effects of changing altitudes on direct solar radiation, such as the well-known phenomenon of significantly increasing amounts of direct solar radiation in high mountain environments (Böhner, 2006).

REMARK 6. *Assuming clear-sky conditions, the direct shortwave solar radiation can be estimated using only a DEM.*

Elevation is, similarly to direct solar radiation, closely correlated with the amount of diffuse solar radiation S_h. The diffuse fraction of the total solar irradiation distinctly increases with decreasing altitudes due to rising contents of aerosol particles, small water droplets and water vapour molecules in the lowest troposphere layers, scattering the solar radiation. The diffuse shortwave radiation (or *diffuse sky light*) again can be obtained either from modelling applications of previously cited radiative transfer models or estimated using empirical approaches. In its simplest form, the diffuse solar radiation income S_h on a horizontal surface at altitude z under clear sky conditions can be estimated by:

$$S_h = 0.5 \cdot \sin\theta \cdot S_C \cdot c \cdot (1 - \tau) \tag{3.27}$$

where the factor 0.5 is used to reduce the total attenuated radiation to its downward flux component (received at the surface from the overlaying celestial hemisphere), and the empirical[4] coefficient $c < 1$, again, has to be calibrated on the

[4] The coefficient c considers the loss of absorbed exo-atmospheric solar energy when passing the atmosphere.

base of available pyranometer measurements of the diffuse irradiance. More detailed physically based formulations for diffuse radiation can be found in Gates (2003) and Perez et al. (1987).

The sum of S_s and S_h on a horizontal surface, the so-called *global radiation* S is an important climate factor, often required for many applications (e.g. for the calculation of potential evapotranspiration rates according to the FAO–Penman–Montieth equation). Nikolov and Zeller (1992) described an empirical model for estimation of average monthly global radiation at an unobstructed horizontal surface S and its diffuse component S_h as a function of latitude, elevation and average monthly data for ambient temperature, relative humidity and total precipitation. This approach has been tested for global radiation against average monthly data from 69 meteorological stations throughout the northern hemisphere, including different climatic zones. Test results demonstrated a high accuracy of the model in describing seasonal patterns of solar radiation for each included station from subpolar regions to tropics.

The net longwave radiation L_n effectively falls within the infrared wavelength of 3–300 μm. The main components of the net longwave radiation, the total incoming longwave radiation L_a, and the upward longwave flux L_s can be estimated using (Marks and Dozier, 1979):

$$L_n = L_a - L_s \tag{3.28}$$

$$L_s = \sigma \cdot T_s^4 \tag{3.29}$$

$$L_a = 1.24 \cdot \left(\frac{e}{T_l}\right)^{\frac{1}{7}} \cdot \left(\frac{P_z}{P_0}\right) \cdot \sigma \cdot T_l^4 \tag{3.30}$$

where σ is the Stephan–Bolzmann constant,[5] T_s and T_l are surface and air (screen) temperatures (K), P_z and P_0 are air pressures at altitude z and sea level (hPa) and e is the water vapour (hPa). According to the Stephan–Bolzmann law, the upward longwave flux increases with the fourth power of the absolute surface temperature and thus depends considerably on the nature of the surface.

In Equation (3.28) L_n is simplified, i.e. expressed as the difference between the total incoming longwave radiation L_a emitted from clouds, atmospheric dust, and some gaseous atmospheric constituents (particularly water vapour and carbon dioxide) and the upward longwave flux L_s, emitted from the surface according to its temperature. Note also that, since most natural surfaces absorb nearly all incoming longwave radiation (just like *black bodies*), the small part of L_a reflected by natural surfaces is usually neglected in the longwave radiation balance.

Since all climate variables that indicate or affect the components of the net longwave radiation are closely correlated with altitude, elevation again has to be assessed as an important control on the longwave radiation. For more detailed discussions of relevant atmospheric processes and particularly the role of clouds, please refer to Deacon (1969), Kyle (1991), Häckel (1999), Bendix (2004).

[5] $\sigma = 5.6693 \cdot 10^{-8}\ \mathrm{W\,m^{-2}\,K^{-4}}$.

3.3 Final remarks about modelling topographic radiation

It is clear from the preceding sections that direct estimation of total net shortwave topographic solar radiation (S_n^*) needs to include: (1) estimations of direct (S_s) and diffuse component (S_h) of total net shortwave solar radiation incoming at the land surface (S_n) and (2) calculation of all effects caused by the topography of this surface, for each particular component (as described in Section 3.1).

The first is dominantly influenced by local/regional weather (in the case of calculation for an exact moment) or by local/regional climate (in a case of calculation for average conditions), and can be obtained by use of: (1) site-specific scattering and absorbing properties of the atmosphere and related physically based formulations, (2) site-specific pyranometric measurements, (3) empirical estimates in terms of site-specific climatological variables (e.g. as in the above mentioned approach of Nikolov and Zeller, 1992) or (4) satellite data for the area of interest (see e.g. Gautier and Landsfeld, 1997).

Site-specific data of atmospheric properties, as well as pyranometric measurements are not often available, and they are usually limited in spatio-temporal coverage. In cases where they are available and also sufficient for describing local/regional S_s and S_h fields, use of these actual data can be recommended as the most precise solution for direct estimation of topographic solar radiation and its components.

Climatological variables such as air temperature, precipitation, relative humidity and/or cloudiness are usually more available, but due to the fact that they are also usually limited in spatial coverage, use of these is more appropriate for calculations under averaged (e.g. monthly mean) conditions. For instance, a combination of the Nikolov–Zeller approach with the previously described approach of Antonić (1998) and Antonić et al. (2000) is probably the best solution for direct estimation of topographic solar radiation in cases without any site-specific radiation data, which still results in the environmentally most relevant monthly mean daily values affected by the local/regional climate.

In cases where fine spatial as well as temporal resolution is required (e.g. calculations for specific hours and/or days on large areas with complex topography), an ultimate solution is to use satellite data for estimation of incoming solar radiation. Dubayah and Loechel (1997) demonstrated this possibility, combining the coarse spatial resolution data of *Geostationary Satellite Server imagery*[6] (http://www.goes.noaa.gov) with the fine spatial resolution DEM-based topography, where direct-diffuse partitioning was performed by algorithm of Erbs et al. (1982), elevation correction by formulations of Dubayah and van Katwijk (1992), and topographic correction by use of land-surface parameters presented in Section 3.1.

When none of site-specific atmospheric properties, pyranometric data, suitable climatological variables or if appropriate satellite data are not available, direct estimation of real topographic solar radiation can not be performed. In such cases, a calculation under potential solar radiation conditions could probably be used

[6] Used for estimation of surface solar radiation flux by the method of Gautier and Landsfeld (1997), and spatially averaged in a 50×50 km^2 window.

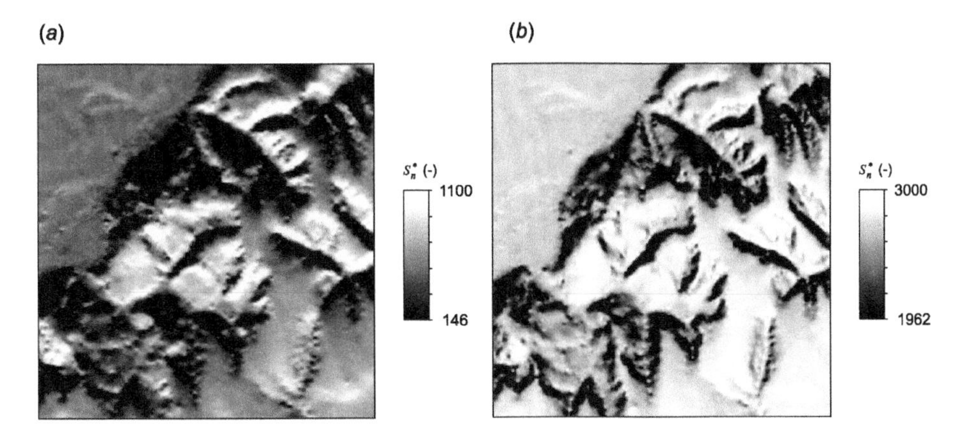

FIGURE 7 Potential net shortwave topographic solar radiation (J cm^{-2} day^{-1}) under clear-sky conditions for Baranja Hill area: (a) — winter solstice; (b) — summer solstice.

instead (see Figure 7 as an example for Baranja Hill, assuming uniform albedo of 0.1 and clear sky conditions), or, probably even better for environmental applications, particular land-surface parameters (such as $K_{s(d)}$, Ψ_S and Ψ_t) can be applied as separate indirect estimators, i.e. inputs to the regression analysis where the contribution of the particular land-surface parameter (as independent variable) to the explanation of the examined spatial variability (of some dependent variable, such as e.g. vegetation or snowmelt pattern) will be obtained *a posteriori*.

Regarding the net *topographic longwave radiation*, probably the most crucial factors in Equations (3.28) and (3.23) are the spatially averaged, longwave radiation $L_{(\text{avg})}$ from the neighbouring visible surface and the outgoing longwave surface radiation L_S. Proper surface temperature values are required in order to estimate longwave fluxes according the Stephan–Bolzmann law given in Equation (3.29). Since spatially extensive, remotely-sensed data (e.g. Landsat) only enable a precise estimation of surface longwave radiation values for the observed date, surface temperatures may have to be approximated empirically by using near-ground air temperatures.

Moreover, if we consider that the longwave radiation income from the atmosphere L_a likewise has to be approximated empirically as a function of air temperature and water vapour [see Equation (3.30)], modelling longwave radiation poses particular requirements for the estimation of these climate factors.

4. TOPOGRAPHIC TEMPERATURE

4.1 Modelling surface temperature

Land-surface parameters discussed so far are physically or trigonometrically based expressions with a clear deterministic relation to the physics of radiation fluxes and radiation geometries. Although there is obvious evidence for multiple orographic effects controlling or affecting the distribution pattern of temperatures

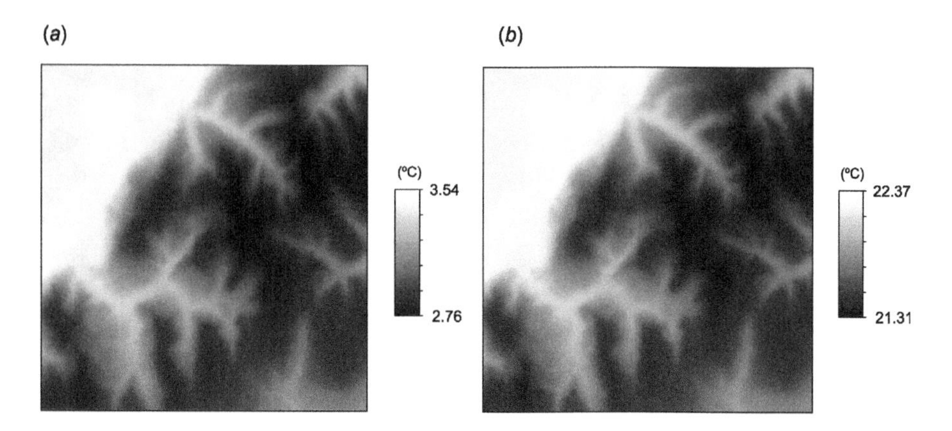

FIGURE 8 Distribution of lower troposphere temperatures (°C) for Baranja Hill area. Sea level temperatures and lapse rates are delineated from NCEP/NCAR reanalysis series (Kalnay et al., 1996): (a) — January; (b) — July.

and the intimately related moisture contents in the near surface layers of the atmosphere, these effects can not be expressed in pure physical terms but require geomorphometric analysis approaches, which represent, or at least approximate the nature of the orographically induced modulation of near-ground atmospheric processes.

Land-surface parameters proposed in this section are still under development and they will require further calibration with field observations. Spatial variations of both temperature and moisture are to a large degree determined by the vertical state of the troposphere and thus, if not affected by inversion layers, decrease with altitude. The long term mean hypsometric temperature gradient, delineated from representative network observations at different elevations or covered in GCM circulation data (Kalnay et al., 1996), mirrors the regional frequency of moist- or dry-adiabatic lapse rates and the occurrences of stable, neutral or unstable vertical troposphere profiles and thus generally varies with macro-climates.

Typical *temperature laps rates*, in the order of −0.4 to −0.8 K/100 m with a characteristic seasonality, are valid for most climates, apart from extreme polar climates, and result in a corresponding temperature distribution pattern, closely related to the surface elevation. Examples of troposphere temperatures are given in Figure 8, delineated from atmospheric fields of the NCEP/NCAR reanalysis series (Kalnay et al., 1996).

Since the atmospheric moisture content decreases exponentially with height and because the saturation vapour pressure is determined by the air temperature, a strict correlation with surface elevation is likewise valid for the spatial distribution pattern of water vapour. On the topo-climatic scale, however, typical residues in the temperature and moisture distribution are due to two major processes: (1) the diurnal differential heating of sloping surfaces and (2) the nocturnal cold air formation and cold air flow.

In mid and higher latitudes exposure-related changes in daily solar radiation income of north and south facing slopes and the resulting differences in heat and moisture exchange control the spatial variation, for instance, in the current soil moisture content, the phenological state, physiognomy of plants and similar. Even the distribution pattern of soil types reflects a differentiation in the long term transient process of Holocene soil formation owing to changing radiation geometries (Böhner, 2006).

Wilson and Gallant (2000, p. 98) suggested a formula to utilise this close relation between shortwave irradiation at sloping surfaces and air temperatures, and estimate *land-surface temperature* (T) by:

$$T = T_b - \frac{\Delta T \cdot (z - z_b)}{1000} + C \cdot \left(S - \frac{1}{S}\right) \cdot \left(1 - \frac{\mathrm{LAI}}{\mathrm{LAI}_{\mathrm{max}}}\right) \tag{4.1}$$

where z is elevation at grid location, z_b is the elevation of the reference climatic station, T_b is the temperature at the reference station, ΔT is the temperature gradient (e.g. 6.5 °C per 1000 m), C is an empirical constant (e.g. 1 °C), S is the net shortwave radiation, LAI is the leaf area index at the grid cell and $\mathrm{LAI}_{\mathrm{max}}$ is the maximum leaf area index. In this case, a map of LAI is used to adjust for the vegetation cover (higher cover, lower temperatures) and a map of S is used to adjust for the relative exposition (lower shortwave radiation, lower temperatures).

Apart from this obvious omnipresent topo-climatic differentiation between shady north and sunny south facing slopes, there is also a significant asymmetry in the components of the diurnal energy balance of western and eastern slopes. Even if we assume a symmetrical distribution of solar radiation with almost identical daily radiation totals on western and eastern slopes, the diurnal shift in the *bowen ratio* with a higher fraction of latent heat flux in the morning hours, when the ground surface is still moist, and an increasing transfer of sensible heat in the afternoon results in a relative heat surplus on western slopes, most obviously shown in the favoured south to west sloping stands of sensitive crops such as grapes.

A proper estimation of this asymmetrical heating of the surface layer requires the use of physically-based modelling approaches, integrating high-resolution temporal radiation and top-soil moisture models to simulate the diurnal course of the Earth's energy budget and its components. However, a rather simple approximation of the anisotropic diurnal heat (H_α) distribution may be obtained by:

$$H_\alpha = \cos(\alpha_{\mathrm{max}} - \alpha) \cdot \arctan(\beta) \tag{4.2}$$

where α_{max} defines the aspect with the maximum total heat surplus, α is the slope aspect and β is the slope angle. Figure 9 shows the resulting distribution of this anisotropy parameter for an α_{max} angle of 202.5° (SSW) in accordance with the soil mapping guidelines of the German soil surveys (Boden Ag, 1994).

> REMARK 7. *Topographic temperature is the consequence of two major processes: (1) the diurnal differential heating of sloping surfaces and (2) the nocturnal cold air formation and cold air flow.*

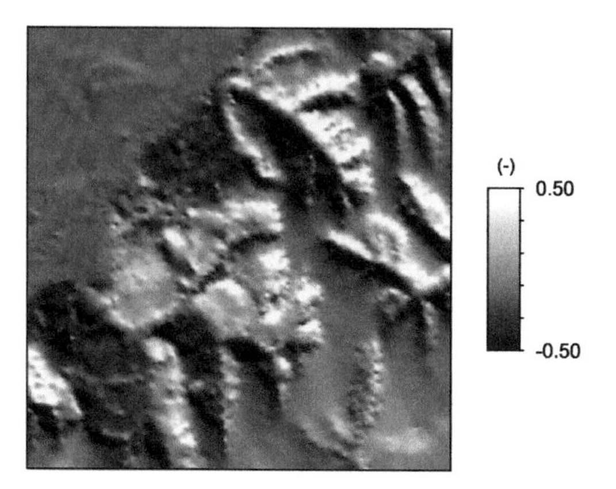

FIGURE 9 Diurnal anisotropic heating for Baranja Hill area. $\alpha_{\max} = 202.5°$ (SSW).

4.2 Modelling cold air flow

The second previously-mentioned process, the formation of cold air due to radiative heat loss of the ground surface and the resulting radiative transfer of sensible heat from the near surface layer to the ground is a typical phenomenon in cloud free calm nights. In sloping settings, the force of gravity causes cold, and thus denser air to flow downhill along gorges and valleys towards hollows or basins, quite similar to the flow of water. While in gently undulating terrains, the movement of cold air proceeds slowly with hardly noticeable speeds of usually less than 1 m/sec, in mountainous regions with steep sloping surfaces and deep valleys, pulsating cold air currents or even avalanches of cold air are a frequently-occurring phenomena (Deacon, 1969). In mountain-rimmed basins such as the broad basins of Central and High Asia, stagnating air throughout the winter months even leads to the formation of huge, high-reaching cold air domes and persistent inversion layers over the basins (Böhner, 2006; Lydolph, 1977).

REMARK 8. *Depth of a sink can be used as an indirect estimator of temperature conditions in the sink, as well as estimator of air humidity, soil depth or duration of flood stagnation.*

The course and frequency of cold air formation and *cold air flow* varies with the nature and roughness of the underlying ground and the topological structure of the surface. However, if we simply assume a sloping terrain with isotropic surface properties, completely homogeneous in terms of vegetation cover and soil moisture content, the amount of cold air flow is solely determined by the shape of the terrain.

The most simple example of the influence of land surface on cold air flow is the temperature inversion effect that occurs in the sinks,[7] which is conditioned by the confluence of the colder and heavier air in the sink. The magnitude of this effect

[7] Such geomorphological features frequently occur in the karst areas [see Figure 6(b) for illustration].

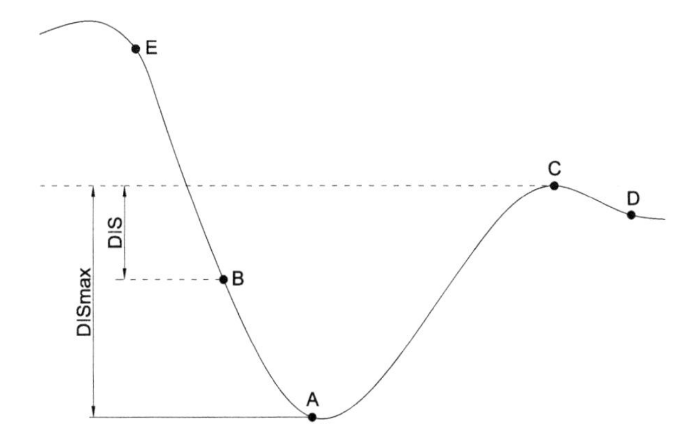

FIGURE 10 Schematic sink cross-section. Bold line represents land surface. Point A (sink bottom) has the largest depth in sink (DIS_{max}), point B has the depth in sink DIS, point C is the lowest point of sink brink (pour point) and it has zero depth in sink, as well as points D and E which are out of the sink. Reprinted from Antonić et al. (2001a). With permission from Elsevier.

is mostly correlated with the total *depth of the sink*, i.e. the vertical distance between the lowest point of the sink brink (pour point) and the point of sink bottom (see Figure 10). Antonić et al. (2001a) show that standard procedures of removing sinks from DEM (usually considered as errors in DEM that have to be corrected by the *filling*; see also Section 2.8 Chapter 4) can also be used for the mapping of depth in sinks (DIS on Figure 10). This simple indirect parameter (calculated as the difference between *corrected* and original DEM) can be considered as an indirect estimator of temperature conditions in the sink, as well as an indirect estimator of some other environmental variables potentially connected with sink depth — air humidity, soil depth influenced by soil erosion and sedimentation, or duration of flood stagnation in the microdepressions of lowland areas.

Other influences of land-surface shape on cold air flow are more complex, and particularly related to the area of the cold air contributing catchment. Consequently, the DEM-based upslope catchment area (see Section 4 in Chapter 7) is frequently suggested as a suitable approach to the terrain parameterisation of cold air flow.

Despite certain analogues with the gravity forced down slope flow of water and cold air, the momentum and dynamics of cold air currents distinctly differ from the way the much denser agent water flows. Particularly in broad valleys, the cold air distribution is not limited to channel lines as indicated by the pattern of DEM derived catchment area sizes. It disperses and normally covers the entire valley ground, depending on the volume of the nocturnal produced cold air. In order to enable a better representation of cold air dispersion in broader plain areas, an iterative slope dependent modification of the catchment area size is suggested:

$$C_m = C_{\max}\left(\frac{1}{10}\right)^{\beta\cdot\exp(10^{\beta})} \tag{4.3}$$

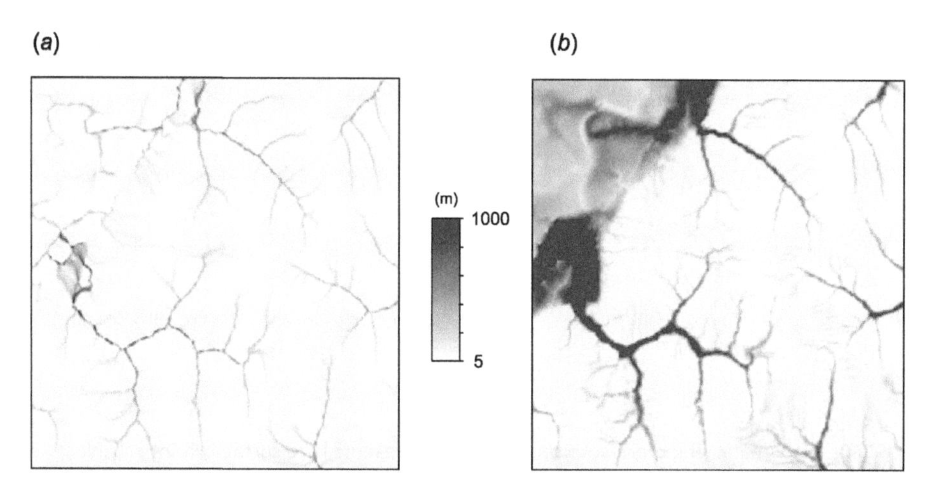

FIGURE 11 Cold air contribution upslope area for Baranja Hill (square roots of the upslope areas sizes): (a) — multiple flow method; (b) — SAGA method. Displayed using a logarithmic stretch.

for:

$$C < C_{\max}\left(\frac{1}{10}\right)^{\beta\cdot\exp(10^{\beta})} \tag{4.4}$$

where β is the slope angle, $C_{\max}$ is the maximum DEM catchment area in a 3×3 moving window environment and C_m is the modified catchment area, computed according to the multiple flow direction method of Freeman (1991). This algorithm is, for example, implemented in the SAGA GIS, originally developed as an adjusted terrain wetness index that should better represent soil moisture distribution in broad plain areas with rather homogeneous orohydrological conditions (Böhner and Köthe, 2003; Böhner, 2006). Compare the resulting spatial pattern of C and C_m in Figure 11(a) and (b).

Catchment area parameters prove suitable to approximate the *flow path of cold air* and the *size of the cold air contributing upslope area*. However, a DEM-based representation of spatially discrete topo-climatic settings like warm belts at slopes or persistent inversion layers — both phenomena are closely related to the course and frequency of cold air formation and cold air flow — require a more sophisticated parameterisation of the relative position of a point (a grid cell) within a sloping surface.

> REMARK 9. *Some hydrological modelling functions used in geomorphometry can also be applied to modelling cold air flow to provide relative estimates of meteorological conditions.*

An important parameter in this context, the *altitude above channel lines*, is a particularly valid measure to estimate potential inversion heights for valley settings which often experience late night or persistent wintertime temperature inversions. The calculation of this land-surface parameter first of all requires a reasonable

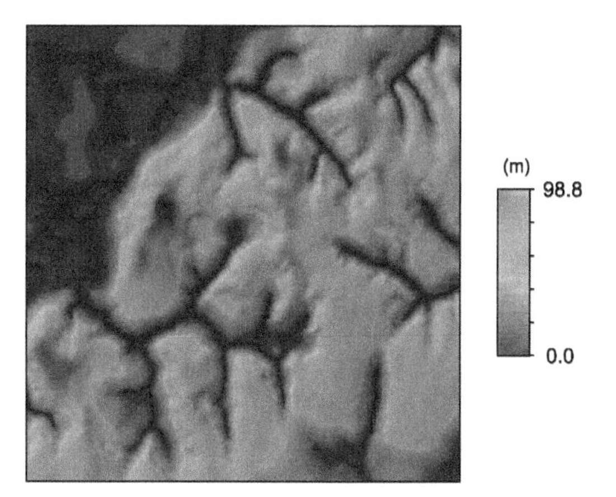

FIGURE 12 Altitude above channel lines. (See page 717 in Colour Plate Section at the back of the book.)

channel network grid, in order to assign the base 0 m elevation to those grid cells, indicating channel lines (see also Section 6 in Chapter 7 for methods to extract drainage lines).

The channel network elements in Figure 12 were initialised with a catchment area threshold of 100,000 m^2. In this example, the catchment areas were computed, using the SAGA GIS single flow direction method, which, differing from the often cited deterministic 8 algorithm, considers a terrain convergence index as a basic morphometric criterion to define overland flow paths (Böhner et al., 2002). Once the channel network grid is identified, vertical distances to the channel lines can be calculated for each grid cell using, again, the single flow direction method.

Methods which rely on overland flow paths, however, produce abruptly changing values at the watersheds and thus distinctly limit the usability of these relative altitudes for further applications. To overcome this disadvantage, a rather simple but efficient iterative procedure proved suitable during the SAGA GIS development, which delineates the altitude above channel lines z_c directly from elevation differences in a moving 3×3 grid cell window. The iterative approximation of z_c is done by:

$$z_c = z_0 - \bar{z}_8 + \bar{z}_8^* \tag{4.5}$$

where z_0 is the elevation z of the centred grid cell, $\bar{z}_8$ is the arithmetic mean of the 8 neighbouring elevations and $\bar{z}_8^*$ is the corresponding mean value of the approximated altitudes above channel network in the neighbourhood at a certain iteration step. A sufficient number of iterations — each performed with a constant 0 m base elevation at channel lines and recalculated z_c values out of the channel network — finally leads to nearly stable results and, thus, can be aborted if the maximum change of z_c between two iterations remains below a predefined threshold. The resulting elevation pattern in Figure 12 was reached after 1604 iterations (using an abort-threshold of 0.1 m).

The altitude above channel lines is suggested as a suitable land-surface parameter (indirect estimator) for climate regionalisation applications in case of sharply-shaped alpine land surface with frequent formations of temperature inversions. In rather shallow low mountain ranges, instead, not only valleys and hollows but also elevated cold air producing expanses are comparatively cold areas, whilst mid slopes remain relatively warm throughout the night. The regionalisation of these warmer slope settings, most familiar known as the *thermal belt at slopes*, requires a land-surface parameter, which integrates the vertical distances to channel lines and crest lines as well. However, the presupposed delineation of discrete topological segments and particularly the DEM-based definition of crest lines is a crucial task and may need a case-wise approximation for different test sites or different DEM-domains in order to obtain a geomorphological consistent representation.

The following section describes an attempt towards a purely continuous estimation of the altitude above drain culmination z_{dm} and altitude below summit culmination z_{sm} without using any basic discrete entities such as channel or crest lines (for more details see Böhner, 2005). In a first step, relative altitudes are designated as the difference of a grid cells altitude z_0 [or the inverted altitude z_0^* in Equation (4.7)] and the weighted mean of the upslope altitudes z_i [or the inverted upslope altitudes z_i^* in Equation (4.7)] each weighted by the reciprocal square root of its catchment area size C_i or C_i^* respectively:

$$z_{sm} = \frac{\sum_{i=1}^{n} \frac{z_i}{C_i^{0.5}}}{\sum_{i=1}^{n} \frac{1}{C_i^{0.5}}} - z_0 \tag{4.6}$$

$$z_{dm} = -1 \cdot \left[\frac{\sum_{i=1}^{n} \frac{z_i^*}{C_i^{*0.5}}}{\sum_{i=1}^{n} \frac{1}{C_i^{*0.5}}} - z_0^* \right] \tag{4.7}$$

where index m in z_{dm} and z_{sm} denotes the subsequent application of the slope dependent modification as already introduced in Equations (4.3) and (4.4). Based on these two parameters, we can also derive the normalised altitude z_n:

$$z_n = -0.5 \cdot \left[1 + \frac{z_{dm} - z_{sm}}{z_{dm} + z_{sm}} \right] \tag{4.8}$$

which integrates both attributes, using the well-known normalisation form of the NDVI (Normalised Difference Vegetation Index) but stretches the values from 0 for bottom positions to 1 for summit positions [Figure 13(a)]. If we simply assume the mid slopes to be the warmest settings, we can derive the indirect estimator z_m by:

$$z_m = \left[\frac{z_{dm} - z_{sm}}{z_{dm} + z_{sm}} \right]^2 \tag{4.9}$$

which alternatively assigns mid-slope positions with 0 whilst maximum relative vertical distances to the mid slope in valley or crest directions are assigned with 1 [Figure 13(b)].

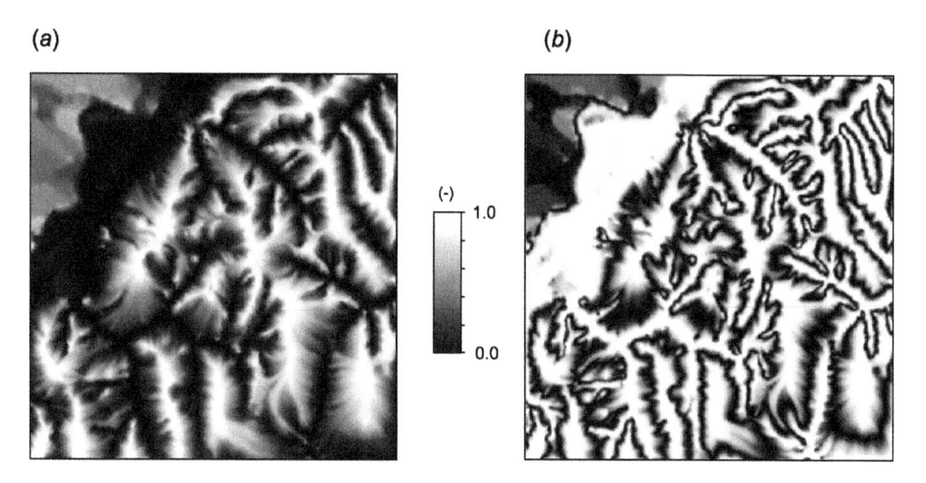

FIGURE 13 Relative altitude: (a) normalised altitude; (b) mid-slope position.

5. TOPOGRAPHIC PRECIPITATION

5.1 Modelling rainfall

The spatio-temporal dynamics of cloud formation and precipitation are likewise significantly affected by the land surface. However, this relationship is much more complex than the previously discussed effects, owing to the alternation of thermally and dynamically induced processes affecting cloud development and precipitation. If we again start with the elevation as a primary topo-climatic control, a global overview reveals a general relation of precipitation regimes and *vertical precipitation gradients*. In the convective regimes of the tropics, precipitation amounts commonly increase till the condensation level at 1000 to 1500 m above the ground surface while the exponentially decreasing air moisture content in the mid to upper troposphere results in a corresponding drying above the condensation level of tropical convection cluster systems (convection type of the vertical precipitation distribution; see also Weischet, 1995).

Likewise, negative lapse rates typically occur in the extreme dry polar climates. Whilst in the mid latitudes and in the subtropics (less pronounced), the frequent or even prevalent high reaching advection of moisture bearing air at fronts leads to increasing precipitation amounts of high mountain ranges such as the Alps (advection type of the vertical precipitation distribution; Weischet, 1995). The reduced precipitation amounts at lower settings are firstly due to the transpiration of rain drops when falling through non-saturated, lower-air levels (Lauer and Bendix, 2004). Moreover, the vertical precipitation gradient in high mountain ranges is often strengthened owing to the diurnal formation of autochthonous upslope breezes, which intensify cloud and shower formation in upper slope positions whilst the subsiding branch of these autochthonous local circulation systems along the valley axis leads to cloud dissolution and a corresponding reduction of rainfall rates in the valley bottoms.

In subtropical and tropical high mountain ranges like the Himalayas or the Bolivian Andes, the thermally induced daytime circulation can be even evident in the physiognomic characteristics of the vegetation, ranging from semi-desert vegetation in the interior dry valleys up to formation of humid forests at upper slopes (Troll, 1952; Schweinfurth, 1956). Besides the DEM elevation itself, the previously defined altitude above channel lines is one sufficient opportunity to represent these strengthened vertical precipitation gradients in steep high mountain environments. In cases of a sparsely and less representative distributed network of meteorological stations, precipitation lapse rates are masked by the predominant topographic effects of nonlinear sharply defined precipitation regimes at different settings.

REMARK 10. *The most common topographic effects on the rainfall are: (1) uplift of moist air currents on the windward side of a mountain range and (2) the intimately related rain shadow effect on leeward settings induced by the blockage of moisture-bearing air.*

Orographic precipitation, caused by the uplift of moist air currents at the windward side of a mountain range or the intimately related rain shadow effect at leeward settings induced by the blockage of moisture-bearing air, are most common effects which place particular demands on DEM-based parameterisation methods. A most frequently used land-surface parameter in this context is the DEM aspect. One often cited example is the statistical-topographic *PRISM approach* (Parameter-elevation Regression on Independent Slopes Model), which divides the land surface into topographic facets of eight exposures (N, NE, E, ..., NW), delineated at six different spatial scales to accommodate varying orographic complexity (Daly et al., 2002). The identification of major topographic orientations supports the computation of optimised station weights for the regression-based delineation of precipitation gradients from network observations.

REMARK 11. *Snow cover pattern can be estimated using solar radiation (thermic gradient), exposition to the winter wind direction (terrain orientation), slope and catchment area (accumulation and decumulation of the snow).*

Based on the assumption, that the uplift of moist air at windward slopes and the resulting precipitation pattern is associated with the increasing angular slope of moisture distributing trajectories, the following equations for the windward horizon parameter function H_W [Equation (5.1)] and the leeward horizon parameter function H_L [Equation (5.2)] are suggested as simple parameterisations of topographically determined effects on flow currents:

$$H_W = \frac{\sum_{i=1}^{n} \frac{1}{d_{WHi}} \cdot \tan^{-1}\left(\frac{d_{WZi}}{d_{WHi}^{0.5}}\right)}{\sum_{i=1}^{n} \frac{1}{d_{LHi}}} + \frac{\sum_{i=1}^{n} \frac{1}{d_{LHi}} \cdot \tan^{-1}\left(\frac{d_{LZi}}{d_{LHi}^{0.5}}\right)}{\sum_{i=1}^{n} \frac{1}{d_{LHi}}} \tag{5.1}$$

for $d_{LZi} > 0$:

$$H_L = \frac{\sum_{i=1}^{n} \frac{1}{\ln(d_{LHi})} \cdot \tan^{-1}\left(\frac{d_{LZi}}{d_{LHi}^{0.5}}\right)}{\sum_{i=1}^{n} \frac{1}{\ln(d_{LHi})}} \tag{5.2}$$

where d_{WHi} and d_{WZi} are the horizontal and vertical distances to the grid-cells in wind direction and d_{LHi} and d_{LZi} are the corresponding vertical distances in opposite (leeward) directions. Böhner (2006) used these parameters to *clean* network observations from topographic effects, when estimating vertical precipitation lapse rates in central and high mountain Asia. More sophisticated physically-based models, simulating precipitation distribution at different horizontal resolution are discussed later in Chapter 26.

5.2 Modelling snow cower pattern

Snow cover pattern (represented by duration, snow cover height or accumulation potential) can be also considered as a climatic variable which is influenced by topographic variables. The most important land-surface parameters are elevation and topographic solar radiation, which control general and local thermic gradients connected with the melting of snow. The land-surface aspect can also be considered as an additional variable which has an impact on snow cover, regarding the influence of terrain orientation to the prevailing winter winds. The impact of slope, curvature and catchment area, which are connected with accumulation and decumulation of snow on the surface, also can not be neglected.

An example of an intuitively constructed relation between land-surface parameters and snow cover pattern is *snow potential index* (SNOW) proposed by Brown and Bara (1994) as an indirect estimator of snow accumulation. It can be calculated as:

$$\text{SNOW} = \alpha_r \cdot C_{rv} \cdot \frac{z - z_{\min}}{z_{\text{range}}} \tag{5.3}$$

where α_r is relative land-surface aspect, i.e. absolute value (°) of angle distance from the terrain aspect α to the azimuth of the prevailing winter wind direction (see also Section 6), C_{rv} is unitless curvature and z is elevation. A higher value of this described index means leeward direction (in the sense of declination of prevailing winter winds), concave land surface and higher elevation. A major disadvantage of Brown's SNOW lies in the fact that it always has zero value if oriented to the prevailing winter winds (windward positions), regardless of elevation and curvature. However, this index illustrates the possibility of logical (intuitive) construction of land-surface parameters in cases when exact understanding of topographic influences on the target dependent variable is missing.

Real topographic influence on snow cover spatial distribution is very complex in its essence due to the large number of interactions between particular topographic variables and, moreover, due to the additional impact of topography on spatial distribution of soil and vegetation variability which also influence snow cover patterns (see also e.g. Walsh et al., 1994). A consequence of this complexity is that exact spatial modelling of snow cover patterns can hardly to be generalised, and it is better to be oriented to the examination of local relationships in particular areas of interest.

An illustrative example of such a local approach is work of Tappeiner et al. (2001) which described DEM-based modelling of direct estimators of *number of snow cover days* in one valley in the central eastern Alps (cca 2 km^2 in the altitudinal range from 1200 to 2350 m), using artificial neural networks as a modelling tool. These empirical models were developed on the basis of snow cover data collected by the 2-year photographic terrestrial remote sensing (using temporal resolution of 1 day and spatial resolution of cca 1 m). Land-surface parameters used as independent variables were elevation, slope and aspect, topographic solar radiation during the winter, number of days with air temperature $<0°$, and forested area (binary data). Tappeiner et al. (2001) showed that the best empirical model explained 81% of spatial variability of annual mean number of snow cover days for two years of observations. It has to be emphasised that the spatial pattern of the achieved model error was not randomly distributed, than it seems to be visually correlated with topographic variability, which probably implies additional impacts of topography not captured by the land-surface parameters used in the model development. Nevertheless, these results suggest that analogous local empirical models could probably be developed for the other areas, with different exact relations between topography represented by the suitable land-surface parameters, other potential independent variables (vegetational, climatological, lithological) and snow cover pattern.

6. TOPOGRAPHIC EXPOSURE TO WIND

Wind is usually defined as movement of air, conditioned by the spatially heterogeneous pattern of air pressure. Wind field near the ground, described with values of wind direction and velocity for each point of examined area, is significantly influenced by the land surface, especially in the areas of rugged relief. This is the reason why data about land surface are usually involved in direct dynamic modelling of wind field near the ground (see also Chapter 26), e.g. for the purposes of weather forecast or for the estimation of wind power potential. This kind of modelling is computationally intensive, with the need for significant amounts of input data.

In cases where the wind field is an important variable in a study (e.g. predicting spatial distributions of air pollutants soil concentrations or vegetation types in mountain regions), but dynamic modelling can not be performed due to the lack of data (and/or dynamic modelling resources), it is possible to use indirect land-surface parameters presented in this section, which estimate topographic exposure to the unknown (but presumed) wind flux (see Figure 14).

The simplest parameter which is a potential estimator of topographic exposure to wind is angle distance from the azimuth of wind direction α_r (i.e. *relative terrain aspect*; RTA on Figure 14). Relative terrain aspect takes into account land-surface orientation only, neglecting the influence of distant terrain (in the sense of shelterness) as well as the influence of slope. It can be generally defined as the absolute

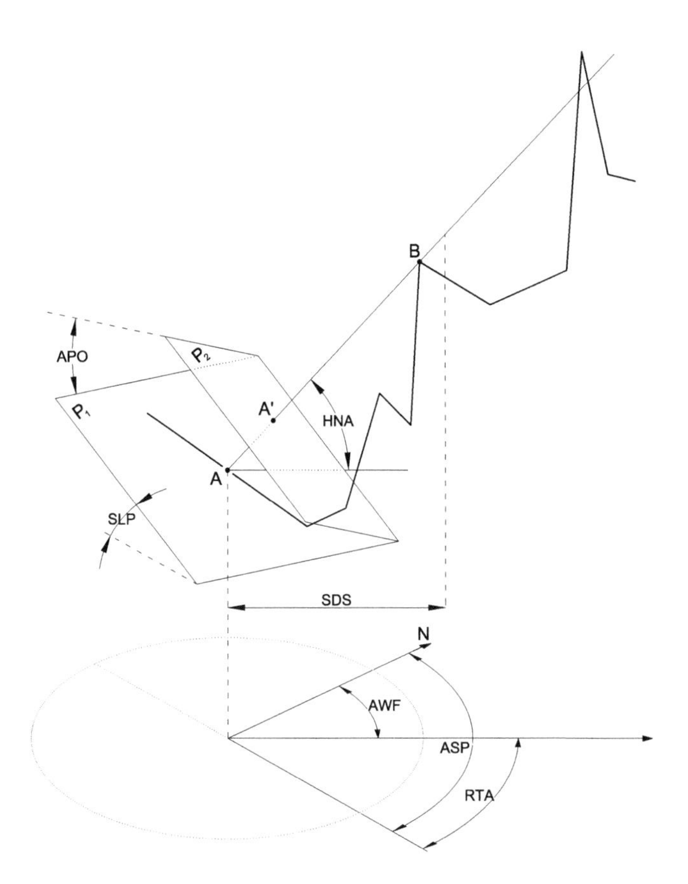

FIGURE 14 Schematic presentation of estimators of topographic exposure to wind. Bold line represents land surface. N indicates north. Terrain around the point A has the slope (SLP) and aspect (ASP). Relative terrain aspect (RTA) is absolute angle distance between the terrain aspect and the azimuth of wind flux (AWF). Horizon angle (HNA) of point A for given azimuth of wind flux is determined by the point B under the chosen search distance (SDS) — note that a change of search distance may change horizon angle. The angle between a plane which locally represents terrain (P_1) and plane orthogonal to the wind (P_2) is denoted by APO (the point A′ is a projection of the point A on the orthogonal plane).

value of the difference between aspect and wind direction, respecting the circularity of both. Consequently, α_r always varies between 0° (strict windward positions) and 180° (strict leeward positions). Alternatively, it can be also expressed as $\cos \alpha_r$ which varies between 1 (windward) and −1 (leeward). In this sense, *northness* presented in Section 2.1 of Chapter 6 can be also understood as relative terrain aspect to the north wind.

The second parameter potentially useful for estimation of topographic exposure to wind is the horizon angle φ in the wind direction (HNA on Figure 14), assuming that higher horizon in this direction implies that the land surface is more sheltered from wind (and neglecting the influence of slope and aspect). Use of a larger search distance in the horizon angle calculation for the purpose of topographic solar radiation modelling (compare Figure 3), usually means a more precise result (with exception of applications on very large areas where the Earth's curvature can not be ignored). In contrast, use of a too large search distance in the context of exposure to wind ignores the adaptability of wind streams to the land surface. Consequently, application of this land-surface parameter requires very careful selection of the search distance, or, even better, use of various search distances and choosing the optimal distance after comparison of results obtained in relation to the examined problem.

REMARK 12. *Exposure to wind can be estimated using the land-surface parameters frequently used in modelling topographic solar radiation — terrain aspect, illumination angle, horizon angle and sky view factor.*

Combination of the two identified land-surface parameters together with accounting for the influence of slope β can be made by use of the following expression (see also Figure 14):

$$\cos \gamma_W = \sin \varphi \cdot \cos \beta + \cos \varphi \cdot \sin \beta \cdot \cos \alpha_r \quad (6.1)$$

where γ_W denotes the angle between a plane orthogonal to the wind and plane which locally represents land surface. The same expression (but with sun elevation and azimuth instead of horizon angle and relative terrain aspect, respectively) is also used for calculation of solar illumination angle γ, as a part of direct solar radiation modelling. In cases where horizon angle (or search distance) is set to zero, this expression estimates land-surface exposure to the horizontal component of wind flux from a given direction, otherwise it is presumed that wind flux is *sloped* (and determined with horizon angle).

All the land-surface parameters suggested here as potential estimators of land-surface exposure to the wind can be applied in two general ways: (1) by using topographic exposure to wind from a dominant (prevailing) direction selected *a priori*, most frequently on the basis of a local wind rose extracted from data collected at the nearest meteorological station,[8] or (2) by using various topographic exposures to wind from different hypothetic directions (e.g. for each 45°). The direction that shows the highest power to explain a spatial variability of the targeted phenomenon can then be used as the most suitable direction.

A typical example of the second approach is the work of Antonić and Legović (1999) in the National Park Risnjak, Croatia. In this research, previously-described land-surface parameters (estimators of topographic exposure to wind) were applied in order to estimate the direction of an unknown air pollution source and to explain spatial variability of heavy metal soil concentrations, resulting in very interpretable outputs. Namely, correlations between heavy metal soil concentrations and land-surface parameters were maximised exactly for these hypothetical directions where major air pollution sources existed and a regression model with selected land-surface parameters as independent variables was able to explain a large part of spatial variability of heavy metal soil concentrations.

[8] One example of this approach is shown in Section 5.2, where relative terrain aspect is used for calculation of snow potential index.

The sky view factor Ψ_S could be also considered to be used as an indirect estimator of overall topographic exposure to the wind (Ψ_{ote}). For this purpose, Equation (3.18) could probably be adjusted to include wind frequency and/or velocity from some respective direction (i.e. wind rose data) as weights within the integration around a full circle. Alternatively, on the areas (and/or for the time periods) where (when) wind directional variability can be ignored (or is not known), the portion of the hemisphere unobstructed by terrain in the total hemisphere could be used as an indirect estimator of overall exposure, following the expression:

$$\Psi_{ote} = \frac{1}{N} \cdot \sum_{i=1}^{N} (1 - \sin \varphi_i) \tag{6.2}$$

7. SUMMARY POINTS

The land-surface parameters, discussed in this chapter are either physically proved expressions or measures suitable to parameterise the multitude of orographically induced or affected atmospheric processes in the boundary layer. Land surface is one major control within the scope of atmosphere surface interactions, however, land use pattern, type and state of natural or managed vegetation cover likewise significantly affect the topo-climatic settings.

Relevant surface properties, such as the meteorological roughness and vegetation/land cover, require consideration of the nature of underlying ground and its influence on the regional climate modelling applications. With the increasing availability of huge amounts of fine-resolution remotely sensed data, a challenge for geomorphometrists is to consistently assimilate and delineate surface parameters specific to climatic-modelling applications. One important, but still crucial, task in this context is the parametrisation of the cold/fresh air production of different surfaces and the approximation of corresponding cold air currents in complex orographic/land cover settings. Geomorphometric measures may contribute to this problem by integrating surface properties using complex catchment-related parameters.

Further methodical tasks such as the realisation of dynamic climate modelling require the development of advanced General Circulation Model (GCM) data assimilation schemes capable of integrating dynamical circulation variables from GCM output. This task however goes far beyond the opportunities of current GIS and thus requires ongoing and intensified measurements in the filed of Dynamic GIS development, supported by suitable programming environments and capable of running modularly-organised climate modelling chains.

IMPORTANT SOURCES

Böhner, J., 2006. General climatic controls and topoclimatic variations in Central and High Asia. Boreas 35 (2), 279–295.

Bendix, J., 2004. Geländeklimatologie. Gebrüder Bornträger Verlagsbuchhandlung, Stuttgart, 282 pp.

Antonić, O., Marki, A., Križan, J., 2000. A global model for monthly mean hourly direct solar radiation. Ecological Modelling 129 (2–3), 113–118.
Dubayah, R., Rich, P.M., 1995. Topographic solar radiation models for GIS. International Journal of Geographical Information Systems 9 (4), 405–419.
Nikolov, N.T., Zeller, K.F., 1992. A solar radiation algorithm for ecosystem dynamic models. Ecological Modelling 61, 149–168.
Oke, T.R., 1988. Boundary Layer Climates, 2nd edition. Taylor & Francis, London, New York, 435 pp.
Deacon, E.L., 1969. Physical processes near the surface of the earth. In: Flohn, H. (Ed.), World Survey of Climatology, vol. 2. Elsevier, Amsterdam, London, New York, pp. 39–104.
Geiger, R., 1969. Topoclimates. In: Flohn, H. (Ed.), World Survey of Climatology, vol. 2. Elsevier, Amsterdam, London, New York, pp. 105–138.

CHAPTER 9

Landforms and Landform Elements in Geomorphometry

R.A. MacMillan and **P.A. Shary**

why are landforms important and why classify them? · difference between general and specific geomorphology · definitions of landforms and landform elements · issues of scale and perception · evolution of automated landform classification techniques · landform elements as defined by Troeh, Dikau, Pennock, Shary and others · automated and manual extraction and classification of repeating landform types · the search window problem and its implications

1. GEOMORPHOLOGY, LANDFORMS AND GEOMORPHOMETRY

1.1 Basic principles

The term *geometry* means literally 'land (surface) measurement', but has been applied mostly to artificial or smooth mathematical surfaces, such as spheres or cubes. *Geomorphometry* returns to the original meaning of geometry as a science devoted directly to quantitative analysis of the Earth's surface. Geomorphometry has previously been considered as a sub-discipline of geomorphology but is now often regarded as a separate discipline in its own right (see Chapter 1). Most classifications of landforms are based either implicitly or explicitly on consideration of how the gravitational field interacts with the land surface with gravity governing surface flow and flow in turn modifying surface forms (Shary, 1995).

Evans (1972, p. 18) recognised two different approaches to geomorphometry, which he termed *specific geomorphometry* and *general geomorphometry* (see Chapter 1). *Specific geomorphometry* applies to and describes discrete landforms such as an esker, drumlin, sand dune or volcano. It can involve arbitrary decisions and subjectivity in the quantification of its concepts. *General geomorphometry* applies to and describes the continuous land surface. It provides a basis for the quantitative comparison of even qualitatively different landscapes and it can adapt methods of surface analysis used outside geomorphology. Dehn et al. (2001)

Developments in Soil Science, Volume 33 © 2009 Elsevier B.V.
ISSN 0166-2481, DOI: 10.1016/S0166-2481(08)00009-3.

observed that landforms were described mainly in two different ways (i) based solely on their geometry or (ii) based on semantics used to express and capture subjective conceptual mental models.

REMARK 1. *A landform is a physical feature of the Earth's surface having a characteristic, recognisable shape and produced by natural causes.*

1.2 Landforms: definitions

A *landform* is defined as "*any physical feature of the Earth's surface having a characteristic, recognisable shape*" (Bates and Jackson, 2005). A subjective semantic definition of landform consistent with specific geomorphometry is "*a terrain unit created by natural*[1] *processes in such a way that it may be recognised and described in terms of typical attributes where ever it may occur*" (Lobeck, 1939; Weaver, 1965; Hammond, 1965; Leighty, 2001). Many geomorphologists, however, prefer a definition that includes recognition of artificial landforms such as quarries, waste heaps and similar (personal communication, I.S. Evans, 2007). A geometrical definition of landforms, consistent with general geomorphometry, would focus on objective consideration of surface shape or form only.

A *landform type* consists of a characteristic pattern of terrain that exhibits a defined variation in size, scale and shape of geomorphic features and occurs in a recognisable contextual position relative to adjacent geomorphic features. A landform type typically repeats one or more full cycles of waveform variation from crest to valley (in regional approaches). A landform type is distinguished by its dimensions (length, width and height) and by the statistical frequency of its principal geomorphic attributes. These include the length, gradient and frequency distribution of its slopes, the frequency of inflections or reversals in slope, the magnitude of its internal relief, the degree of incision of drainage channels and the hydrological order of those stream channels and also any distinguishing considerations of shape or orientation (e.g. long and narrow versus short and round or preferred versus not preferred orientations). By landforms, we usually think of discrete geomorphic/geometric units (e.g. watershed, talus apron, alluvial fan, dune, cirque, drumlin, crater, etc.), whereas the less crisply defined landform types (plateau surface, dune & drumlin fields, a karst landscape, etc.) commonly are collections of such landforms plus '*connecting tissue*'. Repeated landform types form *land systems* or landscapes (Zinck and Valenzuela, 1990; Brabyn, 1997).

Landform types (Dikau et al., 1995) have also been referred to as *relief forms* (Dikau, 1989), *mesoform associations* (Dikau, 1989) and *landform patterns* (Speight, 1974). Examples of landform types include *plains, hills, mountains* and *valleys*. Plains, hills and valleys can be observed at multiple scales. In geography, these names are used for larger landscapes dominated by one landform type.

A *landform element* is a sub-component of a *landform type* at the level immediately below, and hierarchical to landform type (see also Section 1.3 in Chapter 2).

[1] Today, we also consider artificial landforms such as quarries, waste heaps and similar.

Landform elements may be conceptualised as consisting of portions of a landform type that are relatively homogeneous with respect to *shape* (profile and plan curvature), *steepness* (gradient), *orientation* or exposure (aspect or solar radiation), *moisture regime* and *relative landform position* (e.g. upper, mid or lower).

> REMARK 2. *A landform element is a sub-component of a landform type that can be characterise mainly by its morphology (shape, steepness, orientation, moisture regime, etc.).*

Dikau (1989) differentiates form elements with homogeneous plan and profile curvature from even more homogeneous form facets that have homogeneous gradient, aspect and curvature. Shary (1995) and Shary et al. (2005) proposed an objective, local, scale-specific classification of elemental landform features based entirely on consideration of signs of curvatures. This classification was described as "predictable" in the sense that the proportion of an area occupied by each class can be calculated in advance for any terrain (see also Section 2). It can be argued that any landform element that can be further sub-divided into smaller and more homogeneous entities is not technically an elemental form but the concept of a landform element has achieved widespread use in spite of this contradiction.

1.3 Why are landforms important and why classify them?

The Earth's surface is structured into landforms as a result of the cumulative influence of geomorphic, geological, hydrological, ecological and soil forming processes that have acted on it over time. Landforms are therefore widely recognised as natural objects that partition the Earth's surface into fundamental spatial entities. Landform entities differ from one another in terms of characteristics such as shape, size, orientation, relief and contextual position. They also differ in terms of the physical processes that were involved in their formation and that continue to operate within them at the present time (Etzelmüller and Sulebak, 2000). Some authors (Tomer and Anderson, 1995; Clarke, 1988; Shary et al., 2002) have argued that land surface should be thought of as composed of both *deterministic* and *noisy* components (see Section 1.4 in Chapter 2) with geomorphometry concerning itself with the task of extracting deterministic landforms from the noisy land surface.

Landforms define boundary conditions for processes operative in the fields of geomorphology, hydrology, ecology, pedology and others (Dikau, 1989; Dikau et al., 1995; Pike, 1995, 2000a; Dehn et al., 2001). Landform size and shape are interpreted as direct indicators of the processes understood to have produced the landforms. Evans (2003) demonstrated convincingly that the size and spacing of landforms clustered around characteristic scales or limits related to either process thresholds or space available. The surface shape of landforms has consistently been used to interpret or infer hillslope forming processes such as erosion and denudation (from convexities), accumulation and deposition or geomorphic processes such as alluvial, aeolian or glacial deposition.

Surface shape affects the accumulation of surface flow and consequently of surface deposits through two accumulation mechanisms (Figure 1). The first mech-

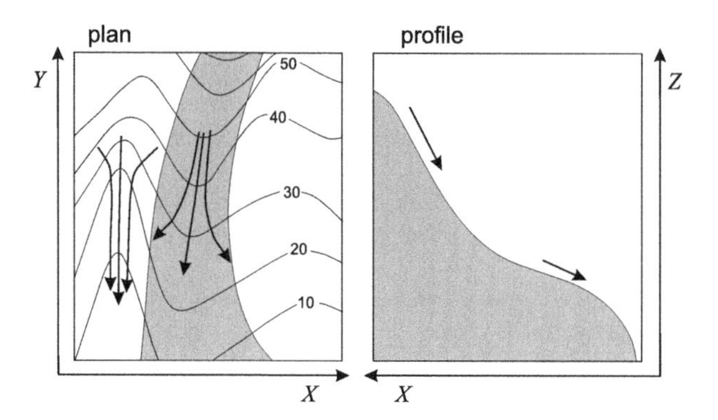

FIGURE 1 Illustration of two fundamental accumulation mechanisms related to surface shape. Reprinted from Shary et al. (2002). With permission from Elsevier.

anism reflects divergence in flow in response to local (or more broadly computed) horizontal convexity and convergence in flow in response to horizontal concavity, resulting in accumulation in areas of horizontal concavity. The second mechanism reflects relative deceleration of flow in the downslope direction as influenced by changes in profile curvature (or optionally some other land-surface parameters) from broadly convex to concave.

Many efforts to automate landform classification have made use of this inferred relationship between surface curvature and erosional or depositional processes (see for example Pennock et al., 1987). Qualitatively described measures of contextual position in the landscape have also been used to interpret and infer geomorphic and hydrological processes (e.g. Ruhe, 1960). Contextual measures have included absolute and relative horizontal and vertical distance to ridge lines or channels (Skidmore et al., 1991) and position in the landscape relative to the order of the nearest stream channel into which a hillslope was drained (Leighty, 2001; Schmidt et al., 1998; Schmidt and Dikau, 1999).

Speight (1974) indicated that "local geometry of surface" and "relative position" are different things. Calculations of local geometry (e.g. slope steepness) consider only a restricted portion of the terrain in a local window. Calculation of relative position requires consideration of extended portions of the land surface (e.g. depressions and hills) within a search window of varying extent. Shary (1995) differentiated these as *local* and *regional* land surface parameters respectively. Both accumulation mechanisms shown in Figure 1 can be described as either local or regional (Shary et al., 2002).

In general, local surface form and local to regional measures of context have been interpreted to infer processes that control the spatial distribution and redistribution of materials and energy in the landscape. Inferences based on process–form relationships are widely used to make further inferences pertaining to expected site conditions with respect to surface and subsurface moisture regimes, types, texture and stability of unconsolidated sediments, and kinds and degree of pedogenic development.

1.4 Issues of scale and perception

The Earth's surface is continuous in most[2] locations and natural terrain features vary across a full range of scales and sizes. Although Evans (2003) has shown that there are limits to the fractal behaviour of landscapes, the concept of fractal dimensions, first formalised by Mandelbrot (1967), leads to the realisation that "*what you see depends largely upon how closely you look*". What one perceives in observing and classifying terrain is therefore dependant upon a combination of the size or extent of the area viewed and the level of detail of the displayed surface as controlled by the horizontal and vertical resolution of the elevation data used to portray it. If one zooms in very closely to view and classify terrain at a horizontal and vertical resolution of mm to cm, one will perceive and classify objects at the scale of, for example, individual furrows and ridges in a cultivated field while likely not being able to recognise or differentiate whether the furrows occur on undulating bottomlands, level tablelands or a sloping hillside. Conversely, if one zooms out to a continental scale and views a terrain surface depicted using a grid resolution of 500 to 1000 m, only the largest and most prominent macro-scale features of the Earth's surface are captured and portrayed.

Most applied classifications of landforms are specific to a particular scale or narrow range of scales. These tend to treat variation in the land surface that occurs below a given scale as random noise to be ignored or removed and to regard variation at a particular scale or over a limited range of scales as signal to be recognised and interpreted. Landscapes and landforms have been widely recognised to occur across a hierarchy of scales and sizes (Hammond, 1964; Meijerink, 1988; Mulla, 1988; Pike, 1988; Weibel and DeLotto, 1988; Dikau, 1989, 1990; Guzzetti and Reichenbach, 1994; Suryana and de Hoop, 1994; Fels and Matson, 1996; Lloyd and Atkinson, 1998; Schmidt et al., 1998; Etzelmüller and Sulebak, 2000; Leighty, 2001; Lucieer et al., 2003). Etzelmüller and Sulebak (2000) observed that "*the geomorphic system must be viewed in its complex, hierarchical context*" and that "*every system consists of an array of smaller, lower-level, systems and is, at the same time, part of a sequence of ever larger, higher-level, systems*".

While some characterisations of landforms are essentially scale-free[3] and recognise the same invariant spatial entities regardless of scale, or at least across a wide range of sufficiently large scales (Shary et al., 2005; Wood, 1996), most consider that specific landforms have characteristic dimensions over which they occur (Evans, 2003; Dikau, 1990) and most either implicitly or explicitly recognise a hierarchy with different types and sizes of landforms occurring at different scales. Dikau (1990), following Kugler (1964), illustrated how different landforms have been conceptualised to occur over different scales (Figure 2). In this hierarchical conceptualisation, relief units ranged from relatively homogeneous form facets to more complex associations or patterns that consisted of assemblages of lower level forms. It was considered that simpler form units (e.g. form facets) could be

[2] Exceptions include cliffs and cavern entrances. '*Before*' and '*after*' landslide surfaces at the same location always are encoded as separate datasets.

[3] See Shary et al. (2002) for a definition and discussion of scale-free topographic features and land surface parameters. Their definition of scale-free recognises that parameters and features have a limit value as grid mesh tends to zero. Shary et al. (2005) show, both theoretically and empirically, that properly introduced regional land surface parameters (e.g. hills or depressions) may be scale-free.

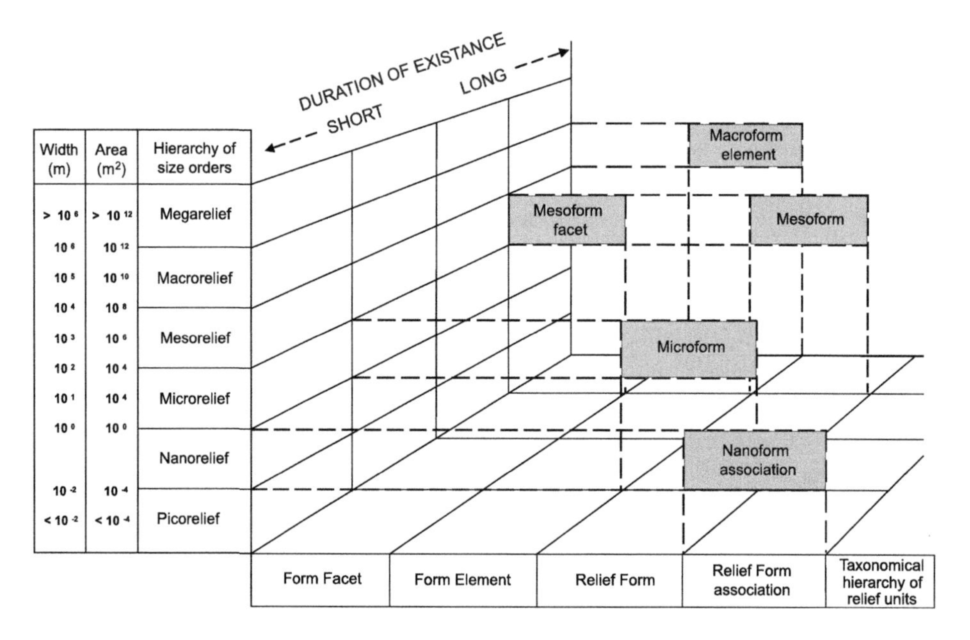

FIGURE 2 Taxonomical hierarchy of geomorphological landforms. Reprinted from Dikau (1990).

described at micro scales and below by inherent shape-based attributes such as curvature, gradient and aspect but that more complex form associations or patterns occurring at coarser scales required a change in defining attributes to include more synoptic descriptors such as *grain, roughness, attribute variability* or *ridge density* (Dikau, 1990).

> REMARK 3. *Classifications of landforms are commonly specific to a particular scale or narrow range of scales.*

Most efforts to classify landforms correspond to the micro (1 to 100 m) and meso (100 m to 10 km) terrain extents of Dikau (1990). Most also typically focus on classifying objects that correspond to one, or perhaps two, levels of complexity on the proposed range of form facets to form associations.

The horizontal and vertical resolution of the elevation data used to portray a terrain surface has a significant influence on the level of detail and the accuracy of portrayal of surface features and on the values of land-surface parameters that are computed from a DEM. Numerous authors have examined the effects of grid resolution on the value and accuracy of land-surface parameters and objects derived from elevation data sets of differing resolutions, usually horizontal resolution with an associated implied vertical resolution (Zhang and Montgomery, 1994; Florinsky, 1998; Jones, 1998; Wilson et al., 2000; Thompson et al., 2001; Tang et al., 2002; Zhou and Liu, 2004; Kienzle, 2004; Waren et al., 2004; Raaflaub and Collins, 2006).

Shary et al. (2002) provided a thorough theoretical examination of the effects of grid resolution on land-surface parameters and objects. They demonstrated that the local variables of slope gradient, aspect and curvatures are very sensitive to grid resolution, with second derivatives (curvatures) more sensitive than first derivatives (slope and aspect). Mean slope gradient tends towards zero as horizontal grid resolution gets larger and increases as horizontal resolution approaches zero. Thus, there is no single true or fixed value for local land-surface parameters such as slope or curvature at a point but rather a whole range of values that are dependant upon the horizontal and vertical resolution (and window size; see for example Figure 4 in Chapter 14) of the grid used to compute the land-surface parameters and objects, as well as the choice of algorithm used to compute them.

There is no single *best* resolution at which to compute local land-surface parameters to portray and classify terrain (Hengl, 2006). The resolution selected needs to be appropriate for capturing and describing the surface features of interest for a particular application. Some approaches explicitly investigate variation over a range of scales by computing land-surface parameters for DEM pyramids of increasing spatial resolution (Wood, 1996). Most applications have operated on a single DEM of a single fixed resolution. By default these applications have made implicit assumptions about the size and scale of the features that would be recognised and classified. For many applications that involve interpreting natural landscapes, a consensus appears to have emerged that regards horizontal resolutions of 5–10 m and vertical resolutions of <0.5 m as optimal for describing local surface form in a manner that is consistent with how micro to meso forms at the level of abstraction of individual hillslopes or parts of hillslopes have tended to be perceived and appreciated by human interpreters (Zhang and Montgomery, 1994; Kienzle, 2004).

Ideally, for each application, we should first test out predictive efficiency for various DEM resolutions and neighbourhood sizes, and then objectively derive the most suitable resolution and search size. For example, Florinsky et al. (2002) used the plots of correlation coefficient versus various resolutions to decide about the optimal grid resolution. M.P. Smith et al. (2006) examined a combination of grid resolutions and neighbourhood windows to suggest the optimal combination (see also Section 2.4 in Chapter 2).

2. APPROACHES TO LANDFORM CLASSIFICATION

Automated classification of landforms almost always represents an attempt to replicate some previously conceived system of manual landform classification and mapping. We briefly consider here some general aspects of manual approaches to landform classification that are relevant to efforts to develop and apply automated approaches. The system of Gauss (1828) recognised four field-invariant geometrical forms defined by signs of total Gaussian and mean curvatures (Figure 3). The system of Troeh (1964, 1965) partitioned land surfaces into four gravity-specific classes intended to recognise the two relative accumulation mechanisms based also on consideration of signs of tangential and profile curvatures (Figure 4).

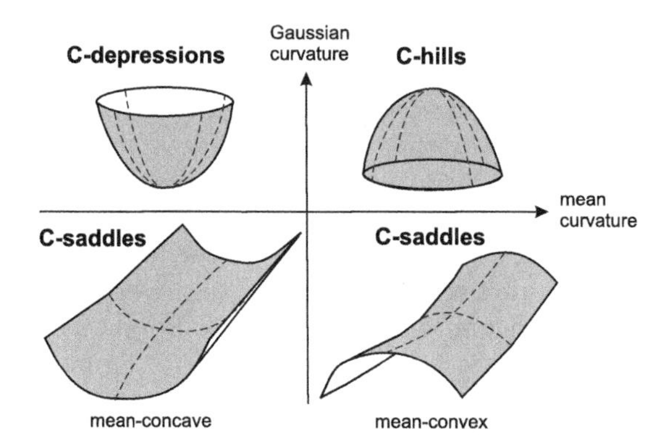

FIGURE 3 Illustration of the four landform classes defined by Gauss (1828) based on total Gaussian and mean curvature. Reprinted from Shary et al. (2005).

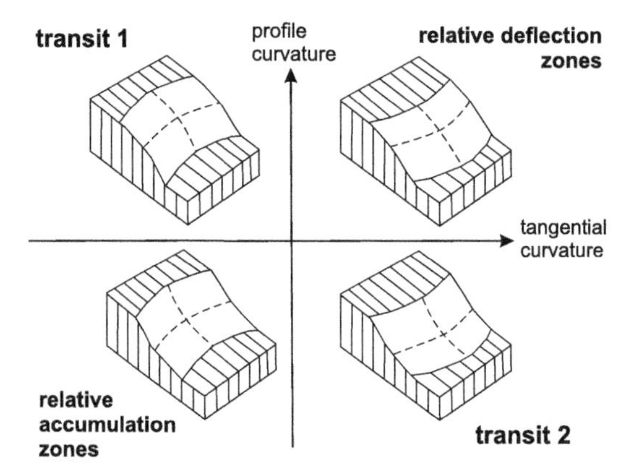

FIGURE 4 Illustration of the four landform classes defined by Troeh (1964, 1965) based on tangential and profile curvatures. Reprinted from Shary et al. (2005).

These systems can be applied to any surface at any scale to produce similar results.

Many subjective *manual systems of landform classification* have been proposed and extensively applied. Examples of widely applied systems include those of Fenneman (1938), Veatch (1935), Hammond (1954, 1964) for the USA, the Australian classification system of Speight (1974), Speight (1990), the SOTER[4] Global Soil and Terrain Database (van Engelen and Ting-tiang, 1995), the ITC system of geomorphic mapping (Meijerink, 1988) and the geo-pedological approach by Zink (Hengl and Rossiter, 2003). Review of such systems is not possible within the constraints of this chapter however a few general observations are relevant. Manual systems of geomorphic classification are invariably hierarchical and implement a sub-division of land surfaces into successively more

[4] http://www.fao.org/ag/aGL/agll/soter.stm.

narrowly described (and typically homogeneous) forms at successively finer scales.

Manual hierarchical systems tend to implement the hierarchies using top–down, divisive approaches in preference to bottom–up agglomerative approaches. Most manual systems invoke semantic models that attempt to capture concepts deemed important by the classifier using subjectively formulated differentiating criteria. Manual classification systems tend be synoptic and synthetic and to require simultaneous consideration and synthesis of multiple differentiating criteria, with different criteria used to differentiate entities at different scales and even under different conditions at the same scale.

REMARK 4. *Many landforms can be delineated manually using photo-interpretation to assess their form, size, scale, adjacency, surface roughness, hydrological and contextual position and geological origin.*

Primary considerations in differentiating landforms at different scales are *local surface form* or shape, *landform size* in horizontal and vertical dimensions, local to regional context, *patterns of cyclic repetition of landform shapes* as exhibited by topographic grain, *topological relationships* such as adjacency, connectivity and relative position, and *hydrological relationships* such as absolute or relative horizontal or vertical *distance to channels*, divides or water tables or *position in the hydrological network* relative to stream channel order (Berry and Marble, 1968, pp. 35–41).

Classifications based on local land-surface parameters (such as curvatures) can be considered to be predictable, while most landform classifications based on regional land-surface parameters and objects should be considered as terrain-specific (Shary et al., 2005). A predictable classification does not mean that landform patterns can be predicted, but rather that the probability of occurrence of a given landform type can be calculated in advance (Shary, 1995). For example, Gauss' saddles (mean-concave and mean-convex, Figure 3) cumulatively occupy 2/3 of any terrain, as do Troeh's relative accumulation and deflection zones.

Argialas and Miliaresis (1997) identified a need to incorporate considerations of spatial reasoning into representation of landform classification knowledge in recognition that rules based solely on consideration of a landform's inherent pattern elements were incomplete. Spatial reasoning was captured mainly in terms of defining the context of individual landform elements. Argialas and Miliaresis (1997) and Leighty (2001) recognised that expert analysts make use of *a priori* physiographic information to focus their search for the correct identification of the landform at a site.

An expert takes into account the *regional context*, the *physiographic context*, the *geomorphic process context* and other forms of context to arrive at an interpretation of a landform (Argialas and Miliaresis, 1997). In most manual systems of landform classification an expert who has familiarity with theoretical concepts applicable for differentiating landforms generally, and who may also possess specific familiarity with local landform types and arrangements, interprets available information about the land surface to partition it into spatial entities that separate and describe different landform classes. This is most commonly accomplished through manual

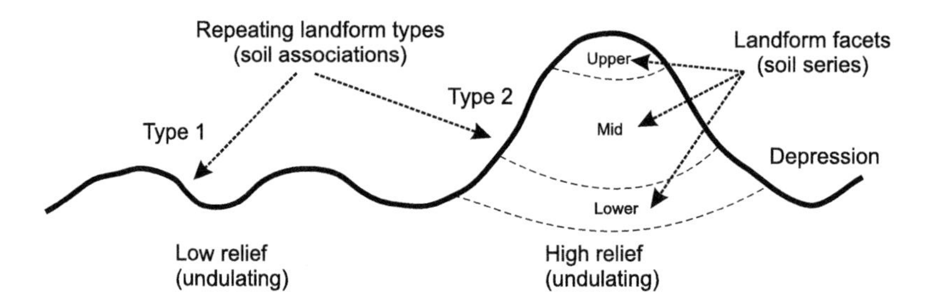

FIGURE 5 Illustration of conceptual differences between repeating landform types and landform facets.

visual examination of stereo air photos to interpret and delineate different patterns that can be observed in three dimensions on the stereo photos (Rossiter and Hengl, 2002).

Several significant early examples of landform classification used topographic contour maps as the primary consideration for identifying and delineating landform entities (Hammond, 1954, 1964). More recently, landforms have been delineated manually on-screen against 2D and 3D backdrops that use various combinations of derivatives of digital elevation models (DEMs) or digital imagery, or both, to support this identification and delineation of landform entities (Hengl and Rossiter, 2003).

2.1 Evolution of automated landform classification concepts and methods

Both manual and automated approaches to landform classification have tended to target recognition of classes that develop at one or more specific levels in multilevel hierarchies of landform entities. Speight (1974, 1990) proposed a two level descriptive procedure for a systematic, parametric description of landforms into landform patterns and landform elements (Klingseisen, 2004). Following the work of Kugler (1964), Dikau (1989) conceptualised a similar hierarchy of entities of increasing size and morphological complexity referred to as form facets, form elements, relief forms and relief associations or patterns. The USDA Geomorphic Classification System recognises two hierarchical components termed landform and element landform (Haskins et al., 1998).

Zinck and Valenzuela (1990) proposed four levels of landscape, relief, lithology and landform. The SOTER Global Soil and Terrain Database (van Engelen and Ting-tiang, 1995) recognised a nested hierarchy of mapping units distinguished principally on the basis of physiographic criteria. Three hierarchical classes of entities of terrain, terrain component and soil component were identified. Most efforts to automatically classify landforms have targeted their classification efforts at entities that are approximately equivalent to one, and only occasionally both, of the two main hierarchical levels of landform patterns or landform facets (Figure 5).

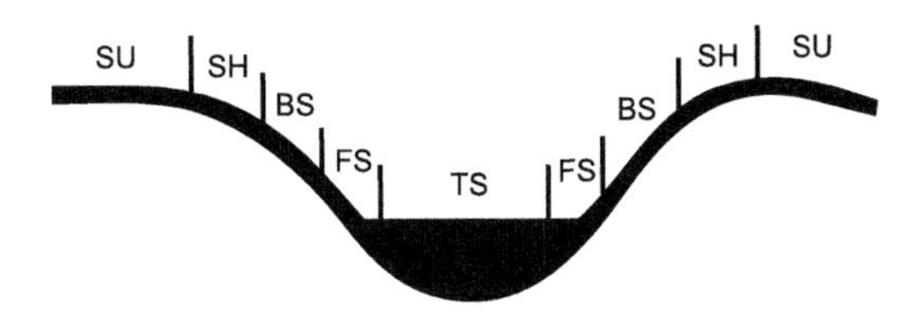

FIGURE 6 Illustration of conceptualisation of geomorphic units (SU — summit, SH — shoulder, BS — back-slope, FS — foot-slope, TS — toe-slope) along a hillslope catena by Ruhe and Walker (1968). Reprinted from Ventura and Irvin (2000). © 2000 John Wiley & Sons Limited. Reproduced with permission.

3. EXTRACTING AND CLASSIFYING SPECIFIC LANDFORM ELEMENTS

Automated extraction of specific landform elements from a DEM typically acts at more detailed level of abstraction and at a larger scale than classification of repeating landform types. Typically, the classification of landform elements involves segmentation of individual hillslopes into more or less homogeneous classes or facets along a catenary sequence or toposequence from ridge crest to valley bottom following concepts outlined by Milne (1935) and elaborated by Ruhe and Walker (1968) and Huggett (1975) (Figure 6). For examples see Dikau (1989), Pennock et al. (1994), Fels and Matson (1996), Irvin et al. (1997), Zhu (1997), MacMillan et al. (2000), Etzelmüller and Sulebak (2000), Bui and Moran (2001), Burrough et al. (2001), Bathgate and Duram (2003) and Drăguţ and Blaschke (2006).

3.1 Basic principles of automated recognition of landform elements

Many of the earliest efforts to partition landforms into landform elements from DEMs were based exclusively on analysis of local surface shape. Automated recognition of surface specific points (Peucker and Douglas, 1975; Collins, 1975) was used to differentiate grid cells into *pits* (all neighbours higher), *peaks* (all neighbours lower), *channels* (neighbours on two opposite sides higher), *ridges* (neighbours on two opposite sides lower), *passes* (neighbours on two opposite sides higher and on the orthogonal sides lower) and *plains* (no prominent curvatures defining distinct shapes). The capabilities of this approach are best illustrated by Wood (1996) who demonstrated how the procedures could be applied to extract and classify these morphological objects across a hierarchy of scales by applying them within calculation windows of ever increasing dimensions. Herrington and Pellegrini (2000) applied similar procedures with similar results.

Several widely applied approaches to automated classification of landform elements are based on consideration of local surface shape as measured by signs or values of curvatures. Curvature values are typically computed within a 3×3 window, but authors such as Dikau (1989), Wood (1996) and M.P. Smith et al. (2006) have demonstrated clear advantages to computing curvatures within a series of larger neighbourhood analysis windows. Shary et al. (2005) cite and review

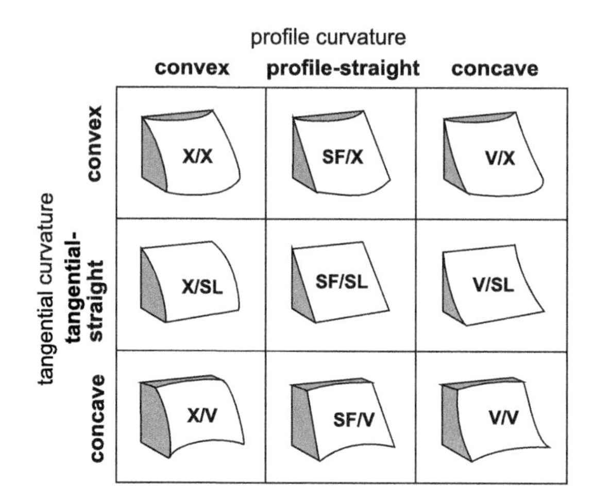

FIGURE 7 Illustration of modified Dikau (1989) classification of form elements based on the profile and tangential (across slope) curvatures. The elements have been further classified as positive or negative based on the radius of curvatures (>600 or <600 m). Shary and Stepanov (1991) provide convincing arguments for replacing plan curvature with tangential curvature in this classification of form elements.

the very early use of curvatures by Gauss (1828) (Figure 3) and much later by Troeh (1965) (Figure 4) to classify form elements. The initial concepts of Richter (1962) and Troeh (1964, 1965) have been adapted and applied by others including Huggett (1975) and Dikau (1989) (Figure 7).

Shary and Stepanov (1991) proposed that tangential curvature be used to replace plan curvature in this classification of form elements. Tangential and profile curvatures are both curvatures of normal sections and they both exhibit similar statistics of distribution whereas plan curvature exhibits markedly different statistics of distribution (SD as much as 10× greater than profile curvature). Plan curvature has therefore been argued to be less stable and less suitable for describing across-slope curvature than tangential curvature. A modified version of the schema of Dikau (1989) is shown in Figure 7. This differs from Troeh's landform classification (Figure 4) only in Dikau's introduction of *straight* slopes that have a curvature radius greater than 600 m.

Dikau's (1989) choice of curvatures to describe straight slopes can be clarified as follows. Imagine a hill that has a form of hemisphere of radius R. For this hill, curvature of any normal section is $1/R$, so that PROFC $=$ TANGC $= 1/R$ at any point of the hill. Contour lines are circles of radius r, so that PLANC $= 1/r$ varies from $1/R$ at the hill boundary to infinity at the top of the hill. Dikau's concept was to define straight slopes using land-surface shape that is described by curvatures of normal sections. According to his concept, the hill's curvature is negligible when $R > 600$ m (i.e. PROFC $< 1/600$ and TANGC $< 1/600$), but his choice of PLANC $= 1/r$ is based on contour shapes instead of surface shape thus making the meaning arbitrary, because r is always small near the top of the hill even for

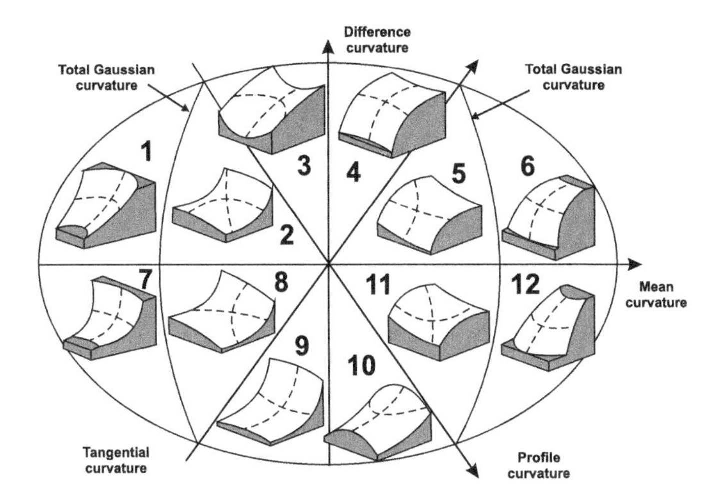

FIGURE 8 Illustration of Shary's complete system of classification according to signs of tangential, profile, mean, difference and total Gaussian curvatures. Reprinted from Shary et al. (2005).

very large R (hundreds of kilometres) when the hill's surface is almost plane and is clearly straight.

An expanded and formalised concept to define a complete system of classification of curvatures that includes the previous classifications of Gauss (1828) and Troeh (1965) as sub-sets (Figure 8) has been presented by Shary et al. (2005).

Many subjective classifications of slope profiles have been proposed that were intended to build upon conceptual classifications of hillslopes into summits, shoulders, backslopes, footslopes and toeslopes as proposed by Ruhe (1960) (Figure 6), or the nine unit classification of hillslopes proposed by Dalrymple et al. (1968) and Conacher and Dalrymple (1977) (Table 1), or the ten types of topographic landform positions described by Speight (1990) (Table 2).

In discussing the detailed qualitative profile description of Dalrymple et al. (1968), Huggett (1975) noted that even small changes in plan may result in strong changes of soil properties. In addition to considering curvature, these conceptual classes either implicitly or explicitly reference slope gradient and relative slope position along a conceptualised toposequence from divide to channel. Automated classifications that have attempted to apply these concepts have consistently recognised a need to consider additional measures of slope gradient and relative landform position in addition to just curvatures.

> REMARK 5. *Recognition of surface specific points and lines defined by pits, peaks, passes, pales, ridge lines and saddle lines is fundamental to most efforts to recognise and classify landforms. These key points establish the size, scale and contextual position of individual landforms and landform elements.*

Currently, no quantitative definitions exist to distinguish between open and closed depressions. Martz and de Jong (1988) described an algorithm that iden-

TABLE 1 Conceptual landform units defined by Conacher and Dalrymple (1977)

Land-surface unit	Characteristics
1 *Interfluve*	Interfluve with predominant pedomorphologic processes caused by vertical (up and down) soil–water movements; 0–1° slope gradient
2 *Seepage slope*	Upland area where responses to mechanical and chemical eluviation by lateral subsurface soil–water movements predominate
3 *Convex creep slope*	Convex slope element where soil creep is the predominant process, producing lateral movement of soil materials
4 *Fall face*	Areas with gradients greater than 45° characterised by the process of fall and rockslide
5 *Transportational midslope*	Inclined surfaces with 1–45° gradients and responses to transport of large amounts of material downslope by flow, slump, slide, erosion and cultivation
6 *Colluvial footslope*	Concave areas with responses to colluvial redeposition from upslope
7 *Alluvial toeslope*	Areas with responses to redeposition from upvalley alluvial materials; 0–4° gradient
8 *Channel wall*	A channel wall distinguished by lateral corrosion by stream action
9 *Channel bed*	A stream channel bed with transportation of material downvalley by stream action as the predominant process

After Ventura and Irvin (2000).

tified and described *closed* depressions, but no corresponding algorithm for unambiguously identifying *open* depressions is currently known. The importance of both open and closed depressions has long been recognised in soil science. For example, Neustruev (1930) described essential soil changes in open depressions that he termed *semi-depressions*. Shary et al. (2002) noted that open depressions are gravity-independent depressions. Shary et al. (2005) showed depressions and hills to be regional, terrain-specific landforms that were scale-free, in terms of their advanced definition of scale-free. The original classification efforts of Pennock et al. (1987) explicitly assumed that surface form, as described by curvature, could be directly related to surface processes and to relative landform position (Table 3). Thus, strong profile convexity was assumed to be indicative of upper, water-shedding slope positions, strong profile concavity was associated with lower, water-receiving landform positions and planar surfaces were associated with backslopes or flat areas (Figure 9). However, this pattern is not always adhered to and there are many instances where convex-concave patterns repeat over short distances along a longer hill slope (reflecting micro-topography).

TABLE 2 Morphological type (topographic position) classes of Speight (1990)

Name	Definitions of Speight (1990)
1 *Crest*	Area high in the landscape, having a positive plan and/or profile curvature
2 *Depression* (open, closed)	Area low in the landscape, having a negative plan and/or profile curvature; closed: local elevation minimum, open: extents at same or lower level
3 *Flat*	Areas having slope <3%
4 *Slope*	Planar element with an average slope >1%, sub-classed by relative position
5 *Simple slope*	Adjacent below a crest or flat and adjacent above a flat or depression
6 *Upper slope*	Adjacent below a crest or flat but not adjacent above a flat or depression
7 *Mid-slope*	Not adjacent below a crest or flat and not adjacent above a flat or depression
8 *Lower slope*	Not adjacent below a crest or flat but adjacent above a flat or depression
9 *Hillock*	Compound element where short slope elements meet at a narrow crest <40 m
10 *Ridge*	Compound element where short slope elements meet at a narrow crest >40 m

After Klingseisen (2004).

TABLE 3 Pennock's (1987) original classification criteria for landform elements

Landform element	Profile curvature (rad/100 m)	Plan curvature (rad/100 m)	Slope gradient (°)
Convergent Footslope (CFS)	<−0.10	<0.00	>3.0
Divergent Footslope (DFS)	<−0.10	>0.00	>3.0
Convergent Shoulder (CSH)	>0.10	<0.00	>3.0
Divergent Shoulder (DSH)	>0.10	>0.00	>3.0
Convergent Backslope (CBS)	>−0.10, <0.10	<0.00	>3.0
Divergent Backslope (DBS)	>−0.10, <0.10	>0.00	>3.0
Level (L)	any	any	<3.0

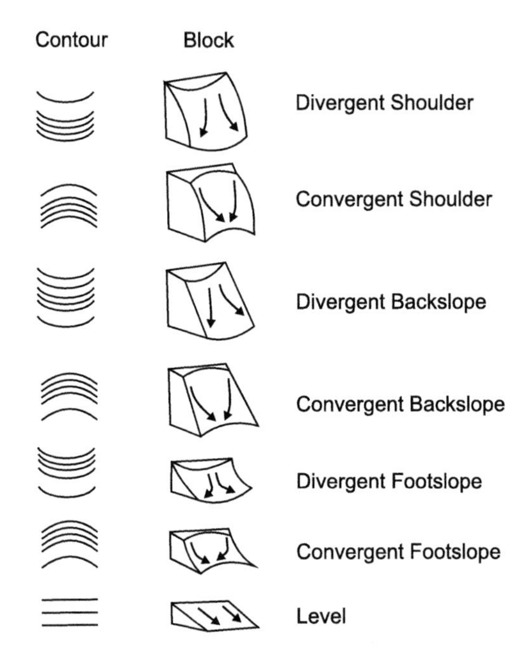

FIGURE 9 Illustration of the original classification of landform elements of Pennock et al. (1987) based on considerations of profile and plan curvatures and slope gradient. Reprinted from Ventura and Irvin (2000). © 2000 John Wiley & Sons Limited. Reproduced with permission.

Irvin et al. (1997) made similar assumptions and developed a similar classification. Due to the presence of multiple scales of topographic variation in natural landscapes, it is not uncommon to encounter areas with strong local profile concavity in upper slopes and strong local profile convexity in lower slopes. Pennock et al. (1994) subsequently recognised limitations associated with using curvature as the sole predictor of relative landform position and tried several alternatives, ultimately adopting use of the regional variable of specific dispersal area to differentiate upper level from depressional landform elements (Pennock and Corré, 2000; Pennock, 2003).

Note that the value of 3.0-degree gradient selected by Pennock et al. (1987) is subjective, and other values may be more suitable for other terrains. Pennock and Corré (2000), Pennock (2003) studied glacial terrain in the Canadian Prairies and selected a critical threshold based on consideration of the gradient at which soil erosion by water was considered to initiate. Dikau (1989) also concluded that consideration of curvatures alone (i.e., predictable landform classifications) was insufficient to permit recognition of conceptual landform elements whose definitions referenced their vertical position in the landscape through use of terms such as crest or valley.

Dikau (1989) proposed and described several regional measures that could be used to infer relative slope position including horizontal distances to channels and drainage divides, the area of drainage basins above each surface point and the

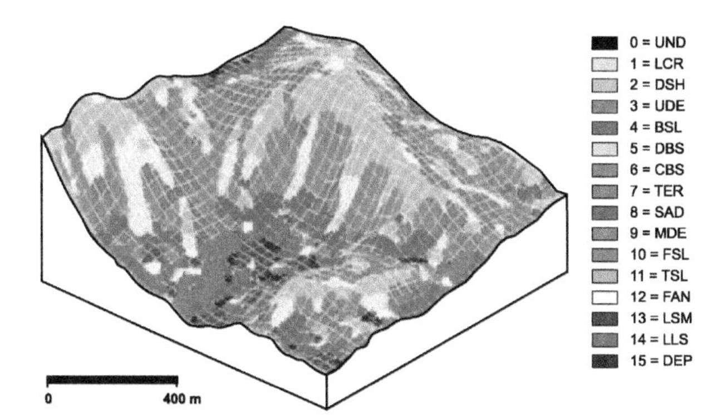

FIGURE 10 Illustration of landform elements extracted from land-surface parameters: 64 ha site in Alberta, Canada. See further Section 2 in Chapter 24. (See page 718 in Colour Plate Section at the back of the book.)

elevation difference of surface points above a channel or valley floor. MacMillan et al. (2000) modified Pennock's original rules to include consideration of relative slope position measured in terms of elevation of a point above a defined channel or sink relative to the total elevation difference from channel to divide or pit to peak (Figure 10).

> REMARK 6. *Landform elements can be extracted automatically by using land-surface parameters such as slope, curvatures, catchment area, distance to streams, peaks and depression depth.*

In a similar vein, Shary et al. (2005) described calculation of regional measures of *maximal catchment area* and its inverse, *maximal dispersion area*, and *depression depth*, and its inverse, *hill height* and showed how hills, depressions and saddles could be defined that subdivide any terrain into three non-overlapping types of regional forms with precisely defined boundaries. These regional measures were able to define the contextual position of points in a landform relative to concepts such as upper or lower, crest or valley.

Automated calculation of relative landform position, as defined qualitatively by Speight (1990), has been applied by Klingseisen (2004) using methods proposed by Skidmore (1990) and Coops et al. (1998). Skidmore clearly related his calculation of relative slope position to implied processes of hillslope (and soil) formation (Figure 11). Upper slope positions were regarded as zones of active removal and transportation of materials while lower slope positions were clearly identified as zones of deposition and accumulation of materials (Skidmore et al., 1991). Other approaches that have adopted calculation and consideration of one or more measures of relative landform position include those of Fels and Matson (1996), Franklin (1987), Franklin and Peddle (1987) and Giles and Franklin (1998). The concept of landform elements defined on the basis of consideration of a combination of measures of local surface shape (convexity/concavity) plus relative slope position is illustrated in Figure 10 using an example drawn from MacMillan et al. (2000) (see also Chapter 24).

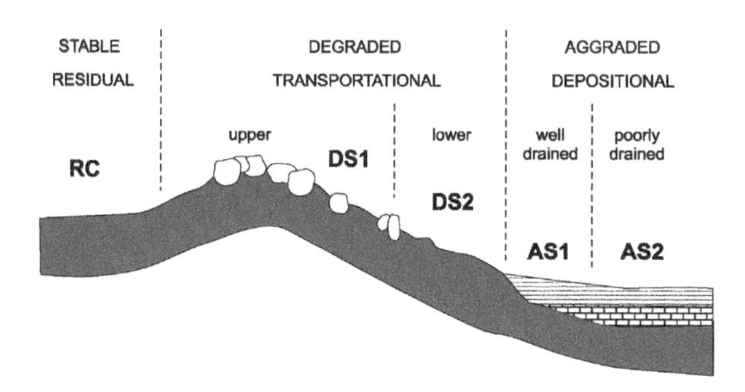

FIGURE 11 Illustration of the inferred relationship between slope position and geomorphic processes inherent in the approach of Skidmore et al. (1991). Reprinted with permission from Taylor & Francis Group.

Basic principles of hillslope analysis recognise that slope profiles may be characterised by changes, or breaks, in slope along recognisable lines of inflection (Pitty, 1969) or other discontinuities. Minár and Evans (2008) propose using what they term singular lines and points defined by local extrema, lines of inflection and other discontinuities to define boundaries between elementary landform segments (for a model surface).

Extraction and classification of landform elements can be improved by including procedures that automatically locate such singular lines and points along hillslope profiles from divide to channel. Giles and Franklin (1998) described a procedure for partitioning two-dimensional slope profiles into geomorphological objects, which they called slope units. Slope units were defined as *a section of a two-dimensional downslope profile having relatively homogeneous form, process, and lithology with upper and lower boundaries located at breaks of slope*. Giles and Franklin (1998) only implemented their approach for two-dimensional cross sections and did not extend it into the third dimension to exhaustively partition complete DEMs into non-overlapping functional areas. The approach of *walking* down a slope from crest to channel while explicitly identifying the locations of major breaks in slope and of using these major slope breaks to classify hillslopes into hillslope elements has considerable potential that has not yet been fully exploited for automated landform segmentation. Lloyd and Atkinson (1998) suggested combination of DEM segmentation and analysis of variograms for extracted landform units to optimise sampling strategies. Geostatistics in general has still a lot more to offer to objective stratification of topography.

4. EXTRACTING AND CLASSIFYING REPEATING LANDFORM TYPES

Automated classification of repeating types of landforms typically acts at a higher level of abstraction and at a broader scale than classification of landform elements (Figure 5). Generally, it involves delineation and classification of regions or areas conceptualised at the level of hills, plains or valleys that are characterised by repeating patterns of cyclical variation in size, scale, relief, morphology and landform context within a defined neighbourhood. For examples consult Hammond (1954, 1964), Pike (1988), Dikau et al. (1991, 1995), Guzzetti and Reichenbach (1994), Bayramin (2000), Brabyn (1998).

Based on Wood and Snell's (1960) terrain index, Pike (1988) defined the geometric signature as *a set of measures that describes topographic form well enough to distinguish among geomorphically disparate landscapes*. All measures used by Pike (1988) to differentiate landslide-prone landscapes in northern California were produced by analysis of terrain derivatives computed from gridded digital elevation data sets (DEMs). The procedures proposed by Pike (1988) for automatically classifying landform types involved passing a moving window of fixed dimensions (21×21 cells) over a raster elevation data set and computing statistical distributions of morphological measures within the window area (these are local land-surface parameters, see Section 2 in Chapter 6). In his 1988 example, he computed up to 75 different variables at each of 33 locations of the sampling window within the study area. These variables represented statistics (mean, minimum, maximum, variance, cumulative frequency) of slope angle computed for a constant horizontal length (30 m) and multiples of it, statistics of curvature in profile and plan the same constant and multiple lengths, statistics of slope between slope reversals and statistics of range and variance in altitude within the sampling window.

REMARK 7. *A geometric signature is a set of measures that describes topographic form well enough to distinguish among geomorphically disparate landscapes.*

The significant contribution of Pike's (1988) geometric signature was advancement of the concept that a series of terrain derivatives and their statistics computed within a moving window of fixed dimensions could be used to uniquely identify different types of terrain. Pike's early (1988) implementation of the concept was limited by the small number of locations sampled and by his failure to propose a set of classes for a generic classification of landform types (e.g. rolling, dunes, undulating) nor did he identify which variables would be required to implement and apply such a classification successfully. Iwahashi and Pike (2007) recently extracted 16 (unsupervised) classes of land-surface form from the world SRTM 1-km DEM by an iterative nested-means algorithm and a geometric signature that requires only three land-surface parameters — slope gradient, local convexity, and spatial *intricacy* ('*texture*').

4.1 The Hammond/Dikau method

The method of Dikau et al. (1991, 1995) follows Hammond (1954, 1964) in recognising four classes of proportion of gentle slopes, six classes of relative relief and four classes of profile type (Table 4) which, when combined, lead to 96 possible landform sub-classes (Brabyn, 1998). These are commonly re-grouped into 24 landform classes and five major landform types of Plains, Tablelands, Plains with Hills or Mountains, Open Hills and Mountains, and Hills and Mountains (Table 5), after Bayramin (2000).

TABLE 4 Classification criteria of the Dikau et al. (1991) method

Distribution of gentle slopes	Local relief	Profile type
(A) More than 80% of the area is gently sloping	(1) 0–30 m	(a) More than 75% of gentle slope is lowland
(B) 50–80% of the area is gently sloping	(2) 30–91 m	(b) 50–75% of gentle slope is lowland
(C) 20–50% of the area is gently sloping	(3) 91–152 m	(c) 50–75% of gentle slope is upland
(D) Less than 20% of the area is gently sloping	(4) 152–305 m (5) 305–915 m (6) >915 m	(d) More than 75% gentle slope is upland

TABLE 5 Classes and subclasses of the Dikau method (Bayramin, 2000)

Landform type	Landform class	Landform subclass code
Plains (PLA)	Flat or nearly flat	A1a, A1b, A1c, A1d
	Smooth plains with some local relief	A2a, A2b, A2c, A2d
	Irregular plains with low relief	B1a, B1b, B1c, B1d
	Irregular plains with moderate relief	B2a, B2b, B2c, B2d
Tablelands (TAB)	Table lands with moderate relief	A3c, A3d, B3c, B3d
	Table lands with considerable relief	A4c, A4d, B4c, B4d
	Table lands with high relief	A5c, A5d, B5c, B5d
	Table lands with very high relief	A6c, A6d, B6c, B6d
Plains with Hills or Mountains (PHM)	Plains with hills	A3a, A3b, B3a, B3b
	Plains with high hills	A4a, A4b, B4a, B4b
	Plains with low mountains	A5a, A5b, B5a, B5b
	Plains with high mountains	A6a, A6b, B6a, B6b
Open Hills and Mountains (OPM)	Open very low hills	C1a, C1b, C1c, C1d
	Open low hills	C2a, C2b, C2c, C2d
	Open moderate hills	C3a, C3b, C3c, C3d
	Open high hills	C4a, C4b, C4c, C4d
	Open low mountains	C5a, C5b, C5c, C5d
	Open high mountains	C6a, C6b, C6c, C6d
Hills and Mountains (HMO)	Very low hills	D1a, D1b, D1c, D1d
	Low hills	D2a, D2b, D2c, D2d
	Moderate hills	D3a, D3b, D3c, D3d
	High hills	D4a, D4b, D4c, D4d
	Low mountains	D5a, D5b, D5c, D5d
	High mountains	D6a, D6b, D6c, D6d

Automated procedures for implementing Hammond's (1954) manual system of landform classification, developed by Dikau et al. (1991, 1995), have been widely adopted and recognised by many as a *de facto* standard for automated classification of subjectively defined repeating landform types Brabyn (1998), Bayramin (2000). The method of Dikau et al. (1991, 1995) computes the slope gradient within a 3×3 window centred with horizontal cell dimensions of 200 by 200 m. A large window of fixed dimensions (9.8 by 9.8 km) is then passed over the entire grid and calculations are made at each grid location of the percentage of all cells within the window that are classified as flat (given as $<8\%$) versus sloping. Brabyn (1998) accomplished this task by reclassifying all grid cells with a slope percent $<8\%$ to a value of 100 and all cells with slope $\geqslant 8\%$ to a value of 0. He then computed the focal mean within a circular window with a radius of 5600 m in preference to the 9.8×9.8 km rectangular window used by Dikau et al. (1991). The result of this (local) focal mean filter for each grid cell gives the proportion of grid cells within the window that have a slope less than 8%.

The relative relief within each window is then computed for each grid cell as the maximum minus minimum elevation within the window area. This also can be computed in most GIS systems by passing a focal filter over the matrix and computing the maximum, minimum and range of values within the window. The third input required by the Dikau method is a computation of profile type, defined as the proportion of flat or gently sloping terrain that occurs in upland versus lowland areas. Each centre cell of the moving window is first examined to ascertain whether it lies in the upper or lower half of the landscape within the moving window. Finally, the proportion of flat cells in upland and lowland areas is computed and used to differentiate areas with flat tablelands from areas with flat bottomlands. For a detailed implementation please refer to Brabyn (1998).

Brabyn (1997, 1998) noted two problems with implementation of the methods. The first was a pattern of progressive zonation that developed in transition zones between areas of high and low relief. The second was that the methods classified areas with quite different macro landform into the same class. Guzzetti and Reichenbach (1994) also identified concerns with imprecise boundary locations and mixed classification and tried to address them by developing and applying a different approach that combined automated classification of DEM data with manual interpretation and digitizing of clear, sharp final boundaries.

4.2 The search window problem

Most efforts to classify repeating landform types have approached the problem in the following way:

- Go to each grid cell in a raster array (usually a DEM);
- Determine the optimum horizontal dimensions for a search window centred at each cell within which to compute diagnostic spatial statistics for each cell;
- Compute a variety of statistics that summarise the variation in key morphological properties within a search window of fixed or variable size centred at each cell;

- Apply classification rules to classify each grid cell into a landform type based on consideration of the statistics of variation in morphological properties within the window;
- Agglomerate clusters of adjacent cells with identical landform type classifications to define the spatial extent of different landform types.

A key challenge for this typical approach has been the problem of how to best establish the *extent of the search window* within which to compute diagnostic morphometric statistics. Several different methods have been suggested for varying the dimensions of moving windows used to derive land-surface parameters so that they match the variation in spatial texture of the land surface. The simplest approach has been to visually review the topography for an area and manually assign fixed search window dimensions that approximate the broad-scale spatial texture of the entire area or different search window dimensions to different parts of an area (Pike, 1988; Fels and Matson, 1996; Menz and Richters, in press; Bayramin, 2000).

The concept of a search window determined objectively, by broad-scale land-surface texture, originated in Gutersohn's attempt to more accurately calculate *Reliefenergie* (local, or relative, relief) from contour-map data (Mark, 1975b). His quantitative technique allowed the land surface itself to decide the optimal size of a search window. In each topographic region of interest Gutersohn calculated values of relief for a series of nested squares and plotted relief (Y) against length of the side of each square (X). Relief typically increased steeply with length up to a breakpoint, where the plot levelled off or rose at a much reduced slope; the sampling square in which this breakpoint occurred was designated the optimal spatial extent for calculating relief. The technique was next applied (manually) in the land-surface regionalisation of central Europe by Wood and Snell (1960), who calculated grain on nested circles rather than squares. Wood and Snell were the first to designate the breadth of the sampling area containing the breakover point as topographic grain. Wood (1964, personal communication to R. Pike) later recognised this inflected relief/distance relation as an example of spatial autocorrelation (Waldo Tobler's *First Law of Geography*).

REMARK 8. *It is not possible to select any single fixed dimension for a moving window that will perfectly capture the wavelength of all landform features of interest in any given area.*

Pike et al. (1989) reviewed the concept of topographic grain and automated its cell-by-cell calculation from 30-metre DEMs by nested squares. They also demonstrated that grain and its accompanying value of relative relief are equivalent, respectively, to the range and sill of inflected variograms. Pike et al. (1989) and Guzzetti and Reichenbach (1994) observed that a search window should try to approximate the topographic grain, defined as the longest significant ridge to valley wavelength of topography within a local area (see Figure 5). They concurred with the earlier observation by Wood and Snell (1960) that a window that does not include at least one major ridge and valley is too small and is unlikely to properly represent the dominant local wavelength while a window that is too large will fail to recognise significant smaller local relief features. It is not possible to select any single fixed dimension for a moving window that will perfectly capture the wavelength of all landform features of interest in any given area. This is because landforms are not perfectly uniform (or concentric) and all areas will exhibit some minor to major degree of variation in the wavelength and relief of the landforms they contain.

Several authors since have suggested computing *variograms* within expanding windows centred at every grid cell in a matrix and fitting a model to the variogram to establish the distance over which variation in topography is operative (Franklin et al., 1996; Lloyd and Atkinson, 1998; Kyriakidis et al., 1999; Etzelmüller and Sulebak, 2000; Moran and Bui, 2002). These authors suggested allowing the size of the moving search window to vary in response to the dimensions of the range computed for the semi-variogram. Most acknowledged, however, that computing semi-variograms for every grid cell was prohibitively time consuming and not yet feasible.[5]

Moran and Bui (2002) adopted a compromise approach that involved computing the slope at the origin for a local variogram centred at each grid cell but solved for only a single fixed radius of 1 km about each point. They interpreted this slope to vary the dimensions of their search window within a specified maximum to minimum range. *Fourier and wavelet transforms* (Nogami, 1995; Lucieer et al., 2003), gray level *co-occurrence matrices* (Haralick et al., 1964; Leighty, 2001) and *local binary pattern operators* (Lucieer et al., 2003) have also been proposed as methods for assessing the size and range of local texture that could be used to vary the size of moving windows within which to compute the statistical distributions of terrain characteristics.

To account for the spatial extent of the landform features, Drăguţ and Blaschke (2006) employed *multiscale image segmentation* algorithms in combination with fuzzy classification to automatically extract landform elements (homogeneous spatial objects) for two contrasting terrains. In their segmentation algorithm, the 'scale parameter' was defined as the maximum change in the total heterogeneity that may occur when merging two image objects.

4.3 Addressing the search window problem

An alternative solution to the search window problem is to extract, characterise and classify spatial objects equivalent in concept to individual hillslopes or individual peaks that explicitly capture one full ridge to valley or valley to valley wavelength in the landscape as follows:

- Use flow modelling to delineate local catchments for down slope flow (pit sheds) and repeat for notional upslope flow (peak sheds);
- Compute the unique intersection of drainage sheds and source sheds to define individual hill sheds (or hillslopes);
- Compute a full suite of morphological measures for every grid cell in a raster DEM (e.g. gradient, slope length, aspect, curvatures, etc.);

[5] The three small example maps in Pike et al. (1989), required 8 to 50 hours on contemporary Unix workstations.

FIGURE 12 Illustration of possibilities and problems with using hillslopes as basic spatial entities for classifying repeating landform types. See text for detailed discussion. (See page 718 in Colour Plate Section at the back of the book.)

- Compute summary statistics (means, medians, min, max, range, frequency distributions) for each morphological measure within each unique spatial entity (peak shed, pit shed or hill shed);
- Use these area statistics as inputs to procedures to classify each unique spatial entity.

Several authors have hinted at using hydrological spatial entities, such as local catchments or individual hillslopes, as the basic spatial entities for classifying landform types but none have fully developed the proposals (Dehn et al., 2001). Schmidt et al. (1998) recognised the similarity of hydrological and geomorphological classifications and identified catchments, landform units and form elements as hydrological areal objects that could be characterised and classified. Band (1989) and Wood (1996) both described procedures for extracting channels and ridges that, when intersected, defined a spatial framework of individual hillslopes. Neither proposed using these hillslopes as spatial objects for classifying landform types.

Shary et al. (2002) argue that, a significant unresolved problem is that we have algorithms to describe relative position within a basin (Martz and de Jong, 1988; Freeman, 1991), but we have no physically based, substantiated algorithms to describe slope profiles. The topographic index (also known as topographic wetness index) of Beven and Kirkby (1979) might be considered as a measure of relative slope position that takes into account both plan and profile contributions, but the profile contribution in this approach, was deemed by Shary et al. (2002) to be essentially local by 1/slope. One may propose that introduction of new land-surface parameters for physically based, or substantiated terrain-specific regional profile description, is of great importance for further development of objective landform description at the regional level.

Although conceptually attractive, use of individual hillslopes as basic spatial entities for classifying repeating landform types does present some problems and limitations.

In Figure 12, individual hillslopes are outlined using thick yellow lines. Most of the hillslopes consist of two, three or more different hillslope segments characterised by significantly different slope gradients. The thick black lines identify the locations where changes in slope gradient in the down-slope direction are deemed significant. It is not clear which of the illustrated breaks in slope is the most significant, or indeed whether any one is consistently more important than the others. Although the hillslope entities do identify the extent of one full ridge to channel wavelength of topographic variation, it becomes obvious in reviewing Figure 12 that a single individual hillslope can enclose areas that would be considered by any reasonable human interpreter to consist of very different landforms consistent with concepts such hill or mountain versus valley or basin.

The semantic difficulties of precisely defining where a mountain or hill ends and where its associated valley begins have been discussed in detail by Smith and Mark (2006), Mark and Smith (2004) and Fisher et al. (2005), who highlighted the vagueness and imprecision with which the transition from an entity conceptualised as a mountain to one conceptualised as a valley occurs. Solutions to the problem of differentiating hills from valleys have been proposed by Shary et al. (2005) who demonstrated that the largest closed contour line that completely encloses a hill or peak can be taken to define the boundary between a hill and a valley.

The problem of further subdividing individual hillslopes is well illustrated in Figure 12 as are limitations that arise from the delineation of partial or fragmented hillslopes. So, while use of hillslopes as basic spatial entities that exactly capture the full extent of one full cycle of topographic variation from crest to channel is conceptually attractive, in practice, limitations are encountered that may render this approach untenable for isolating and classifying individual repeating landform types. More work needs to be done before automated classification of repeating landform types can make effective use of hillslopes, or portions of hillslopes, as basic spatial entities which can be classified as specific landform types.

5. IMPLEMENTING EXTRACTION OF LANDFORMS

Virtually all procedures for developing and applying automated (or manual) landform classifications can be thought of as involving a similar set of basic activities. McBratney et al. (2003) identified seven main stages in developing and applying rules for automated production of classed soil maps. We outline here a modification to five steps of a three step organisation of procedures for classifying landforms originally described by Weibel and DeLotto (1988):

- Establishment of the spatial object(s) of classification;
- Specification and computation of the input variables;
- Extraction or creation of the classification rules;
- Application of the classification rules;
- Evaluation and assessment of accuracy.

Different applications of automated landform classification are initially distinguished by differences in the types and scales of the identified spatial objects of classification. Examples of automated landform classification are subsequently distinguished mainly in terms of the variables selected to support the classification or in the methods used to create and apply either objective or subjective classification rules.

All methods of automatically predicting output classes of geomorphic spatial entities are based on identifying and developing rules for establishing predictive relationships between input variables, or statistics of input variables, and desired output classes. A key step in any approach to automated classification is therefore to identify and create, or obtain, a collection of suitable input variables in digital format.

Rules for classifying terrain entities can only be created after the size, scale and nature of the terrain objects of interest have been specified and after the terrain derivatives or input variables required to effect a classification have been identified and computed. Efforts to develop rules to replicate or approximate subjective systems of landform classification usually do require some means of uncovering effective classification rules. Unsupervised, supervised and knowledge-based (heuristic) approaches have been applied to automatically extract and classify subjectively-defined landform entities.

REMARK 9. *Extraction of landform types and elements from DEMs commonly consist of: (a) preparation of the legend, (b) preparation of the LSPs (inputs), (c) creation of the rules, (d) extraction of landforms and (e) assessment of accuracy.*

Classification and mapping systems devised by humans often represent the end products of assimilation and mental analysis of large volumes of field observations and other data by local experts to create mental or conceptual models of the rules believed to govern soil-landform (or ecological-landform) spatial relationships. Such existing expert heuristic knowledge can be directly expressed using simple Boolean rules (Pennock et al., 1987) or using fuzzy heuristic rules that attempt to allow for vagueness in subjective heuristic knowledge (MacMillan et al., 2000).

Various approaches for automatically extracting rules for classifying areal objects, such as landform entities, are reviewed in Section 2.3 of Chapter 19 and further demonstrated in Chapter 22.

6. SUMMARY POINTS

For practical reasons, procedures for automated extraction and classification of landforms can be distinguished into those that attempt to recognise and classify repeating types of landforms and those that attempt to partition landforms into landform elements along a toposequence from divide to channel.

It is also important to distinguish between objective classification approaches that are primarily based on consideration of surface shape (curvatures) and that are typically applicable across all scales and subjective classifications that are typically applicable at a specific scale or range of scales that tend to use a variety of local and regional (contextual) land-surface parameters as inputs to the classification. Shary et al. (2005) put great emphasis on the need to explicitly differentiate between predictable and terrain-specific, and also between scale-specific and scale-free, landform classifications. They contend that recognition of these distinctions needs further research, especially if we want to be able to choose which approaches are most appropriate for particular landform classification needs.

We can conceptualise landform types as mainly consisting of *waveform features* that exhibit entire repeating cycles of variation in morphological properties such as slope gradient, slope lengths, relief, curvatures and moisture regime. These cyclic patterns can only be identified and characterised by analysing the distribution of variation in morphological attributes within neighbourhoods defined by windows of appropriate dimensions and shape, since morphological variables computed for any given cell describe only a small portion of the total cyclic variation that characterises a repeating landform type. Thus, procedures for automated classification of landform types mainly involve consideration of measures of texture and context that apply to regions within a window of suitable shape and dimensions. Automated classification of repeating landform types has therefore mainly relied upon morphological measures that describe statistics of variation in morphological properties within neighbourhoods or windows selected to encompass at least one full cycle of variation in topography from crest to trough.

Landform types provide information on the size and scale of landform features and how this size and scale might affect the amounts of energy available for geomorphic, pedogenic and hydrological processes. Landform types provide context that can be used to inform and improve the further sub-division of the landscape into landform elements.

Landform elements (except for local ones) are almost universally conceptualised as sub-divisions of hillslopes into segments or facets of relatively uniform shape, orientation, gradient and landform position along a toposequence from crest or divide to channel or valley.

Landform elements have been classified based solely on consideration of their local surface shape, on consideration of a combination of surface shape and slope gradient and on consideration of a combination of surface shape, slope gradient and contextual measures of relative landform position. Automated classification of landform elements has to date mainly relied upon consideration of the values of local and regional land-surface parameters computed for single cells and not for collections of cells within fixed windows.

Future advances in automated classification of landform elements are likely to be linked to improved abilities to explicitly partition hillslopes into hillslope segments through automated recognition of the locations of major slope breaks at points of inflection along hillslope profiles. Hillslope elements are conceptually linked to differences in hydrological, geomorphological and hillslope forming processes that operate at the level of an individual hillslope. Effective classification of hillslopes into hillslope elements can therefore support association of each hillslope element with an inferred hydrological and geomorphological regime that can be interpreted to predict the most likely assemblages of soils or ecological site types that might develop under such conditions.

Procedures for automatically extracting and classifying landform types and landform elements also differ in terms of the kinds of classification methods applied to extract the entities. Repeating landform types have mainly been classified using Boolean rules based on expert knowledge and Heuristic beliefs. Classification of landform elements has been achieved using a wide variety of classification methods including knowledge-based heuristic approaches, supervised classification, and unsupervised classification.

Human-devised conceptual classifications of repeating landform types tend to be far richer, subtler and more complex than any equivalent automated classifications that have been achieved to date. A great deal of progress will have to be made before automated classifications begin to approach the level of detail and subtlety of manually interpreted classifications. Automated classification of landform elements is closer to achieving effective recognition of concepts equivalent to those recognised by manual interpretation and classification of hillslopes into hillslope components. There are only a limited number of widely recognised conceptual hillslope components and several systems of automated classification of landform elements have come quite close to successfully capturing these concepts.

Despite the limitations noted above, there is considerable excitement surrounding the current state of development of procedures for automated classification of landforms. The results of automated classification of landforms can be of considerable interest and use, in and of themselves. Landform classifications can also establish context and identify fundamental conceptual spatial entities that can be interpreted in terms of the kinds and amounts of soils, ecosystems, vegetation communities, or environmental hazards that might reasonably be assumed to be associated with any given landform setting. While they may not be able to match the subtlety and richness of manually interpreted landform classifications, automated approaches have the advantage of being consistent, repeatable, updatable, and quantifiable. We can anticipate continued rapid improvements in the methods and data sources available to classify landforms automatically and can envisage situations, in the near future, where automated approaches will have become routine and will be consistently selected in preference to methods based on manually-based human interpretation.

IMPORTANT SOURCES

Iwahashi, J., Pike, R.J., 2007. Automated classifications of topography from DEMs by an unsupervised nested-means algorithm and a three-part geometric signature. Geomorphology 86 (3–4), 409–440.

Schmidt, J., Hewitt, A., 2004. Fuzzy land element classification from DTMs based on geometry and terrain position. Geoderma 121 (3–4), 243–256.

Burrough, P.A., van Gaans, P.F.M., MacMillan, R.A., 2000. High resolution landform classification using fuzzy *k*-means. Fuzzy Sets and Systems 113 (1), 37–52.

MacMillan, R.A., Pettapiece, W.W., Nolan, S.C., Goddard, T.W., 2000. A generic procedure for automatically segmenting landforms into landform elements using DEMs, heuristic rules and fuzzy logic. Fuzzy Sets and Systems 113, 81–109.

Lane, S.N., Richards, K.S., Chandler, J.H., 1998. Landform Monitoring, Modelling and Analysis. Wiley, 466 pp.

Wood J., 1996. The geomorphological characterisation of digital elevation models. Ph.D. Thesis. Department of Geography, University of Leicester, Leicester, UK, 185 pp.

Part II

Software

CHAPTER 10

Overview of Software Packages Used in Geomorphometry

J. Wood

software packages considered in this book · the relative benefits of using GIS and geomorphometric software · the availability of the software · approaches available for using each software package · other geomorphometry-connected software packages not listed in the book · the future of geomorphometric software

1. INTRODUCTION

As the first section of this book has demonstrated, there is a rich source of theory and techniques available for those studying the measurement of landscape form. This section considers some of the software that has been developed to implement those techniques and theory. What is considered here is not an exhaustive list of all geomorphometry software, but rather a selection of packages that represent different approaches to analysing surface data. They range from generic Geographic Information Systems to specific domain-focussed application software; from proprietary commercial software to open source solutions; from command-line script-based languages to graphical user interfaces to spatial data. The chapters in this part each introduce a particular software package, provide details on how to get access to the software, demonstrate how the software can be used to perform geomorphometric analysis using the common Baranja Hill dataset, and evaluate the software's strengths and weaknesses.

2. THE SOFTWARE LANDSCAPE

Before considering each software package individually, it is worth reflecting on the choices available to the researcher wishing to use software to perform some geomorphometric exploration and analysis. Eight packages are discussed in this section: ArcGIS, GRASS, ILWIS, LandSerf, MicroDEM, RiverTools, SAGA and TAS. All

Developments in Soil Science, Volume 33 © 2009 Elsevier B.V.
ISSN 0166-2481, DOI: 10.1016/S0166-2481(08)00010-X.

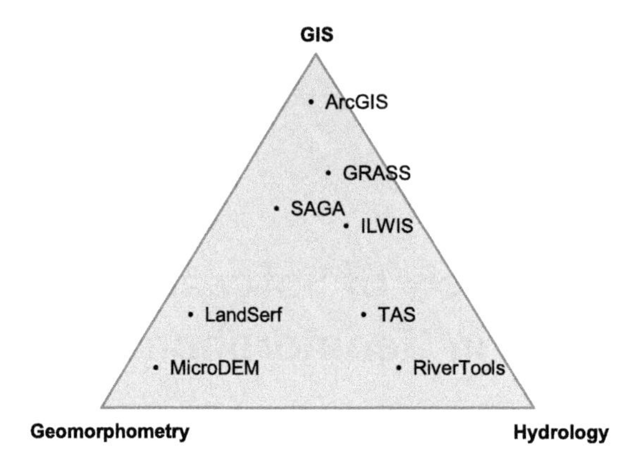

FIGURE 1 Approximate orientation of the software packages considered in this book.

can be considered as examples of *Geographical Information Systems* in that they all process geographically referenced data, and all have some functionality in common (e.g. all 8 packages can import, display and process Digital Elevation Models). They do though vary in the emphasis placed upon generic GIS tasks, geomorphometric analysis and hydrological analysis. Figure 1 summarises these different orientations of the various software packages.

Considering the triangle in a vertical direction, there is a choice to be made between general Geographic Information Systems (e.g. ArcGIS and GRASS) and software more specifically designed for analysis of surface models. Full GIS have the advantage of providing a wide range of functionality, large user communities and substantial support. The more application-specific software provides the advantage of functionality and support more tailored towards geomorphometric applications. The very existence of domain-specific software suggests that while there are considerable advantages in using established GIS, they do not necessarily provide sufficient functionality for all geomorphometric tasks. The chapters in this section provide more detail on the functionality and approach of each of these packages so that an informed choice may be made.

Considering the triangle in a horizontal direction, a distinction can be made between functionality directed towards hydrological analysis and that for more general geomorphometry. In practice there can be considerable overlap here, but the distinction can remain useful in identifying the application areas more commonly associated with the respective packages.

REMARK 1. *All geomorphometry software belong to the domain of GIS. They vary in the emphasis placed upon generic GIS tasks, geomorphometric analysis and hydrological analysis.*

Functionality alone is of course not the only criterion to be considered when choosing a software package. Price, availability and existing expertise all play a part. The packages considered in this book vary from free fully open source software (GRASS, ILWIS and SAGA), through free, but closed source software (LandSerf, MicroDEM and TAS), to commercial packages (ArcGIS, RiverTools).

TABLE 1 Operating systems used by geomorphometry software

	Windows	MacOSX	Linux	Solaris/Unix	Mobile OSs
ArcGIS 9.2	√	–	–	–	√
GRASS 6.2	-	√	√	√	–
ILWIS 3.3	√	–	–	–	–
LandSerf 2.3	√	√	√	√	–
MicroDEM	√	–	–	–	–
RiverTools 3.0	√	√	–	√	–
SAGA 2.0	√	–	√	–	–
TAS 2.09	√	–	–	–	–

Table 1 summarises the platforms upon which these packages can be run. While the table identifies the primary platforms upon which the software can be mounted, several packages also offer networking capabilities (most notably ArcGIS and GRASS) allowing software to be used with any networked platform. Many of the packages can also be used on a range of operating system platforms through the use of Operating System emulators (e.g. Parallels Desktop available for MacOSX allows Mac platforms to run Windows applications; CygWin for Windows allows PCs to run Unix-based software). It is perhaps worth noting, that only one package (ArcGIS) has a version available for mobile platforms. Mounting geomorphometry software on a location aware mobile platform offers new possibilities for field-based geomorphometry.

Finally when surveying the software landscape for geomorphometry it is worth considering the evolution of the packages over time. Figure 2 identifies when each of the packages considered in this book first became publicly available. The earliest packages (ArcGIS and GRASS) have been used for geomorphometric analysis for over 20 years. The early interfaces to both of these packages tended to limit their use to GIS experts. The subsequent emergence of more education-focussed packages in the late 1980s (ILWIS as well as MAP and Idrisi) may have partly overcome this problem, introducing the possibilities of automated terrain analysis to a new generation of researchers. However, the emergence of more geomorphometry focussed software in the 1990s suggests that at the time, existing GIS

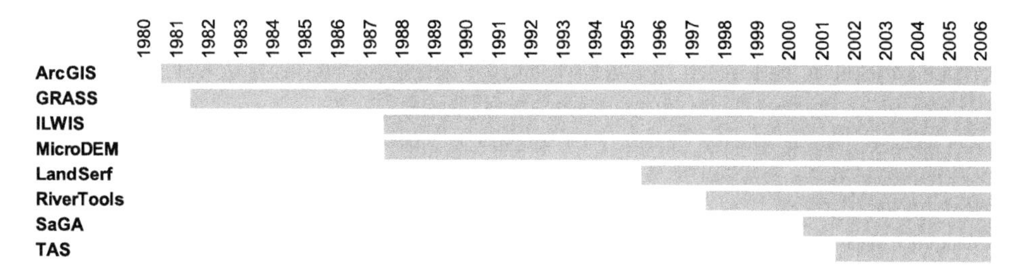

FIGURE 2 Timelines showing public availability of the geomorphometry software considered in this book.

TABLE 2 Interfaces available to the software packages

Package	GUI	Scripting	API
ArcGIS	√	AML, Avenue, Visual Basic, Python	C, C++, Java
GRASS	√	Unix shell, Python, PERL	C
ILWIS	√	ILWIS syntax	–
LandSerf	√	LandScript syntax	Java
MicroDEM	√	–	–
RiverTools	√	IDL	–
SAGA	√	–	C++
TAS	√	–	–

were not sufficiently focussed on the approaches adopted by geomorphometry researchers. If any other trend can be detected in the evolution of these packages it is that there appears to be a tendency for free or open source packages to become increasingly available. This may reflect the close relationship between (non-profit making) academics developing new geomorphometric techniques and the software required to demonstrate them, as well as a philosophy of open-source development that suggests this is a rapid way of building new developments into software.

3. APPROACHES TO USING SOFTWARE

The packages considered in this book offer a range of interfaces with the underlying data and functionality (see Table 2). All offer menu-based interfaces for selecting input, processing and output operations. Some offer command-line interfaces (e.g. GRASS, MicroDEM) or scripting (ArcGIS, GRASS, ILWIS, LandSerf). Some also offer a Application Programming Interfaces (APIs) using common languages such as C, C++, Java and Python (ArcGIS, GRASS, SAGA, LandSerf).

Which approach is most beneficial is going to depend on the expertise of the researcher, the task at hand, and the size of the user-community supporting the software. A well designed graphical user interface, exhibited by most of the packages considered in this book, will allow non-expert users to perform geomorphological analysis fairly rapidly. In some cases, where graphical interaction with surface models is required, that interface is part of the visual functionality of the package (e.g. LandSerf). In other cases, the GUI can be considered a convenience allowing the functionality of the package to be explored easily.

For tasks that are repetitive or that need to be shared between users, the scripting functionality of the software can be advantageous. The very large user-base of ArcGIS and its precursor ArcInfo combined with their long history means that there are many thousands of scripts available in the Arc Macro Language (AML) and Visual Basic to achieve geomorphometric and other spatial tasks. Scripting tends provide more precise control over the functionality of a software package, but without the requirement of learning a low level programming language. In some cases, the scripting languages used (e.g. in GRASS and LandSerf) has a direct correspondence with the menu-based functionality of the software. This can reduce the extra effort required to learn the scripting language. In other cases (e.g. ArcGIS, RiverTools), the scripting language can be used to create graphical user interfaces, effectively allowing the distribution of mini applications between users.

REMARK 2. *A competitive advantage of some geomorphometry software is that they provide scripting possibilities. The scripting language can be used to create graphical user interfaces, effectively allowing the distribution of mini applications between users.*

For researchers interested in implementing novel or recently developed algorithms, the graphical or scripting interfaces to software may not provide sufficient control for the programmer. In such cases, a documented API may be helpful. This provides the lowest level access to the software with a corresponding overhead in learning a programming language. For those already familiar with such languages (e.g. Java and C++), this may not represent much of a problem. The ease with which the API can be used will depend in part on how well it is documented. In the case of open source software (e.g. GRASS, SAGA), the continued development of the software itself is dependent on good quality documentation as well as a community of users familiar with the API.

4. OTHER PACKAGES FOR GEOMORPHOMETRY

A number of other software packages that provide capabilities for geomorphometric analysis have not been included in this book. In this section, we will review some of the more widely known and used packages that are either general GIS environments or specialised packages (or modules) for geomorphometric analysis. A summary review of the packages listed here is given in Table 3. Typically, general GIS/image processing packages such as ENVI/IDL, ERDAS Imagine, gvSIG, MapWindow, IDRISI or PC-Raster are usually limited considering the number of built-in parameters and objects that can be extracted in them. However, these packages usually provide programming environments where algorithms can be implemented, modified and improved by using scripts or batch commands. Packages developed mainly to process DEMs and to serve as stand-alone tools (or plugins) for geomorphometric analysis include: GIS Eco, LandLord, LandMapR, TAPES, TAUDEM, TopoMetrix, Wilbur and WMS.

Probably the most detailed set of stand-alone software tools for geomorphometry come from the Centre for Resource and Environmental Studies in Canberra (CRES[1]). Different modules of it, such as EROS (Wilson and Gallant, 1996) used for erosion modelling or SRAD used for solar radiation modelling, have been especially interesting for environmental applications. CRES also distributes the ANUCLIM (Houlder et al., 2000), which uses meteorological data (points) and DEM

[1] The same Institute developed the TAPES package extensively described by Wilson and Gallant (2000).

TABLE 3 A list of packages for geomorphometric analysis not described in further chapters

Package	Description	WWW domain
ENVI/IDL	A most general package for visualisation and analysis of imagery used in environmental modelling applications. It can be used to generate and visualise DEMs out of stereoscopic images. It is a platform for development of image processing applets.	ittvis.com/envi/
GIS Eco	A stand-alone package written by Peter Shary. It can calculate grids of 18 basic parameters, compute their statistics and run correlation analysis with environmental variables.	giseco.info
gvSIG	Spanish abbreviation that stands for *Generalidad Valencia Sistema de Información Geográfica*. It is an open-source GIS developed for Linux OS by the Valencian Regional Council for Infrastructures and Transportation (CIT). Victor Olaya has been developing Java-based modules for hydrological analysis.	gvsig.org
Idrisi	Consists of over 250 modules for the analysis and display of spatial information, including a module for surface modelling that can be used to extract slope, aspect, hillshading, fractal dimension, etc.	clarklabs.org
LandLord	Developed by Igor Florinsky who does not distribute the program, but is open for collaboration.	iflorinsky.narod.ru
LandMapR	An in-house toolkit (executable files) developed by Robert MacMillan. It can be used for extraction of basic morphometric and hydrological parameters and for extraction of landforms based on the expert-systems.	–
Manifold GIS	A general GIS and map-making package with powerful surface generation and visualisation tools.	manifold.net
PC-Raster	GIS modelling environment specialised in construction of iterative spatio-temporal environmental models. It has a number of built-in functions for extraction of basic morphometric parameters.	pcraster.geo.uu.nl

(*continued on next page*)

TABLE 3 (*continued*)

Package	Description	WWW domain
Surfer	Windows-based application specialised in surface modelling. It includes a large number of gridding techniques, as well as hill-shading and 3D displays.	goldensoftware.com
TAPES	A suite of command-based programs to build hydrologically-correct DEMs and extract a large number of parameters. Run as executables for Windows (ArcGIS), Linux or Sun Solaris OS.	uscgislab.net
TAUDEM	TauDEM (Terrain Analysis Using Digital Elevation Models) is a set of tools distributed as plugins to ESRI ArcGIS and MapWindow GIS. It is specialised for hydrological analysis and extraction of parameters and objects specific to hydrology (contributing area, channel networks, etc.).	hydrology.neng.usu.edu/taudem/
TNTmips	A general GIS and image processing system. It can be used to build and visualise DEMs (a large number of gridding options). Its strongest points are land-surface modelling and building of quality DEMs.	microimages.com
TOPAZ	A package specialised in hydrologic and hydraulic modeling, management of watersheds and ecosystems. It can extract a number of parameters and objects including channel network, watersheds, drainage distances and similar.	grl.ars.usda.gov/topaz/
TopoMetrix	Stand-alone Windows-based application that can extract a number of parameters from a DEM including aspect, slope, terrain shape index, terrain concavity/convexity, catchment area, wetness index, shaded relief, etc.	undersys.com
Wilbur	A stand-alone package developed by Joe Slayton. It can import, generate surfaces and export to a number of formats. It can be also used to generate synthetic terrains, shaded relief, etc.	ridgecrest.ca.us/~jslayton/
WMS	Watershed Modeling System (WMS) is specialized in modeling of watershed hydrology and hydraulics. WMS allows basin delineation, geometric parameter calculations, GIS overlay computations, cross-section extraction from terrain data, etc.	scisoftware.com

to produce estimates of monthly mean climate variables and bio-climatic parameters. Analytical GIS Eco, developed by Peter A. Shary (http://giseco.info), can extract some twenty morphometric land-surface parameters. It can also be used for correlation analysis between point-sampled soil/vegetation properties and land-surface parameters. Somewhat more extensive is LandLord (http://iflorinsky.narod.ru), which offers various operations from DEM interpolation to derivation of gradients, aspect, curvatures, rotor, specific catchment area, topographic and stream power indices (Florinsky et al., 2002). TopoMetrix, distributed by the Understanding Systems OasisGIS (http://undersys.com) can be used to derive aspect, slope, Terrain Shape Index (TSI), terrain concavity/convexity, catchment area, wetness index, shaded relief and other parameters.

TARDEM and TauDEM (used as extensions for MapWindow), developed at the Utah Water Research Laboratory (Tarboton, 1997), can be used for mapping of channel networks and watersheds. WMS, developed by the Environmental Modeling Research Laboratory at the Brigham Young University (EMRL), can be used for automated watershed and sub-basin delineation, geometric parameter computation, hydrologic parameter computation and 3D visualisation of results. Before developing the SAGA GIS (discussed in Chapter 12), Olaf Conrad developed a package he named "Digitales Gelände-Modell" or DiGeM. This windows-based application can be used to derive slope, aspect, curvatures, catchment area, topographic indices, drainage networks and visualise the results in 3D space.

Note that the list of geomorphometry-connected software is probably much longer. We have tried to list all packages that are accessible via web and that we are aware of. If you know a package (module or a script) that should be included in this list, login to geomorphometry.org and upload a full description to our database. You are also welcome to rate your favourite package (module or a script).

5. THE FUTURE OF GEOMORPHOMETRY SOFTWARE

So by considering the software currently available for conducting geomorphometric analysis, can we conclude anything about the future of software? Examining the functionality of the software discussed in this book, one may conclude that there have been relatively few substantial changes in what we do with geomorphometry software in the last 20 years. The techniques for land-surface parametrisation first published two or three decades ago (e.g. measures of slope and curvature, drainage basin and flow magnitude estimation) are still implemented in software used today. Changes have tended to be in the size and quality of datasets that are processed, not necessarily what we do with those data. Algorithms for calculating land-surface parameters may have been refined over the decades (e.g. D8 to D∞ flow direction estimation), but they have not radically changed.

So should we conclude that the future for geomorphometry software is more of the same? There are perhaps some significant recent trends in software and data that suggest this may not be the case.

5.1 New datasets

Probably the most important new development in the field of geomorphometry in the last decade has been the emergence and increasing use of high resolution elevation models produced from sensors such as LiDAR and IFSAR. The datasets produced by these sensors differ from traditional contour interpolated elevation models in two important respects. Firstly, the data volumes of these new datasets are much greater than older coarser resolution elevation models. A LiDAR DEM with a planimetric resolution of 1 metre may typically consist of many tens of thousands of rows and columns. This results in dataset sizes a couple of orders of magnitude greater than those used a decade ago. We might expect the availability and use of such data to increase in the coming years, and that the size of these data is likely to increase at a faster rate than the memory capacity of the computers required to process them. Handling these very large datasets poses challenges to the internal software architecture that must process them in reasonable lengths of time. We might therefore expect the emergence of techniques for streaming, caching and hierarchically organising data for geomorphometric processing. While some of the more generic GIS software that have been used for geomorphometry (e.g. GRASS and ArcGIS) already have such architectures in place, the more specialised software applications may require some re-engineering to be able to handle these new data effectively.

> REMARK 3. *The main challenges for geomorphometry software developers will be: (a) computationally-efficient algorithms that can can process very large high resolution dateasets; (b) algorithms that allow integration of DEM data with RS and other thematic data; (c) algorithms that allow error propagation and assessment of uncertainty.*

The challenge of dealing with large datasets is one that it likely to be addressed by many forms of software, not just those dealing with geomorphometry. However a more geomorphometric-specific problem is introduced by the type of information represented by these new datasets. Remotely sensed elevation models such as those from LiDAR and IFSAR represent surface form at a resolution much finer than many of the geomorphologic and hydrologic models have traditionally considered. For example, the profile curvature of a 1 m patch of terrain will likely represent the results of a very different surface processes to that over a 90 m patch. As these newer datasets start recording surface form at increasingly higher resolutions, geomorphometry software will need to account for the new scales of surface form and the features that are expressed in these models. As resolution increases, so the effects of vegetation, of human influence (buildings, transport infrastructure, etc.), and change over time will become more apparent.

Software will need to be able to process high resolution digital surface models in such a way as to discriminate between surface features (trees, buildings, roads) and the underlying terrain form. This may involve more sophisticated processing of sensor data such as handling LiDAR first and last return signals, or combination with other topographic datasets that identify surface features. There is currently an emerging body of literature that is beginning to address such problems, but on the whole these techniques are not yet implemented in commonly used geomorphometry software.

5.2 Visual approaches to analysis

One of the more noticeable changes in geomorphometry software in recent years has been in the increased use and quality of graphical handling of data. Whereas two decades ago, graphical output may have been considered only as the final cartographic presentation that occurs after some geomorphometric analysis has taken place, more recently software has relied on graphical interaction throughout. As hardware and software becomes more adept at handling large datasets graphically, we might expect this trend to continue. In particular, the use of visualisation for integrating and exploring data and performing analysis interactively. The emerging fields *geoanalytics* and visual data mining are likely to have application in the domain of geomorphometric analysis.

The rapid emergence and uptake of Google Earth may be considered evidence for a trend towards a more visual approach to spatial data handling. Google Earth's sophisticated spatial indexing of very large datasets combined with an open architecture for integrating and customising new data is having a radical effect on many Geographic Information Systems, including those used for geomorphometric processing. One of its biggest impacts is that it has opened up the exploration of spatial data to a much wider non-expert community of users. While that community may not be interested in the details of geomorphometric analysis, there appears to be a demand for exploring the results of such analysis. For example, large scale catastrophic events such as Hurricane Katrina in 2005 and the SE Asian Tsunami of 2004 were represented in Google Earth largely through high resolution aerial photography. While at present, Google Earth is primarily used as a *geo-browser* for exploring spatially referenced data, we might expect this visual approach to be increasingly integrated with analytical processing so that the flood modelling or wind storm modelling results could be explored in a similar fashion.

> REMARK 4. *Google Earth's sophisticated spatial indexing of very large datasets combined with an open architecture for integrating and customising new data is having a radical effect on many Geographic Information Systems.*

IMPORTANT SOURCES

ArcGIS and ESRI software — www.esri.com.
GRASS — grass.itc.it.
ILWIS — www.itc.nl/ilwis.
LandSerf — www.landserf.org.

MicroDEM — www.usna.edu/Users/oceano/pguth/website/microdem.htm.
RiverTools — www.rivertools.com.
SAGA GIS — www.saga-gis.uni-goettingen.de.
TAS — www.sed.manchester.ac.uk/geography/research/tas.

CHAPTER 11

Geomorphometry in ESRI Packages

H.I. Reuter and **A. Nelson**

importing DEM data in ESRI software products · computation of land-surface parameters using menus and scripts in ArcInfo · execution of ESRI scripts · scripts most commonly used to run geomorphological analysis · exporting data to other software

1. GETTING STARTED

Generally speaking, there are two major GIS products from Environmental Systems Research Institute (ESRI): ArcView 3, and the ArcGIS suite. We will give a brief introduction to these two products before discussing DEM import, creation and analysis.

ArcView 3 was released in 1996 and is a moderately powerful but easy to use Desktop GIS package for basic visualisation and data management. It can be used for quite complex GIS operations when specialised extensions — such as Spatial Analyst and 3D-Analyst for raster and tin analysis — and user-developed Avenue scripts are employed. ArcView 3.3 was the last version to be produced in 2002 and although it still available, ESRI support is limited and users are encouraged to use ArcGIS instead. ArcView 3 is still a commonly used platform due to its simplicity, relatively low cost, and the large user base (on the 1st of June, 2008 there were around 4600 scripts on the ESRI ArcScripts website, and about 2000 of these are in Avenue and of the ten most downloaded scripts, eight are written in Avenue).

ArcGIS is a collection of software products, of which ArcInfo is the most powerful and perhaps best-known. ArcInfo Workstation — a predominantly command line driven GIS with powerful scripting capabilities in AML (*Arc Macro Language*) — has been in existence since 1982 on various Unix platforms and since the mid 1990s on MS Windows NT until the final release of version 7 in 1997. The most rel-

Developments in Soil Science, Volume 33 © 2009 Elsevier B.V.
ISSN 0166-2481, DOI: 10.1016/S0166-2481(08)00011-1.

evant extensions of ArcInfo for DEM are the GRID raster modelling environment and the commands for Triangulated Irregular Network analysis (TIN).

> REMARK 1. *ArcGIS is a collection of software products, of which ArcInfo is the most powerful and perhaps most well-known. It contains a GRID raster modelling environment that allows a large number of raster-based GIS operations.*

In 1999, ArcInfo 8 was released as a totally new GIS package with a graphical user interface based on COM objects and hence available only for MS Windows (see Figure 1). By moving to COM objects, users now had the possibility to program or write scripts in a range of languages (Python, Visual Basic, C++, Delphi and Java for example), and it was clear that AML would no longer be supported as a scripting language in this new environment. In 2001, ArcGIS 8.1 was released which essentially combined the ease of use of ArcView 3 with the flexibility and functionality of ArcInfo. This also signalled the end of support for ArcView 3, and the Avenue scripting language. ArcGIS (now in version 9.3) is a scalable framework of several GIS products. The most powerful of the ArcGIS components is called ArcInfo, but confusingly this is different from the command line ArcInfo Workstation which is still available and is often included in the installation of ArcGIS.

The transitions in ESRI GIS products over the last 10 years have meant that there is still a large user base for ArcView 3 (and Avenue), ArcInfo Workstation (and AML) as well as ArcGIS (ArcInfo Desktop in this case), and so we will consider all three packages in the following sections. For clarity ArcView 3 will be referred to

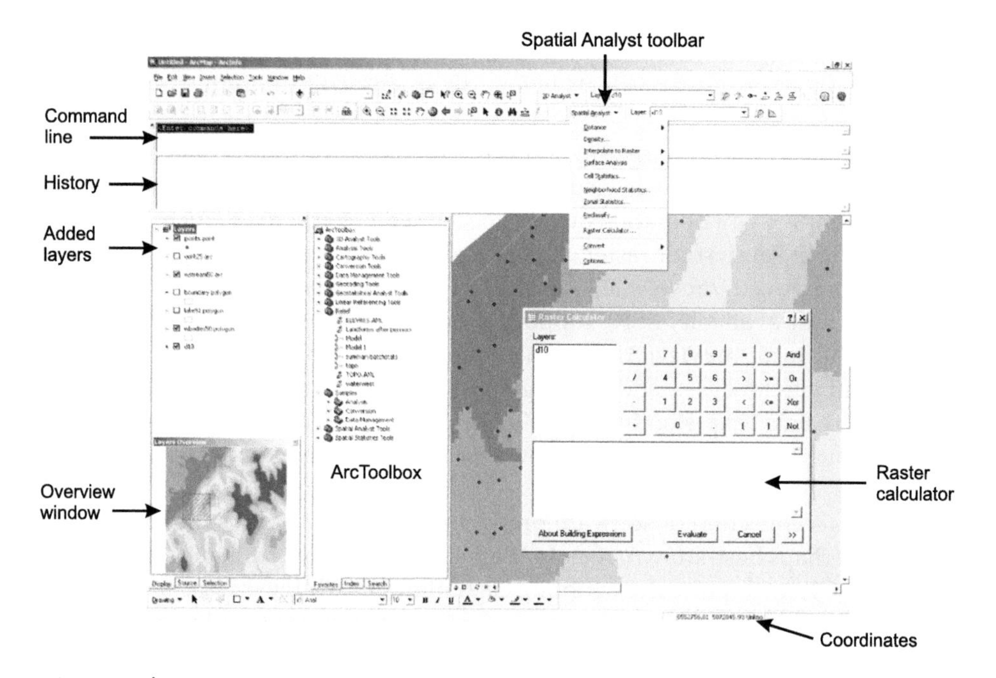

FIGURE 1 The ArcGIS 9.1 main window showing different ways for analysing Surfaces using the Command Line, the Menu and the Toolboxes.

as ArcView, ArcInfo Workstation 9.x as ArcInfo and ArcInfo Desktop 9.x as ArcGIS. Where possible, we will provide examples for geomorphometry workflows for all three packages.

There are two major groups of geomorphometric algorithms in ESRI software that are available in all three packages: *raster-based analysis* (in ArcInfo via GRID and in both ArcView and ArcGIS via the Spatial Analyst Extension) and *Triangulated Irregular Network analysis* (in ArcInfo via TIN commands and in both ArcView and ArcGIS via the 3D Analyst Extension). GRID, Spatial Analyst and 3D Analyst are additional components of the base GIS packages that must be purchased separately, but they are included in most installations. The following section will demonstrate how to:

- import a DEM or other elevation data;
- create a DEM;
- compute land-surface parameters using predefined scripts which can be downloaded from the arcscripts.esri.com or geomorphometry.org websites, and;
- export data into other formats/software.

We assume that you possess a basic familiarity with ESRI products, that the appropriate software extensions (Spatial Analyst/TIN) are available and that you understand how to use a command line interface. If the ideas behind the command line operations are understood, it should be easy to transfer that knowledge to a graphical user interface such as ArcView and ArcGIS.

Comments in the command line code start with "`/*`" as used in the ArcInfo Macro Language (AML). Files developed in AML will have the extension `*.AML`. Within our code examples, file names are in italics.

1.1 Data import

The capabilities for import/export in ESRI products would fill several book chapters. Here we will cover four major data formats as they are standards or *de facto* standards and most GIS packages can handle them[1]:

> `*.e00` — Export files are ASCII text files, which you can even read with a normal text editor, and which can contain any type of information (vector, raster or tabular);
> `*.shp` — ESRI Shapefiles contain vector data with their corresponding attribute data in a dbase file;
> `*.asc` — Raster data represented as ASCII text files;
> `*.flt` — Raster data represented as binary files, with an accompanying header file (*.hdr).

In ArcInfo: The commands to import the Baranja Hill data are shown below.

[1] The upcoming and supported data standards KML, GML GTiff are supported but not discussed.

/ To get help on any command line options, type the command without arguments. Usually the arc command syntax is <command> <input file> <output file> <parameters>, although other extensions may have a different syntax.*

```
Arc: import
Usage: IMPORT <option> <interchange_file> <output>
```

/ Check where we are on the drive with the WORKSPACE command — we need a workspace where we work in. A workspace is a directory where you can store your data. It always contains a sub-directory called INFO, which stores important information about your datasets:*

```
Arc: workspace
Current location: d:\gis\test
```

/ Wrong place — therefore we change. We can abbreviate WORKSPACE to just W.*

```
Arc: w d:\gis\barhill
WARNING: New location is not a workspace.
```

/ There is no workspace here, so we need to create one with the command CREATE-WORKSPACE, which you can abbreviate with CW. Most commands can be abbreviated in this way:*

```
Arc: cw barhill
```

/ Now we want to import an ASCII raster (dem25m.asc) to create a DEM (dem25) in GRID format. It is helpful to name the DEM after the study area and if possible you should include the cell size too. First we should check the command syntax:*

```
Arc: asciigrid
Usage: ASCIIGRID <in_ascii_file> <out_grid>
```

/ Let's execute the command, which will convert dem25.asc into a DEM in GRID format called dem25m:*

```
Arc: asciigrid dem25m.asc dem25m
```

/ Next, we want to import a Shapefile (elevations.shp) into a Coverage (points) — a Coverage is a vector based data model of spatial information, and can contain lines (or arcs), points, polygons, tables and networks:*

```
Arc: shapearc
Usage: SHAPEARC <in_shape_file> <out_cover>
{out_subclass} {DEFAULT | DEFINE}

Arc: shapearc elevations points
6370 Type 1 (POINT) shape records in
D:\GIS\BARHILL\POINTS.
```

/ Now the contour lines:*

```
Arc: shapearc contours contours5k
6370 Type 1 (POINT) shape records in
D:\GIS\BARHILL\CONTOURS5K.
```

Note: If you work with ArcGIS you may not need to convert the Shapefiles into ArcInfo Coverage format since ArcGIS can use the Shapefile format directly in a wider range of methods than ArcInfo. However, we recommended that you convert the ASCII files to Grids and large Shapefiles to Coverages/Personal GeoData Base (PGDB) in order to speed up computation.

In ArcGIS: To import an ASCII or binary format DEM you need to: Load the Spatial Analyst Extension via: *Tools* ↦ *Extensions* ↦ Tick box for spatial analyst; convert data via Arc Toolbox ↦ *Conversion Tools* ↦ *To Raster* ↦ *ASCII to Raster or Float to Raster*. If you want to convert a Shapefile to a Coverage/PGDB then use: *Arc Toolbox* ↦ *Conversion Tools* ↦ *To Coverage* ↦ *Feature Class to Coverage*; *Arc Toolbox* ↦ *Conversion Tools* ↦ *To Geodatabase* ↦ *Feature Class to Geodatabase*.

In ArcView: To import an ASCII or binary format DEM you need to: Load the Spatial Analyst Extension via: *File* ↦ *Extensions* ↦ Tick box for spatial analyst. Then open a view and from the View menu bar ↦ *File* ↦ *Import Data Sources* ↦ *ASCII raster or Binary raster* ↦ *Select File* ↦ Ok. ArcView can read, write and edit Shapefiles but can only read and edit the table attributes of Coverages.

2. DEM PREPARATION

2.1 Generating raster DEMs

Raster DEMs can be created using a variety of models, as explained in Section 2 of Chapter 2. The ESRI products support interpolation methods such as: Spline, Inverse Distance Weighting (IDW), IDW with inverse exponential interpolation, Kriging, Trend, TIN, and TOPOGRID. Topogrid is the only method that supports both point and line data directly. We will describe only two methods as an example: Inverse Distance Weighting (IDW) which will serve as an example for all geostatistical interpolation methods and the TOPOGRID approach.

TOPOGRID is an adaptation of the ANUDEM procedure (Hutchinson, 1989), which creates a hydrologically correct DEM using a multi-resolution iterative finite difference interpolation method. If possible, the TOPOGRID approach should be applied in preference of any other interpolation approach available in ESRI products for creating a DEM (however, some specific routines are provided in this book for LiDAR data; see Section 2.10 in Chapter 4).

> REMARK 2. *Hydrologically correct DEMs can be generated in ArcInfo using the Topogrid procedure.*

Filling in sinks or pits in elevation surfaces is a common processing step in DEM generation. The command <FILL> will fill sinks up to a user defined threshold, where this threshold value needs to be chosen carefully to avoid filling in too much of the drainage system. Please refer to Section 2.8 in Chapter 4 for an in-depth discussion of this problem.

/ First we check which Coverages are available using LISTCOVERAGES or LC command:*

```
Arc: lc
Workspace: D:\GIS\BARHILL Available Coverages
     contours5k        points
```

/ We know from the import section before that the digitised contour lines are in the Coverage contours5k and that the points are in the Coverage points. The ITEMS command lists the data attributes in the Point Attribute Table (PAT/`points.pat`) and the Arc Attribute Table (AAT/`contours.aat`):*

```
Arc: items points.pat
COLUMN ITEM NAME WIDTH OUTPUT TYPE N.DEC ALTERNATE NAME INDEXED

1   AREA          4  12  F  3
5   PERIMETER     4  12  F  3
9   POINTS#       4  5   B  -
13  POINTS -ID    4  5   B  -
17  VALUE         4  5   F  1
```

/ The attributes of polygon and point Coverages are stored in files with a "`.pat`" extension (for Polygon or Point Attribute Table), and line or arc attributes are stored in "`.aat`" (Arc Attribute Table) files.*

```
Arc: items contours5k.aat
COLUMN ITEM NAME WIDTH OUTPUT TYPE N.DEC
          ALTERNATE NAME INDEXED?
1 FNODE#             4  5  B  -    -
...
25 CONTOURS5K -ID    4  5  B  -    -
29 VALUE             4  3  B  -    -
```

/ For both datasets the "VALUE" item in the attribute table contains the elevation information. You can use the LIST command to show the values for each record in the attribute table. Next, we start the GRID environment to generate a DEM using IDW:*

```
Arc: grid
Copyright (C) 1982-2005 Environmental Systems
Research Institute, Inc. All rights reserved.
GRID 9.1 (Thu Mar 3 19:02:07 PST 2005)
```

/ Note that the prompt has changed from Arc: to Grid:. First we check the command syntax for IDW.*

```
Grid: idw
Usage: (F) IDW (<point_cover | point_file>, {spot_item},
{barriers}, {SAMPLE, {num_points}, {max_radius}}, {cellsize},
{xmin, ymin, xmax, ymax})
Usage: (F) IDW (<point_cover | point_file>, {spot_item},
{barriers}, {RADIUS, {radius}, {min_points}}, {cellsize},
{xmin, ymin, xmax, ymax})
```

/ There are a lot of options in the IDW command. We will use the standard values for these options by simply not including them on the command line when we generate the output. Alternatively, you can denote the standard values using the # symbol. In this case <`dem25idw`> is the output, <`points`> is the input file, and <`value`> as identified above is the item in the*

attribute table containing the height data. Note that GRID output file names cannot be longer than 16 characters in length. Here we want to create a 25 m resolution DEM Grid:

```
setcell 25
Grid: dem25idw = idw (points, value)
Running ...100%
```

/ Now, leave grid (Q or QUIT to quit) and check the statistical and spatial parameters of the DEM using the DESCRIBE command:*

```
Grid: q
Arc: describe dem25idw
     Description of Grid dem25idw
Cell Size = 25.000 Data Type: Floating Point
Number of Rows = 156
Number of Columns = 155

       BOUNDARY STATISTICS
Xmin =  6551786.000  Minimum Value =  85.715
Xmax =  6555661.000  Maximum Value =  241.725
Ymin =  5070459.000  Mean =           156.658
Ymax =  5074359.000  Standard D =     43.879
      NO COORDINATE SYSTEM DEFINED
```

/ The statistical parameters look okay. If you see huge negative values, then you probably have some incorrectly specified* `Nodata values` *in your input dataset. Before proceeding with further parameterisation you must define the projection system using PROJECTDEFINE, so that all other results based on the DEM will inherit the same projection, otherwise you will have to define the projection of each output one by one. Now we will create a DEM using the ANUDEM approach. First we want to see some usage information for the TOPOGRID command:*

```
Arc: topogrid
Usage: TOPOGRID <out_grid> <cell_size>
```

/ Create a DEM named* `dem10mtopo` *with a cell size of 10 m:*

```
Arc: topogrid dem10mtopo 10
```

/ The prompt changes from Arc: to Topogrid: to signify that we are in the TOPOGRID environment. Instead of supplying the parameters on a single command line, we will supply them one by one. First, we specify that we will generate the DEM primarily from contours (points will also be used but they will be of less importance in the interpolation).*

```
TopoGrid: datatype contour
```

/ Now we specify the <contour> dataset named <contours>, and that the elevation values are stored in the <value> item:*

```
TopoGrid: contour contours5k value
```

/ Specify the <point> dataset named <points>, and that the elevation values are stored in the <value> item:*

```
TopoGrid: point points value
```

/ To create a hydrologically correct DEM you can (but you don't need to) specify further hydrological input features likes streams and lakes. However, in contrast to the TIN approach you can not directly specify hard and soft breaks. We assume that you know how to import these additional hydrological data layers using the import methods describe above:*

```
TopoGrid: stream river_1
TopoGrid: lake lake_p
```

/* *Now that all the input parameters have been defined, we want to see the sinks and rivers that will be created by the TOPOGRID procedure (this is a type of error diagnosis) by specifying:*

```
TopoGrid: outputs sink_test drain_test
```

/* *We end the process by typing END:*

```
TopoGrid: end
```

/* *For further options within TOPOGRID please refer to the ESRI Help, under the heading Topogrid.*

To create a DEM using IDW in ArcGIS: *Arc Toolbox* ↦ *Spatial Analyst Tools* ↦ *Interpolation* ↦ *IDW* ↦ *Add data* ↦ Ok. For the Topogrid procedure *Arc Toolbox* ↦ *Spatial Analyst Tools* ↦ *Interpolation* ↦ *Topo to Raster*.

To create a DEM using IDW in ArcView: First add a point data set to the view and then from the menu *Surface* ↦ *Interpolate Grid* ↦ *Specify Extent* (Same As View) and Output Grid Cell Size (e.g. 25) ↦ OK ↦ *Method* IDW ↦ Z Value Field is: *Value* ↦ OK. To our knowledge, the TOPOGRID procedure has not been implemented in ArcView.

If the TOPOGRID (ANUDEM) method has been used in the creation of a DEM, then spurious sinks should have been removed already. Still, ESRI products contain other functions to fill sinks. Filling in sinks can take up to three iterations of processing.

/* *First, we find out if there are any sinks in the DEM, and where they are located:*

```
Grid: sink_test = sink (fldir_test)
Grid: sarea_test = watershed (fldir_test, sink (fldir_test))
Grid: sdepth_test = zonalfill (sarea_test, dem10mtopo) -
zonalmin (sarea_test, dem10mtopo)
```

/* *Next we display a map of the sinks and determine the sink depth required to fill the sinks. Several commands on one command line can be separated with a ";" sign:*

```
Grid: display 9999; mape sink_test; gridpaint sink_test;
gridpaint sdepth_test
```

/* *Finally we fill all sinks with a threshold of 10 metres. If no value is specified then all sinks are filled.*

```
Grid: FILL dem10mtopo dem10m_fill SINK 10
```

To identify sinks in a DEM in ArcGIS: *Arc Toolbox* ↦ *Hydrology* ↦ *Sink* ↦ *Add flow direction grid* ↦ Ok; the same location for the Fill command. In ArcView: Load the Spatial Analyst and Hydrologic Modelling Extensions, add DEM to active view *Hydrology* ↦ *Fill*.

2.2 Generating TIN DEMs

TIN DEM generation in ESRI products uses Delaunay Triangulation, which is a proximal method that satisfies the requirement that a circle drawn through the three nodes of a triangle will contain no other point. Application is straightforward for simple cases by defining the input dataset, some tolerances and an output dataset. If features like break lines, ditches or lakes exist in the study area, then some specific coding of attributes is needed to ensure that they appear correctly in the final TIN. We will not go into the details of such specific coding in the case presented below, and users should refer to the documentation in the respective packages (e.g. search for TIN in the help system) for these details.

```
Arc: createtin
Usage: CREATETIN <out_tin> {weed_tolerance}
{proximal_tolerance} {z_factor} {bnd_cover | xmin ymin
xmax ymax} {device}
```

/ We want to create a TIN named* `demtin`. *The # symbol signifies that we will use the DEFAULT values for the optional parameters.*

```
Arc: createtin demtin # # #
```

/ Again, the prompt has changed to signify that we are in the CREATETIN environment. First we input the contour lines.*

```
Createtin: cover
Usage: COVER <in_cover> {POINT | LINE | POLY}
{spot_item} {sftype_item | sftype} {densify_interval}
{logical_expression | select_file} {weed_ tolerance}
```

```
Createtin: cover contours5k LINE value mass
```

/ Now the points.*

```
Createtin: cover points POINT value mass
```

/ We can specify additional datasets such as rivers and lakes by coding them as hardline (specified constant elevation) or softline (varying elevation) breaks. In this case we will not include these features in the TIN, so we just type END.*

```
Createtin: end
```

/ The following lines are just messages reported by the TIN module as it processes the data before returning us to the Arc environment.*

```
Loading arcs from the Coverage CONTOURS5K...
Loading points from the Coverage POINTS...
Computing line spot values... Interpolating... Proximal tolerance
set to 0.000
Removing points within tolerance...
Within tolerance 20. Remaining 7651...
Creating TIN Writing tin data structures...
Creating non-convex hull...
Arc:
```

In **ArcGIS**: First load the 3D Analyst: *Tools* ↦ *Extensions* ↦ Tick box for 3D analyst. Then the TIN creation procedure is: *Arc Toolbox* ↦ *3D Analyst Tools* ↦ *TIN Creation* ↦ *Create TIN* — name is for example `demtin` and projection system (this will create an Empty tin) ↦ again: *Arc Toolbox* ↦ *3D Analyst Tools* ↦ *TIN Creation* ↦ *Edit TIN* ↦ Select available TIN `demtin` as input TIN, add points and contours as input feature class ↦ change `<Height_field>` to value and `<SF_type>` to mass points ↦ OK.

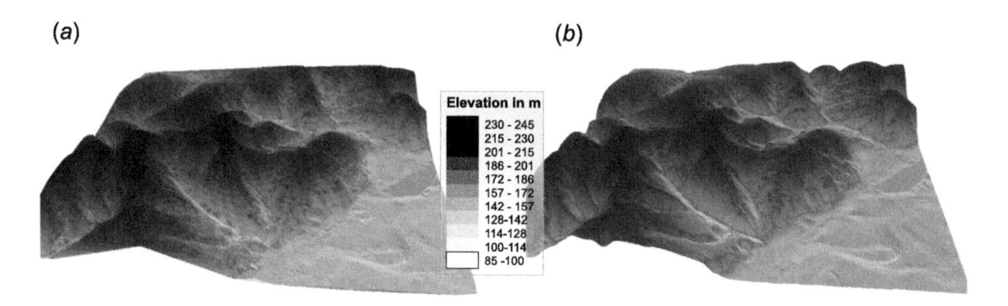

FIGURE 2 Digital Elevation Models generated using (a) the TIN and (b) the TOPOGRID approaches.

In ArcView: Load 3D Analyst extension, add and select contour lines and points to active view, choose *Surface* ↦ *Create Tin*, select points and contours ↦ In the dialog, select contour line theme and specify `<value>` as the elevation value field, change Input *As to Mass Points*, click on the points file, choose *Input as Mass Points*, and specify height source as the elevation value field ↦ OK; specify output file.

You now know how to create DEMs using both raster and TIN approaches as seen in Figure 2.

3. EXTRACTION OF LAND-SURFACE PARAMETERS AND OBJECTS

3.1 Basic principles

First we will explain some simple raster based functions. If you just want to compute land-surface parameters, then jump to the second part of this section. Several primary land-surface parameters are implemented in ESRI products: e.g. slope gradient using the D8-algorithm (the ESRI help file gives Burrough, 1986, as a reference), curvature parameters based on the algorithms of Zevenbergen and Thorne (1987) and flow accumulation using algorithms by Tarboton et al. (1991). The slope in degrees is computed using the D8 method, aspect in degrees as line of steepest descent, and curvature values as the second derivative of the slope. For profile curvature this is the direction of the flowline of a cell, whereas plan curvature is the direction perpendicular to that direction (see Section 2.1 in Chapter 6). The values are given as radians/100 m or 1/100 metre. If you want to test other flow or curvature implementations please refer to page 289. The general syntax for GRID commands is:

```
<outputgrid> = command ( <inputgrid> )
```

/ Let's compute the slope using the D8-method [Figure 3(a)]. You should refer to the DOCELL command if you want to implement your own algorithm:*

```
Grid: slope
Usage: (F) SLOPE (<grid>, DEGREE | PERCENTRISE)
Usage: (F) SLOPE (<grid>, <z_factor> {DEGREE | PERCENTRISE})
Grid:slp25m = slope ( dem25m )
```

```
/* Next we compute the aspect and curvature.
Grid: aspect Usage: (F) ASPECT (<grid>)
Grid:asp25m = aspect( dem25m )
Grid: curvature
Usage: (F) CURVATURE (<grid>, {out_profile_curve},
{out_plan_curve}, {out_slope}, {out_aspect})
Grid:curv25m = curvature ( dem25m , prof25m , plan25m ,
slp25m , asp25m )
```

In ArcGIS go to *Arc Toolbox* ↦ *Spatial Analyst Tools* ↦ *Surface*. Here you can select the operations shown above like Slope and Aspect. For example for Curvature *Arc Toolbox* ↦ *Spatial Analyst Tools* ↦ *Surface* ↦ *Curvature* (input raster name is for example `dem25m`). In ArcView: Add the desired Grids to a View; for Slope and Aspect go directly to: *Surface* ↦ *Slope* or *Surface* ↦ *Aspect*; for curvature parameters: *Analysis* ↦ *Map Calculator* ↦ type in the field.

> REMARK 3. *In ArcInfo, the user can derive the primary parameters, such as catchment area, slope and similar, by using the built-in commands. Secondary parameters can be derived by directly typing the equations to calculate them.*

Secondary land-surface parameters can be computed by just typing the equations in the grid command line using the same syntax structure as for slope and curvature. Secondary topographic attributes are computed from two or more primary attributes. For example, the Topographic Wetness Index (TWI), which quantifies the role of topography for redistributing water in the landscape, is derived using the catchment area map and the slope map. The syntax for these commands is: `<outputgrid> = <inputgrid1> / <inputgrid2>`, e.g. the command for the stream power index (SPI) by Moore et al. (1993a) would be written as [Figure 3(b)]:

```
GRID: SPI25m = flacc25m * TAN ( slp25m DIV DEG )
```

where `flacc25m` is the catchment area and `slp25m` is the slope map in radians.

In ArcGIS go to *Arc Toolbox* ↦ *Spatial Analyst Tools* ↦ *Map Algebra* ↦ *Single Output Map*. Add the commands by clicking on graphically or by typing on the command line:

```
(SPI25m= [flacc25m] * [slp25m].tan)
```

In ArcView: Add the desired Grids to a View ↦ *Analysis* ↦ *Map Calculator* ↦ add the commands by clicking on the available options or by typing the following commands in the Map Calculator text box:

```
([flacc25m] * [slp25m].tan)
```

Furthermore, a series of zonal and neighbourhood commands exist (see Figure 4 in Chapter 1). An example application with these functions might be that we want to know the maximum height in a watershed, which would be a zonal function.

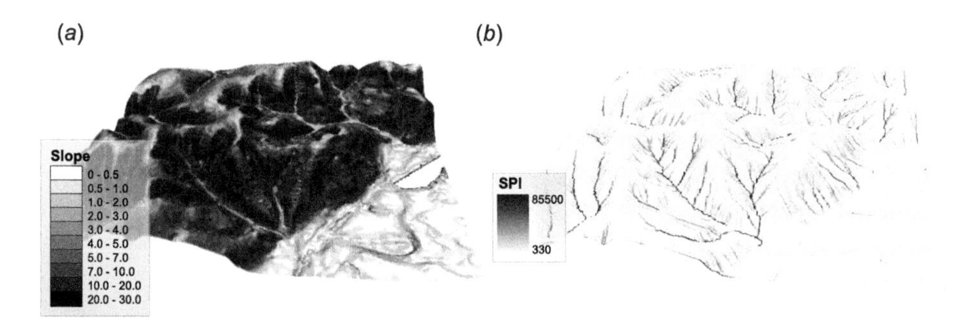

FIGURE 3 (a) Slope and (b) stream power index maps of the Baranja Hill Case study with a resolution of 10 m.

/* *First we look at the usage of the commands:*

```
Grid: zonalmax
Usage: (*) ZONALMAX (<zone_grid>, <value_grid>,
{DATA | NODATA})
```

/* *This means we need a value grid, in our case the elevation (`dem25m`) and a zone grid containing the watersheds (`wshd25m`). If both grids exists, we can execute the following command:*

```
Grid: wshmax25m = zonalmax (wshd25m, dem25m)
```

In ArcGIS go to *Arc Toolbox* ↦ *Spatial Analyst Tools* ↦ *Zonal* ↦ *Zonal Statistics*. Select the desired parameters. In ArcView: Add the desired Grids to a View. Add a Shapefile containing the zones to the View. Be aware that we use a vector file, not a Grid. In the case where only a grid is available you will need to convert it using Menu *Theme* ↦ *Convert to Shapefile*; click on the added Shapefile to make it active. If it is not active then the next command will not work. Secondly, the zone value MUST be an integer or character: Menu *Analysis* ↦ *Summarize Zones*. Pick the field that defines the zones. Pick the grid which contains the data you want to summarise — a table will be created, which contains the statistical information.

Other analysis that uses FOCAL functions involves the calculation of the difference between the minimum and maximum elevation in a five pixel neighbourhood of every cell. Focal function can also be used to filter or smooth DEM data:

```
Grid: focalrange
Usage: (*) FOCALRANGE (<grid>, {DATA | NODATA})
Usage: (*) FOCALRANGE (<grid>, <RECTANGLE>, <width>,
<height>, DATA | NODATA)
```

/* *We need an elevation grid (`dem25m`), the focal neighbourhood is 5 pixels [Figure 4(b)].*

```
Grid: rang25m = focalrange (dem25m, rectangle, 5, 5)
```

In ArcGIS go to *Arc Toolbox* ↦ *Spatial Analyst Tools* ↦ *Neighbourhood* ↦ *Focal Statistics*. Select the desired parameters as shown above. In ArcView: Add the desired Grids to a View; Menu *Surface* ↦ *Neighbourhood Statistics*. Select the desired parameters as shown above.

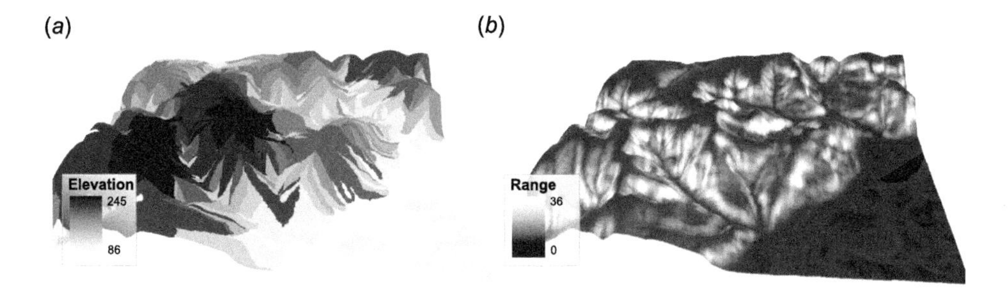

FIGURE 4 (a) The maximum elevation in each watershed identified using a flow accumulation threshold of 100; and (b) the elevation range for a moving window of 5×5 cells of the Baranja Hill Case study with a resolution of 10 m.

If a land-surface parameter does not exist, it can be implemented in a <DOCELL> loop, where the neighbourhood of an investigated cell is specified using an array notation as shown in Figure 4 in Chapter 1. An example would be to compute the sum of the two northern and two southern cells of every neighbour. The following is an example of the notation. Again, a specific notation is used here within the DOCELL environment denoted by a ": :" prompt.

```
Grid: DOCELL
:: sum25m = dem25m (0, -1) + dem25m (0, -2) + dem25m (0, 1) +
dem25m (0, 2)
:: END
```

The hillshade and visibility functions have a common feature in that they relate to the line of sight of a location either from the point of view of the sun (hillshade) or from the point of view of an observer (visibility). The shading can be used to: create an intuitive view of the DEM, highlight errors in a dataset, or create solar radiation budgets as shown below in the script for solar radiation modelling. The hillshade formula needs the azimuth and slope at the local cell. To create a hillshade [Equation (2.14) in Chapter 6] of a DEM we need to type (Figure 5):

```
Grid: hillshade
Usage: (I) HILLSHADE (<grid>, {azimuth}, {altitude},
{ALL | SHADE | SHADOW}, {z_factor})
Grid: hsd25m = hillshade (dem25m, 315, 45)
Computing hillshade...
```

In ArcGIS: *ArcToolbox* ↦ *Spatial Analyst Tools* ↦ *Surface* ↦ *Hillshade*. In ArcView: Add the desired Grids to a View; Menu *Surface* ↦ *Compute Hillshade*.

For visibility, we will demonstrate the application using the following subject: A client has three lookouts (each one is 15 m high) and wants to know the visible ground area (or viewshed) from each lookout. We assume that we have a coverage named `bar_lookoutpt` containing the three lookouts and a DEM:

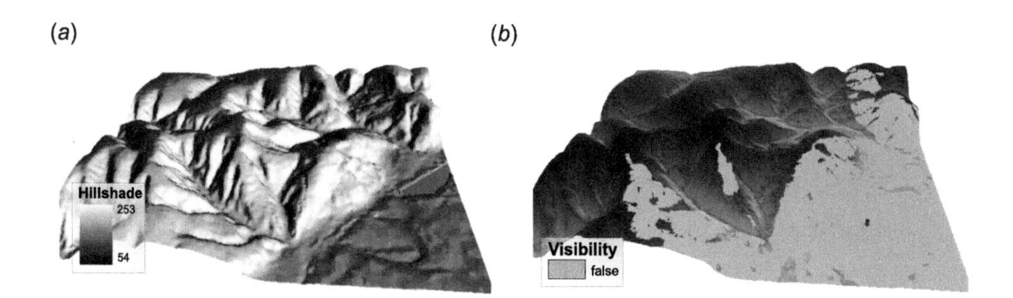

FIGURE 5 (a) Hillshade for azimuth 315 and altitude 45; and (b) visibility analysis for a single lookout point which is 15 m above the surroundings. The area in light gray is not visible from the lookout (e.g. Tower).

/ As we want an offset (the height) to our three points we need to add another item to our lookout dataset. First we make sure that a PAT, a "Point attribute table" exists by BUILDing point topology for our Coverage*

```
Arc: build bar_lookoutpt POINTS
Building points...
```

/ First we add an item to the attribute table, then we start the Tables environment:*

```
Arc: additem
Usage: ADDITEM <in_info_file> <out_info_file> <item_name>
<item_width> <output_width> <item_type> {decimal_ places}
{start_item}
Arc: additem bar_lookoutpt.pat bar_lookoutpt.pat offseta 8 8 i
Adding offseta to bar_lookoutpt.pat to produce
bar_lookoutpt.pat
```

/ Set all lookouts to the same value (15 m):*

```
Arc: tables
Copyright (C) 1982-2005 Environmental Systems Research
Institute, Inc. TABLES 9.1 (Thu Mar 3 19:02:07 PST 2005)
```

/ Select our attribute table (SELECT):*

```
Tables: SELECT lookoutpt.pat 3 Records Selected.
```

/ Use the command CALC to set the value to 15, then Quit TABLES*

```
Tables: CALC offseta = 15
Tables: Q
Leaving TABLES...
```

/ Now change to the Grid environment to compute the visibility for these three points:*

```
Grid: visibility
Usage: (I) VISIBILITY (<grid>, <cover>, {POINT | LINE},
{FREQUENCY | OBSERVERS})
```

```
Grid: visi10m = visibility ( dem10m , bar_lookoutpt, POINT,
GRID , OBSERVERS ) Computing visibility analysis for
bar_lookoutpt...
```

```
Observer 1 at 6553223.000,5072799.000,208.760.
     Observer offset 15.000,
     set near radius (3-D distance) 0.000,
     set far radius (3-D distance) 10000000000.000,
     horizontal angles 0.000->360.000, and
     vertical angles 90.000->-90.000
          with an Object offset of 0.000...
...
Curvature and refraction correction is OFF...
Computing Visibility...
```

/ Now we look at the grid to see the different viewsheds for each lookout point (e.g. Obs1, obs2, obs3)*

In ArcGIS: Viewsheds from points to grid: *ArcToolbox* ↦ *Spatial Analyst Tools* ↦ *Surface* ↦ *Viewshed.* Add the `dem25m` as input raster, the `bar_lookoutpt` as input point files and specify the output raster as `visi10m`. The `z` factor will need to be adjusted only if your height information is in a different unit than your spatial extent (e.g. feet or metres). There is no viewshed analysis included in ArcView, although the functionality exists. Several extensions have been made by users to fill this gap in the functionality, and a search for viewshed avenue scripts on the ArcScripts website will provide several options.

4. ARC SCRIPTS

Now we proceed with the second part of this section showing some examples of scripts for quantitative and qualitative geomorphometry. For ease of computation, several scripts have been developed to compute 28 different land-surface parameters. The script names and the list of land-surface parameters can be seen on the Arc scripts website (http://arcscripts.esri.com). In ArcInfo, the scripts will be executed at the arc or grid prompt using:

```
&run <script name> <parameter_1> ... <parameter_n>
```

In ArcGIS, the scripts can be executed by clicking on the special toolbox *terrain* which is provided at http://arcscripts.esri.com and via geomorphometry.org.

4.1 Grid-based parametrisation

First we will extract some quantitative land-surface parameters. Primary and secondary land-surface parameters, which do not rely on any watershed delineation, can be computed using `topo.aml` with the input of a DEM and a stream flow threshold. In certain landscapes these thresholds for watershed delineation need to be adjusted iteratively, therefore `topowshd.aml` computes parameters which depend on these thresholds.

> REMARK 4. *More sophisticated geomorphometric analysis is possible in ArcInfo by using Arc scripts. So far, ESRI users are the largest GIS community in the world.*

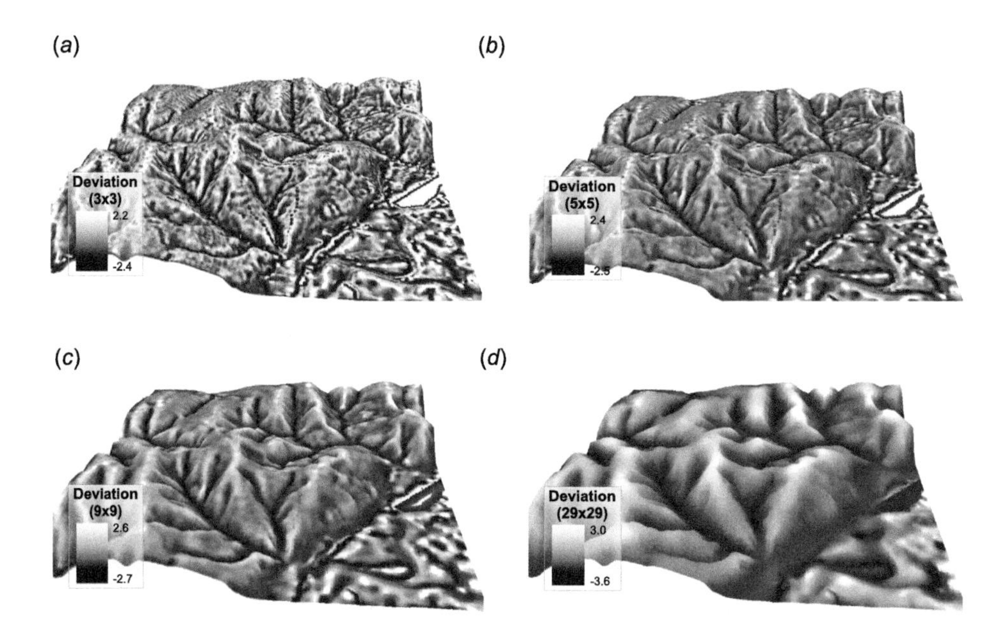

FIGURE 6 Deviation of Elevation in meters for mowing windows with a size of 3 (a), 5 (b), 9 (c) and 29 (d) for the Baranja Hill Case study with a resolution of 10 m.

Secondly, we often need to account for uncertainties and inaccuracies in DEM creation when computing quantitative land-surface parameters from a DEM. A robust procedure to reduce artefacts and errors is to employ a Monte-Carlo simulation approach (see Section 3.2 in Chapter 4). This approach computes the TWI n times and produces the mean and standard deviation TWI of all model runs (Reuter, 2004). The AML will stop if (i) the number of iterations (n) is reached or (ii) the difference between two successive iterations is smaller than a threshold value. The threshold is computed by dividing the standard deviation by n or by specifying it [Figure 7(b)].

Land-surface parameters described by Wilson and Gallant (2000), which are based on neighbouring areas (similar to the *zonalrange* command shown before) can be computed from the `elevres.aml` (Figure 6).

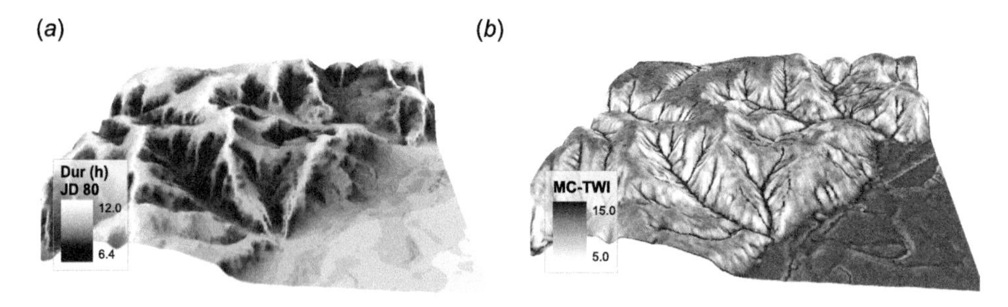

FIGURE 7 (a) Duration of direct solar radiation in hours and (b) a Monte-Carlo simulation of the topographic wetness index for the Baranja Hill Case study with a resolution of 10 m.

Finally, we will demonstrate the calculation of incoming solar radiation [Figure 7(a)] using the `solarflux.aml` (Rich et al., 2002). Other approaches are (1) the more detailed and advanced SRAD model provided by the TAPES-G-suite (Wilson and Gallant, 2000), or (2) the `shortwave.aml` by Kumar et al. (1997), but this is only applicable for time-steps greater than 1 day.

Besides a DEM, the `solarflux.aml` script requires the Julian day (see Section 3.1 in Chapter 8), for which Schaab (2000, p. 259) recommended using three specific days, winter solstice (22.12), summer solstice (21.06) and the spring solstice (21.03). Also, the start and end times should be specified as 4.00 and 22.00 respectively, the time steps (increment) are spaced 12 minutes apart and the transmissivity of the atmosphere will be set to a value of 0.6. Besides that, you will need the location in decimal geographic coordinates (e.g. N45°47′ E18°40′) and the local time meridian.

/ Due to the length of the terrain extensions the DEM name should not exceed 4 characters:*

```
Arc: &run topo
USAGE: topo <DEM> streamflow threshold streamcover
Arc: &run montewi
```

/ Change to GRID.*

```
Arc: grid
Copyright (C) 1982-2005 Environmental Systems Research
Institute, Inc. All rights reserved. GRID 9.1 (Thu Mar 3
19:02:07 PST 2005)
Grid: &run topo dem25m 100
```

/ At this stage we look at the watersheds created by the* `topo.aml`*, if these are not satisfying (too small or too large) then we can re-run the watershed based land-surface parameters:*

```
Arc: &run topowshd
USAGE: topo <DEM> streamflow threshold {streamcover}
```

/ Here we see that the stream network is not detailed enough:*

```
Grid: &run topowshd dem25m 50
```

/ Now we compute the topographic wetness index to characterise the wetness of the landscape. Lets assume that our DEM has an error of 0.15 cm, 50 simulations are a good starting point:*

```
Grid: &run montewi
Usage: MONTEWI <dem> <out-grid> <stdev> <n steps> {break}

Grid: &run montewi dem25m mwi25m 0.15 50 0.001.

Grid: &run elevres
USAGE: elevres <DEM> {cell size}
```

/ Run the analysis for window sizes of 5, 11 and 21 neighbours:*

```
Grid: &run elevres dem25m 5
Grid: &run elevres dem25m 11
Grid: &run elevres dem25m 21
```

/ Solar radiation — here we need several parameters. The Julian day 70, local start time 4, local end time 22, (may change depending on the time of the year), incremental interval 0.12, latitude 47, longitude 18, local time meridian 12, transmissivity 0.6, surface grid dem25.*

```
Grid: &run solarflux
Please enter station file: 9999
```

/ Finally, we want to generate quantitative landforms using McNab's or Bolstad's methods:*

```
Grid: &run landformshape
Usage: LANDFORM <DEM> <OUTGRID> {MCNAB | BOLSTAD}

Grid: &run landformshape dem25m mcnab25m MCNAB
Calculating McNab's Landform Index
Running... 100%
McNab's Landform Index written to dem25mcnab
```

For qualitative geomorphometry we will apply three different landform classification algorithms as examples, which are suitable for this dataset:

- a simple algorithm from Agriculture Canada (MacMillan and Pettapiece, 1997);
- a landform classification for hummocky landscapes by Pennock et al. (1987, 1994) which classifies up to 11 landforms (Figure 8);
- an algorithm by Park et al. (2001).

As we have already computed the input parameters for these algorithms using `topo.aml`, we can go ahead and execute the algorithms straight away. Generally, if the landforms do not satisfy the expectations then the classification parameters will need to be adjusted. This is an iterative process, which depends on the users knowledge of the landscape under investigation. See also Reuter et al. (2006) for one approach to transfer identified classified parameters across a range of different generalisation scales.

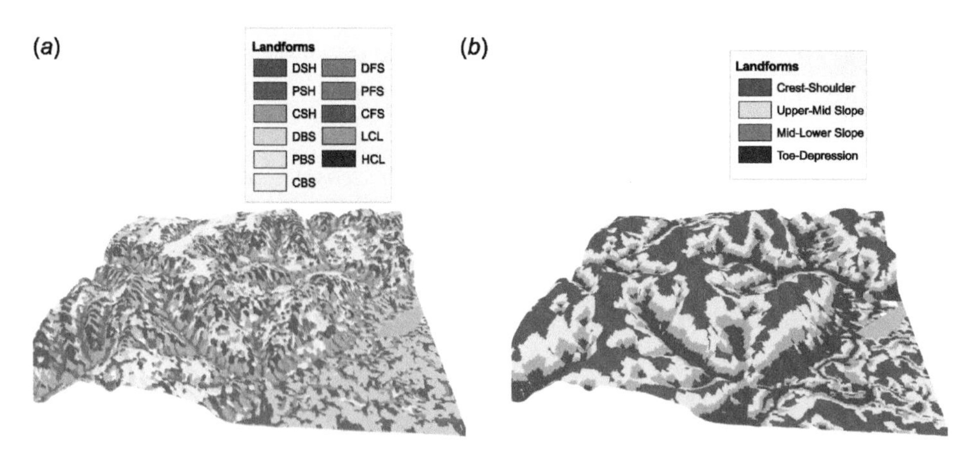

FIGURE 8 Landform classification as shown above using (a) `pennock97.aml` and (b) `simplelfabc.aml` scripts for the Baranja Hill Case study with a resolution of 10 m. (See page 719 in Colour Plate Section at the back of the book.)

```
Grid: &run simplelfabc
USAGE: inputgrid outputgrid method filter slope threshold1
threshold2 threshold3
```

```
Grid: &run simplelfabc dem25m dema25m a
```

/ Note: there are three different methods in simplelfabc. Pennock's original paper used a grid resolution of 10 m*

/ Now lets get Pennock's classification:*

```
Grid: &run landform
USAGE:pennock94 <DEM> <OUTDEM> {method} {...threshold}
{profile} {planform} {slope} {watershedarea} {all/original}
{graphic y/n}
```

/ Add the day to the output DEM name. We start with the default values:*

```
Grid: &run landform dem10m dem10m_lf0301
```

/ If the results are not good, you will have to experiment with the profile, planform and slope thresholds.*

```
Grid: &run landform dem10m dem10m_lf03012006 11 5 0.1 0.1 2.9
```

/ Finally we want to compute the landscape units / land-surface characterisation index of* Park et al. (2001)

```
Grid: &run tci
Usage: tci <dem> {cl_csi} {cl_asi} {cl_ast} {cl_ap}
Grid: &run tci dem10m
```

In ArcGIS: Go to *ArcToolbox* ↦ and execute the desired scripts.

In ArcView: We will provide one example for Landform classification by Schmidt, F.: Download the `topocrop.ave` extension from arcscripts.esri.com or the books website. Copy this to your extension folder[2]; load it: *File* ↦ *Extensions* ↦ tick box front of *Terrain Analysis and Spatial Analyst* ↦ Click Ok; Create a new view; Add a elevation grid using the "plus" button; Make that grid active; Menu *Topocrop* ↦ *Landform Elements 1:5000*; Enter a directory if asked for. It must already exist. Follow the instructions on the screen for reclassifying data. You may need to apply the `landformelements_d.avl` legend to the nine landform elements grid.

4.2 TIN based parametrisation

In contrast to raster based analysis, TIN based analysis is not as advanced in the ESRI products in terms of geomorphometry — slope and aspect (Figure 9) can be computed but landform classifications, watershed delineation and other land-surface parameters are not available using TIN. A workaround is to convert the TIN into a raster dataset and execute the land-surface algorithms there. Still, hill-shading and visibility analysis can be performed. In the following section we will show: (i) how to compute the slope and aspect for a TIN and (ii) how to convert a TIN into a raster:

[2] e.g. c:\esri\av_gis30\ArcView\ext32\.

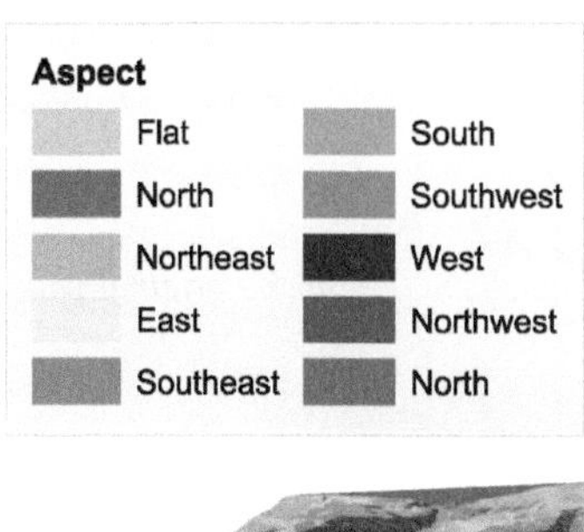

FIGURE 9 Aspect classes calculated for the Baranja Hill DEM TIN. (See page 719 in Colour Plate Section at the back of the book.)

/ Check which tins we can work with, LISTTINS (LT) will list the tins in the Workspace:*

```
Arc: lt
Workspace: D:\GIS\BARHILL
Available TINs
DEMTIN TESTARCGIS
```

/ Use the TIN demtin — check the usage for the slope and aspect computation:*

```
Arc: tinarc
Usage: TINARC <in_tin> <out_cover> {POLY | LINE | POINT |
HULL} {PERCENT | DEGREE} {z_factor} {HILLSHADE} {azimuth}
{altitude}
```

/ The name of the <outcover> is slptin, and we want to compute it for the polygons. zfactor is important if the vertical and horizontal units are different:*

```
Arc: tinarc demtin slptin POLY
```

/ We need more land-surface parameters than the FILTER, VIP, HIGHLOW and TINARC commands provide. So let us convert the TIN to a raster (within the Arcinfo TIN environment, a raster is called a lattice), which allows for many more TPs to be calculated:*

```
Arc: tinlattice
Usage: TINLATTICE <in_tin> <out_lattice> {LINEAR | QUINTIC}
{z_factor} {FLOAT | INT}
Arc: tinlattice demtin dem10mtin
Converting tin demtin to linear lattice dem10mtin...
TIN boundary
```

```
Xmin = 6551798.500 Ymin = 5070471.500
Xmax = 6555639.500 Ymax = 5074356.000
X-extent = 3841.000 Y-extent = 3884.500
Lattice parameter input
Enter lattice origin <xmin> <ymin>:
Enter lattice upper-right corner <xmax> <ymax>:
Enter lattice resolution <n_points>:
Enter distance between lattice mesh points <d>:
10 Default lattice origin (x,y) is (6551798.500, 5070471.500)...
Default Upper-right corner of lattice (x,y) is (6555639.500,
5074356.000)...
Lattice has 385 points in x, 389 points in y...
Spacing between mesh points (d) is (10.000)...
Computing lattice...
```

/* *Now we can perform further analysis with this raster.*

In ArcGIS: *ArcToolbox* ↦ *3D Analyst* ↦ *TIN Surface* ↦ *TIN Slope* / or *TIN Aspect;* For Conversion of a TIN to raster: *ArcToolbox* ↦ *3D Analyst* ↦ *Conversion* ↦ *TIN to Raster*.

4.3 Data export and conversion

Having performed a geomorphometric analysis or even only created a DEM using the TIN based method, a user may want to export the data in order to use it in different software. The export of a grid is performed as follows:

/* *First check which grids (LISTGRIDS or LG) are in the workspace:*

```
Arc: lg
```

```
Workspace: D:\GIS\BARHILL
Available GRIDs
DEM25M TEMPOUT2 TEMPOUT3
```

/* *Check usage for ASCII export and then run it:*

```
Arc: gridascii
Usage: GRIDASCII <in_grid> <out_ascii_file> {item}
Arc: gridascii tempout3 tempout3.asc
Arc: gridfloat
```

/* *Check usage for floating binary grid export, and then run it:*

```
Usage: GRIDFLOAT <in_grid> <out_file> {item}
Arc: gridfloat tempout2 tempou2.flt
```

In ArcGIS: ArcToolbox ↦ ConversionTools ↦ FromRaster ↦ Raster to Ascii or Raster to Float. In ArcView: Create/or open a view ↦ File ↦ Export Data sources ↦ Select Export File Type either ASCII Raster/Binary Raster ↦ Select GRID ↦ provide output file name ↦ Ok.

Lastly, we should mention that the TAPES-G land-surface parametrisation suite can be used in conjunction in ArcInfo or ArcGIS. For further information about this suite please refer to Wilson and Gallant (2000). This suite can use both binary or ASCII grids as inputs. To convert between formats you might use the `tapesg.aml` and `tapestoarc.aml`, found on this books webpage. In ArcGIS, the TAPES-G-ArcGIS and SRAD-ArcGIS scripts have been provided by Hong Chen. To run analysis on your DEM use:

```
Arc: &run tapesg
Usage: tapesg.aml <in_dem> <tapesg_file> <STREAM_TUBE |
SLOPE_WEIGHT> <DEP_LESS_DEM | NO_DEP_LESS_DEM> <FILE | RUN>
Arc: &run tapestoarc
Usage: tapestoarc <tapesfile> {<attribute> <outgrid>}
```

4.4 Modelling applications in ESRI

There are many models of land use, soil properties, hydrology and so on. A good overview of all these data models is available at http://support.esri.com under the section *datamodels*. These documents include examples for almost every type of possible connection between the models and GIS packages. We have selected a couple of examples closely related to land-surface parameters. For example, ArcHydro defines the structure of natural hydrology. From a modelling perspective, the MIKE SHE[3] model family is worth mentioning. Erosion models like AGNPS (Tim and Jolly, 1994), SWAT — soil and water assessment tool (Francisco et al., 2004), land use and landscape changes (Jewitt et al., 2006), urban planning (Stevens et al., 2006) and pesticides models (Sood and Bhagat, 2005) might also be of interest.

5. SUMMARY POINTS AND FUTURE DIRECTION

ESRI has for decades been a key provider of software solutions for analysis, management and visualisation of spatial data. ESRI products are especially powerful in providing support for large DEM databases (e.g. >4 GB) and include a wide variety of land-surface parameter functions. However they lack straightforward implementations of some of the more recent geomorphometric algorithms which need to be created by the user.

Several other land-surface parametrisation packages provide more advanced functionality than the ESRI products themselves. One group of packages uses the grid files as an exchange dataset, which implies a number of import and export operations. The advantage is that the whole GIS overhead is not used, as for example in the TARDEM software developed by Tarboton et al. (1991, 1992), Tarboton (1997), which uses binary and ASCII grids. Other software that is closely linked with the ESRI products and can provide similar (much faster) commands is for example the Terraflow[4] approach (Arge et al., 2003).

[3] See http://www.dhigroup.com/mikeshe/.

[4] Terraflow (http://www.cs.duke.edu/geo*/terraflow/) is a software package for computing flow routing and flow accumulation on massive grid-based terrains. It is based on theoretically optimal algorithms designed using external memory paradigms.

In this chapter, we have covered GUI and command line options for geomorphometric analysis in ESRI products. The user has a choice of high and low level programming languages to interact with these products in order to create new datasets and models. The learning curve can be quite steep, unless the user has prior programming experience. A strong user community is available to provide support for people working on these commercial systems. Several external applications can be used in conjunction with ESRI products, therefore providing a seamless work-flow for geomorphometric analysis in different model systems.

IMPORTANT SOURCES

http://esri.com — Home page for courses, books, data, software.
ArcInfo Help/ArcGIS Help.
ESRI-L@esri.com — Mailing list.
http://arcscripts.esri.com — ESRI scripts, Data models, etc.

CHAPTER 12

Geomorphometry in SAGA

V. Olaya and **O. Conrad**

about SAGA: history, system architecture, license · download and installation · working with SAGA: graphical user interface, data visualisation, module execution, modules overview · DEM preparation: import, creation, pre-processing · deriving lands-surface parameters: morphometry, lighting, hydrology, channels and basins, simulation, non-free modules, further analyses

1. GETTING STARTED

System for Automated Geoscientific Analyses (SAGA) is a full-fledged GIS, and many of its features have some relation with geomorphometry, which makes it an ideal tool for operational work, but also for GIS training purposes. For this reason, we will emphasise the particular characteristics of SAGA and, specially, the relation between some of its features and concepts as presented in previous chapters.

SAGA is GIS software with support for raster and vector data. It includes a large set of geoscientific algorithms, and is especially powerful for the analysis of DEMs. Using SAGA you can calculate most of the land-surface parameters and objects described in the first part of this book, and also you can use some of its additional capabilities to use those land-surface parameters and objects in many different contexts. SAGA is thus a complete tool for many practical applications such as those described on the third part of this text.

SAGA has been under development since 2001 at the University of Göttingen, Germany, with aim of simplifying the implementation of new algorithms for spatial data analysis within a framework that immediately allows their operational application. Therefore, SAGA targets not only the pure user but also the developer of geo-scientific methods. SAGA has its roots in DiGeM, a small program specially designed for the extraction of hydrological land-surface parameters (Conrad, 1998), which explains why SAGA provides quite a large number of functions related to geomorphometry.

In 2004 most of SAGA's source code was published using an Open Source Software (OSS) license. With this step the scientific community has been invited to

Developments in Soil Science, Volume 33 © 2009 Elsevier B.V.
ISSN 0166-2481, DOI: 10.1016/S0166-2481(08)00012-3.

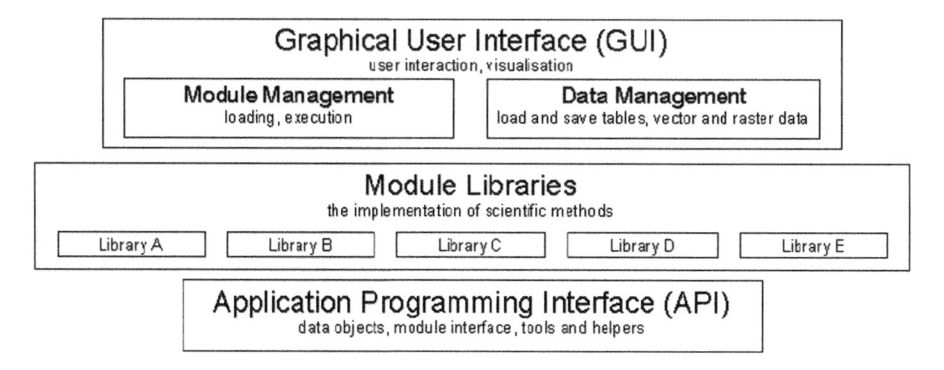

FIGURE 1 SAGA system architecture.

prove the correctness of the implemented algorithms and to participate in their further development. With the release of version 2.0 in 2005, SAGA works under both Windows and Linux operating systems.

In the following text, we will introduce you to SAGA with a strong focus on the analysis and application of DEM data. If you need more information or more detailed descriptions of SAGA functions please consult the SAGA manual that can be accessed from SAGA's website (Olaya, 2004).

To obtain maximum benefit from SAGA, it is crucial to understand how it was designed. SAGA has been designed to be a flexible and useful tool for the geo-scientific community, and a large part of its actual structure is due to that particular aim. Conceptually, the architecture of SAGA consists of three different components (Figure 1):

- The Application Programming Interface (API) provides all the basic functions for performing geographical analysis and is the true 'heart' of SAGA itself.
- A set of modules, which are organised in module libraries, represents the geo-scientific methods.
- The Graphical User Interface (GUI) is the system's front end, through which the user manages data and executes modules.

The GUI and most of the published modules have been put under the GNU General Public License (GPL), which requires programmers to publish derived works also under the GPL or a compatible license, a mechanism called *copyleft*. The API uses the less restrictive Lesser General Public License (LGPL), which permits keeping the modified source codes private. This makes it also possible to distribute a new module as proprietary software.

In addition to the GUI, a second user front end, the SAGA command line interpreter, can be used to execute modules. One of its advantages is the ability to write script files for the automation of complex work-flows, which can then be applied to different data projects. We will not discuss these advanced features here and refer instead to the SAGA manual again. We will neither discuss the API. Although the API is fundamental to the whole system, it is only necessary for the module programmer to know its details. Instead we concentrate on how to use the GUI for data management and visualisation, and also on how to manage and run modules.

Once you have learned how SAGA works, we will use it for the import and preparation of elevation data and will then explain some of the modules that contain methods connected with geomorphometry, presenting a different way of understanding the information given in previous chapters. References to those chapters will be given for each particular module.

1.1 Download and installation

The first step to do when working with SAGA is to download the software. Since February 2004 SAGA has been distributed via SourceForge, a host for many OSS projects. You find the SAGA project homepage at: http://saga-gis.org. Source code, compiled binaries for the different operating systems, demo data, tutorials and manuals can be downloaded from here. It is worth visiting this site frequently to get updated versions with bug fixes and new features. A user forum and more information around SAGA is provided by the accompanying homepage at http://saga-gis.org.

After downloading the appropriate binary distribution, you have to uncompress the downloaded file (dependent on the targeted operating system this is either a zip archive or a tarball) to a folder of your choice. Under Windows you can immediately start SAGA by executing the unzipped file `saga_gui.exe`. Under Linux you have to make SAGA's API library `libsaga_api.so` known to the system first, either by copying it to a standard library location or by adding its location to the searched library paths. Detailed instructions can be found in a *read me* file in the installation folder. To uninstall SAGA simply delete this folder again.

If you have downloaded one of the demo data projects, like the *Forest of Göttingen*, you can immediately start exploring SAGA's capabilities by opening the project file, which can be identified by file extension '`*.sprj`'.

1.2 Working with SAGA

In addition to standard elements like menu, tool and status bars, the GUI has three major control elements: a workspace, an object properties and a message notification window, which are complemented by a varying number of views, which usually show different kinds of data visualisations. The message notification window simply informs the user about actions that have been undertaken. All management tasks regarding modules, data and views can be controlled through the workspace and object windows.

Depending on which object is selected in the workspace window an object specific set of properties is shown in the object window. The workspace has three sub-categories for modules, data and map views. Loaded module libraries are listed with their modules in the modules workspace [Figure 2(a)]. Similarly loaded data appear in the data workspace, sorted by their data type [Figure 2(c)], and created maps can be accessed through the maps workspace. As shortcut to the main menu, right mouse clicks on a workspace object will pop-up a specific context

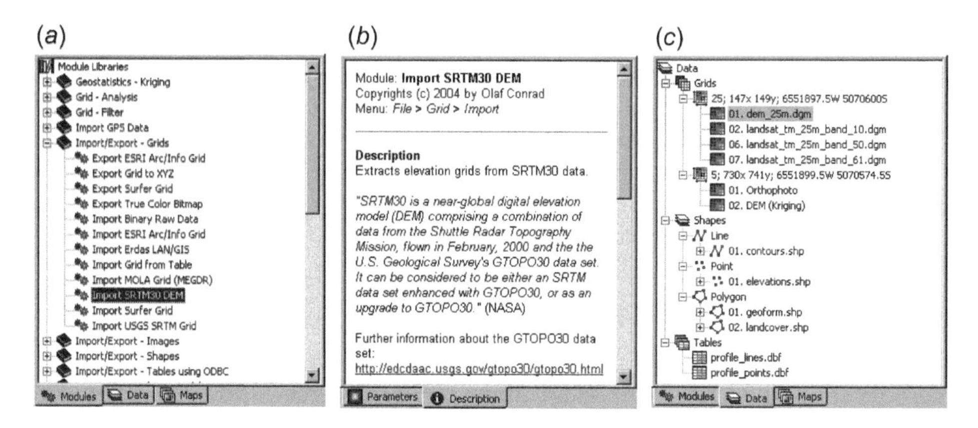

FIGURE 2 SAGA windows: (a) module management, (b) module description and (c) data management.

menu, e.g. to save a data set, to unload a module library or to change the display order of layers in a map view.

The object control provides a *Description* sub-window [Figure 2(b)], that gives information about the selected object, and a *Parameters* sub-window that allows display and modification of data. Other sub-windows appear depending on the object type, e.g. a legend in case of a map. When starting SAGA the first time, all module libraries located in the installation folder will be loaded automatically, which supplies us with all functions that we want to use in the following sections. The data and therefore the maps workspace are still unpopulated and the next step will be to load some data.

SAGA handles tables, vector and raster data and natively supports at least one file format for each data type. It has to be pointed out that SAGA uses *Grid* synonymously for raster structures and refers to vector data as *Shapes*. Table formats can be either tab-spaced text files or DBase files. For vector data the widespread *ESRI Shape File* format is supported. The file access to raster data uses the flexible SAGA raster format, which consists of a separate text file to provide meta information on how to interpret the actual data file.

After loading a data set it appears in the data workspace. Vector data will be sorted by their shape type, either Point, Multi-Point, Line or Polygon, and raster data are categorised by their raster system properties, i.e. the number of columns and rows, cell size, and geographic position. To display a spatial data set in a map, simply double click on it or choose the menu entry *Show* in its associated context menu. Afterwards you can decide whether to create a new map or to add it as new layer to an existing map. The display order of map layers can then be changed in the map workspace.

The most important data display options are related to the colouring, for which you can use lookup tables to manually adjust the value ranges for colour classes, or use a metrical colour classification scheme. One of the display options specific to raster data is *transparency*, which allows using a raster layer for shading effects. Once you have prepared a nice looking set of maps combining a number of data

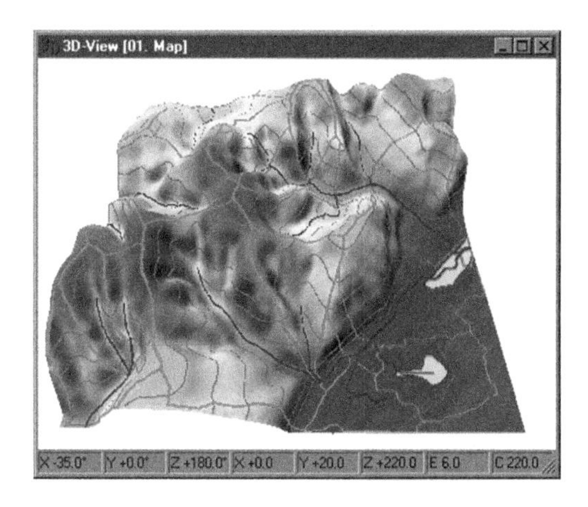

FIGURE 3 A 3D view in SAGA.

sets, you can save all settings in a project file, which can be reopened for further use. Besides maps, several other data visualisations are offered by SAGA, like table views, diagrams, histograms and scatter plots. When appropriate elevation have been loaded, a map can easily be displayed as a 3D view (Figure 3) including the possibility to create animated sequences (*fly through*) and coloured stereo anaglyphs.

Modules can be executed directly by using their associated *Parameters* window. Alternatively we can call a module by its menu entry in SAGA's main menu. The menu entries are hierarchically sorted by the kind of analysis or action they represent. A standard operation when working with DEMs is the calculation of an analytical hillshade model, which is particularly suited for terrain visualisations when combined with other data layers. We find the module *Analytical Hill Shading* at the *Terrain Analysis/Lighting* sub menu. After choosing a module for execution a dialogue will pop up, where the module specific parameters need to be set. Usually at least one obligatory input data set has to be chosen from the loaded data. Here we have to choose the DEM, for which the hill shade calculation shall be performed. Instead of creating a new data set for the results, we can also choose to overwrite the values of an existing one (Figure 4).

Besides settings setting of inputs and outputs, the module will show various options that can be set by the user. For the hill shade calculation we can choose the direction of the light source as well as one of four possible shading methods. After confirming the correct settings by pressing the *Okay* button, the calculation will start. The calculation progress is shown in the status bar and when finished a notification is added to the message window and the newly created data set is added to the data workspace, from where it can be saved to file or added to a map.

Currently SAGA provides about 42 free module libraries with 234 modules, most of them published under the GPL. Not all of these modules are highly sophisticated analysis or modelling tools. Many of them just perform rather simple data operations. The modules cover geostatistics, geomorphometric analysis, im-

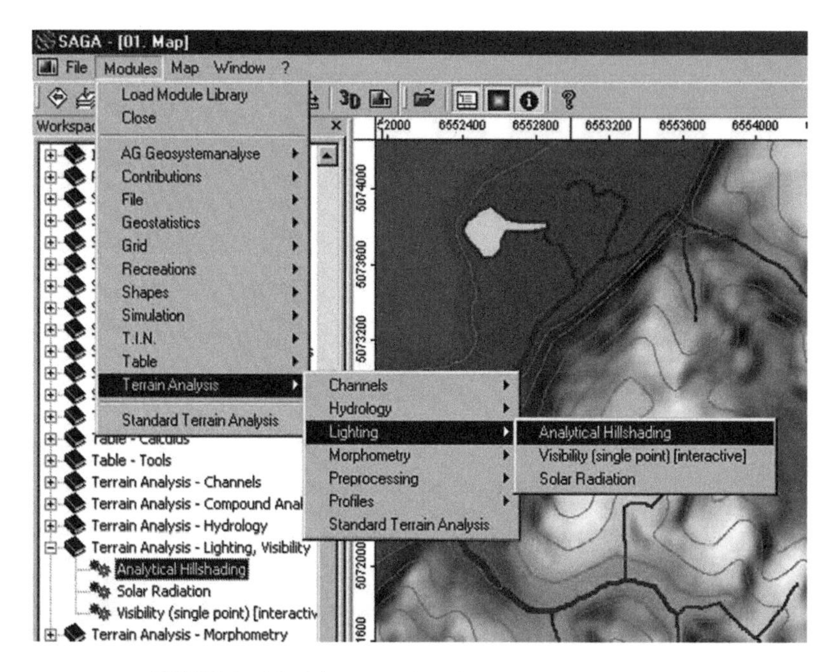

FIGURE 4 Module execution via menu entry in SAGA.

age processing, cartographic projections, and various tools for vector and raster data manipulation.

It is interesting to note that modules, data layers and maps, although connected (modules are executed on data layers, and those are displayed in maps) are completely independent concepts. For instance, you can open a DEM and extract land-surface parameters from it without having to visualise it at all.

2. DEM PREPARATION

Before we continue with geomorphometric analysis, we need to have elevation data in a raster structure loaded in SAGA. Hence we want to know how to load data from various sources, how to derive a raster DEM from point data, and how to prepare a DEM to get best analysis results.

2.1 Import from different sources

Data stored in SAGA's *native* raster file format can immediately be loaded. However, this format is not very widespread, and you are not likely to have your data present in that format. To access data stored in other file formats SAGA provides us with a number of modules for data import and export. You find these modules under the *Modules/Files* menu. To give a practical example, we will see how to incorporate the Baranja Hill layers into SAGA. Open the *Files* $\mapsto$ *Raster* $\mapsto$ *Import* $\mapsto$

Import ESRI ArcInfo Grid module. Click on the button on the right part of its only parameter and you will see a file selection dialogue. Select the DEM25m.asc file containing the DEM. Click on *Okay* to close the parameters window, and you will find the DEM in the *Data* workspace, waiting to be analysed or visualised.

Other modules exist e.g. for the import of SRTM (Shuttle Radar Topography Mission) and MOLA (Mars Orbiter Laser Altimeter) DEM, but the most flexible import tool for raster data uses the *Geospatial Data Abstraction Library* (GDAL), which supports about 40 different file formats. Now let's see how to open other data layers included in the sample set, such as Landsat images. The module that you have to use in this case is *Import Erdas LAN/GIS*, whose parameters window is identical to the one of the *Import ESRI Arc/Info Grid* module, just requiring one file to be selected. If you open the bar_tm.lan file, not just one new layer is added to the data tree, but 8 of them, which represent the different channels of the Landsat TM sensor.

2.2 Creating raster DEM from point samples

Although raster DEM are quite common, these are not always readily available. Particularly when you work in a less investigated area and need a high resolution DEM for further analyses, you probably have to create it by yourself. GPS data or contour lines from digitised topographic maps may then serve as starting point for the DEM creation (see Section 3.2 in Chapter 2). The Baranja Hill dataset includes a vector data file with elevation points, named elevations.shp, which we can load directly into SAGA.

However, this supplies us only with a set of scattered elevation samples and we have to use an interpolation technique to estimate the elevation for each cell of a regular raster. SAGA provides us with a collection of interpolation algorithms:

- Nearest Neighbour takes the value of the nearest observed point.
- Triangulation performs linear interpolation on the triangles, which are defined by applying Delaunay's method to the observed points.
- Inverse Distance calculates the distance weighted average of all observed points within a given search radius.
- Modified Quadratic Shepard is similar to Inverse Distance, but uses a least squares fit for better results.
- Ordinary Kriging is a geo-statistical method based on auto-correlation. It is probably the most sophisticated interpolator, but requires preliminary fitting of the variogram.

We will demonstrate the procedure for the Triangulation, which is a standard technique. After starting the *Triangulation* module you can select the elevation data set as input and the *Attribute* field *Value*, which holds the elevation values. In the dialogue's *Options* section you can change the parameter *Target Dimension* to *User defined*. That way, you will be able to define the exact raster size and extent that you want to continue to work with. After confirming the correct settings by clicking the *OK* button, you are prompted with another dialogue, where you can specify the raster size in the *Grid Size* field, and define the extent whether introducing values

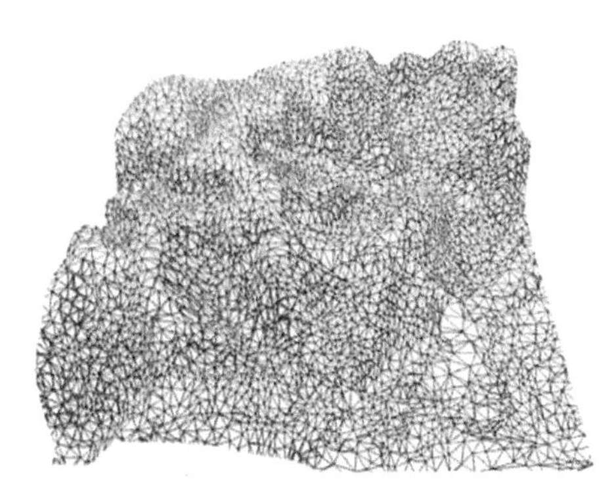
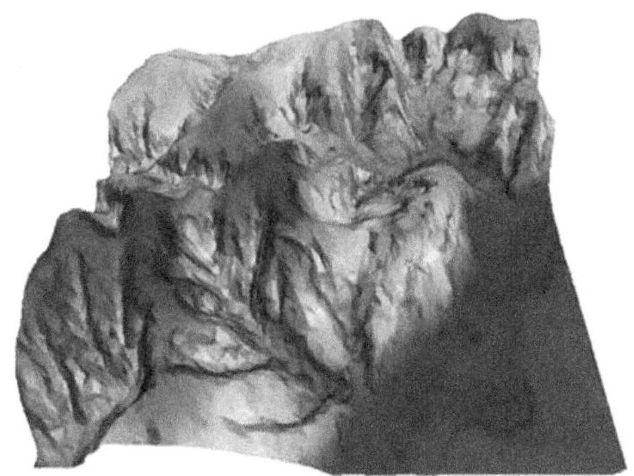

FIGURE 5 Delaunay Triangulation and resulting DEM.

on the correspondent fields or selecting the *Fit Extent* check box, which will cause the module to automatically select the extent according to the boundaries of the vector layer. The resulting layers are shown in Figure 5.

2.3 Further pre-processing

Once a DEM is loaded in SAGA, it might be necessary to do further steps before proceeding with Terrain Analysis. The cartographic projection can be changed for raster as well as for vector data by the use of two alternative cartographic projection libraries, the *GeoTrans* library developed by the National Imagery and Mapping Agency, and the Proj.4 library initiated by the U.S. Geological Survey. You can merge several overlapping or bordering raster tiles or cut a smaller DEM out of a huge one.

In SAGA, data gaps can be solved by combining grids, and grids can be transformed to finer or coarser resolutions using resampling. Several filter algorithms can be used to smooth or sharpen the elevation surface, including special filters, which try to preserve prominent features such as breaks and ridges. Very specific for the pre-processing of DEM are two alternative modules for the removal of closed depressions or sinks, one of them implements the procedure proposed by Planchon and Darboux (2001). When you want to derive water flow dependent Lands Surface Parameters, you should always apply one of these modules first. Otherwise the flow algorithms cannot can not flow continuously (spurious sinks), which can lead to broken streams and artefacts. This happens due to generalisation and other effects.

3. DERIVATION OF LAND-SURFACE PARAMETERS

Once you have loaded your DEM and carried out all preparations on it that are been necessary, you are ready to derive land-surface parameters. In the following we will see the relation between each Terrain Analysis module and the chapter where its fundamentals are described, so you can refer to the latter in case you

FIGURE 6 Convergence Index. (See page 720 in Colour Plate Section at the back of the book.)

need more information. Due to its academic background, where it is of high interest to compare different algorithms to solve one problem, SAGA often offers various ways to calculate many different parameters.

3.1 Morphometric land-surface parameters

Modules of this group analyse and parameterise the shape of the surface. The identification of *Surface Specific Points* makes use of early algorithms for DEM analysis (e.g. Peucker and Douglas, 1975) and classifies the terrain into features like ridges, channels and slopes. *Hypsometric Curves* are particularly useful for the morphometric characterisation of watershed basins (Luo, 2000).

3.2 Lighting

Probably the best known morphometric parameters can be derived with the *Local Morphometry* module, which calculates slope gradient, aspect, and, if supported by the chosen method, also the curvatures. By default the method of Zevenbergen and Thorne (1987) is selected, but you can also choose between those described by Heerdegen and Beran (1982), Tarboton (1997) and others. The *Convergence Index*, proposed by Köthe et al. (1996), uses the aspect values of neighbouring cells to parameterise flow convergence and respectively divergence (Figure 6, described in Conrad, 1998). It is similar to plan curvature, but does not depend on absolute height differences.

Curvature Classification after Dikau (1988) can be performed on plan and profile curvatures. Two other modules calculate the real surface area, as opposed to the projected area, and also a morphometric protection index.

Three modules have a direct relation to illumination and how the terrain influences the spreading of light. *Analytical Hillshading* is commonly used for terrain visualisations as has been pointed out. The standard calculation simply returns the angle, under which light coming from a given direction is reflected by the

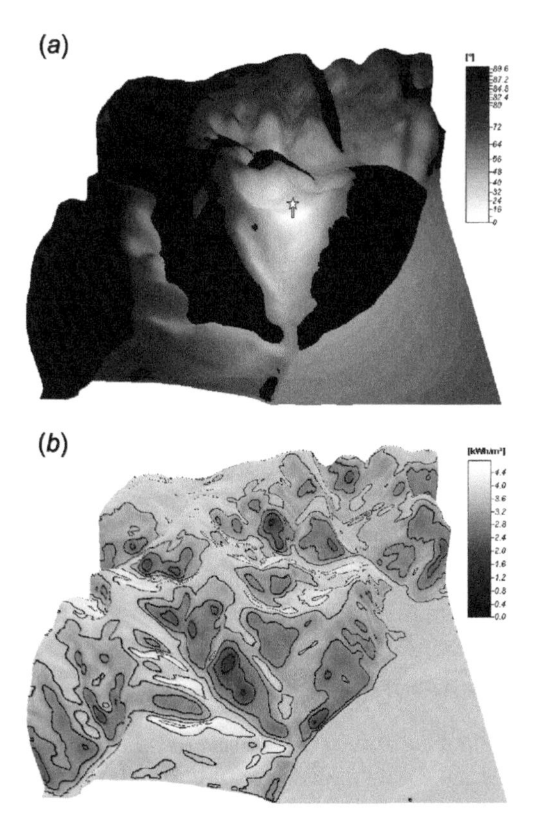

FIGURE 7 (a) Visibility and (b) Solar Radiation.

terrain. This can be combined again with the slope values to emphasise the contrast between hilly and flat areas. With the most advanced option light rays will be traced, so that shadowed areas can be identified. This option is also used by the *Solar Radiation* calculation [Figure 7(b)], where the shading is done for sun's position and the incoming energy is summed for user defined time periods.

Atmospheric effects are taken into account according to the SRAD program of the TAPES-G suite (Wilson and Gallant, 2000). Similarly ray tracing is used in the *Visibility* calculation, an interactive module, where the user can choose by a mouse click on map for which point the visibility analysis shall be executed. The difference is that in this case the light source is not in the far distance, but very close to the terrain. Output is either the visible size of an object, the distance, or the reflectance angle [Figure 7(a)].

3.3 Lands-surface parameters specific to hydrology

If you have compared the results of the different methods for slope and aspect calculation, you have seen that the results do not differ significantly. Due to the nature of the raster structure, this is not the case for calculations based on water flow dis-

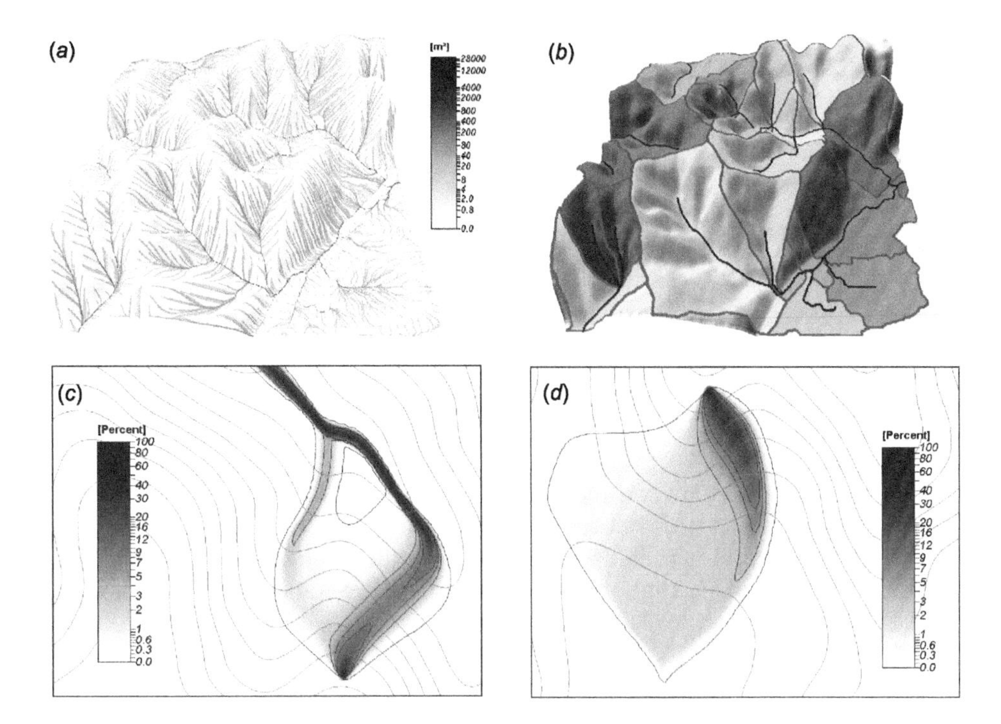

FIGURE 8 Hydrological analysis in SAGA: (a) catchment areas (DEMON, each 100th cell), (b) watershed basins, (c) downslope area (FD8) and (d) upslope area (FD8). (See page 720 in Colour Plate Section at the back of the book.)

tribution models [Equation (3.1) in Chapter 6]. Again, SAGA covers most of the published algorithms for the calculation of catchment areas and related parameters. The *Parallel Processing* and *Recursive Upward Processing* differ only in the way the DEM is processed and give the same results for same flow distribution models. The provided methods include D8 (O'Callaghan and Mark, 1984), D-Infinity (Tarboton, 1997), and FD8 (Freeman, 1991).

The *Flow Tracing* algorithms complement the previously mentioned methods for the *Kinematic Routing Algorithm* (Lea, 1992) and DEMON (Costa-Cabral and Burges, 1994). For a better visualisation of DEMON's flow tube concept, only each 10th cell of each 10th row has been chosen as flow source in Figure 8. Together with the catchment area, associated parameters might optionally be calculated, such as average height, slope, aspect and flow path length.

Most of the other hydrology related modules make use of either D8 or FD8. For example the *Upslope Area* and *Downslope Area*, which determine the hydrologic influence for user defined points or areas [Figure 8(c) and (d)], or the alternative *Flow Path Length*, which accepts additional features for starting a flow path. Among the other related modules, such as *Flow sinuosity*, *Cell Balance*, or *Flow Depth*, maybe the most remarkable one is the one named *Topographic Indices*, which combines catchment areas with slope gradients to indicate soil moisture (TWI) as well as erosion processes (stream power, LS factor, see also Chapter 7).

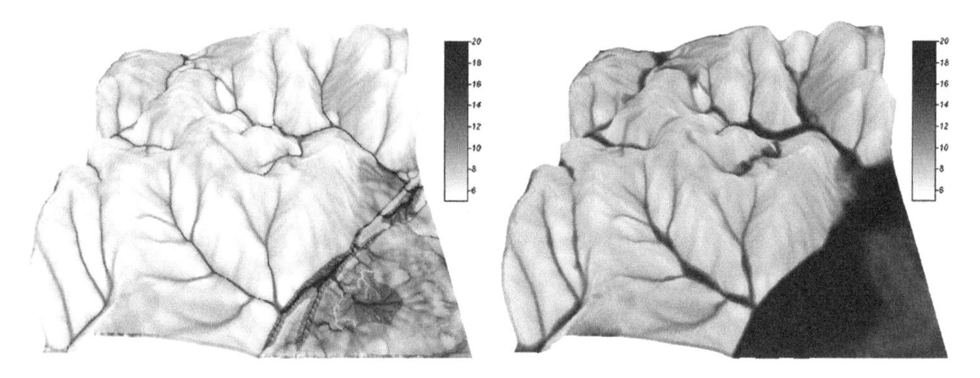

FIGURE 9 The Topographic Wetness Index (left) and the SAGA Wetness Index (right).

The so called *SAGA Wetness Index* (Figure 9) is based on a modified catchment area calculation, which does not consider the flow as very thin film. As a result of this, it predicts for cells situated in valley floors with a small vertical distance to a channel a more realistic, higher potential soil moisture compared to the standard TWI calculation (Böhner et al., 2002).

3.4 Drainage networks and wastershed basins

Drainage or channel networks can be extracted in more than one way using different modules. The most elaborated one is Channel Network, which has various options to control channel origins, density, minimum length and routing. The *Strahler Order* module produces new layers that can be used as initiation grids, yielding different results, sometimes more precise than using e.g. a minimum catchment area as criteria for starting a channel.

Channel networks are generated in raster and vector format. The junctions are stored as special values in the raster and can be directly used to define outlets for the automated derivation of sub-basins (Figure 8). Having a channel network you can calculate the distance of each point to it, either defined by overland flow or to its interpolated base level, which then might be used to estimate e.g. the groundwater influence.

3.5 Hydrological simulations

SAGA is also capable of performing hydrological modelling. For instance, the modules that calculate time to outlet for a defined basin can be used to derive non-synthetic Unit Hydrographs for that basin, using the histogram of time values of the resulting layer. For a more detailed analysis, those same layers can be used as inputs in distributed hydrological models. The TOPMODEL implementation is based on the work of Beven (1997) and is based on the C-port of the Fortran77 sources included in GRASS GIS (see Chapter 17). A predominantly educational module, that is thought of as demonstration of the principles of dynamic computer models, is the Nitrogen distribution model according to Huggett (1993),

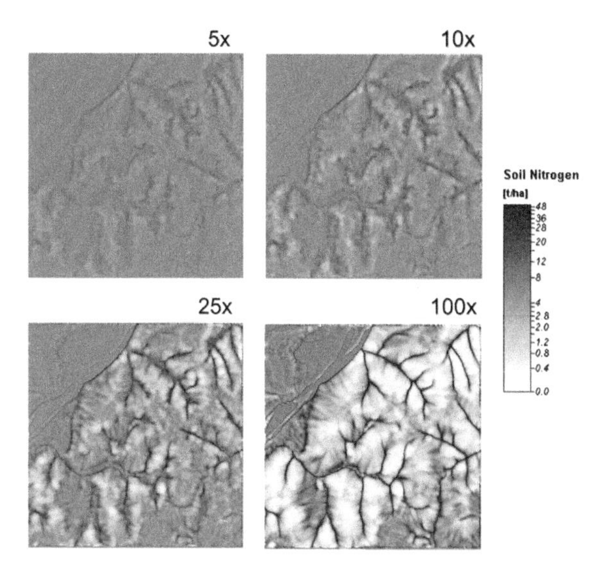

FIGURE 10 Nitrogen distribution simulation.

which simulates the water flow controlled spatial distribution of Soil Nitrogen (Figure 10).

3.6 Commercial modules

As mentioned before, a SAGA module does not have to be *free*. In the following section, several such modules are introduced, because they have a strong relation to geomorphometry. Their theory has already been published, and they are likely to become part of SAGA's OSS distribution in future. *Heights below summit* culminations and *heights above valley floors* respectively (Figure 11) are to some extent similar to the vertical distance to a channel network base level. The advantage is that this land-surface parameter takes only a DEM as input and does not

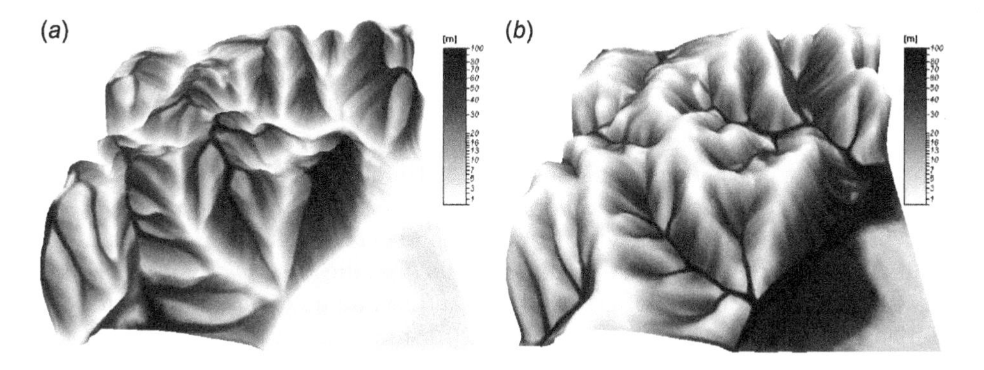

FIGURE 11 (a) Height above valley floors and (b) height below summit culminations.

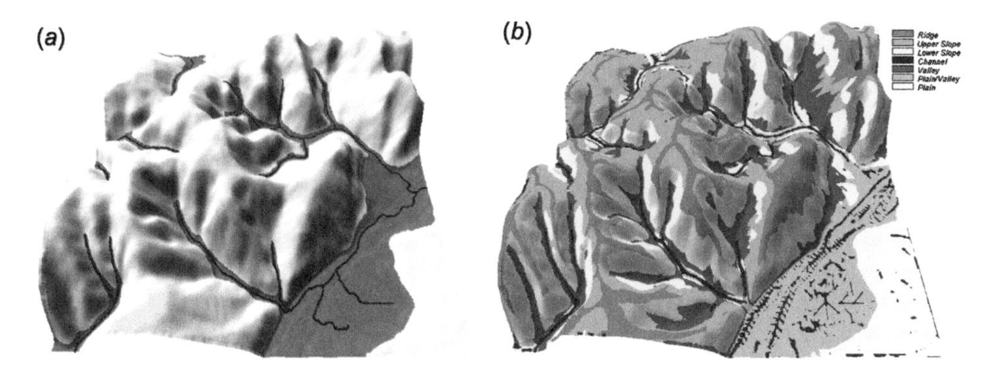

FIGURE 12 (a) Flood plain map calculated using a threshold buffer, (b) terrain classification using Cluster Analysis. (See page 721 in Colour Plate Section at the back of the book.)

depend on arbitrarily dense channel networks. These relative heights have been successfully used for the prediction of soils influenced by solifluction during the pleistocene (Böhner and Selige, 2006).

3.7 Beyond geomorphometric analysis

Being a versatile GIS software, SAGA offers many more methods that do not deal with geomorphometry, but can meaningfully be applied to this subject too. Of course this depends very much on the problem to be solved and the imagination of the investigator. Most relevant but not restricted to DEMs are the modules for profile calculations. Three different profile types can be interactively created with SAGA. Besides simple profiles, where you define a profile line by connecting points, you can derive a flow path profile, where the profile is searched from the initial point down slope according to the D8 method. Swath profiles calculate statistical properties, like mean, minimum, maximum and standard deviation, of the cells lying within a given distance to the chosen profile (Figure 13).

Statistical data analysis can also be used to describe the relation of a point to its neighbourhood, usually determined by a user defined search radius. Wilson and Gallant (2000) describe a number of statistical values for elevation residual analysis, for instance the *value range*, which is a measure of relief energy and *average slope gradient*, and the *percentile*, which is comparable with the curvature. In a similar way Böhner et al. (1997) analyses the variance to get a measure on how representative a cell is for its neighbourhood. The representativeness of altitude can be used to mark summits and floors, while the same concept applied to slope gradient values differentiates between breaks and even areas. Two final examples for alternative calculations shall be given. In the first the *Threshold Buffer* has been used to identify a flood plain, given a DEM and a channel network as input [Figure 12(a)]. The second example in Figure 12(b) shows how *Cluster Analysis* leads to a meaningful terrain classification, when supplying it with well chosen land-surface parameters.

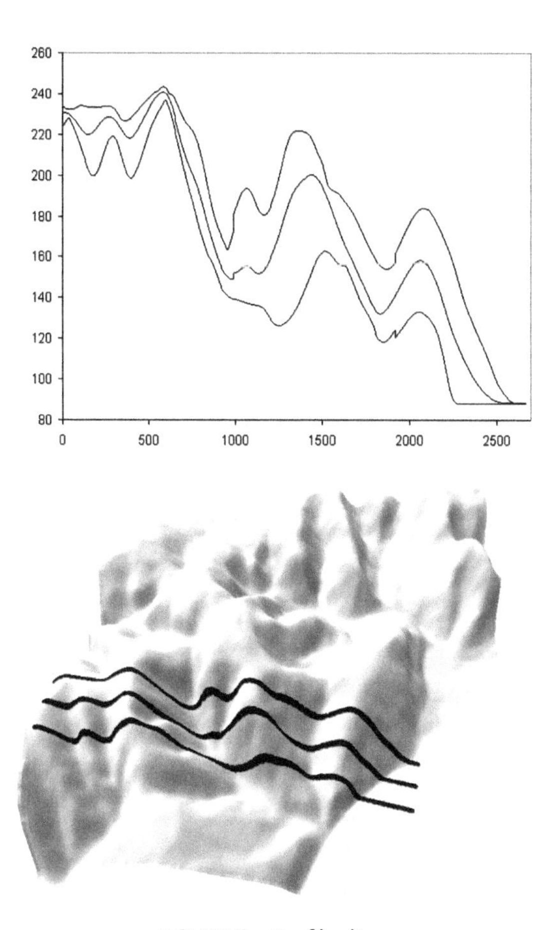

FIGURE 13 Profile diagram.

4. SUMMARY POINTS

Although SAGA has many data management and visualisation features, its true strength remains a comprehensive set of spatial analysis tools with a marked focus on Terrain Analysis. Particularly the Open Source Software philosophy makes the methods transparent to scientists, who, when using commercial software, frequently accept software outputs without an opportunity to improve or validate the underlying algorithms — the so-called outputs from a *black-box*. For users, the system offers an immediate and easy access to a wide range of state of the art methods in spatial analysis and it does this for literally no cost. The free availability and simple installation predestines SAGA for educational purposes, whilst the high performance of sophisticated methods makes it attractive for professional applications.

The easily approachable object oriented API invites every scientist, who has just a basic understanding of programming languages, to choose SAGA as a platform for the implementation of his own models. SAGA is still rapidly evolving and it can be expected that its facilities will increase with a growing community of users and developers.

IMPORTANT SOURCES

Böhner, J., McCloy, K.R., Strobl, J. (Eds.), 2006. SAGA — Analysis and Modelling Applications. Göttinger Geographische Abhandlungen, Heft 115. Verlag Erich Goltze GmbH, Göttingen, 117 pp.
Olaya, V., 2004. A gentle introduction to SAGA GIS. The SAGA User Group e.V., Gottingen, Germany, 208 pp.
http://www.saga-gis.uni-goettingen.de — SAGA homepage.
http://www.geogr.uni-goettingen.de/pg/saga/digem — DiGeM a Program for Digital Terrain Analysis.
http://www.gdal.org — Geospatial Data Abstraction Library.
http://earth-info.nga.mil/GandG/geotrans/ — GeoTrans Geographic Translator.
http://www.remotesensing.org/proj/ — Proj.4 Cartographic Projection Library.

CHAPTER 13

Geomorphometry in ILWIS

T. Hengl, B.H.P. Maathuis and **L. Wang**

first steps in ILWIS · main functionalities — what it can and can't do? how to get support · importing and displaying DEMs · derivation and interpretation of land-surface parameters and objects · use of the hydro-processing module to derive drainage network and delineate catchments · use of ILWIS scripts · strong and weak points of ILWIS

1. ABOUT ILWIS

ILWIS is an acronym for Integrated Land and Water Information System, a stand alone integrated GIS package developed at the International Institute of Geoinformation Science and Earth Observations (ITC), Enschede, Netherlands. ILWIS was originally built for educational purposes and low-cost applications in developing countries. Its development started in 1984 and the first version (DOS version 1.0) was released in 1988. ILWIS 2.0 for Windows was released at the end of 1996, and a more compact and stable version 3.0 (WIN 95) was released by mid 2001. From 2004, ILWIS was distributed solely by ITC as shareware at a nominal price. From July 2007, ILWIS shifted to open source and ITC will not provide support for its further development.

REMARK 1. *ILWIS is an acronym for Integrated Land and Water Information System, a stand-alone GIS and remote sensing package developed at the International Institute of Geoinformation Science and Earth Observations (ITC).*

The most recent version of ILWIS (3.4) offers a range of image processing, vector, raster, geostatistical, statistical, database and similar operations. In addition, a user can create new scripts, adjust the operation menus and even build Visual Basic, Delphi, or C++ applications that will run at top of ILWIS and use its internal functions. In principle, the biggest advantage of ILWIS is that it is a compact package with a diverse vector and raster-based GIS functionality and the biggest disadvantage are bugs and instabilities and necessity to import data to ILWIS format from other more popular GIS packages.

Developments in Soil Science, Volume 33 © 2009 Elsevier B.V.
ISSN 0166-2481, DOI: 10.1016/S0166-2481(08)00013-5. All rights reserved.

1.1 Installing ILWIS

As per July 1st, 2007, ILWIS software is freely available ('as-is' and free of charge) as open source software (binaries and source code) under the 52°North initiative (http://52north.org). The ILWIS binaries are very simple to install. Copy the folder in the downloaded zip file. In this folder there is an `Ilwis30.exe` which is the main executable for ILWIS. Double click this file to start ILWIS.

You will first see the main program window, which can be compared to the ArcGIS catalog. The main program window is, in fact, a file browser which lists all ILWIS operations, objects and supplementary files within a working directory (see Figure 1). The ILWIS Main window consists of a Menu bar, a Standard toolbar, an Object selection toolbar, a Command line, a Catalog, a Status bar and an Operations/Navigator pane with an Operation-tree, an Operation-list and a Navigator. The left pane (Operations/Navigator) is used to browse available operations and directories and the right menu shows available spatial objects and supplementary files.

The user can adjust local settings of ILWIS by entering Preference under the main menu. In addition, the user can adjust also the catalog pane by choosing *View* ↦ *Customize catalog*. This can be very useful if in the same directory we also have GIS layers in different formats. For example, `DEM25m.asc` will not be visible in the catalog until we define `.asc` as external file extension. Note that, although ILWIS provides a possibility to directly write and read from files in external formats, in principle, it is always more efficient to first import all spatial objects to ILWIS format.

> REMARK 2. *There are four basic types of spatial objects in ILWIS: point, segment, polygon and raster maps. Supplementary files include: tables, coordinate systems, scripts, functions, domains, representations, etc.*

1.2 ILWIS operations

ILWIS offers a wide range of vector, raster and database operations that can be often combined together. An overview of possible operations can be seen from the main program window *Help* ↦ *Map and Table calculation* ↦ *Alphabetic overview of operators and functions*. For the purpose of land-surface parametrisation, the most important are the map calculation functions including neighbourhood and filtering operations. A special group of specific land-surface modelling operations is included in the module hydro-processing tools.

Note also that a practical aspect of ILWIS is that, every time a user runs an command from the menu bar or operation tree, ILWIS will record the operation in ILWIS command language. For example, you can import a shape file showing the contour lines from the 1:50,000 map by selecting *File* ↦ *Import* ↦ *ILWIS import* ↦ *Shape file*, which will be shown as:

```
import shape(contours50k.shp, contours50k)
```

on the ILWIS command line. This means that you can now edit this command and run it directly from the command line, instead of manually selecting the operations

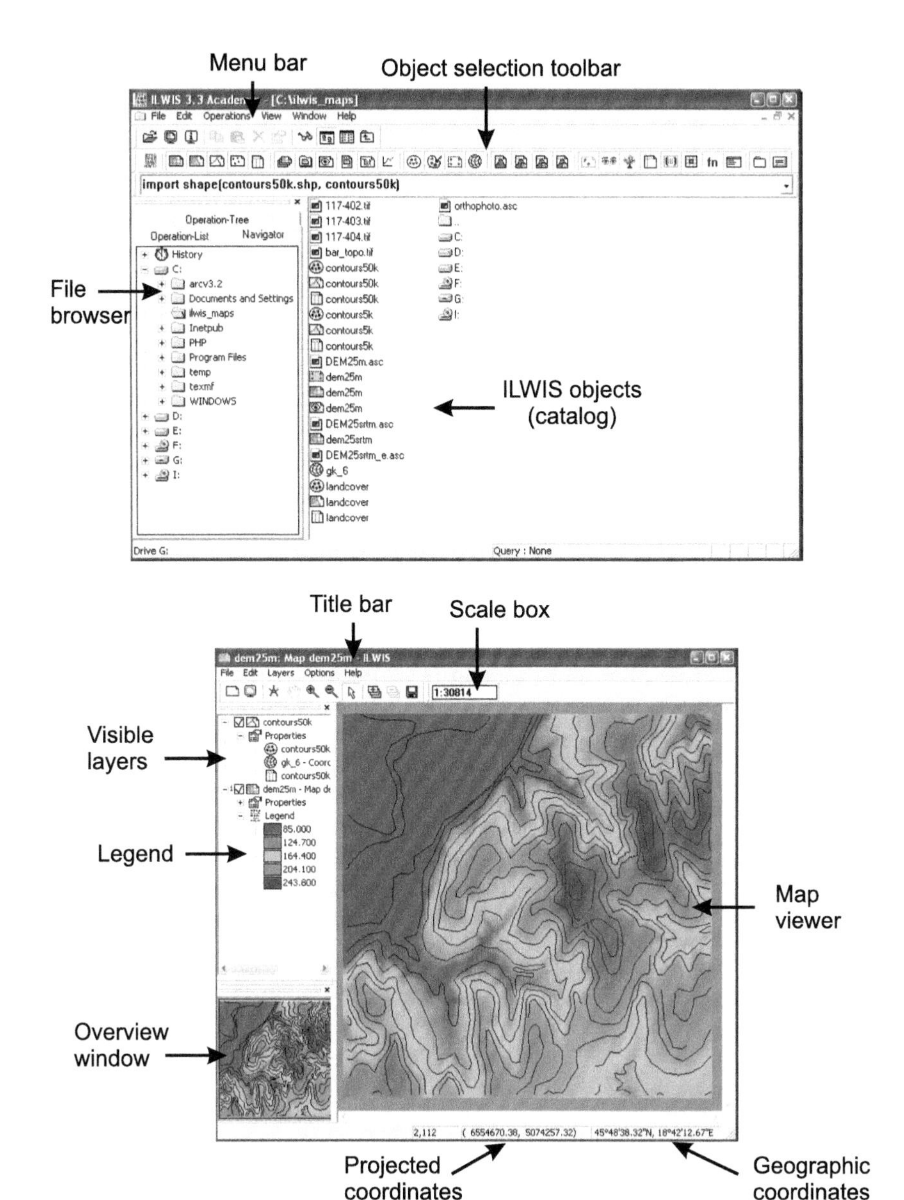

FIGURE 1 The ILWIS main window (above) and map window (below).

from the menu bar. In addition, you can copy such commands into an ILWIS script to enable automation of data analysis (see further Section 3.2).

The most used command in ILWIS is the `iff(condition,then,else)` command, which is often used to make spatial queries. Other commands can be easily understood just from their name. For example command `MapFilter(map.mpr,filter.fil)` will run a kernel filter (`filter.fil`) on an input map (`map.mpr`). Arithmetical operations can be directly done by typing, for example:

```
mapC = mapA + mapB
```

1.3 ILWIS scripts

An ILWIS script gathers a sequenced list of ILWIS commands and expressions with a limited number of open parameters. Detailed instruction on how to create and run a script can be found in ILWIS 3.0 Academic User's Guide, Chapter 12. Parameters in scripts work similar to replaceable parameters in a DOS batch file. Open parameters can be coded using `%1, %2, %3,` up to `%9`. Scripts can be edited using the script tab in ILWIS, which offers editing of both commands, input parameters and their default values. If you have more than 9 variables, then you can create one master script that calls a number of sub-scripts. In that way, the number of parameters can be increased to infinity.

Once you create and save a script, you will see that ILWIS creates two auxiliary files: one `.isf` file which carries the definition of script parameters and an `.isl` file showing the list of commands. Both can be edited outside ILWIS using a text editor. When you have created a script and when you click the *Help* button in the *Run Script* dialog box for the first time, an HTML page will also be automatically generated listing all parameter names of the script and a minimal explanation. This HTML file is stored with the same name and in the same directory as the script and can be edited and modified according to your wishes.

> REMARK 3. *All operations in ILWIS can be run from command line using the ILWIS syntax. List of commands can be combined in ILWIS scripts to automate data processing.*

It is useful to know that remarks and comments within the scripts can be added by using the following commands:

- `rem` or `//` — this is an internal comment. All text on the line after `rem` or `//` is ignored from calculation.
- `begincomment endcomment` environment — has the same functionality as `rem` or `//` commands.
- `message` — this will create a text in message box on your screen. After pressing the OK button in the message box, the script will continue.

Comments and instructions can be fairly important because you can explain calculations and provide references.

After you have built and tested a script, it is advisable to copy it to the ILWIS program folder named `/Scripts/`, so that your script will be available from the operation menu every time you start ILWIS. You can customise the operation menu and operation tree to be able to find these operations much faster (see *ILWIS Help* ↦ *How to customize the Operation-tree, the Operation-list and the Operations menu*). To further customize the Operations menu, the Operation-list and the Operation-tree, advanced users may wish to modify the ILWIS `action.def` file that is located under the ILWIS program folder.

A script can be run by double clicking it from the ILWIS catalog or by typing the run script command in the ILWIS command line, e.g.:

```
run scriptname parameter1 parameter2...
```

2. IMPORTING AND DERIVING DEMS

In the most recent version of ILWIS, you can import GIS layers from a wide range of packages and formats. This is possible due to two built-in translation tools: GeoGateway (see list of supported formats at http://pcigeomatics.com) and GDAL (see list of supported formats at http://gdal.org). Elevation data, prepared as shape files and ESRI ArcInfo ascii grids, can be imported without difficulty. In ILWIS is also possible to import `.hgt` (HeiGhT) blocks, but then a general raster import needs to be used. For example, a command line to import the $1 \times 1°$ SRTM 3 arcsec blocks, which consist of 1201×1201 pixels is:

```
name = map('name.hgt', genras, UseAs, 1201, 0, Int, 2,
SwapBytes)
```

where `name` is the name of the block and `genras` is the general raster map import command.

The following section will explain how to import an existing DEM or derive it from the sampled elevations. First, download the Baranja Hill dataset from geomorphometry.org and save it to a working directory, e.g. `/ilwismaps/` or similar. In this chapter we will work with sampled elevations (contours, height points) digitised from the 1:50,000 topo maps and the 30 m resolution SRTM image. In the case of SRTM DEM, elevations are available at all locations, while in the case of the contour lines, these are just sampled elevations that need to be interpolated to produce a DEM first. Now, import the contour map (`contours50.shp`), point map (`heights50.shp`) with measured heights and a raster mask map (`wbodies.asc`) showing water bodies using the standard import options. Also import the SRTM DEM (`DEM25srtm.asc`), which we will use for further comparison between the DEMs derived from contours and from satellite imagery.

Note that importing a grid file to ILWIS will always create a raster map, a georeference and a coordinate system — you might not need all of these. You can delete redundant coordinate systems and georeferences, but you need to first define in properties of imported maps the replacement grid definition and coordinate system.

2.1 Deriving DEM from sampled elevations

Before you create a DEM from sampled elevations, you need to create a grid definition. Here, you can use either the georeference produced automatically by ILWIS after importing the `DEM25srtm.asc`, or you can create your own grid definition. Use: (*File* ↦ *Create* ↦ *Georeference* ↦ *Corners*) for the output map. By default, we use the following parameters for the grid definition: pixel size of 25 m, and bounding coordinates X, Y (center of pixel): 6551884, 5070562; 6555609, 5074237. This will give you a raster image consisting of 149 rows and 147 columns.

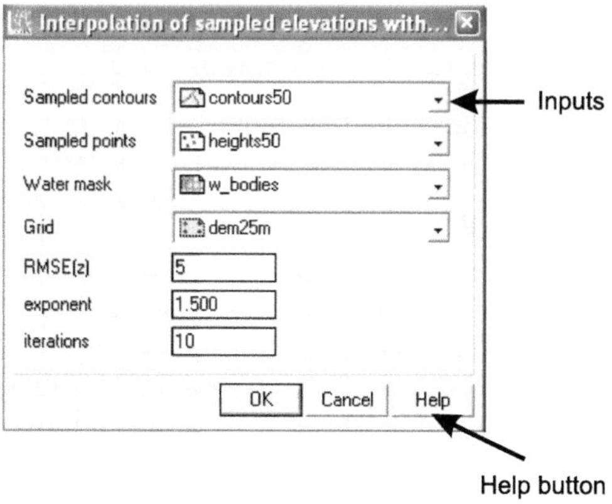

FIGURE 2 Running the DEM interpolation script.

In ILWIS, the default method to interpolate contours is the linear interpolator. The algorithm is described in more detail by Gorte and Koolhoven (1990). This command can be called directly from the *Main Menu* ↦ *Contour interpolation*. A more sophisticated approach is to use the script called `DEM_interpolation`, available from the `geomorphometry.org`. This will interpolate sampled elevations, then detect and filter out the *padi*-terraces and finally adjust elevation for the water bodies. By default, you can run the script using the following command:

```
run DEM_interpolation contours50.mps
heights50.mpp wbodies.mpr dem25m.grf 5 1.5 10
```

where `DEM_interpolation` is the script name, `contours50.mps`, `heights50.mpp` and `wbodies.mpr` are the input maps, `dem25m.grf` is the grid definition, `5` is estimated elevation error, `1.5` is exponent used to adjust for the water bodies and `10` is the maximum number of iterations allowed. A detailed description of the algorithm can be seen by selecting the Help button (Figure 2).

The script works as follows. First the input sampled elevation in segment[1] and point map are rasterized and glued using the target grid:

```
sampled01.mpr = MapRasterizeSegment(contours50.mps, dem25m.grf)
sampled02.mpr = MapRasterizePoint(heights50.mpp, dem25m.grf, 1)
sampled03.mpr = MapGlue(dem25m.grf,
sampled01.mpr, sampled02.mpr, replace)
```

Now we can interpolate the sampled values using:

```
DEM = MapInterpolContour(sampled03.mpr)
```

Of course, the resulting DEM will have many artefacts that will then propagate to land-surface parameters also. We first want to remove the *padi*-terraces, which are absolutely flat areas within the closed contours. These areas can be masked out from the original DEM by using the procedure first suggested by Pilouk and

[1] In ILWIS, segment map is a vector map with no topology, i.e. consisting of only lines.

Tempfli (1992) and further described by Hengl et al. (2004a). First, we need to detect *padi*-terraces using:

```
DEM_TER = iff((nbcnt(DEM#=DEM)>7), ?, DEM)
```

This will detect areas[2] (cut-offs) where more than seven neighbouring pixels have exactly the same elevation and put an undefined pixel "?". Now the medial axes can be detected using the distance operation with the rasterized map of contours:

```
CONT_dist = MapDistance(sampled01.mpr)
MED_AXES{dom=Bool.dom} =
iff((nbcnt(CONT_dist>CONT_dist#)>4), 1, 0)
```

Here the map `MED_AXES` shows detected valley bottoms and ridges, where value "`1`" or "`True`" represents the possible medial axes [Figure 3(b)]. We can attach to these areas some small constant value and then re-interpolate the DEM map. Before we do that, we need to detect which of these medial axes are ridges and which represent bottoms, i.e. which are convex and which concave shapes. Then we can add (concave) or subtract (convex) some arbitrary elevations to the medial axes.

The general shape of the land surface can be detected by using the neighbourhood operation[3]:

```
FORM_tmp{dom=Bool.dom} = iff(DEM>nbavg[2,4,6,8](DEM_TER#), 1,
iff(DEM_TER<nbavg[2,4,6,8](DEM_TER#), 0, ?))
```

The temporary shape map (`FORM_tmp`) needs to be extrapolated using the map iterations to fill the undefined pixels:

```
FORM_ext = MapIter(FORM_tmp.mpr,
iff(isundef(FORM_tmp), nbprd(FORM_tmp#), FORM_tmp))
FORM = MapIter(FORM_ext.mpr, nbprd(FORM_ext#), 5)
```

The last command is used to smooth the `FORM` map and reduce possible artefacts (we recommend at least 10 iterations). The derived map of the general land-surface shape can be seen in Figure 3(c). Finally a constant value (RMSE) is attached to the medial axes [Figure 3(d)] and the remaining undefined pixels are interpolated using linear interpolation:

```
DEM_tmp = iff(MED_AXES=True, iff((FORM=True) AND
(isundef(DEM_TER)), DEM+RMSE,
iff(FORM=False, DEM-RMSE, DEM_TER)), DEM_TER)
DEM_L1.mpr = MapInterpolContour(DEM_tmp.mpr)
```

2.2 Filtering of outliers

When using a DEM derived from satellite or airborne imagery (e.g. SRTM DEM), there can often be many artefacts (single pixels or lines) which are erroneous but

[2] This procedure will also detect true cut-offs, i.e. true lakes and water bodies.

[3] The convex land surfaces (ridges) receive value "`1`" or "`True`" and concave land surfaces (valleys) value "`0`" or "`False`". The *padi*-terrace areas will receive undefined value.

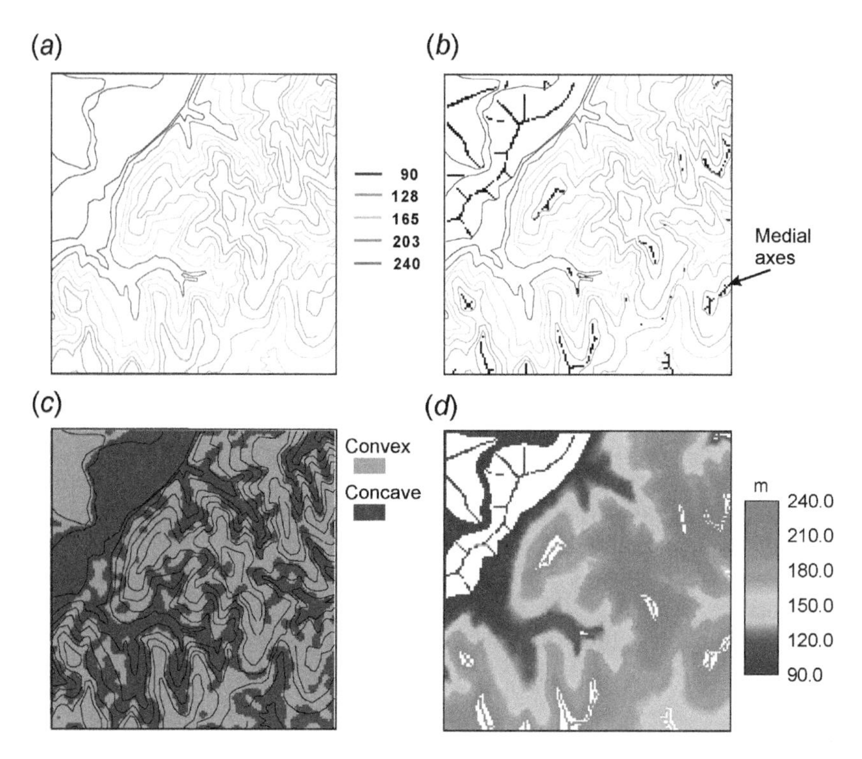

FIGURE 3 Addition of medial axes: (a) original (bulk) contour data; (b) detected medial axes in problematic areas (*padi*-terraces); (c) extrapolated shape of the land surface; and (d) temporary terrace-free map prior to interpolation of the remaining undefined pixels. (See page 722 in Colour Plate Section at the back of the book.)

not easily visible. To detect and filter them, we can apply the statistical procedure explained in Chapter 4. In ILWIS, this can be done using the script called `Filter_outliers`. The script will first predict the central value from the neighbours using the kriging weights calculated in a 5×5 window[4]:

```
Z_PRED = MapFilter(DEM.mpr, zpred.fil, 1)
```

where is the `zpred.fil` is a 5×5 predefined filter matrix calculated for the given covariance function (see Figure 8 in Chapter 4). Because the distances are fixed for the 5×5 window environment, these need to be estimated only once for a given variogram model of elevations. For example, an exponential variogram model with $C_0 = 0$, $C_1 = 1960$ and $R = 1057$ m will give the following weights (see Figure 4 in Chapter 1): $w_A = 0.260$, $w_B = 0.050$, $w_C = -0.025$, $w_D = -0.015$ and $w_E = -0.006$ (see Section 2 in Chapter 4). Now we can calculate the difference between the original and predicted elevation:

```
Z_DIF = DEM - Z_PRED
```

[4] Ideally, one should use a much larger window to filter out the outlier, but this could be computationally demanding for large datasets.

You can display the Z_DIF to see if the values are really normally distributed. The overall average should be 0 and standard deviation should not exceed RMSE(z) as defined for the original dataset. We can then standardise the difference between the predicted and observed value and derive the normal probability of this difference:

```
Z_DIFS = Z_DIF/S_DIF
Z_PROB = (1/sqrt(PI2))*exp(-sq(Z_DIFS)/2)/0.4
```

This probability is then used as the weight function to derive the smoothed DEM:

```
DEM_flt = Z_PROB*DEM + (1-Z_PROB)*Z_PRED
```

where Z_PROB is the normal probability to find a certain value[5] and Z_PRED is the map of elevations predicted from the neighbours.

Note that these filtering steps do not guarantee that all artefacts will be removed (a DEM anyway always carries a measurement error). It is advisable to check the output results and, if needed, digitize extra contours. The above-listed filtering script should also be used with caution because it is possible that also a small number of real features such as small lakes and depressions that can occur naturally will be removed by mistake.

2.3 DEM hydro-preprocessing

A set of built in procedures can be used to prepare the DEM for hydrological processing. These include: (a) removal of sinks, (b) flow optimisation and (c) topological optimisation.

Removal/filling of sinks — Fill sinks operation reduces local depressions (single and multiple pixels). The height value of a single-pixel depression is raised to smallest value of the 8 neighbours of a single-pixel depression and height values of a local depression consisting of multiple pixels are raised to the smallest value of a pixel that is both adjacent to the outlet for the depression and that would discharge into the initial depression. This will ensure that flow direction will be found for every pixel in the map.

ILWIS offers a range of simple procedures to reduce spurious sinks that might not perform equally successful in all areas. More advanced users should consult the work of Lindsay and Creed (2005, 2006).

Note that the flow extraction process allows the occurrences of undefined areas, representing e.g. closed basins, glacial lakes, depressions (sinkholes) within a limestone area or manmade features like reservoirs. These areas are therefore not modified during the fill sink routine. The flow accumulation computation stops at these locations and at a later stage, manually, the topology can be adapted to represent proper flow connection.

> REMARK 4. *ILWIS includes a set of built-in procedures for hydrological processing: removal of sinks, flow optimization and topological optimisation.*

[5] We use the Gaussian function and then simply divide estimated value with the maximum value.

Flow optimisation — This procedure will *'burn'* existing drainage features into a Digital Elevation Model (DEM), so that a subsequent Flow direction operation on the output DEM will better follow the existing drainage pattern. To achieve this, you will need a segment map of the current drainage network and estimate of the smooth drop and sharp drop values. The processed DEM will show (a) gradual drop of (drainage) segments in the output DEM, over a certain distance to the (drainage) segments; (b) gradual raise of (watershed-divide) segments on the output DEM, over a certain distance to the (watershed-divide) segments; (c) additional sharp drop or raise of segments on top of the gradual drop or raise; and (d) simple drop or raise of polygons in the output DEM.

Topological optimisation — Topological consistency can be improved for those areas having undefined DEM values (representing lakes or a reservoir). Before you can perform this operation, you must first prepare a segment map in which the segments connect the inlet(s) of a lake with the outlet(s) of lake (down flow). You can also extract the satellite image based drainage for flat areas and through this manual intervention correct the parallel drainage line occurrences in flat areas. You should first generate a default network, that can be superimposed on a satellite image and then manually adjusted it.

As a result of DEM hydro-preprocessing, you should have a hydrologically consistent raster based elevation representation. Additional modifications might be required as the elevation value assigned to a pixel is an averaged representation only. Furthermore, due to raster resolution in relation to the drainage network or valley width (land-surface discretisation does not allow representation of features smaller than the pixel size) or intrinsic properties of the sensor that acquired the DEM (reflective surface instead of the actual ground surface as is the case with active sensors derived raw elevation models) additional pre-processing is necessary. To overcome the resolution problem, more detailed elevation raster data should be obtained from larger scale aerial photographs or optical stereo satellite images (Aster, SPOT-5 HRS or the ALOS Prism, once operational).

2.4 Visualization of DEMs

In ILWIS, it possible to prepare a 2D visualization of land surface by using a built-in multi illumination angle script. This will produce a colour composite in which the DEM is illuminated from three main directions, the North in red, the North-West in green and the West in blue. The graduated coloured elevation information can be displayed on top of this map as a transparent layer. The example is given in Figure 4.

3. DERIVING PARAMETERS AND OBJECTS

There are two ways to derive LSPs/LSOs in ILWIS: (a) by using built in commands and (b) by building and running scripts. At the moment, ILWIS has only a limited number of built-in land-surface parameterisation algorithms. These are mainly focused on derivation of land-surface parameters related to hydrology.

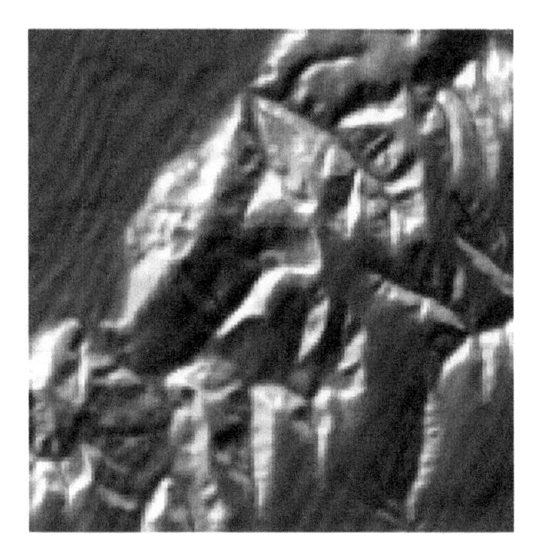

FIGURE 4 Visualization of the DEMs using the multi illuminated angles in ILWIS. (See page 722 in Colour Plate Section at the back of the book.)

3.1 Built-in geomorphometric operations

The version 3.3 of ILWIS has built in hydro-processing module that supports further DEM processing to obtain a full raster and vector based (including topology) schematisation of the (sub-)catchments and drainage network, coupled with additional hydrological relevant parameters (Figure 5). This module is described in detail by Maathuis and Wang (2006) and in ILWIS help files.

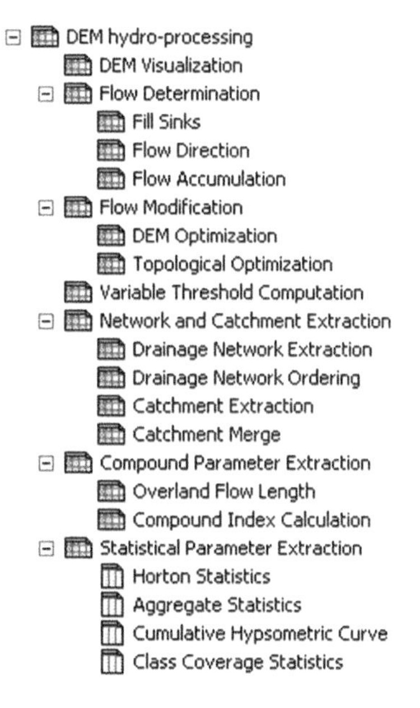

FIGURE 5 ILWIS DEM hydro operations.

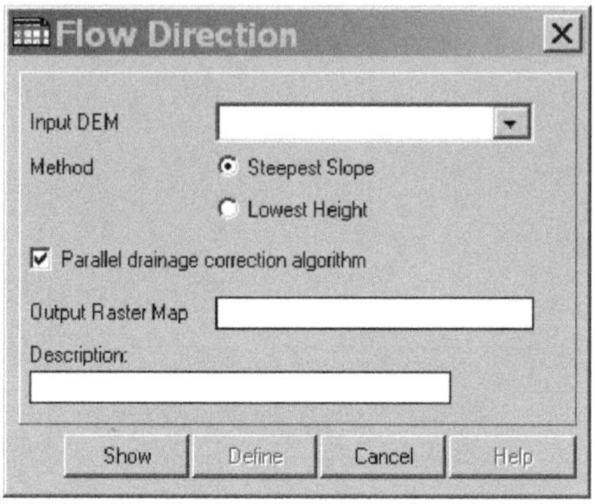

FIGURE 6 ILWIS menu for Flow direction.

The hydro-processing module facilitates various hydrological analysis: extraction of flow direction and flow accumulation, extraction of drainage network and overland flow, extraction of hydrological (flow) indices and statistical parameters. At the moment, only the Deterministic-8 flow model can be used as a built in operation in ILWIS. More advanced models, such as the D-Infinity (Tarboton, 1997) or the Mass Flux algorithm (Section 3.2 in Chapter 7) are under consideration. The raster and vector maps as well as the tables generated in ILWIS can be exported to other formats for incorporation into other software routines.

Flow determination — This module will extract the flow direction (aspect of flow for neighbour pixels — N, NE, etc.) and flow accumulation (cumulative count of the number of pixels that naturally drain into outlets). Optionally a parallel drainage correction algorithm can be incorporated as described by Garbrecht and Martz (1997) to handle the flat areas by imposing an artificial gradient (Figure 6).

Flow direction can be extracted according to the steepest downhill slope between a central pixel and its 8 neighbour pixels[6] or according to the position of the pixel with the smallest elevation.

The current method to address the problem of drainage analysis over flat areas has been implemented in conjunction with the D-8 flow-routine approach. This implementation takes a sink-free DEM as input, and assigns flow direction based on the slope. To the flat areas, where there is no flows (areas with unsolved flow directions), pixels are flagged as *unsolved*. Unsolved directions are resolved by making them to flow toward a neighbour of equal elevation that has a flow direction resolved. The method results in the parallel drainage patterns (as illustrated in Figure 7, left) which requires a lot further manual correction works in order to make a realistic flow.

> REMARK 5. *The true advantage of ILWIS is the possibility of combining vector, raster and database operations together with geomorphometric analysis.*

In the most recent version of ILWIS, a new approach as described by Garbrecht and Martz (1997) is adapted to obtain a more realistic model of flow. This new

[6] To run the multi-direction flow, you can use the script `Flow_indices`, described in Section 3.2.

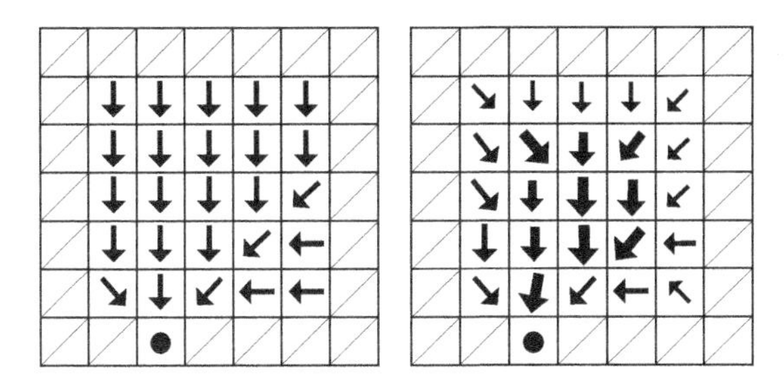

FIGURE 7 Parallel (left) and realistic (right) drainage patterns. The latter is now implemented in the latest version of ILWIS.

approach still makes use of the D-8 flow-routine approach as done in the existing DEM-hydro routine model. Since the D-8 method cannot define the flows over pixels of equal elevations (flat areas), this algorithm will increase the cell elevations of the flat area and produce the desired drainage patterns during the subsequent flow direction assignment processing. This will finally result in the drainage flowing at its outside edge with higher elevations towards lower cells at its centre or middle. Furthermore, the flat area must have at least one outlet so that the down-drainage off the flat area is possible (Figure 7, right).

Note: the adjustment of elevation is applied only to DEM cells within flat areas. No DEM elevations outside flat surfaces are altered. The approach still requires the sink-free DEM data as input as the existing approach done in ILWIS.

> REMARK 6. *Land-surface parameters and objects can be derived by either using built in commands or by building and running scripts.*

In ILWIS, the type of geology and soil can be added to hydrological modelling to emphasises local variability. If a geological or a soil map is available the units of this map can be reclassified to represent flow accumulation threshold values (see ILWIS help). Units with coarse grained sandy soils overlaying deeply weathered sandstones can be assigned higher thresholds compared to thin soils occurring over shales (reflecting the lower permeability and little resistance to erosion).

Drainage network extraction — This operation will extract the basic drainage network (boolean raster map). In this case, a stream threshold (minimum number of pixels that should drain into a pixel examined) need to be defined. Optionally, you can also use the flow direction map and an stream threshold map to allow different thresholds in different parts of the study area. The Flow direction map is used to automatically fill possible gaps between found drainage lines. As a result, you will always obtain a continuous drainage network.

Overland Flow Length — This operation calculates for each pixel the overland distance towards the nearest drainage according to the flow paths available in the Flow Direction map. As input, the Drainage network ordering map need to be derived first. This map shows individual streams within a drainage network and assigns a unique ID to each stream. The short segments can be excluded based on a user defined minimum length threshold. Overland flow length is very useful to quantify the proximity to streams and can be compared with the potential drainage density.

In addition, you can also derive the Flow Length to Outlet and the Flow Path Longitudinal Profile. The output for flow length to outlet will be a value map that will contain for each pixel the down flow distance to the outlet, while outlet pixel will have a value of 0. The Flow Path Longitudinal Profile can derived using the function:

```
outputtabe = TblFlowPathLongitudinalProfile
(LongestFlowPathSegmentMap, SegmentID, Distance, AttributeMap)
```

where `LongestFlowPathSegmentMap` is the input segment map that contains the longest flow path of the catchment, `SegmentID` is the segment ID that you want to use generate the longitudinal profile, `Distance` is the threshold value to obtain the output points at a regular distance along segment and `AttributeMap` is a value map used to obtain the Y column in the output table. The output is a table that will contain 3 columns: X — point ID extracted from the input segment, Y — a value from the input attribute map related to the point and *Coord* — Coordinate of the point:

Flow indices — Based on the DEM and the flow accumulation map, three standard indices can be derived: (a) WTI; (b) SPI; and (c) STI. SPI can be used to identify suitable locations for soil conservation measures to reduce the effect of concentrated surface runoff. STI accounts for the effect of topography on erosion.

In addition to extraction of hydrological LSPs/LSOs, a number of functions are given to provide relevant statistical information of the extracted river and catchment network. The Horton plots show the relationship between Strahler order and total number of Strahler order stream segments for a given order, average length per Strahler order and average catchment area per Strahler order, as well as the bifurcation, channel length and stream area ratio's (by means of a least square regression line). The results can be graphically displayed plotting the Strahler order on the X axis and the number of drainage channels, stream length and stream area on a log transformed Y axis. According to Horton's law the values obtained should[7] plot along a straight line (Chow et al., 1988), which proves that the parameters used for drainage extraction are properly selected. Especially when performing catchment merge operations using Strahler orders, reference to the original Horton plot might be relevant. All extracted catchments can be crossed with e.g. the elevation model and aggregate statistics (mean, minimum, standard deviation, etc.) are computed and appended to the catchment table. Furthermore drainage network and catchment segmentation can be aggregated — merged using different stream orders (e.g. for more generic up scaling purposes) or by user defined drainage outlet locations and the resulting network can be extracted to provide further hydrological model input. Finally, other raster-based layers, e.g. obtained from a soil map or classified satellite image can be crossed with the catchment map and the cross table shows relevant (aggregated) statistical information as well as Horton plots (Chow et al., 1988).

[7] The error in measuring Horton ratio from DEM extracted stream networks is often fairly high. Severe caution is needed when interpreting these values.

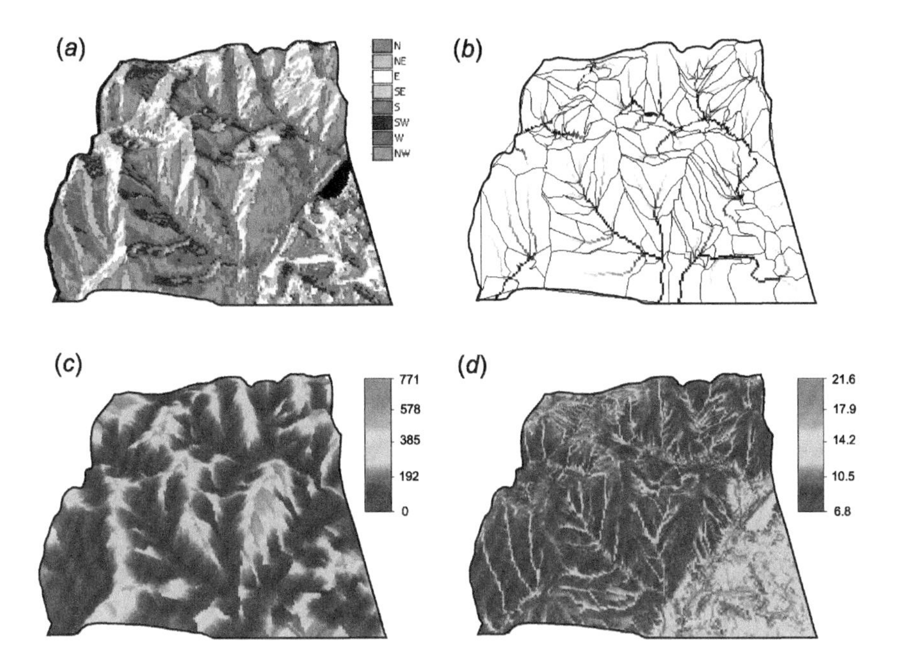

FIGURE 8 Extraction of hydrological parameters and objects using the built-in operations: (a) flow direction, (b) flow accumulation with catchment lines, (c) overland flow length and (d) wetness index. All calculated using the Deterministic-8 algorithm. (See page 723 in Colour Plate Section at the back of the book.)

3.2 Deriving parameters and objects using scripts

In the following section, we will demonstrate how to derive additional number of local morphometric and hydrological parameters using ILWIS scripts. We assume that you have prepared the input DEM using the script described in Section 2.1. Three scripts will be described in more detail:

`LSP_morphometric` can be used to derive local morphometric LSPs: slope in % (SLOPE), aspect (ASPECT), profile curvature (PROFC), planar curvature (PLANC), mean curvature (MEANC), slope-adjusted norhtness (NORTH) and solar insolation for given angles (SOLINS).

`Flow_indices` can be used to derive catchment area and flow indices: topographic wetness index (WTI), stream power index (SPI), sediment transport index (STI) and shape complexity index (SCI), all based on the multiple flow direction algorithm.

`G_landforms` can be used to derive generic landform shapes: channel, ridge, plain (terrace), slope and pit.

`LSP_morphometric` — This script uses the Evens–Young method formulas (see Chapter 6). It differs from similar algorithms in two things. First, the second derivatives (d^2f/dx^2, d^2f/dx^2, $d^2f/dx\,dy$) are smoothed prior to extraction of land-surface parameters in order to produce a more generalized image of land-surface parameters. Second, the local undefined[8] pixels are replaced by iteratively taking the predominant value from the neighbours. The complete script can be seen in Table 1 and the derived land-surface parameters in Figure 9.

TABLE 1 SCRIPT: `LSP_morphometric` — Calculation of local land-surface parameters

REM: Calculation of morphometric land-surface parameters (slope, aspect, curvatures)

```
1  dx.mpr{dom=value.dom;vr=-500.0000:500.0000:0.0001} =
   (%1#[3] + %1#[6] + %1#[9] - %1#[1] - %1#[4] - %1#[7]) /
   (6*pixsize(%1))
2  dy.mpr{dom=value.dom;vr=-500.0000:500.0000:0.0001} =
   (%1#[1] + %1#[2] + %1#[3] - %1#[7] - %1#[8] - %1#[9]) /
   (6*pixsize(%1))
3  //smooth the DEM before deriving second derivatives
4  DEM_s.mpr{dom=value.dom;vr=0.00:5000.00:0.01} =
   %2*(%1#[1] + %1#[2] + %1#[3] + %1#[4] + %1#[6] + %1#[7] +
   %1#[8] + %1#[9])/9 + (1-8*%2/9)*%1#[5]
5  //derive second-order derivates and smooth them to get a more generalised picture
6  d2x_tmp.mpr = (DEM_s#[1] + DEM_s#[3] + DEM_s#[4] +
   DEM_s#[6] + DEM_s#[7] + DEM_s#[9] - 2*(DEM_s#[2] +
   DEM_s#[5] + DEM_s#[8])) / (3*pixsize(DEM_s)^2)
7  d2y_tmp.mpr = (DEM_s#[1] + DEM_s#[2] + DEM_s#[3] +
   DEM_s#[7] + DEM_s#[8] + DEM_s#[9] - 2*(DEM_s#[4] +
   DEM_s#[5] + DEM_s#[6])) / (3*pixsize(DEM_s)^2)
8  dxy_tmp.mpr = (DEM_s#[3] + DEM_s#[7] - DEM_s#[1] -
   DEM_s#[9]) / (4*pixsize(DEM_s)^2)
9  d2x{dom=value.dom;vr =-50.0000:50.0000:0.0001} =
   MapFilter(d2x_tmp, avg3x3)
10 d2y{dom=value.dom;vr=-50.0000:50.0000:0.0001} =
   MapFilter(d2y_tmp, avg3x3)
11 dxy{dom=value.dom;vr=-50.0000:50.0000:0.0001} =
   MapFilter(dxy_tmp, avg3x3)
```

(continued on next page)

[8] This usually happens either due to division by zero or because the mapped values are outside a feasible range.

TABLE 1 *(continued)*

	REM: Calculation of morphometric land-surface parameters (slope, aspect, curvatures)
12	*//derive slope, aspect, curvatures (filter them for undefined values using iterations)*
13	`SLOPE.mpr{dom=value.dom;vr=0.0:5000.0:0.1} = 100*sqrt(dx^2+dy^2)`
14	`ASPCT_tmp = raddeg(atan2(dx,dy)+PI)`
15	`ASPECT{dom=value.dom;vr=0.0:360.0:0.1} = MapIter(ASPCT_tmp.mpr, iff(isundef(ASPCT_tmp), nbprd(ASPCT_tmp#), ASPCT_tmp))`
16	`PLANC_tmp = -(dy2*d2x-2*dx*dy*dxy+dx^2*d2y) / ((dx^2+dy^2)^1.5)*100`
17	`PLANC{dom=value.dom;vr=-50.000:50.000:0.001} = MapIter(PLANC_tmp.mpr, iff(isundef(PLANC_tmp), nbprd(PLANC_tmp#), PLANC_tmp))`
18	`PROFC_tmp = -(dx^2*d2x-2*dx*dy*dxy+dy^2*d2y) / ((dx^2+dy^2)*(1+dx^2+dy^2)^1.5)*100`
19	`PROFC{dom=value.dom;vr=-50.000:50.000:0.001} = MapIter(PROFC_tmp.mpr, iff(isundef(PROFC_tmp), nbprd(PROFC_tmp#), PROFC_tmp))`
20	`MEANC_tmp = -((1+dy^2)*d2x-2*dx*dy*dxy+(1+dx^2)*d2y) / (2*(1+dx^2+dy^2)^1.5)*100`
21	`MEANC{dom=value.dom;vr=-50.000:50.000:0.001} = MapIter(MEANC_tmp.mpr, iff(isundef(MEANC_tmp), nbprd(MEANC_tmp#), MEANC_tmp))`
22	*//delete temporary files* `DEM_s, ????_tmp, d??_tmp`

`Flow_indices` — This script uses the multiple flow direction algorithm described by Quinn et al. (1991). The theory behind is explained in detail in Chapter 7. The derivation of the catchment area consists of four steps.

(1) Generate the slope-lengths for each diagonal and cardinal direction (8 maps) and their sum:

```
S1 = iff(isundef(DEM#[1]) OR (DEM<DEM#[1]), 0,
     (DEM-DEM#[1])/4)
S2 = iff(isundef(DEM#[2]) OR (DEM<DEM#[2]), 0,
     (DEM-DEM#[2])/2)
...
S9 = iff(isundef(DEM#[9]) OR (DEM<DEM#[9]), 0,
     (DEM-DEM#[9])/4)
SSUM = S1 + S2 + S3 + S4 + S6 + S7 + S8 + S9
```

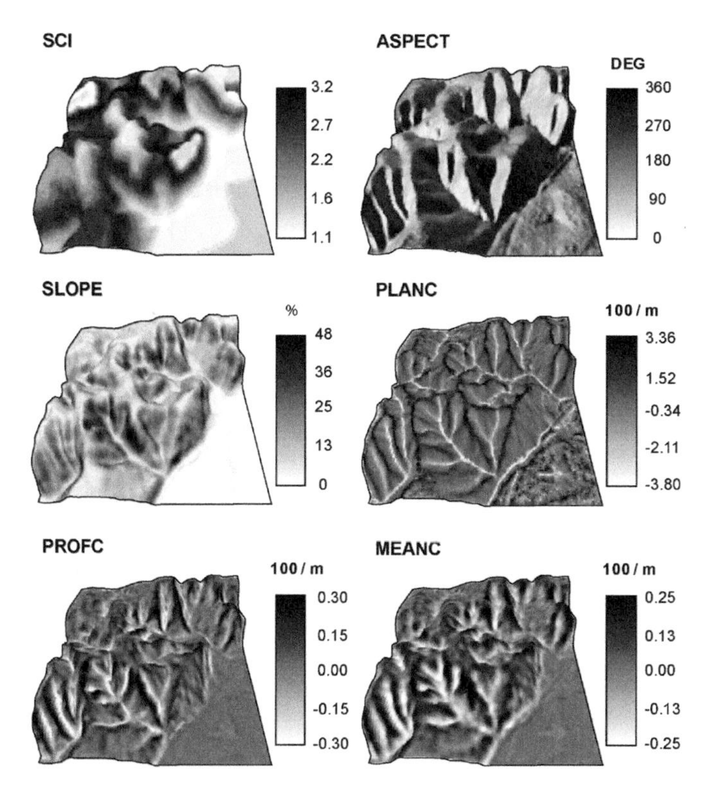

FIGURE 9 SCI (Shape Complexity Index) and other morphometric land-surface parameters derived in ILWIS: ASPECT (0–360°), SLOPE (slope in %), PLANC (plan curvature), PROFC (vertical curvature) and MEANC (mean curvature).

(2) Generate the drainage fraction out of cell for each direction:

```
dA1t = iff(isundef(S1), 0, S1/SSUM)
dA2t = iff(isundef(S2), 0, S2/SSUM)
...
dA9t = iff(isundef(S9), 0, S9/SSUM)
```

(3) Generate drainage fraction into each cell for each direction as a fraction of the contributing cell (Figure 10):

```
dA1 = iff(isundef(dA9t#[1]),0,dA9t#[1])
dA2 = iff(isundef(dA8t#[2]),0,dA8t#[2])
...
dA9 = iff(isundef(dA1t#[9]),0,dA1t#[9])
```

(4) Propagate the total number of contributing cells using n iterations with start map consisting of 1's (Figure 11) and derive the catchment area `CATCH`:

```
start.mpr = iff(isundef(%1), 0, 1)
```

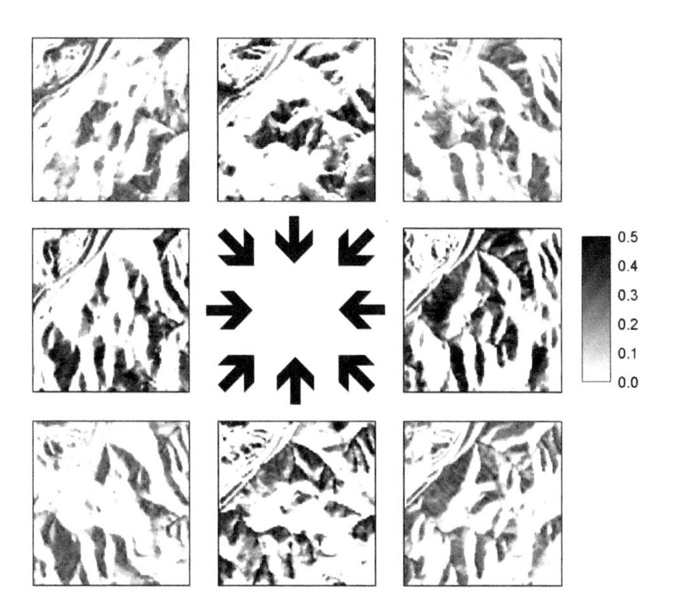

FIGURE 10 Drainage fraction elements in each direction.

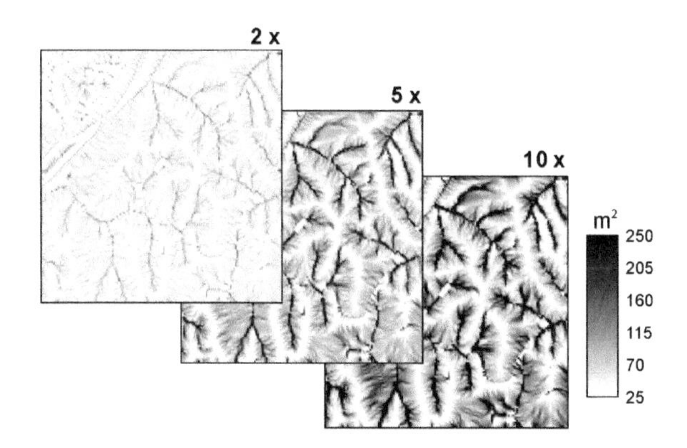

FIGURE 11 Specific catchment area after 2, 5 and 10 iterations.

```
ASUM = MapIter(start.mpr, iff(start<50000, dA1*start#[1]
+ dA2*start# [2] + dA3*start#[3] + dA4*start#[4] + dA6*start#[6]
+ dA7*start#[7] + dA8*start#[8] + dA9*start#[9] + 1, start), %3)
LSUM = pixsize(%1)*(sqrt(2)/4*(iff(dA1>0,1,0)
+ iff(dA3>0,1,0) + iff(dA7>0,1,0) + iff(dA9>0,1,0))
+ 0.5*(iff(dA2>0,1,0) + iff(dA4>0,1,0) + iff(dA6>0,1,0)
+ iff(dA8>0,1,0)))
CATCH_tmp = ASUM*pixarea(DEM)/LSUM
```

The CATCH_tmp map can be iteratively filtered for undefined pixels[9] by taking the predominant value from the surrounding pixels until all zero slopes are replaced:

```
CATCH = MapIterProp(CATCH_tmp.mpr, iff(isundef(CATCH_tmp) and
not(isundef(%2)), nbprd(CATCH_tmp#),CATCH_tmp))
```

This is especially important because in **ILWIS** the undefined pixels will otherwise propagate. This filtering has the effect of creating pools of high TWI in the plain, which is in general realistic.

> REMARK 7. *At the moment, ILWIS scripts are available to derive dozens of morphometric parameters, flow indices using multiple flow direction algorithm and generic landform shapes.*

Each new iteration will propagate flow by a distance equal to the pixel size or the diagonal pixel size. This should be ideally done until only very few downstream pixels are changed with any new calculation, which can be checked by evaluating a difference map of accumulation after n and after $n+1$ iterations. In this case we recommend using at least 100 iterations for flow accumulation. Note that the propagation of the drainage fractions can be time consuming.

After the catchment area has been derived, wetness index (TWI), Stream power index (SPI) and Sediment transport index (STI) can be derived using:

```
TWI = ln(CATCH/SLOPE*100)
SPI = CATCH*SLOPE/100
STI = (CATCH/22.13)^0.6*((sin(ATAN(SLOPE/100)))/0.0896)^1.3
```

G_landforms — channel, ridge, plain (terrace), slope and pit (see also Chapter 22) can be derived using the supervised fuzzy k-means classification. The input maps needed are the slope in % (SLOPE), planar curvature (PLANC) and anisotropic coefficient of variation (ACV), fuzzy exponent and a table with definition of class centres. In this case, the LF_class.tbt table with the definition of classes looks like this:

	SLOPE	PLANC	SCI	SLOPE_STD	PLANC_STD	ACV_STD
channel	5	-2	0.4	5	0.5	0.25
pit	5	-2	0	5	0.5	0.25
plain	0	0	0.2	5	0.5	0.25
ridge	5	2	0.4	5	0.5	0.25
slope	25	0	0.2	5	0.5	0.25
peak	5	2	0	5	0.5	0.25

These are just approximated class centers and variation around the central values (SLOPE_STD, PLANC_STD and ACV_STD) that will probably need to be adjusted from an area to area (see also Figure 8 in Chapter 22). There can be quite some overlap between the pits and streams (see further Section 5.1 in Chapter 22). Other classes seems to be in general easier to distinguish, although there is obviously overlap between streams-plain and ridges-plain.

[9] Division by zero — locations where LSUM = 0.

The script runs as follows. It will first calculate[10] distances from the central value to the attribute band per each class and standardise them according to the standard deviation:

```
t_d11 = abs(%4-TBLVALUE(%1, "SLOPE", 1)) /
TBLVALUE(%1, "SLOPE_ STD", 1)
t_d12 = abs(%5-TBLVALUE(%1, "PLANC", 1)) /
TBLVALUE(%1, "PLANC_STD", 1)
...
t_d63 = abs(%6-TBLVALUE(%1, "ACV", 6))/TBLVALUE(%1, "ACV_STD", 6)
```

where `%1` is `LF_class.tbt` table and `%4`, `%5`, `%6` are SLOPE, PLANC and ACV maps. Then, it will calculate sum's of distances for each class:

```
sum_dc1 = t_d11^2+t_d12^2+t_d13^2
sum_dc2 = t_d21^2+t_d22^2+t_d23^2
...
sum_dc6 = t_d61^2+t_d62^2+t_d63^2
```

and the fuzzy factors per each class:

```
sum_d1 = (sum_dc1)^(-1/(%2-1))
sum_d2 = (sum_dc2)^(-1/(%2-1))
...
sum_d6 = (sum_dc6)^(-1/(%2-1))
sum_d = sum_d1+sum_d2+sum_d3+sum_d4+sum_d5 +sum_d6
```

where `%2` is the fuzzy exponent. Finally, memberhsips for each class as can be derived as `sum_dc/sum_d`:

```
GLF_Channel{dom=Value, vr=0.000:1.000:0.001} = sum_d1/sum_d
GLF_Ridge{dom=Value, vr=0.000:1.000:0.001} = sum_d4/sum_d
GLF_Slope{dom=Value, vr=0.000:1.000:0.001} = sum_d5/sum_d
GLF_Plain{dom=Value, vr=0.000:1.000:0.001} = sum_d3/sum_d
GLF_Pit{dom=Value, vr=0.000:1.000:0.001} = sum_d2/sum_d
GLF_Peak{dom=Value, vr=0.000:1.000:0.001} = sum_d6/sum_d
```

You might also try to classify an area using some other generic landforms, such as *pool* or *"poolness"*, *pass*, *saddle*, etc. These would, of course, require somewhat different clustering of attribute space (see Chapter 9 for more details). The final classification map can be produced by taking the highest membership per cell (Figure 12). In the case of the Baranja Hill dataset, it seems that the most dominant landforms are slopes and plains, while pits occur only in a small portion of the area.

4. SUMMARY POINTS AND FUTURE DIRECTION

ILWIS has many advantages, from which the biggest are the accessibility and richness of GIS operations. For example, next to the elevation data set itself, also information acquired from remote sensing images can be incorporated and up scaling

[10] Note that, in ILWIS, it is possible to run arithmetic operations using raster maps and table values in the same line.

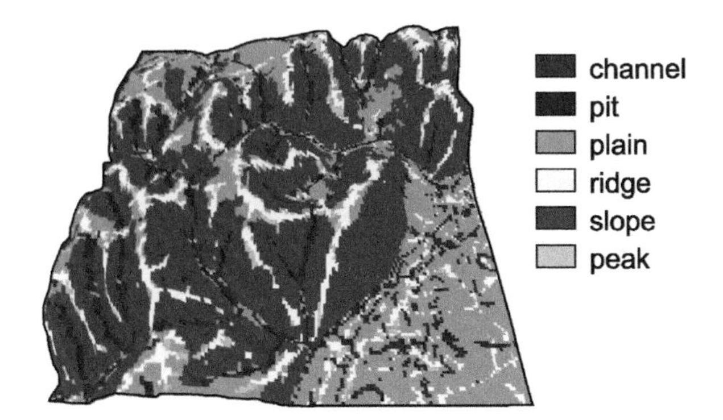

FIGURE 12 Study area classified into the generic landforms. (See page 723 in Colour Plate Section at the back of the book.)

for comparison with data derived from low resolution (meteo) satellites could be facilitated. Relevant features that represent actual topology can also be extracted from satellite images (through screen digitising) and the DEM may be adapted at these locations. It is also possible to improve the assignment of drainage direction over flat surfaces in raster elevation models in order to prevent the occurrence of parallel drainage according to the procedure proposed by Garbrecht and Martz (1997). All this can be achieved because ILWIS already offers a substantial capability for GIS-RS data processing. Also the drainage network and catchment tables generated can be easily linked using common table ID columns and can be exported and incorporated in other packages. The amount of information that can extracted from DEMs is high and can be even extended by building new scripts.

Still, the fact is that the number of ILWIS users is relatively limited to former ITC students and collaborators. There are several probable reasons for this. Number one reason is that the transfer from different packages to ILWIS is still limited. Import/export operations still contains some bugs and can lead to inaccuracies or artefacts in maps. ILWIS needs to import GIS datasets from various popular formats (like ArcInfo ascii, Erdas' `.img` or shape files) to the unpopular ILWIS format which many do not like to do. ILWIS also does not have a website where the users can exchange scripts and user-built modules (compare with ArcGIS, SAGA or GRASS that all have user groups), but only a mailing list.

In addition, the command line is rather user-un-friendly. Unlike in ArcGIS, the user has to already know how are specific functions used and which are input/output parameters. ILWIS will not assist you in running a command directly from the command line or warn you about what is wrong in your command, which usually leads to many tests and trials. Also the neighbourhood operations are fairly limited in ILWIS. For example, unlike ArcInfo DOCELL function, ILWIS is limited to working with 3×3 window environment and further neighbours can not be pin-pointed within an ILWIS scripts.

When displaying multiple raster images, all images need to have the same georeference. Unlike in ArcGIS where the user can overlay literally any GIS layer. On one way, this limitation prevents from creating seamless maps, but does not allow exploration of overlap and position of adjacent maps or maps belonging to different grid definitions. Furthermore, the 3D viewer in ILWIS is practically unusable. Draping large raster images is slow and static, therefore not suggested for large datasets. Similarly, ILWIS is not a professional software to prepare final map layouts.

With its limited support and many known and unknown bugs, ILWIS will continue to be rather a scientific than a commercial product. Still, with its rich computational capabilities can be attractive to users with limited funds interested to learn and modify land-surface parameterisation methods. At least now anybody has a chance to obtain the original code and produce an improved version of the package.

IMPORTANT SOURCES

Maathuis, B.H.P., Wang, L., 2006. Digital elevation model based hydro-processing. Geocarto International 21 (1), 21–26.

Unit Geo Software Development, 2001. ILWIS 3.0 Academic User's Guide. International Institute for Geo-Information Science and Earth Observation (ITC), Enschede, 530 pp.

Unit Geo Software Development, 1999. ILWIS 2.1 Applications Guide. International Institute for Geo-Information Science and Earth Observation (ITC), Enschede, 352 pp.

www.itc.nl/ilwis/ — ILWIS home page.

www.ilwis.org — ILWIS users' home page.

CHAPTER 14

Geomorphometry in LandSerf

J. Wood

LandSerf and its development · installation and running · geomorphometric analysis unique to LandSerf · how to incorporate scale in geomorphometry · mipmapping or level of detail rendering · scripting in LandSerf · using scripting to explore scale signatures

1. INTRODUCTION

LandSerf was first made publicly available in 1996 as a platform for performing scale-based analysis of Digital Elevation Models. Central to its design was the ability to perform *multiscale surface characterisation* (Wood, 1996) where parameters such as slope, curvature and feature type could be measured over a range of spatial scales. This offers the user of the software the opportunity to examine how measurements taken from a land-surface model are dependent on the scale at which they are taken. At that time, the only other software capable of performing multiscale parametrisation was GRASS, using the module `r.param.scale`, also based on Wood (1996).

A secondary design principle of the software was the use of scientific visualisation as a means of exploring the effects of scale on parametrisation through a rich and interactive interface. Subsequent releases of the software have enhanced its visualisation capabilities (for example, 3D real-time flythroughs using OpenGL) and the range of file formats it can import and export. With the addition of vector handling in 1998, attribute tables in 2003, raster and vector overlay in 2004 and map algebra scripting in 2007, LandSerf can be regarded as an example of a Geographic Information System (GIS) specialising in the handling of surface models.

The software is written entirely in Java and can be run on Windows, Linux, Unix and MacOSX platforms. It is freely available from www.landserf.org, along with extensive documentation, an API for Java programmers and a user support forum.

Developments in Soil Science, Volume 33 © 2009 Elsevier B.V.
ISSN 0166-2481, DOI: 10.1016/S0166-2481(08)00014-7.

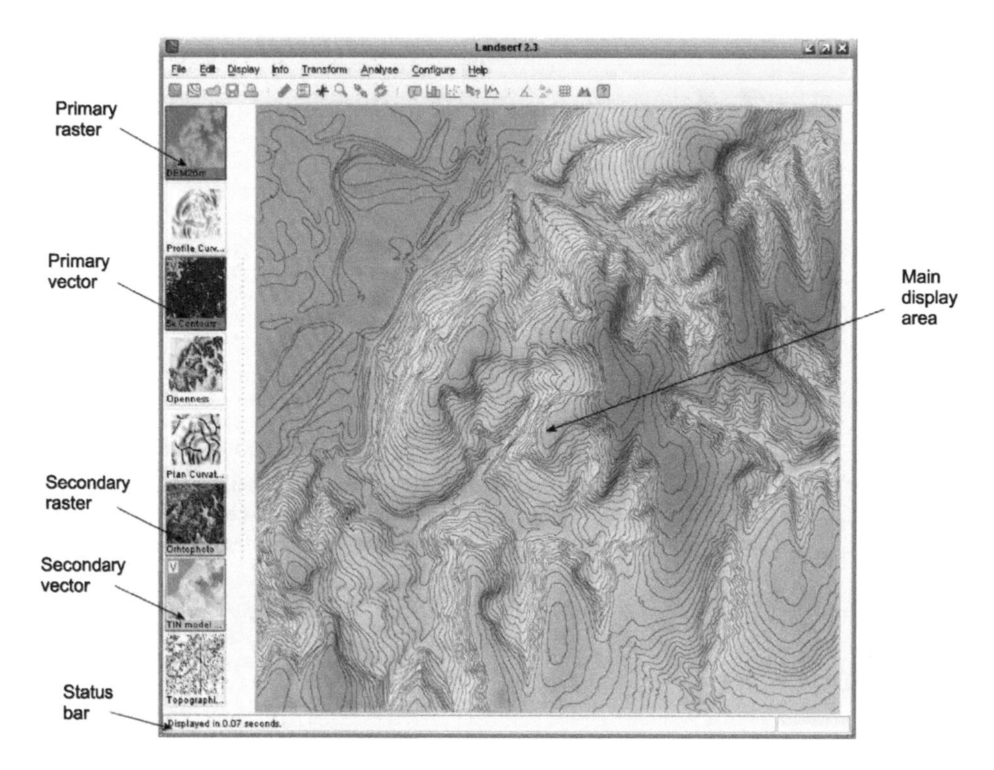

FIGURE 1 Default LandSerf Graphical User Interface showing thumbnail and main views of raster and vector data.

1.1 Getting started with LandSerf

Instructions on how to download and install the latest version of LandSerf can be found at www.landserf.org. The only platform requirement is a working Java Runtime Environment (JRE) which can be downloaded for free from www.javasoft.com.

LandSerf provides three ways with which to interact with spatial data. By default, the main interface provides thumbnail views of all raster and vector maps loaded into LandSerf as well as a larger view of the data being analysed (see Figure 1). The number of maps (raster or vector) shown as thumbnails is limited only by the memory of the platform running the software. To perform analysis or display of any of these maps, a *primary map* is selected from the list of thumbnails by clicking on the relevant map with the mouse. Where analysis or display requires further maps, a *secondary map* can also be selected by right-clicking on a thumbnail. Analysis is performed by selecting operations from the toolbar or menus at the top of the window.

The default presentation of data in LandSerf uses a two-dimensional view of the selected data. When exploring surface models, this view can be enhanced by combining maps with shaded relief representations (as Figure 1) and interactively zooming and panning across the surface. Alternatively, the relationship between

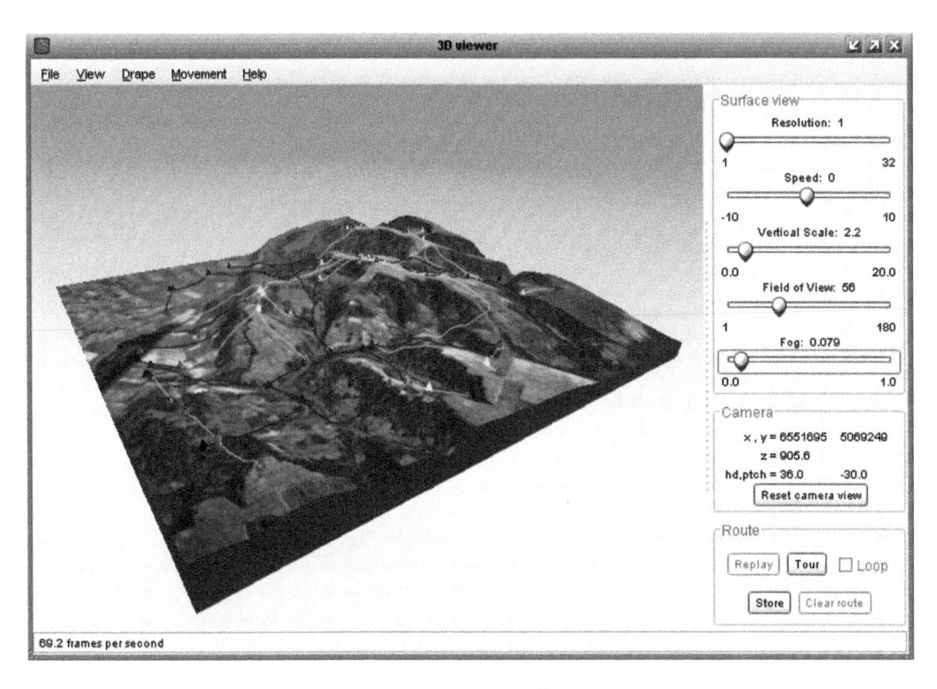

FIGURE 2 LandSerf 3D viewer. The main display area allows interactive '*flythrough*' over a surface while the appearance is controlled via panel to the right. This example shows the Baranja Hill 25 m DEM with orthophoto and metric surface network (Wolf, 2004) draped over the top.

elevation and other data can be explored visually using **LandSerf**'s 3D view (see Figure 2). This view is updated dynamically, based on the current selection of primary and secondary maps and allows interactive navigation over a land surface.

The third form of interface provided by **LandSerf** is via its *LandScript Editor* (see Figure 3). This text-based interface allows analysis to be performed by issuing commands within a script. These commands form part of the language *LandScript* that allow more complex tasks to be represented as a sequence of program instructions. The editor provides simple syntax colouring of keywords, variables and text as well as facilities to aid the debugging and testing of scripts.

Help in using all three interfaces to **LandSerf**, along with tutorials to help getting started and example scripts can be found either via the *Help* menu or online at www.landserf.org.

1.2 The importance of scale in geomorphometry

Central to the design and use of **LandSerf** is the idea that measurements of surface characteristics are dependent on the scale at which they are measured. In this context, scale comprises the spatial extent over which a measurement is taken (also known as the *support* in geostatistical terms — see also Section 2.3 in Chapter 2), and the spatial resolution of sampling within a given extent.

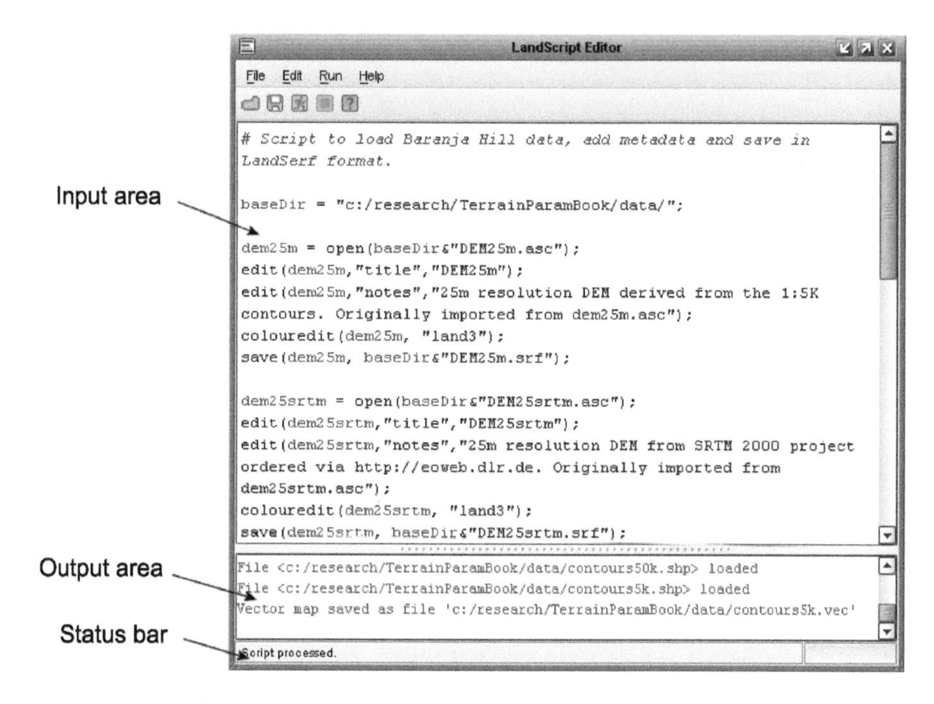

FIGURE 3 LandSerf script view showing syntax highlighted editable script area and output area.

The measurement of land-surface parameters such as slope, aspect and curvature in **LandSerf** uses the widely adopted method of taking first derivatives and partial second derivatives of a bi-quadratic polynomial representing a local patch of a surface (e.g. Evans, 1980). This polynomial expression can be represented in the form:

$$z = ax^2 + by^2 + cxy + dx + ey + f \tag{1.1}$$

where z is the estimate of elevation at any point (x, y) and a to f represent the 6 coefficients that define the quadratic surface. In this respect, **LandSerf** is typical of most packages that derive land-surface parameters from gridded elevation models.

What makes **LandSerf** unique is the way in which the 6 coefficients of the polynomial expression are estimated. Rather than pass a 3×3 local window over a raster grid, a window of any arbitrary size can be selected and the best fitting quadratic surface passing through that window is estimated using least-squares regression. The 6 unknown coefficients are found by solving 6 simultaneous equations using matrix methods. These are further simplified due to the regular spacing of grid cells in a raster and the symmetry of the raster coordinate system in the x and y directions.[1]

The result of this method is the ability for a user to select both the size of window used to derive any land-surface parameters, and the distance decay exponent

[1] See Wood (1996, pp. 92–97) for more detail.

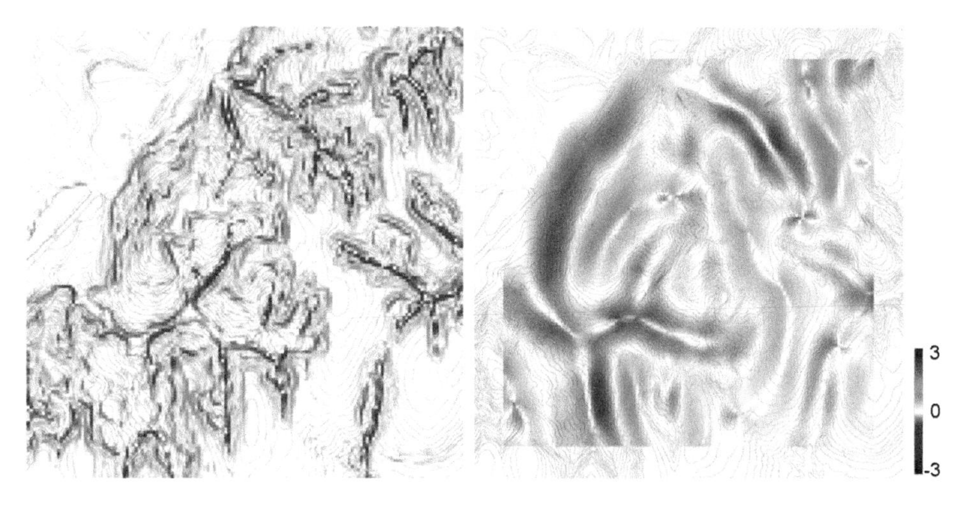

FIGURE 4 Profile curvature (per 100 m) measured over 75 and 625 m spatial extents. (See page 724 in Colour Plate Section at the back of the book.)

that controls the relative importance given to cells at the centre of a window relative to those further from the centre when estimating the quadratic surface. This enables the user to parameterise a surface over spatial extents relevant to their preferred scale of analysis, rather than that implied by the resolution of the raster data they wish to process. This added flexibility can be desirable in that it reduces the possibility of characterising arbitrary artefacts of a DEM, but it increases the complexity and *solution space* of the set of possible derived parameters.

As an example, consider the measurement of *profile curvature* of a surface. Figure 4 shows profile curvature of the same surface (the Baranja Hill 25 m elevation model `dem25m.asc`) with two contrasting spatial extents. As might be expected, measuring curvature at a fine scale, using a 3×3 local window around each raster cell, reveals much more local variation in the surface parameter while the broader scale of analysis (55×55 cell local window) highlights trends in curvature across the surface. The question that then has to be confronted by the geomorphometrist is which scale is most appropriate for analysis? The answer to this question is clearly going to depend to some extent on the nature of the application and scale of features under study by the researcher. This may be already determined, or the researcher may use LandSerf to choose the appropriate scale. Or indeed, it may be that very variation in scale that the researcher is interested in quantifying.

To consider the example in further depth, Figure 5 shows profile curvature of a selected portion of the 5 m Baranja Hill elevation model. Superimposed on the surface are the contour lines from which the elevation model was interpolated. The 15 m profile curvature measures (left of Figure 5) show clear alternating bands of concavity and convexity parallel to the steeper contour lines that are not evident in the 275 m scale measurement (right of Figure 5). Such banding is indicative of a stepped terracing in the surface model.

In this region, the terracing is almost entirely an artifact of the interpolation process rather than a genuine morphological feature at this scale. We can con-

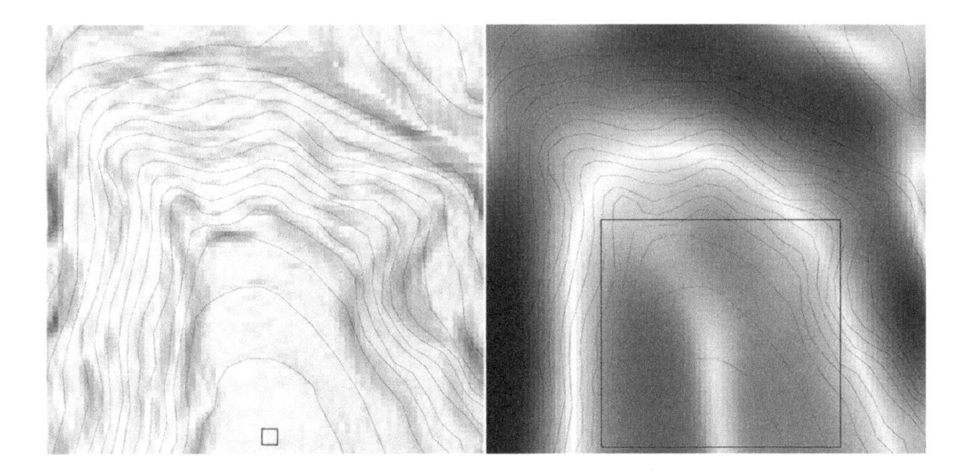

FIGURE 5 Profile curvature (per 100 m) measured from the Baranja Hill 5 m DEM at contrasting spatial scales. The square in the bottom centre of each image represents the size of the window used for processing (15 and 275 m respectively). (See page 724 in Colour Plate Section at the back of the book.)

clude from this that it would be inappropriate to perform much geomorphological analysis using a 3×3 window passed over the 5 m elevation model. Even if we were interested in smaller scale features, it would be wiser to use a slightly larger window size, or a different elevation model.

2. VISUALISATION OF LAND-SURFACE PARAMETERS AND OBJECTS

Once a researcher considers scale as being influential in their measurement and analysis of a surface, the dimensionality of the *solution-space* they are exploring is increased. Somehow, the analysis of a surface must consider variables representing the three dimensions of location, the parameters characterising local surface form (slope, curvature, etc.) and the spatial extent and resolution at which those measurements have been made. While there are a range of multi-variate statistical techniques available for analysing multi-dimensional solution spaces, this problem is also amenable to scientific visualisation. For this reason, LandSerf uses a number of graphical and visualisation techniques to allow the exploration of the relationship between space, scale and morphometry.

2.1 Blended shaded relief

One of the simplest techniques available is to combine visually any surface parameter with the surface from which the measurement was taken. For example Figure 6 shows *plan curvature* of the Baranja Hill 25 m DEM measured using a window size of 15×15 cells. When displayed directly as a coloured image (Figure 6, left image), some indication of variation in curvature is given, with an implied

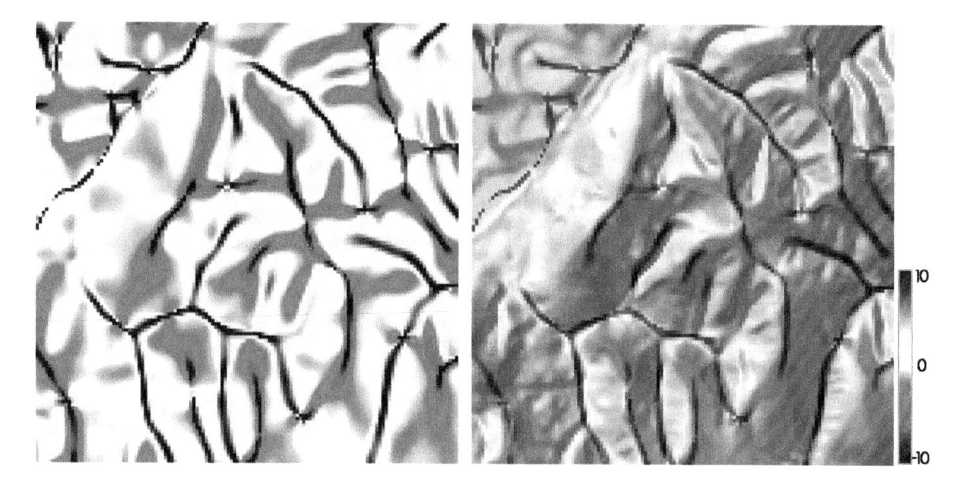

FIGURE 6 Plan curvature (per 100 m) of the Baranja Hill 25 m DEM measured at the 275 m window scale. The image on the left shows only plan curvature. The image on the right shows the same measure but with colour intensity, representing local shaded relief of the underlying surface. (See page 725 in Colour Plate Section at the back of the book.)

relationship to possible landscape features. Ridge and channel lines in particular are emphasised. However the relationship with the land surface is a complex one, and one that is only partially revealed by the image. If, on the other hand, the measurement is combined visually with a shaded relief representation of the surface (Figure 6, right image), more is revealed about this relationship.

In LandSerf, the DEM is selected as the primary raster, plan curvature as the secondary raster, and *Relief* selected from the *Display* menu. The contrast between the relief of the NW corner and the remainder of the study area is highlighted. Also revealed are the smaller scale ridges and valleys (those that appear grey in the figure) that do not result in any significant curvature at the 275 m scale of analysis. Such visual analysis might lead to a refinement of the scale at which analysis is performed.

2.2 Scale signatures

Visual inspection of a combined shaded relief-surface parameter image may help in the exploration of a terrain model, but it is limited in its description of how surface parameters might change with scale. LandSerf allows the variation in scale to be represented explicitly by graphing how a surface parameter varies with window scale. Figure 7 shows examples of how a graph of a surface parameter centred at any one location varies with scale. The x-axis (Figure 7, top) or distance from centre (Figure 7, bottom) represents the local window size used for measurement. The y axis (Figure 7, top) or direction (Figure 7, bottom) represents the surface parameter being measured. Variation in this axis gives a visual indication of how dependent any particular measure is on the scale it which it is taken. This is also summarised numerically as a measure of average and variation below each graph.

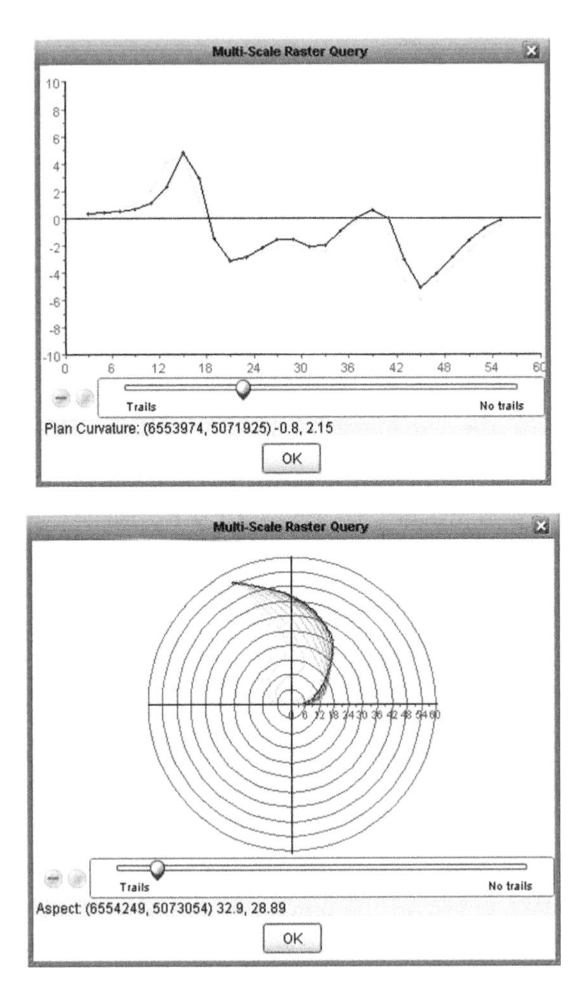

FIGURE 7 Scale signature of plan curvature (top) and aspect (bottom) showing *spatial ghosting* of near neighbours. Spatial extent of measurement in pixels is shown on the horizontal axis at the top, and as distance from the centre at the bottom.

In the case of aspect, circular mean and standard deviation are given. For categorical measures, such as feature classification, mode and entropy are calculated.

Each location on the surface being analysed has its own *scale signature*, and it can be instructive to see how that signature varies in space. By dragging the mouse over the surface model in **LandSerf**, the signature is dynamically updated for the new location under the mouse pointer. In order to aid the visual memory of previously queried locations, *spatial ghosting* shows previous signatures on the same graph that gradually fade as the mouse pointer is moved to new locations. Thus, a visual indication of the 3-dimensional solution space (location, scale, surface parameter) can be used to explore scale-related interactions.

The nature of any one scale signature can be used to identify *characteristic scales* at which a surface parameter is strongest. Since many landscapes will comprise characteristics at many different scales, this method of visual exploration offers an improvement over sampling at a fixed scale. It is considered further in Section 3 below.

2.3 Mipmapping

One of the problems with visually exploring multi-dimensional data spaces is providing an environment that allows relationships between all relevant variables to considered, but without overloading the viewer with too much information. Statistical techniques such as *projection pursuit* (e.g. using principal components analysis) allow data to be collapsed into fewer dimensions. Visual brushing between sets of images (e.g. dynamic updating of scale signatures described above) provides another set of techniques.

Alternatively, maximising the use of visual variables, such as the splitting of colour-space when using blended shaded relief provides another set of possible approaches. However, all of these techniques can require some user-experience and familiarity before they can be used effectively.

An alternative approach provided by **LandSerf**, is to exploit our innate ability to process perspective views in order to reconstruct the 3-dimensional configuration of a surface. By flying an imaginary camera over a surface, a viewer can explore that surface using many of the same cognitive processes they would use when processing the visual field (e.g. Ware, 2004). More importantly for the exploration of scale-related measurements, perspective views allow large and small scale features to be processed simultaneously (Wood et al., 2004). Features in the foreground of a perspective view allow large-scale detailed characteristics to be rendered, while those in the background allow smaller-scale generalised characteristics to be considered.

While simply rendering a draped image as a perspective 3D view affords some scale-specific generalisation of a surface, this does not fully exploit the possibilities of visually exploring the relationship between scale and surface measurements. By using the 3D graphics hardware available in most desktop computers, it is possible to render different surfaces over different parts of a terrain depending on its distance from the imaginary camera. As the viewer moves over a landscape, so the distance dependent rendering is dynamically updated. This process is known as *mipmapping* or *level of detail rendering* (Luebke et al., 2002).

Mipmapping is normally used as a rendering optimisation process to display parts of the surface that are distant from the viewer with less detail than those that are nearer the viewer. **LandSerf** exploits this technique by rendering surface parameters measured at different window sizes at different distances from the viewer. Thus parts of the surface that appear far away from the user might show profile curvature measured using a 55×55 cell window, while those that are near the viewer might show the same parameter measured using a 3×3 window (see Figure 8). By flying to different parts of the surface, or flying towards and away from

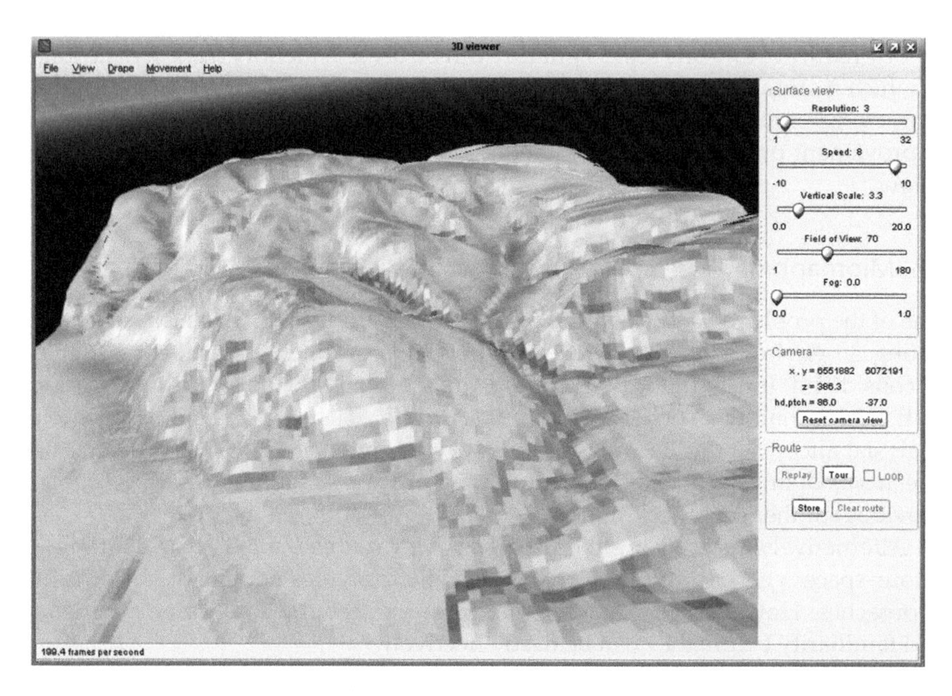

FIGURE 8 Using graphics hardware mipmapping to show multiple scale parameterisations. Here profile curvature measured at the 150 m scale is shown in the foreground ranging to ~2 km on the horizon.

a point on the terrain, an immediate indication of how the surface measurement varies with scale can be given.

3. SCRIPTING WITH LANDSERF

While the benefits of visualisation of landscape models and parametrisation have been demonstrated, especially in an exploratory context, there are occasions when a more systematic and procedural approach is required. By representing process in the form of a script, tasks that need to be repeated or shared between users can be logged in a systematic and reproducible way. Most of **LandSerf**'s functionality can be represented in this form using its own scripting language — *LandScript*. This language contains a series of commands that reproduce those actions otherwise accessible via **LandSerf**'s menus and tool bar, and functions that use *map algebra* (Tomlin, 1990) to perform complex cell-by-cell operations on elevation models and other data.

All map algebra operations can be expressed in the form:

```
newObject = f([Object1], [Object2], [Object3]...);
```

In other words, new spatial objects are created as a function of existing objects. Depending on what is used as input to a map algebra operation, three categories of function are easily scripted with LandScript.

Local operations usually take input from at least two spatial objects. The output for any location is a function of the input objects at that same location. An example of a LandScript local operation might be:

```
errorMap = sqrt((dem1-dem2)^2);
```

which creates a raster containing the root of the squared difference between two elevation models (called `dem1` and `dem2`) for each raster cell. Local operations in LandScript can be created from expressions using common arithmetical and trigonometrical functions. For a comprehensive list functions available, see the documentation at www.landserf.org.

Focal operations usually take input from a single spatial object. The output for any location is a function of the input object at points surrounding the output location. Such functions are often referred to as neighbourhood operations since they process the neighbourhood of a location in order to generate output. LandScript allows neighbouring cells to be identified using a focal modifier in square brackets containing row and column offsets. For example:

```
smoothedDEM = (dem[-1,-1] + dem[-1,0] + dem[1,0]
             + dem[-1,0]  + dem[0,0]  + dem[1,0]
             + dem[1,-1]  + dem[1,0]  + dem[1,1]) / 9;
```

creates a raster where each cell is the average of an input raster's immediate neighbourhood.

Zonal operations are similar to focal operations, but extend the local neighbourhood based on some data-dependent definition of what constitutes a zone. In geomorphometry, zonal operations are commonly used when delineating and characterising drainage basins and other land-surface objects. With LandScript, zonal operations can be implemented using a combination of focal modifiers and iterative or recursive function calls (see example of flow magnitude calculation below).

LandScript can be written in any text editor and run from the command-line using **LandSerf**'s *scriptEngine* command, or can be written and run interactively from within **LandSerf** using the LandScript Editor (*Menu* ↦ *File-Landscript Editor*). The advantage of running from the command-line is in freeing resources that would otherwise be devoted to creating the graphical user interface. This is especially useful when dealing with very large files or memory-hungry operations. The advantage of the built in editor is that it provides coloured syntax highlighting of scripts, identifying commands, functions, variables and text. The following shows a simple example of some LandScript to import the Baranja Hill elevation models and orthophoto, create a new raster containing the elevation differences between the two models, set their colour tables, and save them in **LandSerf**'s native file format.

```
version(1.0);

# Import Baranja Hill DEMs and orthophoto:
baseDir = "c:\data\";
dem25m = open(baseDir & "DEM25m.asc");
dem25srtm = open(baseDir & "DEM25srtm.asc");
```

```
photo = open(baseDir & "orthophoto.asc");

# Calculate difference between the two models:
difference = new(dem25m);
difference = dem25m - dem25srtm;

# Set the colour tables of the rasters and save:
colourEdit(dem25m,"land2");
save(dem25m,baseDir & "dem25m.srf");
colouredit(dem25srtm,"land2");
save(dem25srtm,baseDir & "dem25srtm.srf");
colouredit(photo,"grey1");
save(photo,baseDir & "orthophoto.srf");
colouredit(difference,"diverging1");
save(difference,baseDir & "demDiff.srf");
```

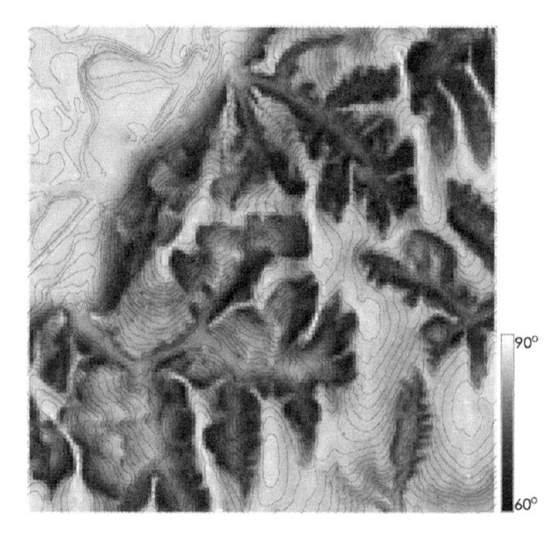

FIGURE 9 Openness measure applied to the Baranja Hill 25 m interpolated DEM. The property was calculated using focal operators in LandScript.

Scripting, such as that shown in the example above can be useful for automating routine processing tasks. It is also useful in defining reproducible algorithmic tasks. For example Yokoyama et al. (2002) proposed a new measure to represent the visual dominance of locations on a landscape based on their local exposure. This requires the calculation of angles along 4 vertical planes for each cell in a DEM — a process that lends itself to a procedural implementation in a language such as LandScript. The script[2] to calculate *openness* (Figure 9) of the Baranja Hill surface is as follows:

```
version(1.0);

baseDir = "c:\data\";
surf = open(baseDir & "DEM25m.asc");
```

[2] The complete script is available via the geomorphometry.org website. Here only an excerpt showing how focal operators are used is shown.

```
openness = new(surf);
DphiL_EW = new(surf);
DphiL_NS = new(surf);
DphiL_NESW = new(surf);
DphiL_NWSE = new(surf);
rad2deg = 180/pi();
res = 25;
diagRes = sqrt(2)*res;
DphiL_EW = 90-rad2deg*max(atan((surf[0,1]-surf)/(1*res)),
                          atan((surf[0,2]-surf)/(2*res)),
                          atan((surf[0,3]-surf)/(3*res)),
                          atan((surf[0,4]-surf)/(4*res)),
                          atan((surf[0,5]-surf)/(5*res)),
                          atan((surf[0,-1]-surf)/(1*res)),
                          atan((surf[0,-2]-surf)/(2*res)),
                          atan((surf[0,-3]-surf)/(3*res)),
                          atan((surf[0,-4]-surf)/(4*res)),
                          atan((surf[0,-5]-surf)/(5*res)));
DphiL_NS = 90-rad2deg*max(atan((surf[1,0]-surf)/1*res)),
...
```

LandScript allows functions to be created and called from within a script. The functions can recursively call themselves, opening up the possibility of map algebra zonal operations. For example the following excerpt from a script to identify the flow magnitude and drainage basins of the Baranja Hill 25 m DEM shows how recursive processing through drainage basins can be implemented. The scale at which this analysis is performed can be controlled by initialising the `windowSize` variable in the script[3]:

```
version(1.0);

# Recursive flow magnitude function:
function calcFlowMag(r,c)
{
  # Check we haven't been here before:
  visitedCell = rvalueat(basins,r,c);
  if (visitedCell == basinID);
  {
     # We have already visited cell during this pass:
     return 0;
  }
  flow = 1;

  # Log this cell as belonging to the drainage basin:
  rvalueat(basins,r,c,basinID);

  # Stop if we have reached the edge:
  if ((r==0) or (c == 0) or
      (r >= numRows-1) or (c >= numCols-1));
```

[3] For full script, see the geomorphometry.org website.

```
    {
        return flow;
    }

    # Look for neighbours that might flow into this cell:
    aspVal = rvalueat(aspect,r-1,c);
    if ((aspVal > 135) and (aspVal <= 215));
    {
        fl = calcFlowMag(r-1,c);
        flow = flow + fl;
    }
    aspVal = rvalueat(aspect,r+1,c);
    if ((aspVal > 305) or
         ((aspVal <= 45) and (aspVal != null())));
    {
        fl = calcFlowMag(r+1,c);
        flow = flow + fl;
    }
    aspVal = rvalueat(aspect,r,c-1);
    if ((aspVal > 45) and (aspVal <= 135));
    {
        fl = calcFlowMag(r,c-1);
        flow = flow + fl;
    }
    aspVal = rvalueat(aspect,r,c+1);
    if ((aspVal > 215) and (aspVal <= 305));
    {
        fl = calcFlowMag(r,c+1);
        flow = flow + fl;
    }
    return flow;
}
```

3.1 Using scripting to explore scale signatures

As a final example of how scripting can be used to explore the scale dependency of land-surface parametrisation, consider the profile convexity scale signatures discussed in Section 2.2. Visual examination of scale signatures can reveal characteristic scales at which a particular surface parameter appears most extreme (peaks and troughs in the scale signature — see Figure 10).

While dynamic exploration of signatures in **LandSerf** can reveal how this characteristic scale varies in space, we can use scripting to identify that characteristic scale and then store the parameter measured at that scale. The result is a pair of maps showing the land-surface parameter and the scale at which measuring that parameter is most extreme (see Figure 11).

The script to do this is given below. The land-surface parameter to investigate can be selected by changing the variable `param`, and the scales over which to investigate variation are set by the variables `minWinSize` and `maxWinSize`:

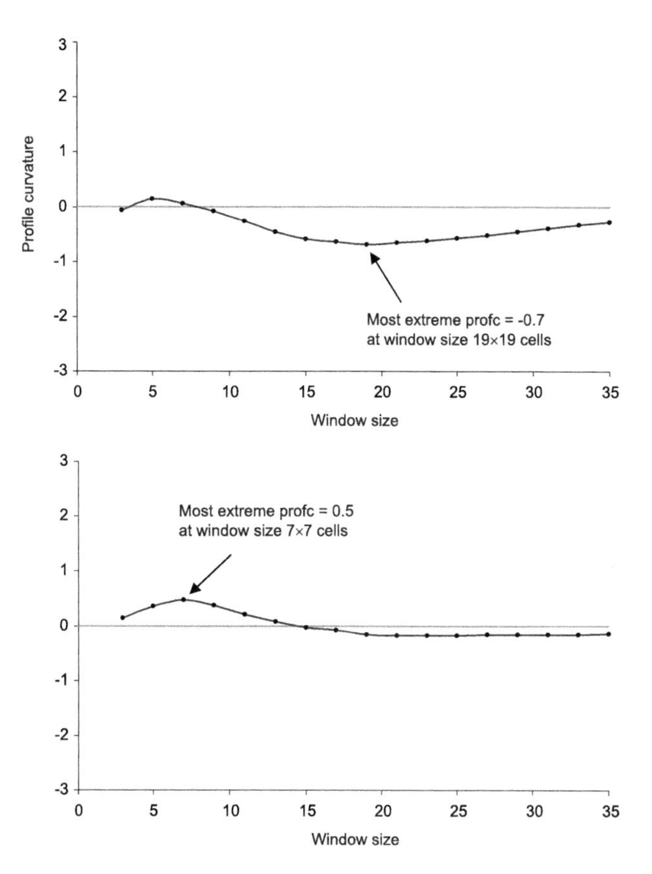

FIGURE 10 Two profile curvature scale signatures for different locations on the Baranja Hill DEM. In each case a *characteristic scale* can be identified at which curvature is strongest.

```
version(1.0);
# Script to measure surface parameter at characteristic scales:

baseDir = "c:\data\";
param = "planc";
maxWinSize = 35;
minWinSize = 3;

dem25m = open(baseDir & "dem25m.srf");
scaleMax = new(dem25m);
scaleMax = minWinSize;

maxParamSurf = surfparam(dem25m,param,minWinSize,1.0);

winSize = minWinSize+2;
maxParam = 0;

while (winSize <= maxWinSize);
{
    # Test if this scale produces extreme parameter value:
```

```
    paramSurf = surfparam(dem25m_, param, winSize,1.0);
    scaleMax = ifelse(abs(paramSurf) > abs(maxParamSurf),
               winSize, scaleMax);
    maxParamSurf=ifelse(abs(paramSurf) > abs(maxParamSurf),
                 paramSurf, maxParamSurf);
    winSize = winSize + 2;
}

# Give the characteristic scale surface a greyscale:
colouredit(scaleMax,"rules",minWinSize&" 255 255
255, "& maxWinSize&" 0 0 0");

# Save the two new surfaces:
save(maxParamSurf,baseDir&"max"&param&".srf");
save(scaleMax,baseDir&"charScale"&param&".srf");
```

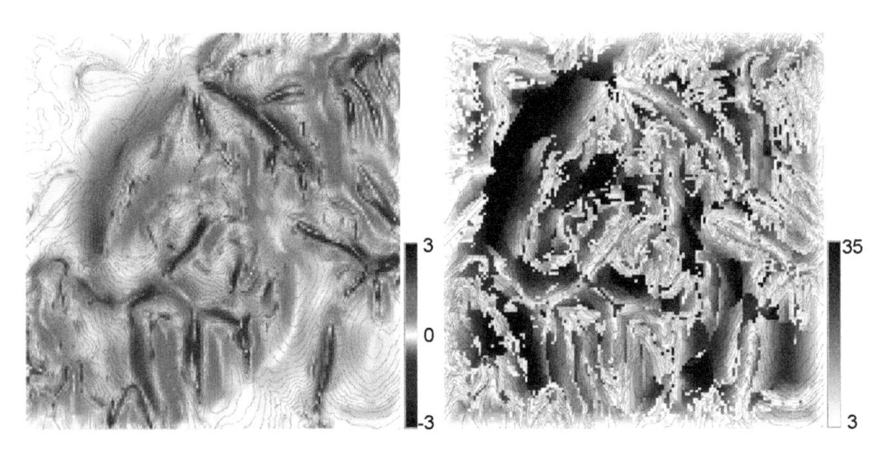

FIGURE 11 Maximum absolute profile curvature (per 100 m) measured over all scales between 75 m and 1.7 km (window sizes 3 to 35). The image to the right shows the window scale (in pixels) at which the most extreme value of profile curvature occurs. (See page 725 in Colour Plate Section at the back of the book.)

4. SUMMARY POINTS AND FUTURE DIRECTION

LandSerf is best suited to geomorphometric analysis where rich visual interaction is considered important and where the effects of scale are to be considered. This chapter has examined three approaches in which LandSerf can be used to consider scale dependencies in land-surface parameters.

Firstly, standard land-surface parameters such as slope and curvature can be measured at any arbitrary scale; a scale determined by setting the local window size over which the parameter is estimated. Secondly, the variation in land-surface parameters with scale can be considered explicitly either by plotting the scale-signature at points over a surface, or by finding the scales at which land-surface parameters are most extreme. Thirdly, variation of land-surface parameters with scale can be explored visually through the use of mipmapping in a dynamic 3D environment.

The strong visual control that underlies the design of LandSerf remains one of its strengths. One of the consequences and weaknesses of this design is that all handling of spatial data is carried out in memory in order to increase the speed of visual interaction. This imposes practical limits on the size of data that can be handled at any one time. Each raster cell is stored as a 32 bit floating point number, so a 1000×1000 cell raster requires 4 MB of heap memory. Combining this with the memory required for display and undoable copies of edited rasters, a size of around 3000×3000 cells per raster is probably the practical limit before performance degradation becomes evident. While disk caching of memory and more recent versions of the Java Virtual Machine can partially overcome this limit, Digital Elevation Models greater than about 6000×6000 pixels become impractical to work with.

It is hoped that as the software is developed, more efficient storage and caching of data will improve the handling of very large datasets. It is anticipated that with the increasing availability of very high resolution elevation models such as those produced by LiDAR, there will be greater need for LandSerf and other geomorphometric software to handle multi-gigabyte datasets.

LandSerf has been publicly available for 10 years and has remained, and will continue to remain, free software. Its non-commercial status and the fact that it is written in Java makes it a package easily accessible to most geomorphometry researchers. The weakness of this model of development is that it does not have the distributor-led support that commercial packages provide. There is however, a large user-base (at the time of writing, over 30,000 copies have been downloaded), and it is hoped that with the recently introduced availability of LandScript, the LandSerf scripting language, this user-community will develop and share scripts to enhance the software.

For those wishing to exercise greater control of the software, there is documentation that provides support for linking it with the Java programming language via the LandSerf API. This requires some Java programming skills, but has the advantage of providing a set of classes for handling surface models and graphical interaction that would otherwise have to be written from scratch.

IMPORTANT SOURCES

Wood, J., 2002. Java Programming for Spatial Sciences. Taylor and Francis, London, 320 pp.
www.landserf.org — Homepage of LandSerf.

CHAPTER 15

Geomorphometry in MicroDEM

P. Guth

MicroDEM and its history · how do I get MicroDEM on my computer? · what can MicroDEM do? · how do I use MicroDEM? · what is terrain organisation? · how is MicroDEM unique?

1. INTRODUCTION

1.1 MicroDEM history and development

MicroDEM grew out of work in the early 1980s to provide computerised terrain analysis for U.S. Army terrain teams in the field. The first operational version was fielded in 1985 for an Apple II computer, although development work had been done on an IMB PC (Guth et al., 1987). MicroDEM was written in Turbo Pascal; the DOS source code was distributed with the program until 1995, and is still available on the web (http://www.usna.edu/Users/oceano/pguth/microdem/source_code/dos/).

In 1995 a Delphi (Object Pascal, the successor to Turbo Pascal) version appeared (http://www.usna.edu/Users/oceano/pguth/website/microdemdown.htm) which is available as freeware without source code. Between January 2003 and May 2008 there have been over 87,000 downloads of the complete program installation, and another 28,000 downloads of an updated version of the executable program.

A forum for discussion of problems with MicroDEM and suggestions for modifications can be found at http://forums.delphiforums.com/microdem/start, with over 4550 messages currently posted.

MicroDEM began with a heavy emphasis on practical application of DEMs, including slope maps, 3D oblique views, line of sight profiles, and viewsheds. It has since become a general purpose GIS, integrating imagery and shape files with DEMs, but it retains a strong emphasis on geology and geomorphometry.

> REMARK 1. *MicroDEM is a GUI program for MS Windows that emphasises geomorphometry but also performs many GIS functions.*

Developments in Soil Science, Volume 33 © 2009 Elsevier B.V.
ISSN 0166-2481, DOI: 10.1016/S0166-2481(08)00015-9.

The target data for MicroDEM was initially the *Digital Terrain Elevation Data* (DTED) from what was then the US Defense Mapping Agency (DMA) and is now the National Geospatial-Intelligence Agency (NGA). Because of the horizontal data spacing in arc seconds, DTED has some unique characteristics that influence the algorithms MicroDEM uses to process data and extract land-surface parameters. The algorithms discussed earlier in this book all considered DEMs with a square grid, with equal x and y spacing in metres, such as a Universal Transverse Mercator (UTM) grid. Some software can only deal with such grids, and must reinterpolate a geographic DEM before using it. MicroDEM has always sought to use DEMs like DTED in their native format, and has adapted all algorithms accordingly.

Guth (2004) discussed differences in line of sight algorithms for geographic and UTM DEMs, and that for small areas geographic grids can be considered to be a rectangular grid with constant but different x and y spacings. Over larger areas, the y spacing will be constant but the x spacing will vary with latitude. Rectangular grids cannot work over large areas because of Earth's curvature, and seamless operation over large areas makes geographic grids attractive. Both major US producers of DEMs now use geographic coordinates for their best data: DTED and Shuttle Radar Topography Mission (SRTM) from NGA, and the National Elevation Dataset (NED) from the US Geological Survey (USGS), and supply free data covering most of the world.

Many of the best medium resolution (about 10–100 m) DEMs now use geographic coordinates, and analysis software should use these in their native format and not require reinterpolation. If reinterpolation is done, it must be suspected as contributing to any anomalies or differences in the resulting analysis.

MicroDEM now reads DEMs in both geographic and rectangular (UTM-like) grids, and can read many other DEMs. A partial list of supported data formats includes: DTED, SRTM in both `DTED` and `.hgt` formats, USGS ASCII and SDTS, Geotiff, `.bil`, `.asc` (ESRI ASCII grid), the United Kingdom's Ordnance Survey Grids, and netCDF. The program has a bias for the formats of the US government mapping agencies, because the formats are openly published and the data freely available. Few other countries freely supply comparable data (Canadian CDED, in USGS ASCII format, is a major exception), and with the SRTM data, the United States took a giant step toward supplying free topographic data for the world.

NGA supplies[1] a number of both raster and vector data sets worldwide, and at least two free sources[2] of Landsat imagery can enhance DEM geomorphometry. MicroDEM can display and integrate all of this data. It can automatically load US data like the Census Bureau's `TIGER` files to show roads and water bodies, or the National Land Cover Dataset (NLCD), which can provide context for geomorphic interpretation of DEMs. This is only possible because the data is freely available in a standard format covering the entire country.

[1] http://geoengine.nga.mil/geospatial/SW_TOOLS/NIMAMUSE/webinter/rast_roam.html.
[2] http://glcf.umiacs.umd.edu, https://zulu.ssc.nasa.gov/mrsid/.

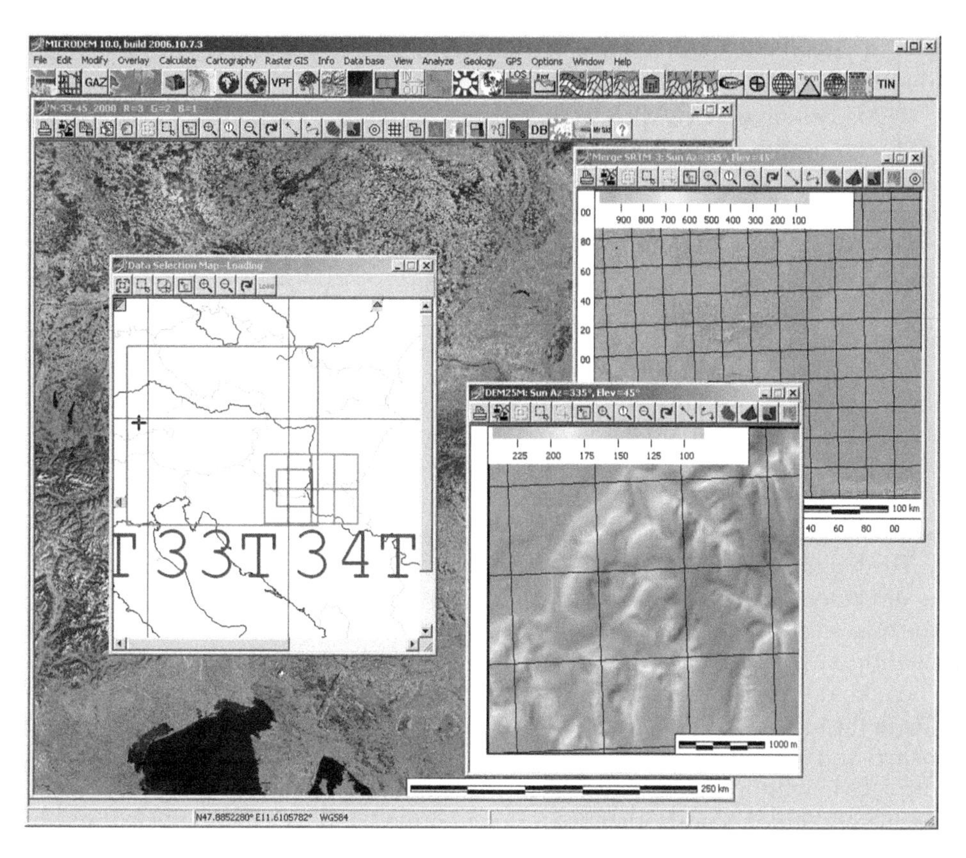

FIGURE 1 The main window of MicroDEM, with standard Windows controls and four active child windows. The centre left window is an index map showing eastern Europe with available Landsat imagery outlined by the large red rectangle, SRTM data shown in green, and the Baranja Hill DEM barely visible at this scale. Selecting the small box in red opened two DEMs, one a merge of 4 SRTM cells, and the satellite image visible in the background. (See page 726 in Colour Plate Section at the back of the book.)

1.2 Getting started

MicroDEM runs on 32 bit versions of Windows, and can also run on 64 bit versions in 32 bit mode. Download MicroDEM and run the installation program which will set up the program, its large integrated help file, and sample data. Open the program from the icon on the start menu, and you will see a splash screen and a standard Windows program (Figure 1). This discussion centers on Version 10.0, Build 2006.12.1.

Options can generally be selected in three ways: using buttons on the toolbars, using the main menu whose choices change with the currently active child window, or by right clicking on the map window to activate a popup menu. The status bar on the bottom of the screen shows the action expected by the program in the leftmost panel, the coordinates and elevation at the mouse cursor in the second panel, and additional information in other panels.

DEMs can be opened directly with the *File, Open DEM* menu choice, or the DEM icon on the main program toolbar. In addition, most data can be opened graphically using an index map, which does not require users to remember cryptic file names or where the files are stored. MicroDEM defines six major categories of digital map data, including *DEMs, bathymetry, DRGs* (digital raster graphics, or scanned maps), *imagery*, and *land cover* like the USGS NLCD.

Each of these occupies a subdirectory under the MAPDATA directory, which can be anywhere on the user's disk drives. In each category, the user can create sub-categories or series, for instance broken down by DEM producer and scale. To help the user manage data, series can contain multiple directories. When selecting DEMs, each series will be displayed in a different colour, and can be turned on or off. Example DEM series include SRTM-3, SRTM-1, USGS-NED-1, LA-LiDAR, or UK-OS. For all these DEMs, users are likely to have a large number of DEMs covering a significant area

The Baranja Hill DEM, with its associated `.asc` and `.prj` files, was placed in the directory: c:/mapdata/indexed_data/dems/misc/dems/, and then indexed. Indexing creates a database with the extent of each DEM, and the user does not have to remember file names or where the files were placed on the hard disk. When the user selects a region on the index map, MicroDEM can determine all the data in the selected region. Multiple DEMs in a single series can be automatically opened and merged on the fly to create a large, seamless DEM. Merging works with DEMs with a regular quadrangle structure and which share the same data spacing and other characteristics. Merging on the fly allows more efficient use of system resources, by combining data sets which cover the area of interest.

Indexing works best with data like the SRTM data sets, USGS data sets, Canadian CDED, or the high resolution state DEMs available in the United States which all have standard extents, data spacing, and format. The complete SRTM-3 data set has abut 15,000 files and requires 35 GB of hard disk storage; with indexing, a DEM covering any desired area can be rapidly opened. MicroDEM can merge almost 400 one degree cells of SRTM-3 data; any larger areas should use SRTM-30. MicroDEM opens a selection map with each DEM, which can be a reflectance, elevation, or contour map. Indexing can also open related data sets in the other categories such as imagery or scanned maps, to provide context for the DEM or merging for visualisation. Figure 1 demonstrates the use of an index map to open the Baranja Hill DEM.

2. GEOMORPHOMETRIC ANALYSIS IN MICRODEM

2.1 Creating geomorphic graphs in MicroDEM

MicroDEM can create a number of statistical graphs for the DEM, either for the entire DEM or a subset currently on the screen. These choices occur on the Analyze menu, available with an active DEM map. Figure 2 shows samples for the Baranja

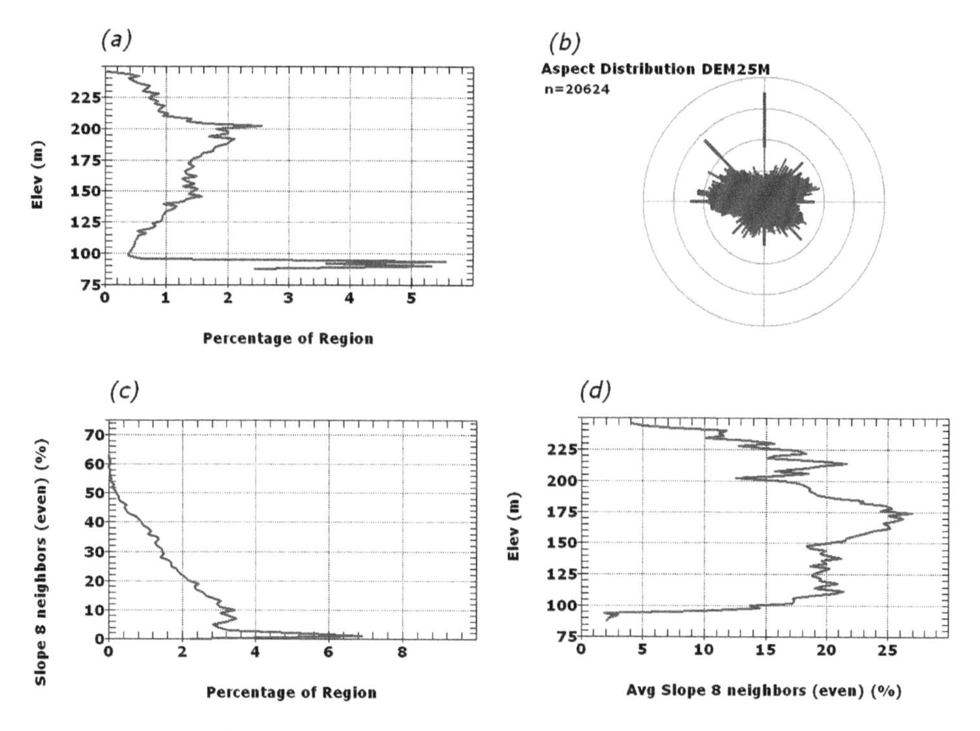

FIGURE 2 Statistical graphs computed for the Baranja Hill DEM: (a) histogram of elevations, (b) rose diagram with aspect distribution, (c) histogram of slope, (d) elevation versus slope, showing that the flattest terrain occurs at the both the lowest and highest elevations.

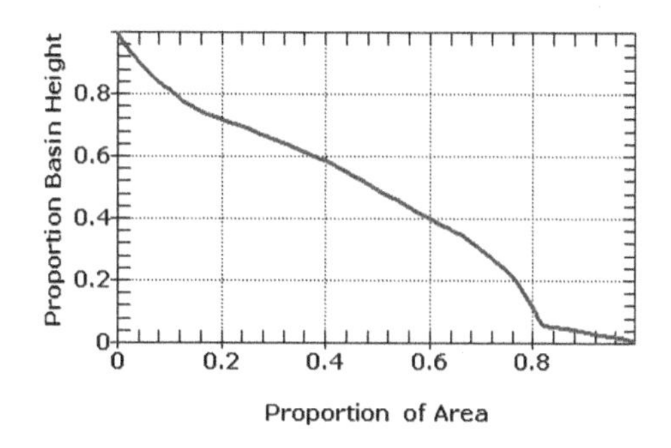

FIGURE 3 Cumulative Strahler curve (Strahler, 1952) for the Baranja Hill DEM, with both elevation range and area normalised to 1.

Hill DEM, including histograms of elevation and slope, a rose diagram of aspects, and a graph showing average slope by elevation.

Figure 3 shows a normalised elevation distribution as suggested by Strahler (1952). Figure 4 shows aspects for the Baranja Hill DEM by slope categories, which demonstrates at least two things. First, the distribution of aspects clearly varies with slope. There are very few SE-facing steep slopes (those over 20%), but a great

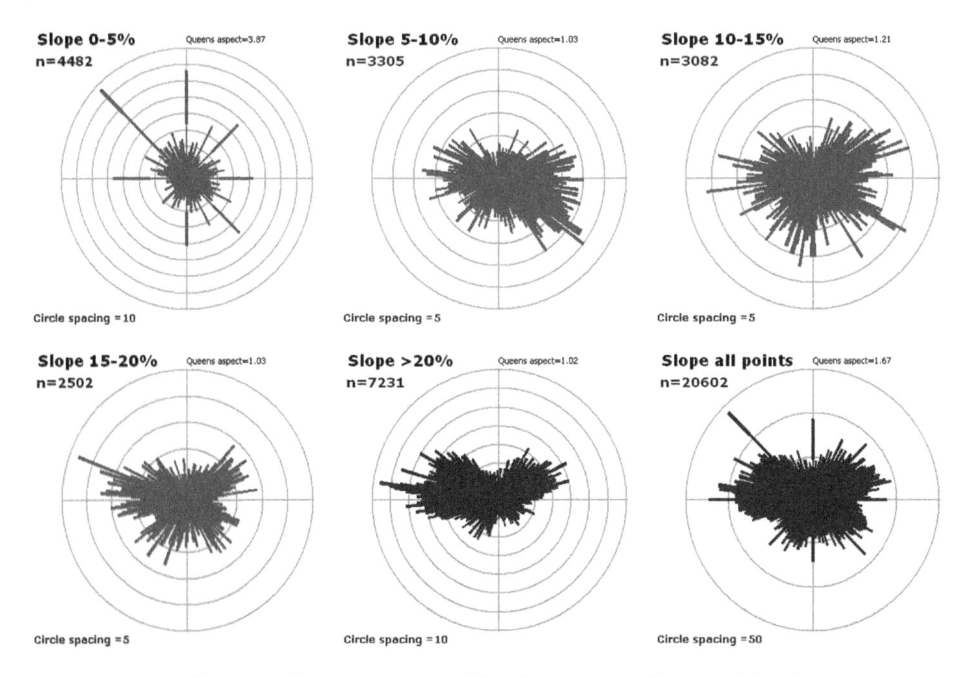

FIGURE 4 Rose diagrams of aspect computed for the Baranja Hill DEM, with 5 slope categories and the entire DEM. In addition to what this says about the landscape, the results for the 0–5% slope category show how the algorithm greatly overestimates aspects in the 8 principal compass directions in flat terrain.

many gentle slope (5–10%). Secondly, the aspect algorithm has performance problems in gently sloping regions, and produces too many aspects in the 8 principal compass directions.

Each aspect rose in Figure 4 has a computed *Queen's Aspect Ratio*, which is the ratio of the number of aspects in each of the principal directions compared to the number that would occur if all 360 directions occurred with equal likelihood. A Queen's aspect ratio of 1 indicates no bias in the DEM and algorithm, and occurs here for slopes over 15%. But for the gentlest slopes, the preferred directions occur almost 4 times too often.

MicroDEM supports 12 different slope algorithms. Guth (1995) demonstrated that six of those algorithms produced highly correlated results, although there were consistent differences and that the definition of slopes at ridge crests and valley floors presents something of a philosophical question "*do you want the gentle slope of the break line, or the very steep orthogonal slope?*" MicroDEM retains the ability to compare slope algorithms, and has added additional algorithms that have been suggested in the literature. Hodgson (1998) and Jones (1998) both confirmed the strong correlation among all the slope algorithms that have been proposed.

Figure 5 shows how aspect distributions for the Baranja Hill DEM vary with the slope algorithm used. Obviously the last four algorithms should not be used because of the extreme quantization of the aspect distribution, but as shown in the first four images of the diagram, clearly the 8 neighbour algorithms outperform

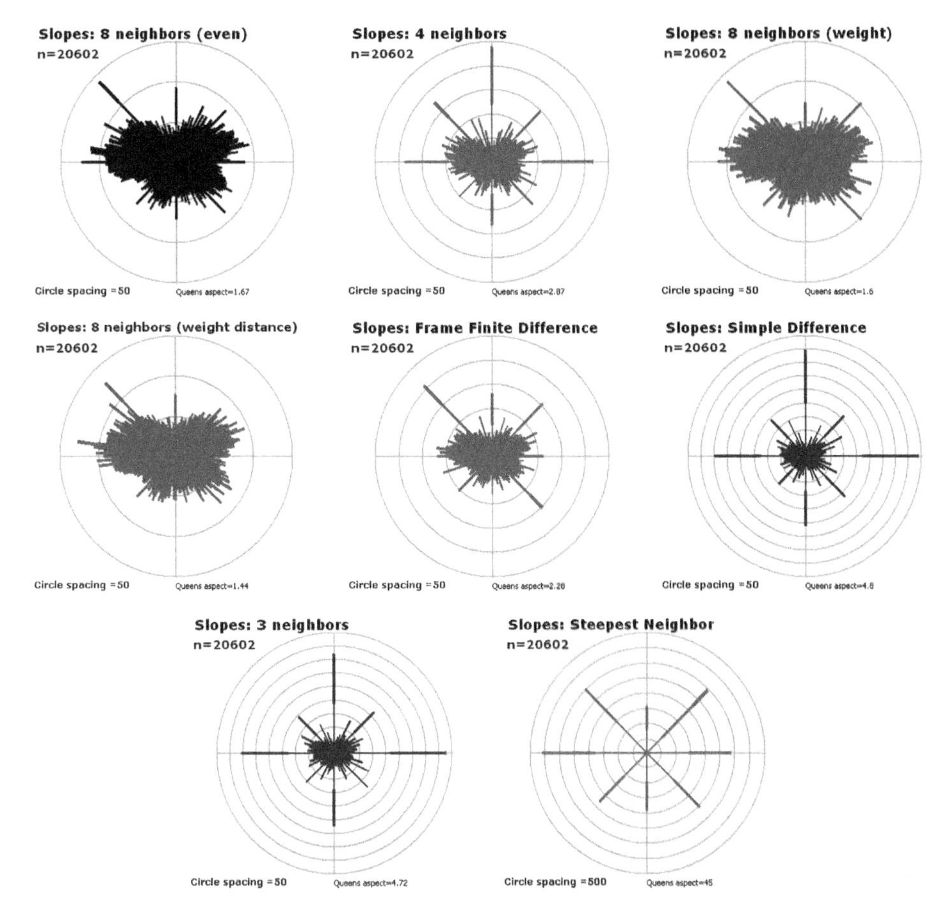

FIGURE 5 Rose diagrams of aspect computed for the Baranja Hill DEM, using eight different slope algorithms. Note that all the algorithms provide too many points in the eight principal compass directions, but that those that use eight neighbours provide a more uniform distribution.

the 4 neighbour algorithm. Because of this effect, and the effect of the algorithm on the moment statistics of the slope distribution, MicroDEM recommends a default setting with an eight neighbour evenly-weighted slope algorithm to produce the most natural slope distributions.

2.2 Deriving land-surface parameters in MicroDEM

Derivative terrain grids for local morphometric land-surface parameters can be created in several ways. First, the display parameter on the selection map can be changed from elevation to several others by either right clicking on the map, or by using an option on the modify menu. This does not affect the original DEM, only its display. Parameters available include elevation, contour, slope, aspect, reflectance, and curvature categories.

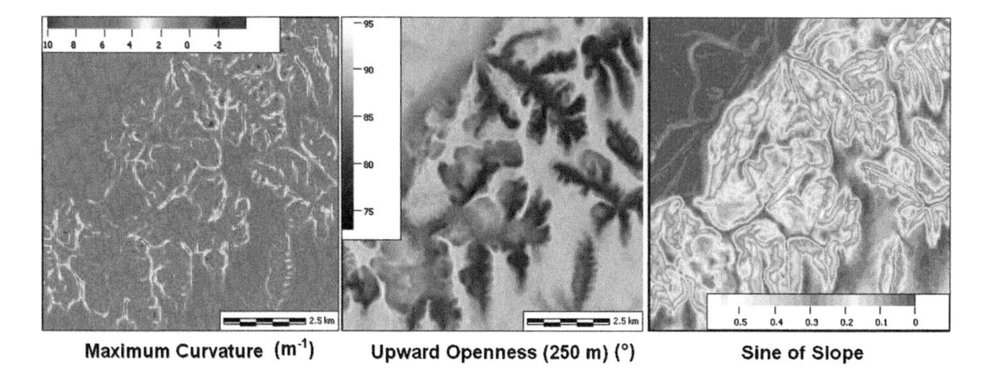

FIGURE 6 Sample maps of land-surface parameters created with MicroDEM. From left to right these show three options for colour coding: a continuous colour scale, a greyscale, and a discrete colour scale. These maps also show the options for placement and orientation of legend and scale bar. (See page 726 in Colour Plate Section at the back of the book.)

A second option creates a new grid with a derived parameter from a larger list of parameters which includes curvature measures, slope (in degrees, percent, or the sine), and aspect. With this new grid operations like moment statistics or filtering can be performed on the derivative data set. Figure 6 shows three standard land-surface parameters, while Figure 7 shows two parameter maps draped on the original DEM.

MicroDEM can also create parameter maps for regional statistics which require a much larger neighbourhood around the point than the typical 8 neighbours used for slope, aspect, and curvature. Examples of these larger neighbourhood parameters include: relief, summit and base level surfaces, and openness. Yokoyama et al. (2002) introduced the concept of openness, and as Figure 8 shows, this correlates strongly with some of the curvature measures.

Because openness uses a larger computation region, it has greater practical value, for instance as a fast predictor for locations that will have good viewsheds.

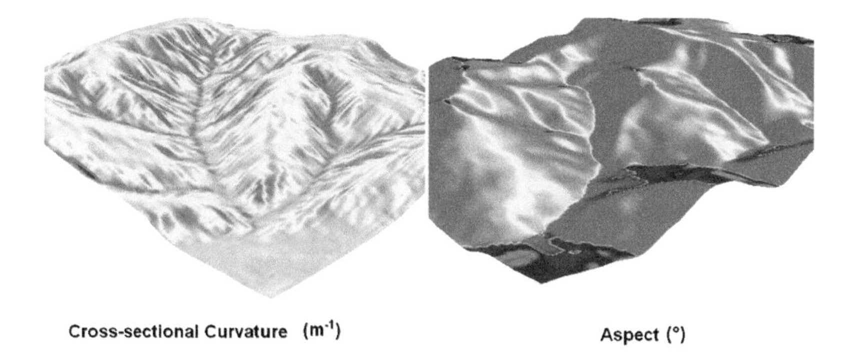

FIGURE 7 Sample land-surface parameters draped on the Baranja Hill DEM. (See page 727 in Colour Plate Section at the back of the book.)

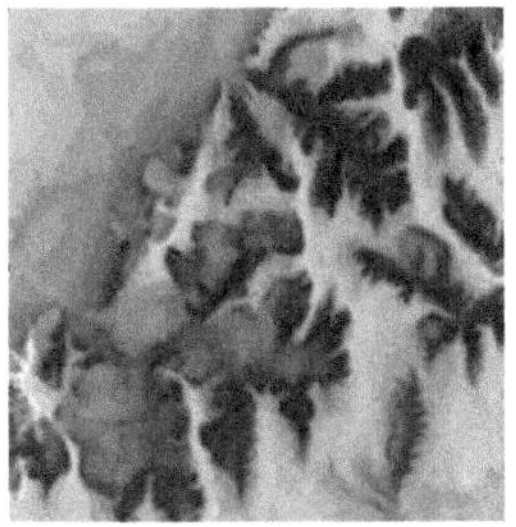

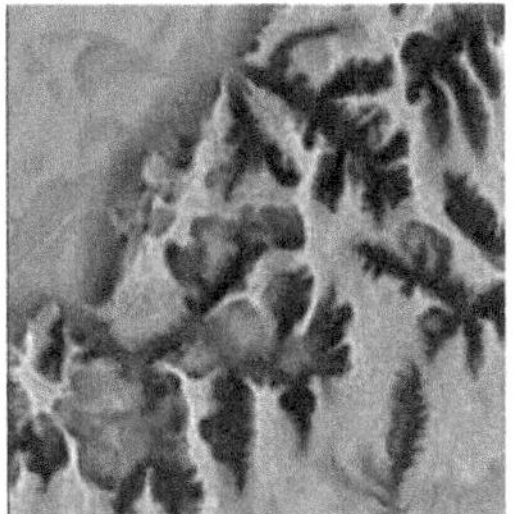

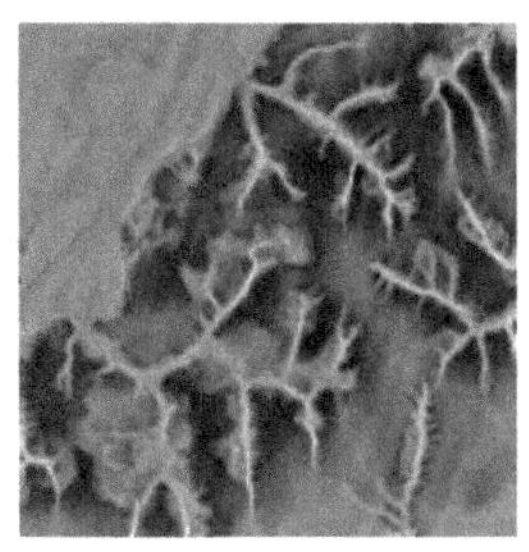

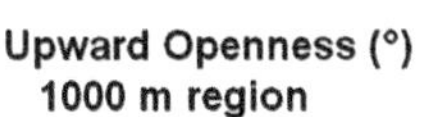

FIGURE 8 Openness maps created with MicroDEM. The maps on the left show how upward openness changes with region size, and the map on the right shows that downward openness is close to a mirror image of upward openness.

Upward openness takes significantly longer to compute than simple land-surface parameters, but is orders of magnitude faster than computing exhaustive viewsheds.

2.3 Terrain organisation

In a series of papers, Guth (2001, 2003) discussed an eigenvector technique to quantify terrain organisation. Drawing on Chapman's (1952) manual method for map analysis and Woodcock's (1977) technique for geologic fabric analysis, the method finds the dominant terrain direction and assigns a numerical score for the degree to which hills and valleys share the same orientation. Terrain organisation requires an analysis region, and results vary with the region size.

REMARK 2. *Terrain organisation quantifies the degree to which ridges and valleys align, and determines the preferred orientation.*

Figure 9(a) shows how the user sets the parameters that control the organisation vectors plotted on the Baranja Hill DEM in Figure 9(b). The length of the line reflects the strength of the organisation parameter in a 400 m region centered on the point, and the vector points in the direction of dominant terrain fabric. Points with a large value of flatness (Woodcock's 1977 definitions of the ratios of the logs of the eigenvalues used flatness rather than steepness) do not have a vector plotted because random noise dominates those regions. The example on scripting in MicroDEM later in Section 2.7 shows how results of terrain organisation vary with the size of the analysis region.

2.4 Regional morphometric land-surface parameters

MicroDEM can compute a series of 30 parameters for a region, with the region size determined by the user. The variables include:

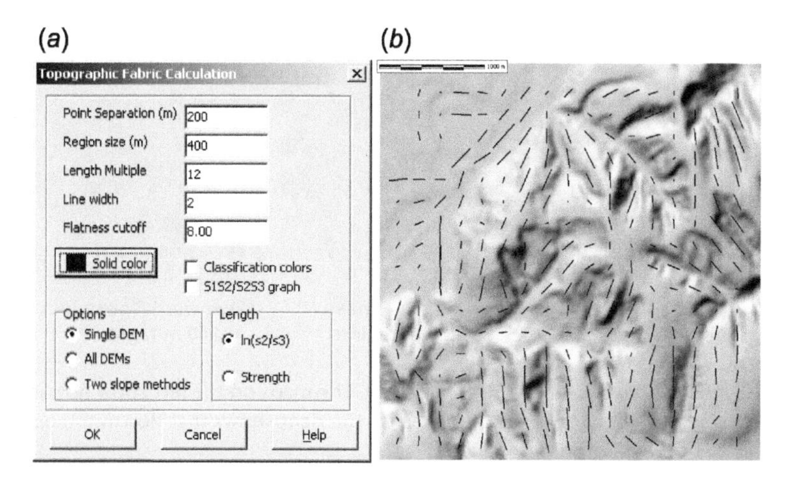

FIGURE 9 Options to create a topographic fabric overlay (a). Point separation, region size, and the flatness cutoff are the key parameters. Reasonable values depend on the DEM spacing, and the nature of the topography-bathymetric DEMs, where abyssal hills show strong organisation, typically require much different values than terrestrial DEMs. Terrain organisation vectors overlaid on the Baranja Hill DEM (b). The length of the lines indicates the organisation in the region, and the vector points in the dominant direction. This computation requires a region size (400 m) and a minimum steepness required to consider the computations valid.

- DEM_AVG, DEM_STD, DEM_SKW, DEM_KRT: the first four moments of the elevation distribution. DEM_STD correlates strongly with slope.
- SLOPE_AVG, SLOPE_STD, SLOPE_SKW, SLOPE_KRT: moments of the slope distribution in percent (100 × rise/run).
- PLANC_AVG, PLANC_STD, PLANC_SKW, PLANC_KRT: moments of the plan curvature distribution.
- PROFC_AVG, PROFC_STD, PROFC_SKW, PROFC_KRT: moments of the profile curvature distribution.
- S1S2, S2S3, FABRICDIR: Computed using logs of the eigenvectors of the surface normal vector distribution. S1S2 measures flatness (a logarithmic inverse of slope), S2S3 measures terrain organisation, and FABRICDIR gives the dominant direction of ridges and valley. Because FABRICDIR measures circular angles, its statistics have anomalies.
- SHAPE, STRENGTH: Fisher et al. (1987) defined these ratios of the logs of the eigenvectors; defined somewhat differently that those used by Woodcock (1977) and Guth (2003).
- RELFR: the relief ratio ($[\bar{z} - z_{\min}]/[z_{\max} - z_{\min}]$) is computed for a region (Pike and Wilson, 1971; Etzelmüller, 2000) and is equivalent to the coefficient of dissection (Klinkenberg and Goodchild, 1992), after Strahler (1952).
- SLOPE_MAX: the largest slope (percent) in the sampling region. While this is largely of value for detecting blunders during DEM creation, it also has geomorphic significance.

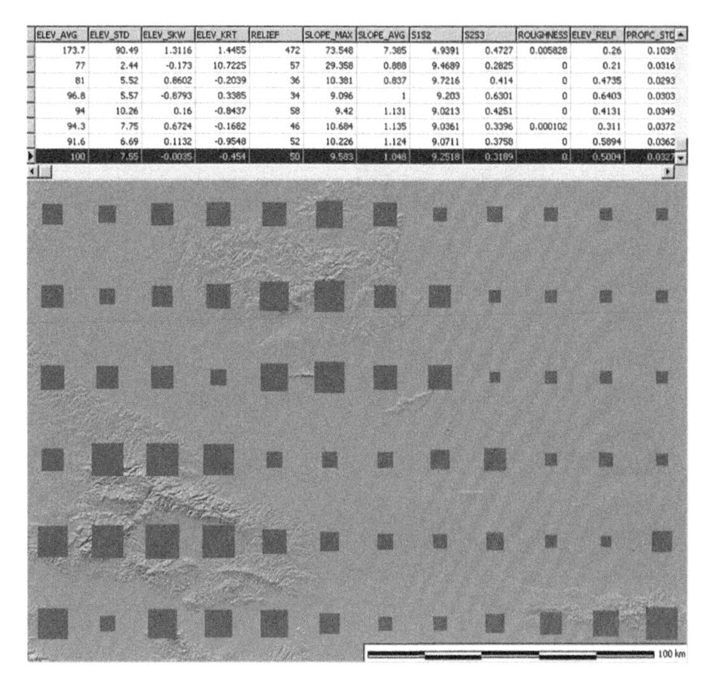

ELEV_AVG	ELEV_STD	ELEV_SKW	ELEV_KRT	RELIEF	SLOPE_MAX	SLOPE_AVG	S1S2	S2S3	ROUGHNESS	ELEV_RELF	PROFC_STD
173.7	90.49	1.3116	1.4455	472	73.548	7.385	4.9391	0.4727	0.005828	0.26	0.1039
77	2.44	-0.173	10.7225	57	29.358	0.888	9.4689	0.2825	0	0.21	0.0316
81	5.52	0.8602	-0.2039	36	10.381	0.837	9.7216	0.414	0	0.4735	0.0293
96.8	5.57	-0.8793	0.3385	34	9.096	1	9.203	0.6301	0	0.6403	0.0303
94	10.26	0.16	-0.8437	58	9.42	1.131	9.0213	0.4251	0	0.4131	0.0349
94.3	7.75	0.6724	-0.1682	46	10.684	1.135	9.0361	0.3396	0.000102	0.311	0.0372
91.6	6.69	0.1132	-0.9548	52	10.226	1.124	9.0711	0.3758	0	0.5894	0.0362
100	7.55	-0.0035	-0.454	50	9.583	1.048	9.2518	0.3189	0	0.5004	0.0327

FIGURE 10 Regional statistics for a 2°×2° block of SRTM data that includes the Baranja Hill DEM. The data base includes 30 parameters for 0.25° analysis regions. The square symbols show the centre of the analysis region, scaled to the maximum slope. The symbols can also be coloured to increase the effectiveness of the map display.

- GAMMA_NS, GAMMA_EW, GAMMA_NESW, GAMMA_NWSE: Nugget variance (C_0) from the variogram (Woodcook et al., 1988a, 1988b). This is a measure of the elevation difference from each point to its nearest neighbour in four directions; smaller values reflect smooth terrain, and high values rougher terrain.
- ROUGHNESS: Measure correlating strongly with slope (Mark, 1975b; Etzelmüller, 2000).
- RELIEF: difference between the highest and lowest elevations within the sampling region (Drummond and Dennis, 1968).
- MISSING: the percentage of holes in the SRTM data. This can be used to filter the results, to avoid looking at statistics where missing data might bias the results.

Figure 10 shows regional statistics from the SRTM data set for the region surrounding the Baranja Hill DEM. The tables shows values for 12 of the parameters, and the map display shows how they can be displayed over the DEM. Many of these parameters actually measure slope, so they might not all be interesting for further applications. Guth (2006) presented a list of the most useful parameters, building on earlier suggestions by Evans (1998) and Pike (2001a).

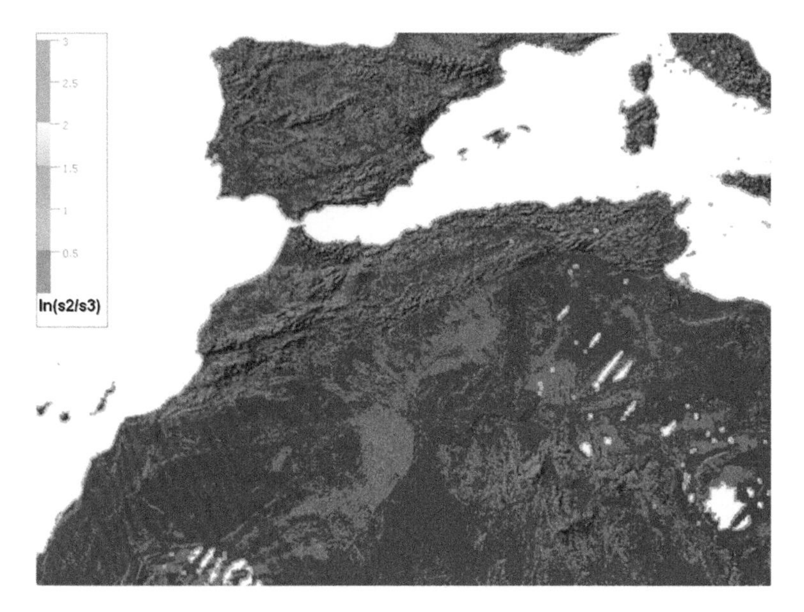

FIGURE 11 Organisation map of North Africa, with colour displaying the degree of organisation (red highly, to blue poorly organized), draped on shaded topography. Note the large void regions where dry sand led to no radar returns. (See page 727 in Colour Plate Section at the back of the book.)

2.5 SRTM atlas-high resolution continental geomorphometry

The 3″ SRTM elevation set has 35 GB of data in 14,277 files covering the Earth's land areas surface between 60° N and 56° S. We divided this data into blocks 2.5′ (arc minutes) on a side, which provides about 7.4 million regions for analysis, which can be considered random sampling areas on a global or continental scale. We masked out the water bodies in the SRTM water mask[3] (Slater et al., 2006), so that we got true terrain statistics without artificial flattening of large lakes and rivers. If there were no holes or water, each block would have 2601 data points, sufficient for robust statistics describing terrain. MicroDEM created grids for 39 parameters, including 5 fractal measures that ultimately proved too noisy for meaningful analysis. Since each DEM took approximately 15 minutes for the computations, we set up a grid of 63 PC's located in 3 college labs to perform the task in two days. Figure 11 shows a detail of one of the maps created, with the values of terrain organisation.

REMARK 3. *MicroDEM has produced an atlas of geomorphic parameters computed from the SRTM data set.*

Figure 12 shows the topography of the North African region with the highest values of terrain organisation due to long, linear sand dunes, as well as examples of three other types of highly organised topography. The SRTM voids limit what

[3] Shuttle Radar Topography Mission Water Body Dataset (http://edc.usgs.gov).

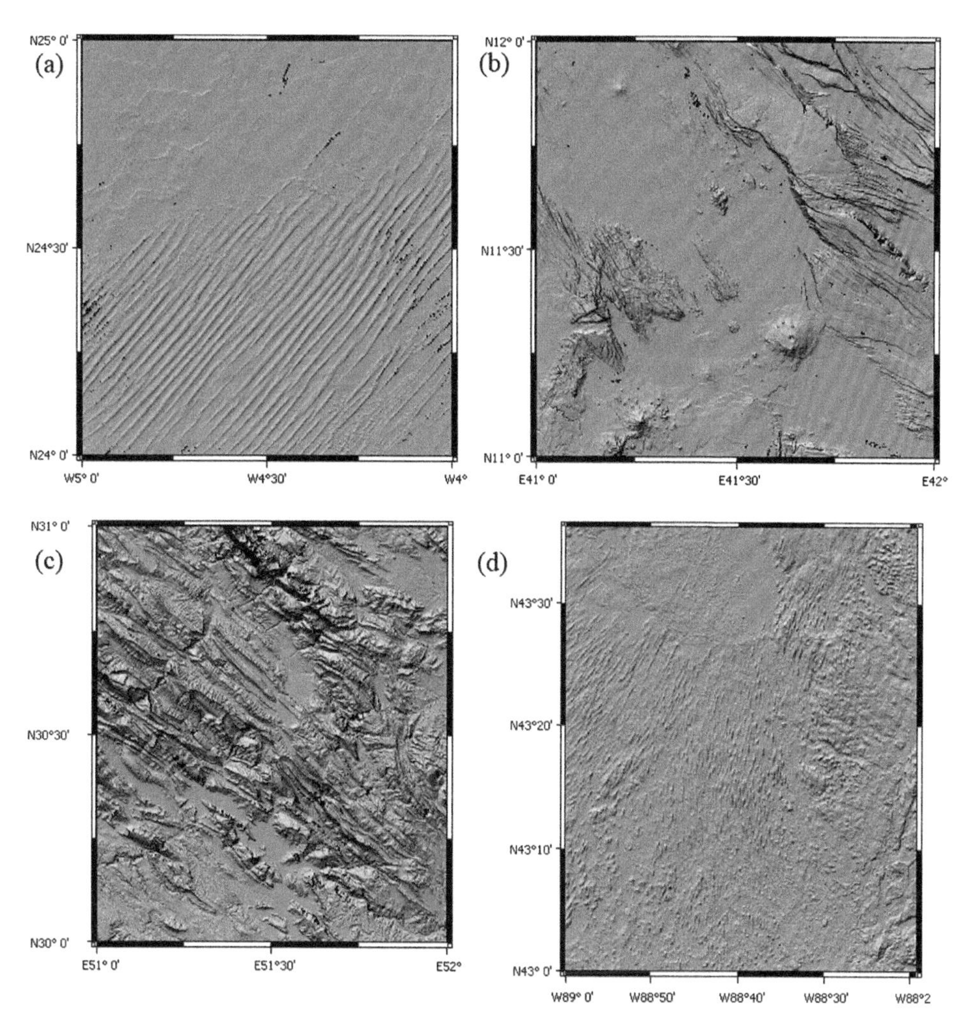

FIGURE 12 Four SRTM data sets shown in shaded reflectance to demonstrate the kinds of highly organized terrain: (a) sand dunes in the Sahara Desert, (b) block faulting in the Afar Triangle of Ethiopia, (c) the folded Zagros Mountains of Iran, (d) glacial drumlins in Wisconsin.

this atlas can do. We investigated whether the SRTM could identify the steepest point or region on Earth. We found 5 points with slopes between 350 and 495% in the SRTM data set, but all 5 are within one posting of a major data void. Since data quality at the edge of a void likely drops, it's unclear how good these point slopes really are. Single extreme points occur in the southern Andes, British Columbia, the Alps, and two are in south central Asia.

We then looked at the average slopes in the 2.5′ analysis regions, and found 20 blocks with an average slope >85% (1 in the Andes, and 19 in central Asia). However, all of these analysis regions were at least 75% holes, so the statistics will be biased. If the holes preferentially occur in steep terrain, the true slopes

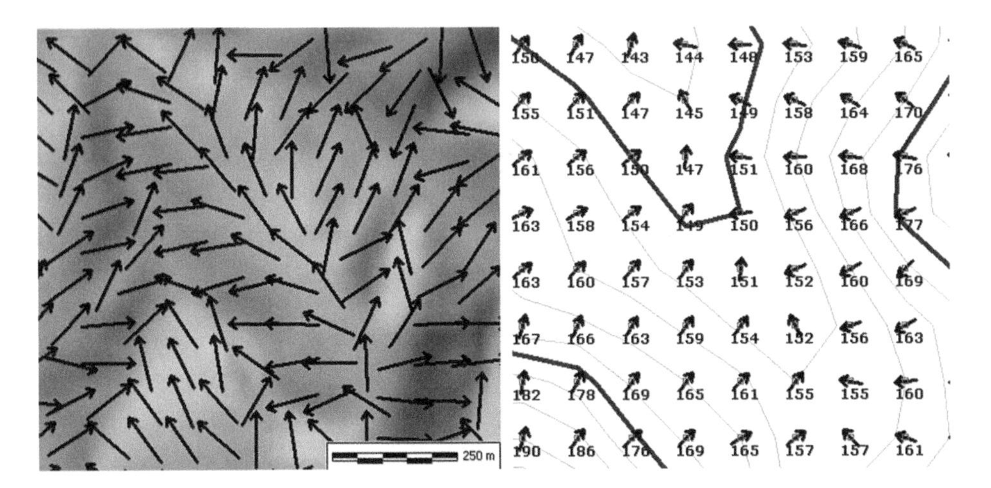

FIGURE 13 Computed drainage vectors for a portion of the Baranja Hill DEM. The option on the left overlays vectors at a user-determined spacing, while the view on right draws a vector on each grid elevation in the DEM and shows contour lines.

might be steeper. Thus, while the SRTM data clearly shows where on Earth very steep terrain occurs, it cannot provide new entries for the Guinness Book of World Records.

2.6 Hydrological modelling

MicroDEM has limited capabilities for hydrological modelling. It will compute and display drainage directions as shown in Figure 13 as an aid to interpreting the topography. For more detailed drainage basin computations, including Strahler stream order and contributing basin areas, MicroDEM has a graphical interface to the DOS version of TARDEM (http://www.engineering.usu.edu/cee/faculty/dtarb/tardem.html) and can display the grids created by TARDEM. MicroDEM can also compute coastal flooding; an animation at http://www.usna.edu/Users/oceano/pguth/website/microdemoutput.htm shows the flooding in downtown Annapolis, Maryland for various levels of storm surge including that from hurricane Isabel.

2.7 Scripting in MicroDEM

MicroDEM was designed as a standard Windows program, with all operations controlled by the graphical user interface or the keyboard. Even the DOS program first described in 1987 used a primitive graphical interface and menus rather than command line scripts (Guth et al., 1987). MicroDEM now has a growing Transmission Control Protocol (*TCP*) interface originally designed for two purposes: (1) to allow other programs to tap into MicroDEM's computation and display capabilities, and (2) allow a web server to access the MicroDEM GIS engine. The interface uses simple ASCII commands, passed by TCP from any computer with a network con-

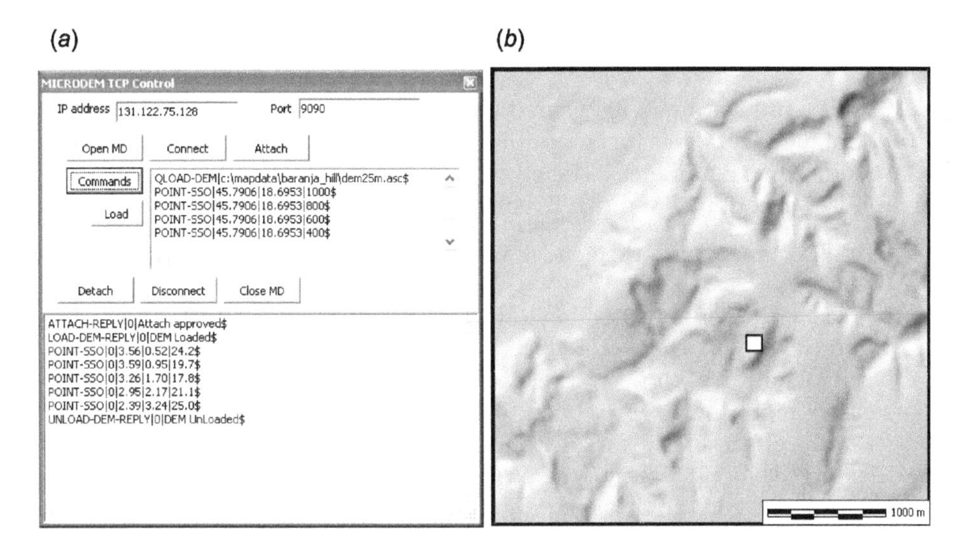

FIGURE 14 Using the TCP interface to run scripts in MicroDEM (a). Scripts can be typed or pasted directly into the upper memo box, or loaded from a saved file. Location (b) for the organisation calculations depicted in (a). Note that this point is on a NNE trending hill and the bottom left memo box shows the computed terrain organisation for this location with five computation regions of increasing size.

nection to MicroDEM. Programs written in C++, Delphi, and Java have been used for this purpose.

The installation for MicroDEM installs a TCP interface program, originally designed for testing the TCP server built into MicroDEM. The program can also be used for scripting geomorphometric or other computations. Figure 14 shows the control program, and a script in the upper memo box. This script loads the Baranja Hill DEM, computes the organisation for a box centered at `N45.7906, E18.6593` and then closes the DEM.

The computation is repeated for box sizes of 1000, 800, 600, 400, and 200 m. The lower memo box shows the replies from MicroDEM, including the computed flatness, the organisation parameter, and the dominant terrain direction. The dominant terrain direction is fairly consistent at about 20°, and corresponds with the location on a NNE trending ridge (Figure 14). This is the last parameter returned by the computations.

Flatness, the first parameter returned by the computations, decreases from 3.36 to 2.39 as the region size decreases and the average steepness increases. The organisation parameter, the second parameter returned, increases from 0.52 to 3.24 as the region size decreases reflecting increasing homogeneity as the smallest region consists only of the single ridge. The program calling MicroDEM would have to interpret the results of the TCP responses, or the user could interpret a text file from the results of the TCP responses. While the TCP interface will probably never include all of the functions available in MicroDEM, it is very easy to add individual operations as desired.

3. SUMMARY POINTS AND FUTURE DIRECTION

MicroDEM is a full featured GIS, geared for geological applications with DEMs. It features unique capabilities for computing terrain organisation, and computing regional geomorphic parameters. The program has been evolving for over 20 years, and promises to continue to grow. Expected major improvements include:

- Using geomorphometric characteristics for predicting good viewshed locations.
- Increasing the options available through scripting with the TCP interface, and the ability to use MicroDEM as a GIS engine for web applications.
- Documentation of slope and related algorithms using geographic DEMs instead of requiring a reprojection to UTM, including options in MicroDEM to show the effects of these algorithms.
- Making additional parts of the program thread safe, and coding more algorithms in parallel, so that the program can utilise the increasing capabilities of multi-CPU and multi-core processors.
- Investigating further applications of the grid to perform massively parallel geomorphometric computations.
- Further investigation of fractal algorithms for classifying Earth's topography in the SRTM 3 second data set.
- Terrain classification, using the clustering and terrain atlas of 30 parameters computed for the SRTM 3 second data set.

IMPORTANT SOURCES

http://www.usna.edu/Users/oceano/pguth/website/microdem.htm — MicroDEM home page.
http://forums.delphiforums.com/microdem/start/ — Delphi MicroDEM forum.

CHAPTER 16

Geomorphometry in TAS GIS

J.B. Lindsay

what is TAS GIS? · who was TAS designed for? · how can you obtain and install the software? · how do you get data into and out of TAS? · what can TAS do? · how do you use TAS to calculate land-surface parameters? · how do you write and execute a script in TAS?

1. GETTING STARTED

1.1 Project history and development

TAS GIS is a stand-alone geographical information system and image processing package that has been designed specifically for geomorphometry applications. The TAS GIS project started in 2002 as part of the author's doctoral research, and was originally called the *Terrain Analysis System*. Early versions of the software were primarily used for DEM pre-processing and some basic analytical functions. Since its inception, however, TAS has grown into a well-equipped GIS with a toolbox capable of advanced modelling of catchment processes (Lindsay, 2005). Although it is powerful software for geomorphometry, TAS is also easy to use, partly owing to its familiar graphical user interface (GUI). This property makes TAS ideally suited to undergraduate and postgraduate education.

TAS was originally developed for the members of the Catchment Research Facilities at the University of Western Ontario to replace a DOS-based land-surface parametrisation program that interfaced with the *RHYSSys hydro-ecological simulation model*. As TAS increased in its spatial analysis and visualisation capabilities, its potential usefulness for a more general audience interested in spatial modelling was obvious. A recent survey of users revealed that approximately 60% of users were members of universities (students and lecturers), with most of the remaining users belonging to government organisations and research institutes. A large

Developments in Soil Science, Volume 33 © 2009 Elsevier B.V.
ISSN 0166-2481, DOI: 10.1016/S0166-2481(08)00016-0.

majority of TAS users claimed research as their intended use, whilst most of the remaining users were interested in the software for educational purposes.

> REMARK 1. *The development of TAS GIS has been driven by two main objectives: the software must satisfy the research needs of scientists while being simple enough in operation to be used for student instruction.*

TAS has been developed using the Visual Basic® 6 programming language. The program has been complied to native code, rather than the slower pseudo-code that many VB programs are distributed with. To further enhance the speed of several TAS functions, the program relies heavily on Windows® Application Programming Interface (API) functions, particularly for graphical operations. One consequence of its VB development is that TAS is limited to operation on IBM PCs running under Microsoft Windows® platforms (i.e. 98, 2000, NT, and XP), unlike LandSerf (Chapter 14) which is platform independent.

Currently, there are no plans to extend usage to other operating systems, partly because of the widespread availability of Windows® emulators. Hardware requirements vary depending on the size of dataset being processed, but the program itself requires approximately 6 MB of RAM and takes 13 MB of disk space. For example, some TAS sub-programs require storing multiple copies, or intermediate steps, of the image in RAM, whilst others only read small blocks of the image into memory.

1.2 Obtaining and installing TAS

TAS is freely available and can be downloaded from the University of Manchester, School of Environment and Development research webpage.[1] At present, the source code is not public domain, unlike GRASS and SAGA (Chapter 12), which are distributed under the GNU General Public License. This is partly because the author wants to retain distribution rights and control over the program's development, although there is interest in fostering collaborations.

TAS does not save property settings to Windows® system files, a feature that greatly simplifies installation of the program. Users simply need to download the TAS main folder to their computer or external drive and the software will execute properly. This characteristic avoids the problems that instructors and students frequently have installing software without administrative rights and makes TAS ideal for instruction. However, when TAS is executed for the first time on a computer, it is necessary to initialise the default settings and the working directory by following these steps:

1. After the TAS main folder has been saved to a computer for the first time, the user must go into the folder and double-click the TAS executable file (`TAS.exe`). The TAS shortcut, which is also contained in the main folder, can be saved to the desktop or quick launch tool bar.

[1] http://sed.manchester.ac.uk/geography/research/tas/.

2. Once in the TAS environment, select *Set Working Directory* under the File menu. Scroll through the directory structure until the directory containing the data to be processed has been found. The Samples folder contained in the TAS main folder serves as the default working directory when the program is loaded onto a computer for the first time. The working directory is the default location for displaying images and all new images that are created are written to this directory. TAS makes several calls to Windows® API functions, and therefore, file names that are longer than 120 characters (including the directory path) can cause the program to error. Thus, if it is necessary to use long file names, it is best if the working directory is high up in the directory structure of the computer (e.g. C:/Baranja_hill/).
3. Select *System Settings*, which is also under the File menu. This window contains several options that affect the way that TAS looks and behaves, including several default display settings. TAS is distributed with numerous system palette files (users can also modify palettes or create custom palettes). The default image palette should be set to an appropriate quantitative palette such as `high_relief` or `soft_earthtones`, which are ideal for displaying DEMs. The default vector palette should be set to a qualitative palette such as `black`.

1.3 First steps in TAS

The TAS GIS environment possesses many of the elements commonly found in Windows® applications, including a menu bar, tool bar, and status bar (Figure 1). Additionally functionality can be accessed through floating tool bars. For example, the *Image Attributes* tool bar, which appears on the left side of the work space (Figure 1) when an image is displayed, is used to query image values, alter the display properties of a displayed image (e.g. the palette, the minimum and maximum displayed values, and hillshading properties), and navigate around a zoomed image. The *Digitise* tool bar is used to create and edit vector data.

DEMs can be created in TAS using either an *inverse-distance to a weight* (IDW) interpolation routine, or a *TIN* algorithm. TAS can be used to pre-process DEMs for hydro-geomorphic analyses (e.g. depression and flat area removal) or more generally using image processing techniques such as filtering. Numerous simple and compound land-surface parameters can be extracted from DEMs and are discussed in greater detail below.

At present, TAS does not contain extensive facilities for modelling climate-related land-surface parameters (e.g. solar radiation indices). The program can extract and perform analyses on stream networks and drainage basins. General GIS analyses (e.g. distance operations, buffering, and clumping) and statistical analyses (e.g. image correlation, semivariogram analysis, and histogram generation) can be performed on images. Most of the analytical functions require raster images although some functions do accept vector coverages. Vector data are generally used to overlay onto raster images to enhance data visualisation and interpretation.

TAS users can get support in a number of ways. The program does have an extensive help function. The TAS website has several resources for learning how

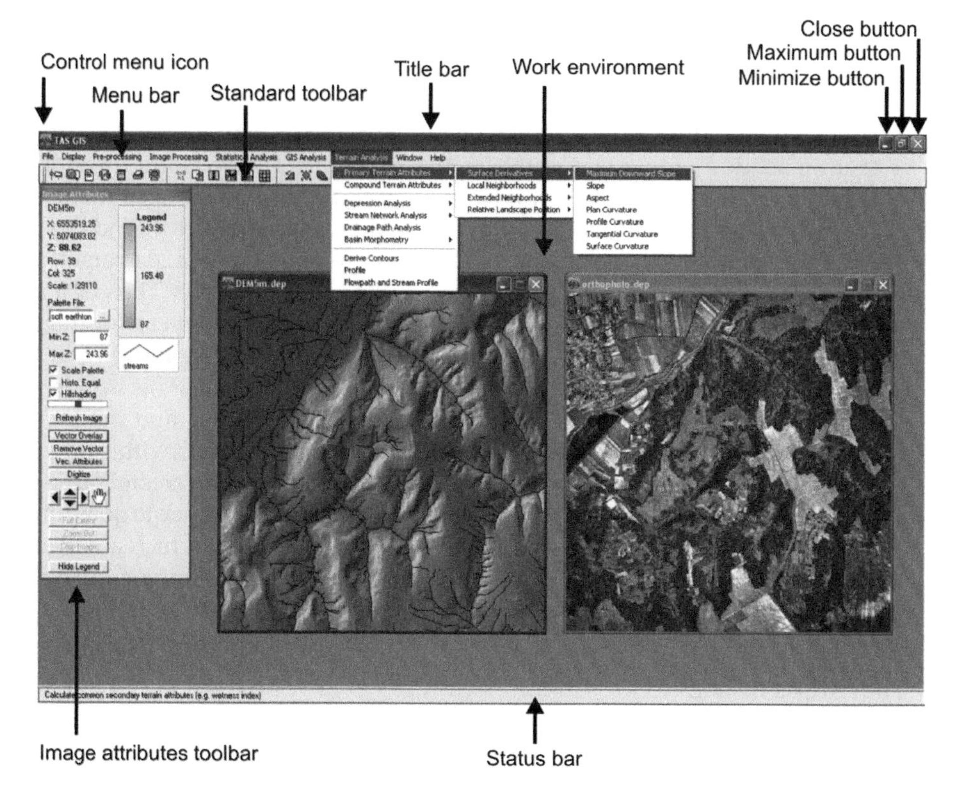

FIGURE 1 The TAS GIS environment.

to use the software, including tutorials, associated data sets, and documentation for the native scripting language. The TAS user forum also provides a means for users to access information about the program, report bugs, provide feedback, and to communicate with the user community. The author receives TAS related email enquires regularly and strives to respond promptly.

2. GEOMORPHOMETRIC ANALYSIS

2.1 Importing and displaying data

DEMs are one the main input data types to TAS but the program does utilise other types of spatial data, including satellite imagery and vector data. The *Import/Export* sub-menu under the File menu offers several facilities for importing and exporting spatial data of various formats. Raster import/export functions include sub-programs to read and write ArcGIS raster formats, IDRISI images, GRASS images (`.asc` format only), Surfer® grids, Ordnance Survey grid files (`.ntf`), device independent bitmaps and the Shuttle Radar Topography Mission (SRTM) DEM format (`.hgt`). TAS reads and writes ArcView Shape files, IDRISI and GRASS ASCII vector files, and delimited XYZ vector point files.

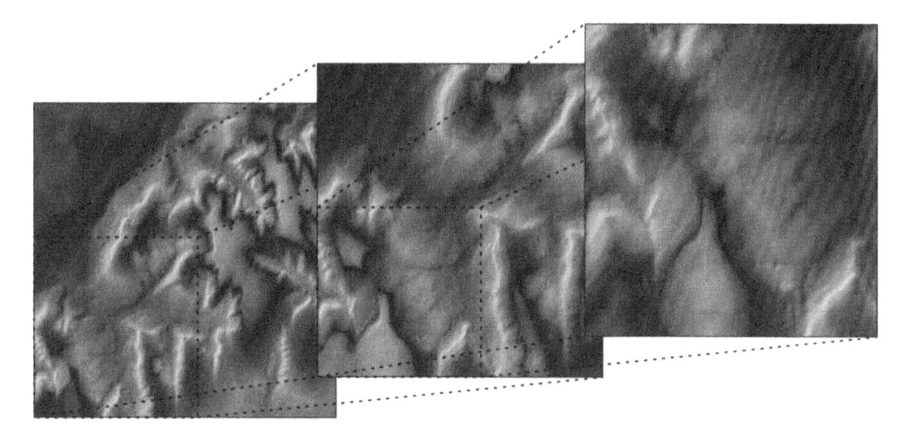

FIGURE 2 TAS can apply a histogram equalisation stretch dynamically as an image is zoomed into. (See page 728 in Colour Plate Section at the back of the book.)

Image data are contained in `.tas` files which are formatted as simple grids (north to south rows and west to east columns) of byte, integer, or single-precision floating point data. Data are stored in the little endian byte order. Image meta-data are contained in separate header (`.dep`) files. When images are created by TAS the program automatically finds the data format that requires the least disk storage. For example, an extracted stream network image is Boolean and is therefore saved in a byte format, whilst a precise DEM may be saved in a floating-point format. Users may also convert between image data formats if the default format is unsuitable. Unlike ArcGIS (Chapter 11) and SAGA (Chapter 12), TAS does not currently accommodate no-data values for raster images. Instead, most sub-programs allow users to specify a mask image to force the program to ignore grid cells beyond the area of interest.

TAS vector files (`.vtr`) contain both meta-data and coverage data within a single file. The co-ordinates of the bounding rectangle of a vector coverage are stored in double-precision format and attribute information is stored in single-precision format. Point and line node co-ordinates are stored in single-precision format and are relative to the minimum X and Y co-ordinates contained in the file header (double-precision). This provides a means of storing precise co-ordinates in a format that requires less disk storage. TAS vector files can contain points, lines, and polygons within the same file. At present, there are no means of storing 'donut-hole' vectors in a TAS vector format.

DEMs can be displayed using several system palettes which are specifically intended for visualising elevation data. Users can either choose to use a linear stretch, scaled between user-defined minimum and maximum values, or a histogram equalisation stretch. Histogram equalisation is applied dynamically, such that when the image is zoomed into, the palette is re-scaled to the displayed data only (Figure 2). This is a useful option for visual interpretation of DEMs, particularly in high-relief areas. For example, notice how ridge-like artifacts, likely to result from the interpolation process, become apparent along rounded hilltops as the Baranja Hill DEM is progressively zoomed into (Figure 2).

To zoom into a displayed image, the user must press the left mouse button and hold the button down while moving the cursor. A dashed white box is drawn on the image outlining the bounding box from where the mouse button was first pressed to the final cursor location. When the left mouse button is released, the image is resized to the bounding box. It is possible to move around the zoomed image by selecting the Pan Tool (the white hand on the Image Attributes tool bar).

Users can refresh the image and zoom out to the original image size by either pressing the right mouse button while the cursor is over the image or the Full Extent button on the *Image Attributes* tool bar. The TAS GUI allows multiple images to be displayed simultaneously, which greatly facilitates visual analysis of multiple parameters. Zooming and navigation operations can be linked between multiple displayed images. Displayed images can also be combined with shaded-relief images to enhance visualisation of terrain. In these composite-relief models, variations in colour correspond to the displayed attribute and tonal variations visualise hill shading (Figure 2). Currently, there are no facilities for 2.5-D visualisation of terrain or fly-though capabilities in TAS. Users that require 2.5-D visualisation may wish to use LandSerf (Chapter 14) or SAGA (Chapter 12) both of which possess extensive DEM visualisation capabilities. The focus of TAS's development has largely emphasised terrain analysis, with visualisation being secondary.

Graphical output (i.e. displayed images with vector overlays) can be saved as Windows® meta-files (`.wmf`), which can be read by most graphics packages and word-processing programs. TAS is not a cartographic package and can not create a cartographically correct map output. Instead, users must import TAS meta-files into a graphics package, or other GIS such as ArcGIS (Chapter 11), for further cartographic editing.

2.2 Deriving land-surface parameters and objects

Several of TAS's algorithms are recursive. These algorithms are generally very efficient, but can encounter problems with larger sized DEMs possessing very long flow-paths because they rely on stack memory. Many of these algorithms perform pre-processing, or have options to perform pre-processing, to ensure that stack memory is not exceeded. Most, although not all, of the TAS analytical algorithms are RAM intensive sub-programs. These sub-programs store one or more images in memory rather than continually reading from and writing to the hard disk. This allows for quicker running operations but does restrict the size of DEM that can be processed, depending on the available memory of the user's computer.

Users are encouraged to analyse data with the smallest possible spatial extent, i.e. to crop DEMs to the extent of the study basin. TAS's *Crop To Object* sub-program can be useful for eliminating unnecessary data beyond an area of interest.

2.2.1 DEM pre-processing

TAS GIS contains an extensive toolbox for the processing and analysis of digital elevation data. There are several sub-programs for DEM pre-processing, including algorithms for removing topographic depressions and flat areas by filling, breaching, the impact reduction approach (Lindsay and Creed, 2005), and selectively

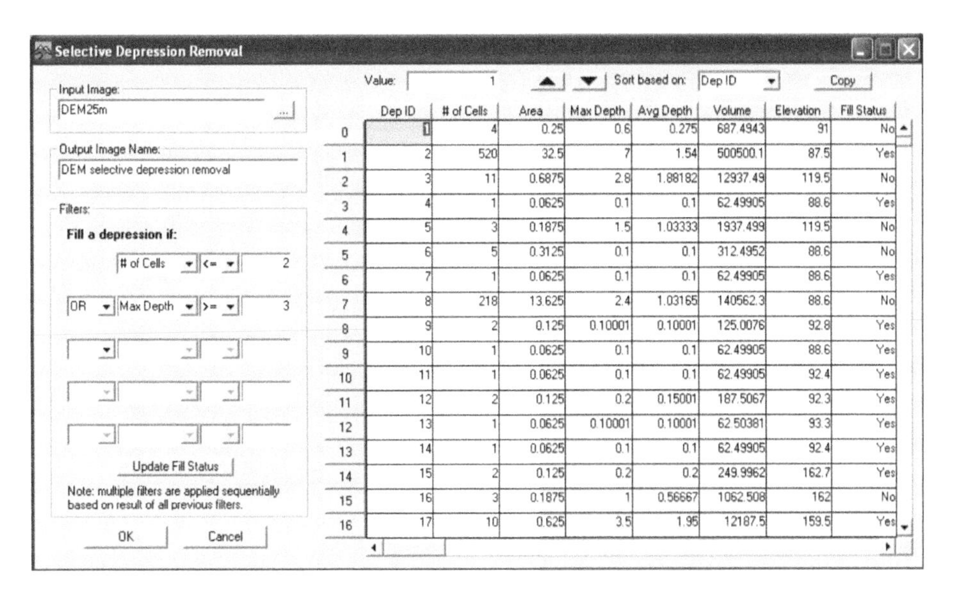

FIGURE 3 The selective depression removal dialog box.

filling based on depression characteristics. Figure 3 shows TAS' selective depression removal dialog box applied to the Baranja Hill 25 m SRTM DEM.

Users are able to review the morphometrics associated with individual depressions and to selectively fill depressions based on thresholds in the number of cells, depression area, volume, maximum or average depth, or elevation. The SRTM DEM contains 41 depressions, many of which are likely to be artifacts. Nonetheless, at least one of the depressions can be confirmed by the presence of marshland in the 1:5000 topomap. The selective depression removal algorithm provides a means of removing depressions that are clearly artifacts whilst retaining actual topographic depressions, which can significantly affect hydrological processes in a region. Other DEM pre-processing operations in TAS include *cropping, burning* streams, and modifying individual or groups of grid cell elevations. An image's datum can be changed and co-ordinate transformations can also be performed.

2.2.2 Deriving land-surface parameters

After a DEM has been satisfactorily pre-processed for the specific application, it is possible to derive numerous simple and compound (i.e. primary and secondary) land-surface parameters. TAS can be used to calculate surface derivatives (e.g. slope, aspect, and curvatures), indices related to local neighbourhoods (e.g. flow direction and number of upslope neighbours) and extended neighbourhoods (e.g. mean upslope elevation and viewsheds), relative landscape position (e.g. elevation relative to local peaks and pits), and compound indices (e.g. wetness index and relative stream power index). Each of the land-surface parameters can be accessed from the *Primary Terrain Attributes* and *Compound Terrain Attributes* sub-menus of the Terrain Analysis menu. Figure 4 shows several parameters that have been derived from the Baranja Hill 25 m SRTM DEM.

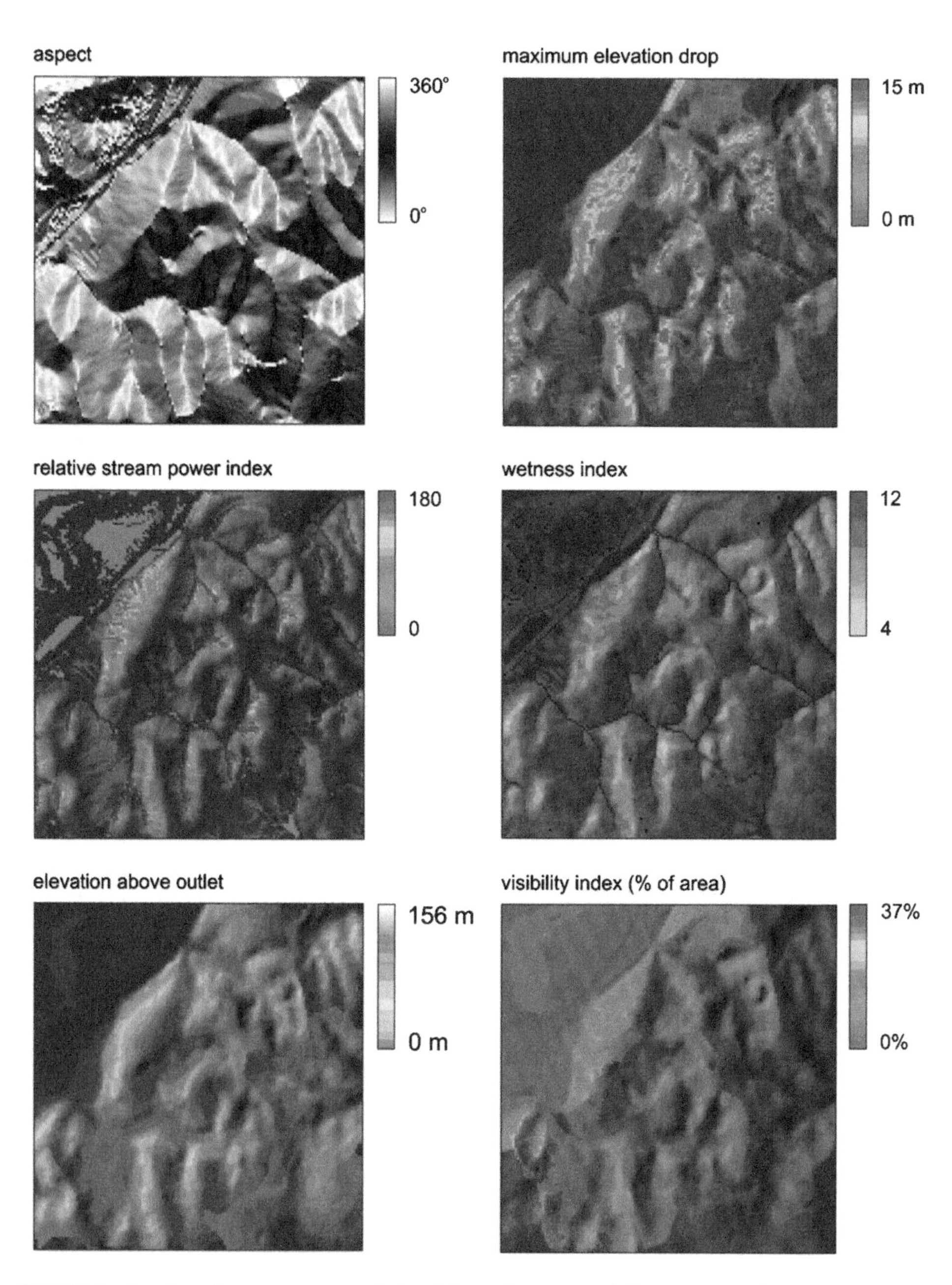

FIGURE 4 Land-surface parameters derived from the Baranja hill SRTM DEM. (See page 729 in Colour Plate Section at the back of the book.)

2.2.3 Land-surface parameters related to flow-paths and stream networks

All land-surface parameters related to flow-paths and stream networks require a DEM that has been pre-processed to remove artifact depressions and flat areas. Although TAS can calculate flow direction and flow accumulation (upslope area) using one of seven flow algorithms, most of the functions that involve tracing flow-paths to calculate land-surface parameters (e.g. downslope flow-path length and watershed delineation) use the steepest descent (O'Callaghan and Mark, 1984), or D8, flow algorithm. This is because many functions assume that there is a unique flow-path connected to each grid cell in a DEM. Flow divergence is not permitted in these cases. For example, it is assumed that there is only one value of downslope flow-path length for each grid cell.

The alternative flow algorithms that are available in TAS (e.g. D∞, FD8, and ADRA[2]) are generally used to calculate more complex land-surface parameters (e.g. the wetness and stream power indices) as inputs to environmental simulation models.

Because stream network analysis algorithms (e.g. Strahler stream ordering) require flow-path tracing, each of these algorithms also use the D8 flow algorithm to route downstream. In TAS, stream networks are single-cell wide raster networks, and therefore, there is one unique flow-path connecting each point in the network to the outlet. Each of the stream network analysis algorithms require a pre-processed DEM and a DEM-extracted stream network as inputs. The DEM is used for routing, with the D8 flow direction grid calculated internally, and the stream image is used as a mask.

Most of the stream network analysis algorithms travel downstream from *channel heads*, passing through each link and bifurcation in the network until an outlet node is finally reached. Channel heads are identified as stream grid cells with no inflowing cells belonging to the stream network. Bifurcations in the network are identified as cells with more than one inflowing stream cell. Figure 5 shows the results of several of TAS' stream network analysis algorithms applied to a network derived from the Baranja Hill 25 m SRTM DEM. The DEM was pre-processed to remove artifact topographic depressions and flat areas using the *Fill all depressions* sub-program (located in the Remove Depressions sub-menu of the Pre-processing menu). This sub-program is capable of simultaneously enforcing flow on flat areas.

The *main channel* algorithm identifies the main channel for each stream network in an area by identifying which link has the largest contributing area at bifurcations (Figure 5). Thus, it assumes that contributing area can be used as a surrogate for discharge, a common assumption in the field of geomorphometry.

In addition to spatial outputs, such as those displayed in Figure 5, TAS can calculate numerical stream network morphometrics, which are out in a textual or chart form. For example, the number of interior and exterior *stream links*, *Horton ratios* (i.e. the bifurcation, length, area, and slope ratios), *drainage density*, and the *network width* function can each be estimated for stream networks.

[2] Adjustable Dispersion Routing Algorithm — see Lindsay (2003) for more info.

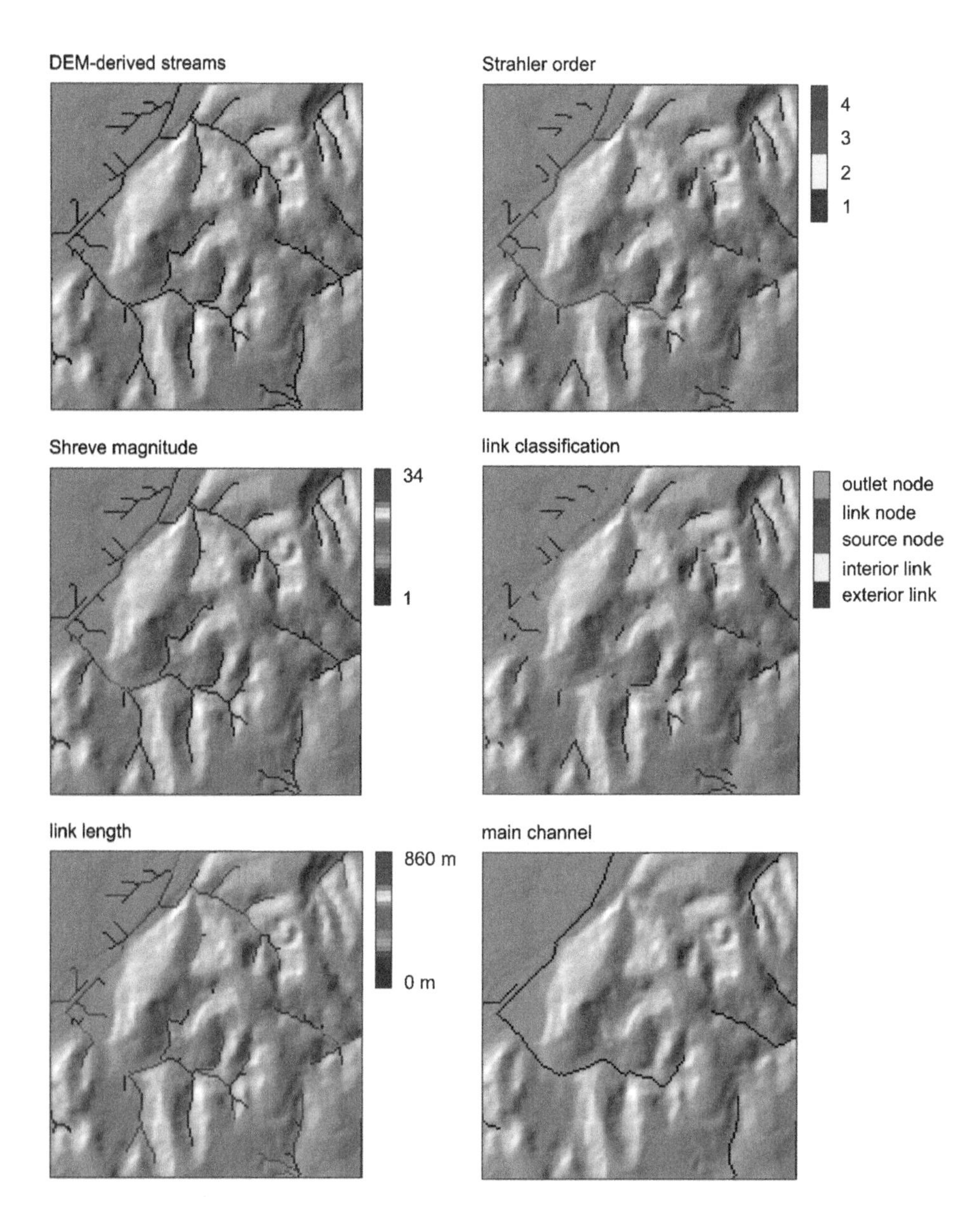

FIGURE 5 Stream morphometrics calculated for a stream network derived from the Baranja Hill DEM. (See page 730 in Colour Plate Section at the back of the book.)

2.2.4 Extracting watersheds and basin morphometrics

TAS possesses a sophisticated sub-program for delineating watersheds, accessed from the Extended Neighbourhoods sub-menu. The user must specify a pre-processed DEM (i.e. artifact depressions and flats removed) and provide points of interest for which to extract watersheds.

Watersheds can be mapped based on user-defined co-ordinates, digitised points, or a seed point image. A stream network image can be used to identify *sub-basins* (areas draining to each link in the network), hillslopes (areas draining to either side of a link), *Strahler order basins*, and *Shreve magnitude basins* (Figure 6). Additionally, users can partition a landscape into a collection of basins of a similar user-defined size in a way that minimises the variation in basin areas (i.e. *isobasins*).

Users are also able to calculate 14 common basin shape and relief indices including the *form factor, basin shape, length-area, circularity ratio, elongation ratio, lemniscate ratio, maximum relief, divide-averaged relief, relief ratio*, and *relative relief*. Each of these shape and relief indices can be calculated using the *Shape and Relief Indices* sub-program located within the Basin Morphometry sub-menu of the Terrain Analysis menu. Additionally, it is possible to perform a hypsometric (i.e. area-relief) analysis and to calculate the hypsometric integral of a basin.

2.2.5 Landform classification in TAS

TAS can perform automated landform classification using the crisp classification scheme of Pennock et al. (1987) (Figure 7). Each of the seven classes used in this scheme are entirely based on measures of local slope and curvature. As such, the method is most appropriate for use with smooth DEMs. Thus, in the example shown in Figure 7 the Baranja Hill 5 m DEM was filtered using a 21×21 mean filter before applying the Pennock classification scheme. The appropriate size of the low-pass filter used to smooth the DEM is dependent on the degree of generalisation in the landform classification that is desired, which is actually an issue of relevant scale. Additionally, it is possible to apply user-defined fuzzy classification schemes based on measures of relative landscape position and other land-surface parameters.

2.3 The Raster Calculator and scripting in TAS

In the lower left-hand side of TAS' Raster Calculator there is a listbox that contains the names of several functions (Figure 8). These are the same operations that are called when functions are accessed through the menu structure of the TAS GUI. When a function name is selected from the Raster Calculator listbox, text appears in the box occupying the bottom of the Raster Calculator (Figure 8). This text describes the syntax that is used to call the selected function using the Raster Calculator.

Each function's syntax follows the pattern:

```
KEYWORD(parameter1, parameter2, parameter3...)
```

in which the function's keyword is typed in capital letters followed by a series of parameters in brackets. For example, the syntax for the function that removes short

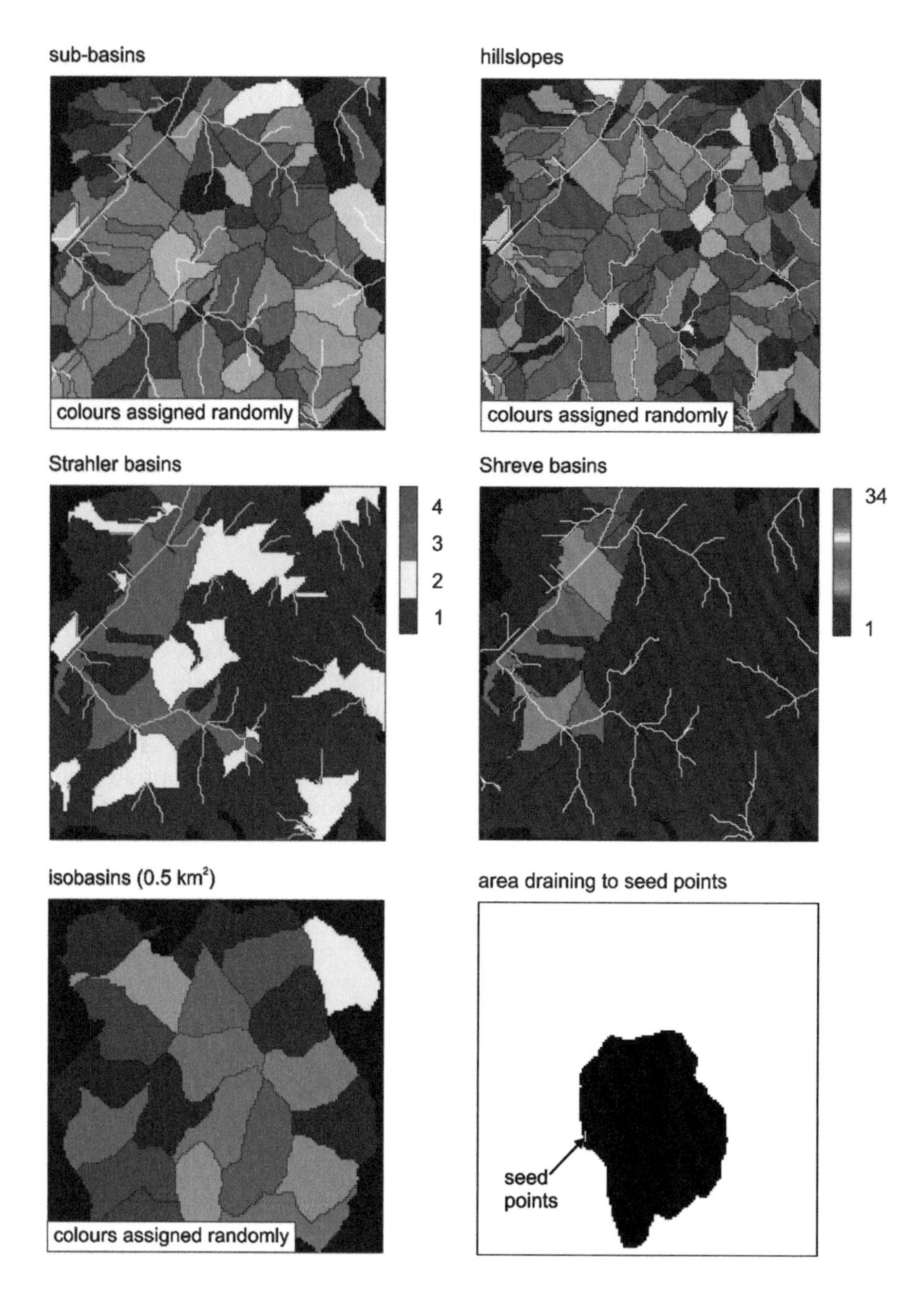

FIGURE 6 Various means of extracting watersheds for the Baranja Hill DEM. (See page 731 in Colour Plate Section at the back of the book.)

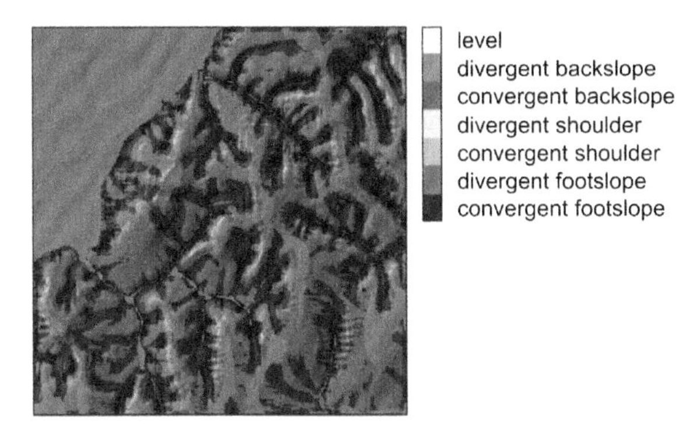

FIGURE 7 Automated landform classification of the Baranja Hill 25 m SRTM DEM, based on the crisp classification scheme of Pennock et al. (1987). The DEM was pre-processed by running a 21×21 mean filter to remove fine-scale topographic variation. (See page 732 in Colour Plate Section at the back of the book.)

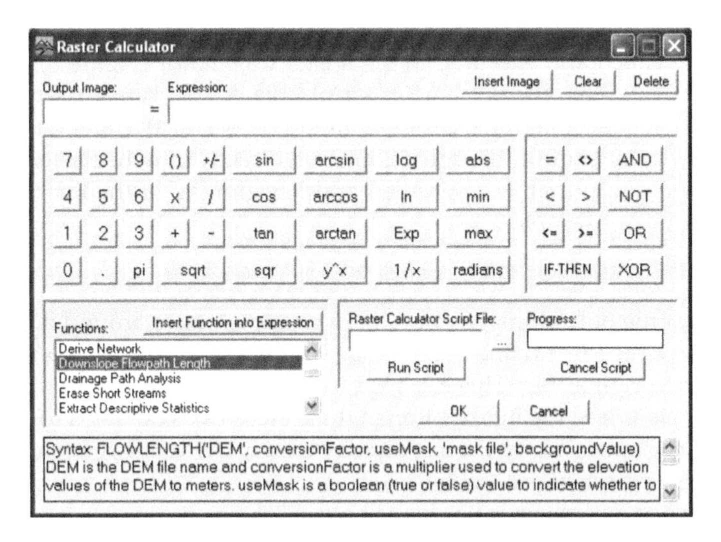

FIGURE 8 TAS' Raster Calculator.

streams from a drainage network, a task commonly performed for cartographic reasons, is:

```
ERASESTREAMS('streamImage', 'DEM', conversionFactor, minLen)
```

Generally, use of spatial analysis functions in the Raster Calculator follows the same conventions as other mathematical or logical operations. Thus, image names, such as `streamImage` in the above example, are always enclosed by apostrophes and must be located in the working directory. Output images are always saved in the working directory. Parameters are separated by commas. The *Syntax Box* on the Raster Calculator gives a description of each of the parameters for the selected function and also provides one or more examples of usage.

It may not be immediately obvious why a user would want to use the Raster Calculator to call a function rather than accessing the corresponding sub-program through TAS' menu structure. It can however be considerably quicker to insert a function and a few parameters into the Raster Calculator than to find the relevant sub-program through the menu structure and enter all of the required information into the dialog box. This is particularly true when several function must be performed in series, i.e., if there are several intermediate steps before arriving at the final answer.

When a lengthy procedure must be performed, scripts can be used such that the Raster Calculator executes each step consecutively without the user's input, i.e. in a batch mode. Scripting is useful when a procedure must be executed again in the future, perhaps in a slightly modified form, e.g. changing an input file name or a parameter value. Scripts enable users to automate complex, repetitive, time-consuming, and common tasks. A TAS script file is a text file with an `.rcs` extension. Script files can be written in any text editor, including TAS' text editor, although they must be saved with the `.rcs` extension.

Scripts are called and executed in the Raster Calculator (Figure 8). Comments are preceded by the characters `//` in TAS scripts. Blank lines can be used to separate blocks of similar code, making it easier to interpret a script at a later date. Each line in a script works the same as though it is entered directly into the Raster Calculator, except that the output name is specified at the beginning of the line followed by an equals sign. For example:

```
New DEM=FILTER('Old DEM',mean,5)
```

Notice that the output image does not have apostrophes around it. The output image can have the same name as an image specified in the script line; TAS simply overwrites the original file. This can be a useful property when there are several intermediate steps and the information in those steps does not need to be retained. If an output file specified in a script already exists, TAS will overwrite it without warning when the script is executed.

The following script example shows how a TAS script can be used to calculate complex parameters, in this example a multi-scale landscape position index:

```
//This script calculates a multi-scale landscape position index:
DEM='DEM5m'*1
//Renames the DEM so the script can be easily reused with a different DEM,
min=Filter('DEM',minimum,11,circular)
//Performs an 11×11 minimum filter on DEM,
max=Filter('DEM',maximum,11,circular)
//Performs an 11×11 maximum filter on DEM,
relief='max'-'min'
relief=if('relief'=0,(-1),'relief')
//Ensures there is no division by zero,
EPR 11x11=('DEM'-'min')/('relief')*100
min=Filter('DEM',minimum,101,circular)
```

```
//Performs a 101×101 minimum filter on DEM,
max=Filter('DEM',maximum,101,circular)
//Performs a 101×101 maximum filter on DEM,
relief='max'-'min'
relief=if('relief'=0,(-1),'relief')
//Ensures there is no division by zero,
EPR 101x101=('DEM'-'min')/('relief')*100
//This next block reclasses the EPR images into high, medium, and low local positions.
temp1=RECLASS('EPR 101x101',UDC,10,0,33,20,33,66,30,66,101)
//Classes are 10, 20 & 30
temp2=RECLASS('EPR 11x11',UDC,1,0,33,2,33,66,3,66,101)
//Classes are 1, 2 & 3
Relief Index='temp1'+'temp2'
//Sums the two reclassed images
```

The first two main blocks of the script calculate the Elevation as a Percentage of local Relief (EPR) at two different scales (i.e. using an 11×11 filter and then a 101×101 filter). In the next block of the script, the local and meso-scale EPR images (EPR 11×11 and EPR 101×101) are each reclassed into low (0–33%), medium (33–66%), and high (66–100%) classes of landscape position. Class values are assigned such that when the reclassed images are finally summed in the last line of the script, the information at the local and meso-scale is preserved. Figure 9 shows the two EPR images as well as the final output of this script.

TAS scripts are also very useful for assessing the uncertainty in land-surface parameters and other DEM-derived stream and basin geomorphometry. The following script uses the Monte-Carlo method, specifically an unconditional simulation, to assess uncertainty in the boundaries of the area draining to a small group of seed points in the Baranja Hill 25 m SRTM DEM:

```
//This script assesses the uncertainty in watershed boundaries due to elevation error:
DEM='DEM25m'*1
//Renames the DEM so the script can be easily reused with a different DEM,
//Initialise some images for later use,
counter='DEM'*0+1
watershed total='DEM'*0
counter='counter'+1
random field=RANDOM('DEM',uniform,0,1,0)
//Creates a random field,
temp=FILTER('random field',gaussian,15,circular)
//Increases the spatial autocorrelation,
random field=RESCALETOCDF('temp',normal_0_5)
//Ensures the field has a normal spatial distribution with a mean of 0 and SD of 5 m,
new DEM='DEM'+'random field'
new DEM filled=DEPFILL('new DEM',1,true)
temp=WATERSHED('new DEM filled',1,'seed point')
watershed total='watershed total'+'temp'
watershed prob='watershed total'/'counter'
REPEAT 999 TIMES
counter='counter'+1
random field=RANDOM('DEM',uniform,0,1,0)
```

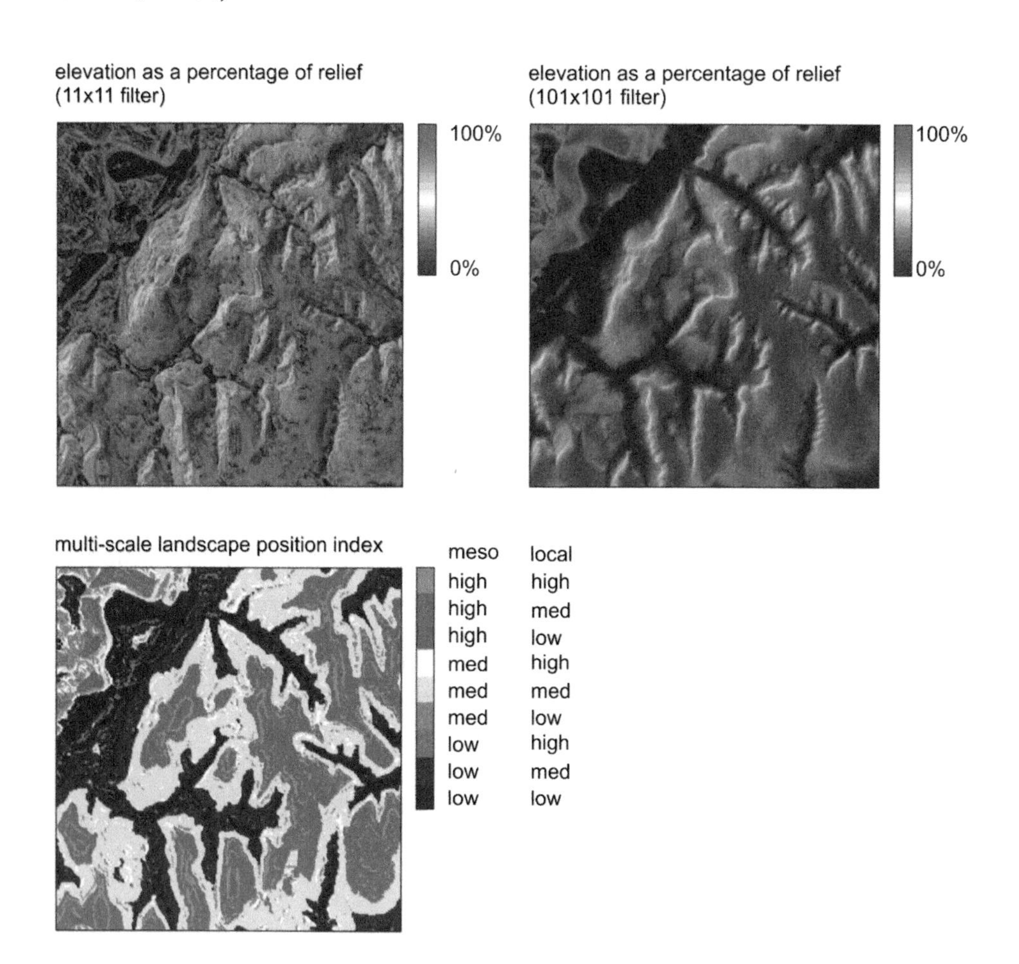

FIGURE 9 Elevation as a percentage of local relief (EPR) calculated using an 11×11 (a) and a 101×101 (b) filter and a multi-scale landscape position index (c). Images have been derived from the sample script applied to the Baranja Hill 25 m SRTM DEM. (See page 732 in Colour Plate Section at the back of the book.)

```
temp=FILTER('random field',gaussian,15,circular)
random field=RESCALETOCDF('temp',normal_0_5)
new DEM='DEM'+'random field'
new DEM filled=DEPFILL('new DEM',1,true)
temp=WATERSHED('new DEM filled',1,'seed point')
watershed total='watershed total'+'temp'
temp='watershed total'/'counter'
temp2=IF(MAD('temp','watershed prob',savedTextAppend,simulation
results)<=0.0001,stopScript,continueScript)
watershed prob='watershed total'/'counter'
END REPEAT
```

The second last block of the script represents the first iteration of the simulation. The first portion of this block is dedicated to creating a random field, with

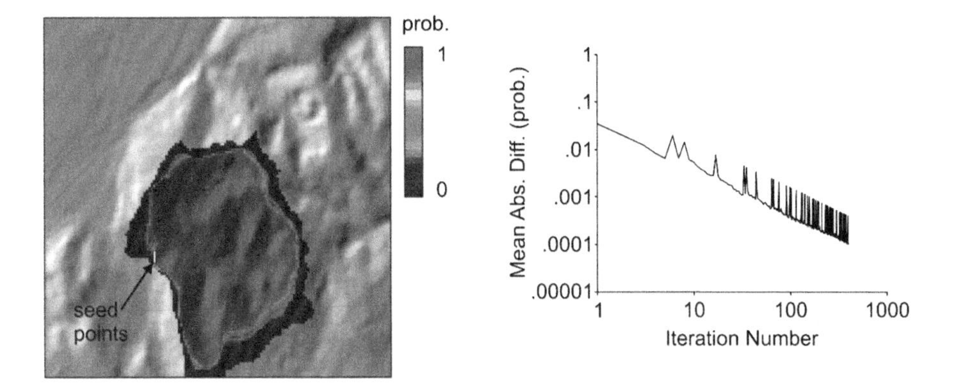

FIGURE 10 Results of a Monte-Carlo uncertainty analysis of the watershed area of a group of seed points in the Baranja Hill 25 m SRTM DEM. (See page 733 in Colour Plate Section at the back of the book.)

the desired spatial properties and statistical distribution. This random field is then added to the original DEM to create a new realisation of the elevation surface. Note that this is not the only means of producing a DEM realisation but it is an efficient method.

In the final portion of this script block, the watershed of the specified group of seed points is defined for the DEM realisation. The last block of the script performs the same tasks as the previous block. However, it is enclosed in a `REPEAT` structure that iterates the procedure another 999 times. The line (which should appear on a single line in the script file):

```
temp2=IF(MAD('temp', 'watershed prob', savedTextAppend,
simulation results) <= 0.0001, stopScript, continueScript)
```

acts as a stopping condition for the simulation. That is, the simulation will continue for `999` iterations unless the mean absolute difference (`MAD`) between the watershed probability image in the current iteration (`temp`) and the previous iteration (`watershed prob`) is less than some small, pre-defined value (`0.0001`). If this condition occurs, the simulation is deemed to have stabilised and the `REPEAT` procedure is stopped prematurely using the stopScript keyword.

This statement also records the value of the `MAD` for each iteration in a text file called `simulation results.txt`. This file can be useful for evaluating the stabilisation of the simulation. Figure 10 shows the image of uncertainty in watershed delineation for the above simulation procedure. The simulation ended after 397 iterations, after which point the probability image was deemed to have stabilised.

Unfortunately, not all of TAS' functions are accessible through scripting in the current version of the software, although a substantial number of functions are available. The author is currently working to resolve this issue. Additionally, TAS' scripting language will eventually be supplemented by a graphical spatial modeler of the kind that are found in other GIS packages (e.g. ArcGIS's ModelBuilder and IDRISI's Macro Modeler).

3. SUMMARY

TAS is a freely available GIS for Windows® operating systems that has been developed to analyse digital elevation data. The software is predominantly used for academic research and education in universities worldwide. Interest in TAS as a research tool reflects its substantial capabilities for modelling catchment processes. The software contains a very extensive lists of land-surface parameters, beyond that which is generally available in commercial GIS packages. Several stream network and basin morphometrics can also be calculated from DEM-extracted streams and watersheds. Table 1 lists the primary and secondary land-surface parameters that can be extracted from DEMs using TAS. Many of these parameters are not commonly available in GIS packages and several are only available in TAS. References are given for algorithm descriptions where appropriate.

TABLE 1 Land-surface parameters derived by TAS GIS

Land-surface parameter	Description
Slope	Slope gradient (Zevenbergen and Thorne, 1987).
Maximum downward slope	Slope to the lowest neighbour in a 3×3 window (Burrough and McDonnell, 1998).
Aspect	Direction of max downward gradient (Zevenbergen and Thorne, 1987).
Plan curvature	Along-slope curvature (Zevenbergen and Thorne, 1987).
Profile curvature	Downslope curvature (Zevenbergen and Thorne, 1987).
Tangential curvature	Curvature in an inclined plane (Mitášová and Hofierka, 1993).
Surface curvature index	Index of total curvature within a group of grid cells (Blaszczynski, 1997).
Shaded relief	Hill shaded image (Horn, 1981).
Flow direction	Direction of flow calculated using one of seven flow algorithms.
Number of downslope neighbours	Number of neighbours in a 3×3 window that are of lower elevation.
Number of upslope neighbours	Number of neighbours in a 3×3 window that are of higher elevation.
Number of inflowing cells	Number of neighbours in a 3×3 window that flow into the centre cell.
Maximum downslope elevation change	Maximum elevation drop to a neighbour in a 3×3 window.

(continued on next page)

TABLE 1 *(continued)*

Land-surface parameter	Description
Average downslope elevation change	Average elevation drop to a neighbour in a 3×3 window.
Local elevation percentile	Elevation percentile within a user-specified window.
Difference from mean elevation	Difference from mean elevation in a 5×5 window.
Standard deviation of elevation	Standard deviation of elevations in a 5×5 window.
Valley bottoms	Grid cells for which the steepest downslope neighbour has a steepest upslope that points back to the grid cell. This will only occur along narrow valley bottoms.
Catchment/dispersal area	Spatial pattern of contributing/dispersal area derived either the D8, Rho8, FD8, FRho8, D8, FD8-Quinn, or ADRA flow algorithm.
Watershed	Areas draining to digitised/inputted points. Also sub-basins draining to each link in a network, Strahler or Shreve basins, hillslopes, and isobasins.
Downslope flowpath length	Downslope distance along flowpath to outlet.
Maximum upslope flowpath length	Maximum flowpath distance to ridge.
Minimum upslope flowpath length	Minimum flowpath distance to ridge.
Average upslope flowpath length	Average flowpath distance to ridge.
Average upslope elevation	Average elevation of contributing area.
Average upslope slope	Average slope of contributing area.
Edge contamination	Cells that are connected to an edge cell by a flowpath.
Drainage network	A vector network indicated the flowpath, based on the D8 algorithm, for each grid cell in a DEM.
Viewshed	Area that can be seen from a point (Franklin and Ray, 1994).
Visibility index	Spatial pattern of the number of visible cells (Franklin and Ray, 1994).
Elevation above pit cell	Absolute elevation above nearest downslope depression cell.
Elevation relative to peaks and pits	Elevation relative to nearest upslope peak and downslope depression cell.

(continued on next page)

TABLE 1 *(continued)*

Land-surface parameter	Description
Elevation relative to minimum and maximum	Elevation relative to DEM minimum and maximum.
Elevation relative to channel and divide	Elevation relative to nearest channel and divide cells.
Elevation above target cell	Absolute elevation above nearest downslope target cell.
Wetness (topographic) index	Beven and Kirkby (1979) TOPMODEL index.
Stream power index	Spatial pattern of stream power (Moore et al., 1991a).
Sediment transport capacity	Spatial pattern of erosion/deposition (Moore et al., 1991a).
Mass accumulation	Mass routed throughout a basin with loading and efficiency terms.
Network wetness index	Minimum downslope wetness index along flowpaths (Lane et al., 2004) and/or areas of possible overland flow re-infiltration.
Landform classification	Landform classes based on the Pennock et al. (1987) scheme.
Depth in sink	Elevation difference between the surface and depression outlet elevation (Antonić et al., 2001a).

TAS is easy to install and simple to use, which makes it ideal for educational purposes and for use by non-experts. Development of TAS has been directed largely by feedback provided by the user community and is ongoing. Most of the bugs that occurred in early versions of TAS have been fixed in later versions, however, as with most complex programs, some will always persist. Future development plans include enhanced support for vectors, incorporation of attribute tables, in-process help, support for no-data values, 2.5-D terrain visualisation, and increased documentation and user support.

IMPORTANT SOURCES

Lindsay, J.B., 2005. The Terrain Analysis System: a tool for hydro-geomorphic applications. Hydrological Processes 19 (5), 1123–1130.

Lindsay, J.B., 2003. TAS Tutorial. The University of Manchester, Manchester, 42 pp.

http://sed.manchester.ac.uk/geography/research/tas/ — TAS home page.

CHAPTER 17

Geomorphometry in GRASS GIS

J. Hofierka, H. Mitášová and **M. Neteler**

how to set-up GRASS GIS · computing DEMs from various data sources · local and regional land-surface parameters · land-surface modelling and applications · DEM quality analysis · GRASS command examples with online database

1. GETTING STARTED

GRASS (Geographic Resources Analysis Support System) is a general-purpose Geographic Information System (GIS) for the management, processing, analysis, modelling and visualisation of many types of georeferenced data. It is Open Source software released under GNU General Public License (GPL, see http://www.gnu.org) and as such it provides a complete access to its source code written in ANSI C programming language. The main component of the development and software maintenance is built on top of highly automated web-based infrastructure sponsored by OSGeo Foundation, http://grass.osgeo.org in Trento, Italy with numerous worldwide mirror sites. This chapter is based on GRASS 6.2 version available for all commonly used operating systems. It includes 2D raster and 3D voxel data support, a new topological 2D/3D vector engine and capabilities for vector network analysis. Attributes are managed in a SQL-based DBMS.

1.1 Installing and running GRASS GIS

Complete information about GRASS GIS features, software installation and usage can be obtained from the GRASS homepage (http://grass.osgeo.org). Neteler and Mitášová (2008) provide detailed information about the use of GRASS including land-surface modelling and analysis, and various tutorials in several languages are available at http://grass.osgeo.org/gdp/tutorials.php. GRASS binaries for different architectures, source code, as well as the user's and programmer's manuals can be downloaded from the GRASS homepage: http://grass.osgeo.org/download/.

Developments in Soil Science, Volume 33 © 2009 Elsevier B.V.
ISSN 0166-2481, DOI: 10.1016/S0166-2481(08)00017-2.

TABLE 1 GRASS commands naming convention

Prefix	Functional group	Example command
d.*	display, query	d.what.rast
r.*	2D raster	r.watershed
r3.*	3D raster (voxel)	r3.mapcalc
i.*	imagery	i.rectify
v.*	2D/3D vector	v.net
g.*	general	g.remove
ps.*	postscript maps	ps.map
db.*	database	db.select

The easiest way to learn GRASS is to start with an existing, ready-to-use data set. Several are available at the GRASS web site.[1] At geomorphometry.org, we provide the GRASS database for the Baranja Hill dataset used in this chapter along with the shell script file containing all GRASS commands used to produce the figures shown here and perform the described analysis.

GRASS data are stored in a directory referred to as *database* (also called `GISDBASE`), in our case the directory `grassdata`. Within this *database*, the projects are organised by *locations* (subdirectories of the *database*): the provided data set is therefore a *location* called `baranja`. It is important to know that each *location* is defined by its coordinate system, map projection and geographical boundaries. Each *location* can have several *mapsets* (subdirectories of the *location*) that are used to subdivide the project into different topics, subregions, or as workspaces for individual team members. Each *mapset* includes subdirectories for raster and vector data, attribute data and a working (current) spatial extent definition file `WIND`; all these subdirectories and files are hidden from the user. When defining a new *location*, GRASS automatically creates a special *mapset* called `PERMANENT` which is used to store the core data, default spatial extent and coordinate system definitions.

GRASS has over 350 modules, so it is helpful to get familiar with its naming convention — it is very intuitive, as shown in Table 1. The prefix indicates the functional group (type of operation that the command performs), the word after the dot describes what the command does or what type of data it works with.

After downloading and installing GRASS and the test data set, GRASS 6.2 can be launched from either the menu or from a terminal window by typing:

```
grass62
```

A GRASS startup window will open as shown in Figure 1. The path to the *database* goes into the first field, then `baranja` is selected as *location* and `topobook` as *mapset*. After clicking the *Enter GRASS* button at bottom left, basic information about the GRASS version and help access appears followed by the GRASS shell prompt and the GUI display manager (`gis.m`, see Figure 2). GRASS commands

[1] http://grass.osgeo.org/download/data.php.

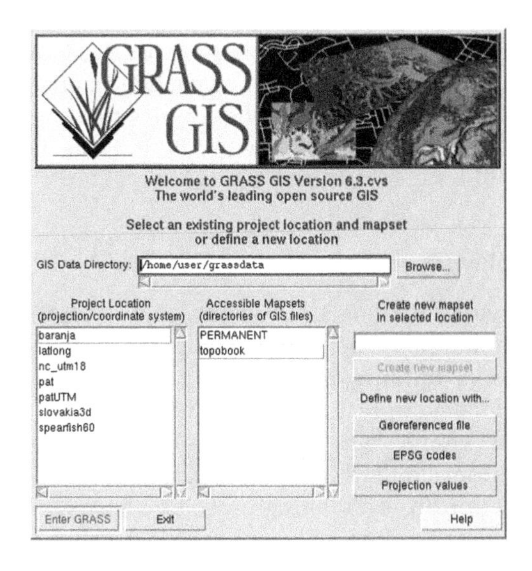

FIGURE 1 GRASS 6 startup screen with selection of database, location and mapset.

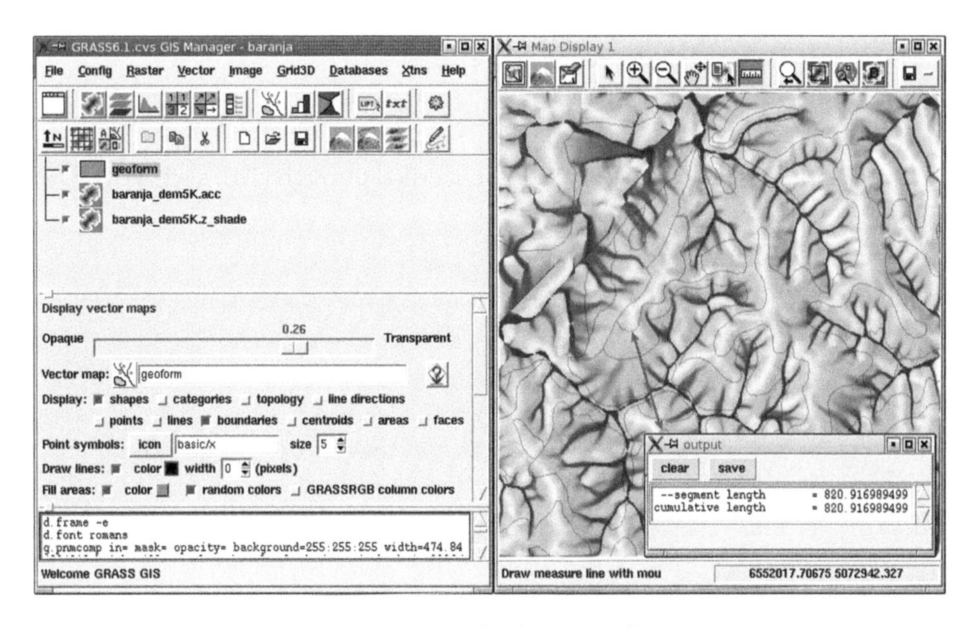

FIGURE 2 GRASS 6 graphical user interface `gis.m`.

can be entered either via GUI, or by typing the commands directly in the UNIX shell window. Execution of tasks that require a sequence of GRASS commands or operating system procedures can be automated by writing a shell script; a feature that is similar to ARC Macro Language or Avenue of ESRI GIS products.

GRASS includes a set of scripts, which behave like standard GRASS modules. In the next sections we provide a couple of examples that can be used directly in GRASS via the UNIX shell. Lines starting with a '#' indicate a comment that is not interpreted by the shell.

1.2 Importing, displaying and computing DEMs

Grid-based DEMs in various formats can be imported using the `r.in.gdal` command (refer to its manual page for the list of supported formats). Elevation data represented by digitised contours or measured points can be imported using the `v.in.ogr` command that supports numerous vector formats while `v.in.ascii` is used for data given as an ASCII list of (x, y, z) coordinates. Very dense ASCII point data, such as those acquired by LiDAR, can be directly converted to raster using `r.in.xyz` that performs a binning procedure based on different statistical measures (min, max, mean, range, etc.). For example, the data used in this book can be imported as follows:

```
# import contours from a SHAPE file
v.in.ogr -o dsn=contours5K.shp output=contours5K
# import raster DEM in Arc ASCII GRID format
r.in.arc input=DEM25m.asc output=DEM25m
# import Landsat imagery in LAN format
r.in.gdal -o input=bar_tm.lan output=bar_tm
```

Grid-based DEMs can be displayed as 2D raster maps and as 3D views (we use here command line but viewing is best handled through GUI, such as `gis.m`; `nviz`):

```
# zoom to raster map
g.region rast=DEM25m

# display 2D raster DEM
d.mon x0
d.rast DEM25m

# display shaded 2D raster DEM
r.shaded.relief DEM25m
d.rast DEM25m_shaded
d.his h=DEM25m i=DEM25m_shaded

# display 3D views
nviz elev=DEM25m vect=contours5K
```

In addition to the internal display tools, GRASS data can be viewed using external programs such as `QGIS` (http://www.qgis.org, see Figure 3) for 2D maps and `paraview` (http://www.paraview.org) for 3D visualisation. `QGIS` can read

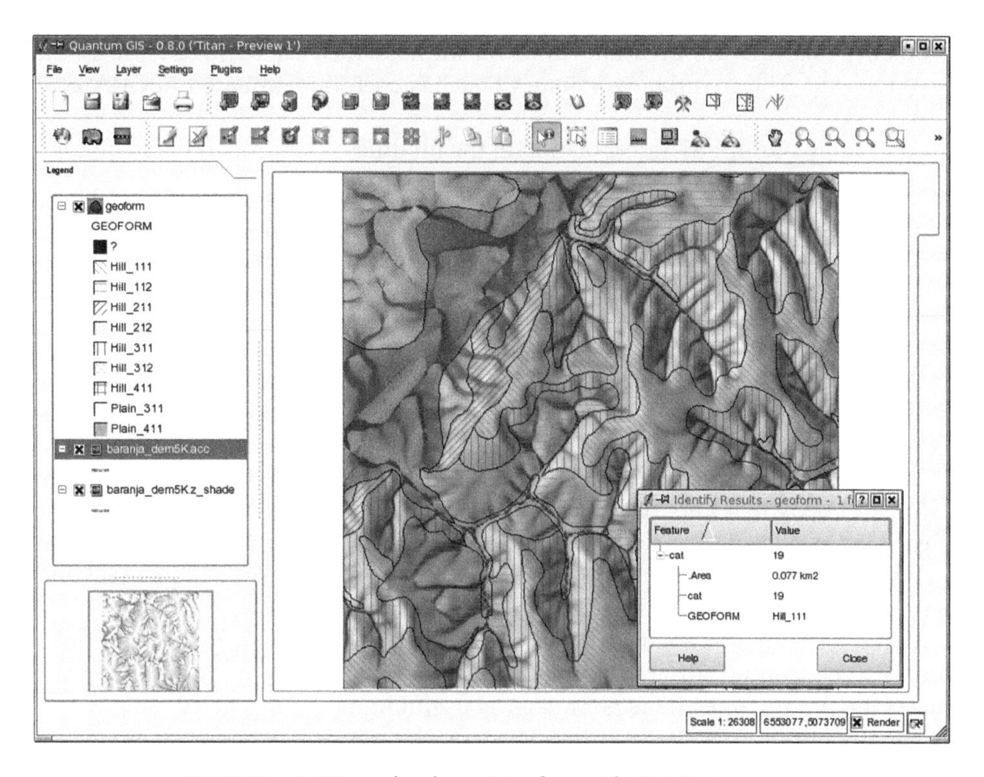

FIGURE 3 QGIS graphical user interface with GRASS 6 support.

the GRASS raster and vector data directly. Its GRASS plugin also offers a toolbox, providing GUI access to important GRASS commands for data analysis. The `r.out.vtk`, `r3.out.vtk` and `v.out.vtk` commands are used to export raster and vector data into VTK format readable by `paraview`.

1.3 Computing a DEM from contours or point data

If the data are given as contours or points we need to use spatial interpolation to first create a grid DEM. GRASS includes several modules for gridding irregularly spaced point or contour/isoline data. A simple inverse distance weighted interpolation is implemented as a `v.surf.idw` module. While this basic method is easy to use, it is not particularly suitable for elevation surfaces (Mitas and Mitášová, 1999). To compute DEM from rasterised contours, `r.surf.contour` can be used.

A more sophisticated interpolation method is based on the variational approach represented by *Regularised Spline with Tension* (RST, see more details in Mitas and Mitášová, 1999; Neteler and Mitášová, 2008). From the viewpoint of geomorphometric analysis, it is important that this interpolation function is differentiable to all orders (Mitášová et al., 1995). Using this property, topographic parameters can be computed simultaneously with interpolation. RST has been explicitly defined by Mitášová et al. (1995) up to four dimensions; for geomorphometric analysis the bivariate function implemented as `v.surf.rst` is the most relevant. The trivariate version of RST has been implemented in GRASS as `v.vol.rst`.

The behaviour of RST interpolation in the modules is controlled by the following parameters:

- *tension*;
- *smoothing*;
- *anisotropy*;
- *minimum and maximum distance between points*.

The parameters can be selected empirically, based on the knowledge of the modelled phenomenon and function, or automatically, by minimisation of the predictive error estimated by a cross-validation procedure (Hofierka et al., 2002).

The tension parameter controls the behaviour of the resulting surface — from a stiff steel plate to a thin, flexible membrane. Using a high tension, the influence of each point is limited to a relatively short distance, while with very low tension each point has a long range of influence. The RST method is scale dependent and the tension works as a rescaling parameter (Neteler and Mitášová, 2008).

Using the smoothing parameter, the RST behaves like an approximation function, i.e. the resulting surface does not pass through the given points, but approximates the input values. This parameter is useful in modelling noisy data, where higher smoothing can filter out the noise, or alternatively, when a phenomenon needs to be modelled at a lower level of detail. Tension and smoothing parameter are linked; lower tension automatically leads to increased smoothing (for more details see Mitášová et al., 2005).

The anisotropy parameter can be used for interpolation of spatially asymmetric data. Anisotropy is defined by orientation of the perpendicular axes characterising the anisotropy and a scaling ratio of the perpendicular axes (a ratio of axes sizes). These parameters scale distances (i.e. the value of tension) in the two perpendicular directions that should fit the spatial pattern of the anisotropic phenomenon.

Minimum and maximum distances between points control the number of points that are actually used in interpolation after reading the input data. The minimum distance allows the user to eliminate the points that are so close to each other that they can be considered identical for the given DEM resolution. The maximum distance can be used only for vector lines with a constant elevation value (e.g. contours) and it allows the user to automatically add points on the contour line if the points are farther apart than the given maximum. The distance parameters also influence the effect of the tension parameter, because tension works as a distance-scaling factor. Therefore, the tension can be set with or without spatial normalisation. The density of data does not affect the normalised tension parameter.

The RST method is appropriate for interpolation of various types of data — irregular data using, for example, `v.surf.rst`, regular (existing grid-based DEM) using `r.resamp.rst`, or alternatively converting grid data to points (`r.to.vect` or `r.random`) that can be re-interpolated using `v.surf.rst`.

2. DERIVING LAND-SURFACE PARAMETERS

GRASS includes an extensive set of modules for deriving land-surface parameters (shown here in Figures 4–20 for the Baranja Hill data set) and performing spatial analysis that involves elevation data.

2.1 Local parameters: slope, aspect, curvatures and derivatives

Topographic (or geomorphometric) analysis provides tools to compute a set of parameters that represent geometrical properties of the land surface. Local parameters describe land-surface properties both at a point and in its immediate surroundings. They can be computed based on the principles of differential geometry using partial derivatives of the mathematical function representing the surface. Local approximation methods are usually applied to estimate derivatives on a regular grid. A surface defined by the given grid point and its 3×3 neighbourhood is approximated by a second-order polynomial, and partial derivatives for the given centre grid point are computed using one of the common finite difference equations — e.g. the method of Horn (1981). This approach works well for smooth and non-flat areas. However, for high resolution data representing relatively flat areas with small differences in elevations or noisy surfaces, the small neighbourhood may not be sufficient to adequately capture the geometry of land-surface features. Also the approximation needs to be modified to estimate derivatives for grid cells on edges of the study area, where the complete 3×3 neighbourhood is not available.

A more general approach to estimation of partial derivatives is to use a differentiable function for DEM interpolation. Then the local surface parameters can be computed using an explicit form of the function derivatives, usually simultaneously with interpolation. However, this task is not trivial because the interpolation function must, at the same time, fulfil several important conditions necessary for reliable land-surface modelling.

GRASS provides both approaches for deriving land-surface parameters. The modules based on RST perform simultaneous interpolation and computation of partial derivatives including the following local land-surface parameters defined in Mitášová and Hofierka (1993):

- *slope* (steepest slope angle, a magnitude of gradient);
- *aspect* (slope orientation, direction of gradient, steepest slope direction, flow direction);
- *profile curvature* (surface curvature in the direction of gradient);
- *tangential curvature* (surface curvature in the direction of contour tangent);
- *mean curvature* (an average of the two principal curvatures).

Alternatively, a user can output first- and second-order partial derivatives instead of land-surface parameters and use them to compute additional maps, such as slope or curvature in any given direction. Plan curvature (contour curvature) can be derived from the tangential curvature and the sine of the slope angle (Mitášová and Hofierka, 1993).

It is important to note that land-surface parameters (especially curvatures based on second-order derivatives) are very sensitive to the quality of interpolation process. For example, interpolation from contours may lead to a false pattern of waves along the contours that can be visible only on a map of profile curvature. This is caused by a very heterogeneous distribution of input data — distances between points on the contours are relatively small, while they are large between the contours. These artifacts can be minimised by tuning the RST parameters (Neteler and Mitášová, 2008). The increase of minimal distance between points to a value that reflects an average distance between contours will reduce the heterogeneity of data density (points that are too close to each other will be removed from the interpolation). The decrease of tension and increase of smoothing will lead to a smoother surface with filtered-out small land-surface variations.

Using the RST method with properly set parameters, the DEM and land-surface parameters can be computed using a single command as follows:

```
v.surf.rst input=contours5K elev=b_dem5K.z
        slope=b_dem5K.s aspect=b_dem5K.a
        pcurv=b_dem5K.pc tcurv=b_dem5K.tc
        mcurv=b_dem5K.mc devi=b_dem5K_dev
        dmin=7.5 dmax=300
        tension=20 smooth=0.5
        zcolumn=VALUE
```

where `input` is the name of the vector data file with contours or elevation data points, `elev`, `slope`, `aspect`, `pcurv`, `tcurv`, `mcurv` are the output DEM and local parameters maps including profile, tangential, and mean curvatures, `devi` is the output deviations file that provides deviations of the resulting surface for each given point, `dmin`, `dmax` are the minimum and maximum distance between points (see explanation in the previous section), `tension`, `smooth` are RST function parameters, also explained in the previous section, and `zcolumn` is the name of the column if the elevation is stored as an attribute rather than as a z-coordinate. The resulting maps are shown in Figures 4–8. The resolution of the resulting maps can be set using `g.region` command before the calculation. In our example, we have used 5 m resolution, based on the data point density and size of the features represented by the given contours.

Computation of curvatures from densely sampled or noisy data, such as LiDAR and SRTM, poses a different type of challenge. Without adequate smoothing, the curvatures will reflect the noise rather than the land-surface features. Figures 9 and 10 show profile curvature maps computed from the original SRTM data using the `r.slope.aspect` command described below and re-interpolated with smoothing using `r.resamp.rst` command, respectively.

Quality of interpolation can be assessed using deviations between the interpolated surface and the given data that are stored in a deviation file. These interpolation errors can be evaluated by statistical measures (e.g., root mean squared error, mean absolute error, etc.). The resulting elevation surface is often a compromise between minimisation of predictive and interpolation errors and the application purpose of the DEM. For example, environmental applications usually require smoother surfaces, while technical applications prefer interpolation accuracy.

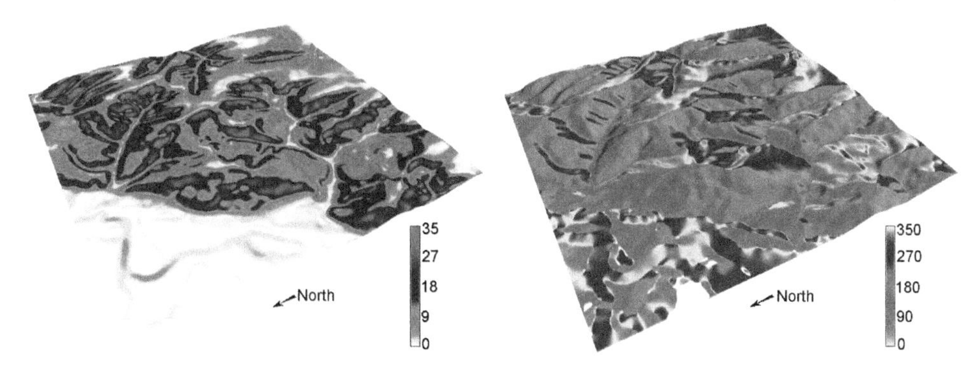

FIGURE 4 Slope steepness [°]. (See page 734 in Colour Plate Section at the back of the book.)

FIGURE 5 Aspect [°]. (See page 734 in Colour Plate Section at the back of the book.)

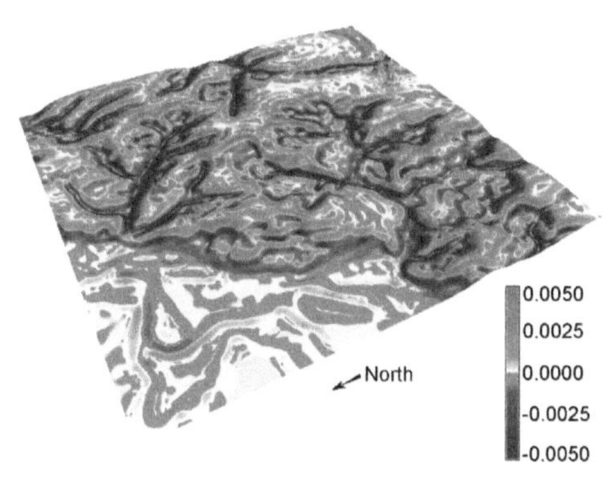

FIGURE 6 Profile curvature [m^{-1}]. (See page 734 in Colour Plate Section at the back of the book.)

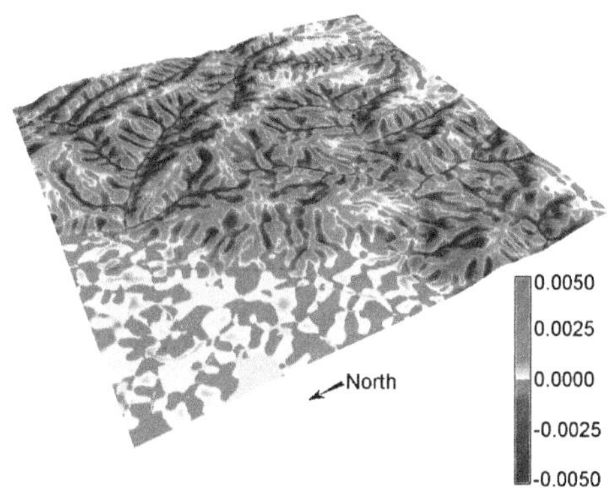

FIGURE 7 Tangential curvature [m^{-1}]. (See page 734 in Colour Plate Section at the back of the book.)

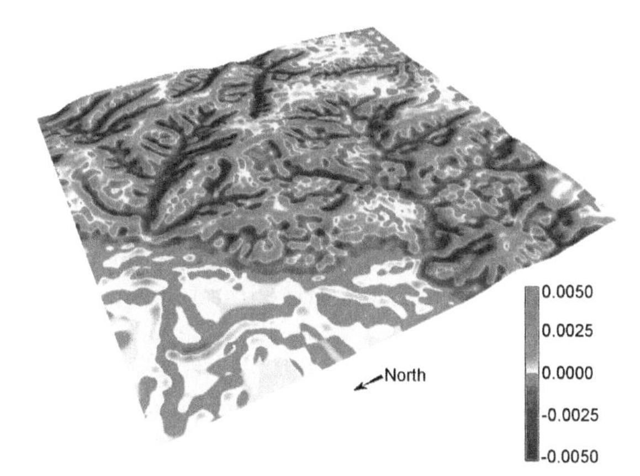

FIGURE 8 Mean curvature [m^{-1}]. (See page 734 in Colour Plate Section at the back of the book.)

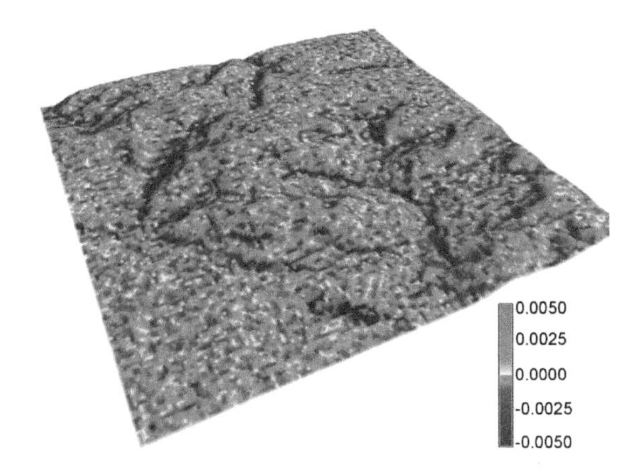

FIGURE 9 Profile curvature [m^{-1}] computed directly from SRTM data using `r.slope.aspect`. (See page 735 in Colour Plate Section at the back of the book.)

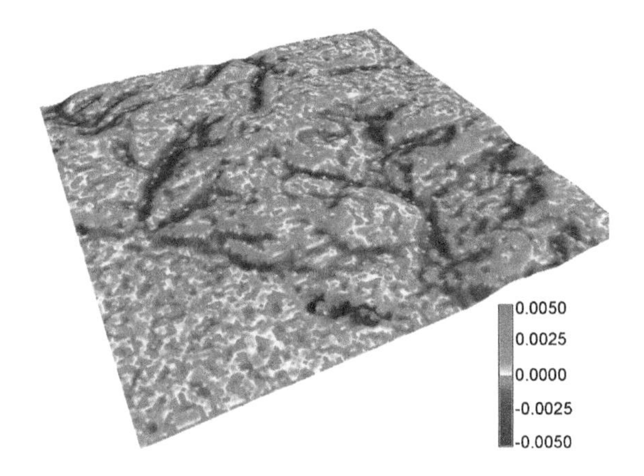

FIGURE 10 Profile curvature [m^{-1}] from smoothed SRTM data using `r.resamp.rst`. (See page 735 in Colour Plate Section at the back of the book.)

If re-interpolation of an existing DEM is not necessary, the local polynomial approximation method implemented in `r.slope.aspect` module can be used to compute local land-surface parameters. Mathematical definitions of the local parameters are identical to the RST modules mentioned above. In `r.slope.aspect` the following second-order polynomial approximation is used:

$$z(x, y) = a_0 + a_1 \cdot x + a_2 \cdot y + a_3 \cdot x \cdot y + a_4 \cdot x^2 + a_5 \cdot y^2 \quad (2.1)$$

By fitting this polynomial to 9 grid points (3×3 array), we can derive the coefficients of this polynomial using weighted least squares. First order partial derivatives are derived using Horn's formula (Horn, 1981; Neteler and Mitášová,

2008):

$$f_x = \frac{(z_7 - z_9) + (2z_4 - 2z_6) + (z_1 - z_3)}{8 \cdot \Delta x} \quad (2.2)$$

$$f_y = \frac{(z_7 - z_1) + (2z_8 - 2z_2) + (z_9 - z_3)}{8 \cdot \Delta y} \quad (2.3)$$

and the second order derivatives are as follows (Neteler and Mitášová, 2008):

$$f_{xx} = \frac{z_1 - 2z_2 + z_3 + 4z_4 - 8z_5 + 4z_6 + z_7 - 2z_8 + z_9}{6 \cdot (\Delta x)^2} \quad (2.4)$$

$$f_{yy} = \frac{z_1 + 4z_2 + z_3 - 2z_4 - 8z_5 - 2z_6 + z_7 + 4z_8 + z_9}{6 \cdot (\Delta y)^2} \quad (2.5)$$

$$f_{xy} = \frac{(z_7 - z_9) - (z_1 - z_3)}{4 \cdot \Delta x \Delta y} \quad (2.6)$$

where $z_3 = z_{i+1,j+1}$, $z_5 = z_{i,j}$, $z_7 = z_{i-1,j-1}$ are elevation values at row i column j, Δx is the east-west grid spacing and Δy is the north–south grid spacing (resolution). Computation of a similar set of parameters as in our RST example, but this time from a raster DEM using `r.slope.aspect`, is performed as follows:

```
r.slope.aspect elevation=b_dem5K.z
        slope=b_dem5Kr.s aspect=b_dem5Kr.a
        pcurv=b_dem5Kr.pc tcurv=b_dem5Kr.tc
```

For an additional example see Section 2.4.

2.2 Regional land-surface parameters

Many landscape processes are influenced by land-surface properties and, at the same time, change the land-surface geometry. Mass and energy flows transport water, air, sediment particles, heat, sound, gases and aerosols within and between landscape elements. Mass flows are influenced by the local land-surface parameters, as well as by landscape configuration that reflects broad-scale geometry of the terrain. The magnitude of the transporting agent (e.g. water) affects its carrying capacity or defines the occurrence of specific phenomena such as floods or gullying. It is often related to the spatial extent of the land surface from which the mass is accumulating while moving downslope. Thus, the movement can be traced by flowlines and currents. GRASS has several modules for computing regional parameters that can be used for analysis of mass flows over the land surface.

2.3 Flow parameters and watersheds

Topography has a profound influence on mass and energy fluxes in the landscape and is often a major factor in many geospatial models and applications. GRASS provides many tools for watershed and water flow analysis. Flow parameters are derived by flow tracing algorithms that approximate the route of water or other liquid over the surface represented by a DEM. Flow routing is based on

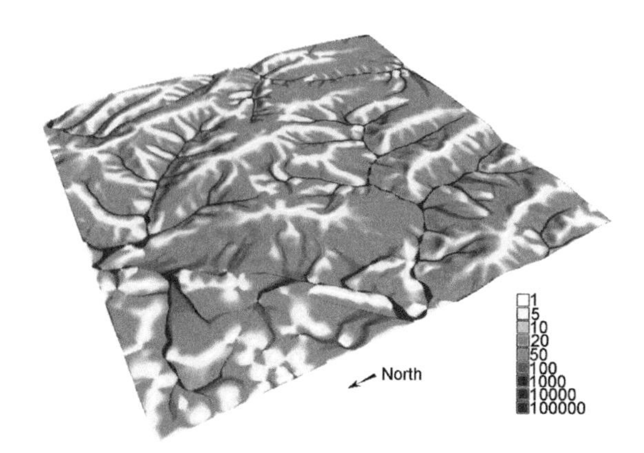

FIGURE 11 Flow accumulation [-] generated by `r.terraflow`. (See page 735 in Colour Plate Section at the back of the book.)

flow-lines — curved lines of descent perpendicular to contours in the direction indicated by aspect. The following basic flow parameters can be computed using GRASS (Neteler and Mitášová, 2008):

- *flow accumulation;*
- *upslope contributing area;*
- *stream network;*
- *watershed (basin) boundaries;*
- *flowpath length.*

Numerous algorithms have been developed for flow routing, based on the approach for estimation of the steepest slope direction and water movement to the downslope cells. In GRASS, the following algorithms have been implemented:

- single flow direction to eight neighbouring cells (SFD, D8) moves flow into a single downslope cell (`r.watershed`);
- single flow to any direction ($D\infty$) or vector-grid approach (`r.flow`);
- multiple flow direction (MFD) to two or more downslope directions (`r.terraflow`, `r.topmodel`);
- 2D water movement simulation based on overland flow differential equations (`r.sim.water`).

The single flow direction approach has the disadvantage that it discretises the flow into only one of eight possible directions. Therefore it produces artificial straight-line patterns especially in areas of flat terrain and on convex landforms with dispersed water flow. SFD is useful for stream network extraction where a single cell representation is needed. Multiple flow routing has the disadvantage that the flow from a cell is dispersed to all neighbours of lower elevation, resulting in a more diffuse flow of water, especially in valleys where concentrated water flow occurs. However, the resulting water flow accumulation surface is smoother, thus more appropriate for further differential geometry analysis (Figure 11).

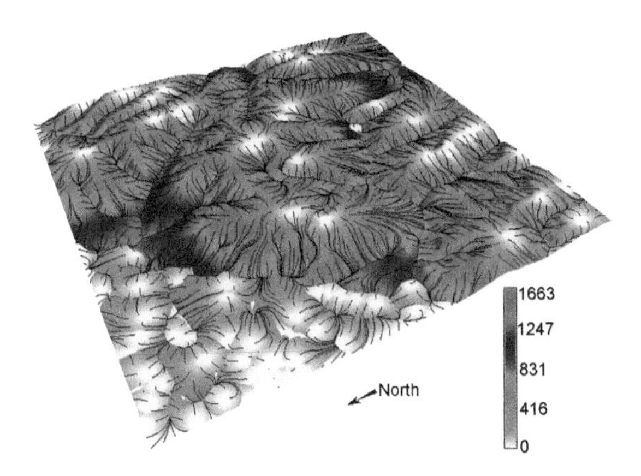

FIGURE 12 Flowpath lengths [m] and flowlines generated by `r.flow`. (See page 736 in Colour Plate Section at the back of the book.)

Many flow routing algorithms are negatively influenced by DEMs of poor quality. Numerous local depressions in valleys or flat areas interrupt the flow-tracing algorithms and create incorrect patterns of flow accumulation (upslope contributing areas), stream networks, flowpath lengths and of other flow parameters. Modules `r.fill.dir` and `r.carve` can be used to remove depressions (sinks) and lakes on DEMs. However, these depression-filling algorithms also introduce positional errors, create artificial features (e.g. flats leading to parallel streams) so that the flow parameters then do not fit with values of other land-surface parameters computed from the original DEM. In GRASS, however, `r.watershed` does not require prior filling of depressions to produce continuous flow accumulation maps, stream networks and other hydrologic parameters, as it uses the least-cost search algorithm to traverse the elevation surface to the outlet. In applications with a new type of DEMs, for example, based on LiDAR or radar-based surveys, this often leads to more accurate results compared to the traditional methods of depression removal (Kinner et al., 2005).

The choice of the module and operations depends on the application. For example, `r.flow` stops flow tracing on flat areas and depressions, so it is more suitable for estimation of flow on hillslopes, smaller watersheds, or DEMs without pits or flat areas. Flowlines, flow accumulation and flowpath lengths for hillslopes in the test region can be computed as follows (Figure 12):

```
r.flow elev=b_dem5K.z aspin=b_dem5K.a skip=15
       flout=b_dem5K_fl dsout=b_dem5K.dd
r.flow -u elev=b_dem5K.z aspin=b_dem5K.a
       lgout=b_dem5K.ul
```

Stream networks and watershed boundaries can be extracted more effectively with `r.watershed`:

```
r.watershed b_dem5K.z accum=b_dem5K.acc
         thresh=10000 basin=b_dem5K.bas
         stream=b_dem5K.st drainage=b_dem5K.dir
```

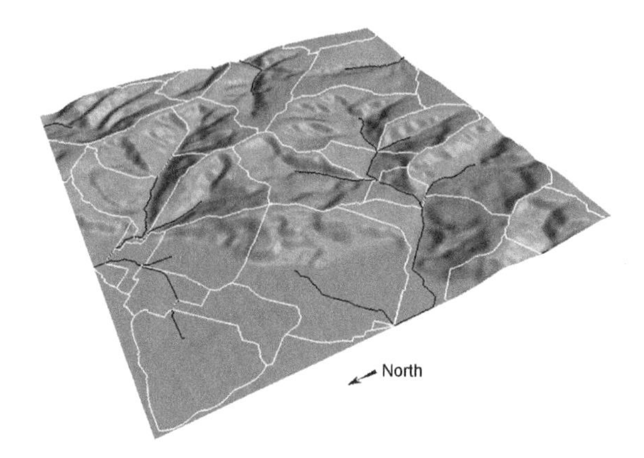

FIGURE 13 Stream network (black lines) and watershed boundaries (white lines) extracted using `r.watershed`.

Note that a more detailed stream network can be extracted from the accumulation map using `r.mapcalc` and a lower threshold than the one defined in the above command. It is often useful to convert the results of `r.watershed` into a vectorised stream network and watershed boundaries (Figure 13). First, the raster representation of the stream network is thinned into a single cell width using `r.thin` and then it is converted to vector lines using `r.to.vect`; watershed areas (basins) can be converted directly, without thinning:

```
r.thin b_dem5K.st out=b_dem5K.st.thin
r.to.vect -s b_dem5K.st.thin out=b_dem5K_st
r.to.vect -s b_dem5K.bas out=b_dem5K_bas
             feature=area
d.vect b_dem5K_st col=blue
d.vect b_dem5K_bas type=boundary
```

If a watershed (contributing area) draining to a given outlet `n,e` is needed, the flow direction map `b_dem5K.dir` generated by `r.watershed` can be used as input for `r.water.outlet`, for example:

```
r.water.outlet drainage=b_dem5K.dir
         basin=b_dem5K.basne easting=6552738
         northing=5071763
```

Specifically designed for handling of very large DEMs (thousands of rows and columns) is `r.terraflow` (Arge et al., 2003, Figure 11); here we show its application to computation of an MFD-flow map and the topographic wetness index:

```
r.terraflow elev=b_dem5K.z filled=b_dem5Kt.fil
           dir=b_dem5Kt.dir swat=b_dem5Kt.swat
           acc=b_dem5Kt.acc tci=b_dem5Kt.tci
```

A wide range of additional parameters can be computed using JGRASS (http://www.jgrass.org/); a Java based GIS built on top of GRASS that includes tools for the HORTON machine (Rigon et al., 2006).

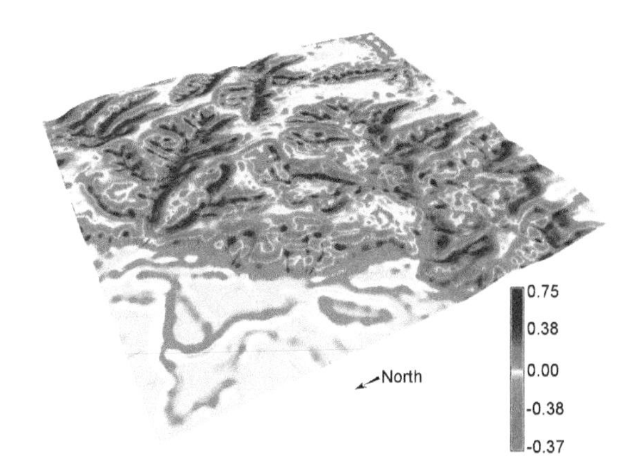

FIGURE 14 Topographic soil erosion index [-]. (See page 736 in Colour Plate Section at the back of the book.)

2.4 Land-surface analysis for modelling

Land-surface parameters play an important role in representing landscape processes. For example, Mitas and Mitášová (1998) have explicitly shown how profile and tangential curvatures influence soil erosion. However, there are applications that may require user-defined or otherwise specific land-surface parameters, or just partial derivatives. To compute specific land-surface parameters not included as outputs of existing modules, such as directional derivatives or flow divergence, a map algebra module `r.mapcalc` and partial derivatives computed by the RST modules or `r.slope.aspect` can be used. The following example demonstrates shell-scripting, map algebra and various land-surface parameters for calculating a topographic index of soil erosion and deposition using the modified LS factor for the Universal Soil Loss Equation; see also the Unit Stream Power-based Erosion Deposition (USPED) model in Mitášová and Mitas (2001). This topographic potential is expressed by a dimensionless index (Figure 14) calculated as a divergence of sediment flow transport capacity:

```
# modified LS factor
# using upslope contributing areas
r.mapcalc "flowtopo = 1.4 *
     exp((upslope_area/cell_size)/22.13,0.4)
     * exp(sin(slope)/0.0896,1.3)"

# sediment transport in x and y directions
r.mapcalc "flowtopo.dx=flowtopo * cos(aspect)"
r.mapcalc "flowtopo.dy=flowtopo * sin(aspect)"

# partial derivatives for sediment transport
r.slope.aspect elev=flowtopo.dx dx=qs.dx
r.slope.aspect elev=flowtopo.dy dy=qs.dy

# topographic potential for
# net erosion and deposition
```

```
r.mapcalc "topoindex = qs.dx + qs.dy"
```

Because the distribution of topographic potential values is skewed, it is helpful to replace a default colour table with a user-defined colour table by the `r.colors` command. This will reveal spatial differences in the values when displaying the resulting map by the `nviz` or `d.rast` commands:

```
r.colors topoindex col=rules
> -1500000 100 0 100
> -10 magenta
> -0.5 red
> -0.1 orange
> -0.01 yellow
> 0 200 255 200
> 0.01 cyan
> 0.1 aqua
> 0.5 blue
> 10 0 0 100
> 1500000 black
> end
```

The first value in every row represents a topoindex value to which a specific colour is attributed either by colour name, or by an RGB triplet. Yellow through red hues represent erosion while blue shades are used for deposition.

Flow parameters represent the potential of relief to generate overland water flow. These parameters do not take into account infiltration or land cover. Therefore, topographical indexes derived from these parameters often represent a steady-state situation or maximal values of overland flow, assuming uniform soil and land cover properties. At landscape scale, a uniform steady-state overland flow is a rare phenomenon occurring only during extreme rainfall events. Therefore resulting patterns of net erosion and deposition based on upslope contributing areas may contradict field observations.

Water and sediment flows are spatial and dynamic phenomena described by complex differential equations that are usually solved by approximation methods. The recently developed `r.sim` group of modules uses the Monte Carlo path sampling method to simulate spatial, dynamic landscape processes. The module `r.sim.water` simulates overland water flow (Figure 15), while `r.sim.sediment` produces sediment flow and erosion/deposition maps based on the Water Erosion Prediction Project (WEPP) theory (Mitas and Mitášová, 1998). The following example shows the application of `r.sim.water` module for the Baranja Hill data set using derived land-surface parameters and uniform *ad hoc* rainfall, soil and land cover properties:

```
r.sim.water -t elevin=b_dem5K.z
dxin=b_dem5K.dx dyin=b_dem5K.dy
rain=b_dem5K.rain infil=b_dem5K.infil
manin=b_dem5K.manning disch=b_dem5K.disch
nwalk=1000000 niter=2400 outiter=200
```

The output of these modules can be in the form of a time-series of maps showing evolution of the modelled phenomenon (available at geomorphometry.org).

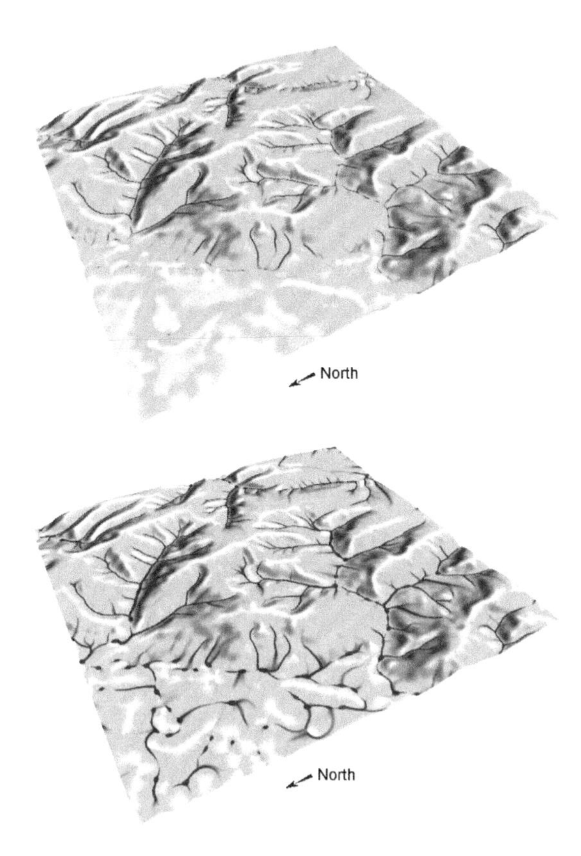

FIGURE 15 Overland water flow simulated by `r.sim.water` after 200 (above) and 2400 (below) seconds.

2.5 Landforms

The GRASS module `r.param.scale` extracts basic land-surface features from a DEM, such as peaks, ridges, passes, channels, pits and plains. This module is based on the work by Wood (1996). It uses a multi-scale approach by fitting a bivariate quadratic polynomial to a given window size using least squares. This module is a predecessor to the system described in Chapter 14 (e.g. Figure 11). In the following example (Figure 16), main land-surface features were identified using a 15×15 processing window:

```
r.param.scale in=b_dem5K.z out=b_dem5K.param
              param=feature size=15
```

2.6 Ray-tracing parameters

Solar radiation influences many landscape processes and is a source of renewable energy of interest to many researchers, energy companies, governments and consumers. GRASS provides two modules related to solar radiation: `r.sunmask`

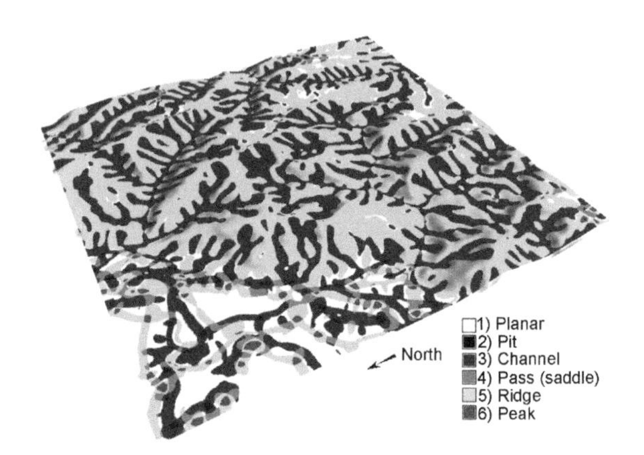

FIGURE 16 Basic land-surface features extracted using `r.param.scale`. (See page 736 in Colour Plate Section at the back of the book.)

calculates a Sun position and shadows map for specified time and Earth position using the SOLPOS2 algorithm from National Renewable Energy Laboratory, and `r.sun` calculates all three components of solar irradiance/radiation (beam, diffuse and reflected) for clear-skies as well as overcast conditions (Šúri and Hofierka, 2004). The clear-sky solar radiation model is based on the work undertaken for development of the European Solar Radiation Atlas (Scharmer and Greif, 2000; Rigollier et al., 2000). The model works in two modes. The irradiance mode is selected by setting a local time parameter; the output values are in W/m^2. By omitting the time parameter, the radiation model is selected; output values are in Wh/m^2.

The model requires only a few mandatory input parameters such as elevation above sea level, slope and aspect of the terrain, day number and, optionally, a local solar time. The other input parameters are either internally computed (solar declination) or the values can be overridden by explicitly defined settings to fit specific user needs: Linke atmospheric turbidity, ground albedo, beam and diffuse components of clear-sky index, time step used for calculation of all-day radiation from sunrise to sunset. Overcast irradiance/radiation are calculated from clear-sky raster maps by the application of a factor parameterising the attenuation of cloud cover (clear-sky index). The clear-sky global solar radiation for Baranja Hill data set, March 21 (spring equinox) has been calculated using `r.sun` as a sum of beam, diffuse and reflected radiation. The shadowing effects of relief were taken into account (Figure 17).

In practical applications related to evaluation of available solar radiation within a specific period of day or year we can use a shell script and the `r.mapcalc` command to compute a sum of available radiation values. Viewshed analysis can be performed using `r.los` that generates a raster map output in which the cells that are visible from a user-specified observer location are marked with integer values that represent the vertical angle (in degrees) required to see those cells (viewshed). A map showing visible areas (in blue) from the position of a man

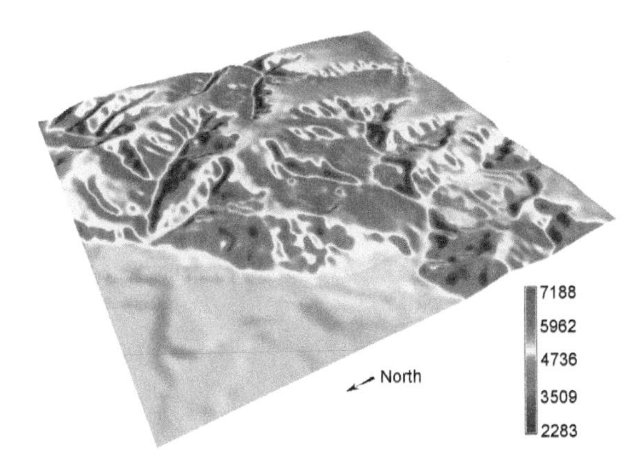

FIGURE 17 Global solar radiation for spring equinox [Wh/m^2]. (See page 737 in Colour Plate Section at the back of the book.)

standing on the hill crest depicted by a black dot in Figure 18 can be computed as follows:

```
r.los b_dem5K.z out=b_dem5K.los
      coor=6553202,5071538
```

An improved viewshed analysis program is available as GRASS extension.[2] Shaded relief maps enhance the perception of terrain represented by a DEM. In GRASS, they are generated using the `r.shaded.relief` module with parameters defining the sun position (sun altitude and azimuth) and vertical scaling

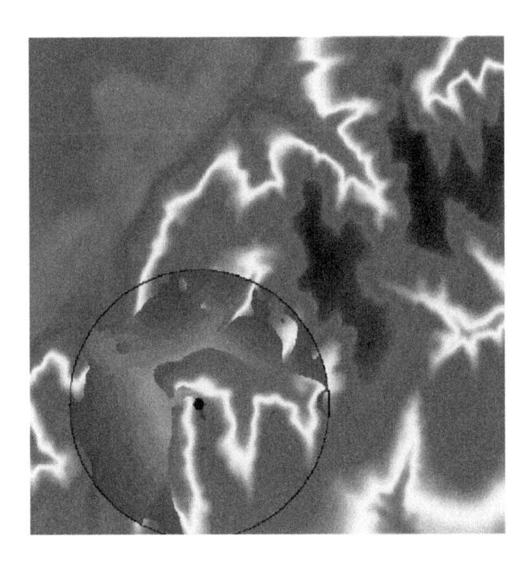

FIGURE 18 Visibility analysis using `r.los`. (See page 737 in Colour Plate Section at the back of the book.)

[2] http://www.uni-kiel.de/ufg/ufg_BerDucke.htm.

FIGURE 19 Random fractal surface generated by `r.surf.fractal`. (See page 738 in Colour Plate Section at the back of the book.)

(z-exaggeration). This shaded map can be used to transform colours of other thematic map using the IHS colour model. The resulting shaded, coloured map, displayed by command `d.his` provides enhanced perception of terrain and better orientation especially in hilly areas (see example in the displaying DEMs section).

2.7 Fractal surfaces

The concept of fractals has attracted the attention of scientists in many fields, including geomorphometry. According to many studies, most real land surfaces have a fractal dimension in the range of 2.2–2.6. However, Wood (1996) notes that landscapes usually do not possess a single fractal dimension, but a variety of values that change with scale. The concept of fractal surfaces and fractal dimension can be employed to generate synthetic, natural-looking surfaces with controllable topographic variation. There are numerous methods of generating fractal surfaces, but the one adopted in `r.surf.fractal` module uses the spectral synthesis approach described by Saupe (1988).

This technique involves selecting scaled (Gaussian) random Fourier coefficients and performing the inverse Fourier transform. It has the advantage over the more common midpoint displacement methods which produce characteristic artifacts at distances 2^n units away from a local origin (Voss, 1988). Wood (1996) has modified this technique so that multiple surfaces may be realised with only selected Fourier coefficients in the form of intermediate layers showing the buildup of different spectral coefficients. The result is that the scale of fractal behaviour may be controlled as well as the fractal dimension itself. In the example for the Baranja region (Figure 19) we have used the `r.surf.fractal` module with the fractal dimension set to 2.05:

```
r.surf.fractal out=b.fractal d=2.05
```

Other fractal-related modules are `r.surf.gauss` and `r.surf.random`. The module `r.surf.gauss` generates a fractal surface based on a Gaussian random number generator whose mean and standard deviation can be set by the user. The module `r.surf.random` uses a different type of random number generator and uniform random deviates whose range can be expressed by the user.

2.8 Summary parameters and profiles

GRASS provides various tools for querying and summarising maps of land-surface parameters. For example, the module `r.report` can be used to create a frequency distribution of map values in the form of a table containing category numbers, labels and (optionally) area sizes in units selected by a user. The command `r.stats` calculates the area present in each of the map categories. Alternatively, `d.histogram` can be used to visualise a distribution of the values in the form of a bar or pie chart. Polar diagrams can be used for displaying distributions of aspect values by the `d.polar` module. If the polar diagram does not reach the outer circle, no data (NULL) cells were found in the map. The vector in the diagram indicates the prevalent direction and vector length the share of this direction in the frequency distribution of aspect values.

The aspect map for the Baranja Hill DEM with a spatial resolution of 25 m and derived from the 1:5000 contours [Figure 20(a)] shows dominant spikes in the polar diagram [Figure 20(d)] indicating a suboptimal land-surface representation in DEM25m. The aspect map of the DEM25-SRTM [Figure 20(c)] does not show dominant spikes but mostly regular spikes representing relatively homogeneous noise typical for RADAR data [Figure 20(d)]. Finally, the aspect computed simultaneously with DEM interpolation from the Baranja Hill contour lines using `v.surf.rst` [Figure 20(b)] is relatively smooth and does not show any significant spikes [Figure 20(d)]. Lengths of the average direction vectors in the diagram are very short which indicates that DEMs for this region show no prevalent aspect direction.

The area of a surface represented by a raster map is provided by `r.surf.area` which calculates both the area of the horizontal plane for the given region and an area of the 3D surface estimated as a sum of triangle areas created by splitting each rectangular cell by a diagonal. More complex analysis is available in `r.univar` and `r.statistics`. The `r.univar` module calculates univariate statistics that includes the number of counted cells, minimum and maximum cell values, arithmetic mean, variance, standard deviation and coefficient of variation. The `r.statistics` module also calculates mode, median, average deviation, skewness and kurtosis. Using the `r.neighbors` module, a *local* statistics based on the values of neighbouring cells defined by a window size around the central cell can be computed. Available statistics include minimum, maximum, average, mode, median, standard deviation, sum, variance, diversity and inter-dispersion. Sophisticated statistics and spatial analysis are available via GRASS interface with the R statistical data analysis language (http://cran.r-project.org/).

Land-surface analysis often requires querying map values at a specific location. This can be done in GRASS either interactively with the mouse, or by a com-

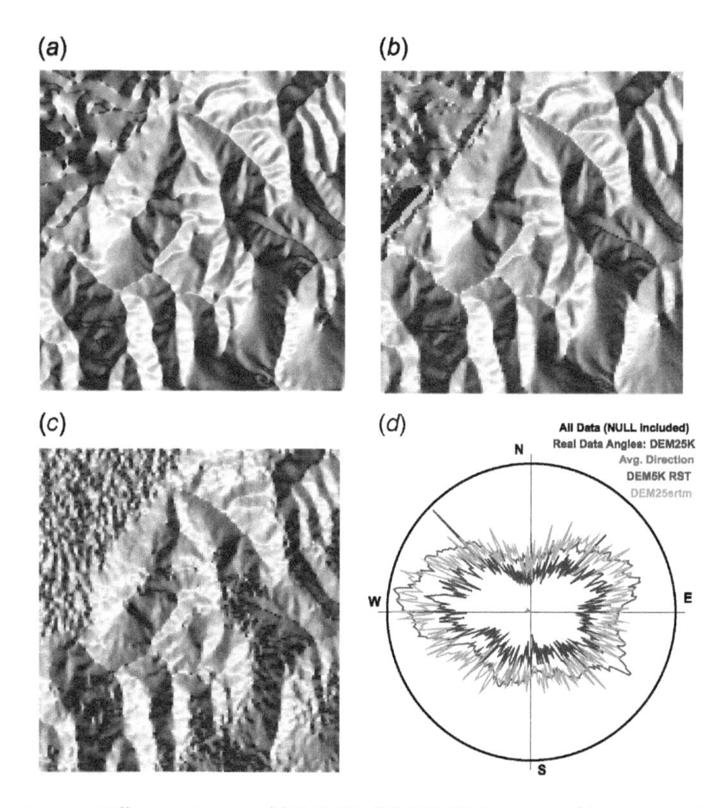

FIGURE 20 Baranja Hill aspect maps: (a) DEM25, (b) DEM5K (generated by `v.surf.rst`), (c) DEM25-SRTM, and (d) a combined polar diagram of all aspect maps from `d.polar`. (See page 738 in Colour Plate Section at the back of the book.)

mand with coordinates defining the location. The simplest command for interactive querying by mouse is `d.what.rast`. To generate profiles, a user can run `d.profile`. It allows one to interactively draw profiles over the terrain by mouse within the GRASS monitor.

Non-interactive query can be performed at specific points defined by coordinates (`r.what`) or along the user-defined profile (`r.profile` and `r.transect`). Similar query commands are available for vector maps as well.

2.9 Volume parameters

Land surface is a 2-dimensional contact between different landscape components (atmosphere vs. lithosphere, or hydrosphere vs. lithosphere). As such, it often represents the surface of a 3D object. To compute the volume of the object, the summary parameter `r.volume` can be used, for example, to estimate the amount of earth that must be excavated for a construction project.

Many landscape phenomena can be investigated using differential geometry tools extended to three dimensions (Hofierka and Zlocha, 1993). GRASS provides several tools for 3-dimensional (volume) modelling. For example, tri-variate Reg-

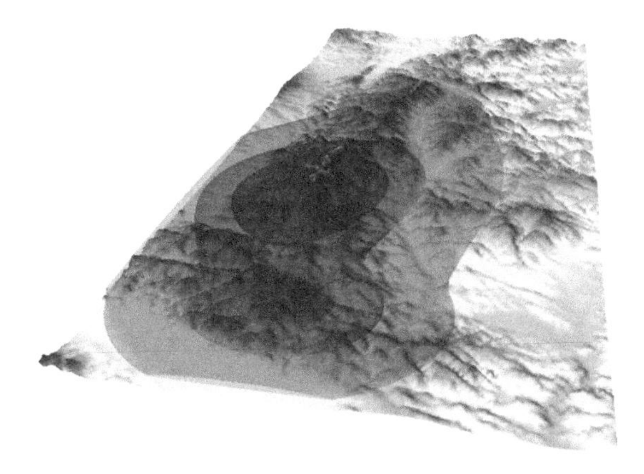

FIGURE 21 Volume interpolation and isosurface visualisation of precipitation (isosurfaces of 1100, 1200, 1250 mm/year are shown) using `v.vol.rst`. (See page 739 in Colour Plate Section at the back of the book.)

ularised Spline with Tension is implemented in `v.vol.rst` for spatial interpolation of volume data. `v.vol.rst` has similar properties and parameters as the bi-variate version of RST, so the principles described in the Introduction section are applicable here as well. Similarly to the bi-variate version, tri-variate RST can compute a number of geometric parameters related to the gradient and curvatures of the volume model: magnitude and direction of gradient, directional change of gradient, *Gauss–Kronecker* and mean curvatures. Mathematical definitions and explanation of volume parameters can be found in Hofierka and Zlocha (1993) and Neteler and Mitášová (2008).

Moreover, tri-variate interpolation can be helpful in spatial characterisation of natural phenomena influenced by land surface. For example, Hofierka et al. (2002) present an application of tri-variate RST in precipitation modelling. Elevation, aspect, slope, or other land-surface parameters can be incorporated in the tri-variate interpolation as a third variable. The approach requires 3D data (x, y, z, w) and a raster DEM. The phenomenon is modelled by tri-variate interpolation. Then, phenomenon values on the land surface are computed by intersection of the volume model with the land surface represented by a DEM. The volumetric visualisation of the precipitation volume model using `nviz` is presented in Figure 21.

3. LIMITATIONS OF GRASS

Although GRASS has rather comprehensive geomorphometry tools it is by no means complete. For example, support for TIN-based land-surface modelling and analysis, often used in engineering applications, is very limited. Also, modelling of terrain with faults and breaklines, although possible, is rather cumbersome as it requires additional pre- and post-processing. Some help is available in `r.surf.nnbathy`, which employs a natural neighbour interpolation library (http://www.marine.csiro.au/~sakov/) and supports interpolation with breaklines. It is provided as an add-on module at GRASS Wiki site (http://grass.osgeo.org/wiki/). The error of prediction can be analysed using a simple comparison of estimated and true values or using more sophisticated cross-validation. GRASS is currently evolving rather rapidly based on the needs of its developers, therefore new capabilities not included here could have happen during the production of this book. The most recent capabilities can be checked at the official GRASS web site.

4. SUMMARY POINTS AND FUTURE DIRECTION

GRASS is a mature, fully-featured open-source GIS capable of a broad spectrum of spatial calculations in geomorphometry. The ANSI C source code provides a comprehensive suite of modules and UNIX-shell scripts to manipulate DEMs, extract a variety of land-surface parameters and objects, and analyse hydro-geomorphological phenomena in both 2D and 3D. Surface-form data can be imported as grid DEMs, digitised contours, or as scattered point-measurements of elevation. Considerable automation has been built into the system, which features a graphical user interface and is readily available through a web-based infrastructure. The 6.2 version of GRASS illustrated in this chapter is available for all commonly used operating systems.

Advances in mapping technologies, especially the rapid evolution of airborne and ground-based laser scanning as well as satellite and airborne radar interferometry are bringing significant changes to geomorphic analysis. The point densities now exceed the level of detail required for most applications and DEMs with resolutions of 3 m and better are becoming common even for large areas. The high mapping efficiency makes repeated mapping at relatively short time intervals feasible, resulting in multi-temporal DEMs. These developments require new concepts and approaches in geomorphometry. In response, GRASS modules are being further enhanced to accommodate very large data sets produced by the new mapping technologies; new tools are added, for example, for efficient handling of very dense elevation or bathymetry data, hierarchical watershed analysis and quantification of land-surface change.

IMPORTANT SOURCES

http://grass.osgeo.org — The GRASS website.
http://www.jgrass.org — JGRASS.
http://skagit.meas.ncsu.edu/~helena/gmslab/viz/sinter.html — Multidimensional Spatial Interpolation in GRASS GIS.
http://skagit.meas.ncsu.edu/~helena/gmslab/viz/erosion.html — Land-surface analysis and applications.
http://skagit.meas.ncsu.edu/~helena/publwork/Gisc00/astart.html — Path sampling modelling.
http://www.cs.duke.edu/geo*/terraflow/ — Terraflow.
http://re.jrc.cec.eu.int/pvgis/ — PVGIS and solar radiation modelling using GIS.

CHAPTER 18

Geomorphometry in RiverTools

S.D. Peckham

history and development of RiverTools · preparing a DEM for your study area · kinds of information that can be extracted using RiverTools and DEMs · special visualisation tools in RiverTools · what makes the RiverTools software unique?

1. GETTING STARTED

RiverTools is a software toolkit with a user-friendly, point-and-click interface that was specifically designed for working with DEMs and extracting hydrologic information from them. As explained in previous chapters, there is a lot of useful information that can be extracted from DEMs since topography exerts a major control on hydrologic fluxes, visibility, solar irradiation, biological communities, accessibility and many human activities. RiverTools has been commercially available since 1998, is well-tested and has been continually improved over the years in response to the release of new elevation data sets and algorithms and ongoing feedback from a global community of users. All algorithms balance work between available RAM and efficient I/O to files to ensure good performance even on very large DEMs (i.e. 400 million pixels or more). RiverTools is a product of Rivix LLC (www.rivix.com) and is available for Windows, Mac OS X and Solaris.

RiverTools 3.0 comes with an installation CD and sample data CD but the installer can also be downloaded from www.rivertools.com. It uses the industry-standard InstallShield installer and is therefore easy to install or uninstall. The HTML-based help system and user's guide includes a set of illustrated tutorials, a glossary, step-by-step explanations of how to perform many common tasks, a description of each dialog and a set of executive summaries for major DEM data sets and formats. All of the RiverTools file formats are nonproprietary and are explained in detail in an appendix to the user's guide. In addition, each dialog has a Help button at the bottom that jumps directly to the relevant section of the user's guide.

Developments in Soil Science, Volume 33 © 2009 Elsevier B.V.
ISSN 0166-2481, DOI: 10.1016/S0166-2481(08)00018-4. All rights reserved.

The purpose of this chapter is to provide an overview of what RiverTools can do and how it can be used to rapidly perform a variety of tasks with elevation data. Section 1.1 explains the layout of the RiverTools menus and dialogs. Section 2 briefly discusses GIS issues such as ellipsoids and map projections. Section 3 introduces some tools in the Prepare menu that simplify the task of preparing a DEM that spans a given area of interest. Section 4 discusses how dialogs in the Extract menu can be used to extract various grid layers and masks from a DEM. Section 5 highlights some of the visualisation tools in the Display menu and Section 5.1 introduces some of the Interactive Window Tools that can be used to query and interact with an image.

1.1 The RiverTools menu and dialogs

RiverTools 3.0 can be started by double-clicking on a shortcut icon or by selecting it from a list of programs in the Windows Start menu. After a startup image is displayed, the Main Window appears with a set of pull-down menus across the top labeled: *File, Prepare, Extract, Display, Analyze, Window, User* and *Help*. Each pull-down menu contains numerous entries, and sometimes cascading menus with additional entries. Selecting one of these entries usually opens a point-and-click dialog that can be used to change various settings for the selected task. Buttons labeled *Start, Help* and *Close* are located at the bottom of most dialogs. Clicking on the Start button begins the task with the current settings. Clicking on the Help button opens a browser window to a context-specific help page and clicking on a Close or Cancel button dismisses the dialog.

The *File menu* contains tools for opening data sets, importing and exporting data in many different formats, and for changing and/or saving various program settings and preferences. The *Prepare menu* contains a collection of tools that can be used at the beginning of a project to prepare a DEM for further analysis, such as mosaicking and sub-setting tiles, replacing bad values, uncompressing files and changing DEM attributes such as elevation units, byte order, orientation and data type. The *Extract menu* contains a large set of tools for extracting new grid layers (e.g. slope, curvature and contributing area), vectors (e.g. channels and basin boundaries) and masks (e.g. lakes and basins) from a DEM or a previously extracted grid layer. The *Display menu* has a collection of different visualisation tools such as density plots, contour plots, shaded relief, surface plots, river network maps, multi-layer plots and many more. Images can be displayed with any of 17 different map projections or without a map projection.

There is also an extensive set of Interactive Window Tools that makes it easy to query and zoom into these images to extract additional information. The *Analyze menu* has a number of tools for analysing and plotting terrain and watershed attributes that have been measured with the extraction tools. Graphics windows can be managed with a set of tools in the *Window menu* and RiverTools can be extended by users with plug-ins that appear in the *User menu*.

2. ADVANCED GIS FUNCTIONALITY

2.1 Fixed-angle and fixed-length grid cells

Virtually all elevation data providers distribute raster DEMs in one of two basic forms. In the *geographic* or *fixed-angle* form, the underlying grid mesh is defined by lines of latitude and longitude on the surface of a chosen ellipsoid model and each grid cell spans a fixed angular distance such as 3 arcsec. Lines of constant latitude (parallels) and lines of constant longitude (meridians) always intersect at right angles. However, since the meridians intersect at the poles, the distance between two meridians depends on which parallel that you measure along. This distance varies with the cosine of the latitude and is largest at the equator and zero at the poles. So while each grid cell spans a fixed angle, its width is a function of its latitude. The fixed-angle type of DEM is the most common and is used for all global or near-global elevation data sets such as SRTM, USGS 1-Degree, NED, DTED, GLOBE, ETOPO2, GTOPO30, MOLA and many others.

The second basic type of raster DEM is the *"fixed-length"* form, where both the east-west and north-south dimensions of each grid cell span a fixed distance such as 30 metres. This type of DEM is commonly used for high-resolution elevation data that spans a small geographic extent so that the Earth's surface can be treated as essentially planar. They are almost always created using a Transverse Mercator projection such as Universal Transverse Mercator (UTM). Examples include USGS 7.5-Minute quad DEMs, most LiDAR DEMs and many state and municipal DEMs. When mosaicked to cover large regions, fixed-length DEMs suffer from distortion and lead to inaccurate calculations of lengths, slopes, curvatures and contributing areas.

2.2 Ellipsoids and projections

Unlike most GIS programs, RiverTools always takes the latitude-dependence of grid cell dimensions into account when computing any type of length, slope or area in a *geographic* or fixed-angle DEM. It does this by integrating directly on the ellipsoid model that was used to create the DEM. In addition, when measuring *straight-line* distance between any two points on an ellipsoid, the highly accurate Sodano algorithm is used (Sodano, 1965). Other GIS programs project the fixed-angle elevation data with a fixed-length map projection such as UTM and then compute all length, slope and area measurements in the projected and therefore distorted DEM.

In RiverTools, various properties of the DEM such as its pixel geometry (fixed-angle or fixed-length), number of rows and columns and bounding box can be viewed (and edited if necessary) with the View DEM Info dialog in the File menu. When working with a fixed-angle DEM, the user should set the ellipsoid model to the one that was used in the creation of the original DEM data. This is done by opening the Set Preferences dialog in the File menu and selecting the Planet Info panel. A list of 51 built-in ellipsoid models for Earth are provided in a droplist, as well as information for several other planets and moons. The ellipsoid models that were used to create several of the major DEM data sets is provided in the RiverTools documentation. Most modern DEM data sets and all GPS units now use the WGS84 ellipsoid model and this is the default. Since maps and images are necessarily two-dimensional, RiverTools also offers 17 different map projections for display purposes via the Map Projection Info dialog in the Display menu.

3. PREPARING DEMS FOR A STUDY AREA

3.1 Importing DEMs

Since elevation and bathymetric data is distributed in many different data formats, the first step when working with DEMs is to import the data, that is, to convert it to the format that is used by the analysis software. The DEM formats that can currently be imported include: ARC BIL, ARC FLT, ENVI Raster, Flat Binary, SDTS Raster Profile (USGS), USGS Standard ASCII, CDED, DTED Level 0, 1 or 2, GeoTIFF, NOAA/NOS EEZ Bathymetry, GMT Raster (netCDF), GRD98 Raster, ASTER, MOLA (for Mars), SRTM, ARC Gridded ASCII, Gridded ASCII, and Irregular XYZ ASCII. While some DEMs simply store the elevations as numbers in text (or ASCII) files, this is an extremely inefficient format, both in terms of the size of the data files and the time required for any type of processing. Because of this, elevation data providers and commercial software developers usually use a binary data format as their native format and then provide a query tool such as the Value Zoom tool in RiverTools for viewing DEM and grid values.

A simple, efficient and commonly used format consists of storing elevation values as binary numbers with 2, 4 or 8 bytes devoted to each number, depending on whether the DEM data type is integer (2 bytes), long integer (4 bytes), floating point (4 bytes) or double-precision (8 bytes). The numbers are written to the binary file row by row, starting with the top (usually northernmost) row — this is referred to as *row major format*. The size of the binary file is then simply the product of the number of columns, the number of rows and the number of bytes used per elevation value. All of the descriptive or georeferencing information for the DEM, such as the number of rows and columns, pixel dimensions, data type, byte order, bounding box coordinates and so on is then stored in a separate text file with the same filename prefix as the binary data file and a standard three-letter extension. This basic format is used by ARC BIL, ARC FLT, ENVI Raster, MOLA, SRTM, RTG and many others. Many of the other common formats, such as SDTS Raster, GeoTIFF and netCDF also store the elevation data in binary, row major format but add descriptive header information into the same file, either before or after the data.

To import a DEM into RiverTools, you choose Import DEM from the File menu and then select the format of the DEM you want to import. If the format is one that is a special-case of the RiverTools Grid (RTG) format (listed above), then the binary data file can be used directly and only a RiverTools Information (RTI) file needs to be created. You can import many DEMs that have the same format as a batch job by entering a *"matching wildcard"* (an asterisk) in both the input and output filename boxes. For example, to import all of the SRTM tiles in a given directory or folder that start with "`N30`", you can type "`N30*.hgt`" into both filename boxes.

Elevation data is sometimes distributed as irregularly-spaced XYZ triples in a multi-column text file. **RiverTools** has an import tool for gridding this type of elevation data. In the current version, Delaunay triangulation is used but in the next release six additional gridding algorithms will be added.

3.2 Mosaicking DEM tiles

The second step in preparing a DEM that spans a given area of interest is to mosaic many individual tiles to create a seamless DEM for the area. These tiles are typically of uniform size and are distributed by DEM providers in separate files. For example, SRTM tiles span a region on the Earth's surface that is one degree of latitude by one degree of longitude and have dimensions of either 1201×1201 (3 arcsec grid cells) or 3601×3601 (1 arcsec grid cells).

To mosaic or subset DEM tiles in **RiverTools**, you first choose Patch RTG DEMs from the Prepare menu. This opens an Add/Remove dialog that makes it easy to add each of the tiles that you wish to mosaic to a list [Figure 1(a)]. Tiles can be viewed individually by clicking on the filename for the tile and then on the Preview button. Similarly, their georeferencing information can be viewed by clicking on the *View Infofile* button. Tiles with incompatible georeferencing information may sometimes need to be preprocessed in some way (e.g. units converted from feet to metres or subsampled to have the same grid cell size) and this can easily be done with the Convert Grid dialog in the Prepare menu.

The file selection dialog that is used to add tiles to the list provides a filtering option for showing only the files with names that match a specified pattern. This dialog also allows multiple files to be selected at once by holding down the shift key while selecting files. If these two features are used, even large numbers of tiles can be rapidly added to the list. The *Add/Remove* dialog itself has an Options menu with a Save List entry that allows you to save the current list of tiles to a text file. You can then later select the Use Saved List option to instantly add the saved list of files to the dialog.

Once you have finished adding DEM tiles to the list, you can type a prefix into the dialog for the DEM to be created and then click on the Start button to display the DEM Patching Preview Window [Figure 1(b)]. This shaded relief image in this window shows how all of the tiles fit together. You can then click and drag within the image to select the subregion that is of interest with a "*rubber band box*", or select the entire region spanned by the tiles by clicking the right mouse button. It is usually best to select the smallest rectangular region that encloses the river basin of interest. If you can't discern the basin boundary, you can easily iterate the process a couple of times since everything is automated. The *DEM Patching Preview* window has its own Options menu near the top and begins with the entry *Save New DEM*. A button with the same label is also available just below the image. These are two different ways of doing the same thing, namely to read data from each of the DEM tiles to create a new DEM that spans the selected region. If there are any "*missing tiles*" that intersect the region of interest (perhaps in the ocean) they are

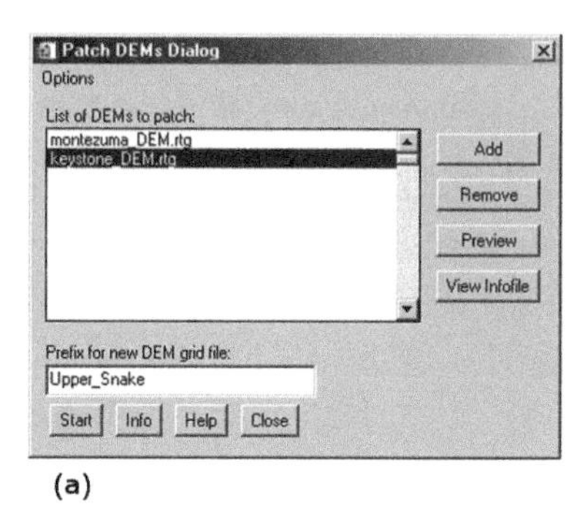

(a)

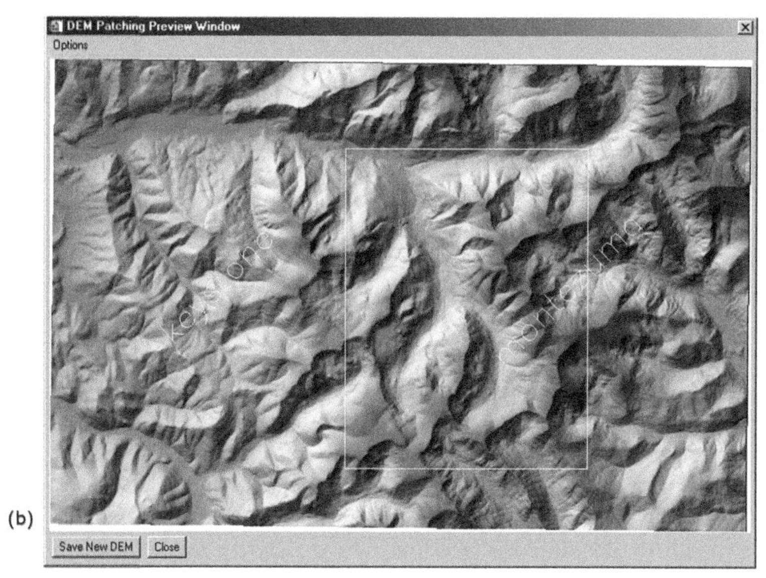

(b)

FIGURE 1 (a) The Patch RTG DEM dialog; (b) The DEM Patching Preview window with subregion selected with a rubber-band box and both tiles labeled with filename prefixes. © 2008 Rivix LLC, used with permission.

automatically filled with nodata values. Other entries in the Options menu allow you to do things like (1) label each tile with its filename, (2) "*burn in*" the rubber band box and labels and (3) save the preview image in any of several common image formats. Once your new DEM has been created, it is automatically selected just as if you had opened it with the Open Data Set dialog in the File menu. You can view its attributes using the *View DEM Info* tool in the File menu.

3.3 Replacing bad values

Sometimes a third step is required to prepare a DEM that spans a region of interest. In SRTM tiles, for example, there are often nodata "*holes*" in high-relief areas that were not in the line of sight of the instrument aboard the Space Shuttle that was

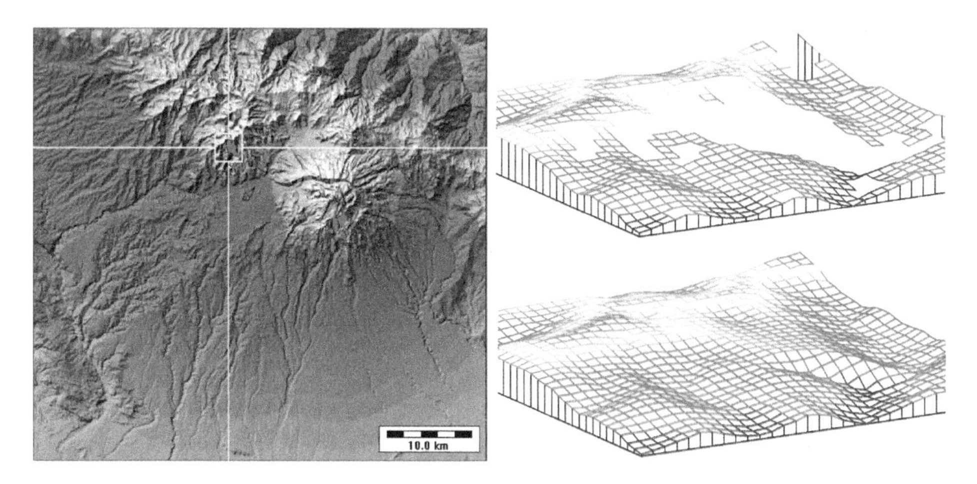

FIGURE 2 A yellow box and crosshairs on a shaded relief image shows the location of a hole (red) in an SRTM DEM for Volcan Baru, Panama. The two images on the right show wire mesh surface plots of the area near the hole, before and after using the Repair Bad Values tool. (See page 740 in Colour Plate Section at the back of the book.) © 2008 Rivix LLC, used with permission.

used to measure the terrain heights. These holes usually span small areas between 1 and 20 grid cells but can be larger. For most types of analysis, these holes must be repaired prior to further processing. RiverTools has a *Replace Bad Values* tool in the Prepare menu that fills these holes with reasonable values by iteratively averaging from the edges of the holes until the hole is filled. The output filename should usually be changed to have a new prefix and the compound extension `_DEM.rtg`. (Figure 2) shows the result of applying this tool to an SRTM DEM for Volcan Baru, in Panama.

4. EXTRACTING LAND-SURFACE PARAMETERS AND OBJECTS FROM DEMS

4.1 Extracting a D8 flow grid

Once you have a DEM for an area of interest, there are a surprising number of additional grid layers, polygons, profiles and other objects that can be extracted with software tools and which are useful for various applications. Some of these were discussed in Chapter 7. Figure 3 shows several land-surface parameters and objects that were extracted for the Baranja Hill case study DEM and which will be discussed throughout this section. A DEM with 5-meter grid cells was created from a source DEM with 25-meter grid cells via bilinear interpolation followed by smoothing with a 5×5 moving window, using the RiverTools Grid Calculator. This smoother DEM was used for creating the images shown except for Figure 3(d).

A D8 flow grid is perhaps the most fundamental grid layer that can be derived from a DEM, as it is a necessary first step before extracting many other objects.

RiverTools makes it easy to create a D8 flow grid and offers multiple options for resolving the ambiguity of flow direction within pits and flats. Choosing Flow Grid (D8) from the Extract menu opens a dialog which shows the available options. The default pit resolution method is *"Fill all depressions"*. In most cases, filling all depressions will produce a satisfactory result since it handles the typically very large number of nested, artificial depressions that occur in DEMs and even provides reasonable flow paths through chains of lakes. However, support for closed

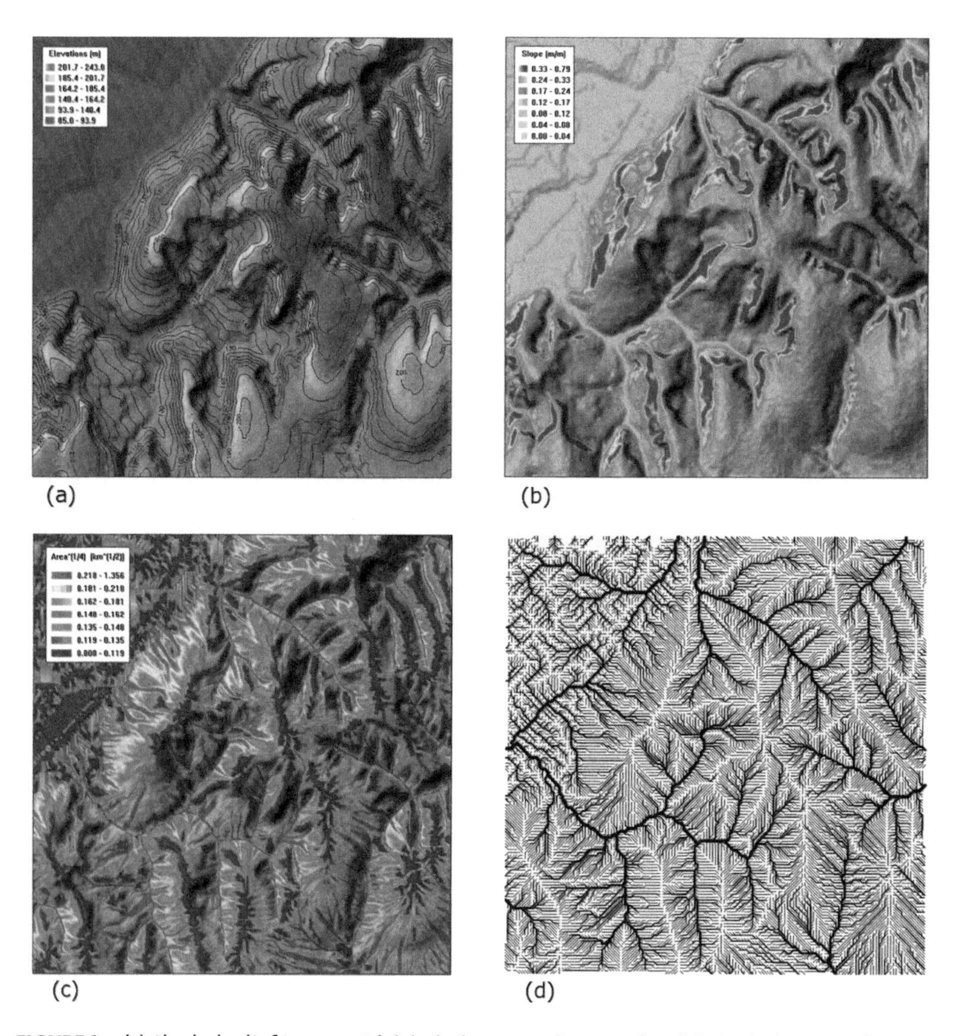

FIGURE 3 (a) Shaded relief image with labeled contour line overlay; (b) Shaded image of a D8 slope grid; (c) Shaded image of a total contributing area grid, extracted using the mass flux method; (d) Drainage pattern obtained by plotting all D8 flow vectors; (e) Watershed subunits with overlaid contours and channels (blue), using a D8 area threshold of 0.025 km^2; (f) Shaded image of plan curvature, extracted using the method of Zevenbergen–Thorne. (See page 741 in Colour Plate Section at the back of the book.)

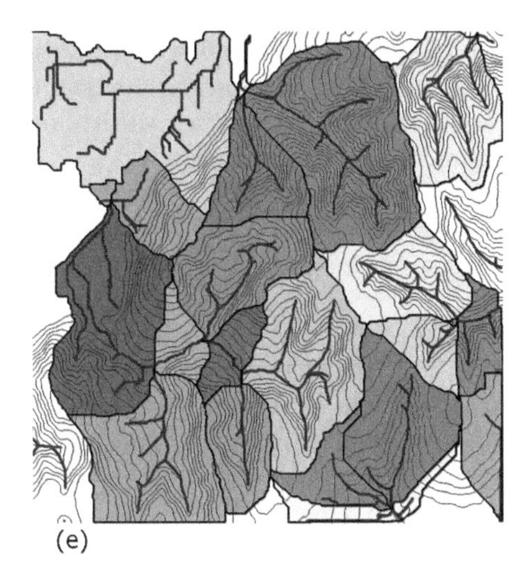

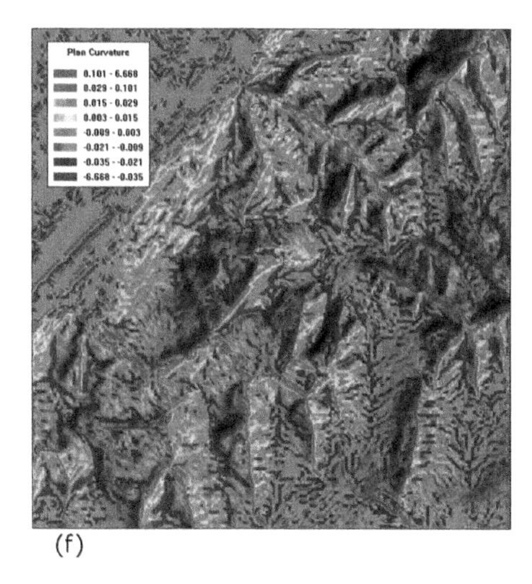

FIGURE 3 (*continued*)

basins is also provided and is necessary for cases where flow paths terminate in the interior of a DEM, such as at sinkholes, land-locked lakes or craters. The default flat resolution method is "*Iterative linking*". As long as the entire boundary of a river basin is contained within the bounding box of the DEM, each of the flat resolution methods will almost always produce flow directions within flat areas of the basin that send water in the right direction, despite the absence of a local elevation gradient (see the discussion of edge effects in Chapter 7).

Within broad, flat valleys, however, the "*iterative linking*" method Jenson (1985, 1991) produces multiple streamlines that flow parallel to one another until there is a bend in the axis of the valley that causes them to merge. The main problem with these parallel flow paths is that the point at which one stream merges into another (the confluence) is often displaced downstream a considerable distance from where it should be. The "*Imposed gradients*" option uses the method published by Garbrecht and Martz (1997) to create a cross-valley elevation gradient in flats and tends to produce a single flow path near the centre of the valley. However, this method sometimes results in two parallel flow paths near the centre of valleys instead of one. The "*Imposed gradients plus*" option was developed by Rivix to merge any parallel flow path pairs (in flats) into a single flow path.

NOTE. Increasing the vertical or horizontal resolution of DEMs does not eliminate artificial pits and flats and can even increase their numbers.

4.2 Extracting and saving a basin outlet

Once you have created a D8 flow grid, there is an easy-to-use graphical tool in RiverTools for precisely selecting which grid cell you want to use as a basin outlet. Choosing Basin Outlet from the Extract menu opens a dialog. Clicking on the dialog's Start button produces an image (shaded relief or density plot) that shows the entire DEM. If you then click within the image window, a streamline from the place where you clicked to the edge of the DEM will be overplotted on the image. You can move the mouse and click again to select and plot another streamline. Some of the streamlines will flow into the main channel of your basin of interest and some will flow into other, disjoint basins. Once you have selected a streamline that flows through the point you wish to use as a basin outlet, you can then use the slider in the dialog to move a red/white indicator along the streamline to your desired basin outlet point.

The precise grid cell coordinates are printed in the *Output Log* window, and you can click on the arrow buttons beside the slider to select any grid cell along the streamline, even if the image dimensions are many times smaller than the DEM dimensions. This graphical tool is designed so that you are sure to select a grid cell for the basin outlet that lies along any streamline that you select, instead of a few pixels to one side or the other. Once you have selected a grid cell as a basin outlet with this two-step graphical process, you simply click on the Save Outlet button in the dialog to save the coordinates in a text file with the extension "`_basin.txt`".

These coordinates identify the watershed that is of interest to you and are used by subsequent processing routines. Additional basic info for the basin will be appended to this file as you complete additional processing steps. By allowing any number of *basin prefixes* in addition to the *data prefix* associated with the DEM filename, RiverTools makes it easy to identify several watersheds in a given DEM and extract information for each of them separately while allowing them to share the same D8 flow grid and other data layers. You can change the basin prefix at any time using the *Change Basin Prefix* dialog in the File menu. This tells RiverTools which watershed you want to work with.

4.3 Extracting a river network

A river network can be viewed as a tree graph with its root at a particular grid cell, the outlet grid cell. The *Extract ↦ RT Treefile* dialog extracts the *"drainage tree"* for the watershed that drains to the outlet grid cell that you selected previously and saved. This is a raster to vector step that builds and saves the topology of the river network and also measures and saves a large number of attributes in a RiverTools vector (RTV) file with compound extension `_tree.rtv`. The *Extract ↦ River Network* dialog can then be used to distinguish between flow vectors on hillslopes and those that correspond to channels in a river network. The flow vectors on the hillslopes are *pruned away* and the remaining stream channels are saved in another RTV file with extension `_links.rtv`, along with numerous attributes. A variety of different pruning methods have been proposed in the literature and each has its own list of pros and cons. Figure 4 shows a river network extracted from SRTM data for the Jing River in China.

RiverTools supports pruning by D8 contributing area, by Horton–Strahler order, or by following each streamline from its starting point on a divide to the first inflection point (transition from convex to concave). In addition, you can use any grid, such as a grid created with the Grid Calculator (via *Extract ↦ Derived Grid*) together with any threshold value to define your own pruning method. The real

FIGURE 4 Jing River in the Loess Plateau of China, extracted from SRTM data with 3-arcsec grid cells.

test of a pruning method is whether the locations of channel heads correspond to their actual locations in the landscape, and this can only be verified by field observations. Montgomery and Dietrich (1989, 1992) provide some guidance on this issue. See Figure 4 in Chapter 7 for additional information on pruning methods.

Once you have completed the *Extract ↦ RT Treefile* and *Extract ↦ River Network* processing steps, you will find that your working directory now contains many additional files with the same basin prefix and different filename extensions. Each of these files contains information that is useful for subsequent analysis. Three of these files end with the compound extensions `_tree.rtv`, `_links.rtv` and `_streams.rtv`. These RTV files contain network topology as well as many measured attributes. For example, the attributes stored in the *stream file* for each Horton–Strahler stream are: upstream end pixel ID, downstream end pixel ID, Strahler order, drainage area, straight-line length, along-channel length, elevation drop, straight-line slope, along-channel slope, total length (of all channels upstream), Shreve magnitude, length of longest channel, relief, network diameter, absolute sinuosity, drainage density, source density, number of links per stream, and number of tributaries of various orders. RTV files and their attributes can also be exported as shapefiles with the *Export Vector ↦ Channels* dialog in the File menu.

4.4 Extracting grids

4.4.1 D8-based Grids

Once you have a D8 flow grid for a DEM, there are a large number of additional grid layers that can be extracted within the D8 framework. RiverTools currently

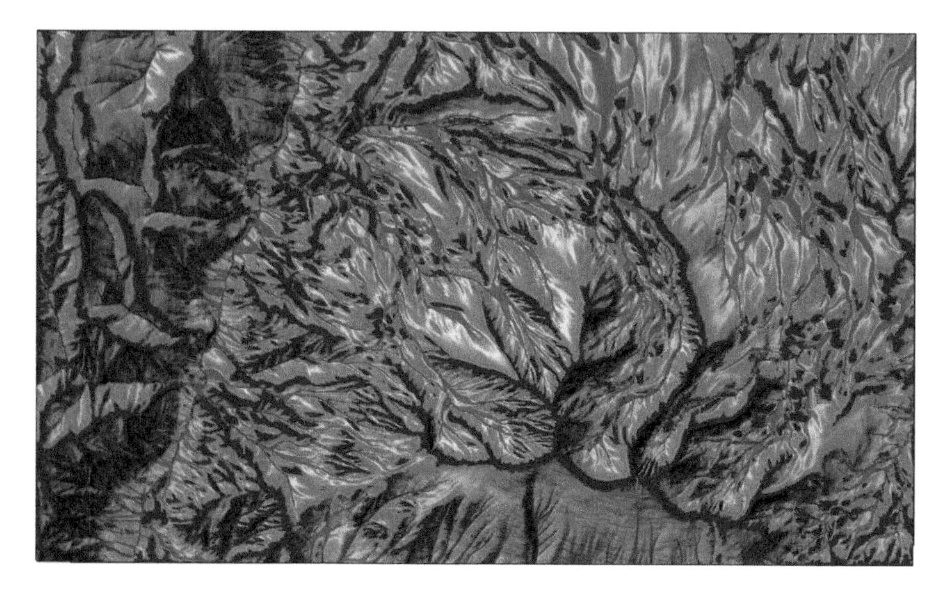

FIGURE 5 A relief-shaded image of a TCA grid for Mt. Sopris, Colorado, that was created using the Mass Flux method. Areas with a large TCA are shown in red while areas with a small TCA value (e.g. ridgelines) are shown in blue and purple. Complex flow paths are clearly visible and results are superior to both the D8 and D-infinity methods. (See page 742 in Colour Plate Section at the back of the book.)

has 14 different options in the *Extract* ↦ *D8-based Grid* menu. D8 area grids and slope grids are perhaps the best-known (see Chapter 7), but many other useful grid layers can be defined and computed, including grids of flow distance, relief, watershed subunits and many others. Each of these derived grids inherits the same georeferencing information as the DEM.

4.4.2 D-Infinity Grids

As explained in Chapter 7, the D-Infinity algorithms introduced by Tarboton (1997) utilise a continuous flow or aspect angle and can capture the geometry of divergent flow by allowing *"flow"* to more than one of the eight neighbouring grid cells. These grids can be computed in RiverTools by selecting options from the *Extract* ↦ *D-Infinity Grid* menu.

4.4.3 Mass Flux Grids

As also explained in Chapter 7, the RiverTools Mass Flux algorithms provide an even better method for capturing the complex geometry of divergent and convergent flow and its effect on total contributing area (TCA) and specific contributing area (SCA). These grids can be computed in RiverTools by selecting options from the *Extract* ↦ *Mass Flux Grid* menu. Figures 5 and 3(c) show examples of contributing area grids computed via this method. Figure 6 shows continuous-angle flow vectors in the vicinity of a channel junction or fork that were extracted using the Mass Flux method and then displayed with one of the interactive window tools.

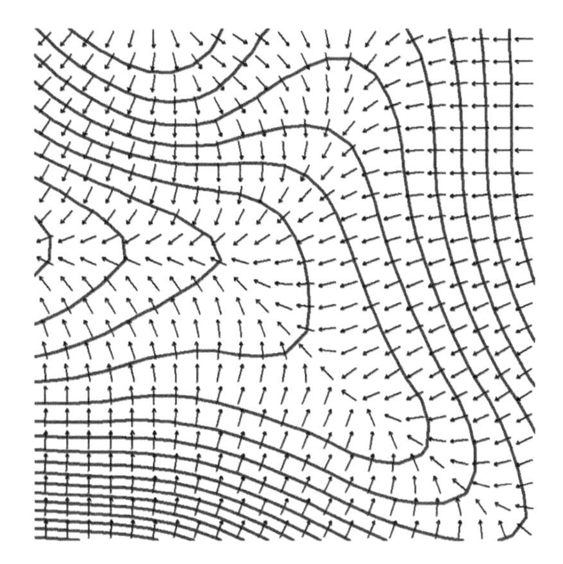

FIGURE 6 Continuous-angle flow vectors in the vicinity of a channel junction or fork, extracted using the Mass Flux method. © 2008 Rivix LLC, used with permission.

4.4.4 Finite Difference Grids

RiverTools can compute many standard morphometric parameters such as slope, aspect, first and second derivatives, and five different types of curvature. It currently does this using the well-known method of Zevenbergen and Thorne (1987) that fits a *partial quartic* surface to the (3×3) neighbourhood of each pixel in the input DEM and saves the resulting grid as a RiverTools Grid (RTG) file. Additional methods are planned for inclusion in the next release. These grids can be computed by selecting options from the *Extract* ↦ *Finite Difference Grid* menu.

4.4.5 Other Derived Grids

The *Extract* ↦ *Derived Grids* menu lists several other tools for creating grids. The most powerful of these is the Grid Calculator that can create a new grid as a function of up to three existing grids without requiring the user to write a script. For example, it can be used to create any type of *wetness index* grid from grids of slope and specific area. The dialog resembles a standard scientific calculator. In addition to the operators shown, any IDL command that operates on 2D arrays (i.e. grids) can be typed into the function text box. The Restricted to RTM tool lets you create grids in which masked values are reassigned to have nodata values. For example, this tool can be used to create a new DEM in which every grid cell that lies outside of a given watershed's boundary is assigned the nodata value.

4.5 Extracting masks or regions of interest

Within grid layers one often wishes to restrict attention or analysis to particular *regions of interest* or *polygons*, such as watersheds, lakes, craters, or places with elevation greater than some value. In order to display or perform any kind of analysis for such a region, we need to know which grid cells are in the region and which are not. This is equivalent to knowing the spatial coordinates of its boundary. A large number of different attributes can be associated with any such polygon, such as its area, perimeter, diameter (maximum distance between any two points on the boundary), average elevation, maximum flow distance or centroid coordinates. RiverTools Mask (RTM) files provide a simple and compact way to store one or more *masked regions* in a file. A complete description of RTM files is given in an appendix to the user's guide.

There are a number of different tools in the *Extract* ↦ *Mask* submenu that can be used to create RTM files. For example, watershed polygons of various kinds can be extracted with the Sub-basin Mask tool, lake polygons can be extracted with the Connected-to-Seed Mask tool, and *threshold* polygons can be extracted with

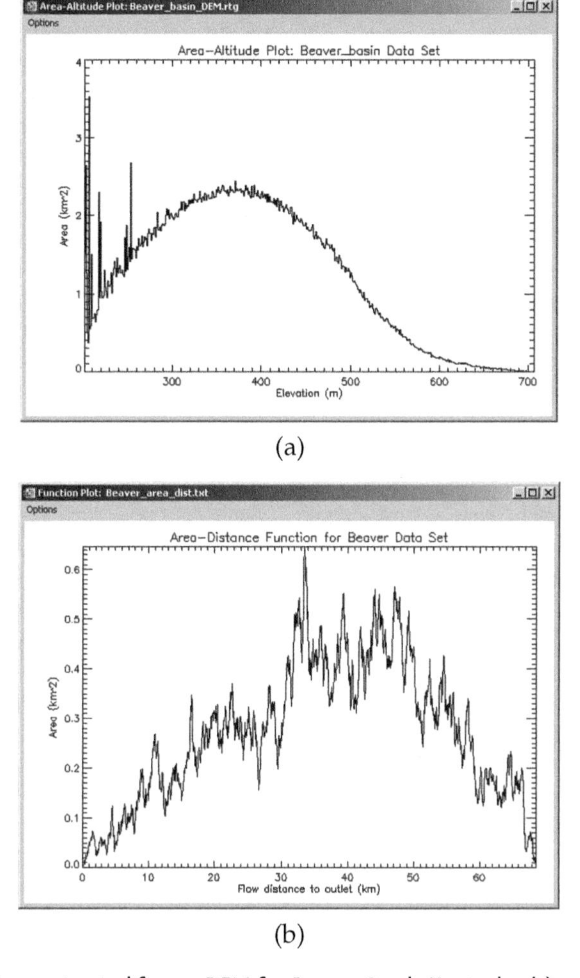

FIGURE 7 Functions extracted from a DEM for Beaver Creek, Kentucky: (a) an area–altitude plot and (b) an area–distance plot.

the Grid Threshold Mask tool. Creative use of these tools can solve a large number of GIS-query problems. RTM files that record the locations of single or multi-pixel pits are created automatically by the *Extract* ↦ *Flow Grid (D8)* tool. A tesselation of watershed *subunits* can be created with the *Extract* ↦ *D8-based Grid* ↦ *Watershed Subunits* tool. RTM files can also be merged by the *Merge Files* tool in the *Prepare* menu. Given an RTM file for a region of interest, the *Export Vector* ↦ *Boundaries* tool in the File menu can create an ESRI shapefile for the polygon and can also compute and save 36 optional attributes (new in the next release).

4.6 Extracting functions

Hypsometric curves or *area–altitude functions* have a long history (Strahler, 1952; Pike and Wilson, 1971; Howard, 1990) and RiverTools can extract this and several other functions from a DEM (Figure 7). The *width function* (Kirkby, 1976; Gupta et al., 1980; Troutman and Karlinger, 1984) and closely related *area–distance function* measure the fraction of a watershed (as number of links or percent area) that is at any given flow distance from the outlet (*Extract* ↦ *Function menu*) and are tied to the instantaneous unit hydrograph concept. The *cumulative area function* (Rigon et al., 1993; Peckham, 1995b) measures the fraction of a watershed that has a contributing area greater than any given value (*Extract* ↦ *Channel Links* ↦ *Link CDF*). *Empirical cumulative distribution functions* (ECDFs) (Peckham, 1995b; Peckham and Gupta, 1999) for ensembles of basins of different Strahler orders have been shown to exhibit statistical self-similarity: *Analyze* ↦ *Strahler streams* ↦ *Stream CDFs*. It has been suggested by Willgoose et al. (2003) that some of these functions can be used together to measure the correspondence between real and simulated landscapes.

5. VISUALISATION TOOLS

RiverTools has a rich set of visualisation tools, many of which are centrally located in the *Display menu*. Each tool provides numerous options which are explained in context-specific help pages, available by clicking on the Help button at the bottom of the dialog. After changing the settings in the dialog, you click on the Start button to create the image. There are too many display tools and options to describe each one in detail here, so the purpose of this section is to provide a high-level overview. Many of the tools have their own colour controls, but colour schemes can also be set globally with the *Set Colors* dialog and saved with the *Set Preferences* dialog. Both of these are launched from the File menu. Most of the images created by tools in the Display menu can be shown with a map projection, and the projection can be configured with the *Map Projection Info* dialog at the bottom of the menu. Menus labelled *Options*, *Tools* and *Info* at the top of image windows provide additional functionality, such as the ability to print an image or save it in any of several popular image formats. The Tools menu contains a large number of Interactive Window Tools that will be highlighted in the next section.

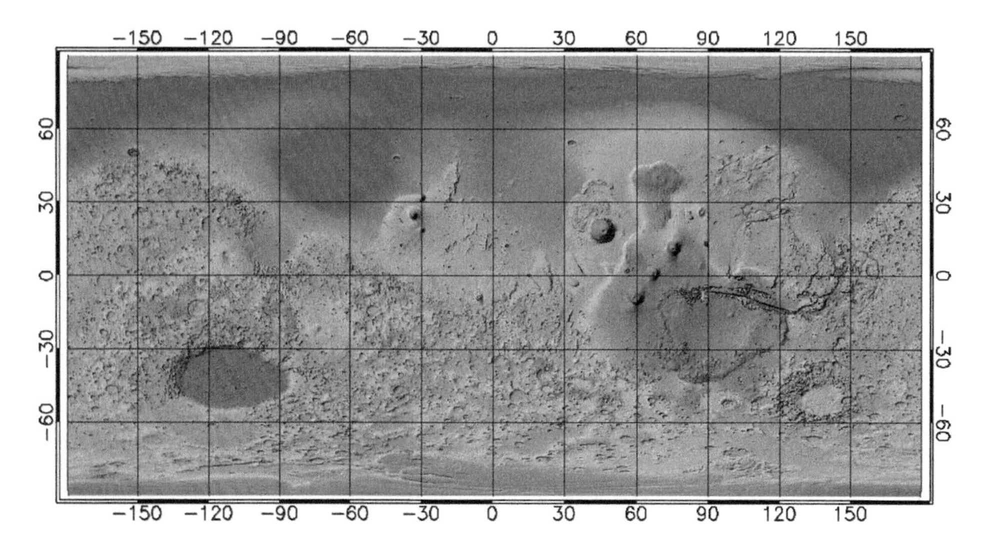

FIGURE 8 High-resolution MOLA (Mars Orbiter Laser Altimeter) DEM displayed in RiverTools: colour shaded relief image for planet Mars shown by the cylindrical equidistant map projection. (See page 743 in Colour Plate Section at the back of the book.)

The *Density Plot* tool creates colour-by-number plots, and offers many different types of contrast-enhancing *'stretches'* including linear, logarithmic, power-law and histogram equalisation. For example, contributing area grids are best viewed with a power-law stretch, due to the fact that there are a small number of grid cells with very large values and a large number with very small values. The *Contour Plot* tool makes it easy to create either standard or filled contour plots (or both as a multi-layer plot) and provides a large number of options such as the ability to control the line style, width and colour of each contour line. Colour shaded relief images with different colour tables and lighting conditions can easily be created with the *Shaded Relief* tool (Figure 8). There is also a tool called *Shaded Aspect* that simply uses D8 flow direction values with special colour tables to visualise DEM texture. A *Masked Region* tool allows you to display the boundaries or interiors of one or more *"mask cells"* or polygons (e.g. basins, pits, lakes, etc.) which are stored in RTM (RiverTools Mask) files with the extension `.rtm`. A related tool is the *ESRI Shapefile* tool which has numerous options for plotting vector data that is stored in a shapefile, including points, polylines and polygons. (Shapefiles may be created from RTV and RTM files with the *Export Vector* ↦ *Channels* and *Export Vector* ↦ *Boundaries* tools in the File menu.) A button labeled *View Attr. Table* at the bottom of this dialog displays a shapefile's attribute table, and the table can be sorted by clicking on column headings. Digital Line Graph (DLG) data in the now-standard SDTS format can be displayed by itself or as a vector overlay with the *DLG–SDTS* tool.

The *Function* tool in the Display menu reads data from a multi-column text file and creates a plot of any two columns. There are several places in RiverTools where data can be saved to a multi-column text file (e.g. longitudinal profiles) and later displayed with this tool. Perspective-view plots for an entire DEM can be displayed with the *Surface Plot* tool as wire-mesh, lego-style or shaded. For larger DEMs, however, better results are obtained with the Surface Zoom window tool which is explained in the next section. Extracted river networks, which are saved in RTV (RiverTools Vector) files can be displayed with the *River Network* tool, or first exported via *File* ↦ *Export Vector* ↦ *Channels* and displayed with the ESRI Shapefile tool. Using the *Multi-Layer Plot* tool, images created by many of the tools in the Display menu can be overlaid, that is, any number of vector plots can be overlaid on any raster image.

One of the most powerful tools in the Display menu is the *Grid Sequence* tool. This tool is for use with RTS (RiverTools Sequence) files, which are a simple extension[1] of the RTG (RiverTools Grid) format. RTS files contain a grid sequence, or grid stack, usually with the same georeferencing as the DEM. Grids in the stack are usually indexed by time and are typically created with a spatially-distributed model that computes how values in every grid cell change over time. For example, a distributed hydrologic model called TopoFlow[2] can be used as a plug-in to RiverTools (see Chapter 25). TopoFlow computes the time evolution of dynamic quantities (e.g. water depth, velocity, discharge, etc.) and can save the resulting sequence of grids as an RTS file. Landscape evolution models also generate grid stacks that show how elevations change over time. This tool can show a grid stack as an animation or save it in the AVI movie format. It allows you to jump to a particular frame, change colours and much more. The Options menu at the top of the dialog has many additional options and there is also a Tools menu that has tools for interactively exploring grid stack data, such as the *Time Profile* and *Animated Profile* tools.

5.1 Interactive window tools

As mentioned previously, image windows that are created with the tools in the Display menu typically have three menus near the top of the window labelled Options, Tools and Info. In RiverTools, the entries in an Options menu represent simple things that you can do to the window, such as resize it, print it, close it or save the image to a file. The entries in a Tools menu represent ways that you can use the mouse and cursor to interact with or query the image. Here again we will simply give a high-level overview of several of these tools, but more information is provided in the user's guide.

The *Line Profile* tool lets you click and drag in an image to draw a transect and then opens another small window to display the elevation values along that transect. Note that this new window has its own Options menu that lets you do things like save the actual profile data to a multi-column text file. The *Channel Profile* tool is similar (Figure 9), except that you click somewhere in the image and then the flow path or streamline from the place where you clicked to the edge of the DEM is overlaid on the image. The elevations (or optionally, the values in any other grid)

[1] All of the RiverTools formats are nonproprietary and are explained in detail in an appendix to the user's guide.
[2] http://instaar.colorado.edu/topoflow/.

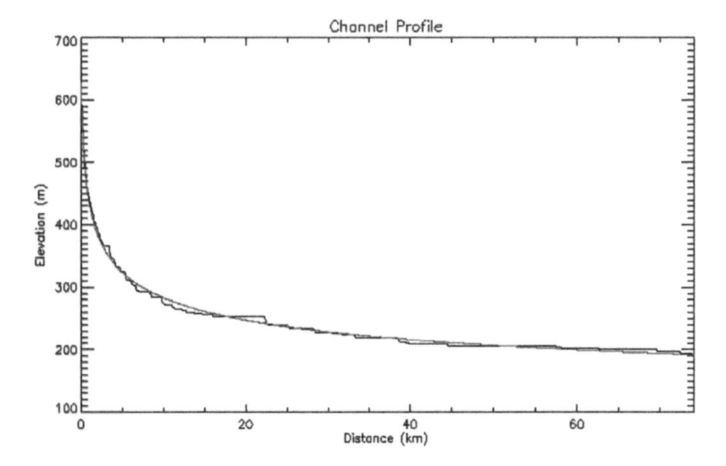

FIGURE 9 Longitudinal profile plot created for a main channel of the Beaver Creek DEM with the Channel Profile tool.

along that streamline are plotted vs. distance along the streamline in another small window. Again, the Options menu of this new window has numerous entries.

The *Reach Info* tool is similar to the Channel Profile tool but opens an additional dialog with sliders that let you graphically select the upstream and downstream endpoints of any reach contained within the streamline and displays various attributes of that reach. If you select *Vector Zoom* from the Tools menu and then click in the image, crosshairs are overlaid on the image and a small window is displayed that shows grid cell boundaries, D8 flow paths and contour lines in the vicinity of where you clicked.

The *Value Zoom* tool is similar but displays actual grid values as numbers and also shows the coordinates of the selected grid cell (Figure 10). This tool has many other capabilities listed in its Options menu, such as the ability to edit grids or jump to specified coordinates. Perspective, wire mesh plots are more effective when applied to smaller regions rather than to entire DEMs, so the *Surface Zoom* tool provides a powerful way to interactively explore a landscape (Figure 11). This

Value Zoom: Beaver_DEM.rtg

Options Info

345	345	342	342	343
333	333	332	334	335
321	320	322	326	328
309	307	310	317	319
296	293	297	304	308

X: 344280.00 Y: 4156140.0

(col, row) = (576, 357)

FIGURE 10 The Value Zoom dialog.

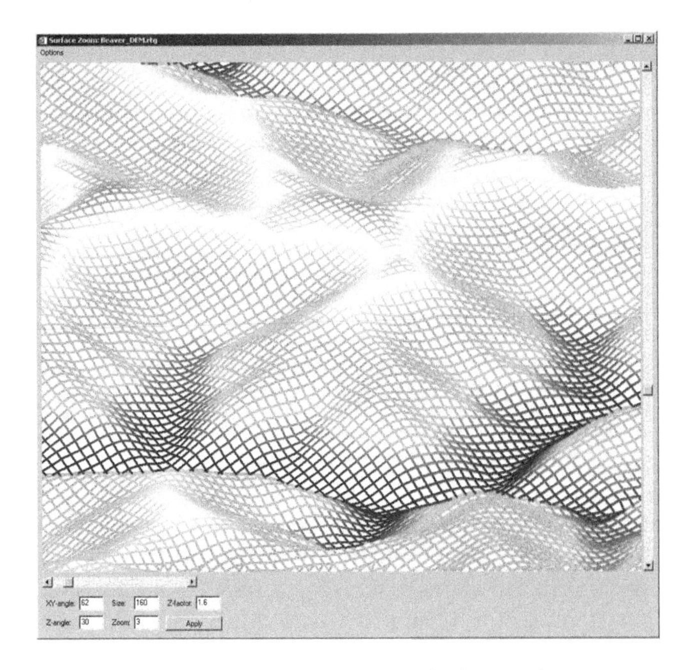

FIGURE 11 The Surface Zoom display window.

tool has many settings at the bottom of the display window and many entries in its Options menu. The *Density Zoom* and *Relief Zoom* tools show density plots (see last section) and shaded relief plots at full resolution for a selected region even though the main image may show the entire area of the DEM at a greatly reduced resolution. All of the Zoom-tools are automatically linked, so that they all update when you move the mouse to another location in the image. The *Add Scale Bar*, *Add Colour Bar*, *Add Text* and *Add Marker* tools can be used to interactively annotate an image prior to saving it to an image file with *Options* ↦ *Save* Window.

Finally, the *Flood Image* tool allows you to change the colour of all pixels below a given elevation to blue, either instantly or as an animation. It is a useful visualisation tool but does not model the dynamics of an actual flood.

6. SUMMARY POINTS

RiverTools is a powerful but easy-to-use toolkit for visualising and extracting information from digital elevation data. It has an intuitive, point-and-click graphical interface, an extensive HTML-based help system and much of the power of a full-featured GIS even though its main focus is on digital elevation data. It also contains state-of-the-art algorithms for computing geomorphometric quantities, such as the new Mass Flux method for computing contributing area. This unique combination of features makes it ideal for teaching courses in hydrology, landscape ecology and geomorphology. RiverTools can import a wide variety of DEM formats as well as vector data in the ESRI shapefile and DLG-SDTS formats. It works well together with other GIS software since it can also export raster data in several common formats (via *File* ↦ *Export Grid*) and vector data in the industry-standard shapefile format (via *File* ↦ *Export Vector*). Publication-quality graphics and posters are easily created and annotated. Many built-in features including a graphical Grid Calculator and support for wildcards in many places where an input filename is required (to allow batch processing) mean that writing scripts is usually not necessary. However, in cases where scripting is required, users have the option to purchase another product called IDL (Interactive Data Language, a product of ITT Visual Information Solutions, www.ittvis.com) that can be used to write extensions to RiverTools. This option provides access to all of the features of the IDL programming language in addition to a large set of documented, low-level RiverTools commands for customisation. Users can also extend RiverTools with free User menu plug-ins, such as a landscape evolution model called Erode and a spatially-distributed hydrologic model called TopoFlow.

RiverTools has been developed and refined over many years around three central themes, namely (1) ease of use, (2) ability to handle very large DEMs (whatever the task) and (3) accuracy of measurements. With regard to ease of use, Rivix has worked with users for many years to develop a user-friendly graphical interface and HTML help system. As for the ability to rapidly extract information from very large DEMs, this has driven the development of advanced algorithms that efficiently distribute the computational workload between available RAM and I/O to files. These types of algorithms are used throughout RiverTools. Finally, RiverTools and MicroDEM may be the only GIS applications that always take the latitude-dependence of pixel geometry into account when working with geographic DEMs. All lengths, slopes and areas are computed by integrating on the surface of the appropriate ellipsoid model to avoid the geometric distortion that is associated with map projections. This feature is especially important when working with DEMs at the regional, continental or global scale.

IMPORTANT SOURCES

Rivix LLC, 2004. RiverTools 3.0 User's Guide. Rivix Limited Liability Company, Broomfield, CO, 218 pp.
http://rivertools.com — RiverTools website.
http://instaar.colorado.edu/topoflow/ — TopoFlow website.
http://www.ittvis.com — ITT Visual Information Solutions.

Part III

Applications

CHAPTER 19

Geomorphometry — A Key to Landscape Mapping and Modelling

T. Hengl and **R.A. MacMillan**

importance of DEMs for mapping natural landscapes · spatial prediction of environmental variables using land-surface parameters and objects · difference between indirect, direct and empirical prediction models · regression-kriging and its properties · implementation of process-based models · expert systems and how do they work? · fuzzy logic and its uses for mapping · evaluation of quality of prediction models

1. IMPORTANCE OF DEMS

The major argument in support of using DEMs for mapping and modelling natural landscapes is the *variety and richness*[1] *of the metrics, measures and objects* that can be derived through automated analysis of elevation data. There are at least 30–50 original univocal land-surface parameters, although the list can be extended to more than 100 land-surface parameters. An updated online gallery of the most used land-surface parameters is available via geomorphometry.org.

DEM data have been also praised for providing continuous coverage for large areas at a relatively *low cost*. Information contained in DEM data tends to be different from, and complimentary to, spectral information contained in airborne and satellite imagery. The relative stability of the terrain surface through time has been widely recognised as a significant advantage (Rowe and Barnes, 1994; McKenzie et al., 2000). Image data primarily capture information about the state of the surface cover at a given instant in time.

Automated analysis of elevation data can consistently and rapidly extract many parameters or object entities that can be treated as direct analogues of the criteria that are used by a manual interpreter to identify and delineate objects in the fields

[1] Although there is almost an immeasurable number of parameters that can be derived from a DEM, many of these represent the same information in slightly altered form, so there are limits to the range of information contained in DEM data. See further Figure 5.

Developments in Soil Science, Volume 33 © 2009 Elsevier B.V.
ISSN 0166-2481, DOI: 10.1016/S0166-2481(08)00019-6.

of soils, ecology, geomorphology and geology. These automatically computed digital outputs provide measures of surface form, context, pattern and texture that can be used as surrogates for the criteria considered in the manual photo interpretation process.

The vast majority of *environmental issues* for which maps of environmental phenomena are prepared tend to include a requirement for interpreting or analysing how water and energy interact with site conditions in response to either man-made or natural influences. Whether we are concerned with crop growth, transport and fate of contaminants, sequestration of carbon, modelling of forest fires, degradation of soils from erosion or salinity, identification of geomorphic hazards, flooding or any of a large number of other issues, a common thread is the ability to track the movement of water through the landscape and to track changes in the status of water and energy in the landscape.

In this chapter, we examine how DEMs can act as affordable sources of information to support mapping and modelling of natural landscapes. Specifically, we provide an overview of many of the ways in which DEM parameters can be used to map various environmental conditions efficiently and consider the capabilities and limitations of different approaches.

1.1 Relief and landscape

Topography is one of the key factors in controlling many of the most significant natural processes of interest to humans. For example, in *soil science*, topography is consistently recognised as the key determinant in the development and functioning of soils at a local or landscape level through reference to the soil–landscape paradigm (Hudson, 2004; McBratney et al., 2003; Grunwald, 2005). Similarly, in ecology, ecological differences are understood to be primarily controlled by changes in topography that produce gradients in moisture, energy and nutrients across the landscape (Davis and Goetz, 1990; Fels and Matson, 1996).

> REMARK 1. *If climate, vegetation and parent material are held constant, information on relief is often sufficient to produce reliable maps of soil, vegetation or ecological units.*

Consider Rowe's (1996) observation that Earth's surface energy/moisture regimes at all scales/sizes are the dynamic driving variables of functional ecosystems at all scales/sizes and that these energy/moisture regimes are primarily controlled by variations in topography. This concept of soil–landform and vegetation–landform relationships has frequently been presented in terms of an equation with five key factors as (Jenny, 1941; McBratney et al., 2003; Grunwald, 2005):

$$S, V = f(c, o, r, p, t) \tag{1.1}$$

where S stands for soil, V for vegetation, c stands for climate, o for organisms (including humans), r is relief, p is parent material or geology and t is time. At regional scales (1–100 km^2), climate and parent material are often relatively homogeneous or are observed to vary in response to topography. At these scales, both vegetation and soils are often observed to exhibit spatial variation that is primarily related to changes in topography and geomorphology. In soil mapping, this soil–landform relationship underpins the so-called *catena concept* (Jenny, 1941).

Where topographic influences are predominant (regional and local scales), the formula from above can be simplified to:

$$S, V = f(r, t) \tag{1.2}$$

This can be interpreted to mean that, if climate, vegetation and parent material are held constant, information on relief of appropriate accuracy should be sufficient to produce correct maps of soil, vegetation or ecological units.

1.2 A review of applications

To list all applications of DEMs in environmental and Earth sciences would require identification of hundreds to thousands of papers in each discipline and is outside the scope of this book. For example, the bibliography of published works on geomorphometry maintained by Pike (2002) contains over 6000 entries. It is neither feasible nor desirable to attempt to replicate such a massive effort in this book. Instead we will focus on the most important reference sources.

The most common groups of applications of geomorphometry in environmental and Earth sciences are:

Geomorphology and geology The field of geomorphology has a long history of analysing digital elevation data to extract and classify geomorphic entities. Weibel and DeLotto (1988), Weibel and Heller (1990) elucidated a framework for automated landform classification using digital elevation data. Pike (1988) introduced the concept of using analysis of digital elevation data to establish what he called a *geometric signature* defined as "*a set of measurements that describe topographic form well enough to distinguish geomorphologically disparate landscapes*". Many early geomorphic studies were concerned with developing procedures for automatically recognising surface specific points identified as pits, peaks, channels, ridges (or divides), passes and the planar hillslope segments that occurred between divides and channels (Peucker and Douglas, 1975; Graff and Usery, 1993; Wood, 1996; Herrington and Pellegrini, 2000). These geomorphic approaches relied upon analysis of local surface shape (convexity/concavity) to differentiate morphological elements. Some shape-based geomorphic models expanded their classifications to differentiate divergent, convergent and planar hillslope components in addition to the pits, peaks, channels and divides (Pennock et al., 1987, 1994; Irvin et al., 1997; Herrington and Pellegrini, 2000; Shary et al., 2005). Subsequent geomorphic research offered suggestions for computing different measures of relative landform position and for including these measures as key inputs to automated procedures for classifying landforms (Franklin, 1987; Skidmore, 1990; Skidmore et al., 1991; Fels and Matson, 1996; Twery et al., 1991; MacMillan et al., 2000). Dikau (1989) and Dikau et al. (1991) developed and applied an automated method for classifying macro landform types from digital elevation data that was based on analysis of variation in topographic measures within areas defined by moving windows. Automated extraction and classification of geomorphic spatial entities has become increasingly sophisticated with recognition of more subtle and complex landform features (Miliaresis and Argialas, 1999; Leighty, 2001; Lucieer et al., 2003; Schmidt and Hewitt, 2004) that incorporate considerations of texture, pattern and context, in addition to shape and relative slope position (see further Chapter 22).

Hydrology The field of hydrology has also made extensive use of automated analysis of elevation data. Many studies have reported methods for simulating surface flow networks using grid-based calculations of cell-to-cell connectivity (Mark, 1975b; O'Callaghan and Mark, 1984; Tarboton et al., 1991). Others have computed flow topology using contour (O'Loughlin, 1986; Moore et al., 1991a) or triangular irregular network (TIN) representations of the topographic surface (Weibel and Heller, 1990). By tracing cell to cell flow to establish flow topology, hydrological researchers have been able to automatically extract a virtually identical set of surface features as those recognised by geomorphic analysis; namely pits, peaks, channels, divides, passes and the hillslopes that occur between divides and channels (Band, 1989). Automated extraction of hydrological spatial entities has also become increasingly sophisticated, with capabilities now offered to extract complex hydrological spatial data models such as the ArcGIS Hydro spatial data model proposed by ESRI (see Chapter 11), RiverTools (see Chapter 18) and TAS packages (see Chapter 16). In addition to automated extraction of hydrological spatial entities, hydrologists investigate rapid and cost-effective mechanisms for estimating the spatial distribution of parameter values for physically-based, deterministic hydrological models (see further Chapter 25).

Soil science Methods that used topographic derivatives to predict the continuous spatial distribution of individual soil properties have been reviewed by Moore et al. (1993a), McBratney et al. (2003) and Bishop and Minasny (2005). A second approach has been to partition the landscape into classes, generally conceptualised as hillslope elements along a topographic sequence, that are typically described as being occupied by a particular soil or range of soils (Pennock et al., 1987; MacMillan et al., 2000; Bui and Moran, 2001; Moran and Bui, 2002). About 80% of automated digital soil mapping applications today are based on the use of DEMs (Bishop and Minasny, 2005).

Vegetation science Ecologists were also quite early to recognise the potential for analysing digital elevation data to quantify environmental gradients and use these to aid in automatically mapping ecological classes. Examples of ecological classification achieved using DEM data are provided by Band (1989), Fels (1994), Burrough et al. (2001). Similarly, derivatives of elevation data have been used to help predict the distribution of tree species in forest classification (Twery et al., 1991; Skidmore et al., 1991; Antonić et al., 2003), help explain spatial patterns of biodiversity (Latimer et al., 2004) and are finding increased use as automated vegetation classification has expanded to rely on additional data sources besides remotely sensed imagery (Paul et al., 2004). A review of applications of DEMs for vegetation mapping can be found in the work of Franklin (1995) and Alexander and Millington (2000).

Climatology and meteorology DEMs are most commonly used to adjust measurements at meteorological stations to local topographic conditions. Two groups of applications are most common today: (a) modelling of soil radiation (Antonić et al., 2000; Donatelli et al., 2006) and (b) modelling of wind flux (McQueen et al., 1995; Chock and Cochran, 2005). In many cases, DEMs are only used to improve interpolation of the climatic variables over regions or continents (Houlder et al., 2000; Lloyd, 2005). In other cases, the objective is to exactly model the processes to create both spatial and temporal predictions of the meteorological/climatic conditions (see further Chapter 26).

2. PREDICTIVE MODELLING OF ENVIRONMENTAL VARIABLES

Relevant and detailed geoinformation[2] is a prerequisite for successful management of natural resources in many applied environmental and geosciences. Until recently, such information has primarily been produced by various types of field surveys, which were then used to create descriptions or maps for entire areas of interest. Because field data collection is often the most expensive part of a survey, survey teams typically visit only a limited number of sampling locations and then, based on the sampled data and statistical and/or mental models, infer conditions for the whole area of interest.

The process of predicting values of a sampled variable for a whole area of interest is called *spatial prediction* or spatial interpolation (Goovaerts, 1997; Webster and Oliver, 2001). With the rapid development of remote sensing and geoinformation science, survey teams have increasingly created their products (geoinformation) using ancillary data sources and computer programs. For example, sampled concentrations of heavy metals can be mapped with higher accuracy/detail if information about the sources of pollution (distance to industrial areas and traffic, map showing the flooding potential or wind exposition) is used to improve spatial prediction.

REMARK 2. *Increasingly the heart of a mapping project is, in fact, the computer program that implements some (geo)statistical algorithm that has shown to be successful in predicting target values.*

Increasingly the heart of a mapping project is, in fact, the computer program that implements some (geo)statistical algorithm that has shown itself to be successful in predicting target values. This leads to the so-called *direct-to-digital* system in which the surveyors only need to prepare their primary survey data, which are then processed in a semi-automated way through data processing wizards. Of course, this does not mean that surveyors are becoming obsolete. On the contrary, surveyors continue to be needed to prepare and collect the input data and to assess the results of spatial prediction. On the other hand, they are less and less

[2] Geoinformation, short for Geographic Information, usually consists of vector or raster maps produced in a GIS that carry information about a location on the Earth's surface. A distinction needs to be made between any raw data that has a spatial reference (geodata) and GIS products (geoinformation), which require no further processing.

involved in the actual delineation of features or derivation of predictions, which is increasingly the role of the predictive models.

In the case of spatial prediction of environmental variables, as in general statistics, we are interested in modelling some (*target variables*) feature or variable of interest using a set of inputs (*predictors*). Behind any statistical analysis is a *statistical model*, which defines inputs, outputs and the computational procedure to derive outputs based on the given inputs (Latimer et al., 2004).

We can distinguish among two major approaches to modelling of reality that vary with respect to the exactness of our understanding and to the amount of the random component in the model:

- *Direct (deterministic) estimation models* Here the assumption is that the outputs are determined by a finite set of inputs and they exactly follow some known physical law. The algorithm (formula) is known and the evolution of the output can be predicted exactly. For example, if we know temperature (in laboratory conditions), we can always calculate the volume of a gas using $V = k \cdot T$ (Charles's law). Note that even formulas from physics will have a small random component, factors that are not accounted for or simply measurement errors, which can be also dealt with statistical techniques. In the case of environmental systems, target variables are the product of dynamic ecological processes (i.e. time-dependent), so that the deterministic models used to predict them are also referred to as *process-based models* (Schoorl and Veldkamp, 2005; Gelfand et al., 2005).
- *Indirect estimation models* If the relationship between the feature of interest and physical environment is so complex[3] that it cannot be modelled exactly, we can employ some kind of the indirect estimator. In this case, we either do not exactly know: (a) the final list of inputs into the model, (b) the rules (formulas) required to derive the output from the inputs and (c) the significance of the random component in the system. So the only possibility is that we try to estimate some basic (additive) model that at least fits our expert knowledge or the actual measurements. In principle, there are two approaches to indirect estimation:
  - *Pure statistical models* In the case of pure statistical modelling, we want to completely rely on the actual measurements and then try to fit the most reasonable mathematical model that can be used to analytically estimate the values of the target variable over the whole study area. Although this sounds like a completely automatic procedure, the analysts have many options to choose whether to use linear or non-linear models, whether to consider spatial position or not, whether to transform or use the original data, whether to consider multicolinearity effects or not, etc.
  - *Expert-based or heuristic models* If we do not posses actual field measurements or if we already have a clear idea about the processes involved, we can employ some empirical rules or algorithms to improve the predictions. As with pure statistical models, we may not exactly know the inputs into

[3] Because either the factors are unknown, or they are too difficult to measure, or the model itself would be too complex for realistic computations.

the model, rules required to clarify predictions or the significance of the random component, however we generally might have a reasonable idea of the conceptual attributes and location in the landscape of the objects of interest. So the challenge is to find some way of identifying and combining relevant input data layers that will result in effective extraction and classification of the output objects of interest.

Note that, in practice, we can also have a combination of deterministic, statistical and expert-based estimation models. For example, one can use a deterministic model to estimate a value of the variable, then use actual measurements to fit a calibration model, analyse the residuals for spatial correlation and eventually combine the statistical fitting and deterministic modelling (Hengl, 2007). Most often, expert-based models are supplemented with the actual measurements, which are then used together with some statistical algorithm (e.g. neural networks) to refine the rules.

2.1 Statistical models

A crucial difference between statistical and deterministic (process-based) models is that, in the case of statistical models, the list of inputs and the coefficients/rules used to derive outputs are unknown and need to be determined by the analyst. However, we first need to ascertain what could be the general relationship between inputs and outputs, which in statistics is referred to as *statistical models* (Chambers and Hastie, 1992; Neter et al., 1996). There are (at least) four groups of statistical models that have been used to make spatial predictions with the help of (ancillary) land-surface parameters (McKenzie and Ryan, 1999; McBratney et al., 2003; Bishop and Minasny, 2005):

Classification-based models Classification models are primarily developed and used when we are dealing with discrete target variables (e.g. land cover or soil types). There is also a difference whether *Boolean* (crisp) or *Fuzzy* (continuous) classification rules are used to create outputs. Outputs from the model fitting process are class boundaries (class centres and standard deviations) or classification rules.

Tree-based models Tree-based models are often easier to interpret when a mix of continuous and discrete variables are used as predictors (Chambers and Hastie, 1992). They are fitted by successively splitting a dataset into increasingly homogeneous groupings. Output from the model fitting process is a *decision tree*, which can then be applied to make predictions of either individual property values or class types for an entire area of interest.

Regression models Regression analysis employs a family of functions called *Generalized Linear Models* (GLMs), which all assume a linear relationship between the inputs and outputs (Neter et al., 1996). Output from the model fitting process is a set of regression coefficients. Regression models can be also used to represent non-linear relationships with the use of *General Additive Models* (GAMs). The relationship between the predictors and targets can be solved using one-step data-fitting or by using iterative data-fitting techniques (neural networks and similar).

Hybrid geostatistical models Hybrid models consider a combination of the techniques listed previously. For example, a hybrid geostatistical model employs both correlation with auxiliary predictors and spatial autocorrelation simultaneously. There are two sub-groups of hybrid geostatistical models: (a) *co-kriging*-based and (b) *regression-kriging*-based (Goovaerts, 1997). Outputs from the model fitting process are regression coefficients and variogram parameters.

Each of the models listed above can be equally applicable for mapping and can exhibit advantages and disadvantages. For example, some advantages of using tree-based regression are that they can handle missing values, can use continuous and categorical predictors, are robust to predictor specification, and make very limited assumptions about the form of the regression model (Henderson et al., 2004). Some disadvantages of regression trees, on the other hand, is that they require large datasets and completely ignore spatial position of the input points.

A statistical technique that is receiving increasing attention by mapping teams is regression-kriging. It is attractive for environmental sciences because it simultaneously takes into account both the spatial location of sampled points and correlation with the predictors. Hence many statisticians consider regression-kriging to be the Best Linear Unbiased Predictor of spatial data (Christensen, 2001, pp. 275–311). An advantage of using GAMs, on the other hand, is that they are able to represent non-linear relationships in the data and therefore fit the actual field observations better. Decision trees are more suited for mixed types of input data and classification-based models are more suited for categorical target variables. There remains much opportunity for development and implementation of even more sophisticated and more generic and robust statistical models.

> REMARK 3. *Each statistical model — classification-based, tree-based, regression-based or hybrid — can be equally applicable for mapping and can exhibit advantages and disadvantages.*

In the following section, we will focus on regression-kriging as one of the most widely used linear statistical prediction models. In order to subsequently explain regression-kriging, we need to explain basic principles of regression analysis first.

As previously mentioned, there is a variety of linear statistical models (GLMs) that can be used for regression analysis. The simplest GLM is the (plain) linear regression with a single predictor and single output:

$$z_i = b_0 + b_1 \cdot q_i \tag{2.1}$$

where z_i is the target variable, b_0 (intercept) and b_1 are the regression coefficients and q_i is the predictor.[4] The coefficients are unknown and can only be determined by using paired observations (i) of both input and output variables. These paired observations can then be fitted so the scatter around the regression line is minimised — the so called *least-squares fitting* (Neter et al., 1996).

[4] Let a set of observations of a target variable z be denoted as $z(s_1), z(s_2), \ldots, z(s_n)$, where $s_i = (x_i, y_i)$ is a location and x_i and y_i are the coordinates (primary locations) in geographical space and n is the number of observations. A discretized study area $\mathbb{A}$ in a grid-based ('raster') GIS, consists of m cells, which can be represented as nodes by their centres, such that $s_i \in \mathbb{A}$ (see also Figure 1).

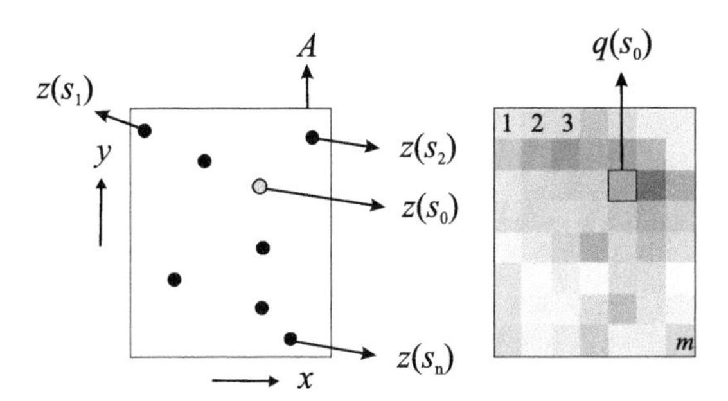

FIGURE 1 Spatial prediction is a process of estimating the values of the target variable z at new locations (s_0), given the sampled observations (s_i) and auxiliary predictor (q).

Regression modelling can also be extended to spatial prediction, so that predictors which are available over entire areas of interest (such as land-surface parameters) can be used to predict the value of a target variable at unvisited locations:

$$\hat{z}(s_0) = \hat{b}_0 + \hat{b}_1 \cdot q(s_0) \tag{2.2}$$

where $\hat{z}(s_0)$ is the estimated value based on the value of the predictor at new location (s_0) and $\hat{b}_0$ and $\hat{b}_1$ are the fitted regression coefficients using the real observations. This technique is often referred to as *environmental correlation* because only the correlation with the predictors is used to predict target variables (McKenzie and Ryan, 1999). The following example shows a regression model from Gessler et al. (1995) derived using 60 field samples (see also Figure 2):

$$\text{solum} = -57.95 + 12.83 \cdot \text{PLANC} + 21.46 \cdot \text{WTI} \tag{2.3}$$

where 'solum' is the solum depth, the target environmental variable. Note that this regression model is valid only for this study area and it would probably give unreliable results if applied to some other study area. This would happen not only because the soils and the soil forming factors would invariably differ between

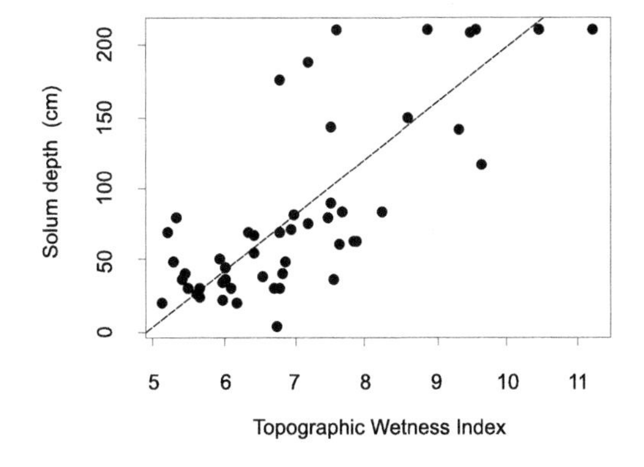

FIGURE 2 A simple example of a regression model used to predict solum depth using only TWI. Reprinted from Gessler et al. (1995). With permission from Taylor & Francis Group.

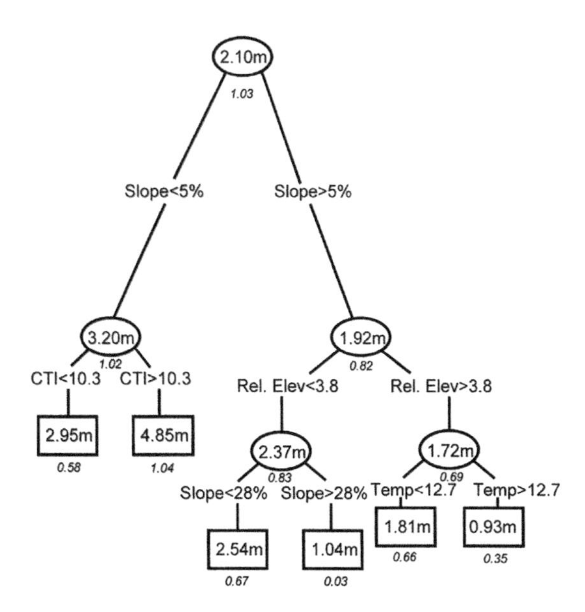

FIGURE 3 Example of a regression tree used to predict soil profile depth using CTI (TWI), relative elevation, SLOPE and temperature. Reprinted from McKenzie and Ryan (1999). With permission from Elsevier.

areas, but also because the land-surface parameters such as PLANC and WTI are relative to scale (grid resolution) and derivation method. Also note that it only makes sense to predict output values using land-surface parameters if the model is statistically significant. In the example presented, the R-square was 0.68, which means that the model accounted for 68% of total variation, which is statistically quite significant.

> REMARK 4. *A procedure where various DEM-based and RS-based parameters are used to explain variation in environmental variables is called environmental correlation.*

Regression analysis can also be combined with tree-based models, which leads to regression trees. This means that many regression models are fitted locally to optimise the data fitting. An example of a regression tree is presented in Figure 3.

A limitation of plain regression modelling as described above is that the spatial location of observations is not considered. Obviously, spatial location of observations plays an important role and should be included in the spatial prediction. The spatial autocorrelation structure can be estimated by plotting the semivariances (differences) between values of variables at pairs of points at various distances. The fitted variogram model showing the change of semivariances in relation to distances between pairs of points can be used to interpolate sampled values based on the spatial similarity structure — which is referred to as kriging (Isaaks and Srivastava, 1989; Goovaerts, 1997; Webster and Oliver, 2001).

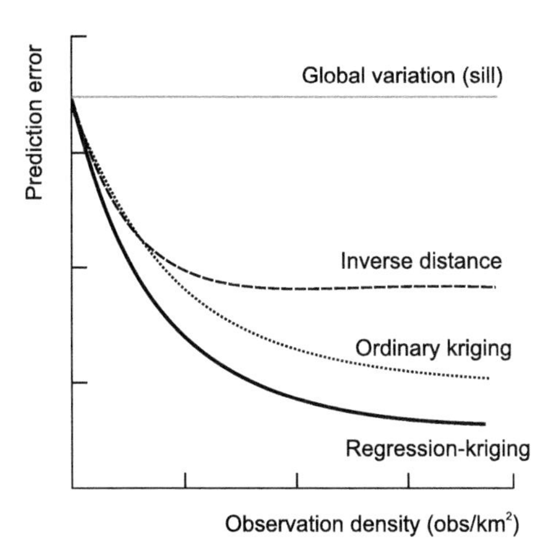

FIGURE 4 Schematised relationship between prediction efficiency and observation density for different interpolation methods. More sophisticated spatial prediction methods have proven to be better predictors (in general, not in all situations), but only to a certain extent. Based on Bregt (1992, p. 49).

The combined influence of both environmental predictors and spatial location of observations is commonly represented using the (additive) universal model of spatial variation:

$$\mathbf{z} = \mathbf{q} \cdot \hat{\beta} + \mathbf{e}, \qquad E(\mathbf{e}) = 0 \tag{2.4}$$

following this model, prediction at a new location can be produced by fitting the regression component and then summing the interpolated residuals back into the final result:

$$\begin{aligned} \hat{z}(s_0) &= \mathbf{q}_0^{\mathrm{T}} \cdot \hat{\beta}_{\mathrm{GLS}} + \lambda_0^{\mathrm{T}} \cdot (\mathbf{z} - \mathbf{q} \cdot \hat{\beta}_{\mathrm{GLS}}) \\ \hat{\beta}_{\mathrm{GLS}} &= \left(\mathbf{q}^{\mathrm{T}} \cdot \mathbf{C}^{-1} \cdot \mathbf{q}\right)^{-1} \cdot \mathbf{q}^{\mathrm{T}} \cdot \mathbf{C}^{-1} \cdot \mathbf{z} \end{aligned} \tag{2.5}$$

where $\mathbf{q}_0^{\mathrm{T}}$ is a vector of predictors at s_0, $\hat{\beta}_{\mathrm{GLS}}$ is a vector of coefficients that are estimated using generalised least squares, λ_0^{T} is a vector of kriging weights for residuals, $\mathbf{z}$ is a vector of n field observations, $\mathbf{C}$ is the $n \times n$ size covariance matrix of residuals and $\mathbf{q}$ is the matrix of predictors at all observed locations ($n \times p + 1$). This technique is known as *regression-kriging* and is equivalent to techniques known as *Universal kriging* and/or *Kriging with external drift*, although, in the literature you will often find that various authors interpret these techniques in different ways (Hengl, 2007).

Note that regression-kriging will, in principle, always fit the observations much better than either pure MLR or pure kriging. In statistical terms, predictors are used to explain the variation in the output signal (target variables), which is measured through the goodness of fit (R-squared). Obviously, if we increase the number of field observations and number and detail of predictors, the amount of explained variation will increase, but only to a certain level (Figure 4).

In fact, one needs to avoid fitting 100% of global variance because we assume that part of the signal is pure noise (measurement error) that we can not explain.

Note also that, although we can extract over 100 land-surface parameters and objects from a DEM, the information contained in a DEM has limits. In fact, many land-surface parameters reveal very similar underlying patterns, resulting in an overlap of information in the land-surface parameters. This is referred to in statistics as the *multicolinearity effect*[5] and can be best determined through Principal Component Analysis or PCA (Tucker and MacCallum, 1997). PCA will reduce or completely eliminate the multicolinearity and reduce the number of predictors.

The extracted PCs will be completely independent and therefore more suitable for regression analysis than the original land-surface parameters. See for example the results of PCA on the Baranja Hill for six land-surface parameters: SLOPE, PROFC, PLANC, TWI, SINS and GWD in Figure 5. In this case, the PC coefficients show that especially SLOPE and TWI are inter-correlated (see also Figure 4 in Chapter 28), probably because SLOPE is used in derivation of TWI. The inter-correlation between predictors is an effect we would like to avoid or reduce because most statistical models assume that the predictors are independent.

PCA can be used to make inferences about the information content of the land-surface parameters — if the amount of variation explained by the first component is high, then there is not much information hidden within the land-surface parameters (Figure 6). In the example given for the Baranja Hill, the first PC explains 49.3% of variation, the second 23.6%, third 12.0% and fourth 9.0%. The PC5 and PC6 explain only 6.2% of variation and it can be seen that they only repeat information from the previous components.[6]

The above example illustrates an advantage of using PCA to reduce the number of input parameters (here from 6 to 4) to ensure a more successful statistical analysis. A problem of using PCA is that the PCs are now compound images that can not be directly interpreted.

2.2 Deterministic process-based modelling

Unlike statistical modelling, in the case of deterministic modelling the formulae to derive environmental variables from inputs are known and do not have to be estimated. The required inputs are also known. This means that we do not actually need to do field sampling to estimate the model, but only to populate or calibrate it. For example, temperature at each location in an area can be determined with reference to the elevation, relative incoming solar radiation and leaf area index (see also Chapter 8):

$$T = T_b - \frac{\Delta T \cdot (z - z_b)}{1000} + C \cdot \left(S - \frac{1}{S}\right) \cdot \left(1 - \frac{\text{LAI}}{\text{LAI}_{\text{max}}}\right) \tag{2.6}$$

[5] This means that the land-surface parameters are inter-correlated and there is an information overlap.

[6] PCA transformation can also be fairly useful to filter out noise or artefacts, which comes out nicely in the higher order PCs.

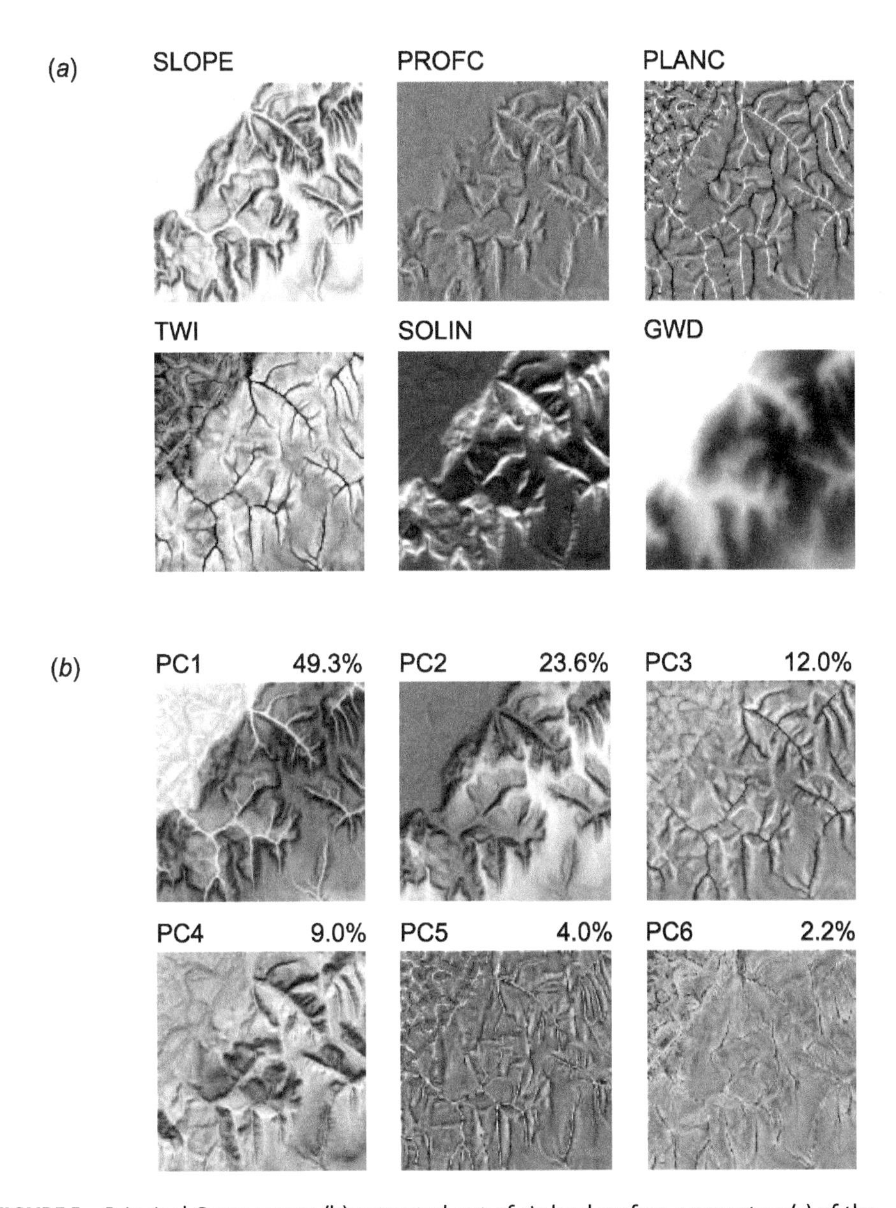

FIGURE 5 Principal Components (b) extracted out of six land-surface parameters (a) of the Baranja Hill. The number above components indicates the percentage of the variance explained by each principal component. The higher order components will typically show the noisy component not visible in the original maps.

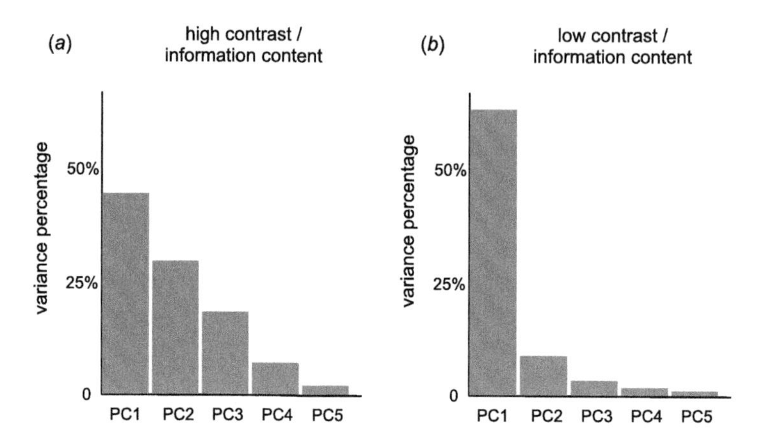

FIGURE 6 Examples of PC plots for the high (a) and low (b) information content.

An example of a more complex, process-based, spatial-temporal model is given by Minasny and McBratney (2001) who developed and tested a mechanistic model to predict soil thickness based on the following model:

$$\begin{aligned}\frac{h_{x,y}^{t+1} - h_{x,y}^{t}}{\Delta t} = & -\frac{\rho_r}{\rho_s} \cdot \frac{\delta e}{\delta t} + \frac{\delta l}{\delta t} + D \cdot \frac{z_{x+1,y}^{t} - 2 \cdot z_{x,y}^{t} + z_{x-1,y}^{t}}{(\Delta x)^2} \\ & + D \cdot \frac{z_{x,y+1}^{t} - 2 \cdot z_{x,y}^{t} + z_{x,y-1}^{t}}{(\Delta y)^2}\end{aligned} \tag{2.7}$$

where $h_{x,y}^{t}$ is the soil thickness at initial time and at position x, y, $h_{x,y}^{t+1}$ is the predicted thickness after some period of time, ρ_r is the density of rock, ρ_s is the density of soil, $\delta e/\delta t$ is the rate of lowering of bedrock surface, $\delta l/\delta t$ is the rate of chemical weathering, D is the erosive diffusivity of material, z is the elevation and Δx, Δy is the pixel size. The rate of lowering of bedrock and the rate of chemical weathering are estimated using:

$$\begin{aligned}\frac{\delta e}{\delta t} &= P_0 \cdot e^{-b \cdot h} \\ \frac{\delta l}{\delta t} &= W_0 \cdot e^{-k_1 \cdot h - k_2 \cdot t}\end{aligned} \tag{2.8}$$

where P_0 is the potential weathering rate of bedrock at $h = 0$, b is the empirical constant, W_0 is the potential chemical weathering rate and k_1, k_2 are the rate constants for soil thickness.

Note that, in fact, the model in Equations (2.7) and (2.8) uses only elevation as input, while all other parameters can be constants. It will simulate evolution of the soil formation and then eventually stabilise after 10,000 years or more (Figure 7). In this system, prediction of soil thickness is a function of time only. Note also that, although Equation (2.7) is rather long, this model is really rudimentary — it does not consider loss of soil by erosion or the impact of vegetation. It simplifies many

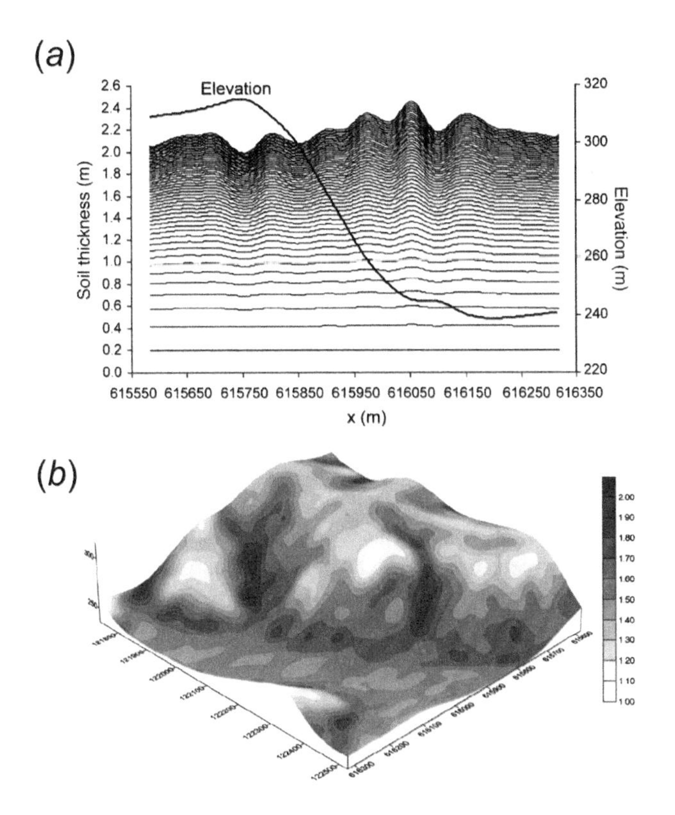

FIGURE 7 Simulated evolution of soil thickness: (a) a cross-section showing the change of soil thickness in relation to relative position in the landscape, (b) soil thickness after 10,000 years. Reprinted from Minasny and McBratney (2001). With permission from Elsevier.

physical and chemical processes and assumes homogeneous and constant conditions. Nevertheless, the final outputs reflect our knowledge about the processes and can help us understand how a landscape functions.

REMARK 5. *The influence of organisms and climate on landscape-formation is often complex and behaves in a non-linear way, so that operational models to simulate evolution of a landscape are still under development.*

An early effort to model processes in the landscape is the *Universal Soil Loss Equation* (USLE) developed by Wischmeier and Smith (1958). USLE takes six inputs to predict potential soil loss by erosion: rainfall erosivemess, soil erodibility, slope length, slope steepness, cropping management techniques, and supporting conservation practices. This can not be consider a process-based model, but rather an empirical estimate of the true physical model. Hydrologists have subsequently worked out more complex process-based landscape models such as the TOPMODEL (Beven et al., 1984; Beven, 1997) used to forecast flood events or Water Erosion Prediction Project (WEPP) model (Flanagan and Nearing, 1995) that is used to predict potential sheet and rill erosion for small watersheds.

Today, the trend is towards modelling landscape evolution in time. More recently, Mitášová et al. (1997) developed virtual soil-scapes that can be visualised as 3D animations. Rosenbloom et al. (2001) and Schoorl and Veldkamp (2005) further extended research in this field. Most of these soil genesis models mainly aim to map the distribution of soil properties based on some mass diffusion model using a DEM as input (Minasny and McBratney, 2001; Rosenbloom et al., 2001), see also the LAPSUS model used in Section 4.2 of Chapter 5. We will purposely not discuss how these mechanistic models work and whether we are really able to model evolution of soils and vegetation using only a few inputs such as a DEM and geological data. The problem is that many environmental processes are as yet poorly understood and many of the inputs are unknown or very poorly known.

Process-based models of complex natural systems such as landscape will have many parameters that need to be identified. For example, in the case of the soil–landscape models, process parameters such as hydraulic conductivity, weathering rates, and also stochastic parameters such as variances and correlations need to be determined. There are enormous challenges. For example, in the case of soil mapping, we need to consider huge state dimensions: for example 40 soil parameters at 5 depths for a 100×100 grid yield 2 million state variables! Moreover, different processes need to be modelled at completely different time and space scales, e.g. podzolization versus event-based erosion (Heuvelink and Webster, 2001). Some of these processes are still poorly understood and many happened rapidly and episodically in the past (e.g. flooding, landslides, movement of glaciers, etc.).

Another issue that complicates process-based modelling is the problem of *scale*. According to Schoorl and Veldkamp (2005, p. 420), a landscape is a system of four dimensions: (1) *length*, (2) *width*, (3) *height* and (4) *time*. Each of these dimensions has a different behaviour at different *scales* of work. Most often, exactly the same models will give completely different results at different scales. However, not only the resolution of DEMs influences the final output, but also the amount of artifacts and vertical precision can seriously propagate inaccurate features to final outputs and result in completely nonrealistic scenarios (Schoorl and Veldkamp, 2005).

Dynamic models of landscape evolution might turn out to be as non-linear and chaotic as long-term weather forecasts (Gleick, 1988). Although the modelling of deposition/accumulation processes and meandering water movement may seem easy, the influence of organisms and climate is often complex and behaves in a non-linear way (Phillips, 1994; Haff, 1996). Moreover, in many cases we will not actually know how the landscape looked in its initial state thousands of years ago. Many believe that such landscape evolution models will need to be calibrated repeatedly to avoid serious divergence between the true and predicted system trajectory (Figure 8). McBratney et al. (2003) believe that it will be a long time before the mechanistic theoretical approach will be truly operational. We can only agree with the statement of Guth (1995, p. 49) who emphasised that DEM users "*should keep their feet on the solid terrain of reality by understanding how the algorithms operate*", before they can claim that their products are accurate and realistic.

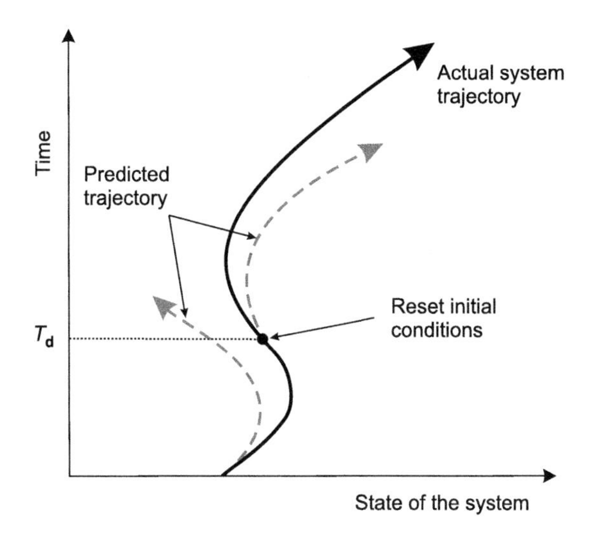

FIGURE 8 Process-based prediction models are usually very sensitive to initial conditions so they need to be calibrated at some divergence time T_d. Reprinted from Haff (1996).

2.3 Expert knowledge-based (heuristic) models

Expert knowledge-based models can also be used to infer environmental conditions or classes based on human understanding of relationships among environmental processes, known controls and resulting outcomes. Expert knowledge is also based on analysis of relationships between observable environmental inputs and those outputs (usually classes) that there is a desire to predict. Such analysis can vary in the degree to which it is systematic, rigorous, empirical or statistically validated. The list of inputs and the coefficients/rules used to derive outputs may also range from completely unknown to imperfectly known to completely known and understood:

Very limited knowledge In many instances, expertise is confined to an expert's ability to correctly identify specific instances or cases of a desired class or outcome. Such knowledge lends itself to analysis using data mining techniques that can uncover relationships between specific cases, as identified by an expert, and the various inputs that are available and considered likely to have some ability to predict the output entity. These data mining techniques determine which inputs and which rules best predict the desired outputs. Examples of data mining techniques for extracting classification rules from example data sets include Bayesian logic, analysis of evidence, spatial co-occurrence analysis, classification and regression tree analysis (CART), neural networks, fuzzy logic, discriminant analysis and maximum likelihood classification. Such models, in fact, are equivalent to the pure statistical models as described in Section 2.1.

Partial knowledge With partial knowledge, an expert may have a general idea of what the objects to be predicted look like, where they typically occur in space and

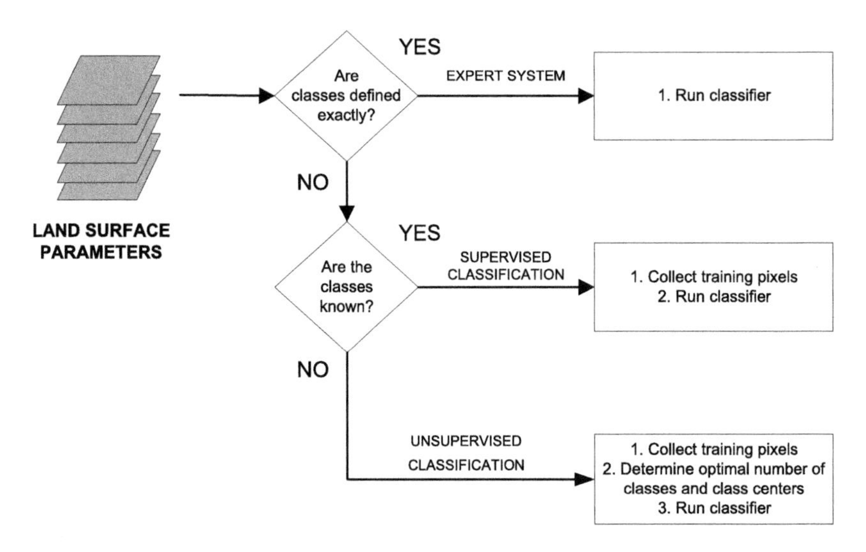

FIGURE 9 A typical strategy for selection of the classification approach.

the main conditions, processes or controls under which they typically develop. Such knowledge lends itself to capture and application using *Boolean logic* or *Fuzzy Semantic models* that can be used to iteratively apply, review and revise knowledge-based rules until such time as the output produced by application of the rules matches the spatial patterns expected for the predicted entity as closely as possible.

Exact knowledge Some forms of expert knowledge may be considered complete and perfect. In these instances the desired outcomes are unambiguous, as are the rules required to recognise the outputs and the inputs required to apply the rules. Such knowledge lends itself to application using Boolean logic to produce clear, crisp entities for which only one spatial expression is correct. An example might be the definition of hillslopes defined by the intersection of complementary divide and channel networks.

Let us first consider the case where human expert knowledge is limited to the ability of a local expert to correctly and consistently assign individual instances to a particular class from among a defined set of classes. The expert may either have no knowledge of the factors and conditions that cause a particular class to occur in a particular location or may have a reasonably good idea of the causal factors but be unable to express that understanding formally and rigorously.

Supervised classification is the most commonly used approach for developing formal, quantitative rules for automatically predicting the spatial distribution of classes of entities of interest given a number of possible predictive input layers and a set of training data (Figure 9). The basic approach of all methods of supervised classification is to first have human interpreters possessed of local expert knowledge identify and locate a series of areas or class instances at which each output class of interest is considered to occur. These instances constitute training data for developing classification rules.

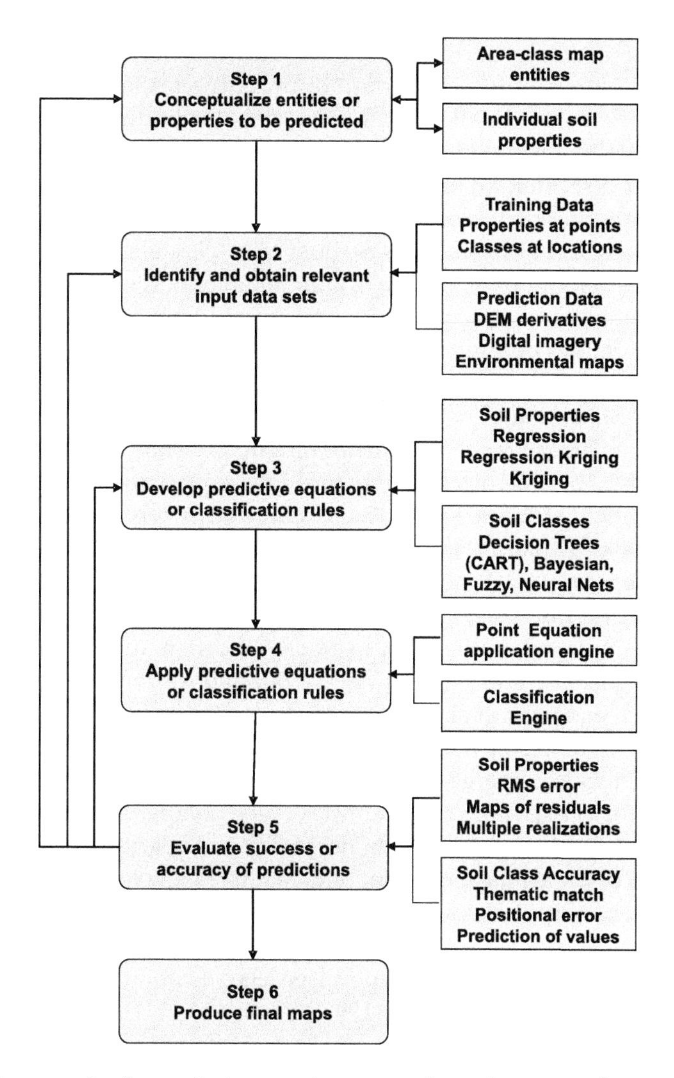

FIGURE 10 An example of a predictive mapping protocols used to map soil mapping units. See also Chapter 24 for more detail.

Any one of a large number of different statistical approaches can be used to analyse spatial relationships between the desired output classes, as identified at training locations, and all available input layers. Some statistical techniques are more suited to analysing continuous input data layers and some are better suited to categorical data while a few handle both types of input data equally well. The intent, in all cases, is to create a series of quantitative rules that will identify which values, or classes, in input layers are most strongly associated with the recognised occurrence of an identified output class or value. They also generally try to identify which input layers are most effective, or useful, in identifying a particular output class.

The success of these various supervised classification approaches is generally evaluated in terms of the proportion of training sites (or of independent test data sets) that are correctly allocated by the classification rules to the class that they were originally assigned to by the expert interpreter. McBratney et al. (2003) identified linear discriminant analysis as perhaps the most widely used classical approach for supervised classification to date. Examples of studies where linear discriminant analysis has been used in supervised classification are provided by Thomas et al. (1999), Dobos et al. (2000), and Hengl and Rossiter (2003).

Classification trees, or decision trees, have found favour for predicting spatial distributions of classes of interest because they require no assumptions about the data, they can deal with non-linearity in input data and they are easier to interpret than GLMs, GAMs or neural networks (McBratney et al., 2003). Tree models use a process known as binary recursive partitioning to develop relationships between a single response variable or class and multiple explanatory variables (McKenzie and Ryan, 1999). Data are successively split into two groups and all possible organisations of explanatory variables into two groups are examined to evaluate the effectiveness of each possible split. Zambon et al. (2006) described four widely applied splitting rules identified as *gini*, class probability, *twoing* and *entropy*. Other examples of studies that used classification and regression trees are provided by Lagacherie and Holmes (1997), Bui and Moran (1999, 2001), Scull et al. (2003, 2005). Henderson et al. (2004) used the Cubist[7] package, that implements regression trees, to map soil variables over Australian continent.

Fuzzy logic has also emerged as a preferred approach for capturing and formalising rules for classifying spatial entities using a supervised approach. It also has no statistical requirements for data normality or linearity and can utilise both continuous and discrete (classed) input data layers. Fuzzy logic associates a fuzzy likelihood of each output class occurring with each value or class on each input map (Figure 11).

Fuzzy logic has been used for supervised classification of soil–landform entities by Zhu and Band (1994), Zhu (1997), Zhu et al. (2001), Carré and Girrard (2002), Boruvka et al. (2002) and Shi et al. (2004). Different methods and equations for computing values for fuzzy similarity of sites to be classified relative to values for reference entities were reviewed by Shi et al. (2005) who provided the following general formula that is applicable to almost all efforts to assess fuzzy similarity:

$$S_{i,j} = \mathop{T}_{t=1}^{n} \left\{ \mathop{P_t}_{v=1}^{p} \left[E_t^v \left(z_{i,j}^v, z_t^v \right) \right] \right\} \tag{2.9}$$

where S_{ij} is the fuzzy membership value at location (i, j) for a specific feature, n is the number of identified typical locations of the feature and p is the number of input data (predictor) layers taken into account, $z_{i,j}^v$ is the value for the vth input attribute at location (i, j) and z_t^v is the corresponding value associated with the tth

[7] See also http://rulequest.com.

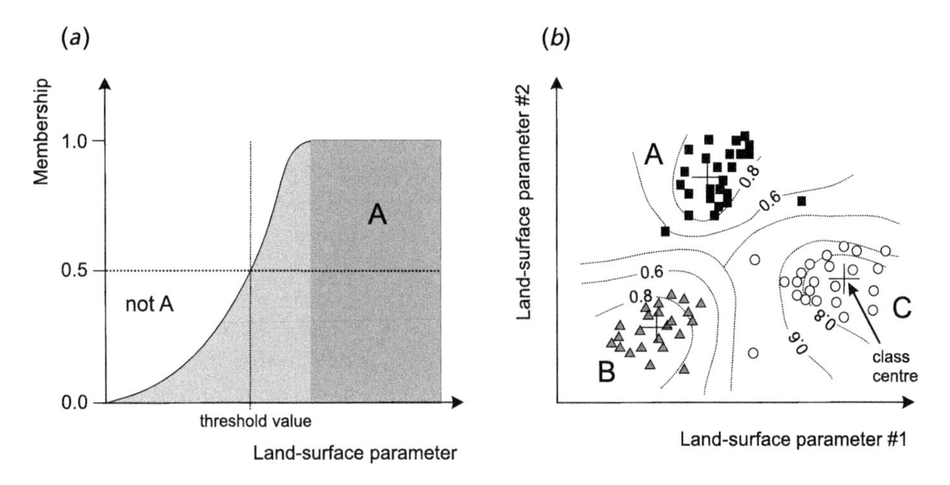

FIGURE 11 Schematic example for derivation of fuzzy memberships using: (a) definition of threshold values and (b) definition of class centres and (see further Section 5.1 in Chapter 22).

typical location, and E represents a function for evaluating the similarity of the vth variable at a particular site relative to the same variable for the reference data.

> REMARK 6. *Expert knowledge is based on semi-subjective analysis of relationships between observable environmental inputs and targeted outputs (usually classes). Such analysis can vary in the degree to which it is systematic, rigorous, empirical or statistically validated.*

The widely-cited *SoLIM method* (Zhu et al., 1997) adopts a limiting factor approach for computing overall similarity P. Here, a fuzzy minimum operator simply selects the smallest similarity value from among all similarity values computed for all attributes for an unclassified entity as the value for overall similarity between that unclassified entity and a reference entity (Shi et al., 2004).

The *Semantic Import model*, as implemented by MacMillan et al. (2000) uses a weighted average method to compute overall similarity of an unclassified site to a reference entity. This is based on the assumption that all input variables should be included in computing the similarity of a site to a reference entity but that some inputs may deserve to be afforded a greater importance or weight than others. The equation used to compute a fuzzy weighted average in the SI model is virtually identical to those used to compute *Bayesian Maximum Entropy* (Aspinall and Veitch, 1993) or to apply *Multi Criteria Evaluation* (MCE) (Eastman and Jin, 1995).

Data mining techniques, such as *Bayesian analysis of evidence* (Aspinall and Veitch, 1993), have also been used to analyse training data to extract knowledge and build[8] rules for classifying spatial entities. This approach analyses patterns of spatial co-occurrence between recognized output classes (in the training data sets) and classes or values in the input data layers to establish both the strength and direction of relationships between input data and the output classes to be predicted.

[8] E.g. the Netica package — http://norsys.com.

This analysis provides an explicit calculation of the conditional probability of occurrence of each output class of interest given each input class on each available input map.

The analysis also supports computation of the relative strength or importance of each input layer in predicting each desired output class. This provides a formal, quantitative mechanism for weighting the different input layers when computing an overall likelihood that each output class of interest will occur given each combination of classes of input layers. Additionally, Bayesian analysis of evidence makes use of a-priori estimates of the proportional extent of each output class to be predicted to constrain the final predictions in such a way that the proportion of each output class that gets predicted (has the highest likelihood value for a particular location) matches the a-priori estimate of the proportion of that class in the area as a whole. Examples of studies that extracted and applied Bayesian expert beliefs are provided by Skidmore et al. (1991), Aspinall and Veitch (1993) and Bui and Moran (1999).

Partial knowledge: expressing and applying inexact heuristic knowledge Let us next consider the case where human experience has led to development of a level of knowledge and understanding that supports expression in terms of semantic statements that relate the spatial distribution of output classes to causal factors that can be approximated by available digital input data layers. This human conceptual understanding has frequently been captured and reported in the form of map legends, field guides, cross sectional or 3D diagrams of soil–landform (or ecological–landform) relationships, edatopic grids (for ecological classifications) and ecological classification keys. These materials record and present beliefs (usually based on empirical analysis of considerable volumes of evidence) about where in the landscape specific classes of soil, ecological, hydrological or geomorphic entities are expected to occur and why. Traditional manual mapping systems make use of this available tacit expert knowledge to guide delineation of map entities and assignment of attributes or classifications to these delineated areas (Arnold, 1988; Northcote, 1984; Swanson, 1990b).

This tacit knowledge has only infrequently been formalised, or made explicit, by recording it as formally expressed semantic or quantitative rules. The Fuzzy Semantic Import (SI) model, as described by Burrough (1989) and applied by Zhu et al. (2001) and MacMillan et al. (2000), provides an ideal platform for capturing local tacit knowledge and systematically quantifying that knowledge by expressing it in terms of fuzzy knowledge-based rules.

REMARK 7. *The key challenge presented to the expert is to identify the range of locations in the landscape over which each particular output class of interest is known or expected to occur.*

For example, we may have already identified that, for a given area, each hillslope is occupied by a characteristic sequence of classes (say A, B, C, D, E) that occur along a catena or topo-sequence from crest to trough. We may be able to state, for example, that entity A almost always occurs along the main, upper portion of ridge crests while entity D almost always occurs in lower to toe slope landform positions, on gentle slopes less than 5% and that it occurs topographically above entity E and below entity C. We may know that the main factors that influence the spatial distribution of these 5 classes are relative values of slope position, moisture regime, slope gradient, aspect or exposure, soil depth, soil texture and perhaps other measures of local context or pattern. It is often feasible to identify and select one or more digital input layers, many of them consisting of land-surface parameters and objects derived from analysis of digital elevation data. For example, an expert may visually select a range of values for land-surface parameters that approximate relative landform position and moisture regime that appear to occupy the same dry, upper portions of the landscape as are associated with hypothetical class A introduced above.

Having an expert associate a range of values of input layers with a particular output class is really not much different than having the same expert select a number of locations to act as reference sites and then using a data mining procedure to establish rules that relate values of an input variable to classes to be predicted. In both cases, the challenge presented to the expert is to identify the range of locations in the landscape over which each particular output class of interest is known or expected to occur. A fuzzy model that expresses the likelihood of a given output class occurring given a particular value (or range of values) of a particular input variable can be constructed using equations given by (Burrough, 1989):

$$s^{v}_{ij,t} = \frac{1}{1 + \left[\frac{(z^{v}_{ij} - z^{v}_{t})}{d}\right]^{2}} \tag{2.10}$$

where $s^{v}_{ij,t}$ is the similarity between the value of the variable v at the unclassified location (i, j), z^{v}_{ij}) and the value of the variable z^{v}_{t} reference location t. In this approach an expert is required to provide two values: (1) the most likely value for the variable of interest for the class of interest (here given as z^{v}_{t}) and (2) a user-selected value for dispersion index (d) that controls the shape of a bell curve centred around this most likely value.

In applications that use actual data from known reference locations, the central, or most likely, value for the variable of interest for the current class of interest is assumed to be given by the value z^{v}_{t} for variable v for each separate case t. The second task required of an expert is to determine the manner in which the various likelihood values for individual input layers for each output class should be analysed, in combination, to estimate the overall likelihood that a given output class occurs at a specific location given a particular combination of input values. If a weighted average approach is used, as with the Semantic Import model, the expert must first decide which input variables will be used to define any given output class and then decide how much weight will be attached to each input variable in computing the overall mean likelihood value according to:

$$S_{ij,t} = \frac{\sum_{v=1}^{p} W^{v}_{t} \cdot S^{v}_{ij,t}}{\sum_{v=1}^{p} W^{v}_{t}} \tag{2.11}$$

where $S_{ij,t}$ is the overall similarity between the unclassified entity at location (i, j) and a reference location at t; $S^{v}_{ij,t}$ in the similarity between the unclassified and reference entity relative to the vth input variable; W_t^v is the weight or importance assigned to the vth input variable.

Heuristic rules created as described above generally represent an initial effort to establish definitive rules for predicting output classes given a particular set of available input variables. It is commonly necessary to go through several iterations of developing knowledge-based rules, applying them to the available input data layers, visualising and evaluating the results and identifying anomalies or errors that suggest where rules need to be revised and how (Qi et al., 2006).

Exact knowledge: knowledge based on proven theory or practice Let us finally consider the case where formalised theoretical principals exist that permit exact recognition of unambiguously defined spatial entities. We put forward an example from hydrology of automated extraction of drainage divides and stream channels from digital elevation data and their subsequent intersection to recognise individual hillslope entities. Drainage divides are recognized theoretically as locations where the direction of flow of surface water diverges with flow on one side of the divide separated from flow on the other side such that the two separate areas contribute flow into different stream channels or at least into different reaches of the same channel. Stream channels are recognizee in locations where surface flow converges and defines a single linear channel.

Typically, the locations of divides and channels can be identified unambiguously and crisp or Boolean logic can be used to extract and classify these hydrological spatial entities, rather than less exact methods such as fuzzy logic. Shary et al. (2005) has described exact and invariant classes of surface forms based on consideration of signs of curvatures. The rules for these classes are the same everywhere and do not require the use of fuzzy methods to accommodate imprecision in their definition. Exact and formal definitions have been proposed for other hydrological entities such as hillslopes and for hydrologically unique partitions of hillslopes into hillslope elements (Speight, 1974; Giles and Franklin, 1998).

These definitions lend themselves to unambiguous recognition of exact locations where continuous land surfaces can be sub-divided into hillslopes and even into components of hillslopes. In such cases, it is un-necessary and likely undesirable to adopt fuzzy classification methods to extract and classify exact objects. Similar exact spatial entities may well exist in other fields such as soils, ecology or forestry.

2.4 Evaluation of spatial prediction models

Evaluation of the accuracy of spatial prediction models is an aspect of mapping that is often forgotten or ignored. McBratney et al. (2003) indicated that "*there has been little work on corroboration of digital soil maps*". Often, accuracy has been reported in terms of the proportions of training data sites that were correctly classified, using the final classification rules, into the class that they were originally designated as. This only tests the internal consistency of the classification rules for the limited subset of training data and should not be considered as a viable assessment of whole map accuracy. Others remove a portion of the total field sample data collected and do not use these data in the preparation or revision of rules. The reserved data are used only to provide an independent assessment of the ability of the rules to predict the correct classes at locations that were not used to create the rules.

According to Li et al. (2005), there are seven criteria that guarantee a successful model:

- *accuracy* — is the output correct or very nearly correct?
- *realism* — is the model based on realistic assumptions?
- *precision* — are the outputs best possible unbiased predictions?
- *robustness* — is the model over-sensitive to the errors and blunders in the data?
- *generality* — is the model applicable to various case studies and scales?
- *fruitfulness* — are the outputs useful and do they help users and decision makers solve problems?
- *simplicity* — is the model the simplest possible model (smallest number of parameters)?

Likewise, in the case of spatial prediction models, we are mainly concern about the quality of final outputs, but we are increasingly concerned about the success of interaction of the users and the model. Some of the criteria listed above cannot really be assessed using analytical techniques. Therefore, in most cases we try to evaluate mainly the accuracy, realism and precision of a technique, then we run a similar analysis on various case studies at different scales and for different environments.

The accuracy of interpolation methods can be evaluated using interpolation and validation sets. The interpolation set is used to derive the sum of squares of residuals (SSE) and *adjusted coefficient of multiple determination* (R_a^2), which describe the goodness of fit:

$$\begin{aligned} R_a^2 &= 1 - \left(\frac{n-1}{n-p}\right) \cdot \frac{\text{SSE}}{\text{SSTO}} \\ &= 1 - \left(\frac{n-1}{n-p}\right) \cdot (1 - R^2) \end{aligned} \tag{2.12}$$

where SSTO is the total sum of squares (Neter et al., 1996), R^2 indicates amount of variance explained by model, whereas R_a^2 adjusts for the number of variables (p) used. In many cases, $R_a^2 \geqslant 0.85$ is already a very satisfactory solution and higher values will typically only mean over-fitting of the data (Park and Vlek, 2002). Note that this number corresponds to the relative prediction error [Equation (2.15)] of $\leqslant 40\%$.

> REMARK 8. *The only way to evaluate the true success of predicting the target variable is to collect additional observations at independent control locations.*

Care needs to be taken when fitting the statistical models — today, complex models and large quantities of predictors can be used so that the model can fit the data almost 100%. But there is a distinction between the goodness of fit and true success of prediction (prediction error at independent validation points). Hence, the only way to evaluate the true success of predicting the target variable is to collect additional separate observations at independent control locations, and to then evaluate the success of predictions at these independent control locations (Rykiel, 1996). The true prediction accuracy can be evaluated by comparing estimated values ($\hat{z}(s_j)$) with actual observations at validation points ($z^*(s_j)$) in order to assess systematic error, calculated as mean prediction error (MPE):

$$\text{MPE} = \frac{1}{l} \cdot \sum_{j=1}^{l} \left[\hat{z}(s_j) - z^*(s_j)\right] \tag{2.13}$$

and accuracy of prediction, calculated as root mean square prediction error (RMSPE):

$$\text{RMSPE} = \sqrt{\frac{1}{l} \cdot \sum_{j=1}^{l} \left[\hat{z}(s_j) - z^*(s_j)\right]^2} \tag{2.14}$$

where l is the number of validation points. In order to compare accuracy of prediction between variables of different type, the RMSPE can be normalised by the total variation:

$$\text{RMSPE}_r = \frac{\text{RMSPE}}{s_z} \tag{2.15}$$

As a rule of thumb, we can consider that a value of RMSPE_r close to 40% means a fairly satisfactory accuracy of prediction. Otherwise, if the value gets above 71%, this means that the model accounted for less than 50% of variability at the validation points.

The overall predictive capabilities of predicting categorical variables (soil or vegetation classes and similar) are commonly assessed using the *Kappa statistics* (Lillesand and Kiefer, 2004), which is a common measure of the accuracy of classification. Kappa statistic is a measure of the difference between the actual agreement between the predictions and ground truth and chance agreement. In remote sensing, a rule of thumb is that the mapping accuracy is successful if kappa $> 80\%$.

Kappa is only a measure of the overall mapping accuracy. In order to see which classes are most problematic, we can also examine the *percentage of correctly classified pixels* per each class:

$$P_c = \frac{\sum_{j=1}^{m} (\hat{C}(s_j) = C(s_j))}{m} \tag{2.16}$$

where P_c is the percentage of correctly classified pixels, $\hat{C}(s_j)$ is the estimated class at validation locations (s_j) and m is total number of control points.

Both the overall measures and partial measures of the success need to be expressed in confidence intervals (Congalton and Green, 1999). The confidence intervals of kappa for different prediction techniques tell us how variable is the success of mapping. A technique which achieves a kappa with a confidence interval 55–95% does not have to be significantly better than a technique with much narrower confidence interval, but lower average kappa e.g. 60–65%.

Another issue is the design of the control surveys. McBratney et al. (2003) recommend adopting a sampling strategy that is designed specifically for corroboration. An almost universal assumption of most efforts to assess the accuracy of classed maps is that the accuracy evaluation should assess whether the correct class has been predicted at specific point locations. Low levels of accuracy may be determined in cases where the size and scale (footprint) of the site locations used to assess classification accuracy are not congruent with the spatial resolution (support) of the input data sets. A field description that applies to a point location with dimensions of less than a few metres on a side is unlikely to compare well with classes predicted using input data layers with dimensions of 10's to 100's of metres. So, it is important to define ground truth sample locations so that they have dimensions that are comparable to the dimensions of the support provided by the input data layers.

3. SUMMARY POINTS

In this chapter, we have reviewed some examples of how automated analysis of DEM data can complement methods that use remote sensing images and field measurement in several scientific disciplines. These examples are by no means comprehensive but rather were selected to be illustrative of some of the many different approaches that can be applied to aid in the production of geoinformation using DEMs as an input. DEMs provide a relatively cheap and easy-to-use information source that has shown benefits for numerous applications such as mapping of landforms, landscape units, vegetation, soils, hydrological entities and modelling of landscapes and landscape-forming processes (see further chapters).

There can be little doubt that the use of digital elevation data, as a key input for the automated production of maps and environmental models of all kinds, has experienced dramatic growth in recent years and that this use will continue to grow. However, there are, as yet, few examples of large national or regional mapping agencies that have adopted automated methods for large scale production of operational maps. Most studies of automated predictive mapping in the disciplines of geomorphology, geology, soils, ecology and hydrology have described efforts to develop, apply and evaluate new concepts for more rapid or improved production of maps of soils, landforms, geological or ecological entities but these concepts have not yet been widely adopted for routine operational use.

We are faced with an ever exploding supply of data, an ever increasing need to process the data to aid in decision making and an inability to effectively manage and process the data using existing manually-intensive methods of analysis and interpretation. Data are generally understood to be the foundation for developing knowledge which in turn leads to improved understanding. We are in danger of becoming data rich and knowledge poor! Our ability to develop and apply knowledge to improve our understanding and decision making has to grow rapidly in order to catch up with our ability to collect raw data itself. It is increasingly necessary that we automate the production of maps (and models) that depict environmental information describing the spatial distribution of soils, landforms, surficial and bedrock geology, hydrological and ecological entities.

New and emerging technologies for mapping or modelling natural landscapes almost universally make explicit (or implicit) use of concepts and scientific knowledge that relate surface form to environmental processes and resulting conditions. These technologies are often rediscovering and applying concepts of soil–landform relationships (catenary sequences), hillslope hydrology, geomorphology or ecological zonation that have been fundamental components of the scientific knowledge of these disciplines for many decades. All of these disciplines have developed conceptual models that elaborate how surface form influences and controls processes such as geomorphic hillslope formation, soil development and evolution of ecosystems and how, in turn, these processes influence the development and evolution of surface form through feedback mechanisms.

For better or for worse, it is expected that the creation of virtually all maps of environmental phenomena will need to embrace and incorporate automated and statistical procedures applied to digital elevation data. It is expected that automated maps will be prepared that portray environmental conditions as both continuously varying values of single variables of interest and as classed maps of discrete spatial entities that are based on partitioning of the topographic surface into landform components. In all cases, automated analysis of DEMs will play a significant role in the production of the resulting maps.

IMPORTANT SOURCES

Schoorl, J.M., Veldkamp, A., 2005. Multiscale soil–landscape process modelling. In: Grunwald, S. (Ed.), Environmental Soil–Landscape Modeling: Geographic Information Technologies and Pedometrics. CRC Press, Boca Raton, FL, pp. 417–435.

Heuvelink, G.B.M., Webster, R., 2001. Modelling soil variation: past, present, and future. Geoderma 100 (3–4), 269–301.

Goovaerts, P., 1997. Geostatistics for Natural Resources Evaluation. Applied Geostatistics. Oxford University Press, New York, 496 pp.

Franklin, J., 1995. Predictive vegetation mapping: geographic modeling of biospatial patterns in relation to environmental gradients. Progress in Physical Geography 19, 474–499.

Bivand, R., Pebesma, E., Rubio, V., 2008. Applied Spatial Data Analysis with R. Use R Series. Springer, Heidelberg, 400 pp.

CHAPTER 20

Soil Mapping Applications

E. Dobos and **T. Hengl**

soils, soil maps, traditional and digital soil mapping techniques · models of soil formation and their implementation · importance of topography for soil formation · soil variables commonly mapped using DEMs · interpolation of sampled profile observations using regression-kriging · interpretation of results of interpolation/simulations · impact of DEM resolution on success of soil mapping · selection of suitable statistical techniques that can be used to map soil variables

1. SOILS AND MAPPING OF SOILS

1.1 Soils and soil resource inventories

Soil plays an important rule in the environment and also in the human life. It is formed in the transition zones of four significant zones of nature — the atmosphere, hydrosphere, lithosphere and biosphere. Soil consists of weathered and unweathered minerals of the underlying rocks (regolith/saprolite), decaying organic matter, living organism, and the pore space filled with gases and liquid solutions. It integrates the four basic spheres — solid, gas, liquid and biosphere — and creates a complex system of processes interfacing these components. Soil is a medium for plant growth, regulator of water supplies, buffer and filter zone for numerous toxic materials deposited from the air or contained by the ground water, recycler of raw material, and habitat and gene reservoir for soil organism. Beyond its ecological functions, soil provides engineering medium to build on and live on. It is a source of raw materials for mining and also a reserve of cultural heritage.

The sustainable and profitable management of this natural resource requires reliable, appropriate information on the soil characteristics influencing its use. Soil parameters are often used for modelling, forecasting or estimating certain environmental processes, e.g. to estimate the environmental risks, for agricultural yield forecasting, carbon stock estimation, or modelling of global warming. Such data is increasingly needed at a fine level of detail. The users of soil information are not

Developments in Soil Science, Volume 33 © 2009 Elsevier B.V.
ISSN 0166-2481, DOI: 10.1016/S0166-2481(08)00020-2.

only interested in summary characteristics of soils, but also in the spatial diversity and variability.

In the last 50 years, soil surveys and mapping institutes were set up in many countries to collect field soil data and create soil resource inventories that can be used to improve the management of soils. The first generation of soil maps has focused on representing distribution of soil variables important for agricultural use (Dent and Young, 1981). This was common for countries where agriculture have dominated the national economy. Starting from the 1970s, the second generation of soil maps has been introduced, and the main purpose of soil maps has shifted from agriculture towards other environmental issues — water management, waste disposal, septic systems, environmental risk assessment. This stage is typical of the industrial countries.

We now live in an era of *digital soil maps*. These maps are not pure soil maps any more. The application-oriented world requires answers for certain questions, like productivity of an area, or resistance against certain human impacts. In this system, soil is only one of the many important sources that need to be considered. Therefore the resulting map is not a pure soil map any more, but a complex representation of the environment.

Results of soil resource inventories are commonly soil maps, which represent spatial distribution of soils and their chemical, physical, or biological characteristics. Here, two types of soil variables can be distinguished: the primary or measured and the secondary or derived soil variables. The *primary soil variables*, e.g. sand, silt or clay content, pH, soil organic matter content, etc., cannot be estimated from other variables. The *secondary soil variables*, such as *soil structure*, *compaction* or *buffering capacity* etc., are estimated from one or more primary soil variables. In that sense, also the *soil types*[1] can be considered to be secondary (categorical) soil variables.

1.2 (Traditional) soil survey techniques

Soils co-evolve with their environment and represent a significant functional part of the landscape. A good soil surveyor can understand the landscape based on its characteristics and can identify the relationships between the soils and the general or specific features of the landscape. The set of rules and relationships, which explain spatial distribution of the soil properties throughout the landscape, is the *soil–landscape model*. An experienced surveyor observes the landscape characteristics, like the landform, geomorphology, vegetation, geology, and then uses that information to estimate the soil variables and delineate the homogeneous soil units in the field (Figure 1). The surveyor then identifies the typological landscape units and selects representative locations for soil sampling and description — so called *soil profiles*.

The number of profiles representing the landscape units depends on the scale. Larger scale needs more observations to describe the soil variability in the de-

[1] Soil types are soils having similar physical, chemical and biological characteristics. Soil types are not as clear entities as species in biology or vegetation communities and almost each country in the world developed their own classification systems. FAO's World Reference Base is the internationally accepted soil classification system (van Engelen and Ting-tiang, 1995; IUSS Working Group WRB, 2006).

FIGURE 1 Soil surveyor uses a stereophoto and a mylar overlay to delineate (presumable) soil bodies also known as soil mapping units.

tail appropriate for the scale. One observation can characterise an area of 1 to 4 hectares at the scale of 1:2000, 10–25 hectares for the 1:10,000, and 25–80 hectares for the 1:25,000 scales. The size of the area depends on the soil diversity of the area as well. The soil profile described in the field, represents the smallest unit of soil, called *pedon*. The pedon is a three dimensional soil body with lateral dimensions large enough to permit the complete study of horizon shapes and relations and commonly ranges from 1 to 10 m^2 in area (Soil Survey Division Staff, 1993).

Soil surveyors attempt to group the contiguous pedons together, which meet certain criteria or have a similar set of characteristics. This grouping of pedons leads finally to a soil delineation, practically a polygon drawn on the landscape and representing an area with (presumably) the same type of soil (Rossiter and Hengl, 2002). Figure 2 shows an example of a soil map with hand-drawn delineations using stereoscopic photo-interpretation (Figure 1). The main assumption of the polygon-based approach to soil mapping is that the polygons are homogeneous with discrete borders between them. As a result, average/representative values can be assigned to the whole soil polygon and the transition between the polygons if often abrupt.

The traditional soil maps show a stratified landscape with discrete units of soils covering certain part of the landscape. This approach serves very well the needs of representing our knowledge and interpreting the spatial distribution of soils over the landscape. However, polygon-based soil-class maps are not of much use for quantitative environmental modelling. This is mainly because the spatial and thematic content of such maps is rather limited — polygons can only represent abrupt changes and large objects and soil-class maps typically show accurately only distribution of soil types, while the distribution of soil properties needs to be inferred.

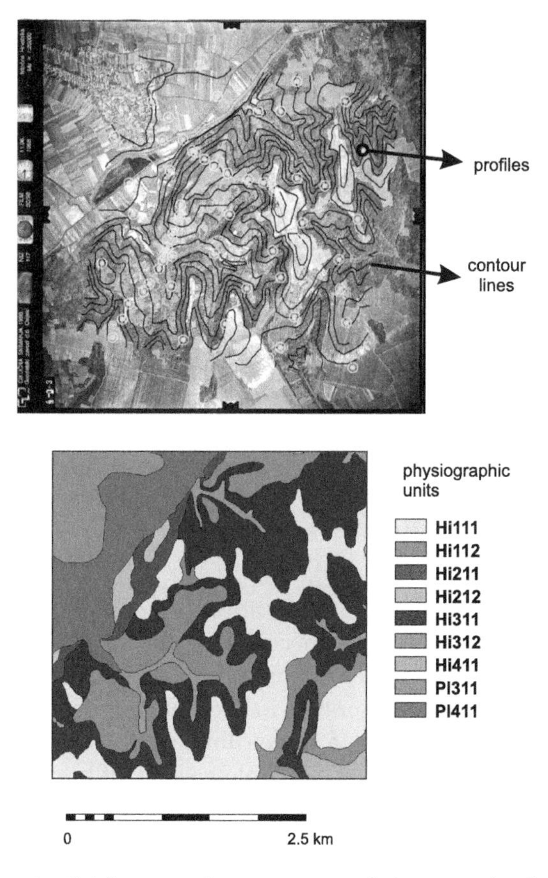

FIGURE 2 A traditional soil delineation drawn on an aerial photo overlain by contour lines (above) and the derived soil map with soil mapping units (below) for Baranja Hill region (Croatia). The lines are delineated manually and points show the location of soil profile observations. (See page 744 in Colour Plate Section at the back of the book.)

1.3 Digital soil mapping techniques

Advances in the raster-based GIS technology and the tremendous amount of remotely sensed and digitally derived data in last few decades have motivated soil mappers to use these resources and try to improve the spatial and semantic detail of the traditional soil maps. The majority of the soil data analysis nowadays happens within a digital environment, i.e. a GIS. The soil information teams are now either digitising the existing soil maps to create vector-based polygon maps inheriting all the limitations of the original choropleth data sources (Burrough and McDonnell, 1998), or deriving new data sources by using soil spatial prediction models (McBratney et al., 2003; Scull et al., 2003; Dobos et al., 2006). Henderson et al. (2004) refer to the latter models as the *point-based spatial prediction models* because the emphasis is now given more to the point (field) samples and analytical soil parameters rather than to the soil delineations or soil classes.

The point-based models usually make use of the raster-based or grid-based GIS structure, which can better represent the continuous nature of the soils. Although there is a technological gap between traditional and digital soil mapping, the two approaches, in fact, do not differ much. Both approaches need input (field) data on soil and covariates characterising the environment where the soil formation takes place. The difference between the two is in the way how the soil information is derived: the traditional models are based on (subjective) mental models in the surveyor's mind, while the digital soil mapping relies on technology and software.

In both cases, field observations are needed to train the models. But there is quite a difference in the processing of the data — digital soil mapping relies on quantitative statistical models; traditional on expert judgement. In addition, digital soil mapping is richer in content because it offers a measure of uncertainty of the prediction models and more possibilities for statistical in-depth analysis of relationship between various variables in the system (Dobos et al., 2006).

REMARK 1. *Digital (quantitative) Soil Mapping relies on use statistical tools in combination with large quantity of predictors, including the DEM-parameters.*

2. TOPOGRAPHY AND SOILS

2.1 The catena concept

Soil and landscape co-evolve and form a very tight soil–landscape relationship (Wysocki et al., 2000). As a result, similar soil populations occur within similar landscape units. Although soil and terrain relationships have been studied intensively, due to their complexity, they are still not fully understood. Many qualitative and subjective rules defining the link between soils and relief have been formulated and used by soil surveyors. Unfortunately, qualitative rules are often difficult to record and share within the soil specialist. The transformation of these rules into quantitative forms, exact equations, helps disseminating this knowledge to a wider audience.

Dokuchaev (1898), a Russian soil scientist, was the first who identified climate, organism, relief or topography, parent material and time as the main factors driving the formation of soils. The soil forming factors have their own spatial distributions and variability, and their site-specific combination defines the soil forming environment and creates a unique niche where certain soil types are formed. Jenny (1941) translated Dokuchaev's theory to the language of mathematics and formulated the most known equation in the soil science. This equation explains the status of a soil variable (S) as a function of climate (c), organism (o), relief (r), parent material (p) and time (t):

$$S = f(c, o, r, p, t) \tag{2.1}$$

Jenny's approach focuses on the prediction of certain soil chemical, physical or biological characteristics on a given location and did not consider the soil as a continuum, where the soil properties at a given location depend on their geographic position and also on the soil properties at neighbouring locations. McBratney et al. (2003) further extended the Jenny's equation and formulated the SCORPAN model:

$$\begin{aligned} S_a &= f(s, c, o, r, p, a, n) \\ S_{cl} &= f(s, c, o, r, p, a, n) \end{aligned} \tag{2.2}$$

where S_a is the estimated soil attribute value and S_{cl} is the estimated soil category, s is related soil property, a is age and n is position. If we consider each soil forming factor as a function of space and time, then the two equations modify to (Grunwald, 2005):

$$\begin{aligned} S_a[x,y,z,\sim t] = f\big(&s[x,y,z,\sim t], c[x,y,z,\sim t], \\ &o[x,y,z,\sim t], r[x,y,z,\sim t], \\ &p[x,y,z,\sim t], a[x,y,z]\big) \end{aligned} \tag{2.3}$$

$$\begin{aligned} S_{cl}[x,y,z,\sim t] = f\big(&s[x,y,z,\sim t], c[x,y,z,\sim t], \\ &o[x,y,z,\sim t], r[x,y,z,\sim t], \\ &p[x,y,z,\sim t], a[x,y,z]\big) \end{aligned} \tag{2.4}$$

Up till now, this equation has been unsolvable, mainly due to the complex nature of these covariates and the lack of data describing them (see also Section 2.1 in Chapter 19). Note also that the soil-environment functions are scale dependent so that different equations need to be developed at different scales, which makes these models even more complex.

In the previous chapter, it has been advocated that, on regional or local scales, distribution of natural soil and vegetation can be explained mainly be the relief factor. Indeed, topography has a great impact on soil formation. The elevation above the sea level and the slope aspect alters and moderates the climatic effect, via changing the rainfall and temperature regime of the area. The slope degree and relief energy drive the intensity of surface runoff, erosion and deposition, infiltration and alter numerous soil properties.

The elevation differences on the plain regions define the depth to the ground water level, which is one of the most significant factors on the development of the soil properties. The shape of the surface, its convex or concave nature, defines the surface drainage network, which defines the lateral transportation of chemicals and physical soil particles. These direct impacts listed above can be complemented with the indirect effects on the other four soil forming factors. Topography modifies the macro-climate and explains the majority of the local variation of rainfall and temperature. Geology and geomorphology are strongly related to topography as well. The combined direct and indirect effects of the topography on soil formation make the topography the most recognised factor with the highest predictive value.

The strong relationship between soils and topography has been recognised early in soil science, and the concept of relative soil-location, also known as *catena*[2]

[2] Meaning *chain* in Latin.

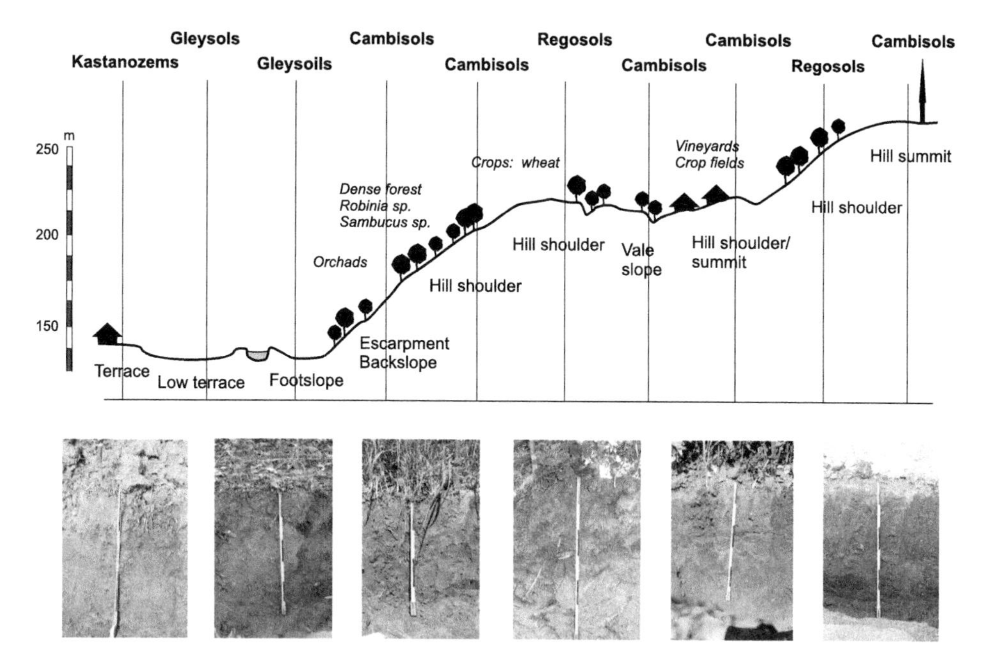

FIGURE 3 Vertical zonation of soils in the Baranja Hill: from deep, drained soils (Kastanozems), to saturated (Gleysoils) and shallow eroded soils (Regosols). (See page 745 in Colour Plate Section at the back of the book.)

or toposequence have been developed. A toposequence of soils can be sampled on a transect — going from a hilltop to the valley bottom — this model can then be used to extrapolate such knowledge over the whole landscape where similar relative positions can be found. An example of a toposequence showing zonation of soils in the Baranja Hill case study can be seen in Figure 3.

The individual land-surface features, like slope or aspect, which are often recognised as leading-forces of the soil formation within a relatively small area, show significant relationship but low predictive value for soil attribute estimation. However, when these land-surface parameters are combined in one model, the predictive value can be significantly improved. An example of this can be an area with relatively small slope steepness and big catchment area. When these two factors coincide under humid climate wetland can be formed. Only the combined effect of the two factors can explain the occurrence of the hydromorphic soils on the area. Such complex nature of relief as a soil forming factor can be quite difficult to represent by using simple linear models.

2.2 DEM as a digital input for soil mapping

According to Bishop and Minasny (2005) and McBratney et al. (2003), in almost 80% of the digital soil mapping projects DEMs are used as the most important data source to run predictions. DEM and land-surface parameters can be used as a digital input for soil mapping in (at least) four ways:

To update existing soil maps Biggs and Slater (1998) characterised the soil landscape with the use of DEM and compared the results with the data derived from existing conventional soil survey. Their derived soil attribute map with a scale of approximately 1:100,000 was used to enhance field validation and increase mapping confidence. Bock et al. (2005) demonstrated that the existing soil mapping units, produced at even relatively detailed scale, can be disaggregated by using DEMs and experts' knowledge.

To extract soil–landscape units or landforms DEMs can also be used to delineate new soil–landscape units to accommodate soil associations. Two potential approaches can be employed to derive soil–landscape units. The first is an automated, clustering based approach (Bathgate and Duram, 2003; Schmidt and Andrew, 2005), when no predefined criteria exist for the terrain classification. In these studies an automated clustering procedure is used to identify meaningful terrain clusters using a set of DEM derivatives. Soil type or soil association information is assigned to the clusters in the second step using an expert knowledge based approach. The second approach of soil–landscape unit delineation is based on existing, predefined, expert-knowledge based terrain classification (MacMillan et al., 2003). Dobos et al. (2005) used elevation, relief intensity, slope and dissection for extraction of SOTER-unit (*SO*il and *TER*rain digital database). Hengl and Rossiter (2003) used the photo-interpretation in typical areas to extrapolate the landform units to the whole area of interest with the help of nine land-surface parameters.

For direct estimation of soil parameters Land-surface parameters can be used to improve prediction of point-sampled soil variables (McKenzie and Ryan, 1999; McBratney et al., 2003). As long as the land-surface parameters show significant correlation with soil parameters, they can be used to predict soil parameters in between the sampling locations. A review of possible prediction techniques is given by Bishop and Minasny (2005).

To optimise the soil sampling strategy Land-surface parameters derived from a DEM can be used to run a representativity study of the sampling scheme, checking whether each combination of all landform classes are well represented among the observations (Minasny and McBratney, 2006). The sampling optimisation algorithms can even be optimised to allocate the points in the feature space so that the prediction error in the whole area of interest is minimised (Brus and Heuvelink, 2007).

> REMARK 2. *DEM parameters are most commonly used to update existing soil maps, to extract soil-landscape units, for spatial prediction and for making new sampling designs.*

Many successful soil mapping applications based on the DEM and DEM-derived data have been implemented for large, medium and small scale mapping. DEMs are most commonly used to map:

Solum and horizon depth The surface and ground water flow potentials determine the amount of available water, which can infiltrate into the soil profile. Among others (like texture), the amount of infiltrating water determines the depth of water penetration and through this, the depth and thickness of certain horizons. Previous models suggested lateral redistribution processes resulting in differential accumulation of carbon and soil mass in convergent and divergent landscape positions. Lateral redistribution of the soluble or physically transportable material also has a significant impact on the changes of the horizon depth along a toposequence. DEMs are often used to estimate the depth of certain horizons, like $CaCO_3$ enriched horizon (Florinsky and Arlashina, 1998; Bell et al., 1994), soil profile depth (McKenzie and Ryan, 1999), A-horizon depth (Gessler et al., 1995; Bell et al., 1994; Moore et al., 1993a). In most of the cases, the reduction in deviance was around 50–60% for the depth estimations.

Soil texture and hydrological properties Land-surface parameters have been used successfully to map topsoil and sub-surface proportions of clay, silt and sand (De Bruin and Stein, 1988; Gobin et al., 2001). This is possible at both continental (Henderson et al., 2004) and very detailed scales (Moore et al., 1993a; Bishop and Minasny, 2005). An extensive evaluation of techniques for mapping of soil texture is given by van Meirvenne and van Cleemput (2005). Land surface defines the way how the water moves through the landscape and transport soil materials in solution or in solid forms. The variables controlling the water flow have the greatest significance in explaining the spatial distribution of numerous soil properties. Soil drainage class is strongly related to the landscape location. Convex surfaces are most likely to be well drained, while concave surfaces, depressions have a higher likelihood of having hydromorphic features. Soil drainage class prediction based on DEM-derived digital variables makes the largest portion of all the DEM-based soil feature estimation (Bell et al., 1992; Thompson et al., 1997; Chaplot et al., 2000; Dobos et al., 2000; Case et al., 2005). The average reduction in deviance is relatively high, a value range of 70–80% can be reached. The most commonly used predictors are SLOPE, curvatures, TWI, flow accumulation and similar.

Soil chemical properties The type and the amount of soil organic matter are strongly related to the presence of water and the lateral redistribution of the surface material by erosion. Both of these phenomena are partially controlled by the topography. Among others TWI, potential drainage density (Dobos et al., 2005), curvature, slope gradient and flow accumulation variables proved to have a significant contribution to the estimation of the depth of A-horizon, soil carbon content (McKenzie and Ryan, 1999; Gessler et al., 2000), soil organic matter content (Moore et al., 1993a), topsoil carbon (Arrouays et al., 1998; Chaplot et al., 2001). The overall predictive values of these models are around 50–70%. Other soil chemical and physical properties estimated by digital land-surface parameters are pH, extractable phosphorus (Moore et al., 1993a; McKenzie and Ryan, 1999), mineral nitrogen, etc. The general impression is that the soil chemical properties are more difficult to estimate using DEMs than the physical properties. This is mainly because the chemical properties are dynamic[3] and are often influenced by several forming factors.

[3] Chemical properties vary not only within a season, but also within few days.

TABLE 1 List of land-surface parameters (supplemented with climatic images, lithology Landsat imagery and land use maps) used to interpolate soil properties over the Australian continent

Land-surface parameters	Mapped soil properties
elevation	pH
deposition path length	Organic carbon
erosion path length	Total phosphorus
relative elevation relief	Extractable phosphorus
slope percent	Total nitrogen
hill slope length	Clay, Silt and Sand %
slope position	Layer (horizon) thickness
river distance	Solum thickness
ridge distance	Bulk density
contributing area	Available water capacity
inverse contributing area	Saturated hydraulic conductivity
transport power in	
transport power out	

Soil taxonomic classes More complex features like soil classification categories were estimated by some authors (Thomas et al., 1999; Dobos et al., 2000; Hengl et al., 2007b). These models were estimated the general distribution of soil types. However, the kappa statistics will rarely exceed 80% because many soil classes are fuzzy and overlapping by definition. Many authors therefore suggest that the classes should be treated as memberships and finally evaluated using the fuzzy-kappa statistics, which is a soft measure of the mapping success (Hengl et al., 2007b).

REMARK 3. *In almost 80% of the digital soil mapping projects DEMs are used as the most important data source.*

An extensive example of how digital soil mapping can be applied to map various soil variables is the one of the Australian Soil Resources Information System (http://audit.ea.gov.au/anra/). In this case, the soil mapping team used a large number of predictors: land-surface parameters, climatic images, lithology, Landsat MSS imagery and land use maps; to map a number of soil variables: textures, soil thickness, pH, OC, etc. (Henderson et al., 2004). To illustrate the computational complexity of this model, we should also mention that there were almost 150,000 soil profiles and over 50 GIS layers as inputs (Table 1). The statistical model applied was regression-trees, which has the advantage of being able to incorporate both continuous and discrete information.

3. CASE STUDY

In the following example we will demonstrate, using the Baranja Hill case study, how to map various soil variables with the help of land-surface parameters as predictors. We will use the technique regression-kriging, explained in detail in Section 2.1 in Chapter 19. For more details about regression-kriging, see also Hengl (2007). The complete script and the input data sets used in this chapter can be obtained from the geomorphometry website.

The inputs to our model are 59 soil profiles, six land-surface parameters and a soil map. All datasets and scripts used in this exercise are available via the geomorphometry.org website. We will focus on how to map two types of soil variables: (1) a continuous soil variable (*solum* in cm) and (2) an indicator soil variable (*occurrence of gleying* — 0 stands for no observation of gleying, 0.5 stands for gleying at depth >60 cm and 1 stands for gleying within 60 cm of soil depth). We have first prepared the land-surface parameters: elevation (DEM), slope gradient in %, (SLOPE), profile curvature (PROFC), plan curvature (PLANC), wetness index (TWI) and slope insolation (SINS), all derived using the scripts in ILWIS (see Chapter 13).

In addition, we will use the (polygon-based) soil map with nine soil mapping units: colluvial footslopes (SMU1), eroded slope (SMU2), floodplain (SMU3), glacis (SMU4), high terrace (SMU5), scarp (SMU6), shoulder (SMU7), summit (SMU8) and valley bottom (SMU9) (see also Figure 2). The list of predictors and target soil variables can be seen in Figure 4. For interpolation, we use the **gstat** package (http://gstat.org) as implemented in the R statistical computing environment (http://r-project.org). This package allows both predictions[4] and simulations using the same regression-kriging model (Hengl, 2007).

The computational procedure is as follows (Figure 5):

1. *Prepare and import predictors*: land-surface parameters and soil map. The soil map needs to be rasterised to the same grid and then converted to indicators.
2. *Match the soil profiles with land-surface parameters and prepare the regression matrix*. Optionally, you can also examine which predictors are the most significant or use factor analysis to reduce the redundancy of the predictors and the negative effects of their inter-correlation (multicollinearity) on the computational accuracy.
3. *Derive the regression residuals, analyse them for spatial autocorrelation and fit the variogram model*. This can also be done in **gstat** using the automated variogram fitting option.
4. *Run interpolations/simulations*.
5. *Visualise and validate the results* using control points.

The command to run regression analysis in R is:

```
> summary(solum.fit <- lm(SOLUM~DEM + SLOPE + PLANC + PROFC
+ TWI + SINS, data=baranja))
```

[4] In gstat and SAGA, regression-kriging is referred to as "*universal kriging*" which is a synonym.

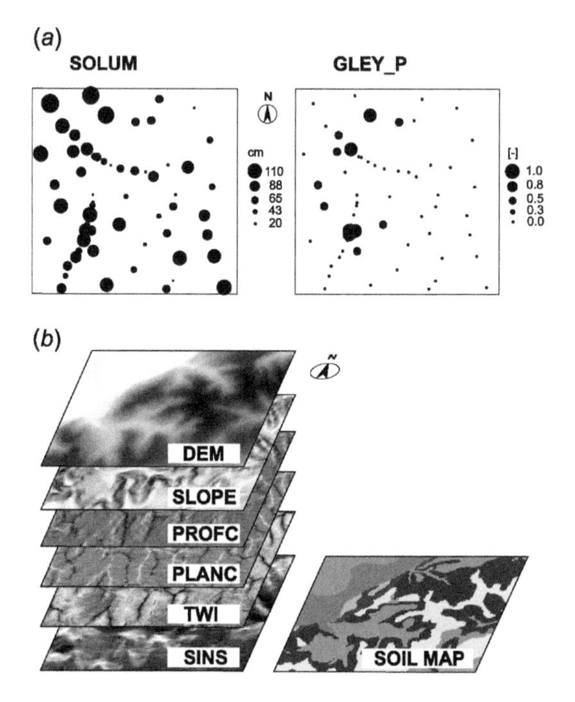

FIGURE 4 The digital soil mapping exercise: (a) sampled soil variables: thickness of soil (SOLUM) and occurrence of gleying properties (GLEY_P); (b) land-surface parameters: DEM, SLOPE, PROFC, PLANC, TWI, SINS; and (c) soil map with nine mapping units.

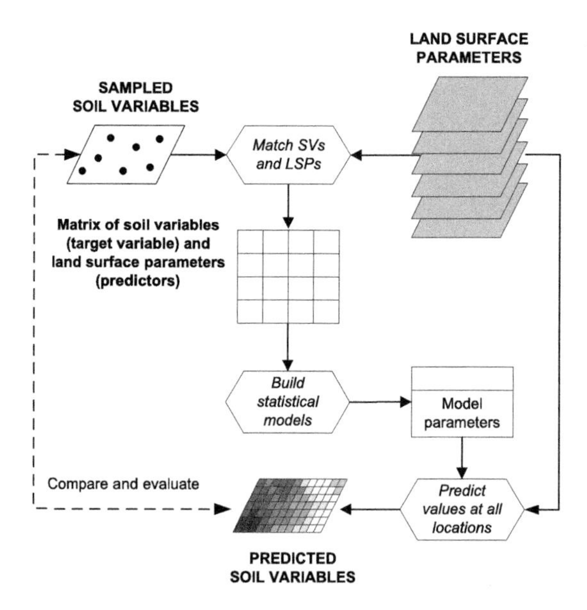

FIGURE 5 General data processing flow used to interpolate soil variables with the help of land-surface parameters.

where `solum.fit` is the output array, `lm` is the linear model fitting function in **R** and `baranja` is the matrix of soil variables crossed with predictors. This will fit the multiple linear regression model and show all summary statistics. In this case, we get a multiple R-squared of 0.46 and the adjusted R-squared of 0.40. The predictors can now be extended to soil mapping units, hence the formula changes to:

```
> summary(solum.fit <- lm(SOLUM~DEM + SLOPE + PLANC + PROFC
+ TWI + SINS + SMU1 + SMU2 + SMU3 + SMU4 + SMU5 + SMU7 + SMU8
+ SMU9, data=baranja))
```

The R-squared is now 0.63 and the adjusted R-squared is 0.51. This means that there is benefit of using also the soil mapping units to map solum, but this improves the R-squared for only 25%. Another possibility to improve R-squared is to run the step-wise regression to select only the predictors that are significant. This can be done using the function `step` in **R**:

```
> solum.step <- step(solum.fit)
> summary(solum.step)
```

In this case, the system has selected only DEM, SLOPE, PROFC, SMU4, SMU5 and SMU9 as significant predictors of SOLUM and DEM, SLOPE, PLANC, SINS, SMU3, SMU7, SMU8 and SMU9 as significant predictors of GLEY_P. Because the number of predictors is much smaller, adjusted R-squared increased to 0.57 (SOLUM), i.e. 0.71 (GLEY_P). Note that the GLEY_P is in fact a binary variable, hence we need to fit the regression model using a GLM:

```
> gleyp.glm = glm(GLEY_P ~ DEM + SINS + SMU3 + SMU5 + SMU9,
> binomial(link=logit), data=baranja)
```

After we have estimated the regression models for both SOLUM and GLEY_P, we need to estimate the variogram for the residuals of this regression model. This can be done in **gstat** using the automated variogram modelling option. Note that even in the case of the automated variogram fitting, an initial variogram needs to be set. We recommend the use 0 value for the initial nugget, the value of global variance in sampled variables (for SOLUM is 533.6 and for GLEY_P is 0.0955) for the initial sill and 1/10 of the largest distance between the points (in this case 4.3 km) as the initial range parameter. In this case study, the number of point pair is relatively low, so the fitting of the variogram is somewhat difficult, both for SOLUM and GLEY_P. Instead, we have fitted the parameters manually. This gave us the following parameters: $C_0 = 161.1$, $C_1 = 56.9$ and $R = 92.0$ (exponential model, Figure 6) for SOLUM and $C_0 = 0.025$, $C_1 = 0.010$ and $R = 148.0$ (exponential model) for GLEY_P.

Once both regression model and variogram parameters are known, we can prepare a **R** script which will implement regression-kriging and give predictions over the whole area of interest [Figure 7(a)]. This is for example a command to predict SOLUM using the **gstat** package:

```
> solum.rk = krige(SOLUM~DEM + SLOPE + PLANC + PROFC + TWI
+ SINS + SMU1 + SMU2 + SMU3 + SMU4 + SMU5 + SMU7 + SMU8 + SMU9,
data=baranja, newdata=maps.grid, model=solum.vgm, nmax=50)
```

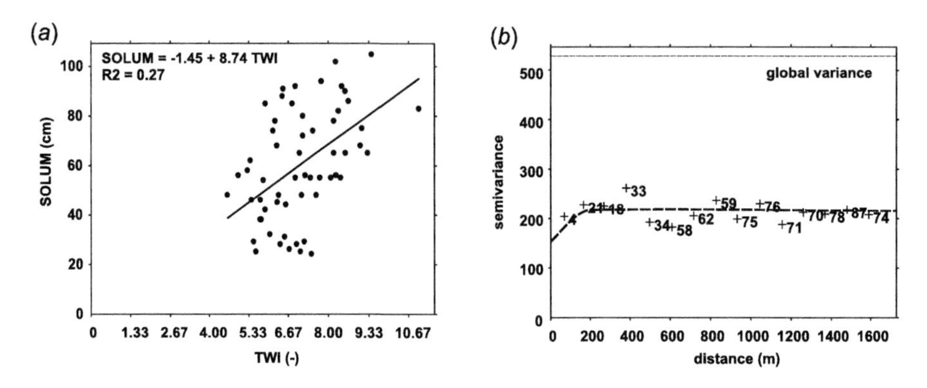

FIGURE 6 (a) Correlation between TWI and SOLUM and (b) variogram model for residuals fitted manually.

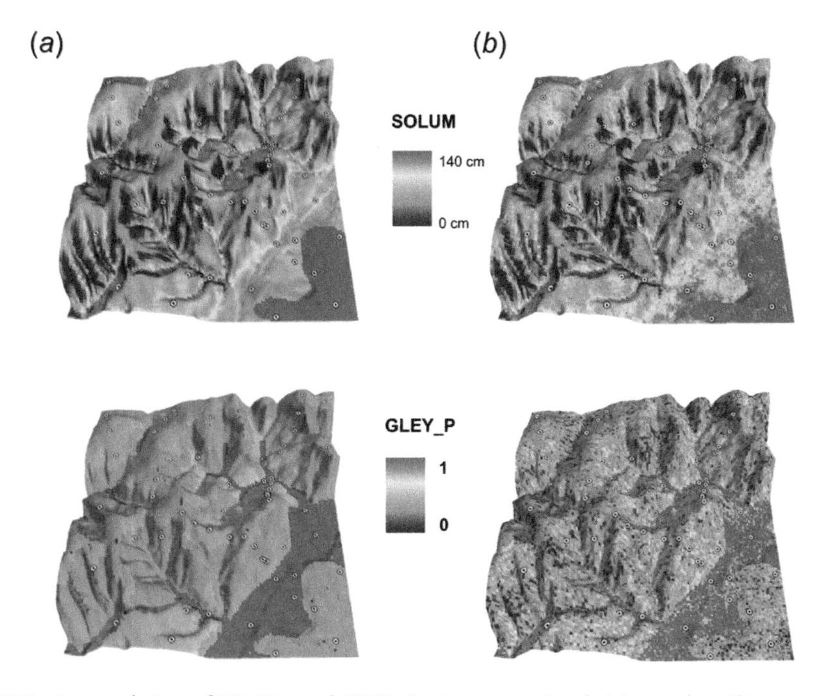

FIGURE 7 Interpolation of SOLUM and GLEY_P using regression-kriging and auxiliary predictors: (a) predictions and (b) simulations.

where `solum.rk` is the output, `krige` is the **gstat** function that runs predictions using a regression model and a fitted variogram of residuals (`solum.vgm= vgm(56.9, "Exp", 92, nugget = 161.1)`) over the grid definition of the map `maps.grid`. In this case, the `maps.grid` dataset has to be a multi-layer map with all predictors combined together. These can be imported using the **rgdal** package.

To produce a more realistic picture of how successful are predictions and how significant is the local variation (nugget), we can also use the conditional Gaussian simulations with the same regression-kriging model [Figure 7(b)]. This can be achieved by adding a parameter `nsim=1` in the **R** code above (Pebesma, 2004). Note the difference between the small-scale variation for the prediction of SOLUM and GLEY_P. In this case, GLEY_P shows much higher amount of small-scale variation than the SOLUM variable.

4. SUMMARY POINTS

Geomorphometry and use of land-surface parameters and objects has shown to be highly beneficial for production of new soil maps or improvement of the existing ones. This is especially true at regional scales and at catchment level, where information on land surface can explain more than 50% of the variability in soil parameters. One should not forget that DEMs are now quite affordable and available globally, which will soon make it an unavoidable input to the soil mapping.

The major issues that are commonly in the focus of digital soil mappers are: which land-surface parameters should be used to map soils? which statistical models should be used to fit the data? which grid resolution should we choose? and how should be uncertainty of the final outputs represented and evaluated? Each of these questions is now addressed down-below.

4.1 Which land-surface parameters to use to map soils?

Not all land-surface parameters are suitable to be used as predictors for all soil variables. If the soil variables are mainly influenced by erosion/deposition processes, one should of course try to employ land-surface parameters that reflect such processes, such as TWI or curvatures. Similarly, if the soil surveyor assumes that the pH in a soil is lower at northern expositions (dark and wet), then modelled incoming solar radiation and elevation might be helpful. One can also calculate statistics to search for the best land-surface parameters having the highest correlation with the soil variable to predict.

> REMARK 4. *Land-surface parameters alone are rarely able to explain the entire spatial variation of soils. To improve such models, it might be wise to supplement land-surface parameters with other data sources, like remotely sensed data, information on geology or land cover.*

Böhner and Selige (2006) believe that, instead of blindly using any possible land-surface parameters to fit variation, we should always try to derive process-descriptive land-surface parameters such as sediment transport indices, mass balance and solifluction parameters. Each soil-process would require design of land-surface parameters that can reflect at least relative impact of relief on the soil formation.

Ideally, we should be trying to build physical process-based models of soil formation and then employ such models for mapping (see Section 2.2 in Chapter 19), but this is still not feasible for many soil processes. Note that, because soils are hidden, mixed and fuzzy bodies, their accurate mapping is often possible only to a certain extent. This also means that land-surface parameters will be successful in explaining the sampled variability of soils only to a certain extent. The R-square of most regression models will rarely exceed 60% (McBratney et al., 2003). Land-surface parameters alone are rarely able to explain the entire spatial variation of soils. To improve such models, it might be wise to supplement land-surface parameters with other data sources, like remotely sensed data, information on geology or land cover (McKenzie and Ryan, 1999; Ryan et al., 2000; Dobos et al., 2000; Bui et al., 2002, 2006).

4.2 Which statistical models to use?

Numerous statistical techniques can be used to handle, process and classify topographic data derivatives (Odeh et al., 1994; De Bruin and Stein, 1988; Lark, 1999; McKenzie and Ryan, 1999). Many of the studies used the land-surface parameter values as direct inputs for regression, or geostatistical procedures, like multilinear regression, logistic regression, regression-trees, regression-kriging, cokriging (Gessler et al., 1995; Moore et al., 1993a; Odeh et al., 1994) or to discrete classification approaches, like maximum likelihood classification (Dobos et al., 2000; Hengl and Rossiter, 2003). Land-surface parameters can be also first preprocessed, classified or transformed and then used as input for statistical or geostatistical methods. Discriminant analysis is often used to enhance the separability of classified soil parameter based on the land-surface parameters (Bell et al., 1992; Sinowski and Auerswald, 1999). Discrete and continuous clustering algorithms (Fuzzy *k*-means) are often employed to create relatively homogeneous landform classes for further use in soil property estimation (Lark, 1999; De Bruin and Stein, 1988).

Bishop and Minasny (2005, Table 7.1) reviewed all possible statistical models used to map soil variables using auxiliary information. They evaluated seven groups of techniques: (1) multiple linear regression, (2) discriminant analysis, (3) *k*-means clustering, (4) Generalized Linear Models (GLMs), (5) Generalized additive models (GAMs), (6) artificial neural networks and (7) classification and regression trees. Each of these groups was evaluated according to various aspects such as: predictive power, ease of use, sensitivity to parsimony, ease of interpretability, handling of mixed data, handling of non-linear relationships, etc. None of the techniques is completely superior to competitors — linear models are easier to use than neural nets but will probably fit the data less successfully. Likewise, GAMs will be more successful with categorical data, but they are more difficult to interpret and might have problems with detecting the parsimony. The choice of model should obviously fit data characteristics (measurement errors, representativity, type of variables), nature of the modelled relationship and user perspective.

4.3 Which grid resolution to use?

One of the most limiting factors of the use of a DEM is its accuracy and spatial resolution. Cell size controls the success of mapping. Various features (e.g. small streams) that are visible at very fine resolutions, will be lost once the resolution increases two or three times. Many soil forming processes happen at large scales and, therefore, soil surveyors are asked to describe soils at 1 m^2 blocks of land or finer. Obviously, not many can afford so detailed DEMs so that the question remains "*Which resolution is good enough?*" (Hengl, 2006).

Numerous authors evaluated the success of spatial prediction models using various resolutions. Ryan et al. (2000) discovered that predictive relationships developed at one scale might not be useful for prediction at different scales. The results of Chaplot et al. (1998) showed an increase in the prediction quality with the decrease of the DEM mesh. Hammer et al. (1995) used slope class maps from soil survey to validate computer-generated slope class maps from 10-metre and 30-metre DEM. They concluded that the GIS-produced maps underestimated the slopes on convexities and overestimated slopes on concavities. The overall accuracy was over 50% for the 10-metre resolution and between 20 and 30% for the 30-metre resolution grids. The majority of the studies were carried out in the field or small watershed scale. Most of the cited research articles on this topic used an original grid spacing of less than 20 m, seven of them used 20–50 m resolutions, while only three used coarser resolutions DEM (100–1000 m). Many of the papers stayed with relatively high resolution DEM to keep the study area small enough to ensure its lithologic and climatic homogeneity.

Thomas et al. (1999) predicted soil classes with parameters derived from relief and geologic materials in a sandstone region of Northeastern France. They could explain more than 70% of the soil class variation in a small catchment area by the nature of geologic substratum and attributes derived from DEM. However, the model predictive potential decreased to 55% after the application to a larger region. The disagreements were due primarily to (1) the existence of superficial deposits not mentioned on the geologic maps, (2) the choice of reference catchments which were not representative of the study area and (3) regional climatic influences which were insufficiently considered during modelling at the local catchment scale. Chaplot et al. (2000) analysed the sensitivity of prediction methods for soil hydromorphy with regard to the resolution of topographical information and additional soil data. From the elevation data they derived the variables of the elevation above the stream bank, the slope gradient, the specific catchment area, and the TWI in resolutions of 10-, 20-, 30- and 50-m. The correlations among these variables and the hydromorphy index were calculated and found to be strong (R-squared up to 0.8). However, the coarser DEM resolution greatly reduced prediction quality.

M.P. Smith et al. (2006) analysed the effect of both the grid resolution and neighbourhood window size on the accuracy of the Soil–Landscape Inference Model. They concluded that various grid resolutions will be suitable for various types of landscape. In areas of less relief, a somewhat coarser resolution (33–48 m) will do the job, while in the areas with higher slopes, one will need to work with somewhat more detailed DEMs (24–36 m). This reflects the idea that the grid resolution needs to be selected to accurately reflect the complexity of a terrain — if the terrain is rather smooth, even a relatively coarse DEM can be used to produce accurate outputs.

Florinsky and Kuryakova (2000) focused specifically on the importance of grid resolution of land-surface parameters on the efficiency of spatial prediction of soil variables. They plotted correlation coefficients versus different grid resolutions and were able to actually detect a cell size with most powerful prediction efficiency. However, the prediction power versus grid resolution graph might give a set of different peaks for different target variables, so that we cannot select a single *'optimal'* grid resolution.

Finally, this grid resolution is then valid only for this study area and its effects might be different outside the area (Hengl, 2006). Zhang and Montgomery (1994) concluded that landscape features were more accurately resolved when cell size decreases, but faithful representation of a land surface by a DEM depends on both cell size and the accuracy and distribution of the original survey data from which the DEM was constructed.

REMARK 5. *The most objective procedure to determine a suitable cell size for soil-landscape modelling is to evaluate predictive efficiency for various cell sizes and then select the one with the best performance.*

It can be concluded, that large scale studies using high resolution DEM (up to 20 m cell size) will focus on land-surface parameters representing actual, site-specific measures of terrain, like, slope, catchment area, TWI, etc. Low resolution DEM used for small scale studies does not represent the actual values of the land-surface parameters, only the overall average values. Therefore, land-surface parameters representing characteristics of a bigger landscape unit are more appropriate sources for small scale mapping.

According to MacMillan (2004), it is unrealistic to expect that elevation data captured on a 90 m grid (the global SRTM DEM) with horizontal positional variation of as much as 90 m and vertical precision of no better then 10–20 m is going to provide an accurate depiction of local (small size) variation in the configuration of the topography. The 90 m DEM data will capture very large features such as major mountains or hills and valleys but it simply cannot resolve minor local variation in topography. Having a point every 90 m means that you cannot reasonably expect to identify and resolve landscape features that are less than about 200 m in length.

Many perturbations of the landscape that have lengths of 10's of metres and vertical relief of 1–10 m exercise significant influence on the variation of soils and soil properties over distances of 10's of metres. Most field assessments of ecological site type changed regularly over distances of 10's of metres (MacMillan et al., 2004). It seems that we should at least use resolutions of about 5 m and a vertical precision of better than 0.5 m to be able to predict variation in soils or soil properties accurately at specific geographic locations.

4.4 How to evaluate the quality of outputs?

One of the major advantages of using quantitative techniques of soil mapping is that one can estimate the direct and propagated uncertainty of the prediction models (see also Chapter 5 and Section 2.4 in Chapter 19). Our experience is that accuracy assessment should always be based on an independent test data set separated from the training data used to calibrate the models (Rykiel, 1996). The best and most commonly used measure of the predictive capabilities is the root mean square prediction error (RMSPE). Categorical values, like soil classes, need different measures such as kappa statistics, fuzzy kappa statistics and confusion indices (Congalton and Green, 1999). The use of this confusion matrix can help the user identifying the major sources of misclassification between the classes and provides the necessary information on how to improve the training setup in order to further increase the accuracy of classification.

IMPORTANT SOURCES

Lagacherie, P., McBartney, A.B., Voltz, M. (Eds.), 2006. Digital Soil Mapping: An Introductory Perspective. Developments in Soil Science, vol. 31. Elsevier, Amsterdam, 350 pp.

Bishop, T.F.A., Minasny, B., 2005. Digital soil-terrain modelling: the predictive potential and uncertainty. In: Grunwald, S. (Ed.), Environmental Soil–Landscape Modeling: Geographic Information Technologies and Pedometrics. CRC Press, Boca Raton, FL, pp. 185–213.

Hengl, T., 2007. A Practical Guide to Geostatistical Mapping of Environmental Variables. EUR 22904 EN Scientific and Technical Research Series. Office for Official Publications of the European Communities, Luxemburg, 143 pp.

McBratney, A.B., Mendonça Santos, M.L., Minasny, B., 2003. On digital soil mapping. Geoderma 117 (1–2), 3–52.

McKenzie, N.J., Gessler, P.E., Ryan, P.J., O'Connell, D.A., 2000. The role of terrain analysis in soil mapping. In: Wilson, J.P., Gallant, J.C. (Eds.), Terrain Analysis: Principles and Applications. Wiley, pp. 245–265.

CHAPTER 21

Vegetation Mapping Applications

S.D. Jelaska

vegetation mapping and its importance · the role of geomorphometry in vegetation mapping · the spatial prediction of vegetation variables using land-surface parameters · statistical prediction models and their use · evaluating mapping accuracy · does data from remote sensing compete or cooperate with land-surface parameters and objects in predicting current vegetation cover? · spatial resolution and statistical methods for mapping vegetation

1. MAPPING VEGETATION

1.1 Why is it important?

Vegetation mapping started with the work of von Humboldt at the very beginning of the 18th century, but did not begin to develop into a profession until more than a century later. Although vegetation (i.e., plant cover of any kind) had been represented on maps for much longer than that, in those distant times, it was mainly shown in coarse thematic resolution, as supplementary information on maps of which the main topics were relief and/or settlements and roads. For example, on 18th and 19th centuries Austrian military maps, vegetation was mapped as forests, pastures, swamps, vineyards and crops. By the 20th century, the development of various hierarchical systems for vegetation classification boosted the creation of maps that focused mainly on vegetation. Especially after the Second World War, this trend gained further support with the development of aerial photography.

Another significant increase in vegetation mapping occurred during the last quarter of the 20th century due to:

- *an increased need for spatially organised data* about the living component of the world. This data was required to inform environmental and nature management, to predict scenarios, to identify and select important areas for nature protection and/or conservation, and to make environmental impact assessments, etc.;

Developments in Soil Science, Volume 33 © 2009 Elsevier B.V.
ISSN 0166-2481, DOI: 10.1016/S0166-2481(08)00021-4.

- *the development of GIS*, as a very efficient way of storing, creating and analysing spatial data. An added attraction is that the capabilities of the system are constantly increasing, while the costs are decreasing;
- *the development of remote-sensing techniques* with ever richer in spatial and spectral detail (further insight into this topic can be found in Alexander and Millington, 2000).

An important, often previously neglected, attribute, that should accompany every vegetation map, is an assessment of the accuracy of the displayed data. This is very often carried out using Kappa statistics (Congalton and Green, 1999), although these have been criticised for being over-used, and that they are not always the best method available (Maclure and Willett, 1987; Feinstein and Cicchetti, 1990). For an overview of rater agreement methods, see e.g. Mun and Von Eye (2004). Besides providing information on the current type of biota[1] at a given area, with the help of well-defined ecological indicator systems (Ellenberg et al., 1992), vegetation maps also provide plenty of information about the prevailing ecological conditions with respect to a number of environmental variables (such as soil acidity, soil-water content, mean air temperature, etc.).

A recent example of soil-parameter prediction using indicator values of current vegetation, mapped using remote-sensing techniques, can be found in Schmidtlein (2005). Furthermore, when mapped at community level, as defined in Braun-Blanquet (1928), vegetation maps provide a good basis for most habitat classifications (Antonić et al., 2005), and for land-cover mapping projects. Data on the spatial distribution of vascular plants can also be very valuable for estimating the overall biodiversity. This was shown by Sætersdal et al. (2003), who demonstrated that vascular plants are a good surrogate group of organisms in biodiversity analyses.

REMARK 1. *Knowing the spatial, and temporal, distribution of vegetation is important because vegetation acts as an identity card — it tells us about the environment and the potential biota under present conditions.*

Nowadays, a thorough understanding of the global changes that are taking place in the environment is a necessity, as is the need to quantify the speed and amount of those changes. Under these circumstances, historical vegetation maps (of various thematic resolutions) have become a very valuable tool in these analyses and estimations. Consequently, there is increased pressure to produce baseline maps of the current situation, so that they can serve as reference for monitoring of future actions, especially in important nature-conservation areas.

There are also initiatives that include large areas, such as *CORINE LAND COVER*[2] (CLC), serviced by the European Environment Agency (http://www.eea.europa.eu). Although some of the CLC's 44 classes of 3-level nomenclature say very little, or nothing, about present vegetation (e.g. 1.1.1. *Continuous urban fabric* or 5.1.1. *Water courses*), some of them give more precise '*green*' information

[1] *Biota* — the animals, plants, fungi and microbes that live, or have lived, in a particular region, during a certain period.
[2] The CORINE (Coordination of Information on the Environment) Programme was established in 1985 by the European Commission, using three main CORINE Inventories (Biotopes, Corinair and Land Cover).

(e.g. 3.1.2. *Coniferous forest* or 3.2.2. *Moors and peatland*). A CLC map, with minimum mapping units of 25 ha, has been prepared, derived from interpretations of satellite images. It shows the land cover of part of Europe, between the 1990s and 2000, and includes a change analysis for the same period. This is a valuable tool and data set for environmental policy makers and for anyone else working in related fields.

1.2 Statistical models in vegetation mapping

Nowadays, statistical models are used in almost all vegetation mapping. Exceptions are local large-scale projects and, for example, in CLC projects for which a methodological prerequisite is that the boundaries of the RS images are delineated manually. In all other cases, the statistical approaches applied are almost as diverse as the vegetation itself. For example, the range of statistical models on disposition is huge, varying from simple univariate linear regressions to very complex models such as Neural Networks (Bishop, 1995), Support Vector Machines (Cristianini and Shawe-Taylor, 2000) or Naïve Bayesian Classifiers (Duda et al., 2000). Overviews of the techniques have been made by Franklin (1995) and Segurado and Araújo (2005), and some direct comparison can also be found in Oksanen and Minchin (2002), Jelaska et al. (2003). A valuable comparison of predictive models used to map distribution of species can be found in Latimer et al. (2004).

Numerous elements can determine which model would be the best to use. These can be objective elements, such as a certain type of variable scale (e.g. nominal, categorical, ordinal), or the size of the input sample and the number of predictors. At the other end of the range, the elements can be purely subjective, such as the researcher's preference for particular methods. However, inevitably, the latter will be limited to those methods that satisfy the conditions dictated by the type and size of the input data. The only rule that can perhaps be pointed out here, is to try to use data sets that are sufficiently large to ensure that a stable model can be built, and that it can be tested on an independent data set. Obtaining a sufficiently large data set, especially when costly and time-consuming field sampling is involved, could be a critical factor.

> REMARK 2. *Statistical methods used in vegetation mapping vary from simple univariate linear regression to neural networks and Bayesian classifiers. Generalised linear models (GLM), classification and regression trees (CART) and general additive models (GAM) are among the most frequently used methods.*

The final combination of predictors and methods will be case-dependent and influenced by five main factors: (1) the density of field observations; (2) the size and character of the (support) data on input vegetation; (3) the availability and quality of auxiliary data, such as remote-sensing images and DEM derivatives; (4) the (thematic and spatial) resolution, i.e. the scale of predictor variables; (5) the capabilities of the GIS and the statistical software, etc.

Among the most frequently used methods in vegetation mapping are: generalized linear models (GLM), classification and regression trees (CART) and general additive models (GAM). These could be combined with ordination (e.g. correspondence) and/or classification (e.g. cluster) analyses (Gottfried et al., 1998; Guisan et al., 1999; Pfeffer et al., 2003; Jelaska et al., 2006). See also Section 2.1 in Chapter 19 for additional information about statistical models.

Geostatistics is only occasionally included in vegetation-mapping projects and papers (e.g. Bolstad et al., 1998; Miller and Franklin, 2002; Pfeffer et al., 2003). Since various interpolation methods deal with continuous variables, when it comes to mapping vegetation classes, i.e. with discrete variables, it is only possible to use those methods indirectly, so it becomes even more complex to apply them. A good theoretical background to this problem can be found in a paper by Gotway and Stroup (1997). Another example can be found in Pfeffer et al. (2003) who employed universal kriging [see Equation (2.5) in Chapter 19] by correlating topographic variables and vegetation scores (specifically, abundance of 147 plant species on 223 plots).

Apart from the problem of nominal scale in vegetation data, Miller and Franklin (2002) found that output pattern is highly dependable on the spatial origin of the sample data set. However, with open-source, user-friendly software packages for calculating spatial statistics, and the more widely accessible they become, the more geostatistics is going to find its place in vegetation mapping.

1.3 The role of geomorphometry in vegetation mapping

Because geomorphometry can be used to describe (and define) the physical environment, the expectation is that it will be possible to use it to explain and model vegetation that is directly dependent on environmental conditions and its spatial characteristics [see also Equation (1.2) in Chapter 19]. In fact, the physical environment has always been used for this purpose, since, only occasionally, entire areas have been completely field surveyed and mapped for their vegetation at that point in time. Depending on thematic and spatial mapping resolutions, and the diversity of the terrain at the time of mapping, mappers use land-surface parameters (elevation belts, aspect, slope, etc.) combined with field observations to create polygons that covered the entire area of interest. When these estimators are not sufficient for estimating the occurrence of a particular type of vegetation, they use land-surface parameters in combination with estimators, such as geology, annual rainfall, and mean temperature.

These conditioned rules can be viewed as simple spatial inference systems, where conditions can be rather trivial: e.g. if the elevation is 350–500 m, then map in mixed oak–beech forest. However, conditions can also be complex: e.g. everywhere in an elevation belt where the soil acidity (pH) is lower than 4, acid beech forest is present, otherwise there is mixed oak-beech forest. In the majority of cases, the mapper has to deal with a combination of conditions. From the schematic distribution of six different vegetation types, represented in Figure 1, several facts can be observed. Vegetation types follow the temperature gradient in both horizontal (i.e. geographical latitude) and vertical directions (i.e. in elevation belts). However, if we use elevation as the sole estimator, we might make a wrong prediction, depending on whether we have input data from, for example, the northern or southern slopes of a mountain. This is because vegetation belts are lower on the

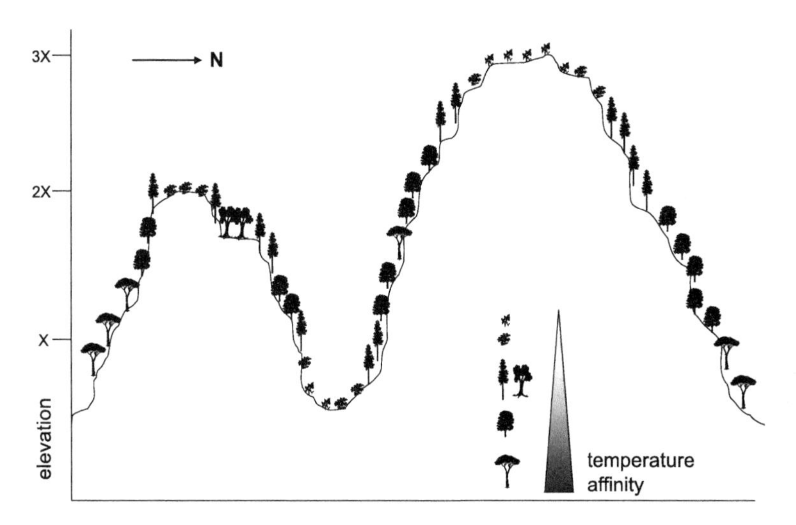

FIGURE 1 A schematic distribution of six types of vegetation, each represented by a different symbol. The base of the temperature affinity triangle denotes an affinity for higher temperatures, and the apex for the lower ones. (The direction of North is shown by the letter "N" and an arrow.)

northern slopes than they are on the southern ones (in the northern hemisphere, that is, and the opposite applies in the southern hemisphere).

Furthermore, vegetation belts occur at higher elevations on larger mountains, therefore we should be careful when extrapolating our models outside the sampled area. Special geomorphometric features such as *sinkholes* (shown somewhat exaggeratedly between two peaks in Figure 1) could cause temperature inversion that would lead to an inversion of the vegetation belts, and these will differ on the northern and southern sides of the sinkhole. The intensity of a slope can be critical for the development of a distinct type of vegetation within the same elevation belt. This is illustrated in Figure 1 by two vegetation types that have the same temperature affinity. Besides the basic land-surface parameters (i.e. DEM, SLOPE, ASPECT) shown and discussed here, other land-surface parameters (e.g. WTI and/or SPI) could also be crucial for certain types of vegetation.

> REMARK 3. *Importance of land-surface parameters in vegetation mapping is case dependent. Thematic resolution of vegetation determines whether elevation, flow accumulation potential or some other parameter will play a crucial role in spatial distribution of a given vegetation type.*

Nowadays, land-surface parameters and objects are not just a set of GIS layers used for shaping and transferring polygons of present vegetation onto a paper map. They are used for constructing very complex statistical models to predict the spatial distribution of vegetation. Table 1 lists several examples of how geomorphometry is applied in mapping vegetation.

TABLE 1 Examples from the literature on the use of geomorphometry in vegetation mapping

Source	Number of input LSPs	Other predictors	Resolution (DEM/thematic)
Gottfried et al. (1998)	18	No	1 m / species & vegetation
del Barrio et al. (1997)	6	Yes (2)	10 m / landscape units
Beck et al. (2005)	11	Yes (12)	20 m / species
Davis and Goetz (1990)	5	Yes (2)	30 m / vegetation types
Sperduto and Congalton (1996)	2	Yes (2)	30 m / species
Franklin (1998)	3	Yes (5)	30 m / species
Guisan et al. (1999)	10	No	30 m / species
Jelaska et al. (2006)	3	Yes (1)	30 m / species
Fischer (1990)	3	Yes (4)	50 m / plant communities

The use of geomorphometry for vegetation mapping applications can be summarised around three points:

- there is no ideal DEM resolution for a thematic resolution of specific types of vegetation, however, most vegetation mapping projects utilise 10–50 m DEMs;
- there are no preferred land-surface parameters that can be used to map vegetation, however, ecological land-surface parameters (climatic and hydrological modelling) are more efficient, in general, for making predictions;
- in most cases, land-surface parameters are used in combination with other parameters — ranging from regolith thickness, substratum characteristics, and parameters derived from remote sensing such as snow cover, water cover, normalised difference vegetative index (*NDVI*), climatic variables, land-use, and leaf-area index.

Another very important role of land-surface parameters in vegetation mapping applications, even if they are not used directly as vegetation predictors, is for the topographic correction of RS images (Riaño et al., 2003; Shepherd and Dymond, 2003; Svoray and Carmel, 2005), especially in hilly and mountainous areas.

The importance of particular land-surface parameters in vegetation mapping is case-dependent. Whether the elevation, flow accumulation potential or another parameter will play a crucial role in the spatial distribution of a given type of vegetation, will depend on the thematic resolution of the vegetation map and with the current diversity of the land-surface parameters.

Land-surface parameters can also be very useful in mapping vegetation that is influenced by human activities, since man adjusts his activities according to existing ecological conditions. For instance, after clear-cutting the forest vegetation from an area, it is more likely that crops will be grown on the flatter terrain, and vineyards (in the case of the Baranja Hill area) on steeper terrain. Similarly, crops will be grown on lower elevations, and the higher elevations will be reserved for pastures.

2. CASE STUDY

In the following section, using the land-surface parameters of the Baranja Hill case study, we will demonstrate how to map the distribution of the presence of a particular plant species (in this case *Robinia pseudoacacia* L. — Black Locust) and the CORINE land-cover categories. Quantitative data of the presence of the Black Locust (step value 0.2) was obtained by field observations. This represents a coverage percentage ranging from 0 (species absent) to 1 (species completely covering the area — i.e. a pure stand of Black Locust plants). We will compare two sets of predictors: (a) land-surface parameters and (b) LANDSAT image bands.

We will use seven land-surface parameters: elevation (DEM), slope (SLOPE), cosine of aspect (NORTHNESS), sine of aspect (EASTNESS), natural logarithm of flow accumulation potential increased by 1 (LNFLOW), profile curvature (PROFC) and plan curvature (PLANC). All these are prepared in an ArcInfo GRID module (see Chapter 11) using the 25 m DEM. The second set consists of eight spectral channels of LANDSAT ETM+ and NDVI (Normalized Difference Vegetative Index).

Both sets were used first separately, and then in combination, which finally gave three sets of predictor variables. We used the STATISTICA program (http://www.statsoft.com) to build predictive models, although operations resembling them are available in R (http://r-project.org) and in similar open-source statistical packages.

2.1 Mapping the distribution of a plant species

Multiple (linear) regression models (MR) can be calculated making exclusive use of independent variables. To select variables that significantly ($p < 0.05$) contribute to explaining the variability of Black Locust data, a stepwise regression was used. The MR models follow the general form:

$$\text{Robinia} = \beta_0 + \beta_1 \cdot q_1 + \beta_2 \cdot q_2 + \cdots + \beta_p \cdot q_p \tag{2.1}$$

or in matrix format:

$$\text{Robinia} = \beta^{\mathbf{T}} \cdot \mathbf{q} \tag{2.2}$$

where 'Robinia' is the coverage of Black Locust, b_0 the intercept and $\beta_1, \beta_2, \ldots, \beta_p$ or $\mathbf{q}$ the coefficients of corresponding predictors $q_1, q_2, \ldots, q_p$ or β, included in the model. After the estimation of the regression coefficient, the spatial predictions can be calculated in the ArcInfo GRID module to produce the coverage of Black Locust over the whole Baranja Hill area (Figure 2).

A disadvantage of the MR is that the predictions may be outside the physical range of the values (in this case <0 and >1). This is obviously erroneous. A better alternative for interpolating the indicator data is to use *Multiple logistic regression* models (Neter et al., 1996):

$$\text{Robinia} = \left[1 + \exp\left(-\beta^{\mathbf{T}} \cdot \mathbf{q}\right)\right]^{-1} \tag{2.3}$$

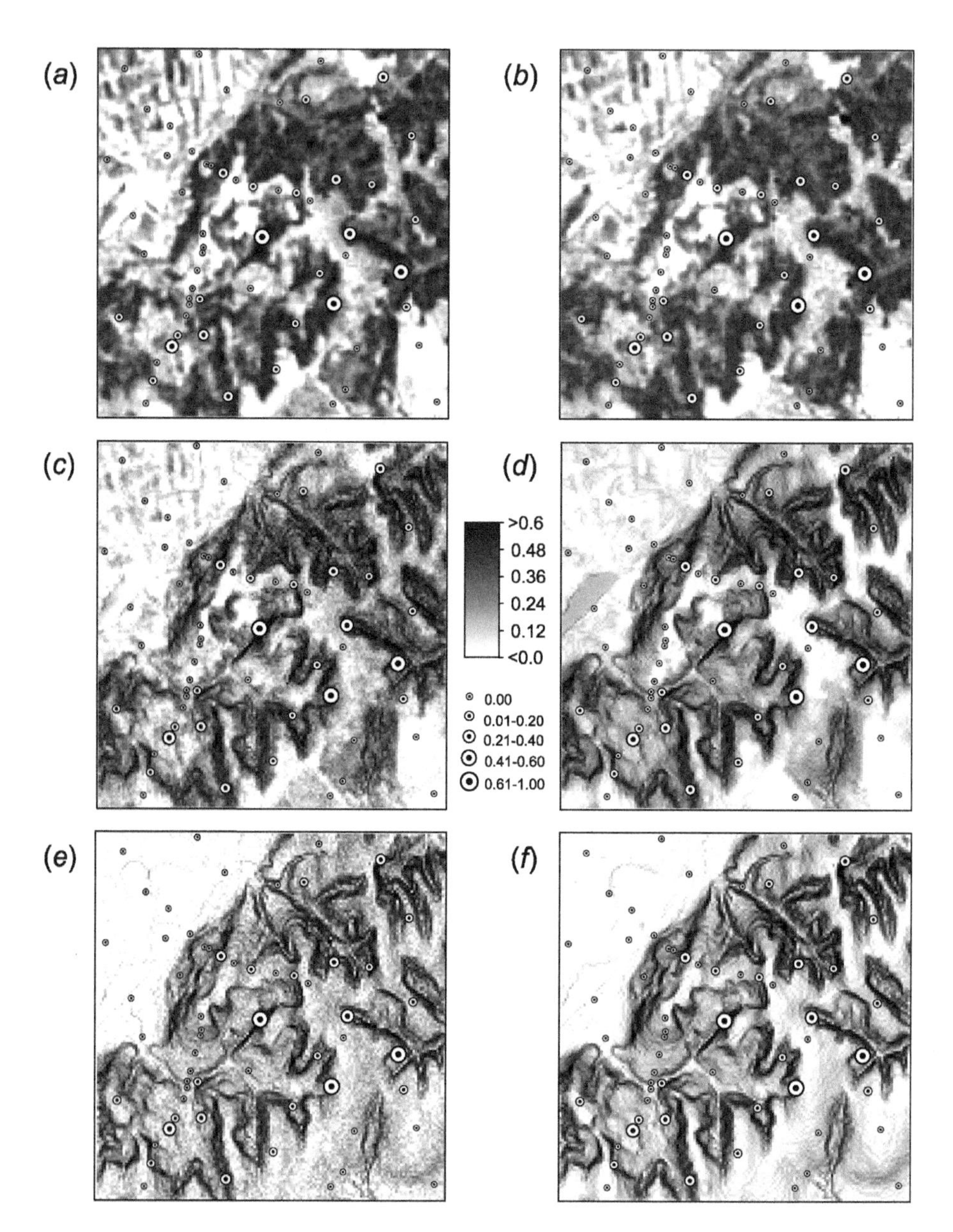

FIGURE 2 Multiple regression models of the percentage of cover of Black Locust on Baranja Hill: on the left, *full regression models* constructed with predictor variables consisting of: (a) RS data only; (c) land-surface parameters plus RS data; (e) land-surface parameters only; and, on the right, *stepwise regression models* constructed using: (b) RS data only, (d) land-surface parameters plus RS data, and (f) land-surface parameters only.

which can easily be linearised if the target variable 'Robinia' is transformed to the logit variable:

$$\text{Robinia}^{+} = \ln\left(\frac{\text{Robinia}}{1 - \text{Robinia}}\right) \tag{2.4}$$

where $0 < \text{Robinia} < 1$. To select just those predictors that contribute significantly ($p < 0.05$) towards explaining the variability of Black Locust data, a stepwise multiple logistic regression can be run, contrary to the full multiple regression that will use all seven land-surface parameters. The two logistic predictive models were also applied to the land-surface parameter grids in the ArcInfo GRID module.

Six MR predictive models of the percentage of coverage (Figure 2) give a similar general pattern for the distribution of Black Locust, with some differences in the north-western corner, whereas models with LANDSAT bands as predictors tend to over-estimate the presence of Black Locust. Over-estimation is also evident in the south-eastern corner, except for those models that use land-surface parameters as predictors. This co-mission is probably due to field sampling that did not cover forests present in that section, as can be seen in the orthophoto of the area (Figure 5).

Models using land-surface parameters seem to have a higher local variability, i.e. a more structured output. The proportions of explained variability for all three sets of predictor variables are similar in the models that use a full set of predictors, to those obtained by stepwise regression. For the Black Locust cover on Baranja Hill, the highest adjusted R-squares were those of models using both LANDSAT channels and land-surface parameters. The value for the full model was 0.57 and for the stepwise model (including SLOPE, PROFC and SC2) was 0.60. The value for the other models was 0.50, with the exception of the stepwise Landsat model that had a value of 0.48 for the predictors SC2, SC3, SC5, and also included NDVI.

The predictors selected from the stepwise regression, using land-surface parameters only, were SLOPE and PROFC. Analysis from the regression model that includes SLOPE, PROFC and SC2, does not reveal the spatial autocorrelation of residuals. Hence geostatistical prediction techniques (e.g. regression-kriging, as used in Chapter 20) are not suitable.

Estimating the accuracy of logistic predictive models (Figure 3) is highly dependent upon the chosen threshold value, since logistic models return values of between 0 and 1 to represent the probability of occurrence of a particular species. Whether we choose 0.2 or 0.8 as the threshold[3] value, it will dramatically affect the outcome of the predicted occurrence on a binary presence/absence level. For the Black Locust distribution, we calculated accuracies of input data for threshold values of 0.4 and 0.6.

A full multiple logistic regression model [see Figure 3(a)] shows a high omission error, or under-estimation, of the occurrence of Black Locust. It only predicted its presence accurately, at just one field point. The stepwise model, that included

[3] A threshold value is a distinct, calculated, probability of the presence or absence of a species. For some very rare species, a smaller threshold value (e.g. 0.3) will produce a more realistic map of the occurrence of that species, but for more dominant species, to prevent over-estimation, higher values (e.g. 0.7 or 0.8) need to be used.

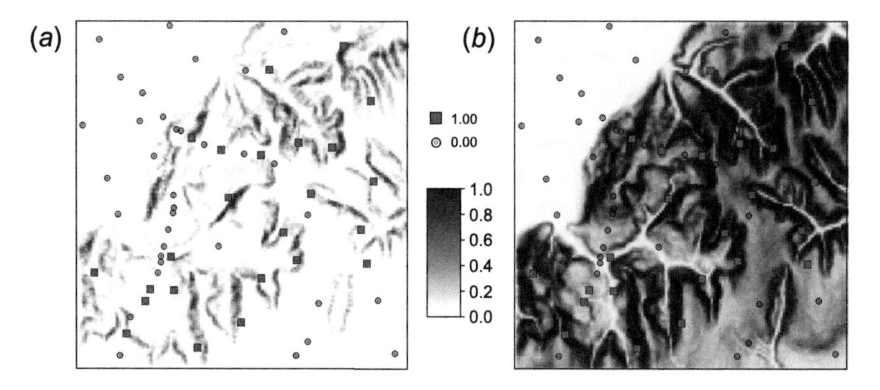

FIGURE 3 Logistic regression models showing the probability of occurrence of Black Locust on Baranja Hill: (a) the full model and (b) stepwise-regression model. Constructed by using land-surface parameters only.

DEM and SLOPE as predictors, has an overall accuracy of 78% for a threshold set at a value of 0.4, and 80% at 0.6, and the latter value gave a higher omission error.

> REMARK 4. *Remote sensing data could be better predictor of main land-cover classes, while land-surface parameters of finer thematic resolution. However this depends upon their spatial resolution.*

2.2 Mapping land-cover classes

Three *classification trees* were constructed, one for each set of predictors, using an exhaustive CART-style search for univariate splits, as the split selection method, with a Gini measure of goodness of fit, and, as a stopping rule, FACT-style direct stopping, using the stopping parameter *Fraction of Object* set at 0.35 in the STATISTICA package. Due to space constraints, only the classification tree constructed using land-surface parameters and Landsat spectral channels is shown in Figure 4. Kappa statistics for all three models are shown in Table 3. The classification tree model that was developed was then run in the GRID module of the ArcInfo software, using a series of nested `IF` statements, on grids containing data about the predictors that had been used:

```
     Grid: IF (sc3<=39.5) map = 311
:: else if (sc3>39.5 & sc2>62.5 & dem>220.5) map = 22
:: else if (sc3>39.5 & sc2>62.5 & dem <=220.5) map = 211
:: else if (sc2<=62.5 & lnflow>1.354 & dem>107) map = 22
:: else if (sc2<=62.5 & lnflow>1.354 & dem<=107) map = 32
:: else if (sc2<=62.5 & lnflow<=1.354 & eastn<=-0.8962)
map = 311
:: else if (sc2<=62.5 & lnflow<=1.354 & eastn>-0.8962 &
curvp > 0.10585) map = 32
:: else map = 24
:: endif
```

TABLE 2 Reclassification scheme of land cover for the Baranja Hill field data, prior to development of the predictive models

Code	Code description	Cases	New code	Reclassification description
211	agricultural land	19	211	agricultural land
221	vineyards	3	22	permanent crops
222	fruit trees and berry plantations	3	22	permanent crops
231	pastures	2	32	non-forest vegetation
242	complex cultivation pattern	2	24	heterogeneous agricultural land
243	land principally occupied by agriculture	4	24	heterogeneous agricultural land
311	broad-leaved forest	21	311	broad-leaved forest
321	natural grasslands	2	32	non-forest vegetation
322	moors and peatland	3	32	non-forest vegetation

where `sc2`, `sc3`, `dem`, `lnflow`, `eastn` and `curvp` represents the predictor grid: second and third spectral channels, elevation, natural logarithm of flow accumulation potential, eastness and planform curvature, respectively. The map represents the output grid, i.e. the predicted land-cover map (Figure 4).

The overall kappa value is the highest for the model that combines Landsat channels and land-surface parameters, and the lowest for the classification tree (land-surface parameters). Landsat predictors, based on Kappa statistics, predicted all the land-cover classes better than the land-surface parameters, with the exception of *non-forest vegetation*, where the kappa values were equal. Combining Landsat channels with land-surface parameters enhanced the predictions of *permanent crops*, *heterogeneous agricultural land* and *broad-leaved forest* compared with the Landsat model, while for *non-irrigated arable land* and *non-forest vegetation*, slightly lower predictions were given.

It is interesting to observe in the classification tree built with land-surface parameters and Landsat channels (Figure 4), that the first two splits were carried out, based on values of the spectral channels that classified the majority of cases in *forest* (code 311) and *agricultural areas* (codes 211 and 22). Further on, land-surface parameters were used to classify most of the occurrences of other types of vegetation cover, representing, in fact, combinations of various types of cover. This suggests that once spectral channels have separated the main classes (e.g. of forest vs. agriculture), land-surface parameters would be more useful for higher thematic resolutions (e.g. pasture vs. meadows).

REMARK 5. *Although DEM has proven to be a powerful input for mapping vegetation, accuracy-assessment should accompany every vegetation map.*

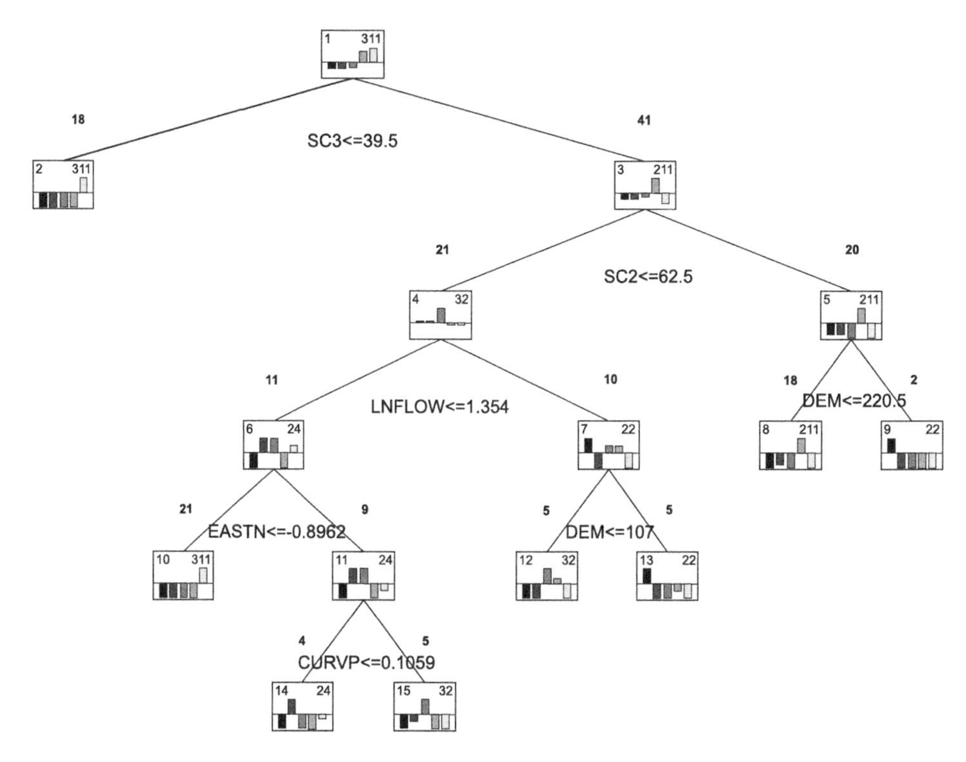

FIGURE 4 A classification tree model for five land-cover classes at Baranja Hill (DEM — elevation; SC2 & SC3 — second and third ETM+ spectral channel; LNFLOW — natural logarithm of flow accumulation potential; EASTN — sine of aspect; CURVP — planform curvature). Numbers above nodes (rectangles) indicate the number of cases sent to that node. The number in the upper-left part of the node is a node number, while that in the upper-right corner denotes the predicted class, i.e. the land-cover unit. In each node, histograms indicate the proportions of cases in each class that occur in that node.

Using split conditions from constructed classification trees through a series of nested `if-then-else` statements, three new grids were calculated in the **ArcInfo** GRID module representing predictive land-cover maps of the Baranja Hill area (Figure 5). *Water bodies* and *inland marshes* could not be predicted with these models, since these two categories were not sampled during the field work. It can be seen from Figure 5 that there are obvious discrepancies between point and polygon data with respect to their land-cover classification, i.e. compared with the polygons, a significant number of point localities have been classified differently. Since point data originate from direct field observations, they are obviously more accurate than the CLC classification at any given point. From that perspective, some information is inevitably lost due to generalisation.

When visually comparing modelled land-cover maps with CLC and point (field) data, the usage of different levels of classification in reference (3rd level) and modelled (2nd and 3rd level) maps should be taken into account. The two classes with the largest number of cases in the field data — *forest* and *agricultural land* —

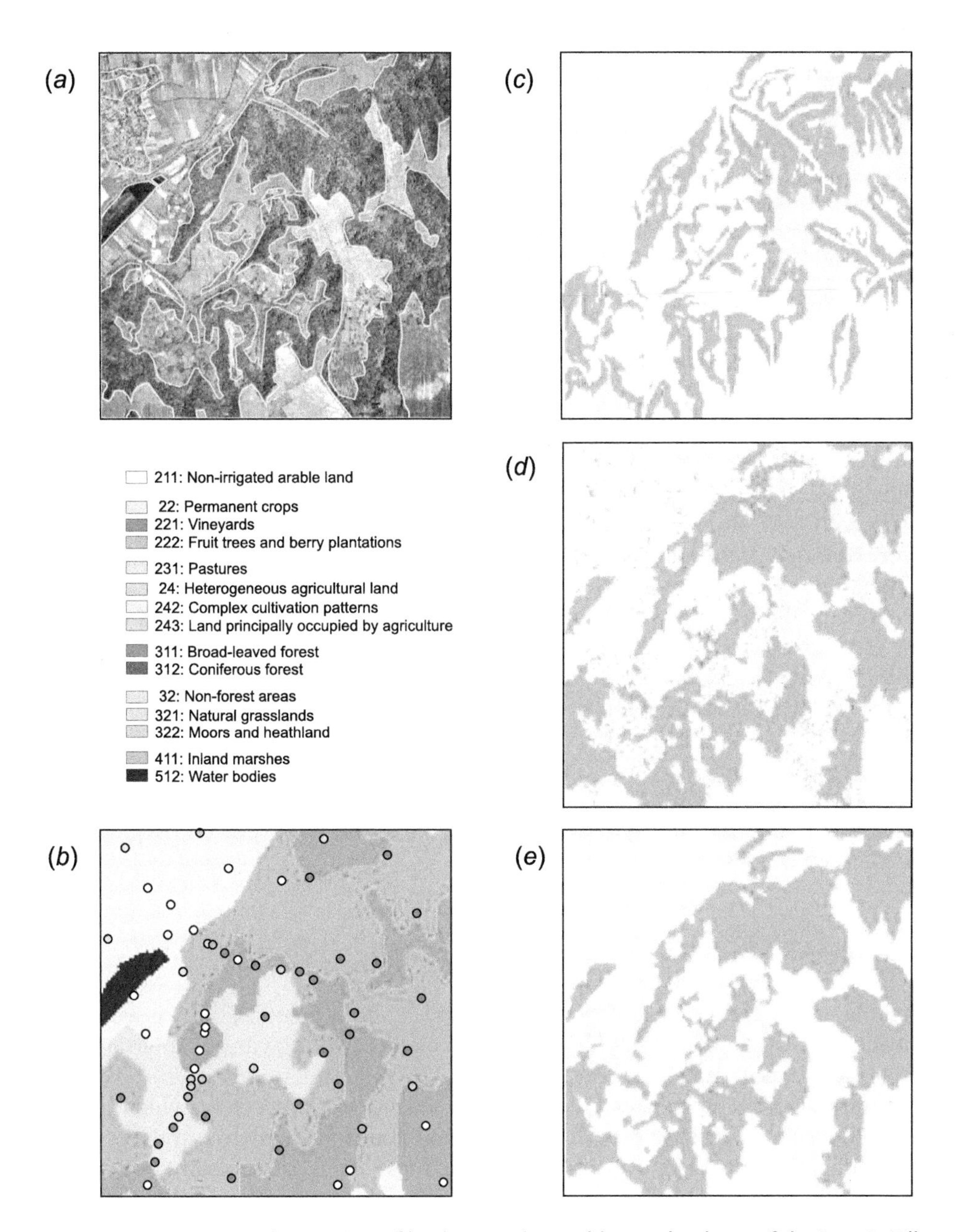

FIGURE 5 An automated extraction of land-cover classes: (a) an orthophoto of the Baranja Hill area, overlaid with manually digitised land-cover areas; (b) land-cover classes from the CLC 2000 Croatia (www.azo.hr) and field observations; (c) the land-cover of the study area, predicted using land-surface parameters only; (d) the land-cover of the study area predicted using land-surface parameters plus RS data; (e) the land-cover of the study area, predicted using RS data only. (See page 746 in Colour Plate Section at the back of the book.)

TABLE 3 Kappa statistics for predicting the land cover classes (CORINE codes) derived using the Baranja Hill field data

Code	Description	Landsat only	LSPs only	Landsat & LSPs
211	agricultural land	0.87	0.70	0.81
22	permanent crops	0.47	0.35	0.91
24	heterogeneous agricultural land	0.51	0.47	0.57
311	broad-leaved forest	0.89	0.67	0.96
32	non-forest vegetation	0.84	0.84	0.80
overall		0.78	0.66	0.84

also had the best predicted resemblance to the reference map. The predicted distribution of *heterogeneous agricultural land* (a second-level CORINE class), mainly follows the pattern of *agricultural areas* in the reference map, in other words, the first-level CORINE class with which it corresponds. *Non-forest vegetation* shows a very different pattern in the predicted models, compared with the reference map, although it has equally high kappa values in all the models (Table 3).

None of the three predicted models gave a satisfactory visual distribution of this class, compared with the reference map. This could be due either to an overestimation of *non-forest vegetation* on the reference map, or because the models did not explain the variability of this class sufficiently, or perhaps it was a combination of both. This would not be surprising, bearing in mind that *non-forest vegetation* was created by merging three classes that had insufficient data sets (Table 2). In conclusion, it can be said that, to achieve a sensible predictive model, input data has to be collated according to thematic resolution and the expected frequency of the mapped classes.

3. SUMMARY POINTS

The role of vegetation maps nowadays is increasing. In practice, there is a serious imbalance between end-users (i.e. policy makers in fulfilling various international conventions, park managements, etc.) and map producers (i.e. experts). Vegetation mappers are currently having to face a *vegetation-data paradox*, which means that there is an increasing need for information about vegetation, but there is not enough field data to support this. On the one hand, there is an instant daily need for updated, spatially organised, vegetation data, but, on the other hand, there are fewer trained biologist-ecologists able to recognise plant entities in the field.

In addition to this, end-users often do not understand, or deliberately neglect to acknowledge that, even though it consumes a lot of time and money, fieldwork is very important for producing a good, i.e. accurate, usable, vegetation map. Fortunately, as two side of the coin, development of GIS and RS did increase pressure on vegetation experts, but also helps them to satisfy the rapidly increasing needs of end-users. Statistical models and land-surface parameters have a very important role to play in vegetation mapping; a role that has at least been partly discussed in this chapter.

There is a wide range of applications of vegetation mapping. At one end, there is, for example, the UNEP/GRID[4] *Global Vegetation map* and at the other end, the paper by Gottfried et al. (1998). The former was derived for the globe as a whole. This Global Vegetation map is based exclusively upon remote-sensing data from the *Advanced Very High Resolution Radiometer* (AVHRR). It has a spatial resolution of approximately 1 km^2 and a thematic resolution of eight general vegetation types, namely: desert, semi-desert, alpine desert,tundra, grassland, deciduous forest, evergreen forest and tropical forest.

In contrast to the Global Vegetation map, Gottfried et al. (1998) mapped 1.7 km^2 of a single Alpine summit in Austria for particular plant species and vegetation communities. The level of association was set at a spatial of resolution of 1 m^2, using land-surface parameters only.

REMARK 6. *Vegetation mapping applications range from maps of whole globe at 1 km^2 resolution, to projects that cover couple of square kilometres with 1 m^2 spatial resolution.*

Between these two *'extreme'* examples, there are dozens of papers dealing with vegetation mapping (e.g. Fischer, 1990; Marshall and Lee, 1994; Krishnaswamy et al., 2004; Jelaska et al., 2005) that differ in method, thematic and spatial resolution, and the size of the area of interest. Vegetation maps can also be used for gaining information about e.g. water supply, soil pH and soil fertility, as in Schmidtlein (2005). This has already been mentioned in Section 1.1, and in e.g. research dealing with creating habitat suitability models for various animal species (Ball et al., 2005).

Nowadays, land-surface parameters and RS data, as well as the software and hardware to manipulate them, are far more readily available and accessible than they were 10–15 years ago. Hence, the main driving factors that will determine how many land-surface parameters and how much RS data will be incorporated into applications of vegetation mapping will be dictated by the thematic and spatial resolution requirements.

Although, today, satellites are equipped with ever more sophisticated sensors, with respect to spatial and thematic (in the sense of the number of spectral channels) resolution, it is expected that land-surface parameters will remain very important predictors at the higher thematic resolutions of vegetation types in small and medium-scale projects (e.g. on particular habitats, protected areas, county levels, etc.). Support for this claim can be found in the work of Jensen et al. (2001), De Colstoun et al. (2003), Dirnbock et al. (2003), Jelaska et al. (2005). In these papers, the more the thematic resolution increases, the more the accuracy achieved in mapping vegetation decreases. By using more complex methods, researchers are continuously searching for ways of improving the accuracy of RS data classification (Carpenter et al., 1999; Krishnaswamy et al., 2004).

[4] UNEP/GRID — United Nations Environment Programme/Global Resource Information Database.

The biggest bottleneck for successful vegetation mapping still seems to be making accurate ground-truth vegetation observations. Remote-sensing data is tending to replace field data. However, at finer thematic resolutions, these are not always sufficiently accurate. The increasing need for data about vegetation, as mentioned in Section 1.1, is not supported enough by real data observed in the field. Future research will inevitably continue to use land-surface parameters and RS data as predictors, but with a focus on achieving better, and more accurate, predictive vegetation maps.

With respect to mapped thematic resolution, and the nature of mapped vegetation, in some cases the classes that are present will have discrete boundaries as for (e.g. crops, pastures and forest), whereas, elsewhere, the boundaries may be continuously changing, as in (e.g. Mediterranean rocky pastures and *garigue*).

An indisputably good method for determining current vegetation types is to use climatic factors. However, the extent of this correlation is largely dependent upon the scale, where land-surface parameters can explain significant degrees of local variability, from the micro-climatic conditions. Land-surface parameters, therefore, can replace climatic data in certain circumstances, and vice-versa. Actually, those two data sets are partly redundant, or mutually predictable in terms of the temperature characteristics, and, when applied to extremely small-scale problems, have certain limitations.

> REMARK 7. *The future of vegetation mapping is in use of advanced statistical methods and new sources of land-surface parameters and remote sensing images. However, ground-truth observations should always remain an irreplaceable data source.*

Where there is a present spatial trend, using geostatistics such as regression-kriging, can enhance the predictive powers of models, in cases where the variables being predicted are represented by real numbers (e.g. the abundance of a particular plant species or vegetation type, or the probability of their occurrence). There is no doubt that we can expect significant improvement in obtaining ever more accurate and precise vegetation maps in various scales and classifications. One of the main sources of data for all phases of vegetation mapping, from optimising the field sampling or running an analysis, to making final predictions, will be land-surface parameters.

IMPORTANT SOURCES

Segurado, P., Araújo, M.B., 2005. An evaluation of methods for modeling species distributions. Journal of Biogeography 31, 1555–1568.

Latimer, A.M., Wu, S., Gelfand, A.E., Silander Jr., J.A., 2004. Building statistical models to analyze species distributions. Ecological Applications 16 (1), 33–50.

Alexander, R., Millington, A.C., 2000. Vegetation Mapping — From Patch to Planet. Wiley, 339 pp.

http://terrestrial.eionet.europa.eu/CLC2000/ — CORINE (Coordination of Information on the Environment) Programme.

http://grid.unep.ch — United Nations Environment Programme/Global Resource Information Database.

CHAPTER 22

Applications in Geomorphology

I.S. Evans, T. Hengl and P. Gorsevski

geomorphological processes and forms · interpretation of distributions of altitude and slope gradient · landsliding and glacial erosion · predicting glacier and snow cover · changing landforms and glaciers · fluvial erosion and flooding · landform extraction · landslide characteristics and susceptibility · Baranja Hill case study · supervised classification with training sets or predefined classes · unsupervised classification · extraction of break lines · other methods · hopes for the future

1. GEOMORPHOLOGICAL MAPPING FROM DEMS

Geomorphology, a science within geology and geography, and closely related to civil engineering, is the study of interactions between climatic, hydrological and tectonic processes at the Earth's surface. It focuses on describing *landforms*, their spatial arrangement, the processes which led to their formation, and their constituent materials. As mentioned previously in Section 1.2 of Chapter 2, there are four main groups of processes important for the formation of landforms: tectonic, erosional/depositional, processes caused by living organisms and processes caused by fall of extra-terrestrial objects. The tectonic processes, such as uplift, faulting, folding, warping and volcanism, are endogenetic, while the others are exogenetic, related to external agents. Very often, landforms relate predominantly to one set of processes and can be classified roughly into these twelve sets: (1) Structural and tectonic, (2) Volcanic, (3) Fluvial, (4) Coastal, (5) Lacustrine, (6) Eolian, (7) Glacial, (8) Periglacial, (9) Mass-wasting, (10) Karst, (11) Submarine and (12) Meteorite impact (Goudie, 2004).

> REMARK 1. *DEMs provide vital numerical input to geomorphology, which focuses on landform description, mapping, and interpretation.*

The morphology of landforms is also very much controlled by the nature of the underlying materials — the mineralogy and structure of the rocks. Thus geo-

Developments in Soil Science, Volume 33 © 2009 Elsevier B.V.
ISSN 0166-2481, DOI: 10.1016/S0166-2481(08)00022-6.

morphology may be defined as the science which studies the nature and history of landforms and the processes that created them. Our focus in this chapter will not be the field of geomorphology as such, but uses of DEMs for geomorphological mapping, i.e. extraction of geological and geomorphological features out of DEMs.

Basic concepts in geomorphology include the magnitude and frequency of processes, spatial scales of landforms and processes, temporal scales (time lags, reaction times, response times, relaxation times) of adjustment, equilibrium and historical inheritance, relations between internal and external processes, and the sediment cascade (Evans et al., 2003). Most of these involve the use of geomorphometric measures, increasingly from DEMs.

The main applications of DEMs in the field of geomorphology, at the beginning of the 21st century, can be roughly grouped to:

- *Visual interpretation of DEMs* — recognition and manual delineation of geomorphologic features;
- *Automated recognition and quantification of geomorphological properties* — extraction and use of morphometric land-surface parameters (slope aspect, curvatures) for geomorphologic analysis and detection of structures;
- *Automated extraction of hydrological/denudational structures* — extraction of drainage networks, valley / ridge lines, recognition of drainage patterns, etc.;
- *Automated extraction of landforms* — extraction of landforms and landform elements using semi-automated or fully-automated algorithms.

As all applications of geomorphometry have some geomorphological dimension, this chapter is highly selective. Thus we will focus mainly on recent papers, which provide references to the earlier work. In addition, we will not consider the visual interpretation of DEMs, because the essence of geomorphometry is the automation of information extraction from DEMs. Other chapters have dealt with drainage network definition and analysis (Chapters 7, 18 and 25), with the use of land-surface parameters in soil mapping (Chapter 20) and ecological mapping (Chapter 24), and with the modelling of mass movement (Chapter 23). Readers interested to find more about the use of DEMs in geomorphological mapping should refer to the three edited books by Rhoads and Thorn (1996), Lane et al. (1998) and Evans et al. (2003).

2. GEOMORPHOMETRY IN THEORIES OF FLUVIAL AND GLACIAL EROSION

2.1 Altitude and slope gradient distributions

The statistical study of altitude and slope gradient distributions has a long history (see also Section 3 in Chapter 1) but has been rejuvenated by the availability of DEMs even on a global scale, leading to novel applications and new interpretations in geomorphology. One concerns the relations between tectonism, isostasy, valley incision and overall surface lowering (denudation). In slope studies, apart from general trends of change with spatial scale, there is a qualitative difference between DEMs at 1 km and coarser horizontal resolution, which blur the representation of valleys, and those at 100 m or finer, which represent slopes at the human scale. Thus progress from GTOPO30 to SRTM and to DEMs with resolutions of 30 m or better permits analysis of slope and curvature (clinometry and more complex studies) and not just altitude (hypsometry), even over broad areas. Such empirical geomorphometric studies are now informing theories of fluvial and glacial erosion in landscape development, especially in mountains.

Most slope gradient distributions are positively skewed, with a few high values and a mode toward the low end of the range. Thus Speight (1971) suggested general use of a logarithmic transformation. However, there is variation between topographic regions (O'Neill and Mark, 1987). Mountains have more symmetrical gradient distributions than lowlands. The square root of sine is a useful compromise, but does not remove skew near either extreme. Using a 90 m resolution DEM, Burbank et al. (1996) calculated slope gradients for six high relief regions on crystalline rocks in the northwest Himalaya, which are being uplifted and eroded at different rates. All were symmetrical with modes around 35°, means of 30 to 34° and few gradients over 60°. This was interpreted as controlled by a common threshold for bedrock landsliding, related to the strength of fractured rock masses. Together with comparable mean altitudes in each non-plateau region, a dynamic equilibrium between uplift, river incision and denudation was inferred.

Likewise in Japan, Katsube and Oguchi (1999) used an approximately 50 m resolution DEM for mountains on igneous and sedimentary rocks in central Honshu. In all three ranges of the Japan Alps the mode of gradients is at 33 to 37° between 1000 and 2800 m, becoming sharper as altitude increases. Mean gradients increase with altitude, to maxima of 32 to 35° above 2000 m. Here too, slopes steeper than 35° fail much more frequently. Relating slope gradient to 1-km relief (range in altitude) for 15 major Japanese ranges, Katsube and Oguchi (2005) found that gradient increased with relief of up to 400 m. Modal gradient is less than mean gradient for relief below 200 m, but greater for relief of 200 to 600 m, again reflecting the influence of limiting gradient [Figure 1(b); see also examples in Evans, 2004, p. 436]. For relief above 600 m gentler slopes are less common and mean gradient catches up with modal gradient. Modal gradient is constant around 35°, a characteristic valley-side slope angle, for relief greater than 400 m.

A broad survey by Wolinsky and Pratson (2005) analysed 30 m mesh DEMs of 28 map sheets in different topographies of the USA, divided into thousands of 690×690 m quadrants. Skewness of the tangent of gradient declined as mean gradient increased, and became negative on average where gradients exceeded 37% (20°). Above 46% (25°), mean skewness levelled out around −0.25. The declining trend applied both generally and within individual map sheets. It was also found for larger quadrats, 2, 4, and 8 times as broad. A process-response model based on a real drainage network reproduces the main trend of declining skew with increasing gradient as the rate of river incision is increased, increasing the importance of slope failure compared with wash and creep. This match supports Wolinsky and Pratson's suggestion that slope gradient frequency distributions contain information about dominant slope processes. Steep slopes are dominated by failure as

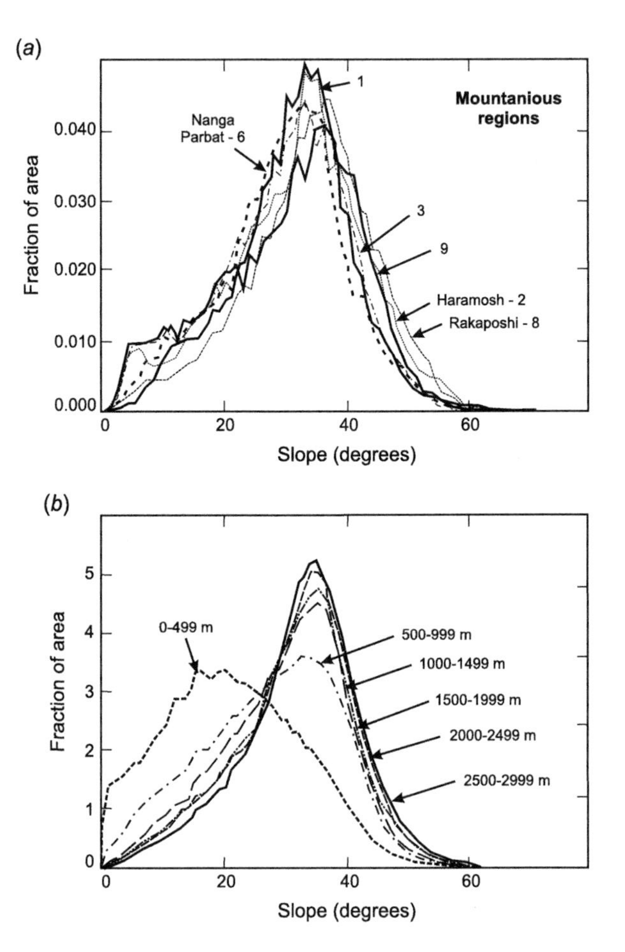

FIGURE 1 Slope gradient frequency distributions for: (a) the northwestern Himalaya, by region (1, 3 and 9 are mountain ranges adjacent to Haramosh), using a 4×4 window on a 90 m DEM (Burbank et al., 1996); and (b) the Northern Japan Alps, by 500 m altitude bins using a 50 m mesh DEM (Katsube and Oguchi, 1999). Reprinted by permission from Macmillan Publishers Ltd.

stability thresholds are exceeded: the gradient mode is near the upper limit, with a tail of lower gradients. On gentler gradients, wash and creep give the reverse distribution shape.

In a more detailed study of coastal mountains in the northwestern USA, Montgomery (2001) found unskewed slope gradient distributions for upland areas of active uplift, and exponential distributions (mode at zero) for depositional topography. In the Oregon Coast Range, upland slope gradients increased with uplift rates, whereas in the Olympic Mountains uplands they tended toward threshold values controlled by rock properties and slope failure. Montgomery (2001) concluded that normal, symmetrical slope gradient distributions are necessary but not sufficient for recognition of bedrock threshold hillslopes. Using classic *hypsometric plots*,[1] Montgomery et al. (2001) contrasted the concave-up plots of the northern Andes, typical of fluvial erosion, with more concave plots for the southern Andes where glacial erosion has reduced the proportion of area at high altitudes, and with somewhat convex plots for the central Andes where aridity kept erosion down and facilitated survival of the *altiplano* at high altitude. An erosion index was derived by summing precipitation over upslope/upstream areas and multiplying by slope gradient. This showed that large-scale climate patterns, together with tectonics, are first-order controls on the topographic development of the Andes.

Erosion or denudation, often measured by sediment yield from river or glacier catchments, is strongly influenced by slope gradient and by discharge: the former tends to increase with relief, the latter with area drained. Thus surface form affects two major factors in erosion. Montgomery and Brandon (2002) found an increase in erosion with the mean slope gradient of 10-km diameter areas in the Olympic Mountains. Mean slope gradient relates linearly to local relief over the same circular area, with R^2 of 0.81. On a global level, erosion rate increases as a power function of local relief, with rates above 1 mm a^{-1} confined to active orogenic belts with local relief in excess of 900 m. Relief is limited by the increasing frequency of landslides on slopes which have reached the threshold of stability, governed by rock strength. Only 5% of the subaerial Earth has 10-km relief in excess of 1000 m, only 2% more than 1500 m, and relief exceeding 2000 m is exceptional.

Around the upper Indus, Brozovic et al. (1997) found that, unrelated to lithology and rate of uplift, slopes are lower around the modal altitude of each subregion, which in turn is a little below present-day *snowlines*. Steepest slopes are above snowline. They went further and suggested that whatever the uplift rate, little mass could rise through the zone of rapid glacial erosion around the snowline. "*In contrast to Tibet, we suggest that in the northwestern Himalaya, the efficiency of surface processes (and, in particular, glaciation) has prevented the mountain range from reaching mean elevations at which the driving forces of tectonism could no longer support it*" (Brozovic et al., 1997, p. 574). This is now termed the '*glacial buzzsaw hypothesis*' and further implies that areas of most rapid (glacial) erosion, on the wet (snowy) sides of mountain ranges, will be areas of most rapid rock exhumation and uplift as isostasy tends to replace the lost mass (Mitchell and Montgomery, 2006a).

Derived geomorphometry from DEMs generated by SPOT 3 panchromatic stereo pairs in tandem with satellite imagery were used by Bishop et al. (2003) to understand landscape denudation and relief production in the high western Himalayas. Greatest meso-scale relief was associated with glaciation at high altitudes and the production of relief decreased at intermediate altitudes with warm-based glaciation. Their research also indicated that scale-dependent topographic analyses were necessary to better address radiation transfer issues and other landscape denudation dynamics.

The most common definition of relief is range of altitude within a grid square, which is highly scale-dependent. Brocklehurst and Whipple (2002) defined *geophysical relief* as the average difference between the land surface and a smooth

[1] Dimensionless plots of cumulated area against altitude.

surface interpolated from heights along a drainage basin perimeter, modified by any internal high summits. This they found appropriate to studies of relief generation by erosion. Although on the east slope of the California Sierra Nevada their glaciated basins had greater relief, this was attributed to their greater areas compared with nearby unglaciated basins. They were, however, able to infer that glacial erosion had been faster than fluvial, and had caused headward extension of the basins. The same authors further found that in the American Rockies, Sierra Nevada and part of New Zealand, mountain glaciation initially increased hypsometric integrals by cirque erosion, but lowering of snowline giving strong erosion by valley glaciers reduced hypsometric integrals by leaving a tail of infrequent high values (Brocklehurst and Whipple, 2004). Evans (1990, pp. 47–48) proposed that mountain glaciation increases the standard deviation of gradient and of profile curvature.

Relief was measured similarly by Korup et al. (2005) in the western Southern Alps of New Zealand, where modal slope gradients are mainly around 38–40° and relate to landsliding. (Note that they used a 25 m grid mesh, finer than Burbank et al.'s 90 m and thus producing steeper gradients.) Above 2700 m, modal values exceed 48° and rock avalanches dominate. They analysed DEMs in terms of both drainage basins and 20 km wide swaths, and mapped topographic depth below the interpolated basin perimeter (major divides) surface: this gave mean basin relief between 510 and 840 m. Altitude distributions (e.g. hypsometric integrals) did not distinguish catchments dominated by glacial or fluvial erosion.

In the Washington Cascades between 47 and 48°N, Mitchell and Montgomery (2006a) analyse trends in mean and maximum altitude in three east–west swathes of DEM each 50 km wide. The highest summits are east of the divide and well east of the zone of greatest uplift. Summit altitudes (excluding volcanoes) rise steadily eastward, as do cirque floor altitudes, and modern glacier median altitudes. Each rises, at 9 to 15 m km^{-1}. Only about 10% of each subdivision rises above the highest cirque floors, and few peaks rise more than 600 m above. Rock exhumation rates suggest vertical erosion of 2 to 5 km in the last 15 Ma, and are greatest 30 to 40 km west of the highest peaks. This is not the pattern expected from fluvial or slope erosion, so Mitchell and Montgomery (2006a) propose a *glacial buzzsaw* of greatest glacial erosion at the average Quaternary glacial equilibrium line represented by the cirque floors, where ice discharge and velocity were greatest. This increased the slope gradients above cirque floors to over 30°, causing slope failure. Thus both vertical and headward erosion in cirques is considered to dominate landscape development at high altitudes.

Uplift rates are greater in the Kyrgyz Range of the Tien Shan, where Oskin and Burbank (2005) use the sub-Cenozoic unconformity to suggest an east–west spatial gradient of uplift and thus of landform development. As mountains are taken above the snowline, glacial erosion both deepens and widens fluvial valleys, increasing local relief as measured from DEMs. Glaciation starts on the north slope where the snowline is some 200 m lower, and has pushed the divide 0.9 to 4.4 km southward. Erosion is localised at the bases of cirque headwalls, and cirque headwall retreat is two to three times the rate of vertical erosion. "*Cirque retreat can effectively bevel across an elevated alpine plateau...*" (Oskin and Burbank,

2005, p. 396). Thus some fairly old ideas on headward and downward erosion by mountain glaciers are now being tested by extensive analyses of hypsometry, relief and gradient from DEMs.

Considerable use has thus been made of hypsometry and *clinometry* — distributions of altitude and of slope gradient. More use could be made of curvature properties of the land surface, and of positional variables. For example Katsube and Oguchi (1999) showed that, while mean profile curvature varies in predictable ways, the standard deviation of profile curvature has an interesting variation with altitude, consistent across the three mountain ranges of central Honshu. High values are found at 200 to 1200 m, declining to minima at 2200 to 2700 m where fluvial topography has fairly straight slopes and there are more points with near-zero curvature. Standard deviations rise again at altitudes affected by glaciation. Standard deviation of profile curvature is high on slope gradients below 20°, and low on those above 40°, with a linear decrease between.

2.2 Snow cover and glacier distribution

Snow cover is important both in agricultural landscapes, in wild landscapes where it affects ecology, and in polar and mountain areas where a surplus of snow over the year is necessary to generate glaciers. Lapen and Martz (1996) showed that in low relief areas, snow depth is related to topographic position rather than to local surface morphology. In mountain areas, variables such as gradient and concavity are expected to be more important. Also as gradient increases (at least up to 30°), slope aspect has a greater effect on mass balance and thus on snow and *glacier distribution* (Evans and Cox, 2005).

The combination of digitised glacier or snowpatch outlines with a DEM permits explanation of glacier distribution in terms of altitude and other land-surface parameters. For the Maladeta massif, the highest in the Pyrenees, Lopez-Moreno et al. (2006) use a binary regression tree and a generalised additive regression model to show that altitude is the most important control of glacier probability, followed by radiation receipt, slope gradient and mean curvature. The probability of glacier cover ranges up to 93% (for altitude over 3021 m, radiation less than 21,721 MJ m^{-2} day^{-1} and slope gradient less than 31.4°).

One problem with this model is that no distinction is made between accumulation areas, where meso-climatic conditions generate surplus snow, and ablation areas. The latter are simply downslope from accumulation areas and suffer net loss of ice, so they are by definition unfavourable to glacier generation. Using a more process-based model, Arrell (2005) attempted specifically to predict areas of net accumulation in mid-latitude mountains, starting from a DEM and climatic station data within the region. She found it was necessary to use different temperature lapse rates for different seasons, and to include a measure of local position.

Radiation is predicted from altitude, slope gradient, aspect and latitude, integrating through the year and with shading taken into account. Results depend sensitively on the vertical lapse rates used and assumptions about surface albedo and its variations: statistical generalisations are necessary and precipitation variations in mountains are notoriously difficult to model. The most difficult aspect to model is the interaction of varying wind fields with the surface, which is basic to the effects of position and shelter. Currently there is too little field control on the models proposed.

3. GEOMORPHOLOGICAL CHANGE

3.1 River channels, coasts, dunes and glaciers

Lane (1998) showed how the study of change in dynamic river channels is improved by construction of *sequential DEMs* (e.g. permitting a 0.1 m contour interval) rather than cross-profiles. A fully 3-D approach can give greater assurance that change has not been missed between profiles, or between surveys over time. The study used TIN-based models and suggested that explicit incorporation of break lines was important in producing difference maps and volumes. The superiority of DEMs to sequences of cross-profiles was demonstrated also by Fuller et al. (2003) in analysis of channel change (both being based on ground survey). Other advantages of DEMs are that they permit changes in channel morphology to be linked to sediment transport processes, with spatially distributed feedbacks between form and process.

> REMARK 2. *LiDAR is especially suitable for geomorphological studies as it allows monitoring of land-surface changes.*

Many recent developments for quantifying and monitoring surface changes rely on the use of IfSAR/SAR or LiDAR survey techniques (see Section 2.2 in Chapter 3) because of their ability to perform accurate topographic mapping including sub-canopy differentiation. But the emphases of scale and resolution are still issues in achieving desired accuracy goals and estimates of surface changes. For example, a study conducted by Hofton and Blair (2002) along the beaches at Assateague Island, MD, USA found that the implementation of such surveying methods should consider resolution and scale of the deformation of features: their results from increasing the size of the laser footprint from 25 to 60 m yielded significant underestimation of the vertical change signal. LiDAR-derived morphometric parameters for five dune systems in England and Wales, conducted by Saye et al. (2005), demonstrated that eroding dunes resulted in narrow and steep beaches while accreting dunes resulted in wide and low-angle beaches. They also found differences in vertical accuracy between vegetation types: on bare sand surfaces the vertical accuracy was ±15 cm while dunes covered by grass and shrubs had a reduced accuracy of ±20–50 cm.

Nagihara et al. (2004) used ground-based laser scanning to map a small sand dune at such high resolution (averaging 10 cm horizontal spacing) that the results of individual wind events could be captured. This permitted the process to be studied and a level of detail to be captured that would have been very difficult with traditional survey techniques. The advantage of minimising foot traffic on dunes is obvious, and although the equipment is expensive and not easy to carry,

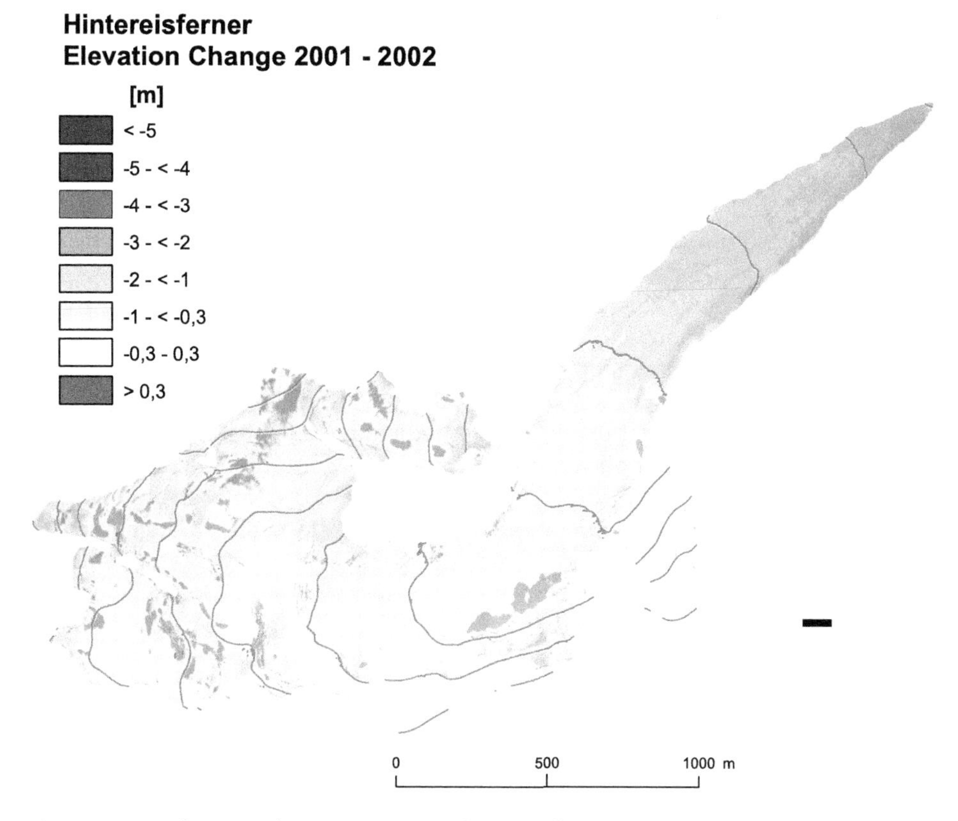

FIGURE 2 Net elevation change on Hintereisferner in the 2001–2002 budget year. Reprinted from Geist and Stötter (2007). Used with permission (http://www.borntraeger-cramer.de). (See page 747 in Colour Plate Section at the back of the book.)

these authors maintain that this approach is suited to landforms between 1 and 100 m across. Airborne LiDAR is recommended for larger landforms.

Geist and Stötter (2007) report on ten airborne laser scanning missions in several years, over several Austrian glaciers totaling 36 km^2 in area. As a relative vertical accuracy of 0.3 m was achieved, it was possible not only to map net change, but to separate winter balances from summer balances by having appropriately timed flights. Both accuracy and surface change are related to slope gradient and aspect. Accuracy in the accumulation area, where because of snow cover there are fewer distinctive points, was greater than from photogrammetry. The build-up of cones of avalanche snow and wind drifts could be mapped (they show as blue areas in Figure 2) and estimation of surface albedo from the return signal intensity is feasible. It remains necessary to measure snow density on the ground, and to allow for the glacier flow field, to obtain the components of glacier mass balance.

Each of these approaches is currently quite expensive, but they point the way to new possibilities of geomorphological and geomorphometric interpretation. At least we can now see, for limited areas, what detail is lost when DEMs of coarser resolution are used and smooth variation is perforce assumed.

3.2 Analysis of erosion

In topographically-based approaches, large-scale fluvial erosion is modelled without a detailed knowledge of stream characteristics by applying variants of stream power indices (Finlayson and Montgomery, 2003). Wilson et al. (2000) applied a sediment transport capacity index to extract changes and map net deposition and net erosion cells, whereas net erosion areas were used to describe the sensitivity to DEM resolution. This sensitivity to net erosion cells was also applied to the Revised *Universal Soil Loss Equation length-slope* (RUSLE-LS) factor using flowpath length and a specific catchment area in the sediment transport capacity index (Moore et al., 1991a; Moore and Wilson, 1992). The findings were that increase in the cell resolution decreases slope gradients and the flowpath routing is significantly decreased, while there is an increase in the specific catchment area which yielded significantly different results in the prediction estimates.

In a similar modelling application, Finlayson and Montgomery (2003) demonstrated that the coefficients used in the stream power indices are sensitive to grid resolution when determined from an analysis of area–slope plots. Their research showed that the stream power per unit area decreased with decrease in DEM resolution from 30 to 900 m in the Olympic mountains, Washington, USA whereas the mean slope of 15 rivers under the study declined by 65% and the mean drainage basin size increased by 14%. This caused a 17% reduction in median main-stem channel length. The reduction in river gradient stems largely from the highly smoothed nature of the 900 m DEM, where each cell is an average of 30×30 cells of the 30 m DEM and is thus an areal value rather than a point value.

A study by Jain et al. (2006) further extended this analysis by using finer DEM and methodological details to determine stream power plots in the upper Hunter River catchment, New South Wales, Australia. Their methodology for deriving stream power profiles used a smoothing, a curve fitting and a theoretical model. The results suggested that different models should be used for different applications related to length of stream profiles because the variability of stream power in headwater reaches is explained by discharge variability while in midstream and downstream reaches it is explained by the high variability in channel gradient.

3.3 Analysis of floods

To predict areas and depths of river or coastal flooding makes the most rigorous demands on DEM accuracy. For example, in the Venezuelan Llanos, J.K. Smith et al. (2006) mapped an area 12.8 km long that varied only between 67 and 70 m altitude. Ground GPS profiles were augmented by air photo interpretation and the effects of dykes could be assessed from a TIN-based DEM accurate to a few centimetres. Flood simulation is not just a matter of flooding to a given contour — flow patterns and the effects of barriers such as dykes and bridges must be reproduced.

Casas et al. (2006) compared seven different DEMs for a reach of the river Ter in NE Spain. The LiDAR model produced an RMSE of 0.3 m for elevation, but 0.85 m for water surface elevation; also, it was the most sensitive to choice of Manning's *n*, the roughness coefficient for flow. However, a study by Carlisle (2005) cautions that the use of a single RMSE value does not fully express the quality of DEMs; it advocates the use of a spatially distributed model of DEM quality. His reasoning is based on the fact that the distribution and scale of elevation error within a DEM are at least partly related to morphometric characteristics of the terrain.

Moreover, high resolution DEMs for large-scale flood risk mapping are still cost-prohibitive or unavailable in many countries. Implementation of the Shuttle Radar Topography Mission (SRTM) X-SAR data, which covers the entire Earth's surface shows that low accuracy may be expected in mountainous terrain primarily due to radar shadow effects, but the overall quality of the data is satisfactory for hydrological applications (Ludwig and Schneider, 2006). For example, Yang and Teller (2005) modelled the history of the Lake of the Woods, Canada, using SRTM data in conjunction with lake bathymetry, isostatic rebound data, and other historical data to reconstruct the paleotopography of the region. Their findings indicated that changes in the extent and the volume of the lake existed through time, and future potential climate changes such as those during the mid-Holocene warming period in the same region will end current overflows and cause the level of the lake to drastically decrease.

4. EXTRACTION OF SPECIFIC LANDFORMS

Landforms have traditionally been defined by interpretation of air photos, preferably in stereo. This is the basis of terrain analysis in the sense used by Way (1973) and van Zuidam (1986). The land systems approach was developed for rapid survey of little-studied areas, for example in Australia, New Guinea and Africa. Adaptation to the possibilities of remote sensing by Townshend (1981) involved introduction of pattern recognition techniques, automating some functions. Mitchell (1991) and Lawrance et al. (1993), for road surveying, set such terrain analysis in the wider context of terrain evaluation.

Most DEM-based work has been applied to the extraction of drainage networks (see Chapter 7). Work on slope runoff processes has found plots of slope gradient versus area drained very useful. For example, Hancock and Hutchinson (2006) mapped drainage heads in the field in Northern Australia and compared them with those generated from a 10 m mesh DEM. They found that despite uniformity in geology, soil and vegetation, the area required for generating concentrated flow was highly variable, with a standard deviation of 890 m^2 around a mean of 480 m^2. Slope–area log–log plots showed that slope was maximal around this drainage area, and declined linearly above the 2000 m^2 threshold at which channel incision began. Coarsening the DEM to 20, 30 and 40 m mesh preserved this linear (power) trend, but lost detail in the diffuse-flow hillslope part of the plot. The channel incision threshold was also well defined by a change in the slope of the cumulative area distribution.

> REMARK 3. *Most DEM-based geomorphology involves semi-automated extraction of surface forms and objects such as watersheds, drainageways, and other break-lines.*

Recent work by Schmidt and Hewitt (2004) and Schmidt and Andrew (2005) has shown how position may be taken into account in recognising landform elements (see also Chapter 9). Summerell et al. (2005) quantified landscape position by an UPNESS index based on the adjacent higher area, including that in different catchments. They used its cumulative distribution function to classify the landscape into four sets: ridge tops and upper slopes; midslopes; lower slopes; and infilled valley/alluvial deposits. This has a number of tunable threshold values and is somewhat resolution-dependent. The approach is intermediate between using upslope catchment area and using altitude percentiles; an alternative might be to combine these in other ways.

Semi-automated extraction of landforms and landform elements is probably one of the most active research areas in the field of geomorphology (Irvin et al., 1997; Burrough et al., 2000; Schmidt and Hewitt, 2004; Fisher et al., 2005). Tribe (1991, 1992b) made a pioneering investigation of automated definition of valley-heads from DEMs, with moderate success. Bue and Stepinski (2006) tested automated extraction of landforms for DEMs of Mars. Drăguţ and Blaschke (2006) showed that fuzzy classification of landform elements can also be combined with image segmentation techniques, so that the classification outputs are spatially continuous. Iwahashi and Pike (2007) recently developed an iterative procedure to extract landforms using slope gradient, local convexity, and surface texture. They applied an unsupervised nested-means algorithm at three scales: (1) 55 m (Shimukappu), (2) 270 m (Japan) and (3) 1 km (whole World[2]) and proposed further development of automated algorithms that can achieve a satisfactory accuracy.

4.1 Morphometric characterisation of landslides

The distinctive morphometric characteristics of topography prone to slope failure have been used to create geometric signatures of shallow landsliding (Pike, 1988) and, emphasising the positions of steep slopes, fingerprints of bedrock landsliding (Densmore and Hovius, 2000). In recent studies such characteristics have been derived through automated geomorphometric analyses that provide basic information about many aspects of landscape functions that are controlled by terrain and have been linked to landslide processes (Gritzner et al., 2001; Gorsevski et al., 2003, 2005, 2006; McKean and Roering, 2004; Glenn et al., 2006). However, the major obstacle still remains a clear understanding of scale-dependent processes associated with the topographical data which is used to represent the relevant scale of morphometric characteristics of landslides. For instance, detailed surface roughness of the moving mass has been measured by vector or aspect dispersion by McKean and Roering (2004) and interpreted, on a given material, as the activity or recency of movement.

REMARK 4. *Geomorphometry, together with airphoto interpretation, can accurately delineate and quantify landslides and other similar features.*

[2] Extracted global map of landforms at 1 km resolution is available at http://gisstar.gsi.go.jp.

A similar study by Glenn et al. (2006) also used LiDAR to examine morphometric characteristics and to differentiate morphological components within two canyon-rim landslides in Southern Idaho, USA. The high resolution topographic data in this investigation conformed to previous results of linking high motion areas of active landslides to high surface roughness and to material types. This roughness is often at length scales below 10 m, and implies that slide masses will seem different as viewed from 25 or 1 m DEMs (variations in approach may be required). The value of roughness measures such as standard deviations of profile curvature, plan curvature and gradient needs investigation. However, such studies cover restricted areas, or single landslides, and the broader applicability of landslide signatures from such studies is uncertain. As yet, landslides are mapped subjectively, rather than being recognised from parameters derived from DEMs.

Other studies that use coarser resolution DEMs (e.g. 30 m) to predict landslide susceptibility for large geographical areas describe processes and morphometric characteristics through mathematical relationships (Gritzner et al., 2001; Gorsevski et al., 2003, 2005, 2006). In this approach, landslide locations are identified from aerial photo interpretation and/or field survey data, and are linked to a spatial database to quantitatively describe water movement, hydrological processes, morphometry, catchment position, and soil-landscape processes. Of course, the accuracy of landslide identification and corresponding characteristics from aerial photographs can vary depending upon the scale of the air photos, the required mapping detail, the landslide size, the photo quality, the season of acquisition, the forest cover density and height, and the skill of the interpreter. Also, other concerns arise from the conversion of inventoried landslides from vector to raster format because derivation of land-surface parameters is raster based and the presence or absence of a landslide is represented as a single grid cell. In modelling applications, often this single cell is assumed to be the initiation area of each landslide (i.e., the area where the main scarp of the landslide occurred) or grid values are assigned based on site-specific estimates and predictions.

Malamud et al. (2004) compared the size–frequency distributions of landslide areas from three major events, related to earthquake, snowmelt and rainfall: each inventory is substantially complete, with between 4000 and 11,000 landslides and coverage down to at least 70 m^2 in area. For areas over 2000 m^2 they find a power-law distribution with an exponent of -2.4. Smaller landslides depart from this trend, and those smaller than 300–700 m^2 have decreasing frequencies. Thus, rather than extrapolating the power trend as some had previously done, they use the three-parameter inverse-gamma distribution, which fits these detailed inventories, to extrapolate in inventories where smaller landslides have been omitted and thus to estimate the total area affected. This is applicable to slide or flow-related landslides, not to rockfalls (where the exponent is -1.1) nor to hyper-concentrated flows. Evans (2003) also found landslides in Japan to be scale-specific, with maximum frequencies for all three dimensions (length, width, and height) (Figure 3). He suggested that landslide scale would be regionally specific, and vary between different classes of landslide.

A study by Gorsevski et al. (2003) suggested that non-road related (NRR) landslides initiated by non-human interaction (spatial locations within non-roaded

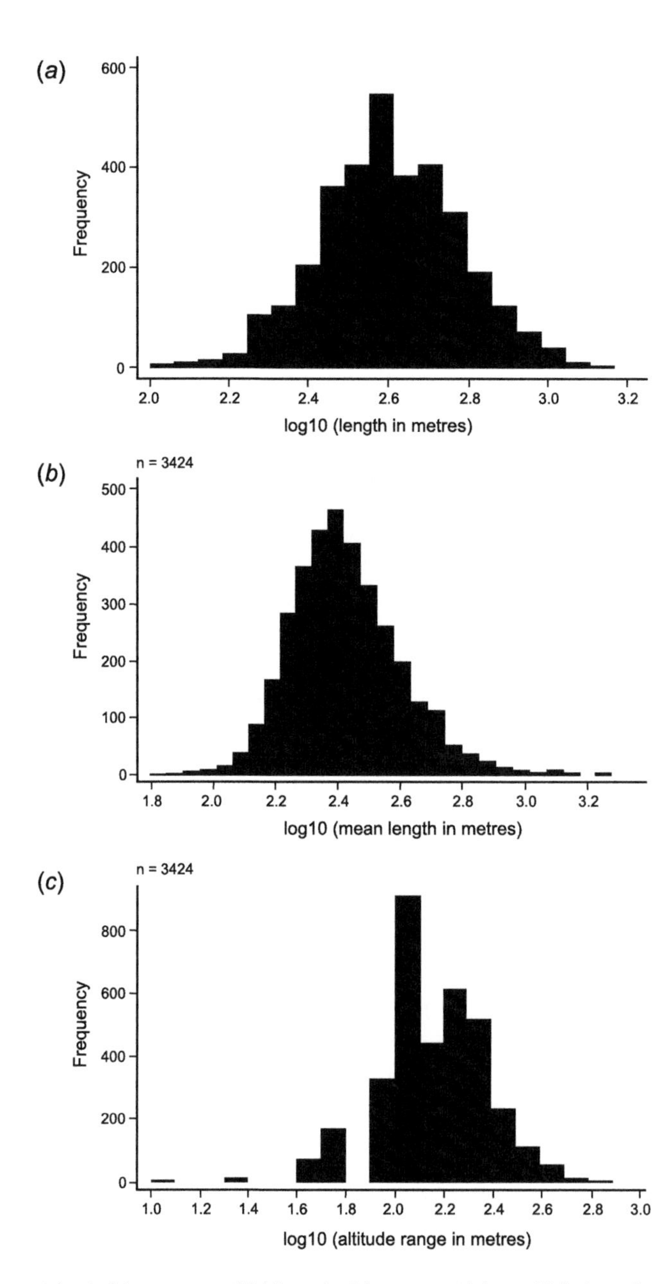

FIGURE 3 Logarithmic histograms of (a) length; (b) mean width; and (c) altitude range of 3424 landslide masses in south-central Japan (after Evans, 2003, p. 68, using data supplied by Sugai and Ohmori).

areas) occur mostly on steep slopes and wet concave drainages, while the road related (RR) landslides initiated by human interaction (spatial locations within road buffers) occur on steep slopes as well as gentle slopes and drier convex areas. The methodology applied was based on fuzzy *k*-means classification that deals with uncertainties in terms of vagueness and incompleteness of known parameters associated with landslides, and with class overlap as well as assigning digital terrain attributes into continuous landform classes. The predictive models were derived from known landslide locations and relationships with the continuous landform classes using the Bayesian approach. In the modelling, a total of six topographic attributes previously used to characterise landform shapes by Irvin et al. (1997) and Burrough et al. (2000) were used to develop the NRR and RR models.

Gorsevski et al. (2005) further extended the previous modelling approach with the aim of representing and managing imprecise landslide delineation information, and dealing with other uncertainties by applying the *Dempster–Shafer (D-S) theory of evidence*. This theory is an extension of the Bayes Theorem, but it is more flexible in the sense that it waives the need for complete knowledge of prior or conditional probabilities before modelling can take place. Additionally the D-S theory introduces the representation of ignorance, which is used to represent the lack of evidence (complete ignorance is represented by 0). For example, absence of a landslide in the database does not necessarily mean that no landslide was present in that location; it may simply suggest that the photo interpreter failed to identify the presence of a landslide.

This modelling approach outputs the uncertainties through consideration of lower and upper probability intervals induced by multi-valued mapping, rather than explicit probability values as with the Fuzzy/Bayesian methodology. Figure 4 illustrates the spatial implementation of D-S theory. Figure 4(a) shows belief function (lower bound of probability function); Figure 4(b) plausibility function (upper bound of probability function); and Figure 4(c) belief interval or uncertainty map (the difference between upper and lower bounds) for the NRR landslides derived from the six sources of evidence (including relationships between landslides and fuzzy *k*-means classes). This confirmed the results of Gorsevski et al. (2003) and suggested that uncertainties are potentially greatest in concave drainages, which may relate to scale-dependent processes and specific topographic attributes that operate on those scales. Landslides and debris flows are discussed further in Chapter 23.

5. CASE STUDY

The following sections demonstrate semi-automated and automated extraction of landforms and some land-surface objects important for geomorphology such as break-lines. We will use the case study Baranja Hill for a manual photo-interpretation of landforms and several land-surface parameters: slope in % (SLOPE), profile curvature (PROFC), tangential curvature (TANGC), topographic

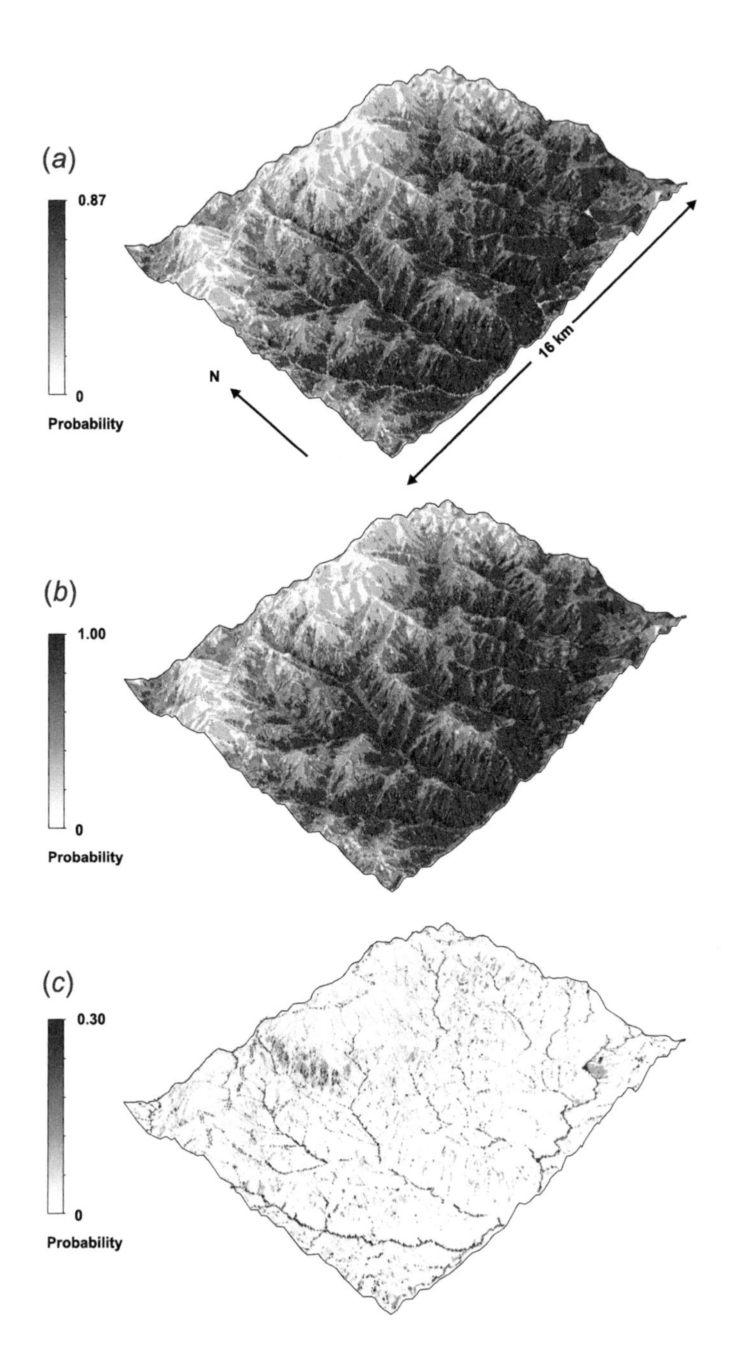

FIGURE 4 Non-road related landslides derived using the Dempster–Shafer concept: (a) belief map; (b) plausibility map; and (c) belief interval or uncertainty map.

wetness index (TWI), solar insolation (SOLIN), ground water depth (GWD[3]) and anisotropic coefficient of variation (ACV[4]). All land-surface parameters were extracted in ILWIS (Chapter 13) and SAGA GIS (Chapter 12).

There are three distinct approaches to extraction of landform classes: expert system, supervised and unsupervised classification (see also Chapter 9). Expert systems do not require training of the classifier and unsupervised classification can be fully automated because the optimal number of classes and their definition can be determined automatically. Each of the three approaches can also consider classes of objects to be defined as either crisp or fuzzy. The crisp objects are homogeneous within their boundaries and a single class absolutely dominates other classes at these locations while the fuzzy objects are without well-defined boundaries between classes and objects can belong to multiple classes described by membership functions. In that sense, the fuzzy definition of classes is just a generalisation of the crisp definition, i.e. the crisp classes are just a special case where a single class absolutely dominates other classes at a given grid node.

Baranja Hill has been mapped extensively over the years so that multi-thematic data (from geological, soil and vegetation surveys) are available. The main geomorphometric features have been photo-interpreted and delineated using the four-level *geo-pedological method* of Zinck and Valenzuela (1990). The first initial legend and the map were then cross-checked in the field to produce the final legend and photo-interpretation map. This map was then orthorectified and digitised to produce the final map of landforms. The full procedure is described in Rossiter and Hengl (2002), Hengl and Rossiter (2003).

Baranja Hill can be divided into two major landscapes (hill land and plain), six relief types and a total of nine landforms: Hi111 (Hill summit), Hi112 (Hill shoulder), Hi211 (Escarpment scarp), Hi212 (Escarpment colluvium), Hi311 (Valley slope), Hi312 (Valley bottom), Hi411 (Glacis slope), Pl311 (High terrace) and P411 (Low terrace) (Figure 5). See also Figure 3 in Chapter 20.

5.1 Supervised classification

The first option to delineate landforms is to rely on the empirical knowledge of a surveyor. In the traditional mapping of landforms, a surveyor tries to reconstruct the main geomorphological processes, recognise the geological material and main structural elements of a landscape and then delineate them using stereoscopic photo-interpretation and field survey. Land-surface parameters can be used to produce a more objective map than the one delineated manually. We can still base processing on a mental concept (i.e. the same legend of landforms), but land-surface parameters can help us to classify the area objectively by using a semi-automated algorithm.

To extract landforms exactly as in our legend (mental model), we first need to prepare a list of input land-surface parameters with the same grid definition, then

[3] Ground Water Depth was estimated as the difference between the DEM and the approximated water table surface, which was interpolated from point data (water table height measurements) using a second order trend function.

[4] See Equation (2.40) in Chapter 6.

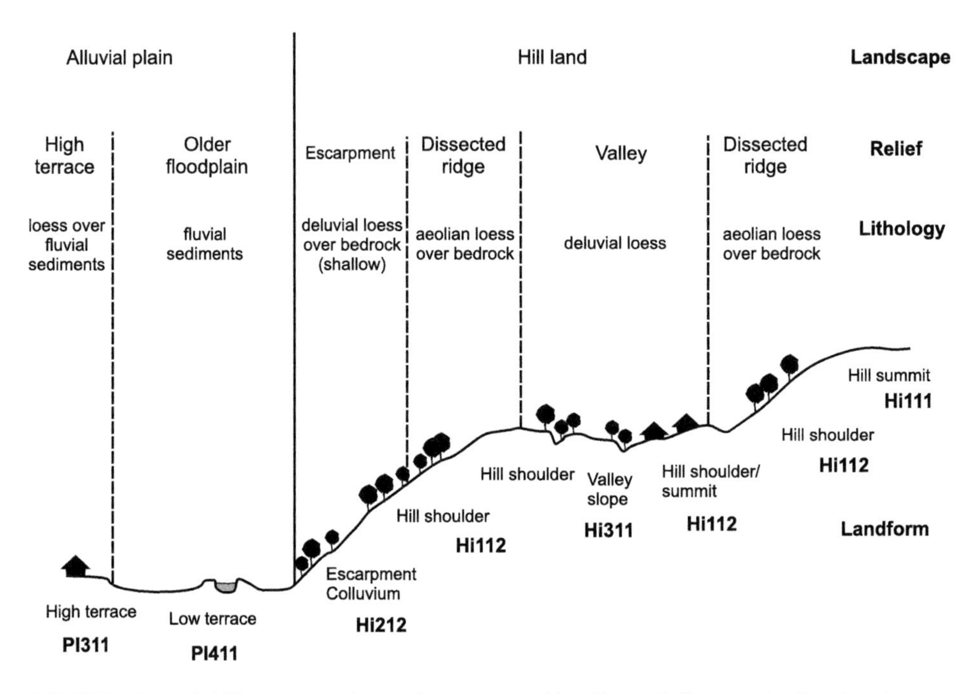

FIGURE 5 Baranja Hill: cross-section and sequence of landforms following the four-level geo-pedological method of Zinck and Valenzuela (1990).

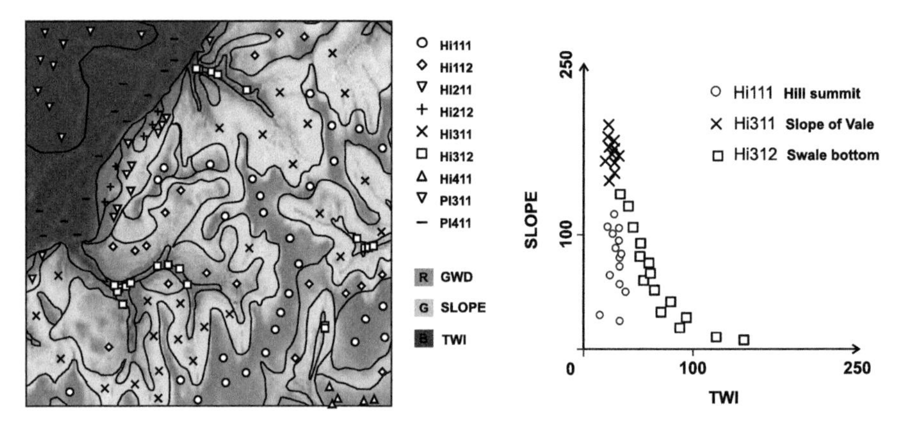

FIGURE 6 Training points displayed in geographical (left) and feature (right) space. The false colour composite (DEM, SLOPE, TWI) can be used to interactively select the most typical locations for each landform class (in this case manually delineated units). The values for TWI and SLOPE in the right plot have been stretched to the 0–255 scale. (See page 747 in Colour Plate Section at the back of the book.)

TABLE 1 Landform elements — description and relation with land-surface parameters

Landform element	Description	SLOPE	PLANC	ACV
Channels (*Valley bottoms*)	Locations of water accumulation and transition; high number of upstream elements and concave shapes	min	min	max
Ridges	Locations of water run-off; lowest upstream contributing area and convex shapes	min	max	max
Slopes	Sloping part with generally higher shape complexity	max	avg	avg
Plains (terraces)	Flat areas of low relief and low shape complexity	min	avg	avg
Pits	Conical concave shapes	avg	min	min
Peaks	Hill-tops, conical convex shapes	avg	max	min

'train' our algorithm by selecting typical[5] locations, and finally run the classifier (see also Figure 9 in Chapter 19). In ILWIS GIS, the processing steps are as follows:

- *Derive the DEM parameters*: SLOPE, PLANC, TANGC, TWI, SOLIN, GWD. All land-surface parameters should be based on grids of consistent resolution.
- *Combine the land-surface parameters* into a map list (land-surface parameters become bands).
- *Create a training dataset* by manually selecting typical locations (at least 20 per class). Use the false colour composite (Figure 6) to pinpoint the most typical locations.
- *Derive the representative* (average) *value* and the variation (standard deviation) for each class centre for each land-surface parameter (see for example Table 1).
- *Run the classifier* and allocate all pixels to classes based on the distances in the feature space.
- *Cross check the classifier* by using kappa statistics (see Section 2.4 in Chapter 19) for the control points.

Note that, ideally, both the training and control points should be field observed locations, but we can also select locations through stereoscopic photo-interpretation. In this case, we have selected about 20 points for each landform class (Figure 6). Note also that we have, on purpose, spread the samples equally around the whole area to get a good coverage and to avoid extrapolation in geographical space. The samples can be visualised using the feature space plots to

[5] Typical locations correspond to the locations of the class centres (gravity centre of point clouds) in the feature space (Shi et al., 2005).

see how distinct are the classes, which classes are the most similar, and which are overlapping (Figure 6).

Feature space[6] is a virtual space representing two or more land-surface parameters (forming a hypercube) where the values of land-surface parameters are used as coordinates. Once the class centres (mean values for each land-surface parameter) and the variation around the class centres (standard deviation for each land-surface parameter) are determined, each grid node in the map can be assigned to a specific class based on discriminant analysis (Gordon, 1981). This procedure is analogous to the supervised classification of Landsat bands for mapping land cover types (Lillesand and Kiefer, 2004).

REMARK 5. *Geomorphological features can be mapped by land-surface parameters in much the same way land cover is mapped by remotely sensed multispectral bands.*

The result of classification and a comparison with manually delineated classes can be seen in Figure 7. Obviously, the supervised classification gives much more local detail than the manual delineation, especially for the classes that are spatially intermingled (mosaicked) with some other landforms. The true advantage of supervised classification of land-surface parameters is that it may be able to replace manual delineation of landforms, as long as the land-surface parameters provide the required level of detail and the geological structure matches the morphological features (Hengl and Rossiter, 2003). The output of classification needs to be filtered for isolated pixels and artefacts (e.g. low terrace at top of a hill). Note that the spatial connection of individual pixels is typically ignored, but can be accounted for if the classification is combined with image-segmentation techniques (Drăguţ and Blaschke, 2006).

The same input parameters (land-surface parameters, class centres and their variation) can be used also to run a continuous classifier such as *fuzzy k-means* (see also Chapter 9). This will result in multiple maps of memberships where each class will be presented as a separate map (Figure 7). The biggest advantage of using memberships is that they can be used to assess confusion between the classes and to detect areas where the confusion between several classes is rather high (Burrough et al., 1997; Hengl et al., 2004b; Shi et al., 2005). We are also able to determine visually which classes are connected with which land-surface parameters — for example, in the case above, class Hi211 (Escarpment scarp) seems to be correlated with aspect (SOLIN parameter) and class Hi312 (Valley bottom) is obviously strongly correlated with the TWI (see Figure 6). This also demonstrates that one might get into problems when using some land-surface parameters over a limited area, because the escarpment here is highly correlated with north-western aspect, which might not apply more widely. Such accidental matches can eventually lead to artefacts and poor mapping accuracy, which are not visible in the final output.

[6] An important difference between geographical space and feature space is that the dimensions of feature space are on different scales. The two are non-linearly related — points that are close in geographical space can be far from each other in feature space and vice versa.

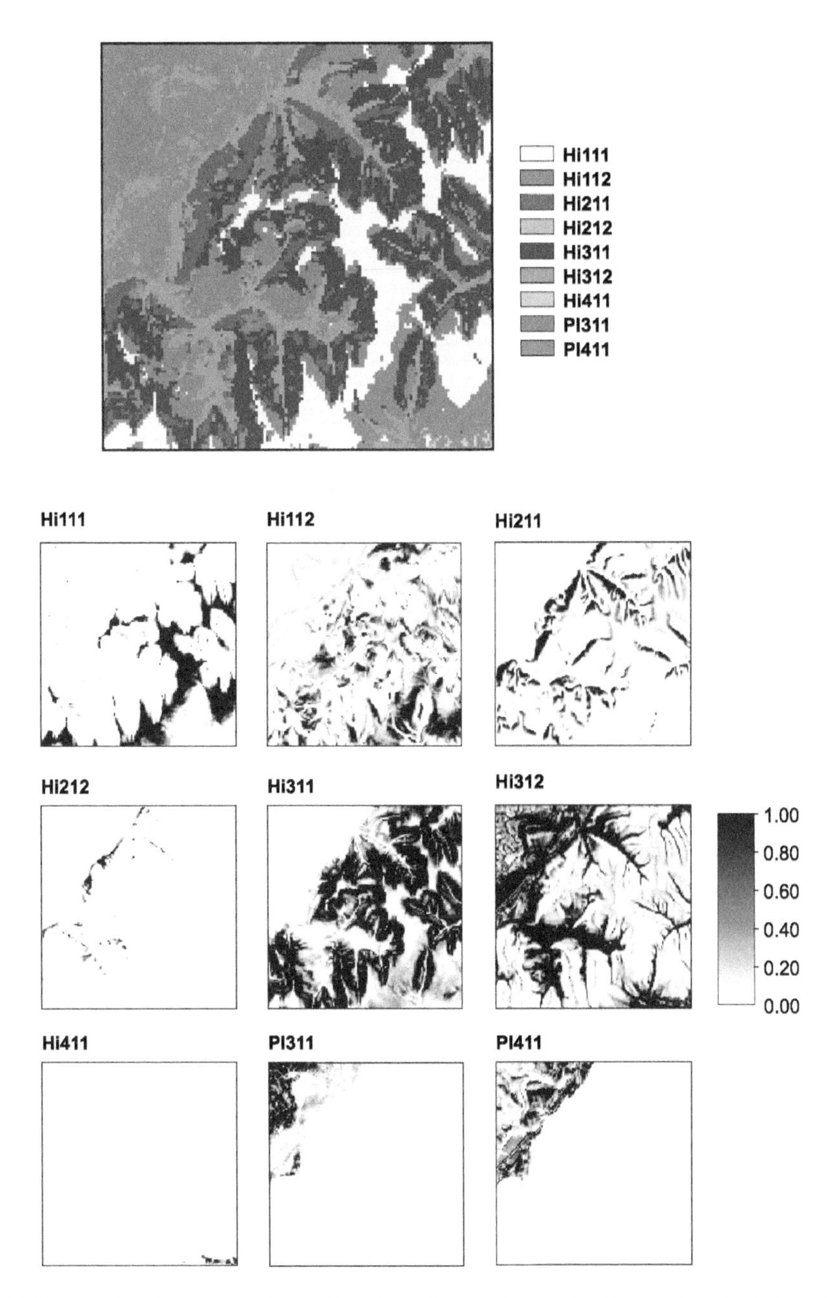

FIGURE 7 Results of supervised classification using maximum likelihood classifier (above) and memberships derived using fuzzy k-means classification (below). Hi111 (Hill summit), Hi112 (Hill shoulder), Hi211 (Escarpment scarp), Hi212 (Escarpment colluvium), Hi311 (Valley slope), Hi312 (Valley bottom), Hi411 (Glacis slope), Pl311 (High terrace) and P411 (Low terrace). Compare with Figure 6. (See page 748 in Colour Plate Section at the back of the book.)

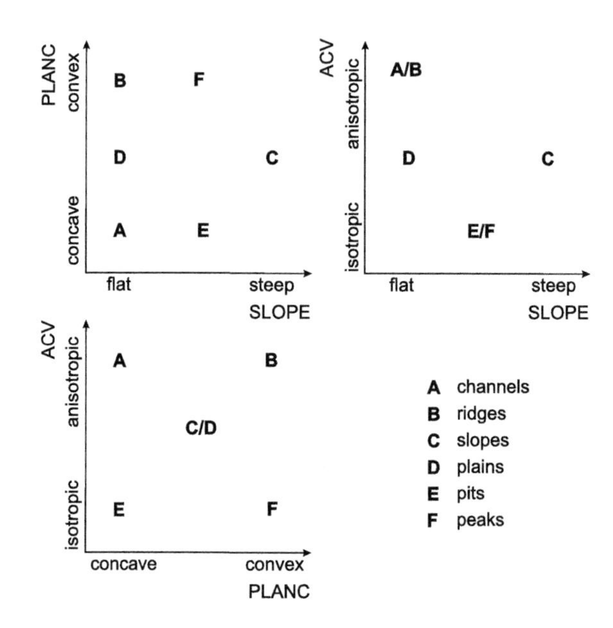

FIGURE 8 Definition of landform elements using slope (SLOPE), plan curvature (PLANC) and anisotropic coefficient of variation (ACV).

Another advantage of using memberships is that they provide continuous predictors to map soils and/or vegetation. On the other hand, fuzzy classification and processing of consequent memberships can be impractical if the number of classes is high or if the landforms are correlated with land-surface parameters in a non-linear way.

The supervised fuzzy *k*-means technique can be run in many image processing packages used to process remote sensing imagery such as Erdas Imagine, PCI Geomatica or ENVI. The example in Figure 7 was produced in ILWIS GIS using an ILWIS script that was prepared just for this case study. The procedure is explained in detail in Hengl et al. (2004b) and the sample script is available from geomorphometry.org. A difficulty of using this algorithm is that each study area will require different inputs (the legend, the land-surface parameters), so that a user needs to redesign the algorithm for each case study.

A variant of supervised classification is to use predefined (generic) classes known as *landform elements, landform facets* (Irvin et al., 1997) or *surface features* (Fisher et al., 2005). As with traditional landform classification systems, here we also have several methodological approaches, although the names might be similar or very similar. In all cases the legend (classes) is universal and can be applied to any study area at any scale. A different issue concerns the total number of generic landform elements, and their exact definitions (see also discussion in Section 4 of Chapter 9).

The example in Figure 9 shows six extracted landform elements (channels, ridges, slopes, plains, pits and peaks) for Baranja Hill. As in the previous example, classes are extracted using a supervised fuzzy *k*-means algorithm based on

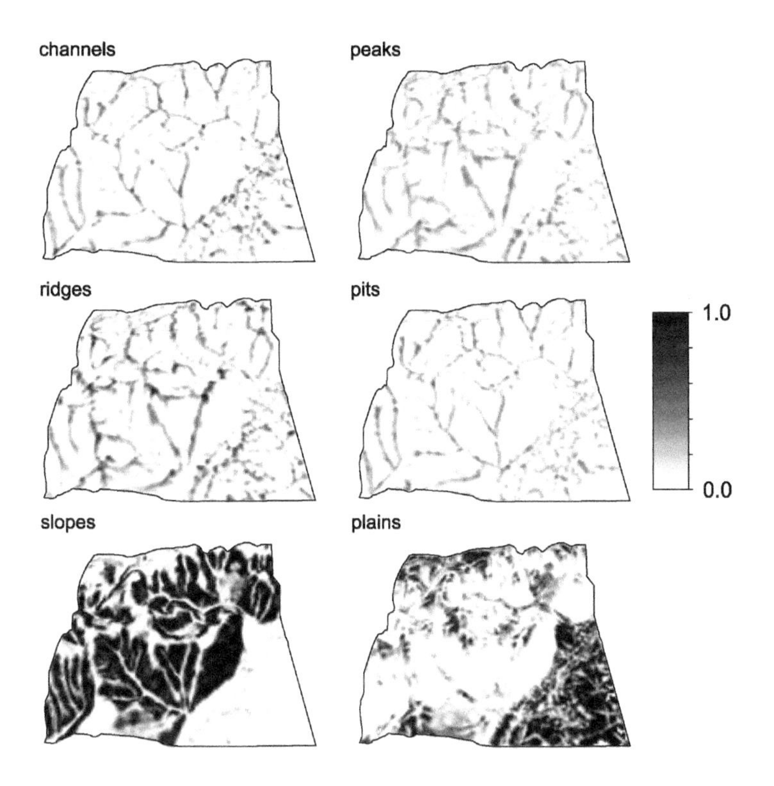

FIGURE 9 Extraction of landform elements: channels, ridges, peaks, slopes, plains and pits. Extracted in ILWIS using the `G_landforms` script.

class centres and variation around them, except that universally applicable classes centres are used (Figure 8, Table 1). Note that this is still a supervised approach because the definition of classes is given prior to their extraction and can be adjusted by the user. The difference between the extraction of landform elements and a fully supervised approach is that a surveyor does not need to sample an area to produce the training data set (central class table). Instead, a user can simply take any DEM and run classification directly.

The landform elements are relative and there can be local channels (streams), ridges, peaks, ... within the broader ridges, terraces, slopes, etc. In this case, no real evaluation of accuracy is possible. The surveyor needs to have a clear concept about the scale of work and then adjusts the definition of the central values in the table. Our experience is that one needs at least 2–3 iterations/adjustments until a satisfactory[7] image is produced. Note that the list of land-surface parameters could also be extended to other more regional land-surface parameters, for example catchment area, etc., so that other landform elements such as *"pool"* (basin) or *"shoulder"* could be introduced. Note also that, because we use a continuous classification, the resulting membership grades for each generic landform

[7] Usually, the maximum information content is achieved if all classes are equally represented. This does, however, mean that the class boundaries are not universal.

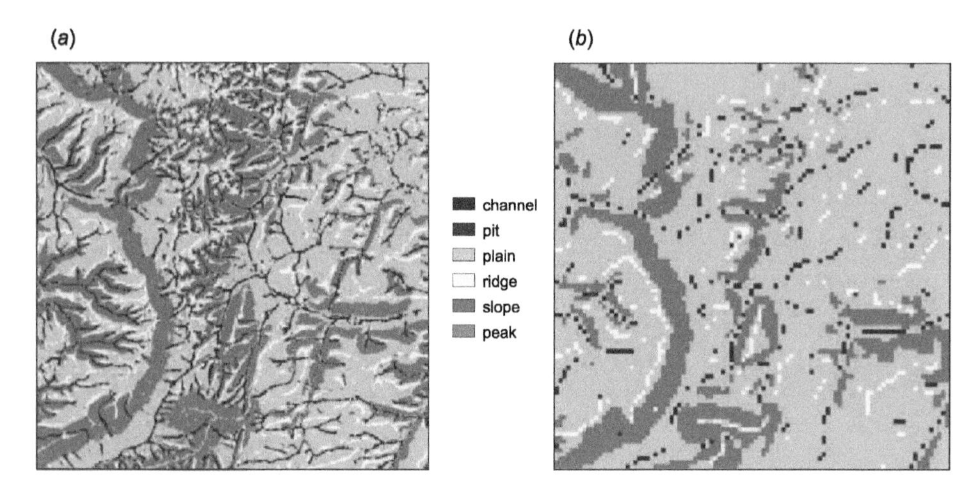

FIGURE 10 Extraction of landform elements for the 10×10 km Ebergötzen study area, Germany using the 25 m DEM (a) and the 90 m SRTM DEM (b). (See page 749 in Colour Plate Section at the back of the book.)

can be considered as special types of land-surface parameters (Fisher et al., 2005): *channel-ness, ridge-ness, slope-ness, terrace-ness, pit-ness, peak-ness*, etc.

Although attractive for its possibility of automation, extraction of landform elements often needs to be adjusted according to the working scale, otherwise it might lead to poor results. Figure 10 illustrates the impact of cell size on the quality of outputs of classification. In the case of the much coarser DEM, landform elements with smoothed values (plains) will over-dominate other classes.

> REMARK 6. *Fuzzy classification excels in extracting landform elements from DEMs; it allows analysis of overlap between similar classes, assessment of overall accuracy, and detection of 'problem' classes.*

Automated extraction of pits, peaks, channels, ridges, passes and plains is also possible in LandSerf by setting up tolerance values for each class (see Chapter 14) or by using ArcInfo scripts (see Chapter 11).

5.2 Unsupervised classification

For unsupervised classification, we do not even have to fix the number of landform classes. This methodology is explained in detail in Irvin et al. (1997) and further refined by Burrough et al. (2000), Schmidt and Hewitt (2004). We will now demonstrate how to achieve it for the Baranja Hill using the program FuZME (http://usyd.edu.au/su/agric/acpa/fkme/), kindly provided by Budiman Minasny of the University of Sydney. For Baranja Hill, we will use the whole area which consists of 149×147 or 21903 grid nodes. The three attribute maps (SLOPE, PLANC, ACV) need first to be transformed to point data sets, then exported as a single dataset where each grid node is attributed with the land-surface parameters. This file can now be opened in FuZME and analysed using the Fuzzy clustering

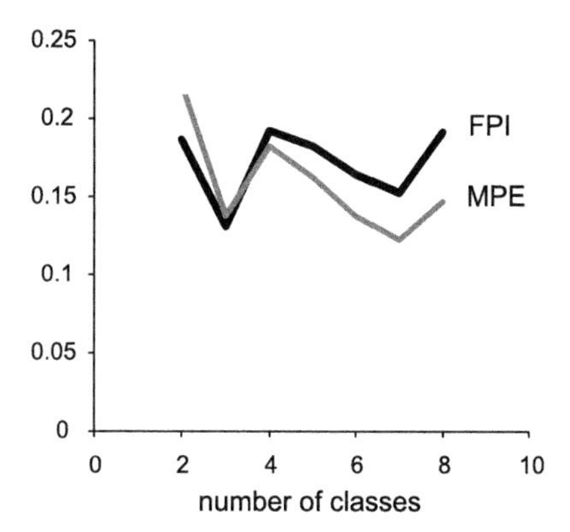

FIGURE 11 Selection of the optimal number of classes for Baranja Hill using Fuzziness Performance Index (FPI) and Modified Partition Entropy (MPE): in this case, either 3 or 7 classes could be a suitable solution.

operation. Optionally, a user can also randomly sample a few thousand locations and then process these in FuZME, which permits repeated testing of the robustness of a classification. After FuZME has determined the class centres, these can be used to classify other grid nodes as explained in Section 5.1.

In FuZME, a user can select the minimum and maximum number of classes, the fuzzy exponent, the maximum number of iterations, and the distance measure in feature space. The program will start by randomly initialising classification and then iteratively allocate points until the objective function is minimised. The same will be repeated for each number of classes (e.g. 2–10), so that a user can then finally decide the optimal number of classes. Two performance measures are commonly used to select the optimal number of classes: Fuzziness Performance Index (FPI) and Modified Partition Entropy[8] (MPE). The optimal number of classes is where both measures reach minima (McBratney and de Gruijter, 1992).

In our case study, the optimal number of classes seems to be 7, although 3 classes could also be satisfactory (Figure 11). The final results of classification with 3 and 7 classes, using diagonal distances, can be seen in Figure 12 (compare also with the results of the supervised fuzzy *k*-means in Figure 9). Note that in the case of unsupervised classification we can not really provide the legend for the final output. We can only try to assign names to extracted classes *a posteriori* by carefully analysing their statistical characteristics (Burrough et al., 2000).

Although attractive for users who want objective estimates, unsupervised extraction of landform elements has not become popular, probably because it can be computationally demanding and because surveyors are not able to interpret the results directly — a class can be well-defined in feature space but what does it signify and how is it to be named?

[8] FPI estimates the degree of fuzziness generated by a specified number of classes and MPE estimates the degree of disorganisation created by a specified number of classes.

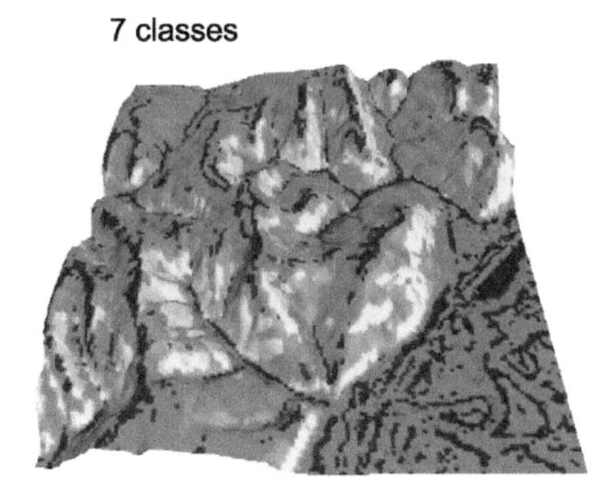

FIGURE 12 Landforms extracted using unsupervised fuzzy k-means classification with 3 and 7 classes in the FuZME package. Because the classification is unsupervised, the legend can be constructed only *a posteriori*. (See page 749 in Colour Plate Section at the back of the book.)

5.3 Extraction of break-lines

Application of geomorphometry for automated extraction of land-surface objects is especially attractive in geomorphology. In the following section we will demonstrate how land-surface parameters can be used to extract break-lines. Break-lines are by definition narrow areas of transitions between significantly different landform elements. In geomorphology these can be clearly seen and delineated on aerial photos: one example is the abrupt change between a hill and a flood plain.

The easiest way to detect break-lines is to derive a slope map and then use the edge enhancement filter to emphasise such abrupt transitions and then create a binary map. A more sophisticated approach is to consider analysis of transition between the landform classes. For example, if the results of classification are fuzzy membership maps, then we can derive a confusion index (see also Figure 11 in Chapter 19), automatically detect areas of very high confusion, and then analyse the width of the transition:

$$\begin{array}{llll} CI > t_{CI} & \text{AND} & h < t_h & \rightarrow \text{break-line} \\ CI > t_{CI} & \text{AND} & h > t_h & \rightarrow \text{extragrades} \end{array}$$

where t_{CI} is the threshold value for the CI, and t_h is the threshold value for the width of the transition. An example of how both break-lines and extragrades can be extracted from the membership maps can be seen in Figure 13. Apart from extracting break-lines, geomorphometry is equally successful in extraction of drainage lines, ridge lines and similar skeleton lines (Chorowicz et al., 1989; Jordan, 2003). More examples of how such features can be extracted can be found in Chapters 7 and 25.

6. SUMMARY POINTS

Geomorphology contains many important applications of geomorphometry; the analysis of DEMs has revolutionised many aspects of geomorphology. It is begin-

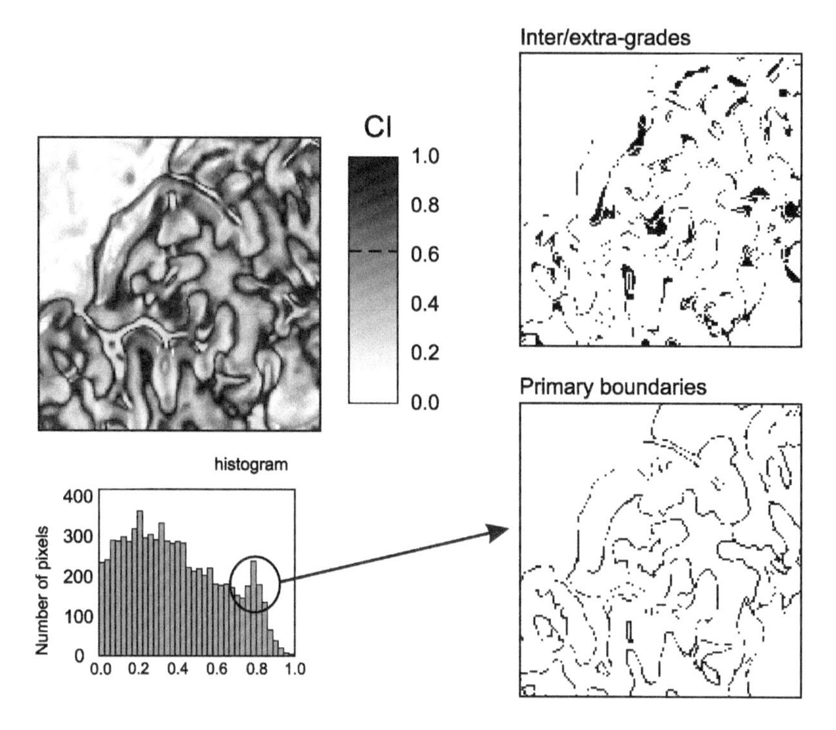

FIGURE 13 Break-lines extracted automatically using the confusion index map. Reprinted from Hengl et al. (2004b). With permission from Taylor & Francis Group.

ning to be taken for granted, and it may not be evident from the title or even the abstract of a paper in geomorphology that a DEM is fundamental to the results obtained. The 'cosmetic' use of a pretty shaded DEM of the study area or its context, often in colour, is even more widespread. The great value of early DEMs was in permitting analysis of surface form over large areas, sampling evenly over space, deriving general trends and crucial differences, and calibrating models. This was at the expense of reduced accuracy compared with precise measurements in the field. However, the use of LiDAR has removed this disadvantage and DEMs are now being constructed at decimetric and centimetric scales. The accuracy of repeat cover permits precise analysis of short-term changes, for example in river channels and glacier mass balance. Integration of DEMs with GIS and modern tools of statistical analysis has hugely increased the analytical power of modern geomorphology.

A major application of geomorphometry in geomorphological studies is the automated extraction of geological/hydrological features and landforms. In Europe, accurate maps of river networks and accurate boundaries of watersheds are becoming increasingly important because watersheds are more and more a basis for regional planning and management (Colombo et al., 2007).

Using a small case study, we have demonstrated various approaches to extraction of predefined, generic and empirically defined landform objects. Many issues are open for discussion and further development. For example, should a user first run unsupervised classification, then build a legend, or vice versa? Should we divide a larger area into markedly different landscapes first, and then run the classifiers? If we run unsupervised classification, should we limit the maximum number of classes and should we try to organise classes in a hierarchy?

Another issue important for successful extraction of landforms is the scale factor. What might be a peak at one scale may be another morphometric class at another (Wood, 1996; Fisher et al., 2005; Iwahashi and Pike, 2007). This makes the extraction of landform elements even trickier. One solution to this problem is to build hierarchical classification systems that refer to exact ranges of grid resolutions. Such systems are already available in traditional geomorphology but are incomplete for the extraction of landform elements. LandSerf, however, can produce hierarchical peak classifications using *relative drop* methods. Another solution is to consider extraction of landform elements only at the most detailed scale and then aggregate the classes instead of aggregating DEM values (Minár and Evans, 2008). The nested-means approach of Iwahashi and Pike (2007) is certainly a development that will progress.

DEMs alone are unlikely to suffice for truly automated and accurate geomorphological mapping at regional or continental scales. DEMs need to be supplemented with geological maps, remote sensing images and similar information that can help analysts distinguish form from material and endogenetic from exogenetic structures (Chorowicz et al., 1995; Jordan, 2003). The combination of DEMs with multi-spectral remote sensing, for example, has proved useful in geological mapping, especially in areas of lower vegetation cover.

6.1 Next steps

This chapter has shown that geomorphometry has gone beyond the experimental stage and produced many substantive results in geomorphology. Applications of DEMs are now almost taken for granted. Although predictions of the future are notoriously unreliable, and the main events (as, in politics, the globally important events of 1989, 1991 and 2001) tend to be unexpected, two trends affecting geomorphometry are very likely to continue. These are the year-by-year increase in computer power, in both speed and storage; and the availability of more DEMs, with improved resolution and accuracy. Freely downloadable data may proliferate, with quality exceeding that of currently expensive data. Applications of geomorphometric analysis are thus likely to become more broadly based and more reliable, with less hand-wringing about data errors and inadequacies. DEM error (Chapter 5) will never cease to be a concern, especially where topographic details are important, but broad-scale studies should be able to rely on the accuracy of data.

Where ideas have been tested with small or regional data sets, we hope that they will in future be tested with large and global data sets. Many studies, however, relate land-surface variables to field observations of either the same or further variables. This may continue to inhibit broad-scale studies of detail, except where remote sensing can be substituted for field observation. Computation should become easier in the near future, so that where vague proposals have gone untested, there will be no excuse to avoid testing against real-world geomorphometric data. There are also possible pitfalls here: editors and reviewers have a responsibility to prevent publication of statistically naïve studies by those who do not understand what goes on inside various computational black boxes. Computer power should not be used to define every possible variation of each attribute, and then condense them into undecipherable indices. More specifically, we expect progress in the near future on the definition of landforms and their elements, and the characterisation of different topographies. These will be compared to those simulated from combinations of processes. Eventually, we may be able to predict the future development of the land surface from its present form and a hypothetical sequence of processes.

IMPORTANT SOURCES

Iwahashi, J., Pike, R.J., 2007. Automated classifications of topography from DEMs by an unsupervised nested-means algorithm and a three-part geometric signature. Geomorphology 86 (3–4), 409–440.

Fisher, P.F., Wood, J., Cheng, T., 2005. Fuzziness and ambiguity in multi-scale analysis of landscape morphometry. In: Petry, F.E., Robinson, V.B., Cobb, M.A. (Eds.), Fuzzy Modeling with Spatial Information for Geographic Problems. Springer-Verlag, Berlin, pp. 209–232.

Gorsevski, P.V., Gessler, P.E., Jankowski, P., 2003. Integrating a fuzzy k-means classification and a Bayesian approach for spatial prediction of landslide hazard. Journal of Geographical Systems 5 (3), 223–251.

Goudie, A.S., 2004. Encyclopedia of Geomorphology, vols. 1 & 2. Routledge, London, 1184 pp.

Burrough, P.A., van Gaans, P.F.M., MacMillan, R.A., 2000. High-resolution landform classification using fuzzy *k*-means. Fuzzy Sets and Systems 113, 37–52.

Lane, S.N., Richards, K.S., Chandler, J.H., 1998. Landform Monitoring, Modelling and Analysis. Wiley, 466 pp.

http://disc.gsfc.nasa.gov/geomorphology/ — Geomorphology from space (online manual).

CHAPTER 23

Modelling Mass Movements and Landslide Susceptibility

S. Gruber, C. Huggel and **R. Pike**

spatial modelling of potential slope instability by geomorphometry · rapid mass-movements in an alpine environment · parameterising single- and multiple-flow-direction models · flow-routing, run-out, and deposition · regional modelling of shallow and deep-seated landsliding in hilly terrain · parameterising a susceptibility model for deep-seated failure · advantages and limitations of DEM-based approaches

1. INTRODUCTION

Mass movement (slope instability or *"failure"*) is an important geo-/ecosystem process in landscapes ranging from gentle hills to steep mountains. *Debris flows, rock falls, deep-seated landslides, snow avalanches,* and other movements also can impact the built environment and threaten human life. Because both the source areas and the downslope paths traced by mass movements are strongly controlled by land-surface form, simple model approaches that rely mainly on topography can aid in understanding these failures and their effects. Advances in GIS technology and the mathematical/statistical tools for modelling and simulation, as well as the increasing availability of DEMs, have led to many applications and a growing literature. The methods and techniques employed range from empirical and heuristic to statistical and physically-based, and each brings its own unique advantages and disadvantages that determine its suitability for a given type of mass movement and application. This chapter samples two of the many approaches that employ geomorphometry. Other methods are referenced as well, and we have sought to explain our examples in sufficient detail for the readers to be able to apply the techniques for themselves.

Developments in Soil Science, Volume 33 © 2009 Elsevier B.V.
ISSN 0166-2481, DOI: 10.1016/S0166-2481(08)00023-8.

2. MODELLING THE PROPAGATION OF MASS: THE AFFECTED AREA AND DEPOSITION

2.1 Background

The methods and case studies on modelling the propagation of mass presented in this section originate from research in steep high-mountain areas. Here, rapid environmental changes — currently even more pronounced due to *accelerating atmospheric warming* — are causing glaciers to retreat and permafrost to thaw (Gruber and Haeberli, 2007). The resulting rapid formation of potential new release areas for mass movements, such as debris flows, and rock or ice avalanches, requires techniques that can quickly evaluate the extent of these potential hazards. This can aid decisions about where detailed investigations, field measurements or monitoring should be carried out, and, in particular, where the need is urgent.

The techniques for this initial assessment should be easily and quickly applicable to large areas. Furthermore, they should require very little input data, because for most high-mountain areas, there is usually little accurate information. The input to the models proposed here consists of only two spatial data fields: (a) location of source areas (starting zones) and the amount of material mobilised in them; and (b) a DEM.

> REMARK 1. *Geomorphometry can aid the assessment of slope instability, especially the reproducible delineation of areas susceptible to future failure.*

Many studies have used geomorphometric models for determining affected areas and the amount of deposition that can be expected. For debris flows or ice avalanches, the potential hazards have been assessed using flow propagation schemes identical or similar to the ones shown here (Huggel et al., 2003, 2004; Salzmann et al., 2004; Noetzli et al., 2006). The *LAHARZ model* (Iverson et al., 1998) propagates lahar flows over a DEM to the point where the depletion of the mass halts the flow. To determine the deposition characteristics, the geomorphological model *LAPSUS* uses a multiple flow-direction algorithm, path length and run-out distance (Claessens et al., 2006) for different types of debris, or for their differing transport, detachment and settlement capacities (Schoorl et al., 2002). Many publications on non-geomorphometry-specific propagation models of debris flows (Hungr, 1995; Gamma, 1999; O'Brien et al., 1993; Bartelt et al., 2005; McArdell et al., 2004) and avalanches (Sampl and Zwinger, 2004) can serve as starting points for further reading.

2.2 Why use parametrisation methods to model the movement of mass?

Instead of representing the physical processes of the event, parametrisation models describe movements of mass in terms of simple parameters. Their great advantage is that they can be applied straightforwardly. When compared with models explicitly representing physical processes, they usually:

- require less input data;
- need less computation time;
- are more easily available; and
- require less technical know-how.

On the other hand, they provide no, or only limited, information regarding the dynamics of flow processes and are often less suitable for application outside their validated domain, compared with models based on physical processes. Which type of model is preferable is thus more a matter of how suitable a model is for a certain purpose, rather than its technical sophistication. Fast parametrisation techniques are often a great asset because they can be applied and employed over large areas. Models based on physical principles or on field measurements and monitoring are usually more appropriate at a later stage, in more detailed studies.

REMARK 2. *DEMs, together with an understanding of landslide processes, enable models of mass-movement to be parameterised over large areas.*

Neither complex models based on physical processes nor simple parametrisation schemes release one from the responsibility of sufficiently understanding the phenomenon (of e.g., a debris flow) that is being modelled and the consequences of the simplifications and errors inherent in whichever model is being used. It is also important to provide a careful interpretation of the results. The methods in this section are described in sufficient detail to be implemented in ArcGIS or in programming languages such as IDL or Matlab.

2.3 Modified single-flow-direction model

2.3.1 The trajectory and the flow-routing component of the model

Single-flow-direction approaches are feasible for mass movements, such as small rock falls or small ice avalanches, which take the steepest descent path along a relatively small corridor. In other processes, such as debris flows or rock avalanches, the flow deviates from the steepest descent path and spreading can be observed. A single-flow-direction path cannot model this process accurately. We therefore introduce a model that makes it possible to calculate multiple flow directions based on single flow direction (D8) and other standard functionalities of ArcInfo™/ArcGIS™. This model, known as *Modified Single Flow Direction* (MSF), was developed by Huggel et al. (2003). It is based on the principle of a mass movement (e.g., a debris flow) being propagated down slope from a specified initiation point. The central flow-line of the mass is assumed to follow the steepest descent path, as calculated by the D8 algorithm.

To account for flow spreading, a diversion function *Fd* has been incorporated into the model. This function allows the flow to divert from the steepest descent path by as much as 45° on both sides. This model is thus better equipped to simulate the different characteristics of processes such as debris flows or larger avalanches when moving along confined channel sections (characterised by converging flow and thus limited spread) and on relatively flat or convex terrain (characterised by diverging flow and thus a greater spread). Once the *areas potentially affected by the mass movement* have been delineated, a function Pq assigns to each grid cell i, the relative probability of it being affected. The probability, described by the function Fr, indicates that the more the flow diverts from the steepest descent path, the greater the resistance. Fr yields a cell value that increases down-valley (i.e. with increasing distance) from the initiation location, and increases laterally (i.e. with increasing flow resistance) at an angle of 45° from the steepest descent path. The ratio between Fr and the horizontal distance L from each cell i to the starting zone represents a functional probability value and, for each cell, there is a probability value of it being affected by the movement of the (ice/rock) mass:

$$Pq_i = \frac{L_i}{Fr_i} \tag{2.1}$$

where Pq_i depicts a qualitative index (for visualisation purposes), rather than a mathematical probability in the strict sense of the term.

2.3.2 The flow-reach component of the model

The flow-reach component corresponds with the approach used by the single-flow-direction propagation model. Accordingly, the H/L ratio is defined as a stopping condition for the trajectory stage of the flow (H is the difference in elevation and L the path length). The trajectory component of the MSF model usually provides the *potential maximum inundation zones of a mass-movement event*. Thus, it indicates which areas are more or less likely to be affected. Consequently, the run-out distance should also be based on a maximum. A reasonable H/L ratio has to be evaluated on the basis of empirical data for the type of mass movement that is being modelled. The fact that large mass movements are often characterised by flow transformations should also be taken into consideration. Debris avalanches, for instance, may change into debris flows and eventually into hyper-concentrated stream flows (Rickenmann, 1999). To accommodate these types of transformations, the run-out length is extended for the more fluid processes.

For debris flows in the Alps, a minimum H/L ratio of 0.19 (corresponding to a slope of 11°) has been found (Rickenmann, 1999; Huggel et al., 2003), but again, larger and more fluid debris flows in other regions may show lower H/L ratios and consequently a larger flow reach. For some mass movements, such as debris flows or debris avalanches, the H/L value can be derived from a relationship between flow volume and H/L (e.g., Rickenmann, 1999).

2.3.3 Implementing the model in ArcGIS

The basic input to the model is the DEM and a starting zone that defines where the mass movement began. Both (called `dem` and `source` here) must be available as **ArcInfo** grids. The starting zone is represented by a number of grid cells. In principle, since there is no implicit consideration of volume, there is no limitation to the number of grid cells that may be selected as the starting location. In practice, however, it is recommendable to use a limited array of cells, because, otherwise, the calculation of the downslope propagation may be too extensive and divergent.

An exception can be mass movements such as rock falls, for which it may be useful to assess multiple-failure zones. Before starting to run the model, the elevation data should be corrected for any topographical sinks (**ArcGIS** command `sink`) that can hinder the propagation of the flow (see also Section 2.8 in Chapter 4).

To implement the model in **ArcGIS** Workstation or **ArcInfo**, it is first necessary to calculate the flow direction from the DEM and using the D8 algorithm:

```
Grid: fldir = flowdirection ( dem )
```

For use later in the model, flow directions need to be converted to values expressed in degrees:

```
Grid: fldir_deg = con(log2(fldir) < 6, (log2(fldir) + 2) * 45,
( log2(fldir) - 6 ) * 45 )
```

The calculation of flow propagation downslope, both along the steepest descent path and in the 45° lateral diversion, is based on the **ArcGIS** command: `pathdistance` (a detailed description of this command can be found in the extensive *help* function provided by **ArcGIS**):

```
Grid: Li = pathdistance( source,#,#,fldir_deg,
"FORWARD ZEROFACTOR=1",#,#,#,start_z,#,int(dem) )
```

with `Li` representing the horizontal distance from the starting location of the flow to each cell potentially affected by it. The function `Fri`, which gives values increasing in horizontal distance from the source location and the lateral flow divergence (increasing flow resistance), is based on a modified `pathdistance` command:

```
Grid: Fri = pathdistance( source,#,#,fldir_deg,
"linear cutangle=90",#,#,#,#,#,dem )
```

As defined above, the index `Pqi` is calculated by:

```
Grid: Pqi = Li / Fri
```

To constrain the flow propagation by the stopping condition, we need to know the H/L ratio at the locations of all the cells:

```
Grid: Hi = start_z - dem,
```

where `Hi` is the vertical distance of the cell `i` from the maximum elevation of the start location `start_z`. The ratio H/L is then derived by:

```
Grid: H_L = Hi / Li.
```

It is essential to define the appropriate `H_L` value, since it defines the run-out of the mass flow, and therefore has to be carefully evaluated for each individual case. As an example, we use here a `H_L` value of 0.19:

```
Grid: H_L_lim = con(H_L >= 0.19, H_L)
```

Finally, to apply the limiting flow-reach condition to the modelling of the complete flow, we use a simple **ArcGIS** function:

```
Grid: Pq_limi = H_L_lim + Pqi - H_L_lim
```

`Pq_limi` is a grid, the cell values of which represent a qualitative index of the probability of being affected by the simulated mass flow which has a stopping condition equivalent to the defined `H_L` value.

2.4 Multiple flow directions

2.4.1 The flow-routing component of the model

Multiple Flow Direction (MFD) methods are well-suited for modelling divergent flow. The method and notation for designating neighbouring cells is described in Section 3.2 of Chapter 7, but some basics are briefly repeated here. Mass M in one cell is propagated by distributing it to its eight neighbours. These are indexed $\mathrm{NB}i$. In classical MFD methods, the fraction of mass d that is propagated into the neighbouring cell $\mathrm{NB}i$ is given by:

$$d_{\mathrm{NB}i} = \frac{\tan(\beta_{\mathrm{NB}i})^v}{\sum_{j=1}^{8}(\tan(\beta_{\mathrm{NB}j})^v)} \tag{2.2}$$

To control excessive dispersion, we introduce an additional feature at this stage: the *draining fractions* $d_{\mathrm{NB}i}$ are corrected to bring them either to zero, or to a value at least larger than a threshold r. This restricts the lateral (sometimes nearly horizontal) propagation of extremely small amounts of mass. First, $d_{\mathrm{NB}i}$ is corrected for small values. This then becomes $c_{\mathrm{NB}i}$:

$$c_{\mathrm{NB}i} = \begin{cases} c_{\mathrm{NB}i} & \text{if } d_{\mathrm{NB}i} \geqslant r \\ 0 & \text{if } d_{\mathrm{NB}i} < r \end{cases} \tag{2.3}$$

The next step is to obtain the corrected draining fractions $cd_{\mathrm{NB}i}$ by bringing the sum over all neighbours to unity in order to preserve mass:

$$cd_{\mathrm{NB}i} = \frac{c_{\mathrm{NB}i}}{\sum_{j=1}^{8} c_{\mathrm{NB}j}}. \tag{2.4}$$

Finally, using $cd_{\mathrm{NB}i}$, the propagation of mass is computed

$$M_{\mathrm{NB}i} = cd_{\mathrm{NB}i} \cdot M \tag{2.5}$$

In the examples presented, we use $r = 0.01$.

2.4.2 The flow-reach component of the model

For the run-out distance approach, we need to determine H/L for each cell. Consequently, not only mass M, but also spatial grids of its source elevation E and accumulated path distance X need to be computed during flow propagation. For each cell, E is the average of the original elevations of each part of mass M that is flowing though that cell, and X is the average travel distance covered by each part of the mass. Both E and X are propagated in a mass-weighted way. This is because H/L is a proxy that contains potential energy (characterised by H) as well as frictional losses (characterised by L). If the source of transported mass for one event originated from a range of elevations, then its proxy for potential energy H should also reflect this distribution. The source elevation of the event (i.e. the product of mass and source elevation) is propagated as ME:

$$ME_{\mathrm{NB}i} = d_{\mathrm{NB}i} \cdot ME \tag{2.6}$$

After flow propagation, to find the mean difference in altitude, H, for each cell, the ME is divided by the mass in each cell M and subtracted from the DEM:

$$H = \mathrm{DEM} - \frac{ME}{M} \tag{2.7}$$

Planimetric path distance is propagated as MX (i.e. the product of mass and the distance it has traveled). Any new *mass–distance* gained is then added to the propagated MX:

$$MX_{\mathrm{NB}i} = cd_{\mathrm{NB}i} \cdot MX + L_{\mathrm{NB}i} \cdot M \cdot cd_{\mathrm{NB}i} \tag{2.8}$$

where the horizontal distance, $L_{\mathrm{NB}2,4,5,7}$, to cardinal neighbours is given by the cell size, and the horizontal distance to diagonal neighbours, $L_{\mathrm{NB}1,3,6,8}$, is given by the cell size multiplied by $\sqrt{2}$. After flow propagation, to find the mean travel distance, MX is divided by the mass in each cell:

$$X = \frac{MX}{M} \tag{2.9}$$

Finally, the overall angle α of the mass movement in each cell is determined as:

$$\alpha = \arctan\left(\frac{H}{X}\right) \tag{2.10}$$

The approach put forward here makes it possible to calculate H/L in a *mass-weighted* way. This can be useful for a number of investigations. Bear in mind that the H/L method originated from field mapping, where both the highest point of a starting zone and the lowest point of a deposit could easily be determined; and where both the distribution of initial mass and the effect of the flow path were unknown.

2.4.3 Deposition

The multiple-flow-direction approach can be used together with a *deposition function* (see Section 3.2 in Chapter 7 and Gruber, 2007). The method used in this deposition function is mass conserving and also allows the depletion and termination of a mass movement to be modelled. When determining the area affected by an event, this offers an alternative to H/L, because it requires finding the deposition parameters rather than the run-out ratio. This is an effective method for resolving differences in topography (e.g., divergent flow on a fan vs. a channelled flow path). However, much more published experience is available for the run-out method, the older approach. Deposition is a data product with an information content that extends beyond the affected area. Because more assumptions need to be made and more parameters need to be determined, it is also more difficult to assess the quality of deposition data. The required input for this method consists of regular grids of maximum deposition $D_{\max}$, elevation z and initial mass I. $D_{\max}$ and I are specified in units of mass or volume per unit area, e.g. $\mathrm{kg\,m^{-2}}$.

Here, *maximum deposition* D_{max} is determined by local characteristics, and these are independent of the volume of mass being transported. Therefore, events of differing magnitude are related to different run-out distances. However, where the events are of equal magnitude, and the path and deposition geometry are variable, the run-out distances will still be different. For instance, deposition in a channel will result in a larger run-out than on a convex fan with a divergent flow. A simple function is used to relate D_{max} to what is assumed to be its most important determinant — the local angle of slope, β:

$$D_{max} = \begin{cases} \left(1 - \frac{\beta}{\beta_{lim}}\right)^{\gamma} \cdot D_{lim} & \text{if } \beta < \beta_{lim} \\ 0 & \text{if } \beta \geqslant \beta_{lim} \end{cases} \tag{2.11}$$

Here, D_{lim} is the *limiting deposition*, i.e. the maximum deposition that would occur on horizontal terrain, the limiting slope β_{lim} denotes the maximum steepness at which mass is deposited, and the exponent γ is used to control the relative importance of steep and gentle slopes. Deposition D in each cell is limited by the local maximum deposition D_{max} and the available mobile mass M:

$$D = \begin{cases} M & \text{if } M < D_{max} \\ D_{max} & \text{if } M \geqslant D_{max} \end{cases} \tag{2.12}$$

where M is the sum of the initial input I and flow received from the neighbouring cells. The only mass that can be drained is the free-flowing mass that has not yet been deposited. The flow F_{NBi} into each neighbour NBi is given by:

$$F_{NBi} = (M - D) \cdot cd_{NBi} \tag{2.13}$$

By computing transport and deposition, *grids of deposition D* and *mobile mass M* expressed, respectively, in units of mass or volume per unit area, can be formed. After computation, the total input I equals the total deposition D. Exceptions to this are the transport of material out of the model domain if not all the relevant deposition areas are included or where there is a loss of mass because sinks, where $M > D_{max}$, were not removed.

2.4.4 Implementation

The multiple-flow-direction propagation scheme, extensions for the mass-weighted determination of flow distance and source elevation, and also the deposition functions are available as IDL source codes. The draining fractions cd_{NB} and the index for accessing the grids from higher to lower elevations are pre-computed and stored for use when making propagation calculations. Iterative sink-filling and correction of horizontal areas (Garbrecht and Martz, 1997) is carried out during the initial phase to prevent the loss of mass in sinks or horizontal areas. During the propagation phase, the algorithm loops from higher to lower elevations through all the cells. For each cell that contains parts of the mass, the algorithm computes the deposition and then, if they also receive parts of the mass, updates the M of neighbouring cells. Grids of deposition D and mobile mass M are computed. The sum of grids D and M describes the amount of mass that has been present in each cell.

FIGURE 1 The track and deposits left by the June 2001 flow of debris that overwhelmed the Swiss village of Täsch. Reproduced by permission of SWISSTOPO (BA081244). (See page 750 in Colour Plate Section at the back of the book.)

2.5 Case study: the Täsch debris flow, 2001

In the following case study, we apply both the MSF and the MTD models to a recent debris-flow event in the village of Täsch (in Valais, Switzerland). It started at Lake Weingarten (3060 m a.s.l.), situated in front of a glacier, but no longer in direct contact with it. The lake is on a large moraine deposit from the Little Ice Age that has a steep slope, with a maximum gradient of 36°. Comprised of loose sediment, the slope is 700 m in length. The section below the moraine, as far down the slope as Täschalp (on the left of the curve to the right, in the flow path of the debris), is characterised by slope angles of about 15 to 20°.

Below this section, and over a short flatter part, the flow path moves into a steep gorge that ends just at the upper edge of the village of Täsch. Like many Alpine settlements, Täsch lies on the debris fan of the torrent from the tributary valley, which thus affords the village protection from floods in the main river valley. The village achieved protection against floods from the tributary torrent by constructing an armoured channel across the village. This structure, however, was designed for flood water that does not contain a significant load of sediment. On 25 June 2001, after a period without significant precipitation, a debris flow rushed down on Täsch, damaging or destroying considerable parts of the village (Figure 1).

Thanks to an alarm given by people who observed the debris flow at Täschalp, there was just enough time to evacuate 150 people in Täsch, but the damage to buildings and other installations amounted to about 12 million EURs (Hegg et

FIGURE 2 The debris-flow deposit of the June 2001 event at the Swiss village of Täsch (photograph by Andreas Kääb).

al., 2002). The reason for this flow was that, due to deposits of ice and snow, the lake had become blocked. This blockage was then overtopped by 6000 to 8000 m^3 of water (Huggel et al., 2003). In the uppermost section, where the sediment was unconsolidated, this body of water eroded 25,000–40,000 m^3 of debris. The combined mass of water and debris rushed down the tributary valley and a small part of the debris was deposited at Täschalp, where a bridge was destroyed. During its passage through the gorge, sediment was probably neither deposited nor mobilised. At the apex of the fan, however, the front of the debris flow surged into the constructed channel. Since the channel was not designed for such heavy loads of sediment, it immediately became obstructed and the flow of debris spread out onto the fan (Figure 2) causing the damage mentioned above. The total volume of debris deposited in Täsch was in the range of 20,000–50,000 m^3 (Huggel et al., 2003).

We used a 25 m DEM (SWISSTOPO DHM25 level 2) to apply mass-propagation models. Three cells at the draining point of Lake Weingarten were selected as starting areas. Figure 3 shows the resulting H/L angles, calculated with the MSF and MTD models. Both results agree well, in terms of their values and in the extent of the flows shown by the models. The large flow-spread in the model, in the uppermost section below the lake, reflects the convex morphology of the moraine complex, which favours flow dispersion. Existing flow channels in the moraine (with cross-sections of about 10–20 m^2) are too small to be adequately represented in the 25 m-gridded DEM. Where the model shows spreading flow on the fan, this is comparable to the dispersion situation found at Täschalp. The modelling was, in fact, very realistic, since debris flows in the past (where the lake had not burst) had often attenuated and spread onto the fan.

Nowadays, because of channelisation to protect buildings and other structures at Täschalp from floods, the channel is confined to the orographic right side. The June 2001 event largely remained confined to the flow channel. In terms of model evaluation, an essential section starts at the apex of the fan, at Täsch, where the

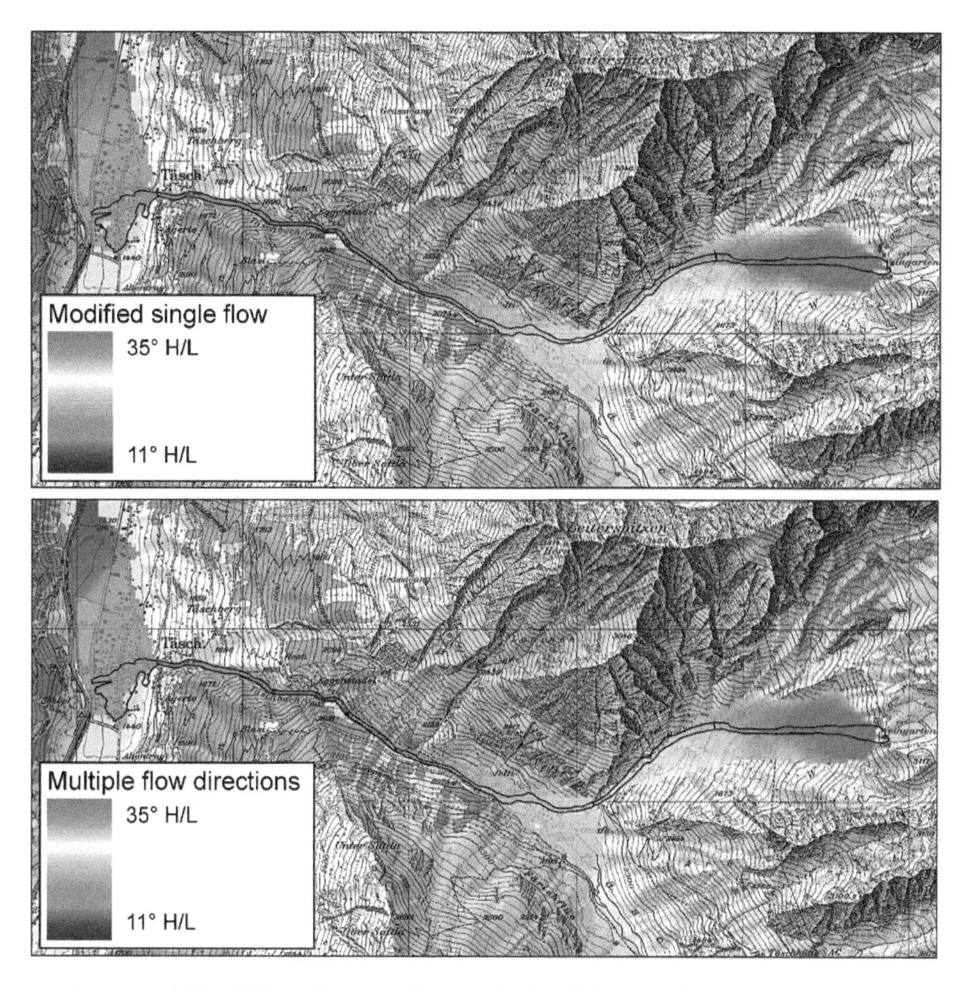

FIGURE 3 Modelling H/L angles using the MSF (top) and the MFD (bottom) models. Map and DEM reproduced by permission of SWISSTOPO (BA081244). (See page 750 in Colour Plate Section at the back of the book.)

model simulates the spread of the debris flow on the fan very well. However, the model is only of limited accuracy, since structures such as buildings, roads or bridges, which significantly influence flow behaviour, are not represented in the DEM. In the model, on the orographic right side below Täsch, the flow disperses widely, which, in the simulation, gives a relatively large affected area. While this may seem to be an error in the model, such points, in reality, may still be critical locations, as it may be the present DEM that causes them to be affected.

For the MFD deposition model, a total flow volume of $M = 50{,}000\ \mathrm{m}^3$ was assigned to source cells. Maximum deposition was defined using $D_{\mathrm{lim}} = 1.5$ m, $\beta_{\mathrm{lim}} = 30°$, and $\gamma = 0.2$. This corresponds to a deposition of up to 1.5 m in the horizontal areas, to deposition starting at slope angles of less than 30°, and to

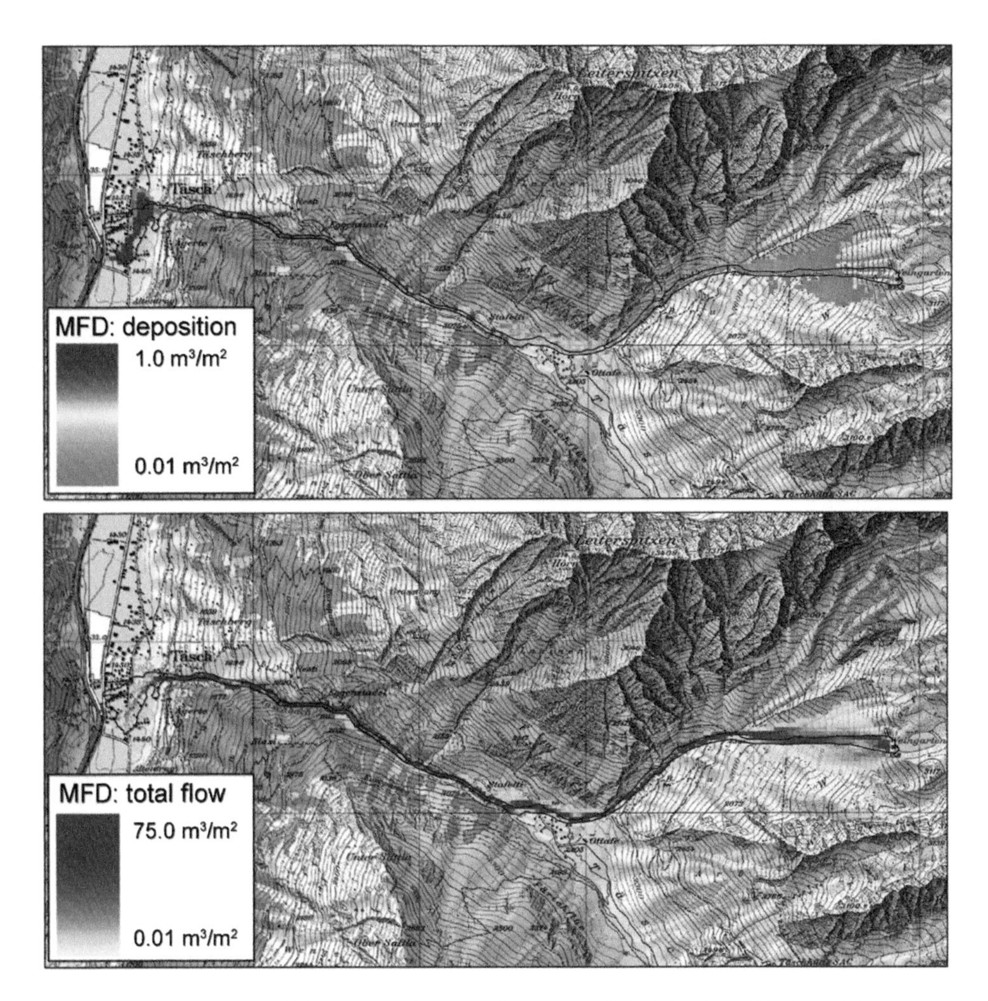

FIGURE 4 Deposition and the total volume of flow as modelled by the MFD deposition approach (map and DEM data reproduced by permission of SWISSTOPO). (See page 751 in Colour Plate Section at the back of the book.)

deposition predominantly on gentle slopes. Deposition and volume have units of length (m) because they are given in unit volume per unit area (m^3/m^2). Figure 4 shows the results of the MFD deposition model. The simulated and observed deposition patterns agree rather well (cf. Figure 2). However, comparison with Figure 3 also reveals that the areas where the flow is widely dispersed, especially at Täschalp and downstream of the village, are much smaller than that shown in the simulation.

The calculations of the H/L angle and the affected area are comparable in both the MSF and the MFD models. The method chosen is based on ease of application: if ArcInfo/ArcGIS are available then MSF is a more practical solution, but if the cost of software is an issue, then the MFD method can be implemented in other packages. The MFD software (including that to simulate deposition), described in this chapter, is provided on the website for this book. It can be run using the IDL™ Virtual Machine, which is available free of charge.

2.6 Important considerations

These techniques are useful for assessing areas affected by diverse movements of masses, or for representing them in other models. However, they should not be used uncritically. In particular, large, fast events, such as dry snow avalanches, are not represented very well by these approaches, because they neglect kinetic energy and the vertical extent of the flow. Slower and smaller events, on the other hand, are represented rather well. In many instances, uncertainties related to the size, location and probability of the event being modelled, together with poor DEM quality, will actually pose more serious limitations than performance of the model itself.

> REMARK 3. *Areas unaffected by debris flow in a model result are not necessarily without hazard; models may guide interpretation but cannot replace experienced judgement.*

3. MODELLING LANDSLIDE SUSCEPTIBILITY

3.1 Background

The increasing losses in life and property from landsliding in steepland areas have become a concern worldwide (Pike et al., 2003). Spatial modelling of the hazard can aid regional planners and other decision-makers in reducing this toll. Turner and Schuster (1996), Pasuto and Schrott (1999), Reichenbach et al. (2002), Chacón and Corominas (2003), Huabin et al. (2005), and Carrara and Pike (2008) are among recent state-of-art summaries. Morphometric analysis of the land surface is now a critical tool in extending our understanding of how topographic form controls slope failure. In fact, of all natural hazards, landsliding is perhaps the one most effectively analysed by GIS and geomorphometry.

In modelling slope failure, it is important to distinguish landslides according to two contrasting sets of environmental circumstances and resulting types of movement: *shallow* (e.g., the rapidly mobilised debris flows; Figure 2) and *deep* (various types of slower-moving slides and flows; Figure 5). The first study (Section 2.5) in this chapter modelled one type of shallow landsliding. The case study presented in this section addresses largely deep-seated landsliding and describes creation of a map that estimates likelihood of this hazard over a broad area; it demonstrates the importance of being able to combine land-surface parameters with categorical spatial information that is not obtainable by processing a DEM.

It is further helpful to distinguish two overarching approaches to modelling of the hazard posed by either deep or shallow failure: one approach treats landslides or their enclosing drainage basins as discrete landforms. Location, dimensions, volume, shape, aspect, and other quantities of individual landslides are correlated

FIGURE 5 Houses in Oakland, California, destroyed or damaged by a deep-seated landslide in 1958 after an unusually rainy winter (Oakland Tribune photo).

with substrate properties, local hydrometeorology, and other physical characteristics to isolate causative factors and model the dynamics and likelihood of failure (Jennings and Siddle, 1998; McKean and Roering, 2004; Glenn et al., 2006). Because this approach to the hazard is designed for landslides as individual landforms, it does not readily lend itself to GIS implementation over the continuous land surface and thus will not be addressed further here.

3.2 Regional landslide modelling

The second, or regional, approach to modelling the landslide hazard involves geomorphometry in a more central role; slope gradient, curvature, aspect, and other quantities computed over a continuous land surface are compared and combined, commonly with non-morphometric data, to identify areas susceptible to landslide activity. GIS technology and the availability of DEMs now enable the approximate severity of the landslide threat to be represented over large areas in the form of a hazard map. Slope-instability mapping has become a veritable *cottage industry*, and hundreds of published studies are available for guidance in modelling the hazard. We caution, nonetheless, that natural-hazard mapping is not a routine point-and-click task to be accomplished rapidly and uncritically, a misleading expectation encouraged by the growing access to DEMs and the user-friendliness of GIS software.

Both shallow and deep-seated failure can be assessed on a regional basis; before presenting the case study of deep-seated landsliding that illustrates this section, we briefly discuss regional mapping of the potential for shallow landsliding, particularly debris flows.

Because many destructive debris flows mobilise within steep concavities, their likelihood depends strongly on land-surface form and thus can be addressed largely by analysis of DEM derivatives (Wieczorek and Naeser, 2000). For example, the often-cited *SHALSTAB* model is based on spatially constant estimates of soil moisture and strength (resistance against shear stress) and reflects water flow-routing controlled by slope gradient and curvature (Dietrich et al., 1993;

Montgomery et al., 1998). The SHALSTAB procedure, based on a coupled steady-state runoff and infinite-slope stability model, creates a map of the relative steady-state precipitation needed to raise soil pore-water pressures to the level where instability is likely. Locations requiring the lowest precipitation for instability (critical rainfall) are assumed to be the most likely to fail. Low-, medium-, and high-hazard categories are assigned empirically from the frequency of actual landslide scars compiled from field observations in each range of critical rainfall on the map. SHALSTAB commonly is implemented with 10 m USGS DEMs, but performance is optimised by using 2 m LiDAR data and a critical-rainfall threshold below a range determined *a-priori* from local experience; these enhancements avoid designating an unduly large area as high hazard. SHALSTAB is not unique; similar GIS-based models, which can be parameterised for soil properties (bulk density, strength, transmissivity), include *SINMAP* (Pack et al., 2001) and *LAPSUS-LS* (Claessens et al., 2005). The latter study also notes some of the effects of DEM resolution. Finally, such non-topographic variables as vegetation type and storm-wind direction are equally important GIS inputs to modelling the location of shallow landsliding (Pike and Sobieszczyk, 2008).

REMARK 4. *Both major types of slope instability — rapid, shallow, landslides and slower-moving deep-seated landslides — can be addressed by geomorphometry.*

3.3 Case study: deep-seated landsliding in Oakland, California

Spatial forecasts of deeper-seated instability are approached somewhat differently although they, too, require an inventory of known failures for proper calibration. A landslide inventory reveals the extent of past movement and thus the probable locus of some future activity within old landslides, but not the likelihood of failure for the much larger area between them. However, existing landslides can be combined with other spatial data to create synthesis maps that show the instability hazard both in and between known landslides (Brunori et al., 1996; Cross, 1998; Rowbotham and Dudycha, 1998; Jennings and Siddle, 1998; Guzzetti et al., 2005; Van Den Eeckhaut et al., 2006). Such a map can be created by many different approaches, ranging from brute-force empiricism to a highly-parameterised physical model of slope instability. Here, we exemplify the principles of regional landslide-hazard mapping by a straightforward GIS model that is easy to understand.

Because a landslide-inventory layer is not available for the Baranja Hill site, we will demonstrate the method for a part of coastal California, where slope hazards are well developed (Figure 6). Pike et al. (2001) mapped the relative likelihood of occurrence — *susceptibility* — for the deeper (e.g., rockslide and earthflow) modes of landsliding in a large tract of diverse geology, topography, and land use centred on the city of Oakland (Figure 7). Described in abbreviated form here, the GIS model is based on the common observation worldwide that deep-seated failure reflects three dominant controls: rock and soil properties, evidence of prior slope-instability, and land-surface form. The resulting 1:50,000-scale susceptibility map of the Oakland study area and a detailed description of the method and input

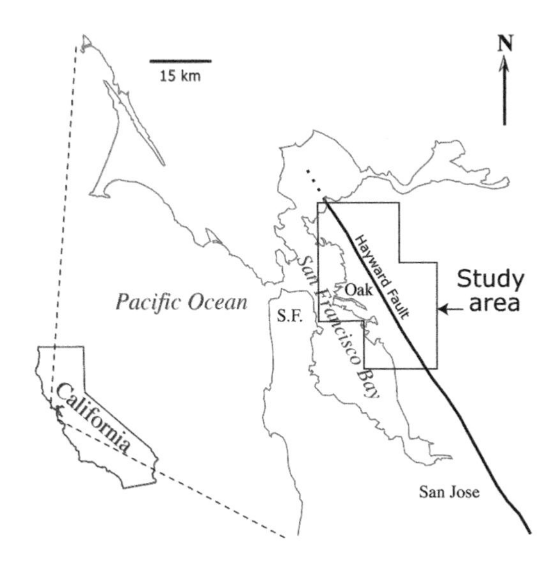

FIGURE 6 Location of the metropolitan Oakland (Oak) study area east of San Francisco (S.F.). The area of the four small representative maps in Figure 7 lies on the Hayward Fault, just east of Oakland.

data are freely available online at http://pubs.usgs.gov/mf/2002/2385/. Here, we illustrate the input data and the model results by four maps showing a small (9 km^2, Figure 7) sample of the larger (872 km^2) area.

Geology The complex geology of metropolitan Oakland is mapped as 120 diverse units, 100 bedrock formations, mostly in the hilly uplands most vulnerable to landsliding, and 20 Cenozoic surficial units in the coastal flatlands (Graymer, 2000). Twenty-five representative units are listed in Table 1 and 21 of these are shown in Figure 7(A). The varied prevalence of landsliding (e.g., mean spatial frequency, SF) with rock type and geologic structure in the Oakland area is well established (Table 1). For example, old to ancient (pre-1970) landslide deposits occupy much of the area underlain by two widespread and comparatively young geologic units that have a high clay content, the Miocene Orinda Formation (SF = 0.28) and the Briones Sandstone (SF = 0.27). Old landslide deposits are far less common in two other important units, the Oakland Conglomerate (SF = 0.01) and the Redwood Canyon Formation (SF = 0.06), both Cretaceous in age.

Prior failure Because the location of past failure is such an important clue to the distribution of future failure, maps that show old landslides as individual polygons [Figure 7(B),(C)] are essential in refining estimates of susceptibility. Brabb et al. (1972) first demonstrated that landslide inventories can be combined numerically with maps of slope gradient and geology to model susceptibility continuously over a large area. Our statistical model incorporates 6700 old landslide deposits (exclusive of debris flows and not distinguishing the degree of failure or triggering mechanism) identified and mapped in the Oakland area by airphoto interpretation (Nilsen, 1975).

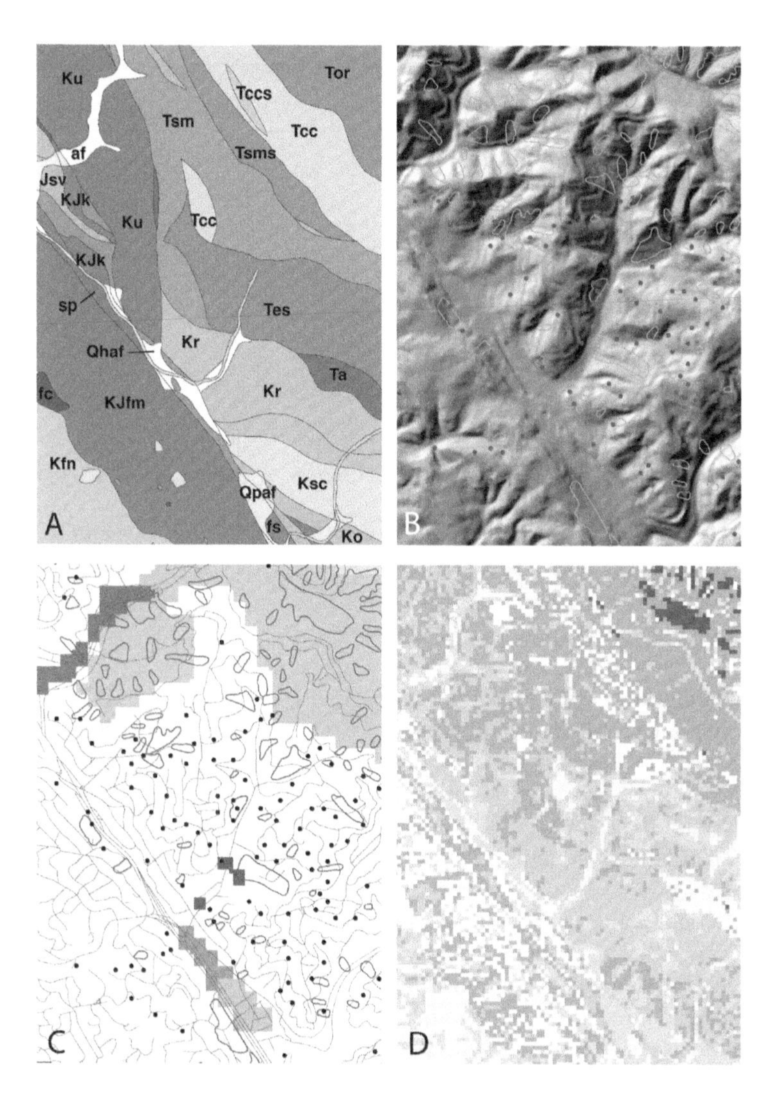

FIGURE 7 Features illustrating preparation of a landslide-susceptibility map for a part of the city of Oakland, California (Pike et al., 2001); the area shown in the four maps is about 2 km across. (A) Geology, showing 21 of the 25 map units in Table 1; the NNW-striking Hayward Fault Zone lies along the eastern edge of unit KJfm. (B) Inventory of old landslide deposits (orange polygons) and locations of post-1967 landslides (red dots) on uplands east of the fault and on gentler terrain to the west; shaded relief is from a 10 m DEM. (C) Old landslide deposits and recent landslides overlain on 1995 land use (100 m resolution): yellow, residential land; green, forest; tan, scrub vegetation; blue, major highway; pink, school; orange, commercial land; brown, public institution; white, vacant and mixed-use land; road net in grey. (D) Values of relative susceptibility at 30-m resolution mapped in eight intervals from low to high as grey, 0.00; purple, 0.01–0.04; blue, 0.05–0.09; green, 0.10–0.19; yellow, 0.20–0.29; light-orange, 0.30–0.39; orange, 0.40–0.54; red, 0.55. Low to moderate values 0.05–0.20 predominate in this 9 km^2 sample of the study area. (See page 752 in Colour Plate Section at the back of the book.)

TABLE 1 Mean spatial frequency (SF ratio) of mapped "pre-1970" landslide deposits for selected geological units (after Graymer, 2000) in metropolitan Oakland; the 21 units accompanied by a symbol appear on the map in Figure 7(A)

Symbol	Geologic map unit	30 m grid cells		
		All	Landslides	Ratio
	Neroly Sandstone (uncertain)	1786	1120	0.63
	Siesta Formation — mudstone	5862	2937	0.50
	unnamed Tertiary sedimentary & volcanic rocks	99,233	35,956	0.36
Tccs	Claremont Chert — interbedded sandstone lens	239	72	0.30
Tor	Orinda Formation	35,166	9682	0.28
	Briones Sandstone — sandstone, siltstone, conglomerate, shell breccia	32,548	8723	0.27
sp	serpentinite — Coast Range ophiolite	3183	720	0.23
KJfm	Franciscan melange (undivided)	12,212	2559	0.21
Tsm	unnamed glauconitic mudstone	3389	438	0.13
Tsms	unnamed glauconitic mudstone — siltstone & sandstone	362	46	0.13
Tcc	Claremont Chert of Graymer (2000)	10,590	1177	0.11
Ksc	Shephard Creek Formation	5675	508	0.09
KJk	Knoxville Formation	8164	663	0.08
Jsv	keratophyre & quartz keratophyre above Ophiolite	15,627	1212	0.08
Ku	Great Valley Sequence — undifferentiated	12,706	965	0.08
Kr	Redwood Canyon Formation	27,503	1697	0.06
Tes	Escobar Sandstone (Eocene)	2513	141	0.06
fs	Franciscan sandstone	3441	109	0.03
Ta	unnamed glauconitic sandstone	163	3	0.02
Qpaf	alluvial fan & fluvial deposits (Pleistocene)	61,867	1010	0.02
Kfn	Franciscan — Novato Quarry terrain	7879	122	0.02
Ko	Oakland Conglomerate	20,921	301	0.01
fc	Franciscan chert	323	1	0.00
af	artificial fill (Historic)	65,934	15	0.00
Qhaf	alluvial fan and fluvial deposits (Holocene)	125,014	254	0.00

Land-surface form The steep upland interior of metropolitan Oakland hosts many old landslides [Figure 7(B)], while its flat coastal lowland has few old landslides (the densely settled area, shown in yellow in Figure 7(C), does have many small recent failures in terrain graded for development). The diagnostic geomorphic features that reveal the presence of deep-seated failure translate into few geomorphometric measures from which hazard maps can be prepared. Slope stability does,

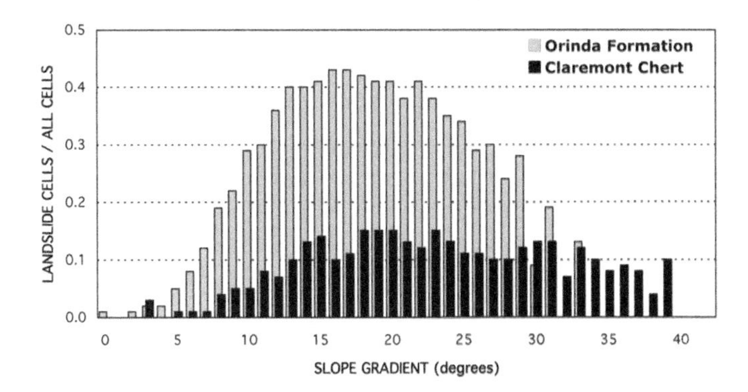

FIGURE 8 Contrast in landslide susceptibility of two geologic units in Oakland, California, shown by spatial frequency of prior failure. Number of 30 m grid cells on old slide deposits/all cells in unit, as a function of slope gradient in 1° intervals. The Claremont Chert (black) is less susceptible than the Orinda Formation (grey). Compare mean values in Table 1.

however, vary importantly with slope gradient; the spatial frequency of landsliding does not increase linearly with gradient for most rock types but rather peaks at intermediate values of slope and declines thereafter (Figure 8). To represent the role of surface geometry in deep-seated landsliding, a slope-gradient value computed from a 30 m DEM was assigned to each digital-map grid square.

The spatial likelihood of future landsliding in metropolitan Oakland was modelled by gridding digital-map databases of geology, landslide deposits, and slope gradient in the ArcInfo GIS at 30 m resolution and combining them statistically by a series of commands programmed as an Arc/Info macro. The resulting index of susceptibility, output as a seven-colour map [Figure 7(D)] (Pike et al., 2001), was computed as a continuous variable over the large (872 km^2) test area at the grid spacing of the DEM. The model further improves upon raw landslide inventories and other types of susceptibility maps by distinguishing, respectively, the degree of susceptibility between and within existing landslide deposits. *Susceptibility* is defined as the spatial frequency of terrain occupied by old landslide deposits (Table 1), adjusted locally by steepness of the topography; the key operational tool is an Info VAT (Value Attribute Table) file, created by the macro for each geologic-map unit, that tabulates the percentage of grid cells that lie on a mapped landside for each one-degree interval of slope gradient (e.g., Figure 8).

Susceptibility S for grid cells located on terrain between the old slide deposits (88% of the study area) is estimated by the ArcInfo macro directly from the (characteristically) bell-shaped distributions of spatial frequency arrayed by slope gradient for each of the 120 geologic-map units. In the Orinda Formation, for example, where 29% of the 30 m × 30 m cells sloping at 10° are located on old landslide deposits (Figure 8), all other cells in the same unit with a slope of 10° are assigned that same susceptibility S, of 0.29. In the less-susceptible Claremont Chert (Figure 8), by contrast, only 5% of the cells in the 10° slope interval lie on mapped slide masses, whereupon an S of 0.05 is assigned to all remaining 10° cells in the

Claremont. Values of S, determined slope interval-by-one-degree-slope interval, are unique to each value of slope gradient in each of the 120 units. Values range from $S = 0.00$ for 300,000 cells in predominantly flat-lying Quaternary units to $S = 0.90$ for 14 cells in the most susceptible (but quite small) hillside formation.

Existing landslide deposits are known to be less stable than the unfailed terrain between them; accordingly, susceptibility within old landslide deposits is refined further as $S_{ls} = S \times$ a multiplier (here 1.33) derived from the relative spatial frequencies of recent (post-1970) failures (here numbering 1192) within and outside old deposits [Figure 7(B),(C)]. Obtaining susceptibility S_{ls} for the much smaller fraction of the Oakland area that is in landslide deposits is more complex. First, raw susceptibility S was calculated for the 116,360 cells within the deposits, by the same procedure as for cells between them. The highest S on landslide masses is 1.00, for 70 scattered cells that occur in 21 different geologic units. To estimate the higher susceptibilities that characterise dormant landslide deposits S_{ls}, these 116,360 values of S were multiplied by a factor a, based on the relative frequency of recent failures in the region:

$$a = \frac{\frac{\#hist_{ls}}{A_{ls}}}{\frac{\#hist_{nls}}{A_{nls}}} \tag{3.1}$$

where $\#hist_{ls}$ and $\#hist_{nls}$ are the numbers of recent failures within and outside old landslide deposits, respectively, and A_{ls} and A_{nls} are the areas (in number of cells) of old deposits and the terrain between them. This correction, $(183/116{,}360)/(1009/852{,}643) = 1.33$, indicates that recent landslides in the Oakland area are about 1/3 more likely to occur within old landslide deposits than on terrain between them. Lacking historic documentation of landsliding for each geologic unit, the 1.33 multiplier is applied uniformly to all 120 units. The highest value of S_{ls} is 1.33, for the same 70 cells mentioned above. The susceptibilities are expressed as decimals rather than percentages.

> REMARK 5. *Some slope-instability problems can be analysed using DEM data alone, but others require non-DEM information.*

All grid-cell values of S and S_{ls}, from zero to 133, were combined to create the map sampled in Figure 7(D); the susceptibility range was divided into seven segments suggested by the shape of the combined frequency distribution (not shown here) and a colour, from grey to red, assigned to each. The strong influence of geology on the resulting map is evident in the good correlation of high susceptibility with the Orinda Formation, unit Tor, Franciscan melange, unit KJfm, and the sandstone lens within the Claremont Chert, unit Tccs (Figure 8); the importance of slope gradient can be seen in the variation in susceptibility within each geologic unit.

Comparison of Figure 7(D) with a 1995 map of land use, Figure 7(C), reveals that 8% of the residential housing in the entire Oakland area, and a substantial 15% in its hilly uplands, occupies terrain where predicted susceptibility exceeds a relatively high 0.30 (compare with the mean values for geologic units in Table 1).

The susceptibility map (Pike et al., 2001) offered an added tool to assist in planning further development and zoning of hillside environments in the greater Oakland metropolitan area; it has been incorporated into the Disaster Mitigation Plan of the adjacent city of Berkeley.

Positive results from two evaluations of the model, not described here, suggest that it is appropriate for wider use. While the model can be applied anywhere its three basic ingredients — geology, prior failures, and slope gradient — exist as digital-map databases, its results could be improved by using more recent and detailed landslide inventories and slope data and by adding parameters that better predict recent failures in developed areas. Further predictive power may reside in such attributes as seismic shaking, distance to the nearest road (a measure of human modification of the landscape), and slope aspect (Pike and Sobieszczyk, 2008). Other, more complex, models of susceptibility to deep-seated landsliding are described in recent papers referenced in this chapter.

> REMARK 6. *In addition to an accurate DEM of an area, an important input to a slope-instability or landslide-susceptibility model is a map of prior failures.*

3.4 Important considerations

Hazard maps created from morphometrically-supported models, regardless of their sophistication, must not be uncritically published or applied to landslide hazards mitigation. Areas of high susceptibility in Figure 7(D), while more likely to fail than locations with low values, also include local occurrences of scattered 30 m cells that are not hazardous. More important for public safety, most low-susceptibility areas on the map are less prone to failure than areas of high value but are not without landslide hazard. Some of these locales slope steeply and are subject to debris flow and other types of failure — small landslides <60 m across, common in the area, were not included in the inventory [Figure 7(B),(C)] on which Figure 7(D) is based. Landslide prediction also remains something of an art, and the locus of much future landsliding cannot be identified with confidence; slopes commonly fail from unanticipated blocking of surface drainage or other consequences of hillside development, as well as from random variation in the operation of landslide triggers and slope processes. Compiling Figure 7(D) at a resolution coarser than 30 m might present a more actualistic picture of some of these uncertainties.

Other potential drawbacks to the approach demonstrated here, as well as to other regional models, include the need for accurate digital-map information on geology, topography, and prior failure over a large area. For example, the landslide inventory used here was not field-checked, and the contour maps from which the 30 m DEM was extracted were surveyed before most of the residential development of the Oakland Hills, thus omitting much cut-and-fill modification of the original land surface. Finally, susceptibility maps show the relative importance of landsliding and thus overall stability but are only a guide to the likelihood of future movement. Where grading for development and construction is contemplated, susceptibility maps can not substitute for a detailed site report by a qualified soil engineering geologist or soils engineer.

Although this chapter mainly treats slope instability due to snowmelt or high rainfall, geomorphometry is equally important to mapping the likelihood of slope failure triggered by earthquake and volcanic eruption. Seismically-induced landslides are a major hazard in all parts of the world where active geological faults coincide with steep topography. Characteristic failures are rock falls and topples as well as debris flows and various translational and rotational slides; portions of large deep-seated landslides may be reactivated. Because the landslide process must be coupled with ground motion resulting from an earthquake, thus requiring linkage of two disparate models, creating a susceptibility map is more complicated than for rainfall-induced landsliding (Jibson et al., 1998; Capolongo et al., 2004; Lee and Evangelista, 2006). For example, parameters of the seismic event may be combined with such site factors as slope gradient, aspect, and curvature; distances from the nearest active fault, road, and drainage; and data on local rock types and moisture content. A hazard map by Miles and Keefer (2001) for seismically-induced landsliding (http://pubs.usgs.gov/mf/2001/2379/) includes the area shown in Figure 7. Finally, lahars are fast-moving highly fluid mudflows of pyroclastic materials and water, commonly derived from glacial ice melted by a volcanic eruption. Their flowpaths and zones of deposition can be modelled by DEMs and menu-driven ARCINFO GRID software[1] (Schilling, 1998; Iverson et al., 1998).

REMARK 7. *Models predicting the location of slope instability and mass-propagation require careful calibration to optimise accuracy.*

4. SUMMARY POINTS

Geomorphometry provides analytical tools and generates DEM-derived data that have revolutionised the spatial modelling of slope failure and mass movements. The resulting large body of findings has advanced our understanding of landslide processes in the natural environment and is being applied to the protection of human life and property. The accompanying literature is growing rapidly in volume and sophistication. An important objective of this research is its inclusion in plans adopted by government officials for land-use zoning and development, at both local and regional scales.

Spatial models of landslide likelihood based on geomorphometry need to be checked for accuracy, because any output map claiming to represent hazard potential carries implications for land-use policy and public safety. All-too-commonly the statistical testing of a landslide model is referred to by the misleading term-of-art *validation*. However, such models of natural systems cannot be *validated* in the True/False sense implied by this term; rather, a model can only be *evaluated* — for its internal goodness-of-fit or suitability for a particular application (Oreskes

[1] Available at http://pubs.er.usgs.gov/usgspubs/ofr/ofr98638/.

et al., 1994; Zaitchik et al., 2003). Such assessments for a spatial model of slope failure usually involve dividing a sample of existing landslides by map location or year of failure (preferred) and running the model separately on both subsets, or by applying the model to nearby landslides in an environmentally similar area. The truest test, of course, remains a future landslide event that generates a fresh population of failures in the same area.

We close this chapter with brief appraisals of the GIS models that illustrate the two general topics highlighted in this chapter, downslope spread of material released in alpine mass-movements and the mapping of landslide susceptibility in hilly topography.

The propagation of mass movements in mountainous areas can be modelled by GIS in many ways; existing approaches include the specific as well as the general and are based on different principles. The parametrisation methods demonstrated here for modelling single and multiple flow-direction routing and deposition have the drawback of neglecting transport dynamics but offer many clear advantages, among them:

- rapid implementation/easy availability;
- operation with minimal input data (e.g., in remote regions); and
- fast computational evaluation that allows investigation of the typically great uncertainties in the model input.

Various regional models of the landslide hazard have been proposed for GIS implementation, and each has its strong and weak points. The multivariate approach described here for mapping susceptibility to deep-seated failure is empirical and straightforward. Shortcomings of the method involve quality and availability of input data rather than robustness of the model itself, which has the following conceptual advantages:

- it can be implemented quickly over a large area, limited only by the extent of the input data;
- method, data, and areal coverage all are 100% quantitative;
- spatial resolution can be as fine as that of the DEM (if high resolution makes sense);
- the susceptibility index is a continuous variable with a range of values within, as well as between, existing landslides;
- the model is *transparent* rather than *black-box*: values of the index can be related directly to field observations; and
- the method is portable; it applies anywhere the necessary data are available.

IMPORTANT SOURCES

Carrara, A., Pike, R.J. (Eds.), 2008. GIS Technology and Models for Assessing Landslide Hazard and Risk. Geomorphology 94 (3–4) 257–507.

Huabin, W., Gangjun, L., Weiya, X., Gonghui, W., 2005. GIS-based landslide hazard assessment. Progress in Physical Geography 29 (4), 548–567.

Chacón, J., Corominas, J. (Eds.), 2003. Special Issue on Landslides and GIS. Natural Hazards 30 (3) 263–499.

Reichenbach, P., Carrara, A., Guzzetti, F. (Eds.), 2002. Assessing and Mapping Landslide Hazards and Risk. Natural Hazards and Earth System Sciences 2. Copernicus Publications, 117 pp. (special issue).
Turner, A.K., Schuster, R.L. (Eds.), 1996. Landslides: Investigation and Mitigation. Transportation Research Board, Special Report 247. National Research Council, Washington, DC, 685 pp.

CHAPTER 24

Automated Predictive Mapping of Ecological Entities

R.A. MacMillan, A. Torregrosa, D. Moon, R. Coupé and **N. Philips**

ecological zones and land classification · the concept of predictive ecosystem mapping (PEM) and its relation to traditional ecological land classification · a case study using the Baranja Hill data set that demonstrates one approach to implementing predictive ecosystem mapping · the benefits and drawbacks of automated predictive ecosystem mapping · developing and applying predictive ecosystem mapping methods · future prospects

1. INTRODUCTION

1.1 Concepts of conventional manual ecological land classification

Ecological land classification is defined as *"a cartographical delineation of distinct ecological areas, identified by their geology, topography, soils, vegetation, climate conditions, living species, water resources, as well as anthropic factors"* (from http://wikipedia.org). The worldwide shift among land and resource agencies towards managing land resources in a more holistic manner has led to a movement that favours delineating landscape units on the basis of their combined *structure* and *function* (Holling, 1992; Bailey, 2002; Omernick, 2004). In ecological land classification, the overall intent is to integrate ecological processes such as fire, flooding, and biotic interactions into spatially defined units. The rationale for moving from producing separate, but related, land-surface interpretations, such as land-cover and vegetation mapping to mapping integrated ecological entities using the principals of ecological land classification (Maybury, 1999; Gallant et al., 2005) has been driven by the needs of management worldwide, to anticipate, and put in place, measures that will provide resilience in the face of temporary, dynamic changes in the landscape.

Ecological classifications are almost universally hierarchical with different *ecological entities* recognised at different scales based on various criteria (Table 1).

Developments in Soil Science, Volume 33 © 2009 Elsevier B.V.
ISSN 0166-2481, DOI: 10.1016/S0166-2481(08)00024-X.

TABLE 1 A comparison of several hierarchical ecological classification systems

Australia	Britain	Canada	USSR	USA
		Ecozone	Zone	
	Land Zone	Ecoprovince		Domain
	Land Region	Ecoregion	Province	Province
	Land District	Eco-district		Section
			Landscape	
Land System	Land System	Eco-section		District
	Land Type	Ecosite	Urochishcha	Landtype Association
Land Unit				
Land Type	Land Phase			Landtype
Site		Eco-element		Landtype Phase
			Facia	Site

Based on Bailey (1981), Klijn and Udo de Haes (1994).

The similarities and differences in management needs for maps of ecological entities have given rise to broadly similar systems of ecological land classification. Paradoxically, however, the various systems have utilised different delineation algorithms, resulting in overlapping mapping concepts and mapped entities. The *Global Earth Observations Ecosystem Mapping Task* is currently trying to eliminate this lack of congruency, by creating a world-wide compilation of ecological entities (see http://earthobservations.org).

The *Canadian System of Ecological Land Classification* (Wiken and Ironside, 1977), and regional modifications thereof, are used in Canada. Land Systems Classification has been widely applied in Australia (Christian and Stewart, 1968). In the U.S., the U.S. Forest Service uses the National Hierarchical Framework of Ecological Units (NHFEU) (Cleland et al., 1997) to classify and map forested and public lands and The Nature Conservancy uses it for setting regional conservation targets (Bow et al., 2005; Nachlinger et al., 2001). At the national level, the *Ecological System Classification* developed by *NatureServe* (Comer et al., 2003) is used by several U.S. programs. These include the GAP Analysis program (http://gapanalysis.nbii.gov) that provides data to support the protection of biological integrity; the U.S. Interagency *LANDFIRE* program (http://www.landfire.gov) that provides support for researching and managing the problem of large landscape fires; and the recently initiated U.S. *Geological Survey National Ecosystem Mapping Initiative* (http://geography.wr.usgs.gov/science/ecosystem.html).

For the U.S. National Water-Quality Assessment Program, an alternative classification based on hydrological-landscape regions (Wolock et al., 2004) is used. The stratification *Biogeographical Provinces of the World* developed for UNESCO's 'Man and the Biosphere' program identifies eight levels of ecological entity, ranging from the largest to the smallest: *eco-zone, eco-province, eco-region, eco-district, eco-section, eco-site, eco-tope, eco-element* (from wikipedia.org). In all of these, ecological units are identified and differentiated based on specified combinations of physical and biological characteristics, which may include climate, geology, geomorphology, soils, hydrology, or potential natural vegetation.

Ecological units at each spatial scale are nested within the broader scales. A crucial concept of ecological land classification is that each of the areas defined is assumed to be relatively stable and not subject to large, sudden changes. Such entities should either remain the same over a certain period of time, or, if they show change, then it should be slow and gradual. Additionally, the pragmatic application of ecosystem concepts recognises that ecosystems are geographically discrete entities; and that their implied levels of resolution more or less coincide with the levels at which management and human intervention typically occur (Sims et al., 1996).

It is beyond the scope of this chapter to review and compare the various systems of ecological land classification. Relatively recent reviews are provided by Thomson et al. (2004), Froude and Beanland (1999), Sims (1992), Russell and Jordan (1992), Klijn and Udo de Haes (1994). The U.S. *National Hierarchical Framework of Ecological Units* (NHFEU) (Cleland et al., 1997) is used here as an example to illustrate the size and scale at which different levels in an hierarchical classification operate and the principal criteria most often used to differentiate ecological units at various scales (Table 2). This example was selected to emphasise the importance placed on landform, topography, slope gradient, aspect and slope position as key criteria for differentiating the lower level units in the system. Most other systems also use topography as a key determinant for identifying ecological units at finer resolutions and use interpretation of recurring topographic patterns to recognise ecological entities at intermediate scales. In the context of this book, the use of topography and derivatives of topography to define ecological units is significant.

1.2 Concepts of predictive ecological mapping

In many jurisdictions, concerns with high costs, slow rates of progress, and uncertain levels of predictive accuracy of ecological maps produced using conventional, manual methods of classification and mapping have resulted in a desire to develop alternative, automated methods of ecological mapping. This has led to the development of the concept of *Predictive Ecosystem Mapping* (PEM). This has been defined by Jones et al. (1999) as *"a computer, GIS and knowledge-based method of stratifying landscapes into ecologically oriented map units based on the overlaying of existing mapped themes and the processing of the resulting attributes by automated inferencing software using a formalised knowledge-base containing ecological-landscape relationships"*. PEM mapping is simply one variant of a larger, worldwide trend towards utilising predictive models, applied to available digital data-sets, to predict the most likely spatial distribution of site conditions, ecological classes or soils rapidly and cost effectively.

Predictive ecological mapping is based on the general understanding that key ecological differences are largely controlled by environmental gradients in moisture, energy and nutrients across the landscape, many of which operate under the influence of topography (see e.g. Aspinall and Veitch, 1993; Band et al., 1991, 1993; Coughlin and Running, 1989; Moore et al., 1993b; Skidmore, 1989b; Skid-

TABLE 2 Design criteria, mapping scales and map unit size of ecological units of the U.S. National Hierarchical Framework of Ecological Units

Ecological unit	Map scale	Size of map unit	Principal design criteria for map units
Domain	1:30M	1,000,000's square miles	Broad climatic groups or zones (e.g. dry, humid, tropical)
Division	1:7.5–1:30M	100,000's square miles	Regional climatic types, vegetational affinities (e.g. prairie or forest), Soil Order
Province	1:5–1:15M	10,000's square miles	Dominant potential natural vegetation, highlands or mountains with complex vertical climate–vegetation–soil zonation
Section	1:3.5–1:7.5M	1,000's square miles	Geomorphological province, geological age, stratigraphy, lithology, regional climatic data, Phases of Soil-Order, Suborders or Great Groups, potential natural vegetation, potential natural communities (PNC)
Subsection	1:250k–1:3.5M	10's to low 1,000's square miles	Geomorphological process, surface geology, lithology, Phases of Soil-Order, Suborders or Great Groups, subregional climatic data, PNC — formation or series
Landtype Association	1:60–1:250k	1,000's to 10,000's acres	Geomorphological process, geological formation, surface geology and elevation, Phases of Soil-Subgroup, families or series, local climate, PNC — series, subseries, plant associations
Landtype	1:24–1:60k	100's–1,000 acres	Landform and topography (elevation, aspect, slope gradient and position), Phases of Soil-Subgroups, families or series, rock type, geomorphological process, PNC — plant associations
Landtype Phase	1:24k or larger	<100 acres	Phases of soil subfamilies or series, landform and slope position, PNC — plant associations or phases

Based on Cleland et al. (1997).

more et al., 1991, 1996; Twery et al., 1991). This supports the assumption that the spatial distribution of different types of ecological classes can be predicted, economically and accurately, by formalising existing ecological knowledge into basic rules. These fundamental rules can then be applied to digital data-sets to predict the most likely ecological classification at any given location.

There are numerous examples of efforts to predict the spatial distribution of ecological units using largely automated procedures. Coughlin and Running (1989) extracted ecological units based on topographically derived spatial entities in combination with information on climate, soil and vegetation. Twery et al. (1991) predicted the spatial distribution of different vegetation-cover types described in terms of the compositions of species. Their analysis was based largely on measures of slope position and aspect derived from a DEM. Similarly, Moore et al. (1991b) applied decision tree analysis to a suite of environmental variables, largely derived from analysing digital elevation data, to model and map ecosystems defined in terms of the distribution of vegetation classes. Skidmore et al. (1991) predicted the spatial pattern of five forest soil–landscape units based on consideration of forest cover, slope gradient, wetness index and a measure of slope position.

Burrough et al. (2001) applied an unsupervised fuzzy *k*-means classification technique to eight topo-climatic attributes computed from a DEM. This enabled them to extract a number of topo-climatic classes automatically; specifically: valley bottoms, drainage channels, lower slopes, ridges, north-facing steep slopes, south-facing steep slopes and lakes. Manis et al. (2001) used a similar set of measures — relative slope position, slope gradient, slope shape and slope aspect — to delineate ecologically predictive landform classes automatically, these being: valley flats; gently sloping toe slopes, bottoms and swales; gently sloping ridges, fans and hills; nearly level terraces and plateaus; very moist steep slopes, moderately moist steep slopes; moderately dry steep slopes; very dry steep slopes; cool aspect scarps, cliffs and canyons; and hot aspect scarps, cliffs and canyons. Blaszczynski (1997) used an analysis of slope gradient and curvatures computed within windows of increasing dimensions to segment landscapes into: convex crests and spurs; concave troughs, side slopes, open and closed basins and sloping or horizontal flats with a view to representing the effects of unique sets of geomorphic processes on ecosystem processes in the landscape. Recently, Barringer et al. (2006) classified areas of steep terrain in New Zealand into landform element classes to support predictive mapping of the distribution of soils in these steep landscapes. These classes were based on consideration of local surface shape in combination with a measure of higher-level landscape context. The list continues. These few selected examples demonstrate how ecological conditions and ecological site types have often been conceptualised and mapped in similar ways using similar methods and roughly similar inputs, many derived by analysing a DEM.

Figure 1 illustrates the conceptual similarities between (a) processes used to infer ecological mapping entities based on consideration of vegetation cover, terrain and topographic features, and (b) processes used to infer the spatial distribution of soil types, using the widely cited CLORPT model (McBratney et al., 2003). In this depiction, information about vegetation cover, where required, usually has to

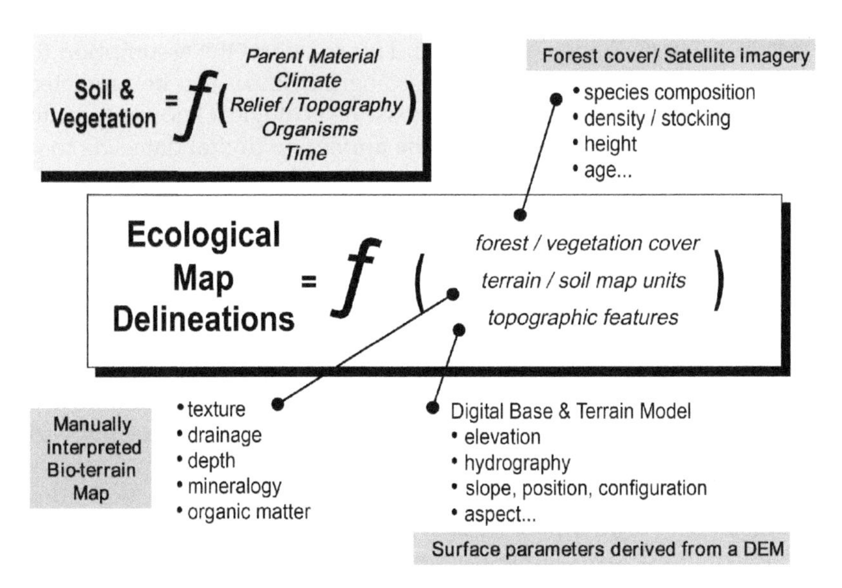

FIGURE 1 A conceptualisation of the predictive ecosystem mapping (PEM) process. Based on Jones et al. (1999).

be obtained either from existing maps of forest cover, or inferred from remotely sensed imagery. Information on soil and terrain properties, such as texture, depth and mineralogy is still often obtained from manually prepared, secondary-source, maps, although conceivably, it could also be obtained by analysing remotely acquired digital data-sets.

Almost all the information required to provide measures of elevation, hydrography, slope position, slope shape, aspect and other topographical attributes can be extracted by analysing a digital elevation model. Predictive ecosystem mapping (PEM) is therefore seen as a very close analogue of predictive soil mapping, which is based on applying the soil–landscape model (see Northcote, 1984; Arnold, 1988; Swanson, 1990a; Zhu et al., 2001; Bui and Moran, 2003; Hengl and Rossiter, 2003; McBratney et al., 2003; Bockheim et al., 2005; Qi et al., 2006). Although ecological site types are often defined more broadly, and are usually described using a different vocabulary, the conceptual and physical entities of both predictive soil mapping and predictive ecological mapping tend to be very similar.

2. CASE STUDY: PREDICTIVE ECOSYSTEM MAPPING

This section describes a case study based on procedures used to produce predictive ecosystem maps (PEM) at a scale of 1:20,000. These maps cover over 3.5 million ha of the 8.5 million ha of forested land in the former Cariboo Forest Region of British Columbia, Canada. PEM maps for the remaining 5.0 million ha were completed in early 2008. For consistency and comparison, the methods and results illustrated here use the small Baranja Hill data-set — the data-set that has been used throughout this book. Data for the actual Cariboo Forest Region study area are only referenced in certain instances, where the size or attributes of the Baranja Hill data are not suitable for effectively illustrating or explaining the procedures or results. Several of the input-data layers used for the Baranja Hill area are purely notional. They were invented to illustrate the methods and concepts actually used in the Cariboo study area.

The *Biogeoclimatic Ecosystem Classification* (BEC) system of site classification (Pojar et al., 1987) was used for this exercise. This system has been officially adopted and widely applied for stratifying and mapping the land resources of the province of British Columbia according to standards specified by the regulatory authorities of that province (Resource Inventory Committee (RIC), 1999, 2000). All digital data-sets for the former Cariboo Forest Region, used to produce PEM maps, were collated and co-registered as regular raster grids with a horizontal resolution of 25 m. These grids matched the finest horizontal resolution that could be supported by the available DEM data. The Baranja Hill data-set used here has a similar 25 m grid resolution. The approximate mapping scale for both data sets is therefore considered to lie within the range: 1:25,000 to 1:50,000.

The landscape-based approach to ecological land classification, illustrated in this chapter, is a form of pattern recognition based on ecological theory. The unit areas delineated represent hypotheses that arise from expert heuristic knowledge of what has been deemed ecologically important for land management. Units defined for mapping are based on local expert knowledge of the interactions and controlling influences of the structural components of ecosystems, as understood by local experts and summarised in *"A Field Guide to Forest Site Identification and Interpretation for the Cariboo Forest Region"* (Steen and Coupé, 1997). This field guide captures knowledge and expertise on how to apply the BEC system of site classification to this local region.

2.1 Methods

The methods described here (see Figure 2) comprise a hybrid of automated, semi-automated and manual procedures that develop and apply heuristic, rule-based conceptual models of ecological–landform and soil–landform relationships in a manner similar to the CLORPT or SCORPAN approaches, as described by McBratney et al. (2003). The procedures attempt to directly parallel, or mimic, the logic and decision-making process followed by local ecological experts.

STEP 1. *Identify and characterise the mapping entities*

The first step in the mapping process is to identify, list and describe the spatial entities that are to be predicted for each unique ecological zone or classification region. In the operational mapping applications carried out to date, this process has made use of a well-developed and clearly documented system of classification, as presented in the Field Guide for the former Cariboo Forest Region (Steen and

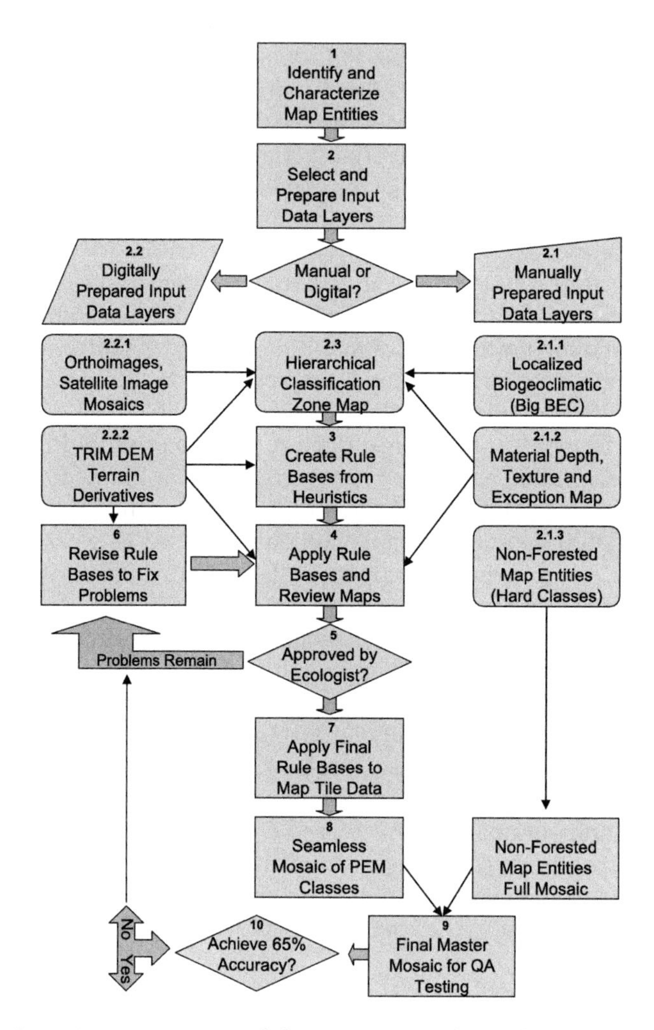

FIGURE 2 A schematic representation of the main steps in the predictive ecosystem mapping (PEM) process.

Coupé, 1997). For the application presented here, classes have been invented to simplify the illustration.

By definition, ecological mapping is hierarchical, with different kinds and types of entities identified at successively finer levels of stratification. In the operational mapping applications in British Columbia considerations of regional climate and vegetation were used to define higher-level, regional entities, referred to as Biogeoclimatic Ecosystem Classification (BEC) zones, subzones and variants.

For illustrative purposes, we assume that the area covered by the small Baranja Hill data-set includes two different ecological subzones characterised by significant differences in both climate and physiography. These two subzones are defined as the flood-plain area and the hill area. Within each of these two main areas, a different set of topographically-controlled ecological sub-classes was identified and defined (Table 3).

TABLE 3 Hypothetical ecological classes defined for the small Baranja Hill data-set

Ecological subzone	Class no.	Class name	Description of the class
Flood Plain	1000	Water	Permanent water (mapped manually)
	1001	Up_Convex	A drier convex surface on upper portions of the flood plain
	1002	Up_Planar	A planar surface on upper portions of the flood plain
	1003	Up_Concave	A wetter concave surface on upper portions of the flood plain
	1021	Mid_Convex	A drier convex surface on mid-slope portions of the flood plain
	1022	Mid_Planar	A planar surface on mid-slope portions of the flood plain
	1023	Mid_Concave	A wetter concave surface on mid-slope portions of the flood plain
	1031	FP_Convex	A drier convex surface on lower portions of the flood plain
	1032	FP_Planar	A planar surface on lower portions of the flood plain
	1033	FP_Concave	A wetter concave surface on lower portions of the flood plain
Hills Medium	2001	Hill_LCR	A relatively level, relatively dry crest along the top of a ridge or hill
	2002	Hill_DSH	A relatively dry and convex divergent shoulder of a ridge or hill
	2003	Hill_USW	A concavity, or swale, with relatively more moisture in an upper landform position
	2004	Hill_PBS	A relatively planar back slope with little across-slope curvature
	2005	Hill_DBS	A slightly drier divergent spur in a mid to back-slope landform position
	2006	Hill_CBS	A convergent swale, or draw, with slightly more moisture in a mid to back-slope position
	2007	Hill_TER	A relatively level step or terrace in a mid to back-slope landform position

(*continued on next page*)

TABLE 3 (*continued*)

Ecological subzone	Class no.	Class name	Description of the class
	2008	Hill_SAD	A mid-slope saddle, convergent in one direction and divergent in the other
	2009	Hill_ MSW	A moister draw, hollow or swale in a mid-slope landform position
	2010	Hill_CFS	A moister convergent foot slope that is concave in the down-slope direction
	2011	Hill_CCH	A very moist convergent lower slope adjacent to a channel or valley bottom
	2012	Hill_PTS	A relatively planar toe slope, not as concave as the convergent foot slope
	2013	Hill_FAN	A lower slope position that is divergent in plan and planar in profile
	2014	Hill_LSM	A drier lower slope position that is convex (shedding) in profile and plan
	2015	Hill_LLS	A lower slope position that has a low slope gradient and low curvatures
Hills Coarse	2024	CRS_PBS	A relatively planar back slope developed on coarse-textured materials
	2025	CRS_DBS	A slightly drier divergent mid slope developed on coarse-textured materials
	2026	CRS_CBS	A slightly moister convergent mid-slope, developed on coarse-textured materials
	2027	CRS_TER	A relatively level mid-slope terrace, developed on coarse-textured materials

These ecological classes illustrate concepts of hierarchical ecological classification for three different classification domains: Flood Plain; Hills — medium; Hills — coarse.

STEP 2. *Select and prepare input-data*

The second step in the mapping process is to select and prepare a number of input-data layers to use in the predictive process. Each input-data layer is selected in the belief that it can act as a surrogate for a particular aspect of a conceptual ecological–landform model. For convenience, we distinguish between manually (Table 4) and automatically (Table 5) and prepared input layers.

The map of localised biogeoclimatic subzones is used to define the spatial extent of higher-level hierarchical entities that are influenced by a particular set of climatic (moisture and energy), geological and physiographic conditions, which, in turn, reveal the presence of a limited and defined sequence of ecological-site types at a finer level of resolution. Only a limited number of ecological-site types are thought to exist within any given biogeoclimatic subzone. These site types are

TABLE 4 Manually interpreted and digitised input layers used in predicting ecological classes for both the small Baranja Hill data-set and in the former Cariboo Forest Region

No.	Name	Description
2.1.1	Localised biogeo-climatic subzones	A manually interpreted and digitised map depicting the extent of higher level ecological subzones within which a characteristic climate and physiography produce a specific, definable, modal vegetation community
2.1.2	Material depth, texture and exception mapping	A manually interpreted and digitised map depicting areas that deviate from the expected conditions with respect to soil depth, texture or moisture status.
2.1.3	Non-forested ecological classes	A manually interpreted and digitised map depicting areas that are thought to be incapable of supporting a forest vegetation community, but which are occupied by non-forested land cover

TABLE 5 Digital input layers extracted from a DEM used in predicting ecological classes for both the small Baranja Hill data-set and in the former Cariboo Forest Region

Abbr.	Description	Concept(s) connected	Reference
`LogQA`	Log of upslope area by multiple descent algorithm	Used mainly as a surrogate for relative landform position	Quinn et al. (1991)
`Qweti`	Quinn Wetness Index	Used as a surrogate for relative moisture regime	Quinn et al. (1991)
`Slope`	Slope gradient in percent	Used as a direct measure of slope steepness	Eyton (1991)
`New_asp`	Values for aspect rotated 45 degrees counterclockwise, so that the 0 aspect value is positioned at 315	Rotated aspect was used to infer degree of exposure, e.g. warm (SW) and cool (NE) aspects	Eyton (1991)
`PctZ2Str`	Percent change in elevation (*z*) of a cell relative to the closest stream and ridge cells to which it is connected	This variable was used as the main measure of local relative landform position or relative relief	MacMillan et al. (2000)
`PctZ2Pit`	Percent change in elevation (*z*) of a cell relative to the closest pit and peak cells to which it is connected	This variable was used as a secondary measure of a more general or regional landform position	MacMillan et al. (2000)

(*continued on next page*)

TABLE 5 *(continued)*

Abbr.	Description	Concept(s) connected	Reference
Z2St	Absolute vertical change in elevation (z) from a cell to the nearest stream channel cell to which it is connected	This variable was used as a measure of absolute vertical change in elevation relative to a local base level at the nearest cell identified as a channel cell	MacMillan et al. (2000)
Plan	Across slope or plan curvature	This variable was used infrequently as an additional indication of convergence or divergence of surface flow	Eyton (1991)
Prof	Down-slope or profile curvature	This variable was used infrequently as an additional indication of acceleration or deceleration of surface flow	Eyton (1991)
L2Wet	The horizontal distance of a cell from the nearest connected cell that is classified as a water body or wetland	This variable was used as a measure of the horizontal distance back from a manually mapped wetland or water body	–
Z2Wet	The vertical distance of a cell above the nearest connected cell that is classified as a water body or wetland	This variable was used as a measure of vertical distance above a manually mapped wetland or water body	–
N2Wet	The buffered horizontal distance of a cell from the nearest cell that was manually mapped as a pasture or meadow	This measure was used on one occasion only to define a buffer around non-forested meadows	–

considered to differ in characteristics and ecological capacity from those in adjacent subzones. Separate classes and separate rules are therefore required for each ecological subzone that is defined.

The map of parent-material depth, texture and exceptions is produced to partition subzones into smaller and more homogeneous spatial entities with an even more restricted range of parent-material depth or texture. Along a toposequence, from crest to channel, one frequently encounters a completely different suite of ecological-site types, depending on whether the parent material is coarse, medium or fine-textured, even though the areas may otherwise be similar. Consequently, it is necessary to subdivide subzones even further, into areas of coarse, medium or fine-textured materials, or deep versus shallow parent materials, and to devise separate rules for each area.

The map of non-forested ecological classes identifies and delineates areas that do not currently have a forest-type land cover and are considered to be incapable of supporting that type of land cover. Most exception classes tend to be rapidly and easily identifiable using manual, visual interpretation of satellite or air-photo imagery. Most exception classes also tend to have clear, hard boundaries, so it is often unnecessary to extract these classes via automated modelling procedures.

The input layers listed in Table 4 are currently prepared using manual, visual interpretation and digitising. In the future, however, this might be done using automated modelling applied to a combination of remotely sensed imagery, secondary-source environmental maps, and terrain derivative input-data layers. For the present, manual preparation of these layers has been found to be faster, more economical and more accurate than any alternative automated methods. This is particularly true in cases, like the one illustrated here, where there are only a very limited number of boundaries or delineations that need to be identified and digitised. There are no efficiencies or improvements in accuracy to be gained by automating procedures that can be done faster, and more reliably, manually.

The only automatically prepared data layers used so far in the PEM procedures applied in the former Cariboo Forest Region of British Columbia have been Landsat7 ETM satellite imagery and BC Terrain Resource Inventory Mapping (TRIM) digital elevation data (Ministry of Environment, Lands and Parks, Surveys and Resource Mapping Branc, 1992). Both of these have been re-sampled to a grid with a horizontal resolution of 25 m. Contrast-stretched digital numbers from bands 1, 2 and 3 of false-colour satellite images have been used to infer the type and density of vegetation cover, where this has been a consideration needed for classifying types of ecological sites. Many different derivatives of the digital elevation data have been computed using a toolkit of in-house programs for analysing DEM data. So far, however, only 12 land-surface parameters have been used for classifying ecological-site types in the former Cariboo Forest Region (Table 5).

The values obtained for several of the land-surface parameters listed in Table 5 are illustrated in Figure 3. The data for all listed LSPs are imported into a GIS and examined visually to assess subjectively whether there appears to be a relationship between any given land-surface parameter and any ecological class to be predicted. Each class is assigned a colour legend and the range of values associated with any colour is adjusted manually until the spatial extent of a legend class, or series of legend classes, appears to match the known or anticipated spatial extent of any ecological class that needs to be predicted. Although this process is subjective, it is not substantively different from manually identifying a large number of locations or training sites at which a given class is known or believed to occur, and then analysing the distribution of those values of an input variable that are common to all the selected training locations.

In British Columbia, the final input map, prepared for a typical PEM project is referred to as a hierarchical classification zone map. This map, described sepa-

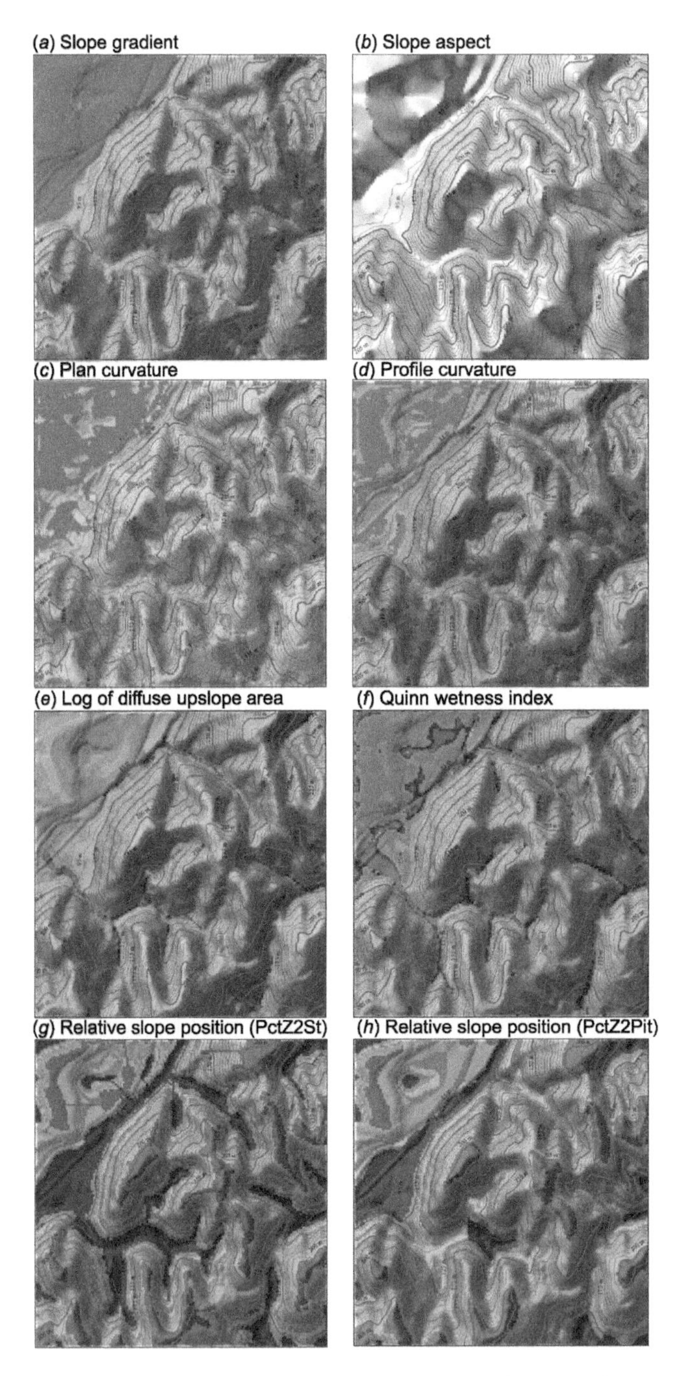

FIGURE 3 An illustration, from the small Baranja Hill data-set, of several of the more frequently used land-surface parameters in the PEM process. (See page 753 in Colour Plate Section at the back of the book.)

rately here, is a combination of several other maps prepared using both manual and automated procedures. It is used to define different domains or zones within the mapping area within which different sets of classification rules apply. The map is prepared by intersecting the manually prepared maps of *localised biogeoclimatic zones* with two other maps. The first of these, extracted from the depth, texture and exception map, shows two or three classes (coarse, medium and fine) of parent-material texture. The other map depicts automatically defined physiographic classes in terms of their maximum local relief, controlling (80th percentile) slope gradient and controlling (80th percentile) slope length.

The map of relief classes permits different knowledge-based rules to be developed for areas in zones which have the same climate and parent-material texture but where there are significant differences in the size, scale and relief of their landforms. This is a necessary capability in the case of the classification system developed for British Columbia, because this system recognises that landforms of differing sizes and scales often exhibit distinctly different *toposequences* of ecological classes from crest to channel, depending upon the size and scale of the individual features (such as hillslopes) of those landforms.

STEP 3. *Create heuristic rule bases*

The process of creating knowledge-based, or heuristic, rule bases is relatively straightforward. As this process has already been described in MacMillan et al. (2007), it will not be repeated in detail here.

In short, each class identified in Step 1 as one that should be predicted is defined using a fuzzy semantic import (SI) model, as described by Burrough (1989) and implemented by MacMillan et al. (2000). Each class of ecological entity is defined as a weighted linear average of a series of defining attributes, where attribute values are computed in terms of *fuzzy membership functions* that relate the value of a parameter (e.g. slope gradient) to the likelihood of that value matching the concept of the class used to define the attribute (e.g. steep slopes).

In building rule bases, the knowledge engineer first has to decide which land-surface parameters, or other digital input values, appear to exhibit a consistent and predictable spatial relationship with the known, or anticipated, pattern of distribution of the class that is being predicted.

The *knowledge engineer* then has to select a value, or, more often, a range of values, for each selected input variable that appears to give the best spatial match with the output class that is being investigated. This range of values is expressed as a fuzzy likelihood membership value, by using one of five fuzzy SI membership functions presented in Burrough (1989). The knowledge engineer then has to assign a relative weight to each attribute used to define each class. For any given attribute, the relative weights place more or less emphasis on the fuzzy likelihood value, thereby making some attributes more important than others when calculating the final overall weighted average for fuzzy membership in the class being predicted.

Any variable, or combination of variables, can be used to define an output class that is being predicted, and any range of values for any selected input variable can

TABLE 6 An example of a rule file used to define fuzzy likelihood models for attributes used in subsequent calculations of fuzzy membership values for ecological classes for the small Baranja Hill data-set

sortorder	file_in	attr_in	class_out	model_no	b	d
1	formfile	PROF	CONVEX_D	4	5.00	2.50
2	formfile	PROF	CONCAVE_D	5	−5.00	2.50
3	formfile	PROF	PLANAR_D	1	0.00	2.50
4	formfile	PLAN	CONVEX_A	4	5.00	2.50
5	formfile	PLAN	CONCAVE_A	5	−5.00	2.50
6	formfile	PLAN	PLANAR_A	1	0.00	2.50
7	formfile	QWETI	HIGH_WI	4	10.00	1.00
8	formfile	QWETI	LOW_WI	5	8.00	1.00
9	formfile	SLOPE	NEAR_LEVEL	5	5.00	1.00
10	formfile	SLOPE	REL_STEEP	4	7.00	1.00
11	formfile	SLOPE	Steep	4	35.00	5.00
12	formfile	SLOPE	SlopeLT20	5	20.00	1.00
13	formfile	NEW_ASP	NE_Aspect	1	90.00	45.00
14	formfile	NEW_ASP	SW_Aspect	1	270.00	45.00
15	relzfile	PCTZ2ST	NEAR_DIV	4	90.00	15.00
16	relzfile	PCTZ2ST	NEAR_HALF	1	50.00	25.00
17	relzfile	PCTZ2ST	NEAR_CHAN	5	10.00	15.00
18	relzfile	PCTZ2PIT	NEAR_PEAK	4	90.00	15.00
19	relzfile	PCTZ2PIT	NEAR_MID	1	50.00	25.00
20	relzfile	PCTZ2PIT	NEAR_PIT	5	5.00	5.00
21	relzfile	Z2PIT	HI_ABOVE	4	2.00	1.00

Parameter model_no tells the program which type of fuzzy model to use, b is a user-assigned value at which an input attribute completely satisfies the conditions for membership in the attribute class, d is a user-assigned value for dispersion that establishes the difference in value from b at which the fuzzy membership value for the attribute class declines to 0.5.

be used to define the likelihood of a parameter value (such as slope) matching the concept of the attribute (e.g. steeply sloping) used to define an ecological class (e.g. steep north-east slope) that is being defined. Examples of fuzzy classification rules applied to the Baranja Hill data-set are given in Tables 6 and 7.

Table 6 is an example of a rule file used to define fuzzy membership values for classifications of individual attributes expressed in terms of ranges of values of specific input variables. The variables chosen are believed to be spatially associated with one or more of the ecological classes that are being predicted. (attr_in) identifies the name of the field in the input file (file_in) that contains the data for the input variable used to define the attribute being classified. The name that the knowledge engineer has elected to give to this attribute classification is (class_out).

The names selected for attribute classes usually give users some idea of the specific morphological (or other) conditions that are believed to be associated with

TABLE 7 An example of a rule file used to establish the overall weighted average fuzzy likelihood models for a set of 16 user-defined ecological classes for the small Baranja Hill data-set

f_name	fuzattr	attrwt	facet_no	f_code
Hill_LCR	NEAR_PEAK	30	1	2001
Hill_LCR	NEAR_DIV	20	1	2001
Hill_LCR	HI_ABOVE	10	1	2001
Hill_LCR	NEAR_LEVEL	20	1	2001
Hill_LCR	PLANAR_D	10	1	2001
Hill_LCR	PLANAR_A	5	1	2001
Hill_LCR	LOW_WI	5	1	2001
Hill_DSH	NEAR_PEAK	30	2	2002
Hill_DSH	NEAR_DIV	20	2	2002
Hill_DSH	HI_ABOVE	10	2	2002
Hill_DSH	REL_STEEP	20	2	2002
Hill_DSH	CONVEX_D	10	2	2002
Hill_DSH	CONVEX_A	5	2	2002
Hill_DSH	LOW_WI	5	2	2002
Hill_USW	NEAR_PEAK	30	3	2003
Hill_USW	NEAR_DIV	20	3	2003
Hill_USW	HI_ABOVE	10	3	2003
Hill_USW	NEAR_LEVEL	10	3	2003
Hill_USW	CONCAVE_D	10	3	2003
Hill_USW	CONCAVE_A	10	3	2003
Hill_USW	HIGH_WI	10	3	2003
Hill_PBS	NEAR_HALF	20	4	2004
Hill_PBS	NEAR_MID	10	4	2004
Hill_PBS	HI_ABOVE	5	4	2004
Hill_PBS	REL_STEEP	20	4	2004
Hill_PBS	PLANAR_D	15	4	2004
Hill_PBS	PLANAR_A	25	4	2004
Hill_PBS	LOW_WI	5	4	2004
Hill_DBS	NEAR_HALF	20	5	2005
Hill_DBS	NEAR_MID	10	5	2005
Hill_DBS	HI_ABOVE	5	5	2005
Hill_DBS	REL_STEEP	20	5	2005
Hill_DBS	CONVEX_A	20	5	2005
Hill_DBS	PLANAR_D	15	5	2005
Hill_DBS	LOW_WI	10	5	2005
Hill_CBS	NEAR_HALF	20	6	2006
Hill_CBS	NEAR_MID	10	6	2006
Hill_CBS	HI_ABOVE	5	6	2006
Hill_CBS	REL_STEEP	20	6	2006
Hill_CBS	CONCAVE_A	20	6	2006
Hill_CBS	PLANAR_D	15	6	2006
Hill_CBS	HIGH_WI	10	6	2006
Hill_TER	NEAR_HALF	20	7	2007
Hill_TER	NEAR_MID	10	7	2007
Hill_TER	HI_ABOVE	5	7	2007
Hill_TER	NEAR_LEVEL	30	7	2007
Hill_TER	PLANAR_D	15	7	2007
Hill_TER	PLANAR_A	20	7	2007
Hill_SAD	NEAR_HALF	20	8	2008
Hill_SAD	NEAR_MID	10	8	2008
Hill_SAD	HI_ABOVE	5	8	2008
Hill_SAD	NEAR_LEVEL	20	8	2008
Hill_SAD	CONCAVE_D	20	8	2008
Hill_SAD	CONVEX_A	20	8	2008
Hill_MSW	NEAR_HALF	20	9	2009
Hill_MSW	NEAR_MID	10	9	2009
Hill_MSW	HI_ABOVE	5	9	2009
Hill_MSW	NEAR_LEVEL	25	9	2009
Hill_MSW	CONCAVE_D	10	9	2009
Hill_MSW	CONCAVE_A	10	9	2009
Hill_MSW	HIGH_WI	20	9	2009
Hill_CFS	NEAR_CHAN	20	10	2010
Hill_CFS	NEAR_PIT	10	10	2010
Hill_CFS	REL_STEEP	10	10	2010
Hill_CFS	CONCAVE_D	20	10	2010
Hill_CFS	CONCAVE_A	20	10	2010
Hill_CFS	PLANAR_A	10	10	2010
Hill_CFS	HIGH_WI	20	10	2010
Hill_CCH	NEAR_CHAN	20	11	2011
Hill_CCH	NEAR_PIT	10	11	2011
Hill_CCH	NEAR_LEVEL	10	11	2011
Hill_CCH	CONCAVE_D	20	11	2011
Hill_CCH	CONCAVE_A	20	11	2011
Hill_CCH	PLANAR_A	10	11	2011
Hill_CCH	HIGH_WI	20	11	2011
Hill_PTS	NEAR_CHAN	20	12	2012
Hill_PTS	NEAR_PIT	10	12	2012
Hill_PTS	REL_STEEP	10	12	2012
Hill_PTS	PLANAR_D	25	12	2012
Hill_PTS	PLANAR_A	25	12	2012
Hill_PTS	HIGH_WI	10	12	2012
Hill_FAN	NEAR_CHAN	20	13	2013
Hill_FAN	NEAR_PIT	10	13	2013
Hill_FAN	REL_STEEP	10	13	2013
Hill_FAN	CONVEX_A	25	13	2013
Hill_FAN	PLANAR_D	25	13	2013
Hill_FAN	LOW_WI	10	13	2013
Hill_LSM	NEAR_DIV	10	14	2014
Hill_LSM	NEAR_CHAN	20	14	2014
Hill_LSM	NEAR_PIT	10	14	2014
Hill_LSM	NEAR_PEAK	10	14	2014
Hill_LSM	REL_STEEP	10	14	2014
Hill_LSM	CONVEX_D	15	14	2014
Hill_LSM	CONVEX_A	15	14	2014
Hill_LSM	LOW_WI	10	14	2014
Hill_LLS	NEAR_CHAN	20	15	2015
Hill_LLS	NEAR_PIT	20	15	2015
Hill_LLS	NEAR_LEVEL	40	15	2015
Hill_LLS	PLANAR_D	5	15	2015
Hill_LLS	PLANAR_A	5	15	2015
Hill_LLS	HIGH_WI	10	15	2015
Hill_DEP	NEAR_CHAN	20	16	2011
Hill_DEP	NEAR_PIT	20	16	2011
Hill_DEP	NEAR_LEVEL	40	16	2011
Hill_DEP	CONCAVE_D	5	16	2011
Hill_DEP	CONCAVE_A	5	16	2011
Hill_DEP	HIGH_WI	10	16	2011

Parameter `fuzzattr` identifies the attribute class previously defined in Table 6, `attrwt` gives the subjective weight assigned to each attribute class by the knowledge engineer.

each attribute class being defined. This makes it much easier for interested users to read and comprehend the rules used to define the overall fuzzy membership values for the ecological classes being predicted. The value selected for `b` establishes the point at which the value for the input variable completely satisfies the requirement for membership in the attribute class being defined. The value selected for `d` is used to establish the range, or cross-over point, at which the likelihood of the value of the input variable (LSP) satisfying the requirement for membership in the attribute class being defined will be 0.5 (or 50%).

For every site that has a particular set of land-surface parameter values for each of the attributes being defined, Table 7 is used to define the weighted combination of fuzzy-attribute membership values for each ecological output classes being considered [see Equation (2.11) in Chapter 19]. Parameter `f_name` identifies the alpha-numeric code for each unique class of spatial entity selected defined for any given area. (`fuzattr`) identifies attributes that the knowledge engineer considers to be definitive for the overall ecological class being predicted. Parameter `attrwt` identifies the relative importance, or weight, of the contribution that each defining attribute is subjectively estimated to make in calculating the overall weighted average fuzzy membership value. This value identifies how likely it is that any given location will belong to any defined class. Parameter `f_code` is used as an integer ID number for each unique output class that is defined by any given set of rules. The same integer ID number can be assigned to several different defined entities. This has been done in cases where the same entity can occur in several different ecological settings, so that the definition of each setting has to be entered separately into the rule file. Parameter `facet_no` is used to sort the rule files, when they are read. This ensures that all rules applying to a defined class will occur together, and will be read sequentially when working through the program.

STEP 4. *Apply rule bases to produce initial maps*

The custom, in-house program that applies the fuzzy semantic import model calculations (**FacetMapR**) first reads Table 6 and then computes the fuzzy membership likelihood value of each attribute class defined in the attribute rule table for each grid location (MacMillan, 2004). The program then reads Table 7 and computes the overall weighted mean fuzzy likelihood value of each grid cell belonging to each of the ecological classes defined in the class rule table.

The program used to compute fuzzy likelihood values for all defined output classes first reads a control file that contains an integer ID number which identifies each unique classification domain or zone using the hierarchical zone classification map prepared in Step 2. A paired set of attribute class (`arule`) and output class (`crule`) rule files is needed for each classification zone. The program computes a value for fuzzy likelihood of occurrence for only the specific classes defined for a given classification zone. Each grid cell is assigned a single integer `ID` number. This identifies the output class with the largest fuzzy likelihood of occurring at that grid cell. Integer `ID` numbers for hard non-forested ecological exception classes are then *'cookie-cut'* into the initial map of predicted ecological classes. Here, they over-ride any predictions made by the fuzzy calculations. A combined map of hard Boolean exception classes and soft fuzzy ecological entity classes is imported into a GIS. In this form, it can be visually assessed and reviewed.

STEP 5. *The initial maps are reviewed by a local expert*

Initial grid maps depicting the spatial extent and pattern of predicted ecological classes are reviewed visually to assess the degree to which the predicted and expected patterns match. The predicted pattern can be compared to either actual site locations classified in the field or, more routinely, to conceptual models formulated by someone who has considerable local experience. Conceptual models are presented using ecological keys, 2D or 3D conceptual landscape profile diagrams, site-features tables or textual descriptions. The predicted pattern is first reviewed to assess the degree to which the predicted classes match conceptual models in terms of a notional crest-to-channel toposequence of entities. Assessments are made of the extent to which the predicted entities follow a logical and expected sequence based on landform position, exposure, moisture conditions and slope gradient.

Next, an assessment is made of whether the total amount or geographical extent of each predicted class matches well with the known or expected extent of each class. This assessment requires someone with considerable local experience to ascertain whether each predicted class occurs in about the geographic locations, and to about the extent, that they would expect; based on their experience. After creating rules for more than fifty different ecological subzones, it has been observed that only a small number of iterations (2–4) have been required to create knowledge-based rules comprehensive enough to predict a relatively reasonable pattern of output classes for most areas.

STEP 6. *Revise the rule bases in order to address problems*

The observations and comments of a local ecological expert have been used to revise and refine the rule bases in order to remove or reduce errors that lead to what the local expert would consider an incorrect prediction of patterns. During each assessment, any classes that have been predicted in an incorrect toposequence order first have to be identified and the classification rules have to be manually adjusted to correct such errors. Once the rules are observed to produce a series of output classes occurring in approximately the desired locations along a toposequence, they are then refined further. This is done by adjusting the ranges associated with the defining attributes to either expand or contract a given output class, or to move the boundaries for a given output class either up or down slope.

The rules can be adjusted by adding one or more attributes to the definition of a specific class, or by deleting rules for specific classes. This is done by modifying the range of values of one of the input variables used to define an attribute class, or by altering the weight placed on any given attribute of an existing output class. A third option is to remove an entire class or to add a definition for a completely new output class. These changes are made by trial and error, using expert judgement. The knowledge engineer, in consultation with the local expert, decides which classes should be revised, removed or added, which attributes should be used for defining each class, and which ranges of input land-surface parameters should be used for defining each class of attributes. It has sometimes been necessary to identify and obtain a new input variable for use in a revised set of rules, or to alter the zone classification of a map by refining a boundary, or by adding an entirely new zone to the classification.

Steps 4 to 6 are repeated until the local ecological expert is satisfied that the predicted output maps correspond as closely as possible with the best available understanding of the actual conceptual and geographical arrangement of ecological classes within any given map area.

STEP 7. *Apply the final rule bases to produce the final predictions*

The local ecological expert acts as a sort of internal quality-control assessor. Knowledge-based rules, and the maps produced by applying them, are not considered final until they have been approved by the local regional ecologist. Once approved, the rules are applied to the assembled input-data layers to produce a predictive ecological map (PEM). This is then subjected to more formal and systematic assessment and evaluation. In many respects, this knowledge-based approach is similar to the definition of classes, using prototype category theory, recently described by Qi et al. (2006).

STEP 8. *Create a seamless mosaic of predicted PEM classes*

Once approval of the rules and predictive maps has been received from the regional ecologist, a complete and seamless mosaic of predicted classes is produced by applying the rules to the layers of input data, which are then joined into a single mosaic.

STEP 9. *'Cookie-cut' the non-forested exception classes into the final map*

All non-forested exception classes are treated as Boolean objects with hard boundaries and a single correct classification. These Boolean non-forested classes are 'cookie cut' into the previously prepared map of predicted ecological classes, covering and displacing all forested ecological classes predicted by using the fuzzy modelling procedures. This step ensures that spatial entities that have clear, hard boundaries retain these boundaries in the final PEM map.

STEP 10. *Send the PEM map for an external assessment of its accuracy*

To date, all predictive ecosystem maps produced for operational mapping projects in British Columbia have been subjected to an arms-length accuracy assessment by an independent third-party contractor. These accuracy assessment procedures do not require the maps to predict accurately the exact class at exact point locations. Rather, they assess the extent to which predictions of the proportions of predicted classes match the proportions of actual classes observed in the field, along randomly-selected, closed, linear traverses.

The intent is to assess the ability of the maps to predict correctly the proportions of ecological classes within small areas that are equivalent to a minimum-sized area for which management decisions are likely to be taken. If the maps fail to achieve a minimum of 65% predictive accuracy, determined according to the approved accuracy assessment protocol, they can be returned to the knowledge engineer for further refinement. So far, this has not been necessary, as all the maps achieved the required minimum level of accuracy of 65% upon first submission.

3. RESULTS

3.1 Results for Baranja Hill

Figure 4 illustrates the kind of results that can be obtained by applying the methods described above. The figure is not meant to present an optimal classification of the area, but rather a relatively simple and comprehensible classification that permits several aspects of the methods to be demonstrated and discussed.

The thick yellow line in Figure 4 illustrates how the Baranja Hill study area was subdivided manually into a hill area and a flood-plain area, with the hill area subdivided further into an area of coarse-textured versus medium-textured parent

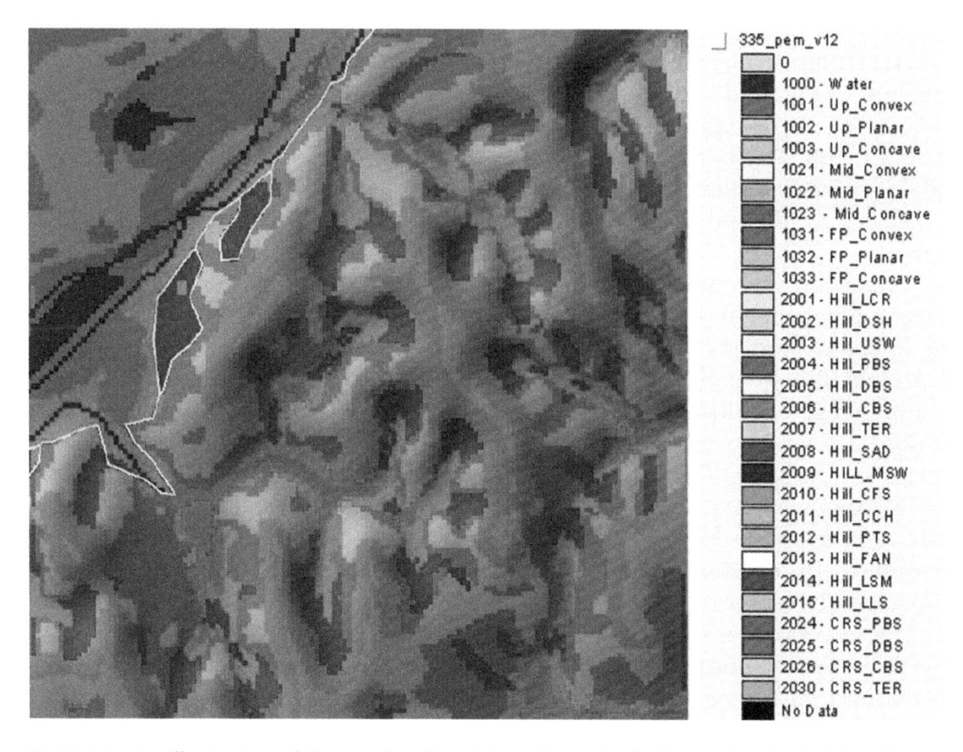

FIGURE 4 An illustration of the results of applying a hypothetical set of ecological–landform classification rules to the small data set from Baranja Hill. See Table 3 for an explanation of legend classes. (See page 754 in Colour Plate Section at the back of the book.)

materials. These partitions defined three different classification zones: flood plain, coarse-textured and medium-textured hilly areas. Three different sets of classification rules were developed and applied in each of these three classification zones.

It is assumed that each of the ecological–landform classes illustrated in Figure 4 represents a unique ecological setting or landscape situation characterised by a particular and describable range of both predictor variable values and on-the-ground environmental conditions. The usual assumption is that soils are thinner and drier in upper landform positions (crests) and on divergent convexities (spurs), and that soils are thicker and wetter in lower landform positions (toe slopes) and concavities (valleys and draws). However, this interpretation does not have to hold for the different entities to remain useful. There may well be instances where, for example, deep organic soils characterised by high levels of moisture could occur on upper crests and plateaus, and where shallow soils characterised by low levels of available moisture might be found in valley bottoms. In such cases, descriptions of the characteristics of each unique entity will differ, but the entities themselves remain useful and valid.

In the hill area, different rules were used to identify different sets of ecological–landform entities, depending upon whether the underlying parent materials were considered to be coarse or medium textured. The rules used to define entities for the medium-textured areas represent a minor modification of a relatively standard set of rules previously used by MacMillan et al. (2000) for classifying generic landform elements. These rules first subdivide hillslopes into three components: upper, mid and lower slope. The upper-slope areas are partitioned into level crests (`Hill_LCR`), divergent shoulders (`Hill_DSH`) and convergent upper swales (`Hill_USW`). It is assumed that the divergent shoulders shed water and exhibit drier conditions and shallower soils than normal, while the convergent swales accumulate surface and subsurface run-off and have moister and deeper soils.

If the relative proportions and extents of these three entities are not considered to be optimal, then it is a relatively simple process to modify the rules by expanding or contracting any one class relative to the others. For example, if one wished to expand the extent of the level-crest class, one would change the rules that establish the range of values of the input variables used to define the attribute classes of *convex down and across* versus *planar down and across*, as well as *steeply sloping* versus relatively gentle. Increasing the range for defining planar attributes will result in expansion of the *level-crest* class at the expense of the *divergent shoulder* and *convergent upper swale* classes. Similarly, increasing the range of slope gradients considered to represent *relatively gentle slopes* versus *relatively steep slopes* would also expand the extent of the *level-crest* class relative to the other two classes.

The mid-slope areas were partitioned into planar back slopes (`Hill_PBS`), divergent back slopes (`Hill_DBS`), convergent back slopes (`Hill_CBS`), relatively level mid-slope terraces (`Hill_TER`), mid-slope saddles (`Hill_SAD`) and mid-slope convergent swales (`Hill_MSW`). Again, the assumption is made that divergent back slopes shed water and exhibit drier conditions and shallower soils, while convergent mid-slopes exhibit wetter conditions and develop deeper soils. Planar back slopes and mid-slope terraces are assumed to exhibit modal or mesic moisture conditions and soil depths. In operational ecological mapping in British Columbia, it has been very common for the local ecological expert to suggest that the rules used to define the modal or mesic class (similar to the *planar back slope* here) should be relaxed or expanded to increase the extent of the modal class relative to the classes used to define drier, divergent mid-slopes and wetter, convergent mid-slopes. It appears that the definition of what constitutes mesic or modal conditions tends to be rather broad, with only the extreme ends of the spectrum considered to represent conditions that are either significantly wetter or drier than normal.

With respect to the knowledge-based rules presented for this example, the extent of the planar back-slope class could be expanded by revising the rules that define the attributes of convex, concave and planar conditions in the across-slope direction. Such revision could involve increasing the range associated with planar conditions and increasing the threshold values that must be attained before a location is considered to be either concave or convex in the across-slope direction. Such decisions can be taken locally for any defined classification zone to tailor the classification to local environmental conditions.

The lower slope areas were partitioned into concave foot slopes (`Hill_CFS`), planar toe slopes (`Hill_PTS`), divergent lower slope fans (`Hill_FAN`), convex lower slope mounds (`Hill_LSM`), level lower flats (`Hill_LLS`) and concave, relatively level, channels (`Hill_CCH`). The concave foot slopes represent areas where surface run-off is assumed to decelerate. The materials carried down-slope in this run-off are deposited, resulting in deeper, wetter soils. The planar toe slopes are assumed to represent even wetter areas that, topographically, usually occur below the concave foot slopes. The lower slope fans are planar in profile and convex in plan, and are assumed to have drier, deeper soils. Lower slope mounds are used to identify areas lower in the landscape that are convex in both profile and plan, and that are assumed to shed water and to be somewhat drier than other parts of the landscape in lower slope areas. Level lower slopes typically occur just outside and above concave channels and are expected to be very wet, but not as wet as the areas classified as concave channels (and depressions).

As discussed above, the knowledge engineer, at the direction of the local ecological expert, may make changes to the rules used to define attribute classes by increasing or decreasing the extent of any one class (e.g. convergent channels) relative to any other (e.g. level lower slopes). Such decisions can be guided by local expert knowledge, or by actual field-sample observations, if available.

The rules used to classify the area of coarse-textured materials in the hill area were simplified to illustrate that rules only need to be developed for classes that are expected to occur within any classification zone. In the case of the coarse-textured area, rules were included only for identifying convergent, divergent and planar back slopes and level mid-slope terraces, because the recognised distribution of coarse-textured materials was limited to the mid-slope portions of the landscape. In operational PEM mapping in British Columbia, it has been quite customary for a more limited number of classes to be identified in coarse areas, as most coarse areas, with the exception of very concave draws and hollows where excess moisture may accumulate, are considered to be drier than the *mesic* conditions encountered elsewhere in that landscape.

The rules used to classify the flood-plain were purely notional and were not particularly realistic or useful. These rules were implemented to show how the lower flood-plain could be partitioned into convex, concave and planar entities, separated from a somewhat elevated terrace. None of the input layers available to support predictions seem capable of extracting a highly effective classification. This was probably because the resolution of the input DEM data was insufficient to capture the subtle local variations in topography necessary, to effectively classify the flood-plain area into meaningfully different components. The results presented for the flood plain were also used to illustrate the concept of cookie-cutting manually mapped hard exception classes into the final map. The areas labelled as open water were extracted from a previously prepared map of permanent water bodies and were 'cookie cut' into the final map.

3.2 Operational ecological mapping in British Columbia, Canada

The methods described and illustrated above have been used by the first author to prepare predictive ecosystem maps (PEM), at a scale of 1:20,000, for more than 3.5 million ha of forested land within the former Cariboo Forest Region of British Columbia, Canada (Figure 5). Mapping of the remaining 5 million ha of this former forest region was completed in early 2008. Several other commercial companies operating in BC have used these procedures and concepts, in a similar fashion, to produce PEM maps for a further 4 million ha. So far, within the former Cariboo Forest Region, heuristic, knowledge-based (KB) rules have been developed and applied for more than 50 different biogeoclimatic subzones. Each subzone contains at least four, and sometimes as many as ten further sub-divisions into classification zones based on considerations of parent-material texture, landform scale and relief, and local climatic conditions (in particular, a high local incidence of frost). As the first author acts as knowledge engineer for the British Columbia operational mapping, he has not been allowed to see, or have any detailed knowledge of, the data collected to assess the accuracy of the final PEM maps. However, some general observations and findings have been passed on via reports prepared by the individuals responsible for collecting and analysing the field accuracy data.

In general, it appears that the PEM maps produced by these procedures have demonstrated a very low capability to correctly predict the exact ecological class at exact point locations along field traverses. This is disappointing, as the PEM maps appear to have captured the main concepts used for defining ecological classes rather well in the mapped areas. This would be even more disappointing, were it not for the fact that no other maps of ecological classes produced for the same areas by any other means have, to date, demonstrated any better ability to predict the exact ecological class correctly, at exact point locations.

Parts of the area for which PEM maps have already been produced have also been mapped at various times in the past, using other methods, including predictive ecosystem mapping and traditional manual methods based on a combination

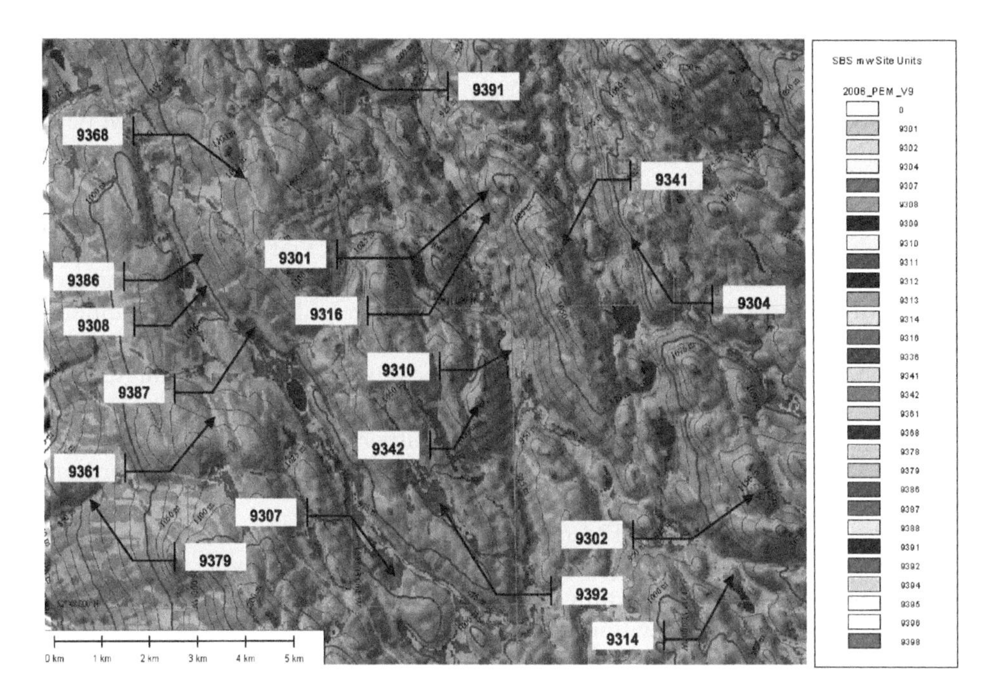

FIGURE 5 Part of a 1:20,000 scale predictive ecosystem map (PEM) produced for an area in the former Cariboo Forest Region of BC, Canada. (See page 754 in Colour Plate Section at the back of the book.)

of air-photo interpretation and field observations. Many of the conventional maps cost between $1.50 CAD and $10.00 CAD per ha to produce and none has been found able to predict either exact ecological classes at exact point locations, or proportions of ecological classes within small areas at a level of accuracy superior to, or even equal to, the PEM maps. Therefore, on a cost-effectiveness basis, the PEM maps, produced at a cost of less than $0.20 CAD per ha using the methods outlined in this chapter, offer a useful alternative.

It is clear that either the rules used for predicting the spatial distribution of ecological classes in British Columbia were incorrect, or the various input layers used to model the spatial pattern of ecological classes were incapable of correctly describing the configuration of the terrain upon which many of the predictive rules were based. Analysis of the traverse data has indicated that, along many field traverses, on-site assessments of the most correct ecological classification have frequently changed over distances of 10 m or less. Since all PEM predictions are based on an analysis of elevation derivatives, and other input data layers, surfaced to a 25 m grid, it would not be reasonable to expect the 25 m grid data to be able to predict changes correctly that occur over distances of 10 m or less. In fact, it is unlikely that the 25 m grid data can support a description of variation in terrain configuration that operates over distances of less than 50 m (2× the grid spacing).

Efforts were made to try interpolating the available elevation data to a 10 m grid and also to produce and analyse a new, custom, DEM with a 5 m horizontal grid spacing to see whether predictions could be improved by using a DEM with a finer horizontal resolution. In both cases, when assessed using the same methods and the same field data-sets that had been used to assess the accuracy of the 25 m PEM maps, the PEM maps using this finer resolution failed to achieve an improved level of predictive accuracy. It was concluded that the finer resolution DEM data-sets were not able to improve the depiction of variation in terrain in any significant manner, compared with the available 25 m DEM data. New methods for producing DEMs, such as LiDAR, may offer the opportunity to procure fine resolution DEMs that truly increase the faithfulness with which the terrain surface is represented. However, none of the finer resolution DEMs investigated to date have been proved to be capable of improving either the spatial accuracy of the PEM predictions, or even the level of accuracy achieved in estimating proportions of ecological classes within small areas.

On the positive side, the PEM maps produced so far for the former Cariboo Forest Region have consistently demonstrated an ability to predict, in excess of the 65% minimum accuracy required by provincial regulators, the proportions of ecological classes observed within small areas of 10–20 ha. This is no small feat, as many ecological maps produced by both conventional manual mapping methods and other PEM modelling approaches have not achieved this level of predictive accuracy. Several such maps, prepared at various times for parts of these same areas, have failed to achieve levels of accuracy in excess of 55–60%. Consequently, the 66–72% levels of accuracy, reported by third-party contractors, for the PEM maps produced by the methods described here, are quite encouraging.

A small test was conducted to assess the ability of personnel charged with collecting the field accuracy data to consistently identify the same ecological classes at exactly the same locations. Four different local ecological experts, working independently, without communicating with each other, were asked to classify four of the 1.5 km long, randomly selected, closed accuracy traverses. It was found that the on-the-ground classifications made by these four experts only had a 65% level of agreement in terms of their estimates of the proportions of specific ecological classes along the total lengths of the traverses. However, when credit was given for ambiguous segments, their level of agreement improved to 70%. There was only a 21% level of agreement among all four ecologists in terms of exact categorical matches at exact, spatially congruent, point locations, while agreement between any two of the ecologists at exact point locations ranged from a low of 23% to a high of 73%. From this small experiment, it seems unreasonable to expect any predictive model to achieve a level of agreement with on-the-ground field assessments that exceeds the level that can be achieved (65–70%) from on-the-ground site examinations made by four different ecologists.

Even if very accurate, very fine-resolution (<5 m horizontal), elevation data were to be obtained, it is believed that it would still be necessary to develop and apply an initial classification based on an analysis of coarser resolution (25–50 m) DEM data in order to establish the contextual knowledge required for interpreting information contained in the finer-resolution data. Thus, in keeping with the hierarchical principals of ecological land classification, it would probably be beneficial to apply and retain a classification that utilises 25–50 m input layers to establish a higher level of classification zones. Within this higher level, rules could be developed for interpreting the effects of short-range, minor variations in topographic and other influences.

4. SUMMARY POINTS

Ecological land classification is a widely practiced activity that partitions space into successively smaller and, hopefully, more homogeneous spatial entities. Statements can then be made about such entities, regarding their important environmental and ecological attributes. Using knowledge of these environmental conditions, management options can be suggested that will help to promote sustainable environmental management for the benefit of both the natural populations and human activities. As practiced conventionally, ecological land classification represents a form of heuristic conceptual modelling. A disadvantage of conventional mapping is that, although systematic, its rules and procedures are not always consistent, replicable or rigorously documented. Manual methods of ecological land classification are usually also costly, time consuming to implement, and prone to error and inconsistency.

The methods presented here demonstrate how traditional manual methods can be extended, and at least partially automated, by developing and applying a combination of Boolean and fuzzy classification rules based on heuristic expert knowledge. The methods capture expert heuristic knowledge in a manner that is relatively easy to understand and interpret. They also cause that knowledge to be applied in a manner that is consistent, replicable, reproducible and testable. The resulting maps can be evaluated to assess the degree to which they provide correct predictions of the occurrence of defined ecological classes at exact point locations, or of the proportions of these defined classes within small test areas equivalent in size to areas for which planning and management decisions are typically made. If improved ecological understanding or improved predictor data-sets become available, the heuristic, knowledge-based rules can also be updated and new and updated maps produced.

The operational PEM maps produced for large areas in British Columbia have so far been unable to predict exact ecological classes at exact point locations correctly, but they have proven capable of providing reasonable estimates of the proportions of specific ecological classes occurring within small areas equivalent to those for which management decisions typically have to be made. Much progress is still required before maps can be produced that will be able to correctly predict all defined ecological classes at all point locations. However, the maps already being produced, that utilise existing methods and predictor data layers, have demonstrated that they can be as accurate as, and often more accurate than, similar maps made by conventional manual methods. They can also be produced at a fraction of the cost.

Automated methods are expected to become the norm for producing all kinds of ecological land classification maps across a variety of scales. In fact, recent applications of automated techniques in Europe,[1] New Zealand (Barringer et al., 2006) and the United States (Qi et al., 2006) provide strong evidence that automated methods, similar to those described here, are being increasingly adopted to support operational mapping of soils, landforms and ecological entities. Methods for achieving automated ecological land classification will continue to improve and will benefit from new sources of predictor data-sets, new procedures for developing and applying predictive models, and improved understanding of predictive ecological-landform relations. However, for the immediate future, it is argued that automated methods utilising existing data and existing predictive tools are already capable of producing maps that improve on those produced using conventional manual methods of air-photo interpretation and field investigation.

ACKNOWLEDGEMENTS

The procedures described in this chapter were developed and refined as part of an ongoing project to complete operational PEM mapping for the entire 8.5 million ha of the former Cariboo Forest Region in British Columbia, Canada. Client representatives for this project included Tim Harding and John Stace-Smith of Tolko Industries Ltd. and Earl Spielman, Al Hicks and Guy Burdikin of West Fraser Mills Ltd. Contractors Timberline Forest Industry Consultants and JMJ Holdings Ltd. produced the manually interpreted exception maps and collected the accuracy assessment field data. Meridian Mapping Ltd. prepared the DEM for the entire area. Funding was received from the BC Forest Investment Account (FIA) and from the, now superseded, Forest Renewal BC (FRBC) account.

IMPORTANT SOURCES

MacMillan, R.A., Moon, D.E., Coupé, R.A., 2007. Automated predictive ecological mapping in a Forest Region of B.C., Canada, 2001–2005. Geoderma 140 (4), 353–373.

Gallant, A.L., Loveland, T.R., Sohl, T.L., Napton, D., 2005. Using an ecoregion framework to analyze land cover and land use dynamics. Environmental Management 34, S89–S110.

MacMillan, R.A., 2003. LandMapR Software Toolkit — C++ Version: Users Manual. LandMapper Environmental Solutions Inc., Edmonton, AB, 110 pp.

Cleland, D.T., Avers, P.E., McNab, W.H., Jensen, M.E., Bailey, R.G., King, T., Russell, W.E., 1997. National hierarchical framework of ecological units. In: Boyce, M.S., Haney, A. (Eds.), Ecosystem Management Applications for Sustainable Forest and Wildlife Resources. Yale University Press, New Haven, CT, pp. 181–200.

Sims, R.A., Corns, I.G.W., Klinka, K., 1996. Global to Local: Ecological Land Classification. Kluwer Academic Publishers, Dordrecht, The Netherlands, 610 pp.

[1] See for example Scilands terrain classification projects available via http://www.scilands.de.

CHAPTER 25

Geomorphometry and Spatial Hydrologic Modelling

S.D. Peckham

how can DEMs be used for spatial hydrologic modelling? · what methods are commonly used to model hydrologic processes in a watershed? · what kinds of preprocessing tools are typically required? · what are some of the key issues in spatial hydrologic modelling?

1. INTRODUCTION

Spatial hydrologic modelling is one of the most important applications of the geomorphometric concepts discussed in this book. The simple fact that flow paths follow the topographic gradient results in an intimate connection between geomorphometry and hydrology, and this connection has driven much of the progress in the field of geomorphometry. It also continues to help drive the development of new technologies for creating high-quality and high-resolution DEMs, such as LiDAR. Like most other types of physically-based models, hydrologic models are built upon the fundamental principle that the mass and momentum of water must be conserved as it moves from place to place, whether it is on the land surface, below the surface or evaporating into the atmosphere. While this sounds like a simple enough idea, it provides a powerful constraint that makes predictive modelling possible. When mass and momentum conservation is similarly applied to sediment, it is possible to create *landscape evolution models* that predict the spatial erosion and deposition of sediment and contaminants.

While hydrologic models have been around for several decades, it is only in the last fifteen years or so that computers have become powerful enough for fully spatial hydrologic models to be of practical use. *Spatially-distributed* hydrologic models treat every grid cell in a DEM as a control volume which must conserve both mass and momentum as water is transported to, from, over and below the land surface. The control volume concept itself is quite simple: what flows in must either flow out through another face or accumulate or be consumed in the interior. Conversely, the amount that flows out during any given time step cannot

Developments in Soil Science, Volume 33 © 2009 Elsevier B.V.
ISSN 0166-2481, DOI: 10.1016/S0166-2481(08)00025-1.

exceed the amount that flows in during that time step plus the amount already stored inside. However, the number of grid cells required to adequately resolve the transport within a river basin, in addition to the small size of the timesteps required for a spatial model to be numerically stable, results in a computational cost that until recently was prohibitive.

REMARK 1. *Since flow paths follow the topographic gradient, there is an intimate connection between geomorphometry and hydrology. Spatial hydrologic models make use of several DEM-derived grids especially grids of slope, aspect (flow direction) and contributing area.*

For a variety of reasons, including the computational cost of fully spatial models and the fact that data required for more advanced models is often unavailable, researchers have invested a great deal of effort into finding ways to simplify the problem. This has resulted in many different types of hydrologic models. For example, *lumped models* employ a small number of *"representative units"* (very large, but carefully-chosen control volumes), with simple methods to route flow between the units. Another strategy for reducing the complexity of hydrologic models is to use concepts such as *hydrologic similarity* to essentially collapse the 2D (or 3D) problem to a 1D problem. For example, TOPMODEL (Beven and Kirkby, 1979) defines a *topographic index* or wetness index and then lumps all grid cells with the same value of this index together under the assumption that they will have the same hydrologic response. Similarly, many models lump together all grid cells with the same elevation (via the *hypsometric curve* or *area–altitude function*) to simplify the problem of computing certain quantities such as snowmelt. All grid cells with a given *flow distance* to a basin outlet can also be lumped together (via the *width function* or *area–distance function*) and this is the main idea behind the concept of the *instantaneous unit hydrograph*. While models such as these can be quite useful and require less input data, they all employ simplifying assumptions that prevent them from addressing general problems of interest. In addition, these assumptions are often difficult to check and are therefore a source of uncertainty. In essence, these types of models gain their speed by mapping many different (albeit similar) 3D flow problems to the same 1D problem in the hope that the lost differences don't matter much. While geomorphometric grids are used to prepare input data for virtually all hydrologic models, fully spatial models make direct use of these grids. For this reason, and in order to limit the scope of the discussion, this chapter will focus on fully-spatial models.

There are now many different spatial hydrologic models available, and their popularity, sophistication and ease-of-use continues to grow with every passing year. A few representative examples of some highly-developed spatial models are: *Mike SHE* (a product of Danish Hydraulics Institute, Denmark), Gridded Surface Subsurface Hydrologic Analysis (*GSSHA*), *CASC2D* (Julien et al., 1995; Ogden and Julien, 2002), *PRMS* (Leavesley et al., 1991), *DHVSM* (Wigmosta et al., 1994) and *TopoFlow*. Rather than attempt to review or compare various models, the main goal of this chapter is to discuss basic concepts that are common to virtually all spatial hydrologic models.

REMARK 2. *Hydrologic processes in a watershed (e.g. snowmelt) may be modelled with either simple methods (e.g. degree-day) or very sophisticated methods (e.g. energy-balance), based partly on the input data that is available.*

It will be seen throughout this chapter that grids of elevation, slope, aspect and contributing area all play fundamental roles in spatial hydrologic modelling. Some of these actually play multiple roles. For example, slope and aspect are needed to determine the velocity of surface (and subsurface) flow, but also determine the amount of solar radiation that is available for evapotranspiration and melting snow. The DEM grid spacing that is required depends on the application, but as a general rule should be sufficient to adequately resolve the local hillslope scale. This scale marks the transition in process dominance from hillslope processes to channel processes. It is typically between 10 and 100 m, but may be larger for arid regions. As a result of the Shuttle Radar Topography Mission (SRTM), DEMs with a grid spacing less than 100 m are now available for much of the Earth. In addition, LiDAR DEMs with a grid spacing less than 10 m can now be purchased from private firms for specific areas. Many of the DEMs produced by government agencies (e.g. the U.S. Geological Survey and Geoscience Australia) now use an algorithm such as ANUDEM (Hutchinson, 1989) to produce *"hydrologically sound"* DEMs which makes them better suited to hydrologic applications (see also Section 3.2 in Chapter 2).

This chapter has been organised as follows. Section 2 discusses several key hydrologic processes and how they are typically incorporated into spatial models. Note that spatial hydrologic models integrate many branches of hydrology and there are many different approaches for modelling any given process, from simple to very complex. It is therefore impossible to give a complete treatment of this subject in this chapter. For a greater level of detail the reader is referred to textbooks and monographs such as Henderson (1966), Eagleson (1970), Freeze and Cherry (1979), Welty et al. (1984), Beven (2000), Dingman (2002), Smith (2002). The goal here is to highlight the most fundamental concepts that are common between spatial models and to show how they incorporate geomorphometric grids. Section 3 discusses scale issues in spatial hydrologic modelling. Section 4 provides a brief discussion of preprocessing tools that are typically needed in order to prepare required input data. Section 5 is a simple case study in which a model called TopoFlow is used to simulate the hydrologic response of a small ungauged watershed in the Baranja Hill case study.

2. SPATIAL HYDROLOGIC MODELLING: PROCESSES AND METHODS

2.1 The control volume concept

Spatially-distributed hydrologic models are based on applying the control volume concept to every grid cell in a digital elevation model (DEM). It is helpful to imagine a box-shaped control volume resting on the land surface such that its top and

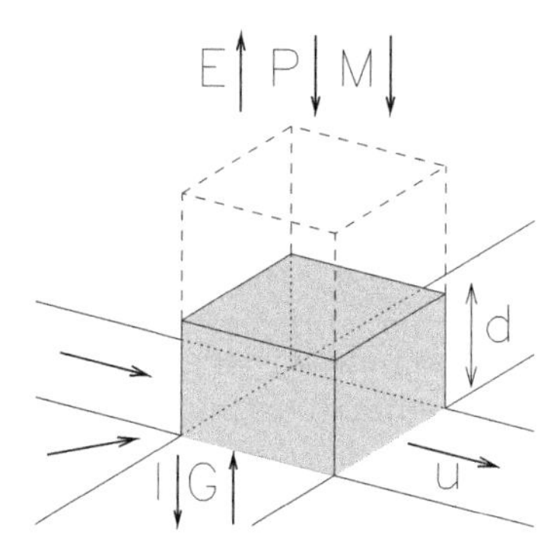

FIGURE 1 A grid-cell control volume resting on the land surface and filled with water to a depth, d. Precipitation, P, snowmelt, M, evapotranspiration, E, infiltration, I, and groundwater seepage, G add or remove water from the top and bottom faces, while surface water flows through the four vertical faces. Overland flow is shown, but a grid cell may instead contain a single, sinuous channel with a width less than the grid spacing.

bottom faces have the x and y dimensions of a DEM grid cell and such that the height of the box is greater than the local water depth (see Figure 1). Water flowing from cell to cell across the land surface flows horizontally through the four vertical faces of this box, according to the D8, D-Infinity or Mass Flux method (see Section 3.2 in Chapter 7). For overland flow, the entire bottom of the box may be wetted and 2D modelling of the flow is possible. For channelised flow, the grid cell dimensions are typically much larger than the channel width, so channel width must be specified as a separate grid, along with an appropriate sinuosity in order to properly compute mass and momentum balance.

Runoff-generating processes can be thought of as "*injecting*" flow vertically through the top face of the box, as in the case of rainfall and snowmelt, or through the bottom of the box, as in the case of seepage from the subsurface as a result of the local water table rising to the surface. Similarly, infiltration and evapotranspiration are vertical flux processes that result in a loss of water through either the bottom or top faces of the box, respectively. If a grid cell contains a channel, then the volume of surface water stored in the box depends on the channel dimensions and water depth, d, otherwise it depends on the grid cell dimensions and water depth.

The net vertical flux into the box may be referred to as the *effective rainrate*, R, and is the runoff that was generated within the box. It is given by the equation:

$$R = (P + M + G) - (E + I) \tag{2.1}$$

where P is the precipitation rate, M is snowmelt rate, G is the rate of subsurface seepage, E is the evapotranspiration rate and I is the infiltration rate.

Each of these six quantities varies both spatially and in time and is therefore stored as a grid of values that change over time. Each also has units of [mm/hr]. Methods for computing these quantities are outlined in the next few subsections of this chapter. Note that the total runoff from the box is not equivalent to the effective rainrate because it consists of the effective rainrate *plus* any amount that flowed horizontally into the box and was not consumed by infiltration or evapotranspiration. Note also that in order to model the details of subsurface flow, it is necessary to work with an additional "*stack*" of boxes that extend down into the subsurface; e.g. there may be one such box for each of several soil layers.

In many models of fluid flow, fluxes through control volume boundaries (e.g. the vertical faces of the box) are not computed directly. Instead, the boundary integrals are converted to integrals over the interior of the box using the well-known *divergence theorem* (Welty et al., 1984). This results in differential vs. integral equations and requires computing first and second-order spatial derivatives between neighbouring cells, typically via finite-difference methods. However, if we assume that flow directions are determined by topography, which is a relatively static quantity, then flow directions between grid cells are fixed and known at the start of a model run. Under these circumstances it is straight-forward and efficient to compute boundary integrals.

2.2 The precipitation process

The precipitation process differs from most of the other hydrologic processes at work in a basin in that the precipitation rate must be specified either from measurements (e.g. radar or rain gauges) or as the result of numerical simulation. All of the other processes are concerned with methods for tracking water that is already in the system as it moves from place to place (e.g. cell to cell or between surface and subsurface). For a small catchment, it may be appropriate to use measured rainrates from a single gauge for all grid cells. For larger catchments and greater realism, however, it is better to use space-time rainfall, which is stored as a grid stack, indexed by time. This grid stack may be created by spatially interpolating data from many different rain gauges. Input data for air temperature (T) is used to determine whether precipitation falls as rain or as snow.

In order to model how temperature decreases with increasing elevation, a grid of elevations can be used together with a lapse rate. If precipitation falls as snow ($T < 0\,^{\circ}\mathrm{C}$), then it can be stored as a grid of snow depths that can change in time. If the snowmelt process is modelled, then snowmelt can contribute runoff to any grid cell that has a nonzero snow depth and an air temperature greater than 0 °C.

2.3 The snowmelt process

In general, the conversion of snow to liquid water is a complex process that involves a detailed exchange of energy in its various forms between the atmosphere and the snowpack. While air temperature is obviously of key importance, numerous other variables affect the meltrate, including the slope and aspect of the topography, wind speed and direction, the heights of roughness elements (e.g. vegetation) and the snow density to name a few. The *Energy-Balance Method* (Marks and Dozier, 1992; Liston, 1995; Zhang et al., 2000) in its various implementations is therefore the most sophisticated method for melting snow, but it is very data intensive. This method consists of numerous equations (see references) and generally makes use of a clear-sky radiation model (see Section 3.1 in Chapter 8; Dozier,

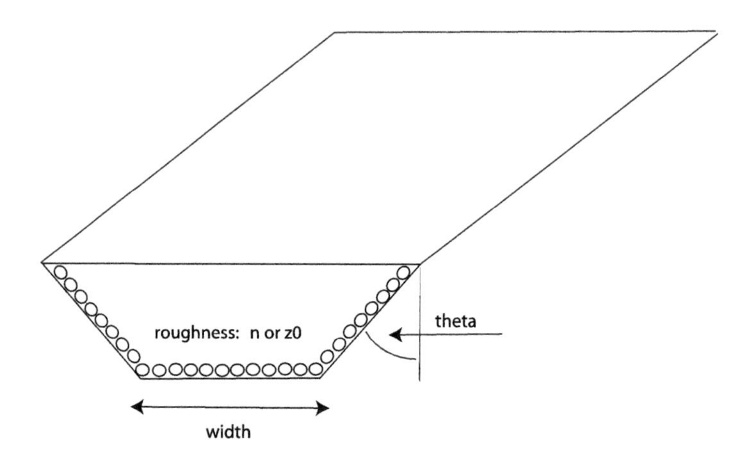

FIGURE 2 A channel with a trapezoidal cross-section and roughness elements that would connect the centres of two DEM grid cells. The cross-section becomes triangular when the bed width is zero and rectangular when the bank angle is zero.

1980, or Dingman, 2002, Appendix E), for modelling the shortwave solar radiation and the *Stephan–Boltzmann law* for modelling the longwave radiation. Most clear-sky radiation models incorporate topographic effects via slope and aspect grids extracted from DEMs.

Since the input data required for energy balance calculations is only available in well-instrumented watersheds, much simpler methods for estimating the rate of snowmelt have been developed such as various forms of the well-known *Degree-Day Method* (Beven, 2000, p. 80). The basic method predicts the meltrate using the simple formula:

$$M = c_0 \cdot \Delta T \tag{2.2}$$

where ΔT is the temperature difference between the air and the snow and c_0 is an empirical coefficient with units of [mm/hr/°C]. In both the Degree-Day and Energy-Balance methods it is possible for any input variable to vary spatially and in time, and many authors suggest that c_0 should vary throughout the melt season. An example comparison of these two methods is given by Bathurst and Cooley (1996). Whatever method is used, the end result is a grid sequence of snowmelt rates, M, that is then used in Equation (2.1).

2.4 The channel flow process

Spatial hydrologic models are based on conservation of mass and momentum, and many of them make direct use of D8 flow direction grids and slope grids to compute the amount of mass and momentum that flows into and out of each grid cell. The grid cell size is generally chosen to be smaller than the hillslope scale and larger than the width of the largest channel (see Section 3). Every grid cell then has one channel associated with it that extends from the centre of the grid cell to the centre of the grid cell that it flows to according to the D8 method. Channelised flow is then modelled as an essentially 1D process (in a treelike network of channels), while recognising that it will be necessary to store additional channel properties for every grid cell such as:

- sinuosity or channel length;
- channel bed width;
- bank angle (if trapezoidal cross sections are used) and;
- a channel roughness parameter.

One method for creating these channel property grids is discussed in Section 4.

The *kinematic wave* method is the simplest method for modelling flow in open channels and is available as an option in virtually all spatial hydrologic models. This method combines mass conservation with the simplest possible treatment of momentum conservation, namely that all terms in the general momentum equation (pressure gradient, local acceleration and convective acceleration) are negligible except the friction and gravity terms. In this case the water surface slope, energy slope and bed slope are all equal. In addition, the balance of gravity against friction (as a shear stress near the bed) results in an equation for depth-averaged flow velocity, u, in terms of the flow depth, d, bed slope (rise over run), S, and a roughness parameter. If the shear stress near the bed is computed using our best theoretical understanding of turbulent boundary layers (Schlicting, 1960), then this balance results in the *law of the wall*:

$$u = (g \cdot R_h \cdot S)^{1/2} \cdot \ln\left(a \cdot \frac{d}{z_0}\right) \cdot \kappa^{-1} \tag{2.3}$$

Here, g is the gravitational constant, R_h is the *hydraulic radius*, given as the ratio of the wetted cross-sectional area and wetted perimeter (units of length), a is an integration constant (given by 0.368 or 0.476, depending on the formulation), z_0 is the *roughness height* (units of length), and $\kappa \approx 0.408$ is von Karman's constant.

Note that the law of the wall is general and is also used by the snowmelt energy-balance models for modelling air flow in the atmospheric boundary layer. However, in the setting of open-channel flow, an alternative known as *Manning's formula* is more often used. Manning's formula, which was determined by fitting a power-law to data gives the depth-averaged velocity as:

$$u = \frac{R_h^{2/3} \cdot S^{1/2}}{n} \tag{2.4}$$

where n is an empirical roughness parameter with the units of $[\mathrm{s/m^{1/3}}]$ required to make the equation dimensionally consistent. Manning's formula agrees very well with the law of the wall as long as the relative roughness, z_0/d is between about 10^{-2} and 10^{-4}. This is the range that is encountered in most open-channel flow problems. Smaller relative roughnesses are typically encountered in the case where wind blows over terrain and vegetation. ASCE Task Force on Friction Factors (1963) provides a good review of the long and interesting history that led to Equations (2.3) and (2.4).

While the kinematic wave method is an approximation, it often yields good results, especially when slopes are steep. The *diffusive wave* method provides a somewhat better approximation by retaining one additional term in the momentum equation, namely the pressure-gradient (water depth derivative) term. In this method, the slope of the free water surface is used instead of the bed slope, and pressure-related (e.g. backwater) effects can be modelled. Note that a general treatment of momentum conservation uses the full *St. Venant equation*, which includes the effects of gravity, friction and pressure-gradients as well as terms for local and convective acceleration. The convective acceleration term corresponds to the net flux of momentum into a given control volume. The most accurate but most computationally demanding approach retains all of the terms in the St. Venant equation and is known as the *dynamic wave* method. Interestingly, the latter two methods create a water-depth gradient and can thereby move water across flat areas (e.g. lakes) in a DEM. These areas have a bed slope of zero and therefore receive a velocity of zero in the kinematic wave method unless they are handled separately in some manner. Whether the kinematic, diffusive or dynamic wave method is used, it is necessary to compute a grid of bed slopes. Given a DEM with sufficient vertical resolution, the bed slope can be computed between each grid cell and its downstream neighbour, as indicated by a D8 flow grid (see Chapter 7).

The D8 flow direction grid indicates the (static) connectivity of the grid cells in a DEM and can therefore be used directly to simplify mass and momentum balance calculations. A D8 flow grid allows fluxes across grid cell boundaries to be computed, which makes it possible to use integral equations instead of differential equations (Welty et al., 1984). In particular, the use of integral equations is simpler because convective acceleration (momentum flux) between cells can be modelled without computing spatial derivatives. Grids for the initial flow depth, d, and velocity, u, are specified, either as all zeros or computed from channel properties and a base-level recharge rate. Given the cross-sectional shape (e.g. trapezoidal) and length, L, of each channel, the volume of water in the channel can be computed as $V = A_c \cdot L$, where A_c is the cross-sectional area. An outgoing discharge, $Q = u \cdot A_c$, can also be computed for every grid cell. For each time step, the change in volume $V(i,t)$ for pixel i can then be computed as:

$$\Delta V(i,t) = \Delta t \cdot \left[R(i,t) \cdot \Delta x \cdot \Delta y - Q(i,t) + \sum_{k \in N} Q(k,t) \right] \tag{2.5}$$

where R is the excess rainrate computed from Equation (2.1), Δx and Δy are the pixel dimensions, $Q(i,t)$ is the outgoing discharge from pixel i at time t, and the summation is over all of the neighbour pixels that have D8-flow into pixel i.

Once Equation (2.5) has been used to update V for each pixel, the grid of flow depths, d, can be updated using the channel geometry grids that give the length, bed width and bank angle of each channel. In the case of the kinematic wave approximation, the grids d and S can then be used to update the grid of flow velocities, u, using either Equation (2.3) or Equation (2.4). For an integral-equation version of the dynamic wave method, the velocity grid, $u(i,t)$, would be incremented by an amount:

$$\begin{aligned}\Delta u(i,t) = \left(\frac{\Delta t}{d(i,t)\cdot A_w}\right)\cdot \Big\{ & u(i,t)\cdot Q(i,t)\cdot (C-1) \\ & + \sum_{k\in N} \big[u(k,t) - u(i,t)\cdot C\big]\cdot Q(k,t) \\ & - u(i,t)\cdot C\cdot R(i,t)\cdot \Delta x\cdot \Delta y \\ & + A_w \cdot \big[g\cdot d(i,t)\cdot S(i,t) - f(i,t)\cdot u^2(i,t)\big]\Big\}\end{aligned} \tag{2.6}$$

where A_w is the wetted surface area of the bed, A_t is the top surface area of the channel and $C = A_w/A_t$. For overland grid cells, $C = 1$, and for channel grid cells $C > 1$. A_w and A_t are computed from the grid of channel lengths, L, and the assumed cross-sectional shape. In the last term, $f \equiv \tau_b/(\rho \cdot u^2)$ is a dimensionless friction factor:

$$f = \left[\frac{\kappa}{\ln\left(a\cdot \frac{d}{z_0}\right)}\right]^2 \tag{2.7}$$

which corresponds to the law of the wall, while $f = g\cdot n^2 \cdot R_h^{-1/3}$ corresponds to Manning's equation. Instead of using the bed slope for S in Equation (2.6), the water surface slope would be computed from the DEM, d and the D8 flow direction grid. As the numerical approach shown here is *explicit*, numerical stability requires a small enough time step such that water cannot flow across any grid cell in less than one time step. If u_m is the maximum velocity, then we require $\Delta t < \Delta x/u_m$ for stability.

2.5 The overland flow process

The fundamental concept of *contributing area* was introduced in previous chapters (see Chapter 7). Grid cells with a sufficiently large contributing area will tend to have higher and more persistent surface fluxes and channelised flow. Conversely, grid cells with small contributing areas will tend to have lower, intermittent fluxes. The intermittent nature of runoff-generating events, and the increased likelihood that small amounts of water will be fully consumed by infiltration or evapotranspiration make it even more likely that grid cells with small contributing areas will have little or no surface flux for much of the time. In addition, the *relative roughness* of the surface (typical height of roughness elements divided by the water depth) is higher for smaller contributing areas so that frictional processes will be more efficient at slowing the flow. Under these circumstances the shear stress[1] on the land-surface will tend to be too small to carve a channel or too infrequent to maintain a channel.

Any surface flux will be as so-called *overland* or *Hortonian* flow and will tend to flow in a sheet that wets the entire bottom surface of a grid cell control volume

[1] Proportional to the square of the flow velocity.

during an event. This flow may be modelled with either a 1D or 2D approach, where the latter method would be required to model flood events that exceed the bankfull channel depth, e.g. a dam break. In this case both channelised and overland flow must be modelled for channel grid cells.

Some models, such as CASC2D (Julien and Saghafian, 1991) have a *retention depth* (surface storage) that must be exceeded before overland flow begins. Note that for sheet flow, the hydraulic radius, R_h is very closely approximated by the flow depth, d. If w is the width of the grid cell projected in the direction of the flow, then the wetted area is given by $w\,d$ and the "*wetted perimeter*" is given by w. It follows that the hydraulic radius is equal to d. It has been found by Eagleson (1970) and many others since that Manning-type equations can be used to compute the flow velocity for overland flow, but that a very large "Manning's n" value of around 0.3 or higher is required, versus a typical value of 0.03 for natural channels.

2.6 The evaporation process

Evaporation is a complex, essentially vertical process that moves water from the Earth's surface and subsurface to the atmosphere. As with the snowmelt process, the most sophisticated approach is based on a full surface energy balance in which topographic effects can be incorporated by including grids of slope and aspect in the solar radiation model. However, since much of the required input data is typically unavailable, a number of simpler models have been proposed. The *Priestley–Taylor* (Priestley and Taylor, 1972; Rouse and Stewart, 1972; Rouse et al., 1977; Zhang et al., 2000) and *Penman–Monteith* models (Beven, 2000; Dingman, 2002) and their variants are two simplified approaches that are used widely. Sumner and Jacobs (2005) provide a comparison of these and other methods.

Whatever method is used, the end result is a grid sequence of evapotranspiration rates, E, that is then used in Equation (2.1). Some distributed hydrologic models have additional routines for modelling the amount of water that is moved from the root zone of the subsurface to the atmosphere by the transpiration of plants. A separate submodel is sometimes used to model the variation of soil temperature with depth, especially for high-latitude applications.

2.7 The infiltration process

The process of infiltration is also primarily vertical, but is arguably the most complex hydrologic process at work in a basin. It has a first-order effect on the hydrologic response of watersheds, and is central to problems involving surface soil moisture. It operates in the *unsaturated zone* between the surface and the water table and represents an interplay between absorption due to capillary action and the force of gravity. A variety of factors make realistic modelling of infiltration difficult, including the nature of boundary conditions at the surface, between soil layers and at the water table (a moving boundary). Variables such as hydraulic conductivity can vary over orders of magnitude in both space and time and the equations are strongly nonlinear.

As pointed out by many authors, including Smith (2002), it is generally not sufficient to simply use spatial averages for input parameters, and best methods for parameter estimation are an active area of research. So-called macropores may be present and must then be modelled separately since they do not conform to the standard notion of a porous medium. Discontinuous permafrost may also be present in high-latitude watersheds. Smith (2002) provides an excellent reference for infiltration theory, ranging from very simple to advanced approaches.

Most spatial hydrologic models use a variant of the *Green–Ampt* or *Smith–Parlange* method for modelling infiltration (Smith, 2002). However, these are simplified approaches that are intended for the relatively simple case where there is:

- a single storm event,
- a single soil layer and
- no water table.

While they can be useful for predicting flood runoff, they are not able to address many other problems of contemporary interest, such as:

(1) redistribution of the soil moisture profile between runoff-producing events,
(2) drying of surface layers due to evaporation at the surface,
(3) rainfall rates less than K_S (saturated hydraulic conductivity),
(4) multiple soil layers with different properties, and
(5) the presence of a dynamic water table.

In order to address these issues and to model surface soil moisture a more sophisticated approach is required.

Infiltration in a porous medium is modelled with four basic quantities which vary spatially throughout the subsurface and with time. The *water content*, θ, is the fraction of a given volume of the porous medium that is occupied by water, and must therefore always be less than the *porosity*, ϕ. In the case of soils, θ represents the *soil moisture*. The *pressure head* (or capillary potential), ψ, is negative in the unsaturated zone and measures the strength of the capillary action. It is zero at the water table and positive below it. The *hydraulic conductivity*, K, has units of velocity and depends on the gravitational constant, the density and viscosity of water and the intrinsic permeability of the porous medium.

Darcy's Law, which serves as a good approximation for both saturated and unsaturated flow, implies that the vertical flow rate, v, is given by:

$$v = -K \cdot \frac{dH}{dz} = K \cdot \left(1 - \frac{d\psi}{dz}\right) \tag{2.8}$$

since $H = \psi - z$ (and z is positive downward). Conservation of mass for this problem takes the form:

$$\frac{\partial \theta}{\partial t} + \frac{\partial v}{\partial z} = J \tag{2.9}$$

where J is an optional source/sink term that can be used to model water extracted by plants. Inserting Equation (2.8) into Equation (2.9) we obtain *Richards' equation*:

$$\frac{\partial \theta}{\partial t} = \frac{\partial}{\partial z}\left[K \cdot \left(\frac{\partial \psi}{\partial z} - 1\right)\right] \tag{2.10}$$

for vertical, one-dimensional unsaturated flow. Many spatial models solve this equation numerically to obtain a profile of soil moisture vs. depth for every grid cell, between the surface and a dynamic water table. However, in order to solve for the four variables, θ, ψ, K and v, two additional equations are required in addition to Equations (2.8) and (2.9). These extra equations have been determined empirically by extensive data analysis and are called *soil characteristic functions*.

The soil characteristic functions most often used are those of Brooks and Corey (1964), van Genuchten (1980) and Smith (1990). Each expresses K and ψ as functions of θ and contains parameters that depend on the porous medium under consideration (e.g. sand, silt, or loam).

The *transitional Brooks–Corey method* combines key advantages of the Brooks–Corey and van Genuchten methods (Smith, 1990, 2002, pp. 18–23). Water content, θ, is first rescaled to define a quantity called the *effective saturation*:

$$\Theta_e = \frac{\theta - \theta_r}{\theta_s - \theta_r} \tag{2.11}$$

that lies between zero and one. Here, θ_s is the *saturated water content* (slightly less than the porosity, ϕ) and θ_r is the *residual water content* (a lower limit that cannot be lowered via pressure gradients). Hydraulic conductivity is then modelled as:

$$K = K_s \cdot \Theta_e^{\epsilon} \tag{2.12}$$

where K_s is the *saturated hydraulic conductivity* (an upper bound on K) and $\epsilon = (2+3\lambda)/\lambda$, where λ is the *pore size distribution parameter*. Pressure head is modelled as:

$$\psi = \psi_B \cdot \left[\Theta_e^{-c/\lambda} - 1\right]^{1/c} - \psi_a \tag{2.13}$$

where ψ_B is the *bubbling pressure* (or *air-entry tension*, ψ_a is a small shift parameter (which may be used to approximate hysteresis or set to zero), c is the *curvature parameter* which determines the shape of the curve near saturation.

Equations (2.8), (2.9), (2.12) and (2.13) provide a very flexible basic framework for modelling 1D infiltration in spatial hydrologic models. The precipitation rate, P, the snowmelt rate, M, and evapotranspiration rate, E, are needed for the upper boundary condition. The vertical flow rate computed at the surface, v_0, determines I in Equation (2.1).

2.8 The subsurface flow process

Once infiltrating water reaches the water table, the hydraulic gradient is such that it typically begins to move horizontally, roughly parallel to the land surface. The water table height may rise or fall depending on whether the net flux is downward (infiltration) or upward (exfiltration, due to evapotranspiration). Darcy's law [Equation (2.8)] continues to hold but $K = K_s$, $\theta = \theta_s \approx \phi$ and $\psi = 0$ at the water table, with hydrostatic conditions ($\psi > 0$) below it. More details on the equations used to model saturated flow are given by Freeze and Cherry (1979).

For shallow subsurface flow, various simplifying assumptions are often applicable, such as (1) the subsurface flow direction is the same as the surface flow direction and (2) the porosity decreases with depth. Under these circumstances the water table height can be modelled as a grid that changes in time, using a control volume below each DEM grid cell that extends from the water table down to an impermeable lower surface (e.g. bedrock layer). Infiltration then adds water just above the water table at a rate determined from Richard's equation and water moves laterally through the vertical faces at a rate determined by Darcy's law. The dynamic position of the water table is compared to the DEM; if it reaches the surface anywhere, then the rate at which water seeps to the surface provides a grid sequence, G, that is used in Equation (2.1). Multiple layers, each with different hydraulic properties and spatially-variable thickness can be modelled, but this increases the computational cost.

2.9 Flow diversions: sinks, sources and canals

Flow diversions are present in many watersheds and may be modelled as another "process". Man-made canals or tunnels are often used to divert flow from one location to another, and usually cannot be resolved by DEMs. They are typically used for irrigation or urban water supplies. Tunnels may even carry flow from one side of a drainage divide to the other. Given the flow rate at the upstream end and other information such as the length of the diversion, these structures can be incorporated into distributed models. Diversions can be modelled by providing a mechanism (outside of the D8 framework) for transferring water between two non-adjacent grid cells. Sources and sinks may be man-made or natural and simply inject or remove flow from a point location at some rate. If the rate is known, their effect can also be modelled. It is increasingly uncommon to find watersheds that are not subject to human influences.

3. SCALE ISSUES IN SPATIAL HYDROLOGIC MODELS

While the preceding sections may give the impression that spatial hydrologic modelling is simply a straight-forward application of known physical laws, this is far from true. Many authors have pointed out that physically-based mathematical models developed and tested at a particular scale (e.g. laboratory or plot) may be inappropriate or at least gross simplifications when applied at much larger scales. In addition, heterogeneity in natural systems (e.g. rainfall, snowpack, vegetation, soil properties) means that some physical parameters appearing in models may vary considerably over distances that are well below the proposed model scale (grid spacing). It is therefore a nontrivial question as to how (or whether) a small number of "*point*" measurements can be used to set the parameters of a distributed model. *Variogram analysis* provides one tool for addressing this problem and seeking a correlation length that may help to select an appropriate model scale. For some model parameters, remote sensing can provide an alternative to using point measurements.

The issue of *upscaling*, or how best to move between the measurement scale, process scale and model scale is very important and presents a major research challenge. A standard approach to this problem that has met with some success is the use of *effective parameters*. The idea is that using a representative value, such as a spatial average, might make it possible to apply a plot-scale mathematical model at the much larger scale of a model grid cell. Unfortunately, the models are usually nonlinear functions of their parameters so a simple spatial average is almost never appropriate. It is well-known in statistics that if X is some model parameter that varies spatially, f is a nonlinear[2] function and $Y = f(X)$ is a computed quantity, then $E[f(X)] \neq f(E[X])$. Here E is the expected value, akin to the spatial average. So, for example, the mean infiltration rate over a model grid cell (and associated net vertical flux) cannot be computed accurately by simply using mean soil properties (e.g. hydraulic conductivity) in Richards' equation.

An interesting variant of the effective parameter approach is to parameterise the subgrid variability of turbulent flow fields by replacing the molecular viscosity in the time-averaged model equations with an *eddy viscosity* that is allowed to vary spatially. This approach is successfully used by many ocean and climate models and may provide conceptual guidance for hydrologic modelers.

When it comes to the channel network and D8 flow between grid cells, upscaling is even more complicated because there is a fairly abrupt change in process dominance at the *hillslope scale* which marks the transition from overland to channelised flow. As seen in Section 6 of Chapter 7, this scale depends on the region and is needed in the pruning step when extracting a river network from a DEM. If the grid spacing is small enough to resolve the local hillslope scale, then it is possible to classify each grid cell as either hillslope or channel. Each channel grid cell will typically contain a single channel with a width that is less than the grid spacing, as well as some "*hillslope area*". Momentum balance can be modelled as long as channel properties such as length and bed width are stored for each grid cell, and the vertical resolution of the DEM is sufficient to compute the bed slope. However, if the grid spacing is larger than the hillslope scale, then a single grid cell may contain a dendritic network vs. a single channel. This is a much more complicated situation, but it may still be possible to get acceptable results by modelling flow in the cell's dendritic network with a single "*effective*" channel, using effective parameters.

> REMARK 3. *Physically-based mathematical models developed and tested at a particular scale (e.g. laboratory or plot) may be inappropriate or at least gross simplifications when applied at much larger scales. The issue of upscaling, or how best to move between the measurement scale, process scale and model scale is very important and represents a major research challenge.*

Using effective parameters and other upscaling methods, researchers have reported successful applications of spatial hydrologic models from the plot scale all the way up to the continental scale. Interestingly, the same model (e.g. MIKE SHE), but with very different parameter settings, can often be used at these two very different scales. While conventional wisdom suggests that traditional, lumped or semi-distributed models are better for large-scale applications, this has been

[2] Anything other than $a \cdot X + b$.

largely for computational reasons and is becoming less of an issue. Note also that a distributed model is similar in many ways to a lumped model when a large grid spacing is used, although a lumped model may subdivide a watershed into a more natural set of linked control volumes.

Although much more work needs to be done on scaling issues, considerable guidance to modelers is available in the literature. Examples of some good general references include Gupta et al. (1986), Blöschl and Sivapalan (1995), Blöschl (1999a) and Beven (2000). References for specific processes include Dagan (1986) (groundwater), Gupta and Waymire (1993) (rainfall), Wood and Lakshmi (1993) (evaporation and energy fluxes), Peckham (1995b) (channel network geometry and dynamics), Woolhiser et al. (1996) (overland flow), Blöschl (1999b) (snow hydrology) and Zhu and Mohanty (2004) (infiltration).

4. PREPROCESSING TOOLS FOR SPATIAL HYDROLOGIC MODELS

As explained in the previous sections, most spatially-distributed hydrologic models make direct use of a DEM and several DEM-derived grids, including a flow direction grid (aspect), a slope grid and a contributing area grid. Extraction of these grids from a DEM with sufficient vertical and spatial resolution is therefore a necessary first step and may require depression filling or burning in streamlines as already explained in detail in previous chapters (e.g. Chapters 4 and 7). But spatially-distributed models require a fair amount of additional information to be specified for every grid cell before any predictions can be made.

Initial conditions are one type of information that is required. Examples of initial conditions include the initial depth of water, the initial depth of snow, the initial water content (throughout the subsurface) and the initial position of the water table. Each of these examples represents the starting value of a dynamic variable that changes in time. *Channel geometry* is another type of required information, but is given by static variables such as length, bed width, bed slope, bed roughness height and bank angle. Each of these must also be specified for every grid cell or corresponding channel segment. *Forcing variables* are yet another type of information that is required and they are often related to weather. Examples include the precipitation rate, air temperature, humidity, cloudiness, wind speed, and clear-sky solar radiation.

Each type of information discussed above can in principle be measured, but it is virtually impossible to measure them for every grid cell in a watershed. As a result of this fact, these types of measurements are typically only available at a few locations (i.e. stations) as a time series, and interpolation methods (such as the inverse distance method) must be used to estimate values at other locations and times. This important task is generally performed by a preprocessing tool, which may or may not be included with the distributed model.

REMARK 4. *A variety of pre- and postprocessing tools are required to support the use of spatial hydrologic models.*

Another important preprocessing step is to assign reasonable values for channel properties to every spatial grid cell. Some spatial hydrologic models provide a preprocessing tool for this purpose. One method for doing this is to parameterise them as best-fit, power-law functions of contributing area. That is, if A denotes a contributing area grid, then a grid of approximate channel widths can be computed via:

$$w = c \cdot (A + b)^p \tag{4.14}$$

where the parameters c, b and p are determined by a best fit to available data. The same approach can be used to create grids of bed roughness values and bank angles. This approximation is motivated by the well-known *empirical equations of hydraulic geometry* (Leopold et al., 1995) that express hydraulic variables as powers of discharge, and discharge as a power of contributing area. Measurements (e.g. channel widths) to determine best-fit parameters may be available at select locations such as gauging stations, or may be estimated using high-resolution, remotely-sensed imagery.

For an initial condition such as flow depth, an iterative scheme (e.g. Newton–Raphson) can be used to find a steady-state solution given the channel geometry and a baseflow recharge rate; this *normal flow* condition provides a reasonable initial condition. Alternately, a spatial model may be "*spun up*" from an initial state where flow depths are zero everywhere and run until a steady-state baseflow is achieved. Similar approaches could be used to estimate the initial position of the water table. Methods for estimating water table height based on wetness indices have also been proposed (see Section 6 in Chapter 7, Section 4.2 in Chapter 8 and Beven, 2000). Any of these approaches may be implemented as a preprocessing tool.

When energy balance methods are used to model snowmelt or evapotranspiration, it is necessary to compute the net amount of shortwave and longwave radiation that is received by each grid cell. As part of this calculation one needs to compute the *clear-sky solar radiation* as a grid stack indexed by time. The concepts behind computing clear-sky radiation are discussed in Section 3.1 of Chapter 8 and are also reviewed by Dingman (2002, Appendix E). The calculation uses celestial mechanics to compute the declination and zenith angle of the sun, as well as the times of local sunrise and sunset. It also takes the slope and aspect of the terrain into account (as grids), along with several additional variables such as surface albedo, humidity, dustiness, cloudiness and optical air mass. A general approach models direct, diffuse and backscattered radiation.

Another useful type of preprocessing tool is a rainfall simulator. One method for simulating space-time rainfall uses the mathematics of *multifractal cascades* (Over and Gupta, 1996) and reproduces many of the space–time scaling properties of convective rainfall.

It should be noted that DEMs with a vertical resolution of one meter do not permit a sufficiently accurate measurement of channel slope using the standard, local methods of geomorphometry. Channel slopes are often between 10^{-2} and 10^{-5}, but for a DEM with vertical and horizontal resolutions of 1 and 10 meters, respectively, the minimum resolvable (nonzero) slope is 0.1. The author has developed an experimental *"profile-smoothing"* algorithm for addressing this issue that is available as a preprocessing tool in the TopoFlow model.

5. CASE STUDY: HYDROLOGIC RESPONSE OF NORTH BASIN, BARANJA HILL

As a simple example of how a spatial hydrologic model can be used to simulate the hydrologic response of a watershed, in this section we will apply the TopoFlow model to a small watershed that drains to the northern edge of the Baranja Hill DEM. This is the largest complete watershed in the Baranja Hill DEM, an area in Eastern Croatia that is used for examples throughout this book.

TopoFlow is a free, community-based, hydrologic model that has been developed by the author and colleagues. The TopoFlow project is an ongoing, open-source, collaborative effort between the author and a group of researchers at the University of Alaska, Fairbanks (L. Hinzman, M. Nolan and B. Bolton). This effort began with the idea of merging two spatial hydrologic models into one and adding a user-friendly, point-and-click interface. One of these models was a D8-based, rainfall-runoff model written by the author, which supported both kinematic and dynamic wave routing, as well as both Manning's formula and the law of the wall for flow resistance. The second model, called ARHYTHM, was written by L. Hinzman and colleagues (Zhang et al., 2000) for the purpose of modelling Arctic watersheds; it therefore contained advanced methods for modelling thermal processes such as snowmelt, evaporation and shallow-subsurface flow. In addition to its graphical user interface, TopoFlow now provides several different methods for modelling infiltration (from Green–Ampt to the 1D Richards' equation) and also has a rich set of preprocessing tools (Figure 3). Examples of such tools include a rainfall simulator, a data interpolation tool, a channel property assignment tool and a clear-sky solar radiation calculator.

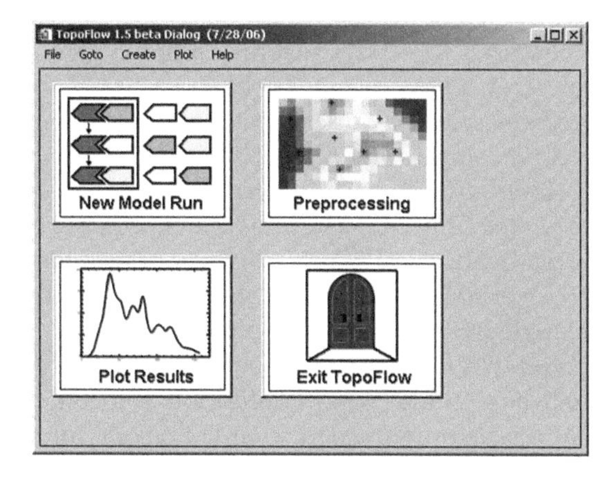

FIGURE 3 The main panel in TopoFlow.

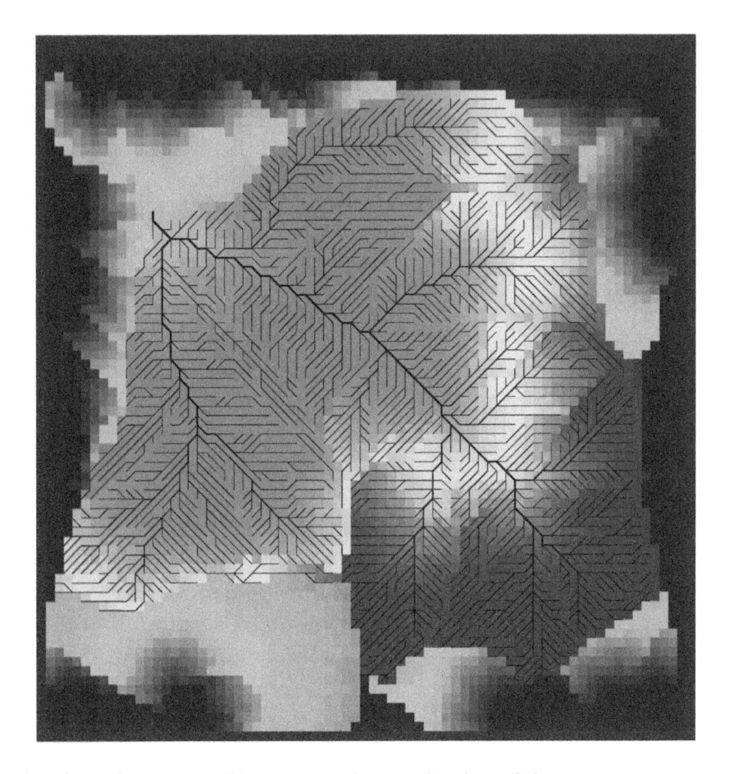

FIGURE 4 Flow lines for the small basin near the north edge of the Baranja DEM, as extracted from a DEM by the D8 method. The flow lines are overlaid on a colour image that shows flow distance to the basin outlet. (See page 755 in Colour Plate Section at the back of the book.)

Before starting **TopoFlow**, **RiverTools** 3.0 (see Chapter 18) was used to clip a small DEM from the Baranja Hill DEM that contained just the north basin. This DEM had only 73 columns and 76 rows, but the same grid spacing of 25 meters. It had minimum and maximum elevations of 85 and 243 meters, respectively. **RiverTools** 3.0 was then used to extract several D8-based grids, including a flow direction grid, a slope grid, a flow distance grid and a contributing area grid. The drainage network above a selected outlet pixel (near the village of Popovac) was also extracted and had a contributing area of 1.84 square kilometers and a fairly large main-channel slope of 0.04 [m/m]. **RiverTools** automatically performs pit-filling when necessary (see Chapter 7) but this was not much of an issue for this DEM because of its relatively steep slopes. Figure 4 shows the D8 flow lines for this small watershed, overlaid on a grid that shows the flow distance to the edge of the bounding rectangle with a rainbow colour scheme.

The **TopoFlow** model was then started as a plug-in from within **RiverTools** 3.0. It can also be started as a stand-alone application using the IDL Virtual Machine, a free tool that can be downloaded from ITT Visual Information Solutions (http://www.ittvis.com/idl/). Figure 5 shows the wizard panel in **TopoFlow** that is used to select which physical processes to model and which method to use

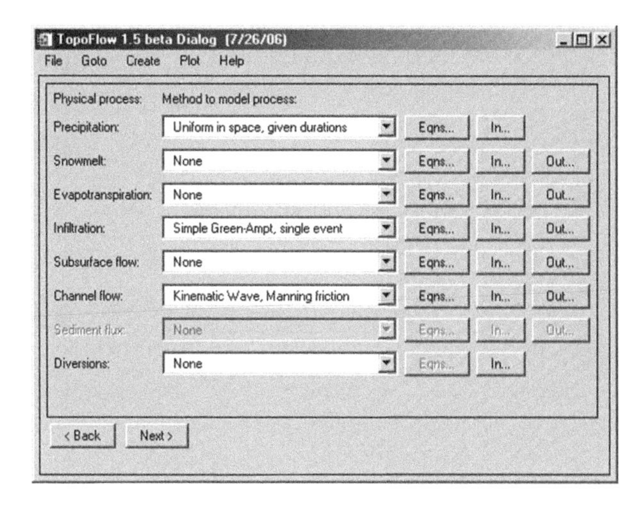

FIGURE 5 A dialog in the TopoFlow model that allows a user to select which method to use (if any) to model each hydrologic process from a droplist of choices. Once a choice has been selected, clicking on the "*In...*" or "*Out...*" buttons opens an additional dialog for entering the parameters required by that method. Clicking on the "*Eqns...*" button displays the set of equations that define the selected method.

for each process. Several methods are provided for modelling each hydrologic process, including both simple (e.g. degree-day, kinematic wave) and sophisticated (e.g. energy balance, dynamic wave) methods. In this example, spatially uniform rainfall with a rate of 100 [mm/hr] and a duration of 4 minutes was selected for the Precipitation process, but gridded rainfall for a fixed duration or space-time rainfall as a grid stack of rainrates and a 1D array of durations could have been used. For the channel flow process, the kinematic wave method with Manning's formula for computing the flow velocity was selected. Clicking on the button labeled "*In...*" in the Channel Flow process row opened the dialog shown in Figure 6.

All of the input dialogs in TopoFlow follow this same basic template; either a scalar value can be entered in the text box or the name of a file that contains a time series, grid or grid sequence. The filenames of the previously extracted D8-based grids for flow direction and slope (from RiverTools) were entered into the top two rows of this dialog. The filenames for Manning's n, channel bed width and channel bank angle as grids were entered in the next three rows. These were created with a preprocessing tool in TopoFlow's Create menu that uses a contributing area grid and power-law formulas to parameterise these quantities.

If available, field measurements can be entered to automatically constrain the power-law parameters, but for this case study default settings were used. This resulted in a largest channel width of 4.1 meters, which may be too large for such a small basin (1.84 km^2). The corresponding value of Manning's n was 0.02, which may similarly be too small. A value of 0.3 was used for overland flow. For this small watershed, a uniform scalar value of 1.0 was used for the channel sinuosity. The initial flow depth was set to 0.0 for all pixels, although TopoFlow has

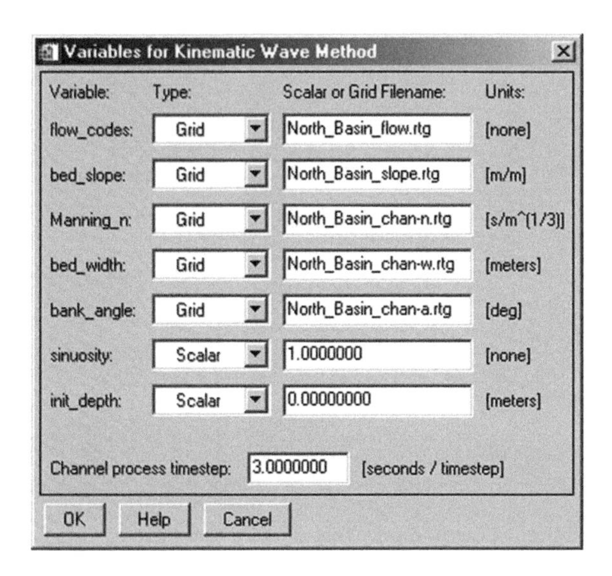

FIGURE 6 The TopoFlow dialog used to enter required input variables for the "*Kinematic Wave, Manning's n*" method of modelling channel flow. Notice that the data type (scalar, time series, grid or grid sequence) of each variable can be selected from a droplist. If the data type is "*Grid*", then a filename is typed into the text box. These names refer to grids that were created with preprocessing tools. Units are always shown at the right edge of the dialog.

another preprocessing tool for computing base-level channel flow depths in terms of an annual recharge rate and the other channel parameters. The channel process timestep at the bottom was set to a value of 3 seconds, as shown. This timestep was automatically estimated by **TopoFlow** as the largest timestep that would provide numerical stability. By clicking on the button labeled "*Out...*" in the Channel Flow process row, the dialog shown in Figure 7 was opened. This dialog allows a user to choose the type of output they want, and for which variables. **TopoFlow** allows user-selected output variables to be saved to files either as a time series (for one or more monitored grid cells) or as a grid stack indexed by time. The check boxes in Figure 7 indicate that a grid stack and a time series (at the basin outlet) should be created for every output variable. A sampling timestep of one minute was selected; this gives a good resolution of the output curves (e.g. *hydrograph*) but is much larger than the channel process timestep of 3 seconds that is required for numerical stability.

Once all of the input variables were set, the model was run with the infiltration process set to None. The resulting hydrograph is shown as the top curve in Figure 10. The "*Simple Green–Ampt, single event*" method was then selected from the droplist of available infiltration process methods. Clicking on the button labeled "*In...*" in the infiltration process row opened the dialog shown in Figure 8. Toward the bottom of this dialog, "*Clay loam*" was selected as the closest standard soil type and the default input variables in the dialog were updated to ones typical of this soil type. The initial value of the soil moisture, shown as `theta_i` was changed from the default of 0.1 to the value 0.35. The infiltration process timestep

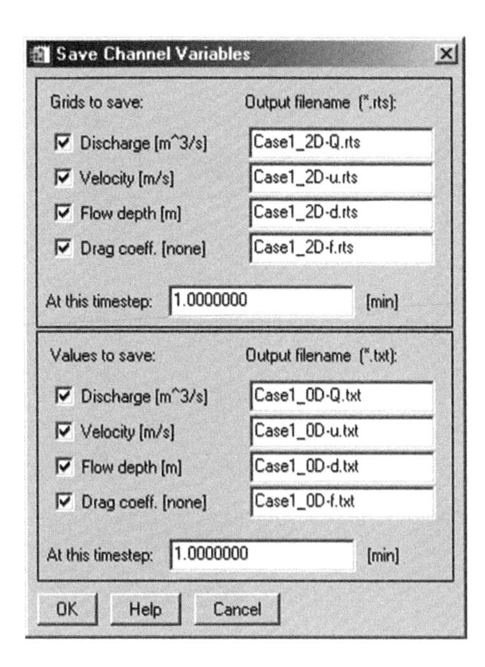

FIGURE 7 The TopoFlow dialog used to choose how model output for the channel flow process is to be saved to files. Any output variable can be saved as either a time series for all monitored grid cells (in a multi-column text file) or as a sequence of grids. The time between saved values can be specified independently of the modelling timesteps.

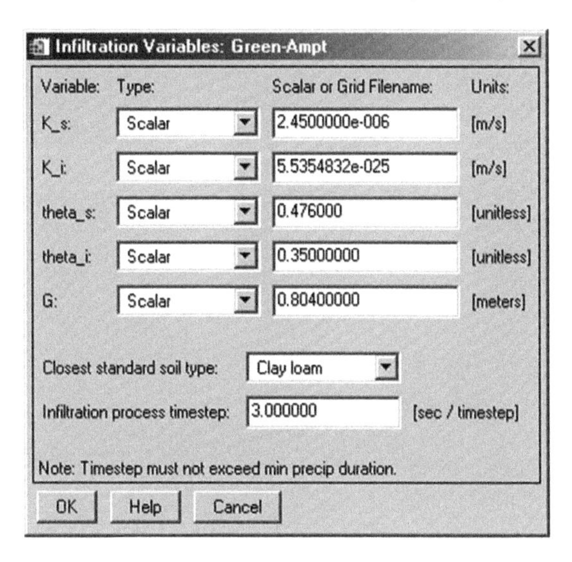

FIGURE 8 The TopoFlow dialog used to enter required input variables for the "*Green–Ampt, single event*" method of modelling infiltration. Here, scalars have been entered for every variable and will be used for all grid cells. Choosing an entry from the "*Closest standard soil type*" droplist changes the input variable defaults accordingly and can be helpful for setting parameters when other information is lacking. This is also useful for educational purposes.

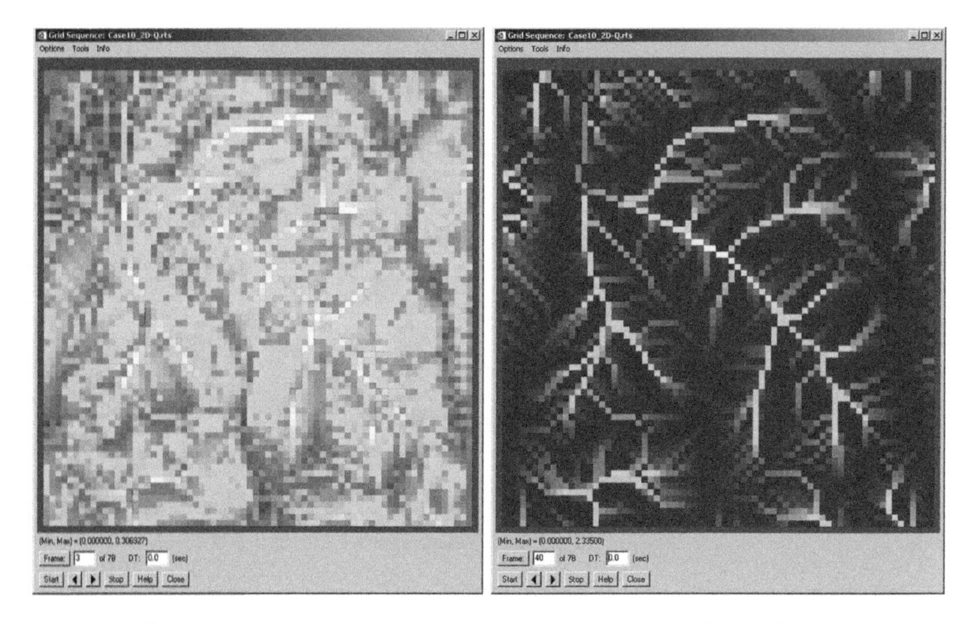

FIGURE 9 The *Display* → *Grid Sequence* dialog in RiverTools 3.0 can be used to view grid stacks as animations or to view/query individual frames. The frame on the left is early in a simulation, and shows flood pulses starting to converge. The frame on the right shows the spatial pattern of discharge well into the storm.

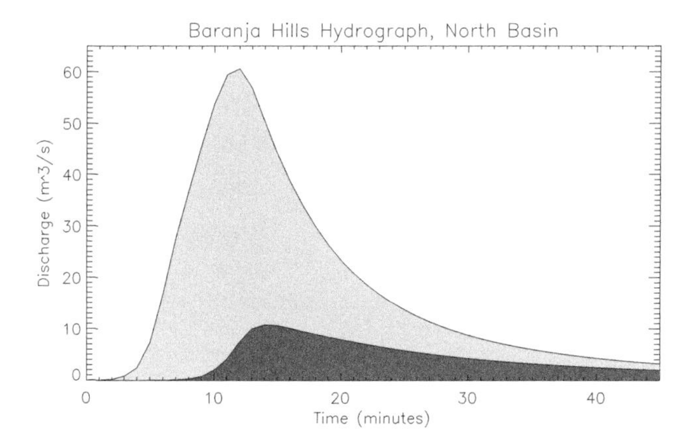

FIGURE 10 Two hydrographs, showing how the hydrologic response of the small basin differs in two simple test cases. Both cases use spatially uniform rainrate, but one also includes the effect of infiltration via the Green–Ampt method.

listed toward the bottom of the dialog was changed to 3.0 seconds per timestep, in order to match[3] the time-step of the channel flow process. When the model was run again with these settings, it produced the hydrograph shown as the bottom curve in Figure 10. It can be seen that, as expected, the inclusion of infiltration resulted in a much smaller peak in the hydrograph and also caused the peak to occur somewhat later. At the end of a model run, any saved time series, such as a hydrograph, can be plotted with the *Plot → Function* option. Similarly, any grid stack can be visualised as a colour animation with the *Plot → RTS File* option. The RTS (RiverTools Sequence) file format is a simple and efficient format for storing a grid stack of data. RTS files may be used to store input data, such as space-time rainfall, or output data, such as space-time discharge or water depth. RiverTools 3.0 (see Chapter 18) has similar but more powerful visualisation and query tools, including the *Display → Function* tool for functions (e.g. hydrographs and profiles), and the *Display → Grid Sequence* tool for grid stacks (see Figure 9). The latter tool allows grid stacks to be viewed frame by frame or saved as AVI movie files. It also has several interactive tools such as (1) a Time Profile tool for instantly extracting a time series of values for any user-selected grid cell and (2) an Animated Profile tool for plotting the movement of flood waves along user-selected channels.

It is important to realise that TopoFlow can perform much more complex simulations without much additional effort at run time. It allows virtually any input variable to any process to be entered as either a scalar (constant in space and time), a time series (constant in space, variable in time), a grid (variable in space, constant in time) or a grid stack (variable in space and time). It can also handle much larger grids than the one used in this case study. Advanced programming strategies including pointers, C-like structures, dynamic data typing and efficient I/O are used throughout TopoFlow for optimal performance and the ability to handle large data sets.

6. SUMMARY POINTS

Spatially-distributed hydrologic models make direct use of many geomorphometric variables. Flow direction or aspect is used to determine connectivity, or how water moves between neighbouring grid cells, and this same flow direction is also commonly used for subsurface flow. Slope is one of the key variables needed to compute flow velocity for both overland and channelised flow. Both slope and aspect are used to compute clear-sky solar radiation, which may then be used by an energy-balance method to model rates of snowmelt and evapotranspiration. Channel lengths (between pixel centres) are used in computing flow resistance. Elevation can be used together with a lapse rate to estimate air temperature. Total contributing area can be used to determine whether overland or channelised flow is dominant in a given grid cell and can also be used together with scaling relationships to set channel geometry variables such as bed width and roughness for every grid cell.

[3] It can often be set to a much larger value (minutes to hours).

One of the main advantages of spatially-distributed hydrologic models over other types of hydrological models is their ability to model the effects of human-induced change such as land use, dams, diversions, stream restoration, contaminant transport, forest fires and global warming. A truly amazing variety of problems can now be addressed with fully-spatial models that run on a standard personal computer. While much work remains in order to resolve issues such as upscaling, these models can be extremely useful if applied with an understanding of their strengths and limitations. Clearly, results do depend on grid spacing, and the greatest uncertainties occur when grid cells are larger than the hillslope scale. For small to medium-sized basins, the problem of upscaling appears to be tractable and significant progress has already been made. Note that many of the problems such as subgrid variability, modelling of momentum loss due to friction and specification of initial conditions are also encountered by fully-spatial climate and ocean models.

REMARK 5. *Spatial hydrologic models can address many types of problems that cannot be addressed with simpler models, such as those that involve the effects of human-induced changes to all or part of a watershed.*

In view of the large number of distributed models now used in hydrology and other fields, there is clearly a growing consensus that their advantages outweigh their disadvantages. A key attraction of physically-based, distributed models is that processes are modelled with parameters that have a physical meaning; note that even an effective parameter may have a well-defined physical meaning. These models also promote an integrated understanding of hydrology, rather than focusing on a particular process and neglecting others. These features combined with their visual appeal makes them very effective educational tools, especially when a variety of different methods are provided for modelling different processes, when any process can easily be turned off and when well-documented source code is made available.

IMPORTANT SOURCES

Rivix LLC, 2004. RiverTools 3.0 User's Guide. Rivix Limited Liability Company, Broomfield, CO, 218 pp.

Peckham, S.D., 2003. Fluvial landscape models and catchment-scale sediment transport. Global and Planetary Change 39 (1), 31–51.

Blöschl, G., 2002. Scale and Scaling in Hydrology — A Framework for Thinking and Analysis. John Wiley, Chichester, 352 pp.

Beven, K.J., 2000. Rainfall-Runoff Modelling: The Primer, 1st edition. John Wiley, New York, 360 pp.

Beven, K.J., 1997. TOPMODEL: a critique. Hydrological Processes 11 (9), 1069–1086.

CHAPTER 26

Applications in Meteorology

S. Emeis and **H.R. Knoche**

meteorological applications of geomorphometry · influence of topography, land form and land use on regional and local weather conditions · importance of weather and climate simulations · case studies where DEMs are used to run climate simulations

1. METEOROLOGY AND TOPOGRAPHY

The discipline *meteorology* comprises the physics and chemistry of the atmosphere and the climatology. Main application products are weather, climate, and air quality analyses and forecasts. We describe the influence which topographical features, especially hills and mountains, land form, land use, and soil type exert on the Earth's atmosphere and how these effects are introduced in numerical models which perform the above-mentioned analyses and forecasts. Case studies will exemplify the influence of topographical information on analyses and forecasts. This chapter will concentrate on the large and regional-scale aspects of topographical influences on atmospheric physics and chemistry. More information about local (topo-climatological) aspects, especially those related to the surface energy balance can be found in Chapter 8.

Due to the high complexity and non-linearity of the dynamics, thermodynamics and chemistry of atmospheric flows, reliable weather, climate, and air quality analyses and predictions cannot be made without numerical modelling. Global, regional and local-scale models for these purposes need — besides correct meteorological input data from ground-based and satellite observations — increasingly precise topographical information from increasingly more detailed DEMs. This information comprises especially the extension, shape, height, and slope of mountains, land form, land use, and soil type. Mountains essentially influence the patterns of atmospheric flow (see further examples and references

Developments in Soil Science, Volume 33 © 2009 Elsevier B.V.
ISSN 0166-2481, DOI: 10.1016/S0166-2481(08)00026-3.

given in Section 1.1) and precipitation patterns (see e.g. James and Houze Jr., 2005; Roe, 2005).

REMARK 1. *Today, high-resolution, regional and local-scale simulation models play an important role especially in local weather forecasts, in the regionalisation of climate predictions, but even in disaster prevention.*

Landform and land use alter due to the surface roughness the friction of atmospheric motions at the lower boundary. Land use (Stohlgren et al., 1998) and soil type (Smirnova et al., 1997; Lynn et al., 1998; Xiu and Pleim, 2001) determine the thermodynamic properties of the lower boundary like e.g. *albedo, emissivity for long-wave radiation, heat capacity, heat conductivity, moisture capacity* and *moisture diffusivity*. They also modify the exchange (emission and deposition) of substances such as moisture and other atmospheric trace gases and pollutants between the atmosphere and the Earth's surface (see e.g. Pleim et al., 2001). Such schemes by which the interaction between the surface and the atmosphere is described in numerical flow and climate models are usually called *soil–vegetation–atmosphere transfer* (SVAT) schemes (see Sellers et al., 1997, for a review).

Today, high-resolution regional and local-scale simulation models play a role in local weather forecasts, in the regionalisation of climate predictions and even in disaster prevention. Today's flood and avalanche predictions are also based on temperature, precipitation, and wind forecasts from local-scale weather prediction models.

1.1 Influence of orography on atmospheric flows and thermodynamics

All energy contained in the Earth's weather and climate system is supplied by the sun through shortwave radiation. Longitudinal and latitudinal gradients in the incoming radiation are the key source for atmospheric motions. This motion energy is finally dissipated by frictional forces and, transformed into heat, it leaves the Earth again as long-wave radiation. In this way the energy budget of the Earth is closed. The major part of the frictional forces appears at the boundary between the atmosphere and the solid Earth.

The atmospheric layer adjacent to the surface is influenced the most by the surface characteristics and is called the atmospheric or *planetary boundary layer*. Boundary layer meteorology constitutes a sub-discipline of its own within the atmospheric sciences (Stull, 1988). It is of utmost importance to describe the characteristics of the Earth's surface correctly in numerical weather, climate, and air quality models (Blumen, 1990). Compared to the Earth's radius even the height of the largest mountains seems negligible. But because more than half of the mass of the Earth's atmosphere is concentrated below 6 km and nearly all weather takes place in the troposphere below 10 to 16 km, the mountains have a considerable influence on meteorological processes. DEMs are thus important input data for these models.

Hills and mountains exert a strong influence on pollutant transport, weather, and climate because the atmosphere is extremely sensitive to vertical displacements (Smith, 1979). First dynamically — because the atmosphere usually has a stable stratification — buoyancy forces will try to return vertically displaced air parcels to their equilibrium level, thus giving rise to wave motions, strong horizontal excursions, and even shooting flows as seen in water flowing over a weir. Second, thermodynamically — because the atmosphere usually contains water vapour — rising motion can rapidly lead to water vapour saturation and subsequently cloud and precipitation formation. Additionally large amounts of latent energy are released during condensation and the freezing of water which further enhance the vertical motions.

A few examples for the topographical influence on weather and climate are given in the following:

- On a global scale or *macro-scale* (see Chapter 8 for exact definitions), large mountain chains influence the overall rotational momentum budget of the atmosphere. These forces exerted by the orography are known as *mountain torque* (Wahr and Oort, 1984). Large north–south-oriented mountain chains like the Rocky Mountains and the Andes modify the general circulation patterns because they favour the formation of troughs and lee cyclones downwind of these mountains, today named Rossby waves (Charney and Eliassen, 1949). Also large west–east-oriented mountains such as the Himalayan Mountains have considerable effects (Deweaver and Nigam, 1995) because they hinder the exchange of polar and tropical air masses and also provoke lee cyclogenesis. Lee cyclogenesis behind the Alps is a well-known phenomenon too (Pichler et al., 1990).
- On a regional scale or *meso-scale*, the air partly moves over the mountains which may result in lee wave formation (Queney, 1948) and *foehn* (Hann, 1866), and partly moves around the mountains which may result in local wind systems known as bora (Smith, 1987) or mistral (Jiang et al., 2003). Mountain-induced perturbations in the surface pressure fields signify sinks for the momentum of the atmospheric flow due to flow blocking and wave formation, an effect called *pressure drag* (Emeis, 1990).
- On all scales from the *macro-scale* down to the *micro-scale*, hills and mountains considerably influence the *areal distribution of precipitation*. This is partly due to the forced rise of air masses when they have to cross mountains; partly it is due to thermally forced convection over the mountain surface which is heated more by the sun's radiation than the air in the surroundings at the same height level. Typically, precipitation increases with increasing elevation. Maximum precipitation intensities are usually found over the windward slopes. Shielding is found in valleys in the interior of larger mountain ranges and areas on the lee side of mountains (Frei and Schär, 1998). Showers and thunderstorms in mountainous areas start to form over mountain summits and crests (Linder et al., 1999) and often travel downstream.
- On a regional and local scale mountains can deflect and channelise the flow. Also on this scale mountains and radiation (see Section 3.1 in Chapter 8 for a detailed description of the interaction between the land surface and solar radiation) interact and mountain-induced wind systems such as *slope winds* and *valley winds* frequently emerge due to differential heating of the topography (see e.g. Defant, 1951). The correct aspect angle and slope of a surface

element with respect to the sun is of great importance for the determination of the local energy balance (Spronken-Smith et al., 2003).

The importance of the influence of topography on air quality, weather, and climate is mirrored in the vast amount of literature dealing with this subject and special experimental efforts. The first experiments were made shortly after the Second World War (Grubišić and Lewis, 2004) in the United States. Recently, in Europe three large meteorological measurement campaigns have focused on the Alps and the Pyrenees: the Alpine Experiment (*ALPEX*) (Kuettner, 1986), the Pyrenees Experiment (*PYREX*) (Bougeault et al., 1997) and the Mesoscale Alpine Program (*MAP*) (Bougeault et al., 2001).

REMARK 2. *Hills and mountains exert a strong influence on pollutant transport, weather, and climate because the atmosphere is extremely sensitive to vertical displacements.*

The results of these campaigns have decisively promoted the simulation of atmospheric flow and thermodynamics over mountainous terrain. Many smaller-scale field campaigns have concentrated on the influence of mountains on air quality, e.g. the Mesolcina Valley Campaign within the EU-funded project "*Vertical Ozone Transports in the Alps*" (*VOTALP*) (Furger et al., 2000). The results of these campaigns have then be verified and extended by numerical modelling studies (e.g. Grell et al., 2000a; Dosio et al., 2001).

1.2 Representation of surface elevation in numerical flow models

Predicting weather and climate requires the solution of a system of prognostic non-linear partial differential equations describing the atmospheric motion, the variation of pressure, temperature and humidity, the formation of clouds and precipitation, and the atmospheric chemistry. These equations are based on budget equations for the atmospheric mass, for momentum, for energy, for several water components (vapour, cloud water, cloud ice, precipitation water and precipitation ice) and for different trace gases, each containing different terms which describe storage, exchange, generation and destruction or dissipation of the respective quantities.

These terms can be either explicit (i.e. a physical or chemical process is described directly in mathematical terms depicting a physical or chemical law) or implicit (only the outcome of a physical process is parameterised by describing it empirically without explicit knowledge of physical or chemical laws). Symbolically the budget equations look as follows (not all terms need to be present in each equation):

$$\frac{\partial a}{\partial t} = F(a) + S(a) + D(a) + C(a) + I(a) \tag{1.1}$$

where a is the density of either momentum, energy, water component, or an atmospheric trace substance. The term on the left-hand side describes the storage of a. On the right-hand side we find the net transport term $F(a)$, a source term $S(a)$, a dissipation term $D(a)$, a conversion term $C(a)$, and an interchange term $I(a)$. Examples for conversions are the phase changes of water, the change from cloud to precipitation water or the exchanges between different forms of energy.

Since the atmosphere cannot be considered separately from its neighbouring systems, exchange processes between the atmospheric system and its surroundings are taken into account by the interchange term $I(a)$. For example, over continents the upper soil layers, a possible snow layer and vegetation have to be considered as relevant adjoining subsystems for which tendency equations, e.g. for soil temperature, soil moisture and snow mass must be formulated too. Here the interchange term e.g. describes the exchange of momentum, energy or substances at the Earth's surface. This is done by *SVAT schemes* (Sellers et al., 1997). A further example for an interchange term is falling precipitation (see also the second example in Section 3).

Weather and climate prediction models are often formulated as *Eulerian Finite Difference Models* (Pielke, 1984). This means that the non-linear system of partial differential equations which govern the flow and chemistry simulated in the atmospheric model are transformed in discrete form and solved in grid cells of a three-dimensional grid with two horizontal coordinates (usually equally-spaced, denoted by x and y) and one vertical coordinate. For the start of a simulation initial values for all grid cells must be specified. During the simulation boundary conditions at all outer boundaries of the grid must be applied to enable the solution of the partial differential equations. This chapter only deals with the boundary conditions at the lower boundary of the atmospheric grid which usually is the interface between the atmosphere and the subsystems soil, snow and vegetation.

In designing a suitable computational grid it is advantageous if the atmosphere is bounded by coordinate surfaces. For the lower boundary this means that the Earth's surface should be a coordinate surface. Over flat terrain a vertical coordinate using fixed constant height levels z (z-coordinates) in a given (not necessarily equal) spacing is sufficient. Here, the lower limit of the atmosphere is a coordinate surface. Over mountainous terrain the lower limit of the atmosphere unfortunately will not be a coordinate surface in z-coordinates. Therefore several coordinate systems have been designed to circumvent this problem. The simplest solution is a height-dependent coordinate as a vertical coordinate:

$$\sigma_z = \frac{z_{\text{top}} - z}{z_{\text{top}} - H(x,y)} \tag{1.2}$$

where z_{top} denotes the chosen height of the top of the model domain and $H(x,y)$ the elevation of the topography. This coordinate assumes the value 0 at the top of the model domain and 1 at the surface. Intentionally z-surfaces are not horizontal in mountainous terrain.

On a large and global scale vertical accelerations in atmospheric flow are small. The atmospheric pressure at a certain height level is then a measure for the mass of the atmosphere above this level (the so-called hydrostatic approximation). Therefore, in hydrostatic weather forecasting, the pressure $p(x,y,z,t)$ is often used as a vertical coordinate and the pressure tendency $\partial p/\partial t(x,y,z,t)$ instead of the vertical velocity $w(x,y,z,t)$. In these equations the density of the air does not appear as an independent variable. Equation (1.2) can then analogously be formulated as:

$$\sigma_p = \frac{p_{\text{top}} - p}{p_{\text{top}} - p_{\text{surf}}} \tag{1.3}$$

with the pressure p_{top} at the chosen top pressure level of the model domain and the time-dependent surface pressure $p_{\text{surf}}(x, y, t)$. The σ_p-surfaces slope like the σ_z-surfaces and they are time-dependent because the surface pressure is time-dependent.

On smaller scales, vertical flow accelerations (atmospheric convection, e.g. thunderstorms) become more important. Meso- and local-scale models therefore cannot be hydrostatic models. Here buoyancy has to be modelled explicitly. In such non-hydrostatic models the pressure is also a function of vertical accelerations. The simple Equation (1.3) is then no longer a practical choice. The frequently used and freely available *"Fifth-Generation Pennsylvania State University/National Center of Atmospheric Research Mesoscale Model"* (*MM5*) (Grell et al., 1994) and its successor the *Weather Research and Forecasting* model WRF (see http://wrf-model.org) are, like many other meso-scale models, based on modified σ_p-coordinates:

$$\sigma_{p_0} = \frac{p_{0,\text{top}} - p_0}{p_{0,\text{top}} - p_{0,\text{surf}}} \tag{1.4}$$

with p_0 as a horizontally homogeneous, temporally constant hydrostatic reference pressure (the true pressure p is $p_0 + p'(t)$). The subscripts 'top' and 'surf' have the same meaning as in Equation (1.3). As the reference pressure p_0 is a function of height z, $p_{0,\text{surf}}$ depends on the height of the surface and the σ_{p_0}-surfaces are again parallel to the Earth's surface near the ground (see Figure 1).

The vertical spacing of the grid-surfaces does not need to be constant. It can be chosen in order to better resolve layers of strong vertical variations. This is very often done for the atmospheric boundary layer adjacent to the Earth's surface (for an example see Figure 1).

Today's global-scale climate models for long runs (decades) use a horizontal spatial resolution of about 200 km, global-scale, short-term models for weather forecasts of a few days use a horizontal resolution of about 60 km. Meso- and local-scale models typically (with a resolution down to about 1 km) use for their simulations the output from global-scale models as initial and boundary conditions. In doing so they are able to simulate weather and climate features with a high spatial resolution for selected regions being guided by the global solution. This technique is often referred to as downscaling or as *regionalisation*.

REMARK 3. *Global-scale climate models for long runs (decades) use a horizontal spatial resolution of about 200 km, global-scale short-term models for weather forecasts of a few days use a horizontal resolution of about 60 km. Meso- and local-scale models use typically resolution down to about 1 km.*

For a specific simulation, the area of interest should — on the one hand — be sufficiently far away from the lateral borders of the *meso*-scale model domain

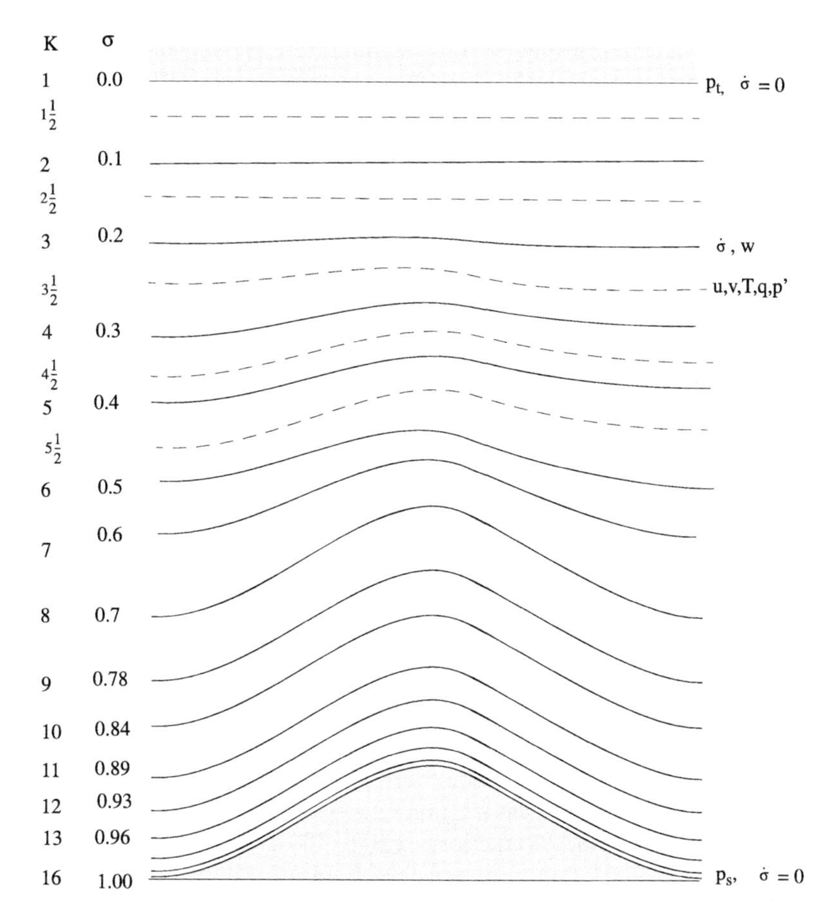

FIGURE 1 Example of σ_p-surfaces over a hill (shaded area). K gives the number of the σ_p-surfaces. The vertical spacing of the surface increases with height. Reprinted from Haagenson et al. (1994). © 1994 University Cooperation for Atmospheric Research.

in order not to be affected by boundary effects.[1] On the other hand the area of interest should be represented with a high horizontal resolution. In MM5 this is achieved by a nesting strategy. The simulation is started on a coarse grid domain with e.g. 27 km grid distance taking the output of a global-scale model as initial and lateral boundary values. Then subsequently finer *nests* with higher horizontal resolution are embedded into the respective coarser grids over that part of the model domain which is of special interest. Each finer grid uses the information from the next coarser grid as initial and boundary values.

The resolution from one nest to the next is usually enhanced by a factor of 3. In a case of three nests the subsequent domains have in our example 9 km, 3 km, and 1 km resolution. Assuming that each of the four grid domains has 60×60

[1] Weather and climate models cover large parts of the Earth's surface. Therefore, modelling with fine-resolution DEMs (e.g. 3 arcsec SRTM DEM) is presently impossible.

horizontal grid cells (this number can be deliberately chosen depending on the available computer resources), the four domains in our example cover areas of 1620×1620 km, 540×540 km, 180×180 km, and 60×60 km.

The nests can be run one after the other (one-way nesting, the information flows from the coarser grid to the finer grid only) or simultaneously (two-way nesting, information exchange is in both directions: from the coarser to the finer grid and vice versa). The preparation of the elevation, land-use/vegetation, land-water mask, and soil type data for the coarse domain and all nested model domains is handled by a pre-processor routine.

The model equations are solved for the above-mentioned grid cells. Thus, the yielded solutions have to be interpreted as averages over a grid cell. Meteorological processes which take place on smaller scales than such grid cells are called *subgrid-scale* processes and cannot be represented explicitly in weather forecast and climate models. Because the effect of such subgrid-scale processes may be important for the large-scale solution (due to the non-linearity of atmospheric motions) they have to be parameterised in meteorological models.

1.3 Representation of surface roughness in numerical flow models

The necessary parametrisation schemes that deal with these subgrid-scale processes need topographical information as input. One important example is the *surface roughness* which decelerates atmospheric flows. Several ways how to derive surface roughness from geometric properties of the surface itself are presented in Section 2.2 of Chapter 6. For the friction between the atmosphere and the Earth's surface another definition is necessary, one which has to be based on the shape of the vertical wind speed profile in the lowest 50 to 100 m of the atmosphere.

Derived from physical requirements the shape of an idealised near-surface wind speed profile $u(z)$ with the height z above the surface is written:

$$u(z) = \frac{u_*}{\kappa} \cdot \ln\left(\frac{z}{z_0}\right) \tag{1.5}$$

with the friction velocity u_* which depends on the wind speed on the top of the atmospheric boundary layer and the thermal stratification of the atmospheric boundary layer, von Kármán's constant $\kappa = 0.4$, and the *aerodynamic surface roughness length* z_0. This Equation (1.5) describes the so-called logarithmic wind profile and z_0 is the height where the logarithmic profile intercepts the z-axis. Although the aerodynamic roughness length z_0 in this profile law is a fixed property of the surface, it has to be derived from measured wind profiles. For flat and horizontally homogeneous surfaces the roughness length is known from a large number of field experiments. It varies from some tenth of a millimetre for a smooth water surface to several metres for the downtown areas of large cities. Grass surfaces have a roughness length of a few centimetres, forests of about 1 m (Stull, 1988, Figure 9.6). These values have to be supplied in the form of look-up tables to numerical flow models (see the following Section 2).

In the case of heterogeneous land use on flat terrain (with single land use patches much smaller than the size of the grid cells of the numerical flow model) one has to determine an *effective roughness length* $z_{0,\mathrm{eff}}$ from the roughness lengths $z_{0,i}$ of the various land use patches. The most simple way would be to put (Taylor, 1987):

$$\ln z_{0,\mathrm{eff}} = \overline{\ln z_{0,i}} \tag{1.6}$$

As also the friction velocity u_* will vary with land use, a more correct method might be (Taylor, 1987):

$$\ln z_{0,\mathrm{eff}} = \frac{\overline{u_* \cdot \ln z_{0,i}}}{\overline{u_*}} \tag{1.7}$$

where $\overline{u_*}$ indicates in both equations the mean over a grid cell. For more details the reader is referred to the detailed discussion in Taylor (1987).

In the case of terrain elevation changes on scales smaller than the grid cells of the numerical flow models one also has to consider orographic parameters such as the slope, height and distance of the hills within one grid cell. The most simple case would be rolling terrain which could be described approximately as a two-dimensional sine-shaped orography with crest height h and horizontal distance L between two adjacent crest lines. If the steepness h/L is below about 0.2 the effective roughness length can be given by (Emeis, 1987):

$$\ln z_{0,\mathrm{eff}} = \ln z_0 \cdot 20 \cdot \frac{h}{L} \tag{1.8}$$

For steeper slopes with flow separation, more complex three-dimensional orographies, and non-neutral thermal stratification of the flow such simple analytical relations are no longer possible. Additionally, in thermally stable stratification the formation of lee waves (also called gravity waves) behind mountain chains becomes possible. This leads to a *gravity wave drag* which — in addition to the surface friction — decelerates the atmospheric flow, too. A review on the parametrisation of these complicated effects can be found in Milton and Wilson (1996).

2. SOURCES FOR TOPOGRAPHIC VARIABLES IN METEOROLOGICAL MODELS

The handling of topographical information in the meso-scale simulation model MM5 (Grell et al., 1994) will be outlined in the following section using a typical example. For all other features of MM5 please refer to the online tutorial at http://www.mmm.ucar.edu/mm5/. The main source for the DEM of MM5 is the GTOPO30 dataset provided by the US Geological Survey Earth Resources Observation and Science EROS service (http://edc.usgs.gov).

The MM5 model system comes with a pre-processor program which horizontally interpolates the regular latitude–longitude topographical information from prescribed height data sets, vegetation and land use data sets, and soil type data sets onto the chosen grid system of the model domain. Three different map projections can be selected (polar stereographic, Lambert conformal, or Mercator projection).

For the interpolation a non-linear 4×4-point interpolation method is applied using a two-dimensional parabolic fit (Guo and Chen, 1994) to obtain the terrain height and the percentages for each vegetation/land use or soil category for each surface grid cell of the model grid. If more than one vegetation/land use category applies for a grid cell the dominant one is chosen by the following majority principle. If the water coverage of this grid cell is determined to be larger than 50%, the category water will be assigned to this grid cell. Otherwise the category with the maximum percentage excluding water will be assigned to the grid cell. Physical parameters needed by the meso-scale simulation model system to describe quantitatively the characteristics of the lower boundary — such as surface roughness, surface albedo for snow-free and snow-covered conditions, soil heat and moisture capacity — are generated from look-up tables when and where they are needed. An additional function of the pre-processor program enables the user to overwrite single interpolated values by user-supplied (corrected) or arbitrary-chosen data. This can be used for corrections as well as for scenario studies. E.g. it is possible to introduce a hypothetical land use via this method for a selected area and then to simulate the effect of this possible land use change on the weather or the climate.

For a multiple-nest simulation the nest domain's values of the topographical parameters at the lateral boundaries of the nests are blended with the parameters valid for these locations in the respective larger (mother) domain.

3. CASE STUDIES

The following three case studies will deal (a) with a large-scale simulation which exemplifies the influence of the Alps as a whole on atmospheric flow dynamics by showing their effect on the track and shape of low pressure systems (cyclones), (b) with a meso-scale simulation and the calculated distribution of precipitation amounts in the Alpine region, and (c) a small-scale simulation presenting local valley and mountain winds and their impact on air quality in complex terrain. The first two case studies are also examples for the information gain which can be achieved by a regionalisation procedure which is designed to downscale coarser forecast and analysis results from a larger-scale or global model to more detailed results for a higher-resolved regional scale.

3.1 Flow modification by the Alps

The first example is taken from a regional climate simulation with the meso-scale model MM5 for Europe (Forkel and Knoche, 2006). The atmospheric lateral boundary conditions were provided by a long-term climate model run with the global model ECHAM4 (Roeckner et al., 1996). The horizontal grid resolution is about 250 km in the global model and 60 km in the regional model. Figure 2 shows the used model orographies for the area covered by the regional simulation.

Because the model orography has to represent mean values averaged over grid cells, the Alps are much lower (only around 1000 m peak height) and much less detailed in the orography used by the global model than in the orography used by

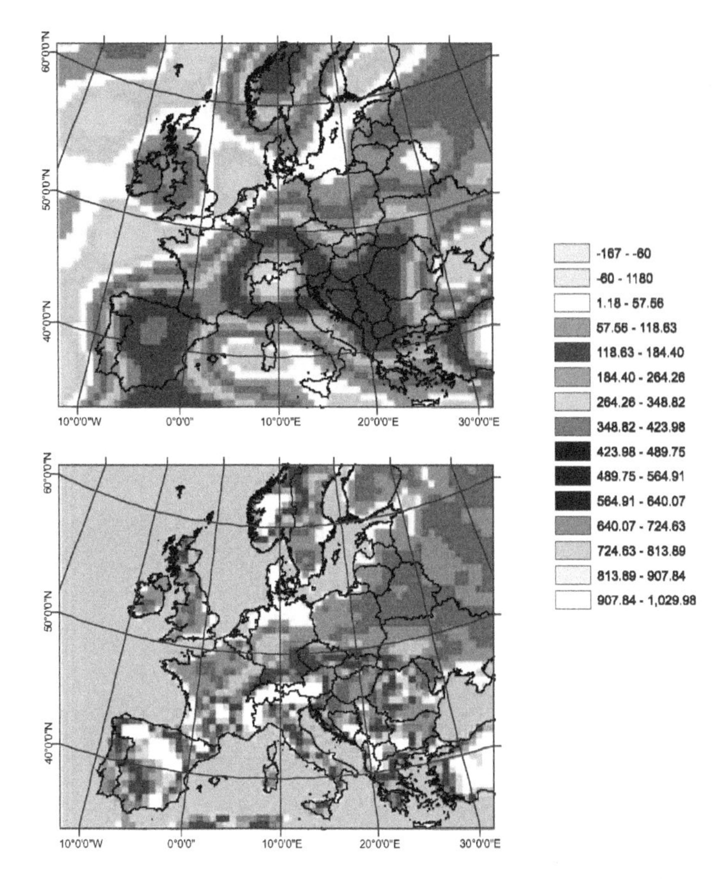

FIGURE 2 Model orography (height in m) for the global model ECHAM4, resolution about 250 km (top), and for the present regionalisation simulation with MM5, resolution 60 km (bottom). (See page 756 in Colour Plate Section at the back of the book.)

the regional model (2160 m peak height). Also the elongated *banana*-shape of the Alps is only discernable in the 60 km-resolution orography used by the regional model. Thus, the barrier effect of the Alps will be much more pronounced in the regional-scale model simulation.

The case study shows the simulation of the movement of a cyclone from West to East over the Alpine Region. Described in Figure 3 is the momentary sea level pressure distribution at noon for a day in April, 1992 as calculated by the global and the regional models. While in the global model the cyclone seems nearly unaffected by the Alps, the enhanced model resolution in the simulation with the regional model shows a stronger impact of the orography. The pressure minimum is slightly deformed and tends to split into a northern part over southern Germany and a southern part over the Gulf of Genua.

This case study makes clear that a sufficient representation of the orography may be decisive for a weather forecast. Especially the simulation and forecast of the above-mentioned lee cyclones (Pichler et al., 1990) depends strongly on the

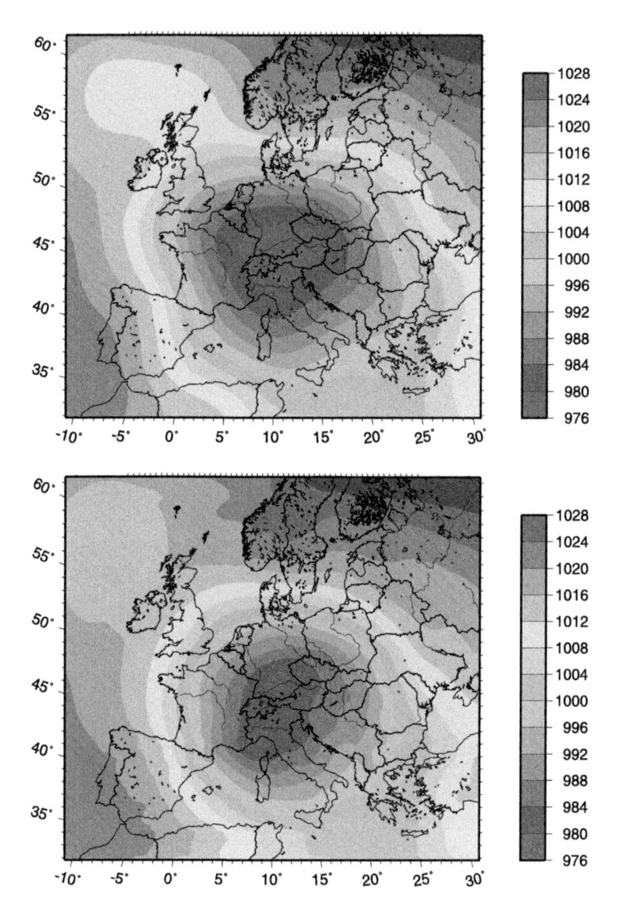

FIGURE 3 Distribution of the sea level pressure in hPa for April 5th 1991, 00 UTC, in the global simulation with ECHAM4 (top) and in the regional simulation with MM5 (bottom). (See page 757 in Colour Plate Section at the back of the book.)

spatial resolution of the orography data and the shape and height of the mountains represented therein.

3.2 Resolution-dependent precipitation amounts in the Alpine region

The second example presents the effect of the chosen horizontal resolution of orographic data on the simulation of the behaviour of the atmosphere by the meso-scale model MM5 (see also Grell et al., 2000b). Here we analyse the annual precipitation amount for the year 1990. One of the most significant effects of orography on weather and climate is the influence of mountains on formation and enhancement of precipitation. The coarse regional simulation is based on global meteorological data (Gibson et al., 1999, ERA15) produced for the years 1979 to 1991 by the global analysis model of the *European Centre for Medium-Range Weather Forecasts* (ECMWF).

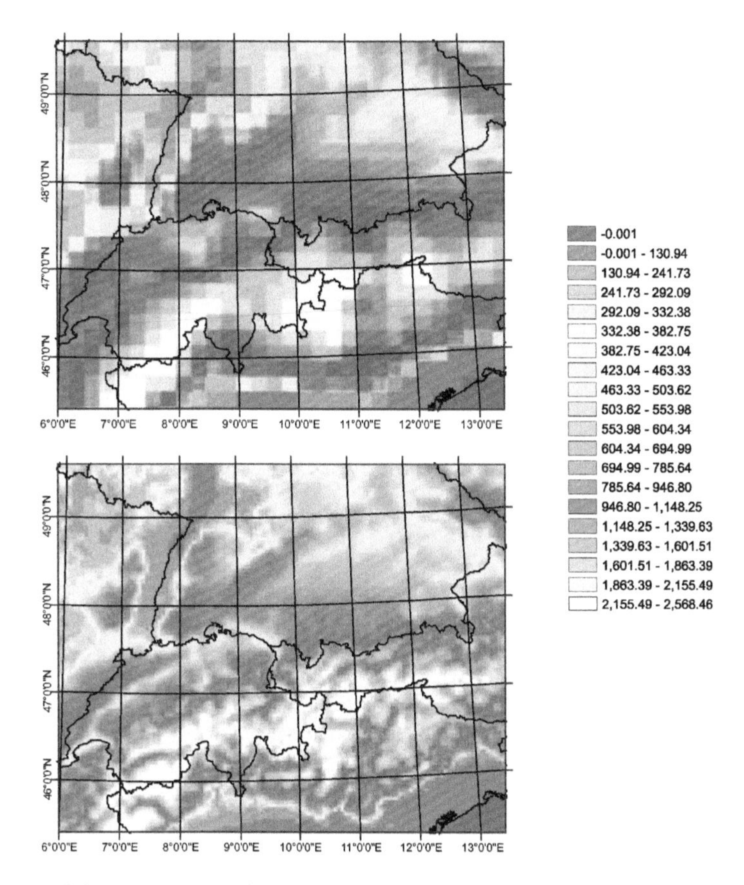

FIGURE 4 Model orographies in the nested regional MM5-simulations: (top) simulation S_1 with a horizontal resolution of 19.2 km and (bottom) simulation S_2 with a resolution of 4.8 km. (See page 758 in Colour Plate Section at the back of the book.)

In two consecutive nesting steps the global fields were downscaled, firstly to a horizontal resolution of 19.2 km (simulation S_1) and then further refined to 4.8 km (simulation S_2). The orographies used in the simulations S_1 and S_2 are shown in Figure 4. Peak height of the Alps in the orography data set for S_1 is 2570 m, for S_2 it is 3220 m. Larger Alpine valleys are only resolved in the finer simulation S_2.

Figure 5 shows the distribution of the annual precipitation sum for the year 1990 as calculated in the two MM5-simulations S_1 and S_2. The precipitation patterns are closely related to the used model orographies. The higher model resolution in the 4.8 km-run leads to more pronounced and more detailed patterns. Comparing both simulations, the S_2-run not only adds additional features in the scale range 4.8 to 19.6 km, but also the precipitation distribution on scales larger than 19.2 km is altered. This can most conveniently be shown by averaging again the output of simulation S_2 over 4×4 grid cells back to the resolution of the simulation S_1, and then taking the difference to simulation S_1.

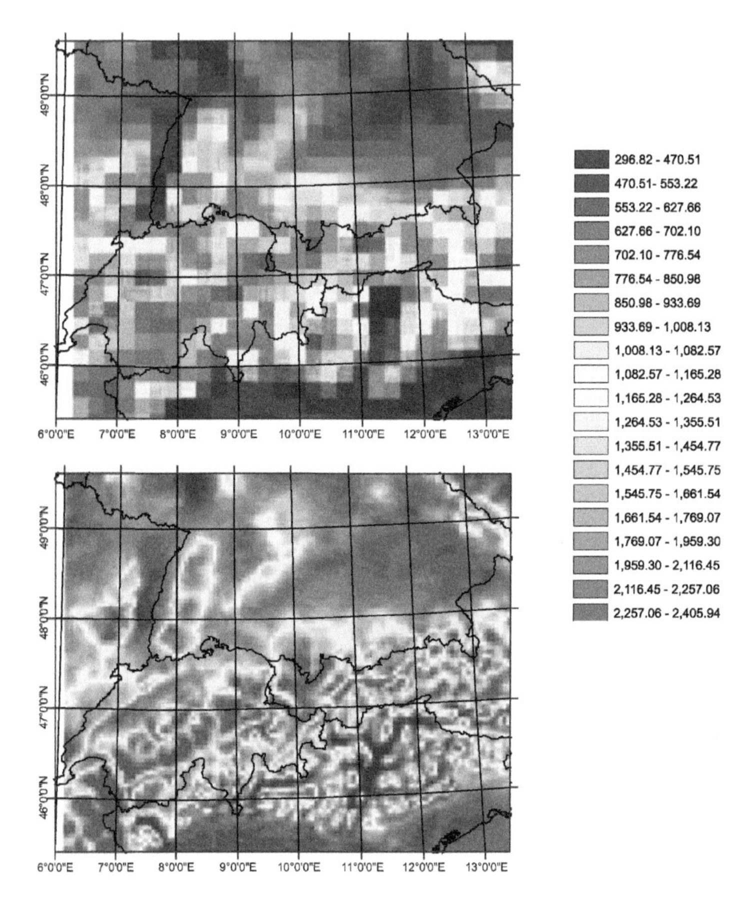

FIGURE 5 Annual precipitation sum (in mm) for the year 1990 as calculated by the MM5-simulations S_1 (top) and S_2 (bottom). (See page 759 in Colour Plate Section at the back of the book.)

The difference field ($S_2 - S_1$) in Figure 6 shows an enhancement of the annual precipitation amount on the windward side of larger mountain chains and a decrease of precipitation amounts on the respective lee sides. This becomes especially clear in the Vosges mountains on the French side of the Rhine river and in the Black Forest.

In comparison to simulation S_1 simulation S_2 shows distinct maxima in the area of the westerly slopes and near the crests. Minima are found in the Rhine valley west of the river Rhine. This is in correspondence with climatological rain distributions (e.g. Swiss Hydrological Atlas) as south-westerly to westerly winds are most frequent in this area.

More extended mountain regions like the Alps show the highest precipitation amounts over the mountain chains at the edges. In the central part, especially in the large valleys (Rhone, Rhine, Inn, Adige) relative minima are observed due to shading effects. In the coarse simulation S_1 these patterns are reproduced only roughly. Figure 5 shows that the higher-resolution simulation S_2 generates more

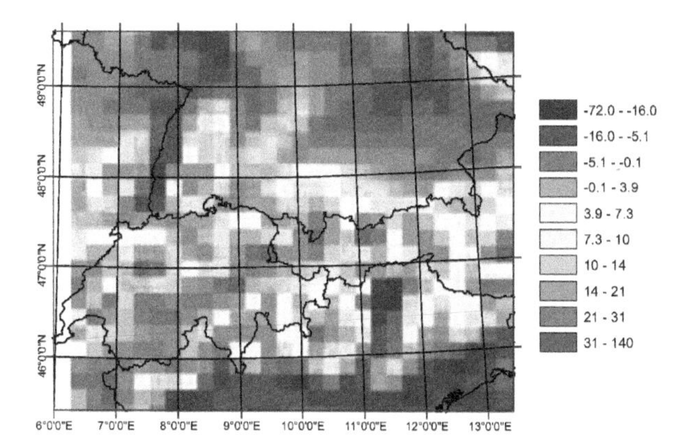

FIGURE 6 Scale reduced difference ($S_2 - S_1$) of the annual precipitation sum (in mm) for the year 1990. (See page 759 in Colour Plate Section at the back of the book.)

precipitation at the edges and significantly less precipitation over the main Alpine divide and in the above-mentioned valleys. By this, the higher-resolution simulation is closer to the climatological distribution (e.g. Frei and Schär, 1998).

This case study demonstrates that a really satisfying simulation of the highly complex spatial distribution of precipitation amounts can only be achieved with spatially high-resolution orographic data. This has great implications for avalanche and flood forecasts, which have to be based on precipitation information generated by regional and local models.

3.3 Orographically generated wind systems and air quality in complex terrain

Small-scale orographically generated wind systems such as valley winds during daytime and mountain winds at night-time have a considerable influence on the air quality in complex terrain. The example taken from Grell et al. (2000a) shows how valley winds during daytime transport high ozone concentrations from the Milan area in the Alpine valleys on the southern flank of the Alps. The simulation has been made with MCCM (see Grell et al., 2000a, for details), a model which consists of the above-mentioned MM5 (Grell et al., 1994) for a computation of the atmospheric dynamics and the RADM2 gas-phase chemical reaction scheme (Stockwell et al., 1990). This has been one of the first studies with a fully coupled dynamics–chemistry model for complex terrain for such a high spatial resolution of the terrain features. Figure 7 shows the chosen domains for the one-way nesting strategy (Section 1.2) which was necessary to resolve explicitly the valleys in the Southern Alpine region in this simulation effort.

To get meteorological input data for domain D_1 and the time period starting on 13 August 00:00 UTC to 19 August 00:00 UTC, 12 hourly *National Centers for Environmental Prediction* (NCEP) global analyses were used as *First guess* meteorological fields. These 12-hourly *First guess* fields were next interpolated to 3-hourly time intervals and enhanced with radiosondes, surface, and ship observations to give

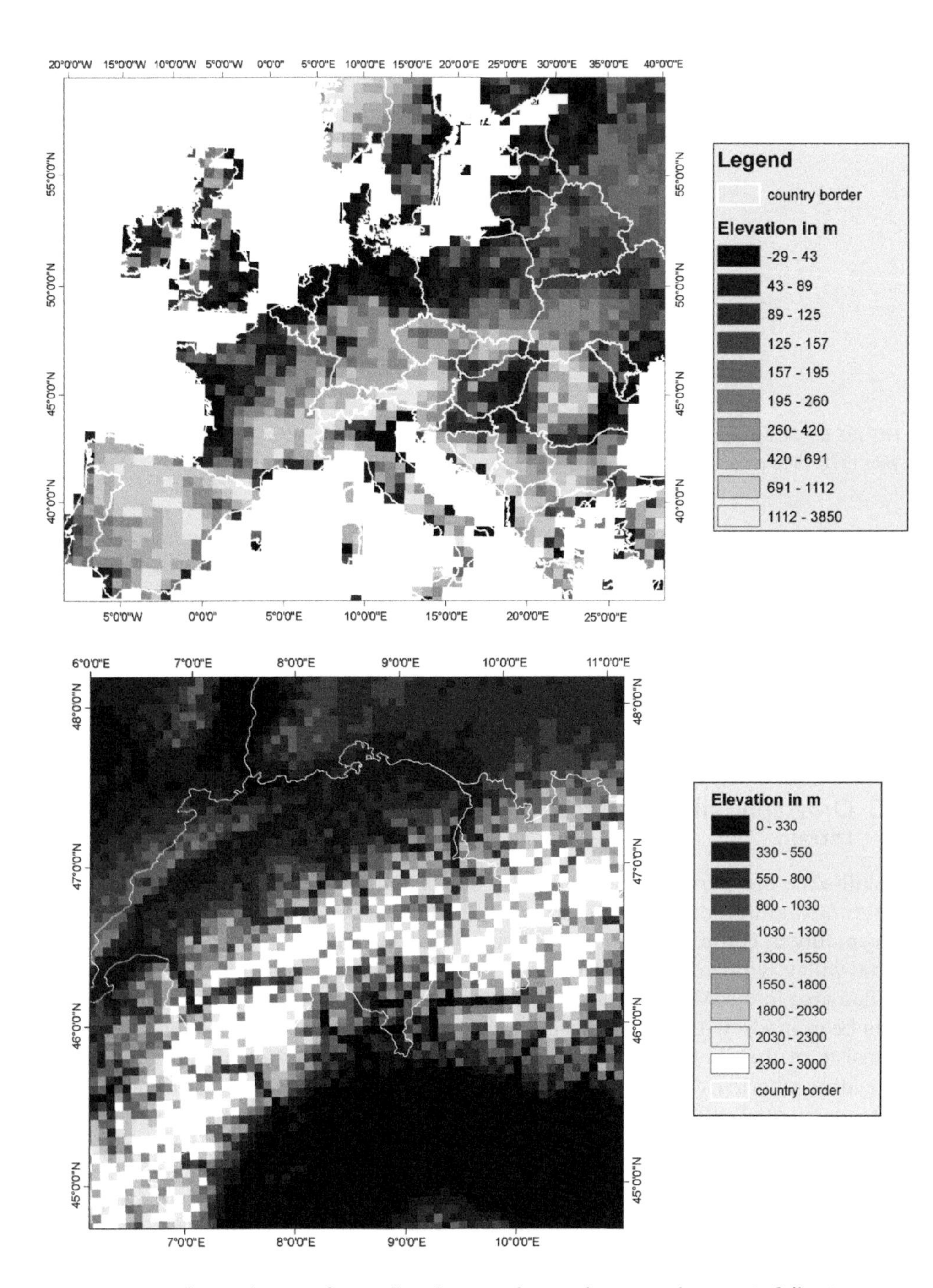

FIGURE 7 Simulation domains for small-scale air quality study in complex terrain following a one-way nesting strategy. The horizontal spatial resolution is 54 km (domain D_1, top frame), 6 km (domain D_2, middle frame), and 1 km (domain D_3, bottom frame). (See page 760 in Colour Plate Section at the back of the book.)

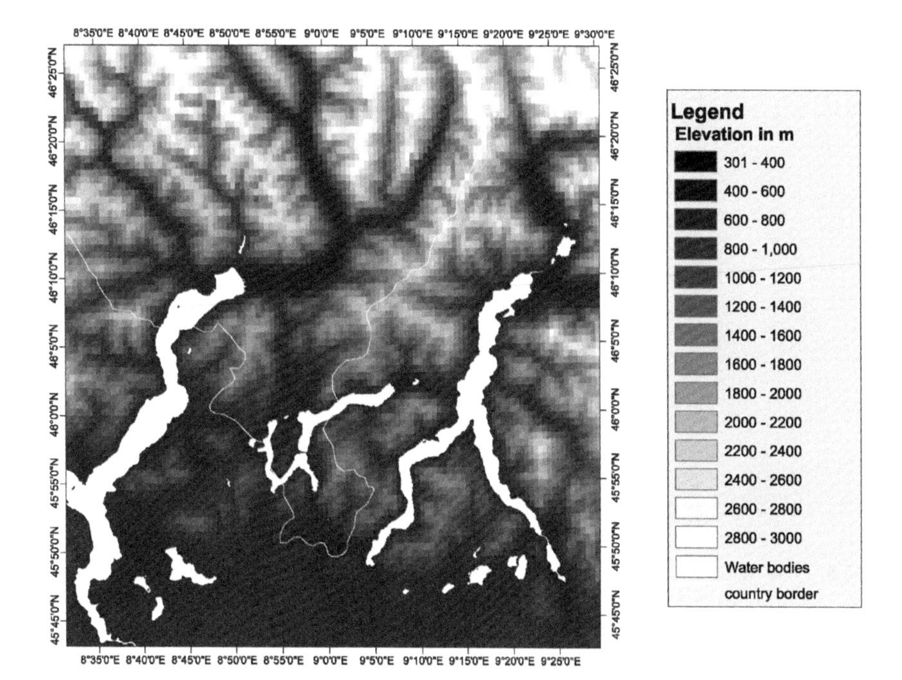

FIGURE 7 (*continued*)

analyses at 3-hourly intervals for the five-day period and domain D_1. Using these observed boundary conditions and analyses, a coupled meteorology/chemistry simulation was performed over the 5-day period. This simulation was then used to provide three-hourly meteorology and chemistry input data on the boundaries of domain D_2. For domain D_2 the model was integrated over a period of three days (starting on 16 August 00:00 UTC). The hourly meteorological and chemical output from D_2 was next interpolated to D_3 with the 1 km horizontal grid. The vertical resolution was identical for all domains.

Domain D_2 was only used here to supply boundary conditions for domain D_3, in order to avoid a too large nesting ratio. Its horizontal grid spacing (6 km) is still much too coarse to resolve any of the important features of the terrain fields. Therefore, we will concentrate on a discussion of the results from the innermost domain D_3. Figure 8 shows a result from this domain D_3 for the evening of August 19, 1996, after the valley wind had blown for more than eight hours. The high ozone concentrations from the Milan area in the lower right of the figure have been carried deep into the Alpine valleys by these valley winds over distances of more than 100 km.

The success of such an air quality study critically depends on the high-resolution representation of the terrain and the correct simulation of the surface energy balance (see Chapter 8) on all differently sloped surfaces within this domain. Deficiencies in the surface energy balance would especially lead to wrong diurnal variations of the surface temperature and this in turn would have a large

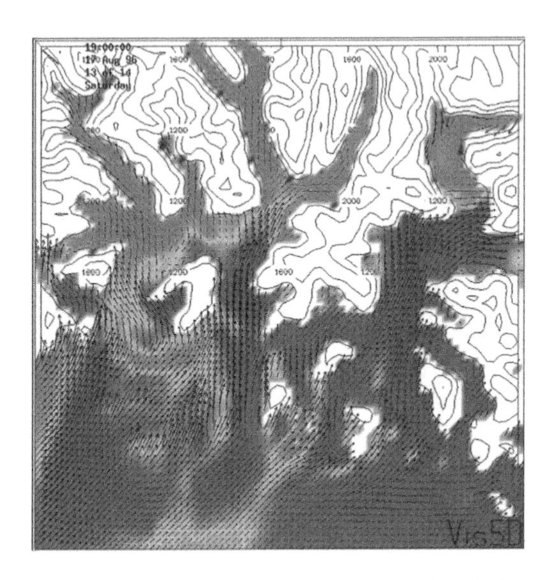

FIGURE 8 The figure shows for August 19, 1996, 1900 GMT, for the whole D_3 (see also Figure 7) the horizontal winds at 1000 m above sea level (arrows, length indicating strength and arrowheads indicating the direction of the flow) and the ozone concentration at the same height (colours, red: high, green: medium, blue: low, white: terrain height higher than 1000 m above sea level). Black lines in white areas give terrain height in 400 m interval (first line: 1200 m above sea level). (See page 761 in Colour Plate Section at the back of the book.)

influence on the formation of the daytime valley wind and night-time mountain wind system, because this wind system is driven by thermal forces.

> REMARK 4. *The success of an air quality study critically depends on the high-resolution representation of the terrain and the correct simulation of the surface energy balance.*

If the surfaces stay too cool, the valley winds start too late or they might not develop at all. If the surfaces become too hot, the wind system would start too early and the valley winds would last too long. If the wind system is not modelled correctly the transport of air pollutants through the valleys would be wrong. The correct modelling not only requires a high-resolution terrain data set, but also correct information on soil type, land use, and vegetation (see Section 2).

4. SUMMARY POINTS

The most relevant topographic data for meteorological applications are: digital elevation data, vegetation and land use data, soil type data, and land-water masks. On the basis of these input data important properties such as surface roughness and the heat capacity of soil and water are specified in air quality, weather, and climate models via look-up tables. Surface roughness and heat capacity are important variables because they govern the exchange fluxes of momentum, heat, moisture, and other trace substances between the different compartments (air, soil, water, ice) in weather and climate models. But it has to be kept in mind that explicit topographic data can be used in numerical flow models only for those features that have the size of at least one grid cell of the applied numerical model. Topographic information on features smaller than one grid cell are conveyed to the simulations only implicitly via the parameterisations in SVAT schemes, i.e. if the size of a grid cell in a numerical flow model is about 5 km by 5 km then all terrain details that are smaller than 5 km are not used explicitly by the model. This means that e.g. valley wind systems in smaller valleys or cold air lakes in small basins are not resolved explicitly in weather forecast models. The effect of such local-scale flow phenomena must be entered indirectly via the SVAT schemes into these models. The size of grid cells in air quality, weather forecast, and climate models are chosen in such a way that the simulations can be done with the available computational resources within a reasonable time period. This restriction is especially important for weather forecasts which have to be ready before the actual weather occurs.

In the near future increasing computer resources will give way to enhanced spatial resolutions in air quality, weather, and climate models. Parameterisations schemes for sub-grid scale processes in these models are going to be adapted to this enhanced resolution. But such higher-resolution models can only show their strength when they have suitable high-quality input data for initial and boundary values available. Therefore, the quality and availability of high-resolution topographic data for meteorological model applications as listed in Section 2 will become of increasingly greater importance in the future. The supplying of high-quality topographic data will eventually lead to better local weather forecasts and to an enhanced *regionalisation* of climate change predictions when the available computational resources allow finer cell sizes. A correct and reliable computation of the exchange fluxes between the different compartments in climate models will make it possible to couple models for the atmosphere, the biosphere, the soil, the ocean, and the ice sheets. This will finally lead to complete *Earth system models* that are expected to become indispensable in the planning of a sustainable future of our planet and its population.

IMPORTANT SOURCES

Blumen, W. (Ed.), 1990. Atmospheric Processes over Complex Terrain. Meteorological Monographs, vol. 23. American Meteorological Society, Boston, MA, 323 pp.

Grell, G.A., Dudhia, J., Stauffer, D.R., 1994. A description of the fifth-generation Penn State/NCAR mesoscale model (MM5). NCAR Technical Note. National Center for Atmospheric Research, Boulder, Colorado, 117 pp.

Pielke, R.A., 1984. Mesoscale Meteorological Modeling. Academic Press, 612 pp.

Smith, R.B., 1979. The influence of mountains on the atmosphere. Advances in Geophysics 21, 87–230.
Stull, R.B., 1988. An Introduction to Boundary Layer Meteorology. Kluwer Academic Publishers, 666 pp.
http://www.mmm.ucar.edu/mm5/ — Pennsylvania State University/National Center for Atmospheric Research.

CHAPTER 27

Applications in Precision Agriculture

H.I. Reuter and **K.-C. Kersebaum**

Precision Agriculture applications · why is geomorphometry important for Precision Agriculture? · what are the current and foreseeable applications? · which land-surface parameters are especially important for Precision Agriculture?

1. INTRODUCTION

Precision farming or Precision Agriculture is an emerging field in agriculture that tries to optimise the in-field variability of biomass production from an economic and ecological perspective. Assessment of spatial variability at local scales makes it possible to more precisely select optimum sowing density, estimate fertiliser demand and other input needs, and to more accurately predict crop yields. In that sense, geomorphometry is also an important source of information that can be used to represent the in-field variability.

Generally, the assumption is made that under heterogeneous soil and field conditions the observed variability in crop yield or quality is mainly attributable to the underlying soil properties. In contrast, under homogeneous soil conditions a homogeneous distribution of yields is expected across a field. However, large variability in the biomass development and grain yield is often observed, even if the management (i.e. fertiliser application) and soil conditions (texture) are relatively homogeneous. In some cases the differences are as much as 6 t/ha for winter wheat under similar soil and management conditions (Dobermann et al., 2004). If it is not simply soil properties that cause the yields or quality to vary, the question arises, what other processes or parameters are contributing to this variability?

Certainly, biomass development across a field can never be totally homogenised. However, Precision Agriculture is intended to (i) reduce uncertainties in decision making, and (ii) to control and manage plant growth variables, e.g. by spatially varying inputs and management practices (Berntsen et al., 2006). This leads directly to two major questions: how can agriculture-relevant parameters be determined in a cost effective and accurate way, and secondly how to then interpret and manage these different sites in a practical way.

Developments in Soil Science, Volume 33 © 2009 Elsevier B.V.
ISSN 0166-2481, DOI: 10.1016/S0166-2481(08)00027-5.

Land-surface parameters mapped at large scales (meso-relief) have been more or less successfully incorporated into Precision Agriculture applications, using continuous and discrete land-surface parameters. As biomass production is an integrative parameter over space and time, land-surface parameters are used together with other data coming from remote sensing, soil profiles, plant sensors and models. Hence land-surface parameters from many chapters of the book can be cross-referenced (see Chapters 20, 21, 25 and 26). In this chapter we mainly focus on biomass parameters, to avoid overlap with soil parameters (Chapter 20) or hydrology, which are of equal importance for Precision Agriculture.

Precision Agriculture has evolved from *farming by soil* (Robert, 1993), through variable application and lately to consumer production issues (product safety). At the same time, that area has also changed from simple yield monitoring systems, and the impressive developments in *vehicle guidance*, to *Decision Support Systems* (DSS). However, to benefit from Precision Agriculture in economic and ecological terms, several requirements must be met (Dobermann et al., 2004): (i) significant spatial variation must exist at the sub-field scale, (ii) crop response to inputs is significant, predictable and not confounded by other factors, (iii) input applications are done accurately, and (iv) the extra cost for Precision Agriculture is kept low. This might be the reason why the implementation of Precisions Agriculture in practice is delayed, and for example DSS do not take into account the highly variable, hardly predictable, and dynamic environment a farmer works in (McBratney et al., 2005).

1.1 Concepts for application of geomorphometry in Precision Agriculture

In the application of geomorphometry in the field of Precision Agriculture two main approaches can be distinguished. The first one delineates practical, agronomically-meaningful management zones that contained distinct plant growth controlling parameters (Pennock et al., 2001; Stevenson, 1996). Several algorithms to delineate management zones (Chang et al., 2003; Ferguson et al., 2003; Fraisse et al., 2001) do not always necessarily include DEMs. Research based on management zones (e.g. see also Chapters 22 and 24) has shown an increase in grain yield (Jowkin and Schoenau, 1998; Manning et al., 2001); an increase in culm height (Vachaud and Chen, 2002); an increase in thousand kernel mass (Ciha, 1982) or for example a decrease in grain protein (Fiez et al., 1994) with increasing curvature character of the landscape. These observed different biomass developments occur due to landscape processes resulting in the redistribution of soil mineral nitrogen or soil moisture, for example (see also Chapter 20). Manning et al. (2001) captured the gross variability at a manageable landscape level by investigating soil mineral nitrogen and soil moisture. Pennock et al. (2001) showed that such differences can also be manipulated to identify the relationship between different fertiliser rates and N_2O emissions. The N_2O emission rates were found to be lowest at shoulder positions, whereas backslope positions showed 5–6 times higher rates compared to shoulder positions across all fertiliser treatments (Pennock et al., 2001).

Landforms combined with other auxiliary information (soil, hydromorphy) have been used by Vosshenrich et al. (2001) for a site-specific tillage operation (shallow tillage of 8–10 cm compared to deep tillage up to 25 cm). The fields showed similar plant development, and similar levels of yields compared to a conventionally tilled site. Additionally, they measured savings in fuel consumption by up to 50%, depending on distribution of soil and landforms at the field sites. However, to be fair, several examples also show no interaction between landforms and biomass parameters, which might be due to scale problems as Dobermann et al. (2004) stated that algorithms applied are sometimes not suitable for the scale of Precision Agriculture.

As the agricultural production system is usually not linear and the farmers need to take informed decisions, different Decision Support Systems are used to model the biomass development in time. Examples, where land-surface parameters are used in one or the other way, are developed for mapping and managing *nutrients* (Khosla et al., 2001), *fertilizers* (Wenkel et al., 2002), *seeding density* (Roth et al., 2001), *seeding variety* (Paz et al., 2001), and *irrigation* (Van Alphen and Stoorvogel, 2000).

The second main approach is based on the use of continuous land-surface parameters. These land-surface parameters are used on the one hand site for the process of creating cost effective production parameters for example a land-surface guided soil and plant sampling (McBratney et al., 2003; Minasny and McBratney, 2006), for soil maps, etc. Another emerging field of applications is the use of continuous land-surface parameters in the prediction of expected crop yields. Wendroth et al. (2003) used the NDVI, crop nitrogen status and surface elevations in bi- and multivariate autoregressive state-space analysis to predict spring barley grain yield, whereby compared to a traditional soil parameter dataset the same prediction quality could be reached. Secondly, continuous land-surface parameters are used in analysis of grain yield in response to land-surface parameters. F.-M. Li et al. (2001), Yang et al. (1998), Kraus and Pfeifer (1998), da Silva and Alexandre (2005) showed inverse relationships between elevation height and grain yield. Elevation might serve here as a proxy for a more complex and more difficult to measure parameter, which is the plant available water over the whole profile. This negative relationship is in line with a recent publication by Berntsen et al. (2006), where elevation height proved to be helpful, and other land-surface parameters not. This focuses on one problem, that no *best* relief attributes can be given to estimate or analyse soil properties or yield components. At different scales, as well as in different landscapes, land-surface parameters used to analyse and estimate soil and plant properties differ strongly. Precision Agriculture can only be successful if adapted regional approaches are used to describe, analyse and manage crop yield and soil variability.

2. PRECISION AGRICULTURE APPLICATIONS

Having shown the variety of different applications of geomorphometry to agriculture production, especially the Precision Agriculture area, we shall now focus on two case studies, where we present examples of how land-surface parameters of different complexity (elevation, solar radiation, topographic wetness index, landforms) have been used in practice. The first example uses a model applied under farm conditions to estimate the biomass development and to predict the amount of nitrogen fertiliser applications in a Precision Agriculture approach. The second example provides results for post harvest analysis to increase our understanding about crop yield development processes using landform classification.

2.1 Case study #1: nitrogen fertilisation and crop yield

To illustrate the processing steps, we will use data from a Precision Agriculture research project called MOSAIK, located at the research farm of the Südzucker Company in Luettewitz-Dreißig in south eastern Germany. The farm is located in the very productive agricultural region Lommatscher Pflege, approximately 50 km South-South-West of Dresden. The soil is derived from loess with an approximate thickness of 25–30 m (Härtel, 1931) and is classified as a Luvisol, with the top 1 m of the soil profile depleted of calcareous material. A detailed LiDAR-DEM was obtained for an area of approximately 200 ha with a spatial resolution of 1 m (0.15 m vertical accuracy). The preprocessing of the LiDAR scan consists of mosaicing, filtering of outliers, smoothing, and resampling. A more in-depth review of the investigation area and methods is given in Reuter (2004).

Precision Agriculture is assumed to provide improved efficiency of applied nutrients combined with lower emissions of agro-chemicals. Nitrogen fertiliser recommendations are usually based on measurements of soil mineral nitrogen content in early spring supported at later stages by measurements of the crop nitrogen status by optic sensors. Both methods are just snapshots of a present situation which do not enlighten the reason for an observed phenomena or the probable future development. Therefore methods are required to estimate the local nitrogen demand considering the spatial variability of soil nitrogen supply and crop yield potentials. Agricultural system models provide a tool to transfer the spatial heterogeneity of time stable soil and terrain attributes which have to be estimated once for a field into the temporal dynamic of the relevant state variables of the soil–crop nitrogen dynamics. We show how a model is used to simulate the spatial variability of crop yield and nitrogen dynamics and to give fertiliser recommendations based on the nutrient deficiency approach under special consideration of geomorphometric information.

The authors used the daily time step *model HERMES* (Kersebaum, 1995) to simulate crop biomass development, soil water balance and nitrogen dynamics on arable fields. The generic type of crop growth model allows the simulation of whole crop rotations considering different crops by external crop parameter files. The model has been applied mainly for cereals, but also for silage maize (Herrmann et al., 2005), sugar beets and potatoes (Kersebaum, 2007). The model uses daily weather data from a local weather station for precipitation, temperature, vapour pressure deficit and global radiation. Basic soil data (texture, soil organic matter, stone content, wetness, groundwater level) were obtained up to a depth of 90 cm. The third group of inputs is related to management: sowing,

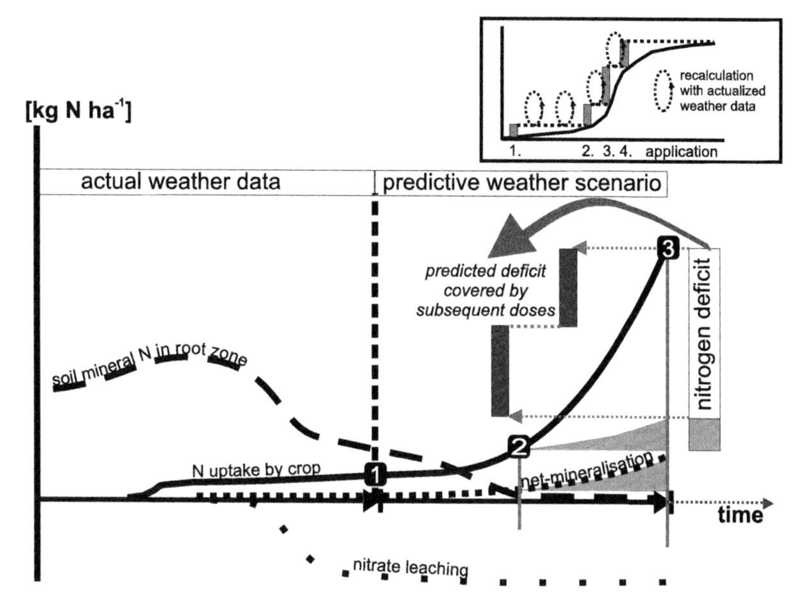

FIGURE 1 Scheme of model derived fertilisation recommendations.

harvest date and amount of fertilisation and irrigation are needed. Additionally, the model can be used to derive fertiliser recommendations based on subsequent model applications throughout the growing season. For this, predictive simulations are performed using typical, site-specific weather scenarios derived from long term data sets. A schematic illustration of this procedure is shown in Figure 1. Further details are described in Kersebaum and Beblik (2001).

In the current version, solar radiation and an adjusted TWI are used to differentiate the model inputs and soil parameters in space. The SRAD model was used to determine the complex interactions of incoming long and short wave radiation between the Earth's surface and the atmosphere (Gallant and Wilson, 1996), which has not been provided in other[1] packages. For a SRAD version with a more modern user interface please refer to http://arcscripts.esri.com.

Irradiance was determined for three days each month with a time step of 12 min. Local weather conditions were used for parametrisation (Reuter et al., 2005). For the simulation of solar radiation a floating grid was derived in ArcInfo using:

/ setting the work environment only to the spatial extent of the field site (SK)*

```
GRID: setwindow SK
```

/ set NODATA values to the minimum value of the DEM*

```
GRID - SRAD glitch GRID: dem1 = con ( isnull (ls1), 170, ls1)
```

/ convert to float file*

[1] SOLARFLUX by Dubayah and Rich (1995), Solar Analyst (Fu and Rich, 2002), Genasys (Kumar et al., 1997).

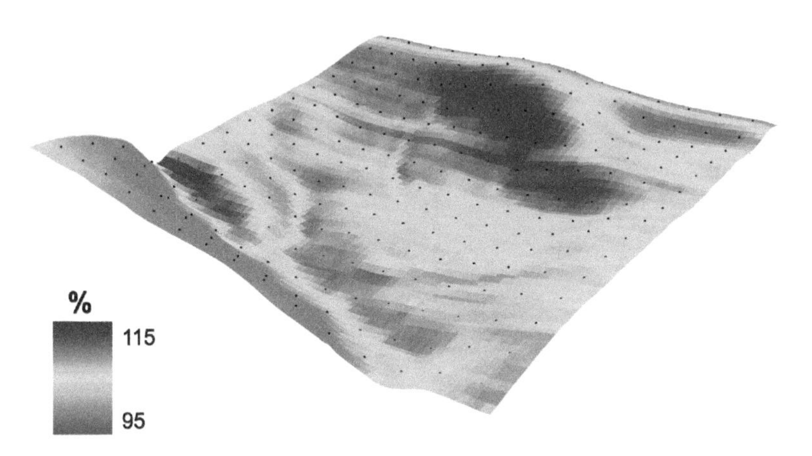

FIGURE 2 Simulated spatial distribution of solar radiation relative to a neighbouring automatic weather station for the field site Sportkomplex. (See page 762 in Colour Plate Section at the back of the book.)

```
GRID: sk.flt = gridfloat (dem1)
```

/ leave ArcInfo and execute srad*

```
srad < sk98sim.txt
```

The SRAD calculation was then performed by providing a text file (`sk98sim.txt`) to SRAD containing the following commands:

sk.bin	Output file name
Y	Specify Binary(Y)/ASCII(n) file output
sk.flt	Input floating grid file
0	Specify grid cell size, if 0 read from header
5	Specify that DEM is ArcInfo Grid
16	Number of sky sections
4	Determine solar constant
1	Lumped transmittance
sk98.srad	Radiation parameter file
Y	Specify Temperature Output
07 1 08 1 10	Time period (Start Day/Month, End Day/Month, Time Interval)

For the preparation of the SRAD radiation file please refer to the SRAD documentation. The binary file so created was re-imported into ArcInfo using the `tapes2arc.aml` and aggregated to represent average conditions for every simulation point as a monthly correction factor. Figure 2 shows the spatial distribution of the irradiation relative to the neighbouring weather station integrated over the year 1998. Compared to the south exposed positions, the locations in the valley received about 25% less irradiation as confirmed by pyranometer measurements at these locations (Reuter et al., 2005).

FIGURE 3 Topographic wetness index calculated for the field site Sportkomplex.

Additionally the elevation model was used to derive the topographic wetness index according to Moore et al. (1993a). The specific catchment area and the slope were derived using the GRID module of ArcInfo using the `topo.aml`. The execution of this command is outlined in Chapter 11). Figure 3 shows the spatial distribution of the TWI across the terrain of the investigation area of 20 ha. The high indices at the footslope areas are confirmed by observations of water saturation or even exfiltration at the surface mainly in early spring. The TWI has been scaled to be used as a hydromorphic modification factor for the field capacity parameter.

Use of the model on the Sportkomplex field study without considering the topography-related modifications was insufficient to reflect the observed spatial variability of grain yield across the field due to a small variability of textural composition and their spatially-related hydrological parameters. However, consideration of the scaled topographic wetness index and the spatial correction of irradiation led to a better spatial pattern of the model results (Figure 4). An analysis of the effects of different spatially variable input data showed that, on this site, the influence of TWI and topographic shading is higher or in the same order of magnitude as textural related variability (Kersebaum et al., 2002).

The result of using TWI as a modifier of the water storage capacity show that the negative effect on grain yield through air shortage is only observed during wet periods, while in dry periods crops may benefit from a better water supply. It is therefore more flexible and realistic, then, to use the TWI as a static correction factor for crop growth.

Using the model within a fertilisation trial on the same field in two years (2000 and 2002) for real-time fertiliser recommendations achieved similar yields in both years, with 40 kg N ha^{-1} less fertilisation compared to a recommendation based on soil mineral nitrogen (N_{min}) in early spring and an online chlorophyll measurement by a Hydro-N-Sensor at later crop development stages (Table 1). More details about the experimental design and the results can be found in Kersebaum et al. (2003).

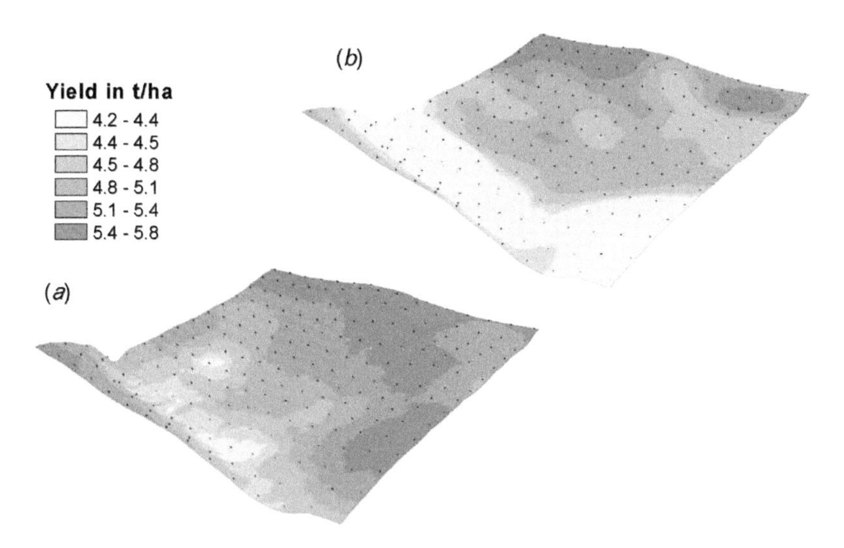

FIGURE 4 Simulated (a) and observed (b) spatial patterns of grain yield of spring barley in 1998 on field Sportkomplex.

TABLE 1 Measured and simulated yields for selected site-specific fertiliser treatments on winter wheat in two years on field Sportkomplex (standard deviation in brackets)

Treatment	Nitrogen fertilisation		Yield (dry matter)	
	Average (min/max) kg N ha^{-1}	combine harvested t ha^{-1}	hand harvested t ha^{-1}	simulated t ha^{-1}
		2000		
Zero	–	4.6[a] (1.56)	–	5.2 (0.46)
N_{min} + N-Sensor site-specific	179 (154/195)	6.9[b] (0.42)	–	6.7 (0.25)
HERMES site-specific	139 (75/157)	6.8[b] (0.74)	–	6.5 (0.61)
		2002		
Zero	–	4.7[a] (0.37)	4.6[a] (0.94)	5.4 (0.11)
N_{min} + N-Sensor site-specific	177 (150/202)	5.7[b] (0.17)	7.6[b] (1.13)	7.8 (0.07)
HERMES site-specific	136 (70/172)	5.7[b] (0.24)	7.9[b] (0.76)	7.8 (0.07)

a,b Grouping according to multiple range test (Nemeny-test with $p < 0.05$).

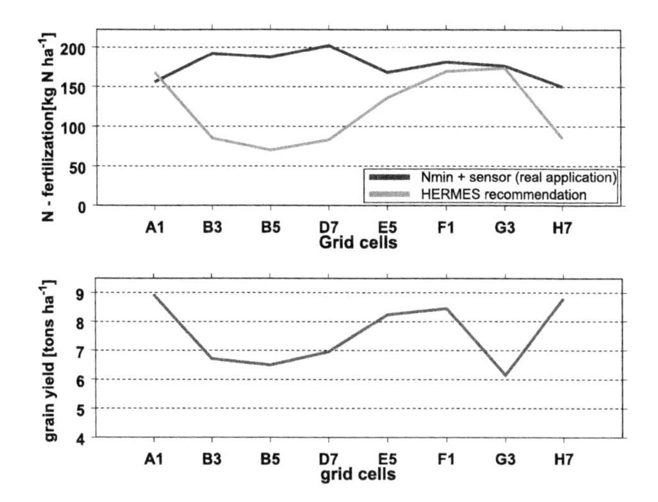

FIGURE 5 Model-based nitrogen fertiliser recommendations in comparison to actually applied fertilisation and observed winter wheat yields at plots of N_{min} + sensor treatment for Sportkomplex in 2002.

Site-specific fertiliser recommendations by the model varied from 75 to 157 kg N ha^{-1} in 2000 and from 70 to 172 kg N ha^{-1} in 2002. They were lowest in some of the footslope areas due to unfavourable wet conditions for crop growth in these years. The question arises, whether the observed low yields at these locations might be the result of the low fertilisation recommendation of the model instead of unfavourable growth conditions. The upper portion of Figure 5 shows the recommendations of the model at those locations where the fertiliser was applied according to the N_{min} + Hydro-N-sensor measurement that were obviously higher at location B3, B5 and D7. These 3 plots were located at footslope areas and, although they received a much higher amount of nitrogen, the observed yields were relatively low as indicated by the lower recommendation of the model. In summary, spatially variable inputs for simulation models derived from geomorphometric information can be used in addition to basic soil information to improve the spatial pattern of simulated crop growth. This has consequences for model-based fertilisation recommendations, leading to a better adaptation of the supply to the demand of the crops.

2.2 Case study #2: post-harvest analysis

Blackmore (2000) and several other authors (Bakhsh et al., 2000; Grenzdörfer and Gebbers, 2001) have shown that temporally stable crop yield patterns are only observed in small sections of agricultural fields. These results are not unexpected, as different grain yield components are influenced by the environmental conditions prevailing throughout the growing season. Various yield components are specifically influenced by the characteristics of different landforms (LF). In other words, each LF unit reflects a characteristic yield development, based on soil, meteorology

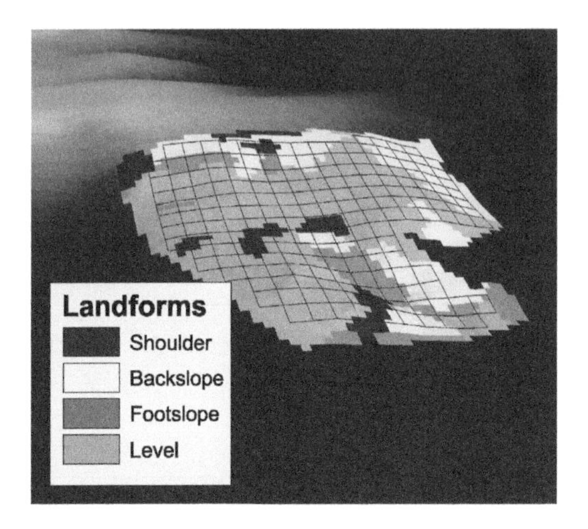

FIGURE 6 Distribution of landforms at the field site Bei Lotte. (See page 762 in Colour Plate Section at the back of the book.)

and management, which may have a different impact from year to year. Therefore, different grain yield development across the observed landform elements would be expected as observed by Manning et al. (2001), Pennock et al. (2001).

The relief parameters slope (SLOPE), profile curvature (PROFC), planform curvature (PLANC), and flow accumulation (TCA) area were created using the `topo.aml` (see Chapter 11), and a landform classification using the algorithm of Pennock et al. (2001) was performed to segment the landscape, including planar areas, into 11 classes (`landform.aml` — see Figure 6). The execution of these commands is presented in Chapter 11. Grain yield was determined by hand harvesting 192 plots of 0.5 m^2 size, spaced 27 m apart, for the years 1999–2001. The harvested material was dried at air temperature, and analysed for number of yielding spikes per m^2 (N_{Spike}), dry mass of spikes in grams (0.5 m^{-2}), the kernel mass in grams, the number of kernels (N_{Kernel}) and the Thousand Kernel Mass (TKM) in grams $(1000 \text{ kernels})^{-1}$. Statistical analysis was performed using classical variance analysis (KS-test, Error-variances, Scheffe, Games–Howell, Kruskal–Wallis test) for the years 1999–2001.

The results of the yield properties within the six diagrams show the landform mean for the respective landform and the field mean for the year (see Figure 7). The N_{Spike} is the first yield component to develop in the growing season [Figure 7(a)]. Shoulder positions contain 60 spikes m^{-2} in 1999 and 50 spikes m^{-2} in 2000 more than the field average. In contrast, N_{Spike} was found to be highest at shoulder positions in 2001. Additionally a slight increasing trend in N_{Spike} for all three subsequent years was found for the landform positions backslope — footslope — level [Figure 7(b)]. Differences in environmental conditions (e.g. precipitation) might be the reason for the year-to-year variation in grain yield within each landform. A reduction in N_{Spike} could be attributed to (I) differences in growth response and (II) to soil nutrients. However, plant nutrients (N, P, K) were found to be sufficient

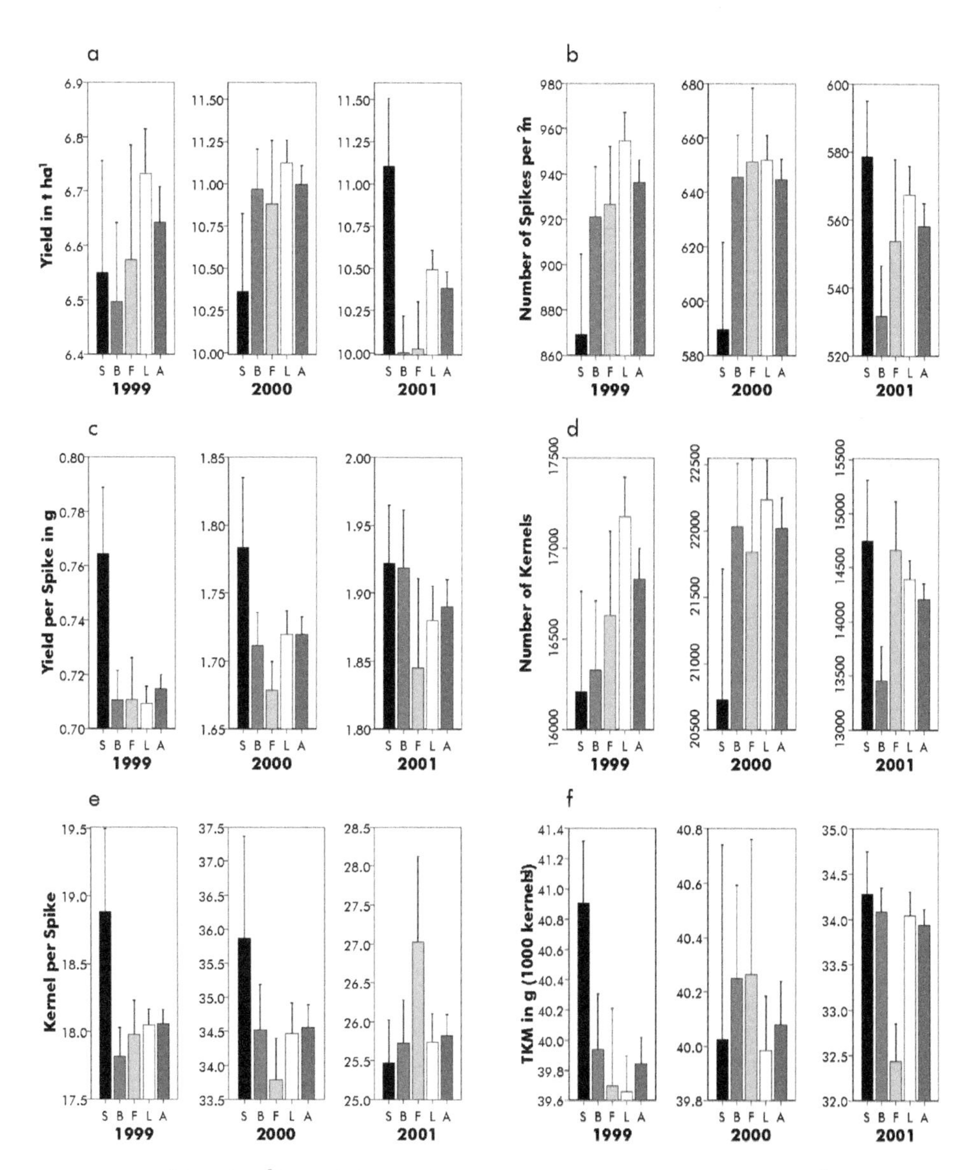

FIGURE 7 Yield in t ha^{-1} (a) and the yield components Number of Spikes per m^{-2} (b), Yield per Spike in g (c), Number of Kernels per m^2 (d), Number of Kernels per spike (e) and TKM in g (1000 kernel)$^{-1}$ (f) for the landforms shoulder ($S, n = 17$), backslope ($B, n = 49$), footslope ($F, n = 16$), level ($L, n = 110$) and the field average ($A, n = 192$). The Y-axis of the diagrams shows, for each year, values of the respective landform and the field average as shown on the X-axis. Error bars show the standard error.

across the field after harvest. This does not imply that there have been sufficient nutrients for plant development at all times during the growing season.

Footslope positions might have been waterlogged due to subsurface flow of water (Freeland et al., 1998) during the moist growing season of 2001 and so caused lack of oxygen in the root zone and therefore a reduction in N_{Spike}. In contrast, plants grown on shoulder positions might have suffered drought stress during the dry growing season of 2000. The second yield property that is developed over time is the number of kernels (N_{Kernel}), which showed decreased values at shoulder positions in 1999 and 2000, whereas in 2001 the largest N_{Kernel} was observed at shoulder [Figure 7(d)]. If the grain yield components N_{Kernel} and N_{Spike} are combined into K_{Spike}, then, a summary of the grain yield development up to that stage can be performed [Figure 7(e)].

In 1999 and 2000, shoulder positions show the largest number of K_{Spike}, additionally for footslope the least is found in 2000. In contrast, during the relatively wet year 2001 the largest K_{Spike} was at the footslope positions and the smallest at the shoulder positions. One of the most important times in yield development processes is the kernel filling phase, where the TKM provides a snapshot of the environmental conditions [Figure 7(f)]. Spring Barley at shoulder positions in 1999 yielded approximately 1 g higher TKM than on other landforms.

In the year 2000 only small differences in TKM were found across landforms, in contrast to the decrease of 1.5 g at footslope compared to all other landforms in 2001. A decrease in TKM was addressed due to increasing N_{Spike} by Darwinkel (1980), as well as to a prolonged time of dryness by Roth et al. (2001). Corresponding results were found in 1999 with an increase in TKM together with a decrease in N_{Spike} at shoulder positions [see Figure 7(f),(b)]. However, results for shoulder positions in 2000 showed no increase, even if N_{Spike} was decreased again. Additionally, TKM was reduced at footslope in the moist year (2001), indicating plant growth stress due to less available water during the grain filling period (Entz and Fowler, 1998). As gravimetric soil moisture after harvest was evenly distributed across all LF, differences in TKM are probably due to the observed lodging (Tripathi et al., 2003) and associated difficulties during hand harvest.

Finally, all yield components discussed so far were combined in the yield itself [Figure 7(a)]. In the moist year 2001 shoulder positions out-yielded all other landforms by 0.75 $t\,ha^{-1}$ probably due to the most favourable growing conditions at this landform. In contrast, in the dry year 2000, shoulder positions are assumed to suffer under soil moisture deficits during the growing season, yielding approximately 0.75 $t\,ha^{-1}$ less than the field average. The year 1999 showed an interesting phenomenon for the observed yield at the shoulder position. Although K_{Spike} and N_{Spike} were least during the development of the spring barley, the plants compensate for this deficiency with an increase in TKM during the grain filling phase; so that any differences in grain yield were small [see Figure 7(d),(b),(f)]. Hence a similar grain yield was reached as that achieved on footslope positions.

To conclude, single yield components reacted differently over three years and for various landforms within the same field. Generally, grain yield differences of up to 0.7 $\mathrm{t\,ha^{-1}}$ at different landforms were observed due to landform-dependent response to observed weather conditions. To some extent, plants were able to compensate differences in yield development by various yield components. In 1999, the environmental conditions were favourable and shoulder positions compensated for possible yield loss. Shoulder positions developed less spikes and further on during the development a higher TKM, therefore compensating for possible yield loss. This effect, for example, did not occur in 2000.

3. SUMMARY POINTS

Precision Agriculture is an approach intended to facilitate management of agricultural fields in an economic and ecologically-efficient way in the temporal and spatial domain. The aim is not to make fields uniform with regard to crop development, but to adjust the scale of management to the scales at which most of the decision uncertainty occurs.

Geomorphometry can support Precision Agriculture in several ways. The first would be as auxiliary variables in decision support systems (e.g. Models) to reduce uncertainties in decision making. In the authors understanding, these variables can be used to help understand the non-linearity of soil and plant processes. However, the challenge is to make these models applicable to farmers, with only a limited amount of input data and calibration, for (i) prior application calculation (if-then scenarios), (ii) real time decision making, and (iii) post harvest analysis. The next challenge would be the generation of soil/hydrological and meteorological information needed for these DSS. This implies the use of soil and plant sensors from various sources and the development of algorithms to create coherent data sets.

Geomorphometry can also support Precision Agriculture via post-harvest analysis to help improve understanding of topography-connected processes, soil and crop growth processes, and crop yield. This analysis can be applied, not only to improve understanding of crop growth and yield, but also to generate improve understanding of spatial variation of diseases, pests, weed growth, crop quality and residuals of applied chemicals and fertilisers.

A last summary point should be made. Precision Agriculture can collect a huge amount of data, similar to geomorphometric analysis. However, to transform this information into management decisions, is a task which requires a lot of expertise, which can be a huge barrier to implementation.

ACKNOWLEDGEMENTS

The technical assistance of numerous people from ZALF Muencheberg and ATB Potsdam is greatly appreciated. Support of this study by German Research Foundation (DFG, Bonn), Suedzucker, Agrocom, and Amazonen-Werke is acknowledged.

IMPORTANT SOURCES

Kersebaum, K.-C., 2007. Modelling nitrogen dynamics in soil–crop systems with HERMES. Nutrient Cycling in Agroecosystems 77 (1), 39–52.

Reuter, H.I., 2004. Spatial crop and soil landscape processes under special consideration of relief information in a loess landscape. Ph.D. Thesis. Universität Hannover, Hannover, 251 pp.

Pennock, D.J., Walley, F., Solohub, M., Si, B., Hnatowich, G., 2001. Topographically controlled yield response of Canola to nitrogen fertiliser. Soil Science Society of America Journal 65 (6), 1838–1845.

CHAPTER 28

The Future of Geomorphometry

P. Gessler, R. Pike, R.A. MacMillan, T. Hengl and **H.I. Reuter**

dynamic geomorphometry · LiDAR and other topographic data · anticipated trends in software tools and methods · prospective applications of geomorphometry · geomorphometric atlas of the World — why? how? when? · last thoughts for the future

1. PEERING INTO A *CRYSTAL BALL*

Whither geomorphometry? When queried, most contributors to this book responded in the spirit of Einstein's memorable "*I never think of the future. It comes soon enough*". The thoughts of those authors who chose to ponder the issue are incorporated in this final chapter. While geomorphometry is now well developed, the next decade will see more routine application of its tools and data: imagine mapping vegetation or rock types in a remote area, carrying an integrated cell phone/video/GPS display, and interactively updating a choice of high-resolution images draped over a DEM showing your location and those of scattered team members!

Geomorphometry will advance in concepts, land-surface measurement, and analytical tools and methods; all are addressed here with an eye to the future. The overarching drivers of progress in geomorphometry, especially in the environmental context, promise to be global warming (http://www.ipcc.ch) and the world's growing population (http://www.prb.org) and how these challenges will be accommodated. Much has been made of how humans are altering the global landscape (Thomas, 1956), but the landscape shapes us as well; understanding its structure and function will be critical to realising a holistic vision of sustainability — and even survival of the human species.

REMARK 1. *Among future drivers of geomorphometry are global warming and a growing world population and how to meet their resulting challenges.*

Developments in Soil Science, Volume 33 © 2009 Elsevier B.V.
ISSN 0166-2481, DOI: 10.1016/S0166-2481(08)00028-7.

2. CONCEPTS

Geomorphometry's unique geometrical-topological approach to representing the landscape will continue to evolve. Its conceptual trajectory, from a barometer hand-carried up a mountain to supercomputer process-modelling based on LiDAR data, surely holds greater promise than just more surface heights at ever-finer resolution manipulated by increasingly complex algorithms. Because the science is still very young (Pike, 1995), the line between a conceptual breakthrough and a mechanistic or technological advance is a narrow one. What fresh new *ideas* might brighten the future of geomorphometry is more elusive than issues of topographic data and their manipulation. One concept worthy of consideration, *dynamic morphometry*, is still in its infancy but already is taking geomorphometry in some interesting new directions.

Nearly 20 years ago a group of Silicon Valley luminaries[1] met not far from Stanford University to plan a *seamlessly scalable system* to dynamically visualise the Earth's surface. Informally dubbed *The Geography Machine*, this raster-structured tool would use stored georeferenced data to zoom into or out from any point on the globe, either normal to the surface or starting at one set of x, y, z coordinates and ending at another, viewing the Earth continually.

> REMARK 2. *Concepts expected to develop in the future include the virtual digital globe and robotic on-the-fly geomorphometry.*

The concept, eventually executed by others, is now realised as the *virtual globe*,[2] e.g. Google Earth (see Figure 9 in Chapter 1; http://earth.google.com). This marks just the beginning; the *virtual globe*[3] concept will grow to unprecedented capabilities and complexity with the delivery of terabytes of data from coming generations of Earth-sensing aircraft and spacecraft. An example of a web-based interface that supports GIS graphical, statistical, and spatial tools for the analysis of planetary datasets is available at: http://astrogeology.usgs.gov/Projects/webgis/.

A more recent conceptual breakthrough combines LiDAR with robotics. *Autonomous on-the-fly morphometry* continually generates point locations to guide its own (moving) laser-ranging platform, unaided by human intervention. Born of the desire to manoeuvre unoccupied vehicles on future battlefields, the need for autonomous driving has led to *probabilistic terrain analysis* (Thrun et al., 2006b) for high-speed cross-country locomotion; one new refinement incorporates a self-supervised roughness estimator to automatically adjust vehicle speed to the land surface. The guiding software employs an error model based on *point clouds* of x, y, z locations and their range from the laser scanner (Figure 1). It achieved operational status in the 2005 DARPA[4] Grand Challenge, when a robotic automobile successfully navigated a 212-km course through the California desert at speeds up to 61 km/h (Thrun et al., 2006a). Might such a concept also be adaptable to, say, guiding agricultural machinery?

[1] Among them computer and artificial-intelligence pioneer Les Earnest and image-processing innovator Irwin Sobel.
[2] See http://en.wikipedia.org/wiki/Virtual_globe/.
[3] But not limited to Earth; see, for example, Figure 3.
[4] The U.S. Defence Advanced Research Projects Agency.

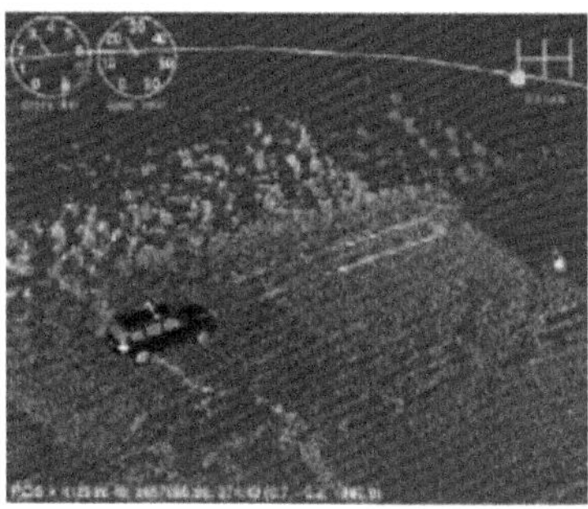

FIGURE 1 Geomorphometry on-the-fly. Laser sensors (top) scan terrain in front of a robotic vehicle as it moves, recording x, y, z location, distance, and time. The resulting data are integrated into 3-D point clouds (bottom) and continuously sorted by the changing relation among the three variables to separate drivable terrain from potential obstacles. Reprinted from Thrun et al. (2006a). (See page 763 in Colour Plate Section at the back of the book.) © 1996 John Wiley & Sons Limited. Reproduced with permission.

These and other prospects for our young science, however, still lack a unifying theory (Pike, 1995) — one either limited to Earth and planetary topography (Schmidt and Dikau, 1999; Rasemann et al., 2004) or including surfaces other than geomorphic (Pike, 2000b). Physicist Wolfgang Pauli's lament "The surface was invented by the devil!" reminds us that the challenge is formidable. One path to a theory of geomorphometry lies through the 150-year-old concept of *Flächenelemente-surface elements* or *elementary forms* (see Figure 7 in Chapter 1). Minár and Evans' (2008) review and new quantitative work in this area, based on Krcho (1973) and others, exemplify recent progress. It remains to be determined how well their unified system of geometric primitives, when developed to accommodate a variety of conditions, will perform on real-world DEMs, and whether the proposed scheme will constitute, or only contribute to, an overarching theory of surface form.

3. DEM DATA: DEMAND AND SUPPLY

Geomorphometry is strongly data-driven. In comparison with ground truth, however, every digital representation of Earth's surface height is in some respect

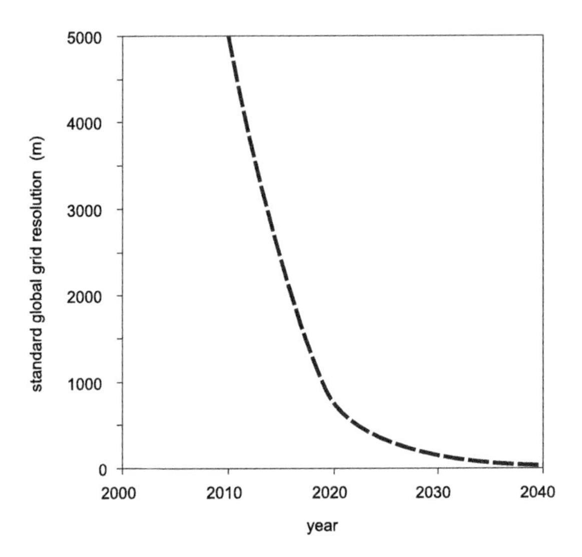

FIGURE 2 Projected increase in standard global DEM resolution over the next 30 years. After Hengl (2006).

flawed. Existing inaccuracies commonly reflect the methods of DEM generation; current DEMs, for example, commonly contain error that originated largely within their source contour maps. Many of these map sheets were compiled under standards that guarantee only a statistical level of quality; locally, their accuracy can be low. However, most contour maps were never intended to provide elevation data of the high density and accuracy required by contemporary geomorphometric analysis. Advanced technologies such as LiDAR do not entirely solve the quality problem, for each new technique brings its own unique shortcomings.[5]

Although the 2000 Shuttle Radar Topographic Mission (SRTM) provided a global DEM that replaces 80% of the worldwide 1-km GTOPO30 DEM, with 90-m coverage for some areas, in locations of dense tree cover its radar did not penetrate the canopy and accurately capture the land surface. SRTM relative vertical accuracy (at the 90% level) averages a significant ± 16 m, which is inadequate for such common applications in morphometry as precision agriculture, floodplain mapping, hillslope hydrology, and modelling debris-flow susceptibility.

The need for better elevation data seems only to be growing, and technological advances continue to accelerate the demand. Among these are computer processing speed and storage capacity, growing coverage of Earth at improved resolution and accuracy, and ease of surface visualisation in both space and time. Today's standard computers work with images of about 1 million pixels; the entire Earth thus may be displayed at a resolution of about 15 km. If Moore's Law holds in this case, the global standard could conceivably improve to 25 m by the year 2040 (Figure 2). Likewise, the next global imaging mission could well provide data substantially finer than the current global 90-m SRTM DEM.

[5] Experienced judgement in preprocessing is essential to reliable LiDAR DEMs (Haugerud and Greenberg, 1998; Haugerud and Harding, 2001).

Web access and freely downloadable DEMs will proliferate, with quality exceeding that of current (often expensive) information. Geomorphometric analysis thus promises to become more widespread and reliable. Demand for DEMs and their derivatives, evident from many different disciplines and applications, will prompt ever higher expectations for rigour and force the global scientific community to commit to an ongoing review of data standards.

4. EXPLOITING LIDAR

The surface detail revealed by geodetic laser scanning, or LiDAR (Carter et al., 2007), is the geomorphometric equivalent of changes wrought in other fields by a more powerful microscope or medical-imaging system. LiDAR-fuelled research, exemplified by Staley et al. (2006), Lashermes et al. (2007), Cavalli et al. (2008), Lehner et al. (2008), and Newell and Clark (2008), is exploding. The technique promises to deliver the long-sought dream of at once broad yet explicit multi-scale surface analysis. Algorithms that efficiently parse and tile multiple datasets will be needed to rapidly generate useful information from the many billions of points; quality control through ground-truth verification for such a volume may be difficult to enforce. As acquisition costs decline and techniques improve for filtering out vegetation and the built environment, LiDAR should routinely provide quality DEMs for studying both the land surface and related phenomena (e.g. biomass, wildfire fuels, etc.). Off-the-shelf large-area coverage is already a reality in a DEM for the entire Canadian province of Alberta, 700,000 km^2 at a 2–4 m footprint.

With the 2004 launch of the *Geoscience Laser Altimeter System* aboard the ICESAT platform (http://icesat.gsfc.nasa.gov) LiDAR went global. Its immense volumes of information will spur research into data structures and algorithms for processing both discrete-return and waveform LiDAR, and GIS software will serve as the means to integrate these data with other geospatial information. The expected outcome is a revolution in how the Earth can be viewed from a variety of perspectives and spatial scales.[6]

> REMARK 3. *Applications software and DEM quality control may struggle to accommodate the LiDAR revolution in mass data-acquisition.*

DEM analysis increasingly includes data from other sources[7]: satellite imagery, gamma radiometrics, thermal imagery, RADAR, sensor networks, and flux towers. LiDAR should enable more rapid correction of distortions in aerial and space imagery, thus reducing the time between data acquisition and product delivery. Remote sensing experts anticipate that imaging and monitoring will make ever more integrated use of DEM data. Geomorphometry, with roots in airphoto interpretation of landforms and photogrammetric generation of DEMs, will be com-

[6] While LiDAR is evolving rapidly and is likely to become the principal source of mass-produced DEMs, traditional photogrammetric and radar systems can be expected to remain in use for the foreseeable future. Soft-copy (workstation) photogrammetry is particularly useful in local applications.

[7] More frequently, remote-sensing platforms are including multiple sensors to acquire several different types of data simultaneously.

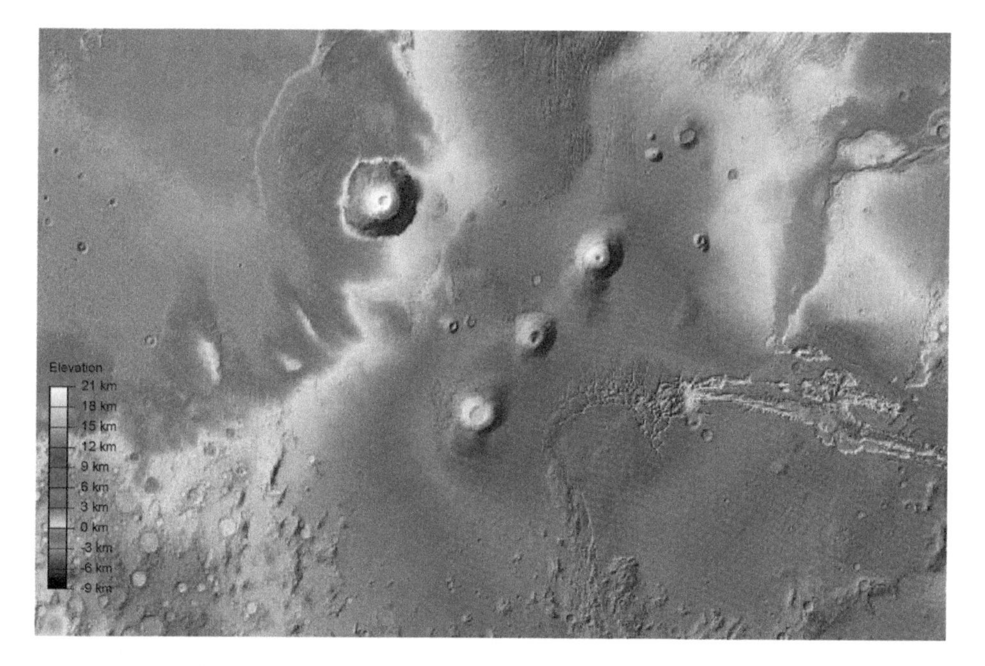

FIGURE 3 Out-of-this-world geomorphometry of volcanoes, canyons and, craters. Shaded relief image of Mars from 40°N to 40°S and 60°W to 180°W colour-coded by elevation, from the MOLA 1/128° DEM (available via http://mars.google.com). Image credit: NASA/JPL/GSFC/Arizona State University. (See page 764 in Colour Plate Section at the back of the book.)

bined with remotely sensed imagery and proximal sensing data for simulation modelling of landscape processes.

Elevation measurements of the Earth's surface acquired by LiDAR and other instruments on board geostationary or orbiting satellites can be expected to provide land-surface models for, variously, daily/monthly/yearly cycles at a variety of grid resolutions (Table 1).

Because LiDAR often shares imaging platforms with multi- and hyper-spectral instruments, more data fusion can be anticipated. The differential GPS capability already incorporated in many devices — cell phones, PDAs, laptops, digital cameras and even wristwatches — will open opportunities for the collection of

TABLE 1 A typical need for topographic data

Spatial & temporal resolution	Daily	Weekly	Monthly	Yearly
Fine			⋆	★
Medium		⋆	★	
Coarse	⋆			

continuous spatio-temporal (x, y, z, t) data. A soil surveyor, for example, could retain a record of each traverse and sample site visited so that field data could be used to test and refine existing models or add observations for locations of low confidence. Geologists already use this approach to map rock-unit contacts in the field.

5. MORPHOMETRIC TOOLS

Many challenges remain to the wider adoption of geomorphometric data, methods, and results in routine mapping and decision-making. The scope of the science, as evidenced by the chapters of this book, suggests the need for research in a number of complementary areas:

- primary elevation measurement and DEM compilation;
- *data structures* for current and new conceptual models — point clouds and honey-comb (hexagon, octagon) patterns;
- algorithm development for both measures and surface objects, including contextual parameters for true *multi-scale characterisation*;
- maintaining *operational ease* of use despite increasing complexity of morphometric procedures;
- temporal analysis and *simulation of landscape processes* and evolution from micro to macro scales;
- tools for static and *dynamic visualisation* of measures and surface objects;
- conceptualisation of both the continuous land surface and discrete or fuzzy object-entities that support development of *knowledge libraries* and *semantic data models*[8];
- interactive Web provision of primary and derivative variables through *multi-scale morphometric atlases*;
- support for increasingly diverse applications that require integration of all aspects of geomorphometry;
- cutting-edge concepts in geomorphometry, many yet undefined, including contributions to geospatial theory.

> REMARK 4. *Applications will increasingly combine land-surface data with information sensed from multiple remote and* in-situ *arrays.*

From the computational perspective, several developments can be anticipated in the coming decade that will improve application of geomorphometry in various fields. These may include:

Faster and friendlier GIS tools Scarcity of user-friendly and computationally efficient tools is the most serious bottleneck in semi-automated mapping. Hybrid techniques for geostatistical estimation, for example, require much prior analysis

[8] Semantic data models are machine readable meta-data, which will promote automated information retrieval, exchange and inter-operability between processes.

and human-moderated selection of parameters. To test multiple prediction techniques, at various resolutions and with a minimum of steps, users require more toolboxes that offer guided analysis via data-processing *wizards*.

Dynamic GIS environments Future understanding of the landscape requires accurate, stable operational models of surface process. Eventually, different thematic groups will work together to develop integrated models of landscape evolution. Such progress faces many computational challenges but promises more accurate and flexible mapping to accommodate a diversity of management scenarios.

More detailed, more accurate DEMs Many case studies have shown that the current *de facto* resolution standard set by the global SRTM DEM is not fine enough for the mapping of soils, vegetation, and similar phenomena. It is difficult to settle on an optimum standard for global modelling at this time, but most current mappers would not be satisfied with resolutions coarser than 25 m.

Online data browsers To many non-GIS professionals, geomorphometry's land-surface measures and objects can appear abstract and operationally confusing; more widely available tools for data browsing could reduce this problem. Such interfaces as Google Earth and World Wind (http://worldwind.arc.nasa.gov) are likely to play an important role in extending geomorphometry to less specialised professionals and the general public.

We anticipate *intelligent* tools that build on what has been learned thus far to provide a *point-and-click* interface for the analysis and classification of morphometric surfaces. A user at the computer screen should be able to define a bounding box over a desired DEM area and have possibilities appear for surface-entity classes and their boundaries that the person could accept, reject or revise interactively to arrive at a final, human-mediated result suited to a particular application at a specific scale. Such tools would be most useful if they accessed *knowledge libraries*, collections of morphometric data, variables, and objects and their interrelations, in a manner that was transparent to the analyst, yet based on sound geomorphometry.

Image segmentation–object recognition by such advanced approaches as wavelet analysis, presently exemplified by e-Cognition® and other software packages, will become more important (Drăguţ and Blaschke, 2006). Abstracting complex surfaces as objects within a broader context is innately human, a trait that will lead to knowledge libraries that store representations of human understanding of morphometry and landforms. Work by Miliaresis (2001), Leighty and Rinker (2003), and others shows that land-surface taxonomy needs to be informed by context and pattern.[9] This requires knowing how some surface forms lie only near, above, or below certain other forms and occur only under particular physical conditions and physiographic settings. Presently the necessary knowledge libraries are only in the planning stage.

[9] A tool for spatial pattern analysis is available at: www.umass.edu/landeco/research/fragstats/fragstats.html.

6. APPROACHES AND OBJECTIVES

The Earth's surface has been quantified in various ways, most of which will persist (Wolf, 2004) despite regular elevation grids having largely replaced spot heights, contours, and topographic profiles. Both raster and vector structures will continue to represent the land surface as square grids, lattices, TINs, polygonal facets, and stream-tube arrays.[10] The two chief types of DEM derivatives are likely to remain dominant: discrete objects comprising landform elements and continuous-surface variables with built-in scale dependency. Because each data structure or object abstraction has advantages and disadvantages, morphometric applications will continue to support this diversity of approaches; a *universal* standard is unlikely to win acceptance.

REMARK 5. *Multi-scale spatio-temporal process modelling and simulation for environmental analysis will require careful integration of DEM derivatives.*

As computational tools improve, vaguely posed research ideas that previously went unexamined may be assessed and developed from real-world geomorphometric data. Approaches to landscape analysis may have been evaluated with local or regional samples, but such testing will be possible with very large, even global, data sets (Figure 6). In so doing, it is well to anticipate potential pitfalls. Among these is publication of statistically naïve work. It is the responsibility of the scientific community to guide (especially new) contributors toward a working knowledge of computational methods, an important role served by this book.

Despite the *objectivity* often claimed for geomorphometry, some *subjectivity* will always guide land-surface analysis. Good judgement is critical in applying digital tools and parsing data sets. Raw computer power must not be used blindly, for example, generating all possible variations of each surface measure and then combining them in meaningless or undecipherable indices. Pike (2001c) cautions that "*new*" geomorphometric parameters rarely are and repeats Evans' (1972) warning that we should not expect development of a *philosopher's-stone* variable that magically summarises many geomorphometric properties in a single measure (most recently, e.g., the fractal dimension D).

A closely related trap for the unwary is parameter redundancy; many variables express much the same attribute of the land surface (most commonly *roughness*) and thus contribute little to a multivariate model or geometric signature except statistical noise. Figure 4 illustrates this problem by the inverse relation between elevation skewness and the hypsometric integral. To guard against these and other failings, statistical testing (e.g. correlation) and a knowledge of geomorphology helps relate individual variables to physical processes in the landscape. This broader background promotes understanding of landscape evolution in a variety of settings, while applying geomorphometry as the analytical tool.

The near future will likely see continued progress in the quantitative definition of landforms, their geometric and topological elements, and the characterisation of contrasting landscapes. Terrain objects need to be compared numerically to those

[10] Despite the declared preference for square-grid DEMs in this book.

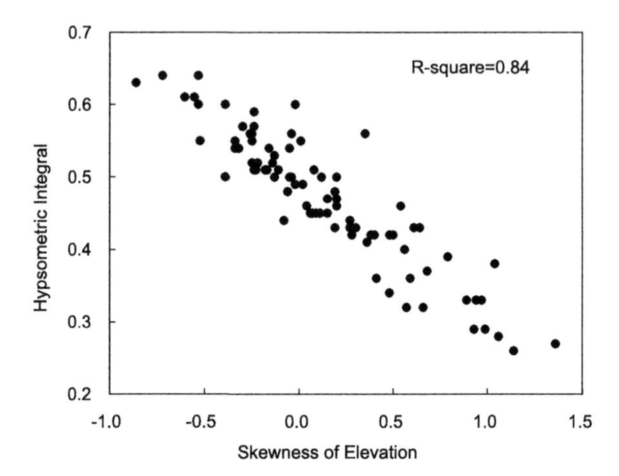

FIGURE 4 Parameter redundancy. An example of why truly "*new*" morphometric measures are few; 80 samples from diverse 1:24,000-scale contour maps (21×21 window on 61-m DEM).

simulated from process models. Modifying existing topography by a hypothetical sequence of natural processes eventually may predict its future configuration.

Multi-scale analyses need to establish and quantify object hierarchies constructed from morphometric primitives. This will facilitate the scaling up or down of mechanistic processes as well as linkage with broader patterns in space and time.

Landform identification will progress at least as much from advanced software as it will from better DEMs. If the array of analytical packages applied in this book is any guide to future software development, today's successes in land surface description and taxonomy will pale beside those of tomorrow.

Improving geomorphometric methods and tools will promote multi-temporal analysis. The evolution of surfaces and objects needs to be modelled through time, as well as at a variety of spatial scales. *Time-lapse estimates* of surface development will be an active area of research, expectedly driven by the need to monitor high-latitude ice and snow cover as well as sea-surface topography. Changes in the global climate are certain to cause weather extremes that will require ever improving and more rapid modelling of water and mass-movement over the Earth's surface. More surface-specific data structures, including the flow-tube concept (Wilson and Gallant, 2000) may better predict the potential magnitude of hydrologic hazards.

7. APPLICATIONS TO COME

The past and present hold keys to the future course of applied geomorphometry; tomorrow's uses of the science most likely will resemble today's, only under greater urgency and with increasing complexity. Onset of widely perceived effects of global warming, coupled with the growing world population, will provoke generation of ever more data to guide stewardship of the planet and improve understanding of its natural processes. Current advances in geomorphology, for example, which are taking every advantage of DEM-based analysis,[11] will only increase in significance.

> REMARK 6. *Tomorrow's geomorphometric applications are likely to resemble today's, only with increasing complexity and under greater urgency.*

The number of management decisions that require topographic information has grown dramatically, along with the volume of supporting DEM data and the need for promptness in its processing. While greater automation in mapping is thus imperative, the necessary procedures have not yet matured to the point of routine application. As new tools and data appear on the market, so will the applications of geomorphometry increase. Four areas immediately come to mind:

Water and energy resources Water could be the oil of the 21st century (Fortune Magazine, Nov. 15, 2000, p. 55). As this resource becomes scarce and more limiting, and urban areas continue to expand, competition for water will intensify and spawn disputes over its ownership and use. Hydro-morphometric analysis will become critical, because the definition of catchments, channel reaches and types, hillslopes, and groundwater zones is vital to assigning jurisdictional responsibility. End users will need to be able to monitor the volume and flux of surface and subsurface water and to forecast changes and future needs in near-real time. Also, geomorphometry is intrinsic to at least two alternative sources of energy; land-surface geometry figures prominently in assessing an area's wind or solar-energy potential (see Mortensen et al., 2006, and Section 6 in Chapter 8).

Environmental management The combined effects of global warming and a burgeoning population will impact such life-sustaining resources as forest cover and agriculture (http://www.fao.org). These must be managed in a responsible and sustainable manner that is recognised as such through official certification. Carbon as a global commodity will need to be carefully accounted for to quantify its relation to the global climate (http://www.ipcc.ch). Mapping, modelling, and monitoring in each of these instances will require detailed geomorphometric data and analysis.

Hazard assessment and mitigation Natural hazards bring another set of unwanted problems (Pike et al., 2003) that will increasingly intrude upon human settlement worldwide, as weather extremes become more common and marginal areas are settled or urbanised. The emergency preparedness and response required for mitigation begins with identifying hazard location, present or potential. The scientific understanding of flood,[12] drought, wildfire, slope instability, volcanic eruption, earthquake, and the like all involve the analysis of elevation data.

Defence requirements Because humankind's ability to settle its rivalries has not kept pace with the lethality and spread of modern weaponry, the need for defence

[11] Including reemergence of orometry as "*neo-orometry*", the DEM-based modelling of mountain ranges and mountain-building processes (Van der Beek and Braun, 1998; Kühni and Pfiffner, 2001; Mitchell and Montgomery, 2006b).
[12] See, for example, potential sea-level rise in Figure 5 and Rowley et al. (2007).

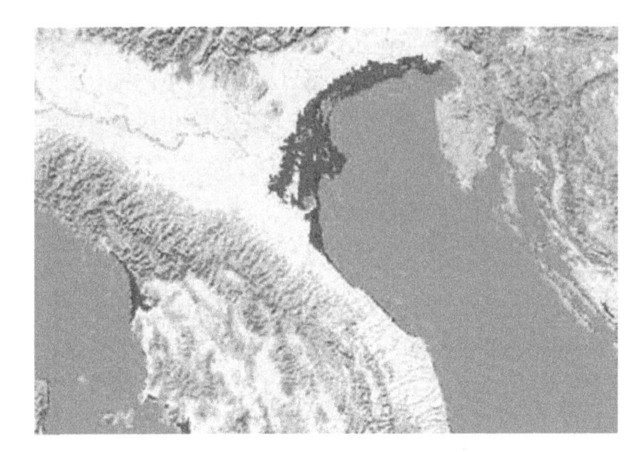

FIGURE 5 Mapping flood risk by geomorphometry. Inundation of southern European coast (in red), given a potential 3-m rise in sea level, estimated from the 1-km GLOBE DEM (http://www.cresis.ku.edu/). Image credit: Center for Remote Sensing of Ice Sheets. (See page 764 in Colour Plate Section at the back of the book.)

against potential foes remains. A military understanding of topography has been essential since antiquity. While current activities must be closed to most of us, such past successes (see, for example, Chapter 1) as the U.S. Corona surveillance program (Cloud, 2002) indicate that not only will geomorphometry continue to aid in keeping the peace but that the eventual declassification of defence technology will be of immense benefit to the science.

8. A GEOMORPHOMETRIC ATLAS OF THE WORLD

This book, through the initiative of the European Commission, has inspired broad collaboration and potential team projects. Individuals or research groups, some in widely scattered locations, need to better collaborate in creating analyses and maps of at least continental scale. It is difficult for a single investigator or even one group to undertake complex geomorphometry (see various chapters in this book) on a global scale. However, cooperative initiatives could develop a *geomorphometric atlas* as an information system to distribute parameters and objects for any area, online and with global coverage (see e.g. Figure 6). Such an atlas could be constructed within a GIS available to the broader research and user community.

> REMARK 7. *Global geomorphometric atlases and knowledge libraries ultimately may provide standardised land-surface parameters and objects via the Web.*

A geomorphometric atlas is particularly useful as a resource of surface measures and objects in support of decision-making and projects over a broad spectrum of applications (Figure 7). One current example is the USGS global 1 km *HYDRO1k* Elevation Derivative Database (http://edc.usgs.gov), which consists

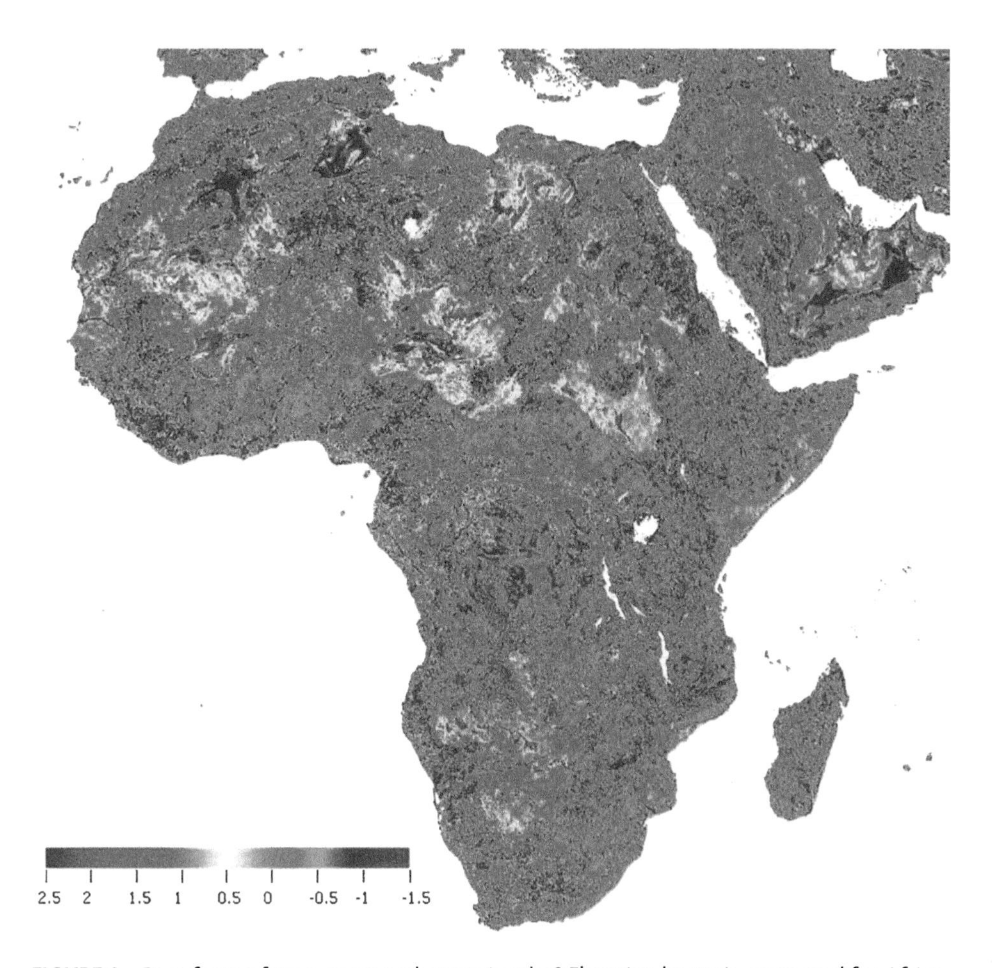

FIGURE 6 Page from a future geomorphometric atlas? Elevation kurtosis computed for Africa from the SRTM 3" data set in regions measuring 2.5'×2.5'. Zero kurtosis indicates a more or less statistically normal (bell-shaped) distribution of elevations; positive kurtosis denotes fewer-than-normal very high and low points, whereas negative kurtosis indicates the opposite — more high and low elevations than normal but fewer points in between. The highest values of kurtosis (red) occur in river deltas like the Nile, Mesopotamia, and isolated dune fields in the Sahara and the Empty Quarter of the Arabian peninsula. The lowest values (blue/purple) occur in fields of uniform linear dunes where most topography is either crest or trough. The SRTM data allow creation of global maps (like this), but their detail can only be appreciated when visualised at continental or finer scales. Courtesy of P.L. Guth. (See page 765 in Colour Plate Section at the back of the book.)

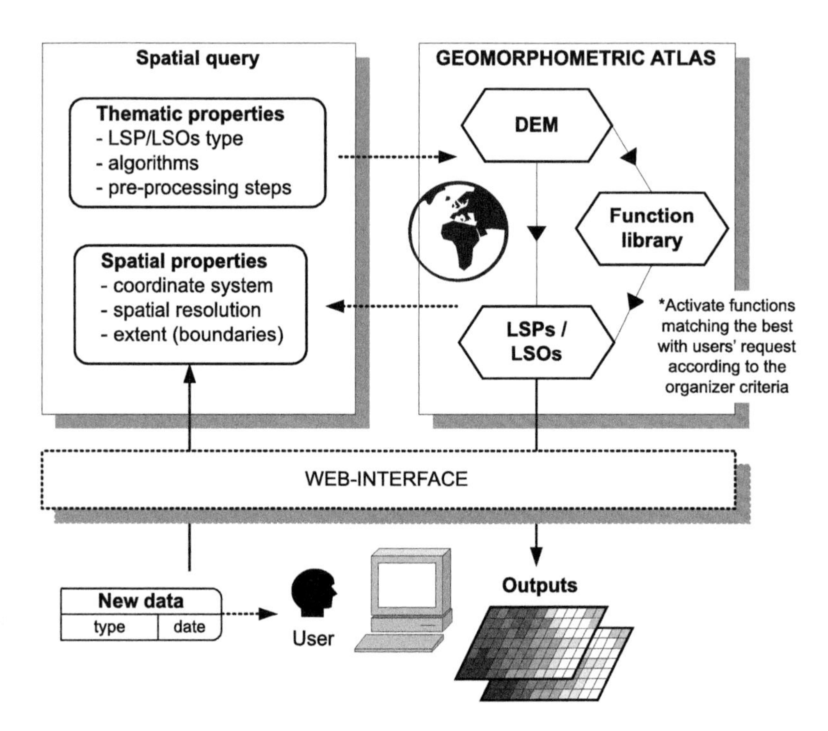

FIGURE 7 Conceptual design for a geomorphometric atlas: a user defines various parameters and then downloads the data. Processing and derivation of surface descriptors is done directly on the server.

of eight raster/vector layers: elevation, WTI, slope, aspect, flow direction, flow accumulation, drainage basins (polygons) and streams (segments). Other countries have similar plans for on-line data browsers from which each user can access the *default*[13] map of geomorphometric features. An example similar to the HYDRO1k project is the current *River and Catchment Database*[14] at intermediate scales of 1:250,000 to 1:500,000 derived from the filtered 250 m SRTM DEM for Europe (Vogt et al., 2003). The list of surface phenomena could be extended to include the most frequently sought parameters and objects (Table 2), whereupon suitable algorithms could be chosen to map them.

To achieve full functionality, a geomorphometric atlas should satisfy the following criteria:

Precision the best estimates of surface measures and objects, derived from the most accurate and current data and algorithms;

Multiscale an input DEM residing within a multi-scale GIS so that it can be available at various resolutions;

[13] Because most land-surface measures and objects are scale dependent, it is often expedient to simply determine an arbitrary scale and declare it the standard.
[14] Available at http://wise2.jrc.it.

TABLE 2 Some potential inputs and outputs for a global geomorphometric atlas

Inputs	Outputs
Global 90-m DEM (SRTM) Global 1–5-m DEM (LiDAR) Global water-mask Vegetation Index & Leaf Area Index Areas under cultivation Protected natural areas Mean annual precipitation Mean annual temperature	Gallery of the chief parameters and objects (basic, hydrological, climatological) 'Memberships' of main generic landforms & eco-units Singular points & break-lines: peaks, closed depressions, ridges, drainages, slope inflections, hydrological nets, watersheds, etc. Propagated uncertainty in mapping various geomorphometric features

Open structure users should be able to modify and adjust spatial queries (see Figure 7) to tailor surface measures and objects to fit specific needs;

Web access the system should be Web accessible and able to deliver the queried data in one of the common or open formats that can be easily implemented in many GIS packages;

Quality the input (DEMs) and algorithms should be continually evaluated and improved (through filtering out artefacts and other errors, improvement of resolution, etc.) and maintained to ensure their fitness for use.

Potential clients for a geomorphometric atlas include climate modelers, regional weather agencies, telecommunications companies, agricultural and forest-product companies, natural-resources agencies, water-supply and waste-management companies, emergency-response agencies, and planning officials at all levels of government. Customers should be able to download a well-thought-out *standard package* of land-surface data and derivatives much like the parameters and objects listed on the geomorphometry.org website. Such a global resource could provide content for regional or metropolitan atlases. Digital atlases should also support the development of knowledge libraries, discussed above, as well as continued testing and comparison of geomorphometric methods by the scientific community.

9. CLOSING REMARKS

Geomorphometry, now a science in its own right, is essential to the understanding of such worldwide concerns as water and food supply, natural hazards, climate change, and sea-level rise. Despite conceptual advances, the emphasis of this largely empirical science is likely to remain on method and technique rather than theory. The field continues to evolve as computing power increases and new instrumentation is developed for sensing the Earth and its interacting processes. Few analytical tools will be more effective in furthering knowledge of the Earth and the activities of its inhabitants than a digital surface model and the methods to manipulate it at varied spatio-temporal scales.

Collecting data on Earth's surface-form at heretofore unprecedented detail and extent will become simpler, more routine, and surprisingly affordable. LiDAR technology will mature, and national agencies and wider collaborative efforts will continue to compile ever-more-detailed global datasets. As the World Wide Web becomes even more essential for interactive analysis, local communities could eventually access geomorphometric atlases over the Internet.

Geomorphometry already can guide vehicles across rough terrain and simulate visual impressions of the landscape through digital models of topography. Future applications will range from saving lives on the battlefield to making it even easier to quantify processes that control Earth's water and bio-geochemical fluxes. Computer models implemented over varied spatial scales will be able to accelerate or slow the passage of time to better comprehend the flow of water, plant growth, changes in air temperature, and the recession of glaciers.

Among the many prospects for geomorphometry are tighter coupling of remote (satellite, airborne, UAV) sensors with in-situ arrays to better capture the dynamics of Earth's moisture and energy balances. Geomorphometry will continue to diversify, from measuring a surface on which to conceptualise geomorphic process, to applying a multi-faceted toolset to observe, portray, and explain Earth's complexities in ways that few could have anticipated thirty years ago. Indeed, these tools are not unique to Earth's surface, but apply to its ocean floor and the surface of Mars or any other planetary body for which digital surface data can be obtained.

BIBLIOGRAPHY

REFERENCES

Aandahl, A.R., 1948. The characterization of slope positions and their influence on total nitrogen content of a few virgin soils of western Iowa. Soil Science Society of America Proceedings 13, 449–454.

Aerts, C.J.H., Heuvelink, G.B.M., Goodchild, M.F., 2003. Accounting for spatial uncertainty in optimization with spatial decision support systems. Transactions in GIS 7, 211–230.

Akima, H., 1978. A method of bivariate interpolation and smooth surface fitting for irregularly distributed data points. ACM Transactions on Mathematical Software 4 (2), 148–159.

Al-Harthi, A.A., 2002. Geohazard assessment of sand dunes between Jeddah and Al-Lith, western Saudi Arabia. Environmental Geology 42 (4), 360–369.

Albani, M., Klinkenberg, B., 2003. A spatial filter for the removal of striping artifacts in digital elevation models. Photogrammetric Engineering and Remote Sensing 69 (7), 755–765.

Alexander, R., Millington, A.C., 2000. Vegetation Mapping — From Patch to Planet. Wiley, 339 pp.

Allder, W.R., Caruso, V.M., Pearsall, R.A., Troup, M.I., 1982. An overview of digital elevation model production at the United States Geological Survey. In: Proceedings of Auto-Carto 5. ASPRS–ACSM, pp. 23–32.

Antonić, O., Kušan, V., Bakran-Petricioli, T., Alegro, A., Gottstein-Matočec, S., Peternel, H., Tkalčec, Z., 2005. Mapping the habitats of The Republic of Croatia (2000.–2004.) — the project overview. Drypis — Journal for Applied Ecology 1 (1), 40 (in Croatian).

Antonić, O., Pernar, N., Jelaska, S.D., 2003. Spatial distribution of main forest soil groups in Croatia as a function of basic pedogenetic factors. Ecological Modelling 170 (2–3), 363–371.

Antonić, O., Hatić, D., Pernar, R., 2001a. DEM-based depth in sink as an environmental estimator. Ecological Modelling 138 (1–3), 247–254.

Antonić, O., Križan, J., Marki, A., Bukovec, D., 2001b. Spatio-temporal interpolation of climatic variables over large region of complex terrain using neural networks. Ecological Modelling 138 (1–3), 255–263.

Antonić, O., Marki, A., Križan, J., 2000. A global model for monthly mean hourly direct solar radiation. Ecological Modelling 129 (2–3), 113–118.

Antonić, O., Legović, T., 1999. Estimating the direction of an unknown air pollution source using a digital elevation model and a sample of deposition. Ecological Modelling 124 (1), 85–95.

Antonić, O., 1998. Modelling daily topographic solar radiation without site-specific hourly radiation data. Ecological Modelling 113 (1–3), 31–40.

Arge, L., Chase, J.S., Halpin, P., Toma, L., Vitter, J.S., Urban, D., Wickremesinghe, R., 2003. Efficient flow computation on massive grid terrain datasets. GeoInformatica 7 (4), 283–313.

Argialas, D.P., Miliaresis, G.C., 1997. Landform spatial knowledge acquisition: identification, conceptualization and representation. In: Proceedings 1997 ASPRS/ASCM Annual Convention and Exposition, Technical Papers. Seattle, WA, April 7–10, 1997. American Society for Photogrammetry and Remote Sensing, American Congress on Surveying and Mapping, Bethesda, MD, pp. 733–740.

Arnold, R.W., 1988. Soil survey: an example of applied research. Soil Survey Horizons, 102–106.

Arrell, K.E., 2005. Predicting glacier accumulation area distributions. Unpublished Ph.D. in Geography. University of Durham, 383 pp.

Arrouays, D., Daroussin, J., Kicin, J.C., Hassika, P., 1998. Improving topsoil carbon storage prediction using a digital elevation model in temperate forest soils of France. Soil Science Society of America Journal 163, 103–108.

ASCE Task Force on Friction Factors, 1963. Friction factors in open channels. Journal of the Hydraulics Division 89, 97–143.
Aspinall, R., Veitch, N., 1993. Habitat mapping from satellite imagery and wildlife survey data using a Bayesian modelling procedure in a GIS. Photogrammetric Engineering and Remote Sensing 59, 537–543.
Axelsson, O., 1999. Processing of laser scanner data — algorithms and applications. ISPRS Journal of Photogrammetry and Remote Sensing 54, 138–147.
Bailey, R.G., 2002. Ecoregion-Based Design for Sustainability. Springer-Verlag, Heidelberg, 223 pp.
Bailey, R.G., 1981. Integrated approaches to classifying land as ecosystems. In: Laban, P. (Ed.), Proceedings of the Workshop on Land Evaluation for Forestry. In: ILRI Publication, vol. 28. ILRI, Wageningen, pp. 95–109.
Bakhsh, A., Jaynes, D.B., Colvin, T.S., Kanwar, R.S., 2000. Spatio-temporal analysis of yield variability for a corn-soybean field in Iowa. Transactions of the ASAE 43 (1), 31–38.
Bakran-Petricioli, T., Antonić, O., Bukovec, D., Petricioli, D., Janeković, I., Križan, J., Kušan, V., Dujmović, S., 2006. Modelling spatial distribution of the Croatian marine benthic habitats. Ecological Modelling 191 (1), 96–105.
Ball, L.C., Doherty, P.F., McDonald, M.W., 2005. An occupancy modeling approach to evaluating a palm springs ground squirrel habitat model. The Journal of Wildlife Management 69 (3), 894–904.
Baltsavias, E.P., 1999a. A comparison between photogrammetry and laser scanning. ISPRS Journal of Photogrammetry and Remote Sensing 54, 83–94.
Baltsavias, E.P., 1999b. Airborne laser scanning: basic relations and formulas. ISPRS Journal of Photogrammetry and Remote Sensing 54, 199–214.
Bamler, R., 1997. Digital terrain models from radar interferometry. In: Fritsch, D., Hobbie, D. (Eds.), Photogrammetrische Woche 1997. Wichmann, Heidelberg, pp. 93–105.
Band, L.E., Patterson, P., Nemani, R., Running, S.W., 1993. Forest ecosystem processes at the watershed scale: incorporating hillslope hydrology. Agricultural and Forest Meteorology 63, 93–126.
Band, L.E., Peterson, D.L., Running, S.W., Couglan, J.C., Lammers, R.B., Dungan, J., Nemani, R., 1991. Forest ecosystem processes at the watershed scale: basis for distributed simulation. Ecological Modelling 56, 151–176.
Band, L.E., 1989. Spatial aggregation of complex terrain. Geographical Analysis 21, 279–293.
Banks, J. (Ed.), 1998. Handbook of Simulation — Principles, Methodology, Advances, Applications, and Practice. Wiley, New York, 864 pp.
Barr, D.R., Mansager, B., 1996. Terrain map resolution. Mathematical and Computer Modelling 23 (1–2), 39–46.
Barringer, J.R.F., Hewitt, A.E., Lynn, I.H., Schmidt, J., 2006. National mapping of landform elements for New Zealand in support of S-map, a New Zealand soils database. In: Zhou, Q., Tang, G. (Eds.), Proceedings of International Symposium on Terrain Analysis and Digital Terrain Modelling. Nanjing Normal University, Nanjing, p. 11.
Barsch, D., 1990. Geomorphology and geoecology. Zeitschrift für Geomorphologie 79, 39–49.
Bartelt, P., McArdell, B.W., Swartz, M., Christen, M., 2005. Evaluation of a two-phase debris flow model. Geophysical Research Abstracts 7, 05948.
Bates, R.L., Jackson, J.A. (Eds.), 2005. Glossary of Geology. 5th edition. American Geological Institute, New York, 800 pp.
Bathgate, J.D., Duram, L.A., 2003. A geographic information systems based landscape classification model to enhance soil survey: a southern Illinois case study. Journal of Soil and Water Conservation 58 (3), 119–127.
Bathurst, J.C., Cooley, K.R., 1996. Use of the SHE hydrological modelling system to investigate basin response to snowmelt at Reynolds Creek, Idaho. Journal of Hydrology 175, 181–211.
Bayramin, I., 2000. Using geographic information system and remote sensing techniques in making pre-soil surveys. In: Proceedings of the International Symposium on Desertification. Soil Science Society of Turkey, pp. 27–33.
Beck, P.S.A., Kalmbach, E., Joly, D., Stien, A., Nilsen, L., 2005. Modelling local distribution of an Arctic dwarf shrub indicates an important role for remote sensing of snow cover. Remote Sensing of Environment 98, 110–121.
Bekker, M.G., 1969. Introduction to Terrain-Vehicle Systems. The University of Michigan Press, Ann Arbor, 846 pp.

Bell, J.C., Cunningham, R.L., Havens, M.W., 1994. Soil drainage class probability mapping using a soil–landscape model. Soil Science Society of America Journal 58, 464–470.
Bell, J.C., Cunningham, R.L., Havens, M.W., 1992. Calibration and validation of a soil–landscape model for predicting soil drainage class. Soil Science Society of America Journal 56, 1860–1866.
Bendix, J., 2004. Geländeklimatologie. Gebrüder Bornträger Verlagsbuchhandlung, Stuttgart, 282 pp.
Berntsen, J., Thomsen, A., Schelde, K., Hansen, O.M., Knudsen, L., Broge, N., Hougaard, H., Hørfarter, R., 2006. Algorithms for sensor-based redistribution of nitrogen fertilizer in winter wheat. Precision Agriculture 7 (2), 65–83.
Berry, B.J.L., Marble, D.F. (Eds.), 1968. Spatial Analysis; A Reader in Statistical Geography. Prentice–Hall, Englewood Cliffs, NJ, 512 pp.
Berry, J.K., 1996. Beyond Mapping: Concepts, Algorithms, and Issues in GIS. Wiley, 246 pp.
Beven, K.J., 2000. Rainfall-Runoff Modelling: The Primer, 1st edition. Wiley, New York, 360 pp.
Beven, K.J., 1997. TOPMODEL: a critique. Hydrological Processes 11 (9), 1069–1086.
Beven, K.J., Kirby, M.J., Schoffield, N., Tagg, A., 1984. Testing a physically based flood forecasting model (TOPMODEL) for three UK catchments. Journal of Hydrology 69, 119–143.
Beven, K.J., Kirkby, M.J., 1979. A physically based, variable contributing area model of basin hydrology. Hydrological Sciences Bulletin 24 (1), 43–69.
Biggs, A., Slater, B., 1998. Using soil landscape and digital elevation models to provide rapid medium scale soil surveys on the Eastern Darling Downs, Queensland. In: Proceedings of the 16th World Congress of Soil Science. Montpellier, France.
Bishop, C., 1995. Neural Networks for Pattern Recognition. Oxford University Press, Oxford, 504 pp.
Bishop, M.P., Shroder Jr., J.F., Colbyb, J.D., 2003. Remote sensing and geomorphometry for studying relief production in high mountains. Geomorphology 55, 345–361.
Bishop, M.P., Shroder Jr., J.F., Hickman, B.L., Copland, L., 1998. Scale-dependent analysis of satellite imagery for characterization of glacier surfaces in the Karakoram Himalaya. Geomorphology 21 (3–4), 217–232.
Bishop, T.F.A., Minasny, B., 2005. Digital soil-terrain modelling: the predictive potential and uncertainty. In: Grunwald, S. (Ed.), Environmental Soil–Landscape Modeling: Geographic Information Technologies and Pedometrics. CRC Press, Boca Raton, FL, pp. 185–213.
Bishop, T.F.A., McBratney, A.B., Whelan, B.M., 2001. Measuring the quality of digital soil maps using information criteria. Geoderma 103 (1–2), 95–111.
Bivand, R., Pebesma, E., Rubio, V., 2008. Applied Spatial Data Analysis with R. Use R Series. Springer, Heidelberg, 400 pp.
Blackmore, S., 2000. The interpretation of trends from multiple yield maps. Computers and Electronics in Agriculture 26 (1), 37–51.
Blaszczynski, J.S., 1997. Landform characterization with geographic information systems. Photogrammetric Engineering and Remote Sensing 63 (2), 183–191.
Blöschl, G., 1999a. Scale and Scaling in Hydrology — A Framework for Thinking and Analysis, 1st edition. Wiley, Chichester, 352 pp.
Blöschl, G., 1999b. Scaling issues in snow hydrology. Hydrological Processes 13, 2149–2175.
Blöschl, G., Sivapalan, M., 1995. Scale issues in hydrological modelling — a review. Hydrological Processes 9, 251–290.
Blow, J., 2000. Terrain rendering at high levels of detail. In: Game Developers Conference Proceedings. CMP Media LLC, San Jose, CA, p. 4.
Blumen, W. (Ed.), 1990. Atmospheric Processes over Complex Terrain. Meteorological Monographs, vol. 23. American Meteorological Society, Boston, MA, 323 pp.
Bock, M., Rossner, G., Wissen, M., Langanke, T., Lang, S., Klug, H., Blaschke, T., Remm, K., Vrščaj, B., 2005. Spatial indicators for nature conservation from European to local scale. Ecological Indicators 5 (4), 322–338.
Bockheim, J.G., Gennadiyev, A.N., Hammer, R.D., Tandarich, J.P., 2005. Historical development of key concepts in pedology. Geoderma 124, 23–36.
Boden Ag, 1994. Bodenkundliche Kartieranleitung, 4. verbesserte Auflage. E. Schweizerbartische Verlagsbuchhandlung, Hannover, 392 pp.
Böhner, J., 2006. General climatic controls and topoclimatic variations in Central and High Asia. Boreas 35 (2), 279–295.

Böhner, J., Selige, T., 2006. Spatial prediction of soil attributes using terrain analysis and climate regionalisation. In: Böhner, J., McCloy, K.R., Strobl, J. (Eds.), SAGA — Analysis and Modelling Applications, vol. 115. Verlag Erich Goltze GmbH, pp. 13–27.
Böhner, J., 2005. Advancements and new approaches in climate spatial prediction and environmental modelling. Arbeitsberichte des Geographischen Instituts der HU zu Berlin 109, 49–90.
Böhner, J., Köthe, R., 2003. Bodenregionalisierung und Prozessmodellierung: Instrumente für den Bodenschutz. Petermanns Geographische Mitteilungen 147, 72–82.
Böhner, J., Köthe, R., Conrad, O., Gross, J., Ringeler, A., Selige, T., 2002. Soil regionalisation by means of terrain analysis and process parameterisation. In: Micheli, E., Nachtergaele, F., Montanarella, L. (Eds.), Soil Classification 2001, EUR 20398 EN. The European Soil Bureau, Joint Research Centre, Ispra, pp. 213–222.
Böhner, J., Köthe, R., Trachinow, C., 1997. Weiterentwicklung der automatischen Reliefanalyse auf der Basis von Digitalen Geländemodellen. Göttinger Geographische Abhandlungen 100, 3–21.
Böhner, J., Pörtge, K.-H., 1997. Strahlungs- und expositionsgesteuerte tagesperiodische Schwankungen des Abflusses in kleinen Einzugsgebieten. Petermanns Geographische Mitteilungen 141, 35–42.
Bolstad, P., 2006. GIS Fundamentals: A First Text on Geographic Information Systems, 2nd edition. Atlas Books, Ashland, OH, 560 pp.
Bolstad, P.V., Swank, W., Vose, J., 1998. Predicting Southern Appalachian overstory vegetation with digital terrain data. Landscape Ecology 13, 271–283.
Bolstad, P.V., Stowe, T., 1994. An evaluation of DEM accuracy: elevation, slope and aspect. Photogrammetric Engineering and Remote Sensing 60, 1327–1332.
Bolstad, P.V., Lillesand, T.M., 1992. Improved classification of forest vegetation in northern Wisconsin through a rule-based combination of soils, terrain, and Landsat TM data. Forest Science 38 (1), 5–20.
Bonner, W.J., Schmall, R.A., 1973. A photometric technique for determining planetary slopes from orbital photographs. Professional Paper 812-A. U.S. Geological Survey, 16 pp.
Borkowski, A., Meier, S., 1994. A procedure for estimating the grid cell size of digital terrain models derived from topographic maps. Geo-Informations-System 7 (1), 2–5.
Boruvka, L., Kozak, J., Nemecek, J., Penizek, V., 2002. New approaches to the exploitation of former soil survey data. In: 17th World Congress of Soil Science. Bangkok, Thailand, August 14–21. IUSS.
Bougeault, P., Binder, P., Buzzi, A., Dirks, R., Houze, J., Kuettner, J., Smith, R.B., Steinacker, R., Volkert, H., 2001. The MAP special observing period. Bulletin of the American Meteorological Society 82, 433–462.
Bougeault, P., Benech, B., Bessemoulin, P., Carissimo, B., Jansa, A., Pelon, J., Petitdidier, M., Richard, E., 1997. PYREX: a summary of findings. Bulletin of the American Meteorological Society 78, 637–650.
Bow, J., Sayre, R., Josse, C., Touval, J., 2005. Modeling nature reserves in Latin America and Caribbean ecological systems throughout South America. Unpublished final report. The Nature Conservancy, Washington, DC.
Brabb, E.E., Pampeyan, E.H., Bonilla, M.G., 1972. Landslide susceptibility in San Mateo County, California. Miscellaneous Field Studies Map MF-360, scale 1:62,500. U.S. Geological Survey, 1 p.
Brabyn, L., 1998. GIS analysis of macro landform. In: Proceedings of the Spatial Information Research Centre's 10th Colloquium. University of Otago, Dunedin, New Zeeland, pp. 35–48.
Brabyn, L.K., 1997. Classification of macro landforms using GIS. ITC Journal 1997 (1), 26–40.
Braun-Blanquet, J., 1928. Pflanzensoziologie. Springer Verlag, Berlin, 423 pp.
Bregt, A.K., 1992. Processing of soil survey data. Ph.D. Thesis. University of Wageningen.
Brock, T.D., 1981. Calculating solar radiation for ecological studies. Ecological Modelling 14, 1–19.
Brocklehurst, S.H., Whipple, K.X., 2004. Hypsometry of glaciated landscapes. Earth Surface Processes and Landforms 29 (7), 907–926.
Brocklehurst, S.H., Whipple, K.X., 2002. Glacial erosion and relief production in the Eastern Sierra Nevada, California. Geomorphology 42, 1–24.
Brooks, R.H., Corey, A.T., 1964. Hydraulic properties of porous media. Hydrology Paper No. 3. Civil Engineering Department, Colorado State University, Fort Collins, CO, 27 pp.
Brown, D.G., Bara, T.J., 1994. Recognition and reduction of systematic error in elevation and derivative surfaces from 7 1/2-minute DEMs. Photogrammetric Engineering and Remote Sensing 60 (2), 189–194.
Brozovic, N., Burbank, D.W., Meigs, A.J., 1997. Climatic limits on landscape development in the northwestern Himalaya. Science 276, 571–574.

Brunori, F., Casagli, N., Fiaschi, S., Garzonio, C.A., Moretti, S., 1996. Landslide hazard mapping in Tuscany, Italy: an example of automatic evaluation. In: Slaymaker, O. (Ed.), Geomorphic Hazards. Wiley, Chichester and New York, pp. 56–67.

Brus, D.J., Heuvelink, G.B.M., 2007. Optimization of sample patterns for universal kriging of environmental variables. Geoderma 138 (1–2), 86–95.

Bue, B.D., Stepinski, T.F., 2006. Automated classification of landforms on Mars. Computers & Geosciences 32 (5), 604–614.

Bui, E., Henderson, B., Moran, C., Johnston, R., 2002. Continental-scale spatial modeling of soil properties. In: 17th World Congress of Soil Science. Bangkok, Thailand. IUSS, p. 1573 (paper no. 1470).

Bui, E.N., Henderson, B.L., Viergever, K., 2006. Knowledge discovery from models of soil properties developed through data mining. Ecological Modelling 191 (3–4), 431–446.

Bui, E.N., Moran, C.J., 2003. A strategy to fill gaps in soil survey over large spatial extents: an example from the Murray–Darling basin of Australia. Geoderma 111 (1–2), 21–44.

Bui, E.M., Moran, C.J., 2001. Disaggregation of polygons of surficial geology and soil maps using spatial modelling and legacy data. Geoderma 103, 79–94.

Bui, E.M., Moran, C.J., 1999. Extracting soil–landscape rules from previous soil surveys. Australian Journal of Soil Research 37, 495–508.

Burbank, D.W., Leland, J., Fielding, E., Anderson, R.S., Brozovic, N., Reid, M.R., Duncan, C., 1996. Bedrock incision, rock uplift and threshold hillslopes in the northwestern Himalayas. Nature 379, 505–510.

Burrough, P.A., van Gaans, P.F.M., Hansen, A.J., 2001. Fuzzy *k*-means classification of topo-climatic data as an aid to forest mapping in the Greater Yellowstone Area, USA. Landscape Ecology 16, 523–546.

Burrough, P.A., van Gaans, P.F.M., MacMillan, R.A., 2000. High resolution landform classification using fuzzy *k*-means. Fuzzy Sets and Systems 113 (1), 37–52.

Burrough, P.A., McDonnell, R.A., 1998. Principles of Geographical Information Systems. Oxford University Press Inc., New York, 333 pp.

Burrough, P.A., van Gaans, P.F.M., van Hootsmans, R., 1997. Continuous classification in soil survey: spatial correlation, confusion and boundaries. In: Fuzzy Sets in Soil Systems. Geoderma 77, 115–135.

Burrough, P.A., 1989. Fuzzy mathematical methods for soil survey and land evaluation. Journal of Soil Science 40, 477–492.

Burrough, P.A., 1986. Principles of Geographic Information Systems for Land Resource Assessment. Monographs on Soil and Resources Survey, vol. 12. Oxford Science Publications, New York, 193 pp.

Burrows, M.T., Robb, L., Nickell, L.A., Hughes, D.J., 2003. Topography as a determinant of search paths of fishes and mobile macrocrustacea on the sediment surface. Journal of Experimental Marine Biology and Ecology 285–286, 235–249.

Cajori, F., 1929. History of determinations of the heights of mountains. Isis (University of Chicago Press) 12 (3), 482–514.

Capolongo, D., Refice, A., Mankelow, J., 2004. Evaluating earthquake-triggered landslide hazard at the basin scale through GIS in the Upper Sele River Valley. Surveys in Geophysics 23 (6), 595–625.

Carlisle, B.H., 2005. Modelling the spatial distribution of DEM error. Transactions in GIS 9, 521–540.

Carlton, D., Tennant, K., 2001. DEM quality assessment. In: Maune, D.F. (Ed.), Digital Elevation Model Technologies and Applications: The DEM Users Manual. American Society for Photogrammetry and Remote Sensing, Bethesda, MD, pp. 395–440.

Carpenter, G.A., Gopal, S., Macomber, S., Martens, S., Woodcock, C.E., Franklin, J., 1999. A neural network method for efficient vegetation mapping. Remote Sensing of Environment 70, 326–338.

Carrara, A., Pike, R.J. (Eds.), 2008. GIS Technology and Models for Assessing Landslide Hazard and Risk. Geomorphology 94 (3–4) 257–507.

Carrara, A., Bitelli, G., Carla, R., 1997. Comparison of techniques for generating digital terrain models from contour lines. International Journal of Geographical Information Science 11, 451–473.

Carré, F., Girrard, M.C., 2002. Quantitative mapping of soil types based on regression kriging of taxonomic distances with landform and land cover attributes. Geoderma 110, 241–263.

Carson, M.A., Kirkby, M.J., 1972. Hillslope Form and Process. Cambridge University Press, Cambridge, 475 pp.

Carter, T.R., Parry, M.L., Harasawa, H., Nishioka, S., 1994. IPCC technical guidelines for assessing climate change impacts and adaptations. Department of Geography, University College London & National Institute for Environment Studies Japan, London, Tsukuba, 59 pp.
Carter, W., Shrestha, R., Slatton, K., 2007. Geodetic laser scanning. Physics Today 60 (12), 41–47.
Casas, A., Benito, G., Thorndycraft, V.R., Rico, M., 2006. The topographic data source of digital terrain models as a key element in the accuracy of hydraulic flood modeling. Earth Surface Processes and Landforms 31 (4), 444–456.
Case, B., Meng, F.R., Arp, P.A., 2005. Digital elevation modelling of soil type and drainage within small forested catchments. Canadian Journal of Soil Science 85 (1), 127–137.
Cavalli, M., Tarolli, P., Marchi, L., Dalla Fontana, G., 2008. The effectiveness of airborne LiDAR data in the recognition of channel-bed morphology. Catena 73 (3), 249–260.
Cayley, A., 1859. On contour and slope lines. The London, Edinburgh and Dublin Philosophical Magazine and Journal of Science Series 4 18 (20), 264–268.
Chacón, J., Corominas, J. (Eds.), 2003. Special Issue on Landslides and GIS. Natural Hazards 30 (3) 263–499.
Chambers, J.M., Hastie, T.J., 1992. Statistical Models in S. Wadsworth & Brooks/Cole, Pacific Grove, CA, 595 pp.
Chang, Y.E., Clay, D.E., Clarson, C.G., Clay, S.A., Malo, D.D., Berg, R., Kleinjan, J., Wiebold, W., 2003. Different techniques to identify management zones impact nitrogen and phosphorus sampling variability. Agronomy Journal 95, 1550–1559.
Chaplot, V., Bernoux, M., Walter, C., Curmi, P., Herpin, U., 2001. Soil carbon storage prediction in temperate forest hydromorphic soils using a morphologic index and digital elevation model. Soil Science Society of America Journal 166, 48–60.
Chaplot, V., Walter, C., Curmi, P., 2000. Improving soil hydromorphy prediction according to DEM resolution and available pedological data. Geoderma 97, 405–422.
Chaplot, V., Walter, C., Curmi, P., 1998. Modeling soil spatial distribution: sensitivity to DEM resolutions and pedological data availability. In: Proceedings of the 16th World Congress of Soil Science. Montpellier, France, p. 354.
Chapman, C.A., 1952. A new quantitative method of topographic analysis. American Journal of Science 250, 428–452.
Charney, J.G., Eliassen, A., 1949. A numerical method for predicting the perturbations of the middle latitude westerlies. Tellus 1, 38–54.
Chock, G.Y., Cochran, L., 2005. Modeling of topographic wind speed effects in Hawaii. Journal of Wind Engineering and Industrial Aerodynamics 93 (8), 623–638.
Chorley, R.J., 1966. The application of statistical methods to geomorphology. In: Dury, G.H. (Ed.), Essays in Geomorphology. Heinemann, London, pp. 275–387.
Chorley, R.J., Dunn, A.J., Beckinsale, R., 1964. The History of the Study of Landforms, vol. 1. Methuen, London, 678 pp.
Chorley, R.J., 1957. Climate and morphometry. Journal of Geology 65, 628–638.
Chorowicz, J., Parrot, J., Taud, H., 1995. Automated pattern-recognition of geomorphic features from DEMs and satellite images. Zeitschrift für Geomorphologie, Supplementband 101, 69–84.
Chorowicz, J., Kim, J., Manoussis, S., Rudant, J., Foin, P., Veillet, I., 1989. A new technique for recognition of geological and geomorphological patterns in digital terrain models. Remote Sensing of Environment 29, 229–239.
Chow, V.T., Maidment, D.R., Mays, L.W., 1988. Applied Hydrology. McGraw–Hill, Singapore, 572 pp.
Chrisman, N., 2006. Charting the Unknown: How Computer Mapping at Harvard Became GIS. ESRI Press, Redlands, CA, 232 pp.
Christensen, R., 2001. Linear Models for Multivariate, Time Series, and Spatial Data, 2nd edition. Springer Text in Statistics. Springer Verlag, New York, 393 pp.
Christian, C.S., Stewart, G.A., 1968. Methodology of integrated surveys. In: Aerial Surveys and Integrated Studies. Toulouse Conference 1964. Proceedings. UNESCO, Paris, pp. 233–280.
Ciha, A.J., 1982. Slope position and grain yield of soft white winter wheat. Agronomy Journal 76, 193–196.
Claessens, L., Lowe, D.J., Hayward, B.W., Schaap, B.F., Schoorl, J.M., Veldkamp, A., 2006. Reconstructing high-magnitude/low-frequency landslide events based on soil redistribution modelling and a Late-Holocene sediment record from New Zealand. Geomorphology 74 (1–4), 29–49.

Claessens, L., Heuvelink, G.B.M., Schoorl, J.M., Veldkamp, A., 2005. DEM resolution effects on shallow landslide hazard and soil redistribution modelling. Earth Surface Processes and Landforms 30 (4), 461–477.
Clark, O., Kok, R., Champigny, P., 1997. Generation of a virtual terrain. Environmental Modelling & Software 12 (2–3), 143–149.
Clarke, J., 1966. Morphometry from maps. In: Dury, G.H. (Ed.), Essays in Geomorphology. Heinemann, London, pp. 235–274.
Clarke, K., 1995. Analytical and Computer Cartography. Prentice–Hall, Englewood Cliffs, NJ, 334 pp.
Clarke, K.C., 1988. Scale-based simulation of topographic relief. The American Cartographer 15 (2), 173–181.
Cleland, D.T., Avers, P.E., McNab, W.H., Jensen, M.E., Bailey, R.G., King, T., Russell, W.E., 1997. National hierarchical framework of ecological units. In: Boyce, M.S., Haney, A. (Eds.), Ecosystem Management Applications for Sustainable Forest and Wildlife Resources. Yale University Press, New Haven, CT, pp. 181–200.
Cloud, J., 2002. American cartographic transformations during the Cold War. Cartography and Geographic Information Science 29 (3), 261–282.
Collins, S.H., 1975. Terrain parameters directly from a digital terrain model. Canadian Surveyor 9, 507–518.
Colombo, R., Vogt, J.V., Soille, P., Paracchini, M.L., de Jager, A., 2007. Deriving river networks and catchments at the European scale from medium resolution digital elevation data. Catena 70 (3), 296–305.
Comer, P., Faber-Langendoen, D., Evans, R., Gawler, S., Josse, C., Kittel, G., Menard, S., Pyne, M., Reid, M., Schulz, K., Snow, K., Teague, J., 2003. Ecological Systems of the United States: A Working Classification of U.S. Terrestrial Systems. NatureServe, Arlington, VA, 83 pp.
Conacher, A.J., Dalrymple, J.B., 1977. The nine-unit landsurface model: an approach to pedogeomorphic research. Geoderma 18 (1–2), 1–154.
Congalton, R.G., Green, K., 1999. Assessing the Accuracy of Remotely Sensed Data: Principles and Practices. Lewis Publishers, Boca Raton, FL, 137 pp.
Conrad, O., 1998. Die Ableitung hydrologisch relevanter Reliefparameter am Beispiel des Einzugsgebiets Linnengrund. Department of Geography Göttingen, Göttingen, 89 pp.
Cook, A.C., Watters, T.R., Robinson, M.S., Spudis, P.D., Bussey, D.J.B., 2000. Lunar polar topography derived from Clementine stereoimages. Journal of Geophysical Research 105 (E5), 12023–12033.
Coops, N.C., Gallant, J.C., Loughhead, A.N., MacKey, B.J., Ryan, P.J., Mullen, I.C., Austin, M.P., 1998. Developing and testing procedures to predict topographic position from Digital Elevation Models (DEMs) for species mapping (Phase 1). Client Report No. 271. Environment Australia, CSIRO Forestry and Forest Products, 56 pp.
Costa-Cabral, M., Burges, S.J., 1994. Digital Elevation Model Networks (DEMON), a model of flow over hillslopes for computation of contributing and dispersal areas. Water Resources Research 30 (6), 1681–1692.
Coughlin, J.C., Running, S.W., 1989. An expert system to aggregate forested landscapes within a geographic information system. Artificial Intelligence Applications in Natural Resource Management 3 (4), 35–43.
Cristianini, N., Shawe-Taylor, J., 2000. An Introduction to Support Vector Machines (and Other Kernel-Based Learning Methods). Cambridge University Press, Cambridge, 204 pp.
Cross, M., 1998. Landslide susceptibility mapping using the matrix assessment approach: a Derbyshire case study. In: Maund, J.G., Eddleston, M. (Eds.), Geohazards in Engineering Geology. In: Engineering Geology Special Publication, vol. 15. Geological Society, London, pp. 247–261.
da Silva, J.R., Alexandre, C., 2005. The spatial variability of irrigated corn yield in relation to field topography. In: Stafford, J.V., Werner, A. (Eds.), Precision Agriculture, vol. 1. Wageningen Academic Publishers, Wageningen, pp. 95–101.
Dagan, G., 1986. Statistical theory of groundwater flow and transport: pore to laboratory, laboratory to formation and formation to regional scale. Water Resources Research 22, 120S–134S.
Dalrymple, J.B., Blong, R.J., Conacher, A.J., 1968. A hypothetical nine unit landsurface model. Zeitschrift für Geomorphologie 12, 60–76.
Daly, C., Gibson, W.P., Taylor, G.H., Johnson, G.L., Pasteris, P., 2002. A knowledge-based approach to the statistical mapping of climate. Climate Research 22, 99–113.

Darwinkel, A., 1980. Ear development and formation of grain yield in winter wheat. Netherlands Journal of Agricultural Science 28, 156–163.
Davis, F.W., Goetz, S., 1990. Modeling vegetation pattern using digital terrain data. Landscape Ecology 4 (1), 69–80.
De Bruin, S., Stein, A., 1988. Soil–landscape modeling using fuzzy c-means clustering of attribute data derived from a Digital Elevation Model (DEM). Geoderma 83, 17–33.
De By, R.A. (Ed.), 2001. Principles of Geographical Information Systems. ITC Educational Textbook Series, vol. 1. ITC, Enschede, 466 pp.
De Colstoun, E.C.B., Story, M.H., Thompson, C., Commisso, K., Smith, T.G., Irons, J.R., 2003. National Park vegetation mapping using multitemporal Landsat 7 data and a decision tree classifier. Remote Sensing of Environment 85, 316–327.
de Dainville, F., 1970. From depth to height. Surveying and Mapping 30 (3), 389–403.
de Martonne, E., 1941. Hypsométrie et morphologie, détermination et interprétation des altitudes moyennes de la France et de ses grandes régions naturelles. Annales de Géographie 50, 241–254.
de Sousa, L., Nery, F., Sousa, R., Matos, J., 2006. Assessing the accuracy of hexagonal versus square tiled grids in preserving DEM surface flow directions. In: Caetano, M., Painho, M. (Eds.), Proceedings of the 7th International Symposium on Spatial Accuracy Assessment in Natural Resources and Environmental Sciences (Accuracy 2006). Instituto Geográphico Português, Lisbon, pp. 191–200.
Deacon, E.L., 1969. Physical processes near the surface of the earth. In: Flohn, H. (Ed.), World Survey of Climatology, vol. 2. Elsevier, Amsterdam, London, New York, pp. 39–104.
Defant, F., 1951. Local winds. In: Malone, T.F. (Ed.), Compendium of Meteorology. American Meteorological Society, Boston, pp. 655–672.
Dehn, M., Gärtner, H., Dikau, R., 2001. Principals of semantic modeling of landform structures. Computers & Geosciences 27, 1011–1013.
del Barrio, G., Alvera, B., Puigdefabregas, J., Diez, C., 1997. Response of high mountain landscape to topographic variables: Central Pyrenees. Landscape Ecology 12 (2), 95–115.
Densmore, A.L., Hovius, N., 2000. Topographic fingerprints of bedrock landslides. Geology 28, 371–374.
Dent, D., Young, A., 1981. Soil Survey and Land Evaluation, vol. xiii. George Allen & Unwin, London, England, 278 pp.
Deren, L., Xiao-Yong, C., 1991. Automatically generating triangulated irregular digital terrain model networks by mathematical morphology. ISPRS Journal of Photogrammetry and Remote Sensing 46 (5), 283–295.
Desmet, P.J.J., 1997. Effects of interpolation errors on the analysis of DEMs. Earth Surface Processes and Landforms 22, 563–580.
Deweaver, E., Nigam, S., 1995. Influence of mountain ranges on the mid-latitude atmospheric response to El Niño events. Nature 378, 706–708.
Dietrich, W.E., Wilson, C.J., Montgomery, D.R., McKean, J., 1993. Analysis of erosion thresholds, channel networks, and landscape morphology using a digital elevation model. Journal of Geology 101, 259–278.
Dikau, R., Brabb, E.E., Mark, R.K., Pike, R.J., 1995. Morphometric Landform Analysis of New Mexico. Zeitschrift für Geomorphologie, Supplementband 101, 109–126.
Dikau, R., Brabb, E.E., Mark, R.M., 1991. Landform classification of New Mexico by computer. Open File Report 91-634. U.S. Geological Survey, 15 pp.
Dikau, R., 1990. Geomorphic landform modeling based on hierarchy theory. In: Brassel, K., Kishimoto, H. (Eds.), Proceedings of the 4th International Symposium on Spatial Data Handling. Department of Geography, University of Zürich, Zürich, Switzerland, pp. 230–239.
Dikau, R., 1989. The application of a digital relief model to landform analysis. In: Raper, J.F. (Ed.), Three Dimensional Applications in Geographical Information Systems. Taylor & Francis, London, pp. 51–77.
Dikau, R., 1988. Entwurf einer geomorphographisch-analytischen Systematik von Reliefeinheiten. Heidelberger Geographische Bausteine, vol. 5, 45 pp.
Dingman, S.L., 2002. Physical Hydrology, 2nd edition. Prentice–Hall, NJ, 250 pp.
Dirnbock, T., Dullinger, S., Gottfried, M., Ginzler, C., Grabherr, G., 2003. Mapping alpine vegetation based on image analysis, topographic variables and Canonical Correspondence Analysis. Applied Vegetation Science 6, 85–96.

Dobermann, A., Blackmore, S., Cook, S.E., Adamchuk, V.I., 2004. Precision farming: challenges and future directions. In: Proceedings of the 4th Internation Crop Science Conference. Cropscience, Brisbane, Australia, 19 pp.
Dobos, E., Carré, F., Hengl, T., Reuter, H.I., Tóth, G., 2006. Digital soil mapping as a support for production of functional maps, EUR 22123 EN. Office for Official Publications of the European Communities, Luxembourg, 68 pp.
Dobos, E., Daroussin, J., Montanarella, L., 2005. An SRTM-based procedure to delineate SOTER Terrain Units on 1:1 and 1:5 million scales, EUR 21571 EN. Office for Official Publications of the European Communities, Luxembourg, 55 pp.
Dobos, E., Micheli, E., Baumgardner, M.F., Biehl, L., Helt, T., 2000. Use of combined digital elevation model and satellite radiometric data for regional soil mapping. Geoderma 97 (3–4), 367–391.
Dokuchaev, V.V., 1898. On soil zones in general, and on the vertical zones specifically. In: Selected Papers 1949, vol. 3. Selhozgiz, Moscow, pp. 322–329 (in Russian).
Donatelli, M., Carlini, L., Bellocchi, G., 2006. A software component for estimating solar radiation. Environmental Modelling & Software 21 (3), 411–416.
Dorninger, P., Jansa, J., Briese, C., 2004. Visualization and topographical analysis of the Mars surface. Planetary and Space Science 52 (1–3), 249–257.
Dosio, A., Emeis, S., Graziani, G., Junkermann, W., Levy, A., 2001. Assessing the meteorological conditions in a deep Alpine valley system by a measuring campaign and simulation with two models during a summer smog episode. Atmospheric Environment 35, 5441–5454.
Dozier, J., Frew, J., 1990. Rapid calculation of terrain parameters for radiation modeling from digital elevation data. IEEE Transactions on Geoscience and Remote Sensing 28, 963–969.
Dozier, J., 1980. A clear-sky spectral solar radiation model for snow-covered mountainous terrain. Water Resources Research 16, 709–718.
Drăguţ, L., Blaschke, T., 2006. Automated classification of landform elements using object-based image analysis. Geomorphology 81, 330–344.
Drummond, R.R., Dennis, H.W., 1968. Qualifying relief terms. Professional Geographer 20, 326–332.
Dubayah, R., Loechel, S., 1997. Modelling topographic solar radiation using GOES data. Journal of Applied Meteorology 36 (2), 141–154.
Dubayah, R., Rich, P.M., 1995. Topographic solar radiation models for GIS. International Journal of Geographical Information Systems 9 (4), 405–419.
Dubayah, R., van Katwijk, V., 1992. The topographic distribution of annual incoming solar radiation in the Rio Grande river basin. Geophysical Research Letters 19, 2231–2234.
Dubayah, R., 1991. Using LOWTRAN7 and field flux measurements in an atmospheric and topographic solar radiation model. In: Proceedings IGARSS 1991, vol. 1. IEEE, Helsinki, Finland, pp. 39–42.
Duda, R.O., Hart, P.E., Stork, D.G., 2000. Pattern Classification, 2nd edition. Wiley, 680 pp.
Eagleson, P.S., 1970. Dynamic Hydrology, 1st edition. McGraw–Hill, New York, 448 pp.
Eastman, J.R., Jin, W.G., 1995. Raster procedures for multicriteria multiobjective decisions. Photogrammetric Engineering and Remote Sensing 61 (5), 539–547.
El-Hakim, S.F., Brenner, C., Roth, G., 1998. A multi-sensor approach to creating accurate virtual environments. ISPRS Journal of Photogrammetry and Remote Sensing 53 (6), 379–391.
Ellenberg, H., Weber, H.E., Dull, R., Wirth, V., Werner, W., Paulissen, D., 1992. Zeigerwerte von Pflanzen in Mitteleuropa. Scripta Geobotanica 18, 1–258.
Emeis, S., 1990. Surface pressure distribution and pressure drag on mountains. Meteorology and Atmospheric Physics 43, 173–185.
Emeis, S., 1987. Pressure drag and effective roughness length with neutral stratification. Boundary-Layer Meteorology 39, 379–401.
Endreny, T.A., Wood, E.F., 2001. Representing elevation uncertainty in runoff modelling and flowpath mapping. Hydrological Processes 15, 2223–2236.
Entz, M.H., Fowler, D.B., 1998. Critical stress periods affecting productivity of no-till winter wheat in western Canada. Agronomy Journal 80, 987–992.
Erbs, D.G., Klein, S.A., Duffie, J.A., 1982. Estimation of the diffuse radiation fraction for hourly, daily and monthly-average global radiation. Solar Energy 28, 293–302.
Etzelmüller, B., 2000. On the quantification of surface changes using grid-based Digital Elevation Models (DEMs). Transactions in GIS 4, 129–143.

Etzelmüller, B., Sulebak, J.S., 2000. Developments in the use of digital elevation models in periglacial geomorphology and glaciology. Physische Geographie 41, 35–58.
Evans, I.S., 2006. Allometric development of glacial cirque form: geological, relief and regional effects on the cirques of Wales. Geomorphology 80 (3–4), 245–266.
Evans, I.S., Cox, N.J., 2005. Global variations of local asymmetry in glacier altitude: separation of north–south and east–west components. Journal of Glaciology 51 (174), 469–482.
Evans, I.S., 2004. Geomorphometry. In: Goudie, A.S. (Ed.), Encyclopedia of Geomorphology. International Association of Geomorphology. Routledge, London, pp. 435–439.
Evans, I.S., 2003. Scale-specific landforms and aspects of the land surface. In: Evans, I.S., Dikau, R., Tokunaga, E., Ohmori, H., Hirano, M. (Eds.), Concepts and Modelling in Geomorphology: International Prospectives. TERRAPUB, Tokyo, pp. 61–84.
Evans, I.S., Dikau, R., Tokunaga, E., Ohmori, H., Hirano, M. (Eds.), 2003. Concepts and Modelling in Geomorphology: International Perspectives. TERRAPUB, Tokyo, 253 pp.
Evans, I.S., Cox, N.J., 1999. Relations between land surface properties: altitude, slope and curvature. In: Hergarten, S., Neugebauer, H.J. (Eds.), Process Modelling and Landform Evolution. Springer Verlag, Berlin, pp. 13–45.
Evans, I.S., 1998. What do terrain statistics really mean? In: Lane, S.N., Richards, K.S., Chandler, J.H. (Eds.), Landform Monitoring, Modelling and Analysis. Wiley, Chichester, UK, pp. 119–138.
Evans, I.S., McClean, C.J., 1995. The land surface is not unifractal: variograms, cirque scale and allometry. Zeitschrift für Geomorphologie, Supplementband 101, 127–147.
Evans, I.S., 1990. General geomorphometry (2.3). In: Goudie, A.S., et al. (Eds.), Geomorphological Techniques. 2nd edition. Unwin Hyman, London, pp. 44–56.
Evans, I.S., 1980. An integrated system of terrain analysis and slope mapping. Zeitschrift für Geomorphologie, Supplementband 36, 274–295.
Evans, I.S., 1979. An integrated system of terrain analysis and slope mapping. Final Report (Report 6) on Grant DA-ERO-591-73-G0040. Statistical characterization of altitude matrices by computer. Department of Geography, University of Durham, 192 pp.
Evans, I.S., Cox, N.J., 1974. Geomorphometry and the operational definition of cirques. Area 6 (2), 150–153.
Evans, I.S., 1972. General geomorphometry, derivatives of altitude, and descriptive statistics. In: Chorley, R.J. (Ed.), Spatial Analysis in Geomorphology. Harper & Row, pp. 17–90.
Evans, J.S., Hudak, A.T., 2007. A multiscale curvature filter for identifying ground returns from discrete return lidar in forested environments. IEEE Transactions on Geoscience and Remote Sensing 45 (4), 1029–1038.
Eyton, J.R., 1991. Rate-of-change maps. Cartography and Geographic Information Systems 18 (2), 87–103.
Fairfield, J., Leymarie, P., 1991. Drainage networks from grid digital elevation models. Water Resources Research 27 (5), 709–717.
Falconer, K., 2003. Fractal Geometry: Mathematical Foundations and Applications, 2nd edition. Wiley, Chichester, 332 pp.
Feinstein, A.R., Cicchetti, D.V., 1990. High agreement but low kappa: I. The problems of two paradoxes. Journal of Clinical Epidemiology 43 (6), 543–549.
Felícisimo, A.M., 1994a. Modelos digitales del terreno. Introducción y aplicaciones en las ciencias ambientales. Pentalfa Ediciones, Oviedo, 222 pp.
Felícisimo, A.M., 1994b. Parametric statistical method for error detection in digital elevation models. ISPRS Journal of Photogrammetry and Remote Sensing 49 (4), 29–33.
Fels, J.E., Matson, K.C., 1996. A cognitively based approach for hydro-geomorphic land classification using digital terrain models. In: Proceedings of the 3rd International Conference/Workshop on Integrating GIS and Environmental Modeling. Santa Fe, NM, January 21–25, 1996. National Centre for Geographic Information and Analysis, Santa Barbara, CA, USA, 6 pp.
Fels, J.E., 1994. Modeling and mapping potential vegetation using digital terrain data: applications in the Ellicott Rock Wilderness of North Carolina, South Carolina, and Georgia. Ph.D. Thesis. North Carolina State University.
Fenneman, N.M., 1938. Physiography of the Eastern United States. McGraw–Hill, New York, 714 pp.
Ferguson, R.B., Lark, R.M., Slater, G.P., 2003. Approaches to management zone definition for use of nitrification inhibitors. Soil Science Society of America Journal 67, 937–947.

Fiez, T.E., Miller, B.C., Pan, W.L., 1994. Winter wheat yield and grain protein across varied landscape positions. Agronomy Journal 86, 1026–1032.
Finlayson, D.P., Montgomery, D.R., 2003. Modeling large-scale fluvial erosion in geographic information systems. Geomorphology 53, 147–164.
Fischer, H.S., 1990. Simulating the distribution of plant communities in an alpine landscape. Coenoses 5 (1), 37–43.
Fisher, N.L., Lewis, T., Embleton, B.J.J., 1987. Statistical Analysis of Spherical Data. Cambridge University Press, Cambridge, 339 pp.
Fisher, P., 1998. Improved modeling of elevation error with geostatistics. GeoInformatica 2 (3), 215–233.
Fisher, P.F., Tate, N.J., 2006. Causes and consequences of error in digital elevation models. Progress in Physical Geography 30 (4), 467–489.
Fisher, P.F., Wood, J., Cheng, T., 2005. Fuzziness and ambiguity in multi-scale analysis of landscape morphometry. In: Petry, F.E., Robinson, V.B., Cobb, M.A. (Eds.), Fuzzy Modeling with Spatial Information for Geographic Problems. Springer-Verlag, Berlin, pp. 209–232.
Flanagan, D.C., Nearing, M.A. (Eds.), 1995. Hillslope Profile and Watershed Model Documentation. NSERL Report #10. USDA–Water Erosion Prediction Project, West Lafayette, IN, 280 pp.
Fleming, M.J., Chapin, F.S., Cramer, W., Hufford, G.L., Serreze, M.C., 2000. Geographic patterns and dynamics of Alaskan climate interpolated from a sparse station record. Global Change Biology 6, 49–58.
Florinsky, I.V., Eilers, R.G., Manning, G., Fuller, L.G., 2002. Prediction of soil properties by digital terrain modelling. Environmental Modelling & Software 17, 295–311.
Florinsky, I.V., Kuryakova, G.A., 2000. Determination of grid size for digital terrain modelling in landscape investigations — exemplified by soil moisture distribution at a micro-scale. International Journal of Geographical Information Science 14 (8), 815–832.
Florinsky, I.V., 1998. Accuracy of local topographic variables derived from digital elevation models. International Journal of Geographical Information Science 12 (1), 47–62.
Florinsky, I.V., Arlashina, H.A., 1998. Quantitative topographic analysis of gilgai soil morphology. Geoderma 82, 359–380.
Fogg, D.A., 1984. Contour to rectangular grid conversion using minimum curvature. Computer Vision, Graphics, and Image Processing 28 (1), 85–91.
Forkel, R., Knoche, R., 2006. Regional climate change and its impact on photooxidant concentration in southern Germany: simulations with a coupled regional climate–chemistry model. Journal of Geophysical Research 111, D12302.
Forstner, W., 1999. 3D-city models: automatic and semiautomatic acquisition methods. In: Fritsch, D., Spiller, R. (Eds.), Photogrammetric Week 99. Wichmann Verlag, pp. 291–303.
Fowler, R.J., 2001. Topographic LiDAR. In: Maune, D.F. (Ed.), Digital Elevation Model Technologies and Applications: The DEM Users Manual. American Society for Photogrammetry and Remote Sensing, MD, pp. 207–236.
Fox, C.G., Hayes, D.E., 1985. Quantitative methods for analyzing the roughness of the seafloor. Reviews of Geophysics and Space Physics 23 (1), 1–48.
Fraisse, C.W., Sudduth, K.A., Kitchen, N.R., 2001. Delineation of site-specific management zones by unsupervised classification of topographic attributes and soil electrical conductivity. Transactions of the ASAE 44 (1), 155–166.
Francisco, O., Valenzuela, M., Srinivasan, R., 2004. ArcGIS–SWAT: a GIS interface for the Soil and Water Assessment Tool (SWAT). In: Critical Transitions in Water and Environmental Resources Management. American Society of Civil Engineers, pp. 1–9.
Franklin, J., 1998. Predicting the distribution of shrub species in southern California from climate and terrain-derived variables. Journal of Vegetation Science 9, 733–748.
Franklin, J., 1995. Predictive vegetation mapping: geographic modeling of biospatial patterns in relation to environmental gradients. Progress in Physical Geography 19, 474–499.
Franklin, S.E., Wulder, M.A., Lavinge, M.B., 1996. Automated derivation of geographic window sizes for use in remote sensing digital image texture analysis. Computers & Geosciences 22 (6), 665–673.
Franklin, S.E., 1987. Geomorphic processing of digital elevation models. Computers & Geosciences 13, 603–609.

Franklin, S.E., Peddle, D.R., 1987. Texture analysis of digital image data using spatial co-occurrence. Computers & Geosciences 12, 195–209.
Franklin, W.R., Ray, C.K., 1994. Higher isn't necessarily better: visibility algorithms and experiments. In: Waugh, T.C., Healey, R.G. (Eds.), Advances in GIS Research: Sixth International Symposium on Spatial Data Handling. Taylor & Francis, Edinburgh, pp. 751–770.
Freeland, R.S., Yoder, R.E., Ammons, J.T., 1998. Mapping shallow underground features that influence site-specific agricultural production. Journal of Applied Geophysics 40 (1), 19–27.
Freeman, T.G., 1991. Calculating catchment area with divergent flow based on a regular grid. Computer and Geosciences 17 (3), 413–422.
Freeze, R.A., Cherry, J.A., 1979. Groundwater, 1st edition. Prentice–Hall, NJ, 604 pp.
Frei, C., Schär, C., 1998. A precipitation climatology of the Alps from high-resolution rain-gauge observations. International Journal of Climatology 18, 873–900.
Froude, V.A., Beanland, R.A., 1999. Review of environmental classification systems and spatial frameworks. Report prepared for the Ministry for the Environment's Environmental Performance Indicators Programme. Pacific Eco-Logic Resource Management Associates, Porirua City.
Fu, P., Rich, P.M., 2002. A geometric solar radiation model and its applications in agriculture and forestry. Computers and Electronics in Agriculture 37 (1), 25–35.
Fujisada, H., Bailey, G.B., Kelly, G.G., Hara, S., Abrams, M.J., 2005. ASTER DEM performance. IEEE Transactions on Geoscience and Remote Sensing 43, 2707–2714.
Fujisada, H., Sakuma, F., Ono, A., Kudo, M., 1999. Design and preflight performance of ASTER instrument protoflight model. IEEE Transactions on Geoscience and Remote Sensing 36, 1152–1160.
Fuller, I., Large, A.R.G., Charlton, M.E., Heritage, G.L., Milan, D.J., 2003. Reach-scale sediment transfers: an evaluation of two morphological budgeting approaches. Earth Surface Processes and Landforms 28 (8), 889–903.
Furger, F., Dommen, J., Graber, W.K., Poggio, L., Prévôt, A., Emeis, S., Trickl, T., Grell, G., Neininger, B., Wotawa, G., 2000. The VOTALP Mesolcina Valley Campaign 1996 — concept, background and some highlights. Atmospheric Environment 34, 1395–1412.
Gallant, A.L., Loveland, T.R., Sohl, T.L., Napton, D., 2005. Using an ecoregion framework to analyze land cover and land use dynamics. Environmental Management 34, S89–S110.
Gallant, J.C., Hutchinson, M.F., 2006. Producing digital elevation models with uncertainty estimates using a multi-scale Kalman filter. In: Caetano, M., Painho, M. (Eds.), Proceedings of the 7th International Symposium on Spatial Accuracy Assessment in Natural Resources and Environmental Sciences (Accuracy 2006). Instituto Geográphico Português, Lisbon, pp. 150–160.
Gallant, J.C., Dowling, T.I., 2003. A multiresolution index of valley bottom flatness for mapping depositional areas. Water Resources Research 39 (12), 1347–1360.
Gallant, J.C., Wilson, J.P., 1996. TAPES-G: a grid-based terrain analysis program for the environmental sciences. Computers & Geosciences 22 (7), 713–722.
Gamma, P., 1999. dfwalk — EinMurgang — Simulationsprogramm zur Gefahren zonierung. Ph.D. Thesis. University of Bern.
Gandoy-Bernasconi, W., Palacios-Velez, O., 1990. Automatic cascade numbering of unit elements in distributed hydrological models. Journal of Hydrology 112, 375–393.
Garbrecht, J., Martz, L.W., 1997. The assignment of drainage direction over flat surfaces in raster digital elevation models. Journal of Hydrology 193 (1–4), 204–213.
Gardiner, V., 1990. Drainage basin morphometry. In: Goudie, A., et al. (Eds.), Geomorphological Techniques. 2nd edition. Unwin Hyman, London, pp. 71–81.
Gates, D.M., 2003. Biophysical Ecology. Dover Publications, New York, 635 pp.
Gauss, K.F., 1828. Disquisitiones generales circa superficies curvas. Commentationes Societatis Regiae Scientiarum Gottingensis 6, 99–146.
Gautier, C., Landsfeld, M., 1997. Surface solar radiation flux and cloud radiative forcing for the Atmospheric Radiation Measurement (ARM) Southern Great Plains (SGP), a satellite, surface observations, and radiative transfer model study. Journal of Atmospheric Sciences 54 (10), 1289–1307.
Geist, T., Stötter, J., 2007. Documentation of glacier surface elevation change with multi-temporal airborne laser scanner data — case study: Hintereisferner and Kesselwandferner, Tyrol, Austria. Zeitschrift für Gletscherkunde und Glazialgeologie 41, 77–106.
Gelfand, A.E., Banerjee, S., Gamerman, D., 2005. Spatial process modelling for univariate and multivariate dynamic spatial data. Environmetrics 16, 1–15.

Gerrard, A.J.W., Robinson, D.A., 1971. Variability in slope measurements. Transactions, Institute of British Geographers 54, 45–54.
Gessler, P.E., Chadwik, O.A., Chamran, F., Althouse, L., Holmes, K., 2000. Modelling soil–landscape and ecosystem properties using terrain attributes. Soil Science Society of America Journal 64, 2046–2056.
Gessler, P.E., Moore, I.D., McKenzie, N.J., Ryan, P.J., 1995. Soil–landscape modelling and spatial prediction of soil attributes. International Journal of Geographical Information Systems 9 (4), 421–432.
Giannoulaki, M., Machias, A., Koutsikopoulos, C., Somarakis, S., 2006. The effect of coastal topography on the spatial structure of anchovy and sardine. ICES Journal of Marine Science 63 (4), 650–662.
Gibson, J.K., Kållberg, P., Uppala, S., Hernandez, A., Nomura, A., Serrano, E., 1999. 1. ERA-15 description. ECMWF re-analysis report series. European Centre for Medium-Range Weather Forecasts, Reading, UK, 74 pp.
Giles, P.T., Franklin, S.E., 1998. An automated approach to the classification of slope units using digital data. Geomorphology 21, 251–264.
Giles, P.T., Franklin, S.E., 1996. Comparison of derivative topographic surfaces of a DEM generated from stereographic SPOT images with field measurements. Photogrammetric Engineering and Remote Sensing 62, 1165–1171.
Gleick, J., 1988. Chaos: Making a New Science. Penguin USA, 368 pp.
Glenn, N.F., Streutker, D.R., Chadwiak, D.J., Thackray, G.D., Dorsch, S.J., 2006. Analysis of LiDAR-derived topographic information for characterizing and differentiating landslide morphology and activity. Geomorphology 73 (1–2), 131–148.
Gloaguen, R., Marpu, P.R., Niemeyer, I., 2007. Automatic extraction of faults and fractal analysis from remote sensing data. Nonlinear Processes in Geophysics 14, 131–138.
Gobin, A., Campling, P., Feyen, J., 2001. Soil–landscape modelling to quantify spatial variability of soil texture. Physics and Chemistry of the Earth Part B: Hydrology, Oceans and Atmosphere 26 (1), 41–45.
Gomes Pereira, L.M., Wicherson, R.J., 1999. Suitability of laser data for deriving geographical information — a case study in the context of management of fluvial zones. ISPRS Journal of Photogrammetry and Remote Sensing 54 (2–3), 105–114.
Goovaerts, P., 1997. Geostatistics for Natural Resources Evaluation. Applied Geostatistics. Oxford University Press, New York, 496 pp.
Gordon, A.D., 1981. Classification: Methods for Exploratory Analysis of Multivariate Data. Chapman and Hall, London, 190 pp.
Gorsevski, P.V., Gessler, P.E., Foltz, R.B., Elliot, W.J., 2006. Spatial prediction of landslide hazard using logistic regression and ROC analysis. Transactions in GIS 10 (3), 395–415.
Gorsevski, P.V., Jankowski, P., Gessler, P.E., 2005. Spatial prediction of landslide hazard using fuzzy k-means and Dempster–Shafer theory. Transactions in GIS 9 (4), 455–474.
Gorsevski, P.V., Gessler, P.E., Jankowski, P., 2003. Integrating a fuzzy k-means classification and a Bayesian approach for spatial prediction of landslide hazard. Journal of Geographical Systems 5 (3), 223–251.
Gorte, B.G.H., Koolhoven, W., 1990. Interpolation between isolines based on the Borgefors distance transform. ITC Journal 1 (3), 245–247.
Gottfried, M., Pauli, H., Grabherr, G., 1998. Prediction of vegetation patterns at the limits of plant life: a new view of the alpine-nival ecotone. Arctic and Alpine Research 30 (3), 207–221.
Gotway, C.A., Stroup, W.W., 1997. A generalized linear model approach to spatial data analysis and prediction. Journal of Agricultural, Biological, and Environmental Statistics 2 (2), 157–178.
Goudie, A.S., 2004. Encyclopedia of Geomorphology, vols. 1 & 2. Routledge, London, 1184 pp.
Graff, L.H., Usery, E.L., 1993. Automated classification of generic terrain features in digital elevation models. Photogrammetric Engineering and Remote Sensing 59, 1404–1407.
Graham, L.C., 1974. Synthetic interferometer radar for topographic mapping. Proceedings of the Institute of Electrical and Electronics Engineers 63, 763–768.
Graymer, R.W., 2000. Geologic map and map database of the Oakland Metropolitan Area, California. Misc. Field Studies Map MF-2342, scale 1:50,000. U.S. Geological Survey, 29 pp.
Grell, G.A., Emeis, S., Stockwell, W.R., Schoenemeyer, T., Forkel, R., Michalakes, J., Knoche, R., Seidl, W., 2000a. Application of a multiscale, coupled MM5/chemistry model to the complex terrain of the VOTALP Valley Campaign. Atmospheric Environment 34, 1435–1453.

Grell, G.A., Schade, L., Knoche, R., Pfeiffer, A., Egger, J., 2000b. Nonhydrostatic climate simulations of precipitation over complex terrain. Journal of Geophysical Research 105, 29595–29608.

Grell, G.A., Dudhia, J., Stauffer, D.R., 1994. A description of the fifth-generation Penn State/NCAR mesoscale model (MM5). NCAR Technical Note. National Center for Atmospheric Research, Boulder, Colorado, 117 pp.

Grender, G.C., 1976. Topo III-a FORTRAN program for terrain analysis. Computers and Geosciences 2 (2), 195–209.

Grenzdörfer, G.J., Gebbers, R.I.B., 2001. Seven years of yield mapping — analysis and possibilities of multi year yield mapping data. In: Grenier, G., Blackmore, S. (Eds.), ECPA 2001 Third European Conference on Precision Agriculture. AGRO Montpellier, Montpellier, pp. 31–36.

Griffin, M.L., Beasley, D.B., Fletcher, J.J., Foster, G.R., 1988. Estimating soil loss on topographically nonuniform field and farm units. Journal of Soil and Water Conservation 43, 326–331.

Griffin, M.W., 1990. Military applications of digital terrain models. In: Petrie, G., Kennie, T.J.M. (Eds.), Terrain Modelling in Surveying and Civil Engineering. Whittles Publishers, Caithness, UK, pp. 277–289.

Gritzner, M., Marcus, A., Aspinall, R., Custer, S., 2001. Assessing landslide potential using GIS, soil wetness modeling and topographic attributes, Payette River, Idaho. Geomorphology 37, 149–165.

Grubb, P.J., 1971. Interpretation of the "Massenerhebung" effect on tropical mountains. Nature 229, 44–45.

Gruber, S., 2007. MTD: a mass-conserving algorithm to parametrize gravitational transport and deposition processes using digital elevation models. Water Resources Research 43, W06412.

Gruber, S., Haeberli, W., 2007. Permafrost in steep bedrock slopes and its temperature-related destabilization following climatic change. Journal of Geophysical Research 112, F02S18.

Grubišić, V., Lewis, J.M., 2004. Sierra Wave project revisited: 50 years later. Bulletin of the American Meteorological Society 85, 1127–1142.

Grunwald, S., 2005. What do we really know about the space–time continuum of soil–landscapes? In: Grunwald, S. (Ed.), Environmental Soil–Landscape Modeling: Geographic Information Technologies and Pedometrics. CRC Press, Boca Raton, FL, pp. 3–36.

Guisan, A., Weiss, S.B., Weiss, A.D., 1999. GLM versus CCA spatial modeling of plant species distribution. Plant Ecology 143, 107–122.

Guo, Y.-R., Chen, S., 1994. Terrain and land use for the fifth-generation Penn State/NCAR mesoscale modelling system (MM5). NCAR Technical Note NCAR/TN-397+IA. NCAR Boulder, CO, 114 pp.

Gupta, V.K., Waymire, E., 1993. A statistical analysis of mesoscale rainfall as a random cascade. Journal of Applied Meteorology 32 (2), 251–267.

Gupta, V.K., Rodríquez-Iturbe, I., Wood, E.F., 1986. Scale Problems in Hydrology, 1st edition. D. Reidel Publications, Dordrecht, Holland, 246 pp.

Gupta, V.K., Waymire, E.C., Wang, C.T., 1980. A representation of an instantaneous unit hydrograph from geomorphology. Water Resources Research 16 (5), 855–862.

Gutersohn, H., 1932. Relief und Flussdichte (Relief & drainage density). Inaugural-Dissertation. Universität Zürich, Zürich, 91 pp.

Guth, P.L., 2006. Geomorphometry from SRTM: comparison to NED. Photogrammetric Engineering and Remote Sensing 72, 269–277.

Guth, P.L., 2004. The geometry of line-of-sight and weapons fan algorithms. In: Caldwell, D.R., Ehlen, J., Harmon, R.S. (Eds.), Studies in Military Geography and Geology. Kluwer, Dordrecht, The Netherlands, pp. 271–285.

Guth, P.L., 2003. Eigenvector analysis of digital elevation models in a GIS: geomorphometry and quality control. In: Evans, I.S., Dikau, R., Tokunaga, E., Ohmori, H., Hirano, M. (Eds.), Concepts and Modelling in Geomorphology: International Perspectives. TERRAPUB, Tokyo, Japan, pp. 199–220.

Guth, P.L., 2001. Quantifying terrain fabric in digital elevation models. In: Ehlen, J., Harmon, R.S. (Eds.), The Environmental Legacy of Military Operations. In: Geological Society of America Reviews in Engineering Geology, vol. 14, pp. 13–25.

Guth, P.L., 1995. Slope and aspect calculations on gridded digital elevation models: examples from a geomorphometric toolbox for personal computers. Zeitschrift für Geomorphologie 101, 31–52.

Guth, P.L., Ressler, E.K., Bacastow, T.S., 1987. Microcomputer program for manipulating large digital terrain models. Computers & Geosciences 13, 209–213.

Guzzetti, F., Reichenbach, P., Cardinali, M., Galli, M., Ardizzone, F., 2005. Probabilistic landslide hazard assessment at the basin scale. Geomorphology 72 (1–4), 272–299.
Guzzetti, F., Reichenbach, P., 1994. Toward the definition of topographic divisions for Italy. Geomorphology 11, 57–75.
Gyalistras, D., Fischlin, A., Riedo, M., 1997. Herleitung stündlicher Wetterszenarien unter zukünftigen Klimabedingungen. In: Fuhrer, J. (Ed.), Klimaänderung und Grünland — eine Modellstudie über die Auswirkungen zukünftiger Klimaveränderungen auf das Dauergrünland in der Schweiz. Hochschulverlag AG, ETH Zürich, pp. 207–263.
Haagenson, P.L., Dudhia, J., Stauffer, D.R., Grell, G.A., 1994. The Penn State/NCAR mesoscale model (MM5) source code documentation. NCAR Technical Note NCAR/TN-392+STR. NCAR, Boulder, CO, 200 pp.
Häckel, H., 1999. Meteorologie. Ulmer Verlag, Stuttgart, 448 pp.
Haff, P.K., 1996. Limitations on predictive modeling in geomorphology. In: Rhoads, B.L., Thorn, C.E. (Eds.), The Scientific Nature of Geomorphology: Proceedings of the 27th Binghamton Symposium in Geomorphology. Wiley, pp. 337–353.
Hammer, R.D., Young, F.J., Wollenhaupt, N.C., Barney, T.L., Haithcoate, T.W., 1995. Slope class maps from soil survey and digital elevation models. Soil Science Society of America Journal 59, 509–519.
Hammond, E.H., 1965. What is a landform? Some further comments. The Professional Geographer 17 (3), 12–13.
Hammond, E.H., 1964. Analysis of properties in land form geography: an application to broad-scale land form mapping. Annals of the Association of American Geographers 54, 11–19.
Hammond, E.H., 1954. Small-scale continental landform maps. Annals of the Association of American Geographers 44, 33–42.
Hancock, P.A., Hutchinson, M.F., 2006. Automatic computation of hierarchical biquadratic smoothing splines with minimum GCV. Computers & Geosciences 32 (6), 834–845.
Hann, J., 1866. Zur Frage über den Ursprung des Föhns. Zeitschrift der Österreichischen Gesellschaft für Meteorologie 1, 257–263.
Haralick, R.M., Shanmugam, K., Dinstein, I., 1964. Textural features for image classification. IEEE Transactions on Systems, Man and Cybernetics 2, 610–621.
Harrison, J.M., Chor-Pang, L., 1996. PC-based two-dimensional discrete Fourier transform programs for terrain analysis. Computers & Geosciences 22 (4), 419–424.
Härtel, F., 1931. Erläuterungen zur Geologischen Karte von Sachsen im Maßstab 1:50000 — Nr. 47 Blatt Lommatzsch. Geologischen Karte von Sachsen. Finanzministerium, Leipzig, 99 pp.
Haskins, D.M., Correll, C.S., Foster, R.A., Chatoian, J.M., Fincher, J.M., Strenger, S., Keys, J.E.J., Maxwell, J.M., King, T., 1998. A Geomorphic Classification System. USDA Forest Service, Washington, DC, 110 pp.
Haugerud, R.A., Harding, D.J., 2001. Some algorithms for virtual deforestation (VDF) of LiDAR topographic survey data. International Archives of Photogrammetry and Remote Sensing 34 (3/4), 211–217.
Haugerud, R.A., Greenberg, H.M., 1998. Recipes for digital cartography — cooking with DEMs. In: Soller, D.R. (Ed.), Digital Mapping Techniques'98—Champagne–Urbana, May 27–30 Workshop, Proceedings. Open-File Report 98-487. U.S. Geological Survey, pp. 119–126.
Heerdegen, R.G., Beran, M.A., 1982. Quantifying source areas through land surface curvature and shape. Journal of Hydrology 57, 359–373.
Hegg, C., Badoux, A., Frick, A., Schmid, F., 2002. Unwetterschiden in der Schweiz im Jahre 2001. Wasser, Energie, Luft 94 (3/4), 99–105.
Henderson, B.L., Bui, E.N., Moran, C.J., Simon, D.A.P., 2004. Australia-wide predictions of soil properties using decision trees. Geoderma 124 (3–4), 383–398.
Henderson, F.M., 1966. Open Channel Flow, 1st edition. Macmillan, New York, 522 pp.
Hengl, T., 2007. A Practical Guide to Geostatistical Mapping of Environmental Variables. EUR 22904 EN Scientific and Technical Research Series. Office for Official Publications of the European Communities, Luxemburg, 143 pp.
Hengl, T., Heuvelink, G.B.M., Rossiter, D.G., 2007a. About regression-kriging: from theory to interpretation of results. Computers & Geosciences 33 (10), 1301–1315.

Hengl, T., Toomanian, N., Reuter, H.I., Malakouti, M.J., 2007b. Methods to interpolate soil categorical variables from profile observations: lessons from Iran. Geoderma 140 (4), 417–427.
Hengl, T., 2006. Finding the right pixel size. Computers & Geosciences 32 (9), 1283–1298.
Hengl, T., Gruber, S., Shrestha, D.P., 2004a. Reduction of errors in digital terrain parameters used in soil–landscape modelling. International Journal of Applied Earth Observation and Geoinformation 5 (2), 97–112.
Hengl, T., Walvoort, D.J.J., Brown, A., Rossiter, D.G., 2004b. A double continuous approach to visualisation and analysis of categorical maps. International Journal of Geographical Information Science 18 (2), 183–202.
Hengl, T., Gruber, S., Shrestha, D.P., 2003. Digital Terrain Analysis in ILWIS. Lecture Notes. International Institute for Geo-Information Science & Earth Observation (ITC), Enschede, 56 pp.
Hengl, T., Rossiter, D.G., 2003. Supervised landform classification to enhance and replace photo-interpretation in semi-detailed soil survey. Soil Science Society of America Journal 67 (5), 1810–1822.
Hengl, T., Bajat, B., Reuter, H.I., Blagojevic, D., 2008. Geostatistical modelling of topography using auxiliary maps. Computers & Geosciences. http://dx.doi.org/10.1016/j.cageo.2008.01.005.
Hensley, S., Munjy, R., Rosen, P., 2001. Interferometric Synthetic Aperture Radar (IFSAR). In: Maune, D.F. (Ed.), Digital Elevation Model Technologies and Applications: The DEM Users Manual. American Society for Photogrammetry and Remote Sensing, MD, pp. 142–206.
Hernández Encinas, A., Hernández Encinas, L., Hoya White, S., Martín del Rey, A., Rodríguez Sanchez, G., 2007. Simulation of forest fire fronts using cellular automata. Advances in Engineering Software 38 (6), 372–378.
Herrington, L., Pellegrini, G., 2000. An advanced shape of country classifier: extraction of surface features from DEMs. In: Parks, B., Clarke, K.M., Crane, M.P. (Eds.), Proceedings of the 4th International Conference on Integrating Geographic Information Systems and Environmental Modeling: Problems, Prospects, and Needs for Research. University of Colorado/USGS/NOAA, Boulder, CO, 9 pp.
Herrmann, A., Kersebaum, K.-C., Taube, F., 2005. Nitrogen fluxes in silage maize production: relationship between nitrogen content at silage maturity and nitrate concentration in soil leachate. Nutrient Cycling in Agroecosystems 73, 59–74.
Hettner, A., 1928. Die Oberflächenformen des Festlandes, 2nd edition. B.G. Teubner Verlag, Stuttgart, 250 pp.
Heuvelink, G.B.M., Webster, R., 2001. Modelling soil variation: past, present, and future. Geoderma 100 (3–4), 269–301.
Heuvelink, G.B.M., Pebesma, E.J., 1999. Spatial aggregation and soil process modelling. Geoderma 89 (1–2), 47–65.
Heuvelink, G.B.M., 1998. Error Propagation in Environmental Modelling with GIS. Taylor & Francis, London, UK, 144 pp.
Hickey, R., Smith, A., Jankowski, P., 1994. Slope length calculations from a DEM within Arc/Info GRID. Computing, Environment and Urban Systems 18 (5), 365–380.
Hill, J., Mehl, W., 2003. Geo- und radiometrische Aufbereitung multi- und hyperspektraler Daten zur Erzeugung langjähriger kalibrierter Zeitreihen. PhotoGrammetrie — Fernerkundung — Geoinformation 1, 7–14.
Hobson, R.D., 1972. Surface roughness in topography: a quantitative approach. In: Chorley, R.J. (Ed.), Spatial Analysis in Geomorphology. Harper & Row, pp. 221–245.
Hodgson, M.E., 1998. Comparison of angles from surface slope/aspect algorithms. Cartography and Geographic Information Systems 25, 173–185.
Hofer, M., Sapiro, G., Wallner, J., 2006. Fair polyline networks for constrained smoothing of digital terrain elevation data. IEEE Transactions on Geoscience and Remote Sensing 44 (10), 2983–2990.
Hofierka, J., Parajka, J., Mitášová, H., Mitas, L., 2002. Multivariate interpolation of precipitation using regularized spline with tension. Transactions in GIS 6, 135–150.
Hofierka, J., Zlocha, M., 1993. Application of surface and volume geometry analysis in geosciences. Geologica Carpathica 44 (1), 94.
Hofton, M., Dubayah, R., Blair, J.B., Rabine, D., 2006. Validation of SRTM elevations over vegetated and non-vegetated terrain using medium-footprint LiDAR. Photogrammetric Engineering and Remote Sensing 72 (3), 279–286.
Hofton, M.A., Blair, J.B., 2002. Laser altimeter return pulse correlation: a method for detecting surface topographic change. Journal of Geodynamics 34, 477–489.

Holling, C.S., 1992. Cross-scale morphology, geometry, and dynamics of ecosystems. Ecological Society of America — Ecological Monographs 62 (4), 447–502.

Holmes, K.W., Chadwick, O.A., Kyriakidis, P.C., 2000. Error in a USGS 30m digital elevation model and its impact on digital terrain modeling. Journal of Hydrology 233, 154–173.

Holmgren, P., 1994. Multiple flow direction algorithms for runoff modeling in grid based elevation models — an empirical-evaluation. Hydrological Processes 8 (4), 327–334.

Hormann, K., 1981. Räumliche Interpolation von Niederschlagswerten. Beiträge zur Hydrologie 8 (2), 5–40.

Hormann, K., 1969. Geomorphologische Kartenanalyse mit Hilfe elektronischer Rechenanlagen. Zeitschrift für Geomorphologie 133 (1), 75–98.

Horn, B.K.P., 1981. Hill shading and the reflectance map. Proceedings IEEE 69 (1), 14–47.

Horton, R.E., 1945. Erosional development of streams and their drainage basins: hydrophysical approach to quantitative morphology. Geological Society of America Bulletin 56, 275–370.

Horton, R.E., 1932. Drainage basin characteristics. Transactions American Geophysical Union 14, 350–361.

Houlder, D., Hutchinson, M., Nix, H., McMahon, J., 2000. ANUDEM User's Guide. Centre for Resource and Environmental Studies, Australian National University, Canberra.

Howard, A.D., 1990. Role of hypsometry and planform in basin hydrologic response. Hydrological Processes 4, 373–385.

Huabin, W., Gangjun, L., Weiya, X., Gonghui, W., 2005. GIS-based landslide hazard assessment. Progress in Physical Geography 29 (4), 548–567.

Hudson, B.D., 2004. The soil survey as a paradigm-based science. Soil Science Society of America Journal 56, 836–841.

Huggel, C., Haeberli, W., Kääb, A., Bieri, D., Richardson, S., 2004. An assessment procedure for glacial hazards in the Swiss Alps. Canadian Geotechnical Journal 41 (6), 1068–1083.

Huggel, C., Kääb, A., Haeberli, W., Krummenacher, B., 2003. Regional-scale GIS-models for assessment of hazards from glacier lake outbursts: evaluation and application in the Swiss Alps. Natural Hazards and Earth System Sciences 3, 647–662.

Huggett, R.J., 1993. Modelling the Human Impact on Nature. Oxford University Press, 202 pp.

Huggett, R.J., 1975. Soil landscape systems: a model of soil genesis. Geoderma 13 (1), 1–22.

Huising, E.J., Gomes-Pereira, L.M., 1998. Errors and accuracy estimates of laser data acquired by various laser scanning systems for topographic applications. ISPRS Journal of Photogrammetry and Remote Sensing 53, 245–261.

Hungr, O., 1995. A model for the runout analysis of rapid flow slides, debris flows and avalanches. Canadian Geotechnical Journal 32 (4), 610–623.

Hunter, G.J., Goodchild, M.F., 1997. Modeling the uncertainity of slope and aspect estimates derived from spatial databases. Geographical Analysis 29, 35–49.

Hutchinson, M.F., Gallant, J.C., 2000. Digital elevation models and representation of terrain shape. In: Wilson, J.P., Gallant, J.C. (Eds.), Terrain Analysis: Principles and Applications. Wiley, pp. 29–50.

Hutchinson, M.F., 1996. A locally adaptive approach to the interpolation of digital elevation models. In: Proceedings of the Third International Conference/Workshop on Integrating GIS and Environmental Modeling. National Center for Geographic Information and Analysis, Santa Barbara, CA, 6 pp.

Hutchinson, M.F., 1995. Interpolating mean rainfall using thin plate smoothing splines. International Journal of Geographical Information Systems 9, 385–403.

Hutchinson, M.F., 1989. A new procedure for gridding elevation and stream line data with automatic removal of spurious pits. Journal of Hydrology 106, 211–232.

Hutchinson, M.F., 1988. Calculation of hydrologically sound digital elevation models. In: Third International Symposium on Spatial Data Handling. Columbus, OH. International Geographical Union, Sydney, pp. 117–133.

Imamura, G., 1937. Past glaciers and the present topography of the Japanese Alps. In: Science Reports of Tokyo Bunrika Daiguku, vol. C.7. Tokyo Bunrika Daiguku, 61 pp.

Imhof, E., 1982. Cartographic Relief Presentation. Walter de Gruyter, Berlin, 389 pp.

Irvin, B.J., Ventura, S.J., Slater, B.K., 1997. Fuzzy and isodata classification of landform elements from digital terrain data in Pleasant Valley, Wisconsin. Geoderma 77, 137–154.

Isaaks, E.H., Srivastava, R.M., 1989. Applied Geostatistics. Oxford University Press, New York, 542 pp.
IUSS Working Group WRB, 2006. World Reference Base for Soil Resources, 2nd edition. World Soil Resources Reports. FAO, Rome, 127 pp.
Iverson, R.M., Schilling, S.P., Vallance, J.W., 1998. Objective delineation of lahar-inundation hazard zones. Geological Society of America Bulletin 110 (8), 972–984.
Iwahashi, J., Pike, R.J., 2007. Automated classifications of topography from DEMs by an unsupervised nested-means algorithm and a three-part geometric signature. Geomorphology 86 (3–4), 409–440.
Jaeger, R.M., Schuring, D.J., 1966. Spectrum analysis of terrain of Mare Cognitum. Journal of Geophysical Research 71 (8), 2023–2028.
Jain, V., Preston, N., Fryirs, K., Brierley, G., 2006. Comparative assessment of three approaches for deriving stream power plots along long profiles in the upper Hunter River catchment, New South Wales, Australia. Geomorphology 74, 297–317.
James, C.N., Houze Jr., R.A., 2005. Modification of precipitation by coastal orography in storms crossing northern California. Monthly Weather Review 133, 3110–3131.
Jarvis, C.H., Stuart, N., 2001. A comparison among strategies for interpolating maximum and minimum daily air temperatures. Part II: The interaction between number of guiding variables and the type of interpolation method. Journal of Applied Meteorology 40, 1075–1084.
Jarvis, R.S., Clifford, N.J., 1990. Specific geomorphometry. In: Goudie, A., et al. (Eds.), Geomorphological Techniques. 2nd edition. Unwin Hyman, London, pp. 63–70.
Jelaska, S.D., Antonić, O., Božić, M., Križan, J., Kušan, V., 2006. Responses of forest herbs to available understory light measured with hemispherical photographs in silver fir-beech forest in Croatia. Ecological Modelling 194 (1–3), 209–218.
Jelaska, S.D., Kušan, V., Peternel, H., Grgurić, Z., Mihulja, A., Major, Z., 2005. Vegetation mapping of "Žumberak — Samoborsko gorje" Nature Park, Croatia, using Landsat 7 and field data. Acta Botanica Croatica 64 (2), 303–311.
Jelaska, S.D., Antonić, O., Nikolić, T., Hršak, V., Plazibat, M., Križan, J., 2003. Estimating plant species occurrence in MTB/64 quadrants as a function of DEM-based variables — a case study for Medvednica Nature Park, Croatia. Ecological Modelling 170 (2–3), 333–343.
Jenness, J.S., 2004. Calculating landscape surface area from digital elevation models. Wildlife Society Bulletin 32 (3), 829–839.
Jennings, P.J., Siddle, H.J., 1998. Use of landslide inventory data to define the spatial location of landslide sites, South Wales, UK. In: Maund, J.G., Eddleston, M. (Eds.), Geohazards in Engineering Geology. In: Special Publications, vol. 15. Geological Society, London, pp. 199–211.
Jenny, H., 1941. Factors of Soil Formation. McGraw–Hill, New York, 288 pp.
Jensen, M.E., Dibenedetto, J.P., Barber, J.A., Montagne, C., Bourgeron, P.S., 2001. Spatial modeling of rangeland potential vegetation environments. Journal of Range Management 54, 528–536.
Jenson, S.K., 1991. Applications of hydrologic information automatically extracted from digital elevation models. Hydrological Processes 5, 31–44.
Jenson, S.K., Domingue, J.O., 1988. Extracting topographic structure from digital elevation data for geographical information system analysis. Photogrammetric Engineering and Remote Sensing 54 (11), 1593–1600.
Jenson, S.K., 1985. Automated derivation of hydrologic basin characteristics from digital elevation model data. In: Proceedings of the Auto-Carto 7, International Symposium on Computer-Assisted Cartography. U.S. Geological Survey, Washington, DC, pp. 301–310.
Jewitt, G.P.W., Garratt, J.A., Calder, I.R., Fuller, L., 2006. Water resources planning and modelling tools for the assessment of land use change in the Luvuvhu Catchment, South Africa. Physics and Chemistry of the Earth, Parts A/B/C 29, 1233–1241.
Jiang, Q., Smith, R.B., Doyle, J., 2003. The nature of the mistral: observations and modelling of two MAP events. Quarterly Journal of the Royal Meteorological Society 129, 857–875.
Jibson, R.W., Harp, E.L., Michael, J.A., 1998. A method for producing digital probabilistic seismic landslide hazard maps — an example from the Los Angeles, California area. Open-File Report 98-113. U.S. Geological Survey, 17 pp.
Jones, K., Meidinger, D., Clark, D., Schultz, F., 1999. Towards the establishment of predictive ecosystem mapping standards: a white paper, 1st approximation. Prepared for Terrestrial Ecosystem Mapping Alternatives Task Force. Resource Inventory Committee (RIC), 88 pp.

Jones, K.H., 1998. A comparison of algorithms used to compute hill slope as a property of the DEM. Computers & Geosciences 24 (4), 315–323.
Jones, N.L., Wright, S.G., Maidment, D.R., 1990. Watershed delineation with triangle-based terrain models. Journal of Hydraulic Engineering — ASCE 116 (10), 1232–1251.
Jordan, G., Meijninger, B.M.L., van Hinsbergen, D.J.J., Meulenkamp, J.E., van Dijk, P.M., 2005. Extraction of morphotectonic features from DEMs: development and applications for study areas in Hungary and NW Greece. International Journal of Applied Earth Observation and Geoinformation 7 (3), 163–182.
Jordan, G., Schott, B., 2005. Application of wavelet analysis to the study of spatial pattern of morphotectonic lineaments in digital terrain models. A case study. Remote Sensing of Environment 94 (1), 31–38.
Jordan, G., 2003. Morphometric analysis and tectonic interpretation of digital terrain data: a case study. Earth Surface Processes and Landforms 28, 807–822.
Jowkin, V., Schoenau, J.J., 1998. Impact of tillage and landscape position on nitrogen availability and yield of spring wheat in the Brown soil zone in southwestern Saskatchewan. Canadian Journal of Soil Science 78 (3), 563–572.
Julien, P.Y., Saghafian, B., Ogden, F.L., 1995. Raster-based hydrologic modeling of spatially-varied surface runoff. Water Resources Bulletin 31 (3), 523–536.
Julien, P.Y., Saghafian, B., 1991. CASC2D User's Manual — A Two Dimensional Watershed Rainfall-Runoff Model. Report CER90-91PYJ-BS-12. Civil Engineering Department, Colorado State University, Fort Collins, CO, 66 pp.
Kalnay, E., Kanamitsu, M., Kistler, R., Collins, W., Deaven, D., Gandin, L., Iredell, M., Saha, S., White, G., Woollen, J., Zhu, Y., Leetmaa, A., Reynolds, R., Chelliah, M., Ebisuzaki, W., Higgins, W., Janowiak, J., Mo, K., Ropelewski, C., Wang, J., Jenne, R., Joseph, D., 1996. The NCEP/NCAR 40-year reanalysis project. Bulletin of the American Meteorological Society 77 (3), 437–471.
Karssenberg, D., De Jong, K., 2005. Dynamic environmental modelling in GIS: 2. Modelling error propagation. International Journal of Geographical Information Science 19, 623–637.
Katsube, K., Oguchi, T. (Eds.), 2005. Terrain relief, local slope and drainage density in Japanese mountains. In: Sixth International Conference on Geomorphology, Abstracts Volume. University of Zaragoza, p. 385.
Katsube, K., Oguchi, T., 1999. Altitudinal changes in slope angle and profile curvature in the Japan Alps: a hypothesis bearing on a characteristic slope angle. Geographical Review of Japan B 72, 63–72.
Katzil, Y., Doztsher, Y., 2000. Height estimation for filling gaps in gridded DTM. Journal of Surveying Engineering 126 (4), 145–162.
Keith, F., Kreider, J.F., 1978. Principles of Solar Engineering. Hemisphere Publishing Corporation, Washington, DC, 778 pp.
Kersebaum, K.-C., 2007. Modelling nitrogen dynamics in soil–crop systems with HERMES. Nutrient Cycling in Agroecosystems 77 (1), 39–52.
Kersebaum, K.-C., Steidl, J., Bauer, O., Piorr, H., 2003. Modelling scenarios to assess the effects of different agricultural management and land use options to reduce diffuse nitrogen pollution into the river Elbe. Physics and Chemistry of the Earth 28 (12–13), 537–545.
Kersebaum, K.-C., Lorenz, K., Reuter, H.I., Wendroth, O., Ahuja, L.R., Ma, L., Howell, T.A., 2002. Modelling crop growth and nitrogen dynamics for advisory purposes regarding spatial variability. In: Agricultural System Models in Field Research and Technology Transfer. CEC Press LLC, Boca Raton, pp. 230–251.
Kersebaum, K.-C., Beblik, A.J., 2001. Performance of a nitrogen dynamics model applied to evaluate agricultural management practices. In: Shaffer, M., Ma, L., Hansen, S. (Eds.), Modeling Carbon and Nitrogen Dynamics for Soil Management. CRC Press, Boca Raton, USA, pp. 551–571.
Kersebaum, K.-C., 1995. Application of a simple management model to simulate water and nitrogen dynamics. Ecological Modelling 81, 145–156.
Khosla, R., Fleming, K., Delgado, J.A., Shaver, T., Wetsfall, D.G., 2001. Use of site specific management zones to improve nitrogen management for precision agriculture. Journal of Soil and Water conservation 57 (6), 515–518.
Kienzle, S., 2004. The effect of DEM raster resolution on first order, second order and compound terrain derivatives. Transactions in GIS 8 (1), 83–111.

King, C.A.M., 1969. Trend surface analysis of Central Pennine erosion surfaces. Transactions, Institute of British Geographers 47, 47–59.

King, D., Bourennane, H., Isambert, M., Macaire, J.J., 1999. Relationship of the presence of a non-calcareous clay–loam horizon to DEM attributes in a gently sloping area. Geoderma 89 (1–2), 95–111.

Kinner, D., Mitášová, H., Harmon, R.S., Toma, L.R.S., 2005. GIS-based stream network analysis for the Chagres river basin, Republic of Panama. In: Harmon, R.S. (Ed.), The Rio Chagres: A Multidisciplinary Profile of a Tropical Watershed. Springer/Kluwer, pp. 83–95.

Kirkby, M.J., 1976. Tests of the random network model, and its application to basin hydrology. Earth Surface Processes 1, 197–212.

Klein, S.A., 1977. Calculation of monthly average insolation on tilted surface. Solar Energy 19, 325–329.

Klijn, F., Udo de Haes, H.A., 1994. A hierarchical approach to ecosystems and its implications for ecological land classification. Landscape Ecology 9 (2), 89–104.

Klingseisen, B., 2004. GIS based generation of topographic attributes for landform classification. Diploma Thesis, Diplom-Ingenieur (FH). University of Applied Sciences, School of Geoinformation, Fachhochschule Technikum Kärnten.

Klinkenberg, B., 1992. Fractals and morphometric measures; is there a relationship? Geomorphology 5 (1/2), 5–20.

Klinkenberg, B., Goodchild, M.F., 1992. The fractal properties of topography: a comparison of methods. Earth Surface Processes and Landforms 17, 217–234.

Kneizys, F.X., Shettle, E.P., Abreu, L.W., Chetwynd, J.H., Anderson, G.P., Gallery, W.O., Selby, J.E.A., Clough, S.A., 1988. Users Guide to LOW-TRAN7. Report AFGL-Tr-88-0177. Air Force Geophysics Laboratory, Bedford, MA, 137 pp.

Knöpfle, W., Strunz, G., Roth, A., 1998. Mosaicking of Digital Elevation Models derived by SAR Interferometry. In: Fritsch, D., Englich, M., Sester, S. (Eds.), The International Archives of Photogrammetry and Remote Sensing, vol. 32/4, ISPRS Commission IV — GIS between Visions and Applications, Stuttgart, pp. 306–313.

Korup, O., Schmidt, J., McSaveney, M.J., 2005. Regional relief characteristics and denudation pattern of the western Southern Alps, New Zealand. Geomorphology 71, 402–423.

Köthe, R., Gehrt, E., Böhner, J., 1996. Automatische Reliefanalyse für geowissenschaftliche Anwendungen — derzeitiger Stand und Weiterentwicklungen des Programms SARA. Arbeitshefte Geologie 1, 31–37.

Kraus, K., Pfeifer, N., 1998. Determination of terrain models in wooded areas with airborne laser scanner data. ISPRS Journal of Photogrammetry and Remote Sensing 53, 193–203.

Krcho, J., 2001. Modelling of Georelief and Its Geometrical Structure Using DTM: Positional and Numerical Accuracy. Q111 Vydavatel'stvo, Bratislava, 336 pp.

Krcho, J., 1983. Teoretická concepcia a interdisciplinarne aplikacie komplexného digitalneho modelu reliéfu pri modelovanï dvojdimenzionalnych poli. Geografický casopis 35 (3), 265–291.

Krcho, J., 1973. Morphometric analysis of relief on the basis of geometric aspect of field theory. Acta Geographica Universitatis Comenianae, Geographico-Physica 1 (1), 7–233.

Krishnaswamy, J., Kiran, M.C., Ganeshaiah, K.N., 2004. Tree model based eco-climatic vegetation classification and fuzzy mapping in diverse tropical deciduous ecosystems using multi-season NDVI. International Journal of Remote Sensing 25, 1185–1205.

Krumbein, W.C., 1959. Trend surface analysis of contour-type maps with irregular control point spacing. Journal of Geophysical Research 64 (7), 823–834.

Kuettner, J.P., 1986. The aim and conduct of ALPEX. Scientific results of the Alpine experiment. Technical Document 108. GARP Publications Series No. 27. World Meteorological Organization, Geneva, pp. 3–14.

Kugler, H., 1964. Die geomorphologische Reliefanalyse als Grundlage großmaßstäbiger geomorphologischer Kartierung. Wissenshaftliche Veröffentlichungen des Deutschen Instituts für Länderkunde N.F. 21/22, 541–655.

Kühni, A., Pfiffner, O.A., 2001. The relief of the Swiss Alps and adjacent areas and its relation to lithology and structure — topographic analysis from 250-m DEM. Geomorphology 41 (4), 285–307.

Kumar, L., Skidmore, A.K., Knowles, E., 1997. Modelling topographic variation in solar radiation in a GIS environment. International Journal of Geographical Information Science 11 (5), 475–497.

Kumler, M.P., 1994. An intensive comparison of triangulated irregular networks (TINs) and digital elevation models (DEMs). Monograph 45. Cartographica 31 (2), 1–99.

Kuo, W.L., Steenhuis, T.S., McCulloch, C.E., Mohler, C.L., Weinstein, D.A., DeGloria, S.D., Swaney, D.P., 1999. Effect of grid size on runoff and soil moisture for a variable-source-area hydrology model. Water Resources Research 35 (11), 3419–3428.

Kyle, T.G., 1991. Atmospheric Transmission, Emission and Scattering. Pergamon, Oxford, New York, 288 pp.

Kyriakidis, P.C., Shortridge, A.M., Goodchild, M.F., 1999. Geostatistics for conflation and accuracy assessment of digital elevation models. International Journal of Geographical Information Science 13 (7), 677–708.

Lagacherie, P., Holmes, S., 1997. Addressing geographical data errors in a classification tree for soil unit predictions. International Journal of Geographical Information Science 11, 183–198.

Lane, S.N., Brookes, C.J., Kirkby, M.J., Holden, J., 2004. A network-index-based version of TOPMODEL for use with high-resolution digital topographic data. Hydrological Processes 18, 191–201.

Lane, S.N., James, T.D., Crowell, M.D., 2000. The application of digital photogrammetry to complex topography for geomorphological research. Photogrammetric Record 16, 793–821.

Lane, S.N., Richards, K.S., Chandler, J.H., 1998. Landform Monitoring, Modelling and Analysis. Wiley, 466 pp.

Lane, S.N., 1998. The use of digital terrain modelling in the understanding of dynamic river channel systems. In: Lane, S.N., Richards, K.S., Chandler, J.H. (Eds.), Landform Monitoring, Modelling and Analysis. Wiley, pp. 311–342.

Lapen, D.R., Martz, L.W., 1996. An investigation of the spatial association between snow depth and topography in a prairie agricultural landscape using digital terrain analysis. Journal of Hydrology 184, 277–298.

Lark, R.M., 1999. Soil–landform relationships within-field scales: an investigation using continuous classification. Geoderma 92, 141–165.

Lashermes, B., Foufoula-Georgiou, E., Dietrich, W., 2007. Channel network extraction from high resolution topography using wavelets. Geophysical Research Letters 34 (23), doi:10.1029/2007GL031140.

Latimer, A.M., Wu, S., Gelfand, A.E., Silander Jr., J.A., 2004. Building statistical models to analyze species distributions. Ecological Applications 16 (1), 33–50.

Lauer, W., Bendix, J., 2004. Klimatologie. Das Geographische Seminar. Westermann, Braunschweig.

Lawrance, C., Byard, R., Beaven, P., 1993. Terrain evaluation manual. HMSO for Transport Research Laboratory, Department of Transport, UK, London, 300 pp.

Lea, N.L., 1992. An aspect driven kinematic routing algorithm. In: Parsons, A.J., Abrahams, A.D. (Eds.), Overland Flow: Hydraulics and Erosion Mechanics. Chapman & Hall, New York, pp. 147–175.

Leavesley, G.H., Lichty, R.W., Troutman, B.M., Saindon, L.G., 1991. Precipitation-Runoff Modeling System (PRMS) Users Manual. Water Resources Investigations Report 83-4238. U.S. Geological Survey, 207 pp.

Lee, J., 1991. Comparison of existing methods for building triangular irregular network models of terrain from grid digital elevation models. International Journal of Geographical Information Systems 5 (3), 267–285.

Lee, J.S., Papathanassiou, K.P., Ainsworth, T.L., Grunes, M.R., Reigber, A., 1998. A new technique for noise filtering of SAR interferometric phase images. IEEE Transactions on Geoscience and Remote Sensing 36 (5), 1456–1465.

Lee, S., Evangelista, D.G., 2006. Earthquake-induced landslide-susceptibility mapping using an artificial neural network. Natural Hazards and Earth System Sciences 6, 687–695.

Lehner, B., Verdin, K., Jarvis, A., 2008. New global hydrography derived from spaceborne elevation data. EOS, Transactions, American Geophysical Union 89 (10), 93–94. http://hydrosheds.cr.usgs.gov.

Leica Geosystems GIS & Mapping, 2003. Leica Photogrammetry Suite OrthoBASE & OrthoBASE Pro User's Guide. Leica Geosystems GIS & Mapping, Atlanta, 481 pp.

Leighty, R.D., Rinker, J.N., 2003. Integration of high-resolution DTED, hyperspectral data, and hypermedia data using the Terrain Analysis System. In: Shen, S.S., Descour, M.R. (Eds.), Algorithms for Multispectral, Hyperspectral, and Ultraspectral Imagery, vol. 4381. SPIE, pp. 253–264.

Leighty, R.D., 2001. Automated IFSAR Terrain Analysis System: Final Report. U.S. Army Aviation & Missile Command, Defense Advanced Research Projects Agency (DoD) Information Sciences Office, Arlington, VA, 59 pp.

Leopold, L.B., Wolman, M.G., Miller, J.P., 1995. Fluvial Processes in Geomorphology. Dover, New York, 522 pp., reprinted from 1964 edition.
Lewis, L.A., 1968. Analysis of surficial landform properties — the regionalization of Indiana into units of morphometric similarity. Proceedings Indiana Academy of Science 78, 317–328.
Li, F.-M., Liu, Y.-L., Li, S.-Q., 2001. Effects of early soil water distribution on the dry matter partition between roots and shoots of winter wheat. Agriculture Water Managment 49, 163–171.
Li, Z., Zhu, Q., Gold, C., 2005. Digital Terrain Modeling: Principles and Methodology. CRC Press, Boca Raton, FL, 319 pp.
Li, Z., 1994. A comparative study of the accuracy of digital terrain models (DTMs) based on various data models. ISPRS Journal of Photogrammetry and Remote Sensing 49, 2–11.
Li, Z., 1992. Variation of the accuracy of digital terrain models with sampling interval. Photogrammetric Record 14, 113–128.
Li, Z., 1988. On the measure of digital terrain model accuracy. Photogrammetric Record 12, 873–877.
Li, X., Baker, B., Dickson, G., 2001. Accuracy of mapping products produced from the STAR-3i airborne IFSAR system. In: Proceedings of the 20th International Cartographic Conference, vol. 2. Beijing, China. ICA/ACI, pp. 1328–1336.
Lillesand, T.M., Kiefer, R.W., 2004. Remote Sensing and Image Interpretation, 5th edition. Wiley, 763 pp.
Linder, W., Schmid, W., Schiesser, H.-H., 1999. Surface winds and development of thunderstorms along southwest-northeast oriented mountain chains. Weather and Forecasting 14, 758–770.
Lindsay, J.B., Creed, I.F., 2006. Distinguishing actual and artefact depressions in digital elevation data. Computers & Geosciences 32 (8), 1192–1204.
Lindsay, J.B., 2005. The Terrain Analysis System: a tool for hydro-geomorphic applications. Hydrological Processes 19 (5), 1123–1130.
Lindsay, J.B., Creed, I.F., 2005. Removal of artifact depressions from DEMs: towards a minimum impact approach. Hydrological Processes 19 (16), 3113–3126.
Lindsay, J.B., 2003. A physically based model for calculating contributing area on hillslopes and along valley bottoms. Water Resources Research 39 (12), 1332–1340.
Liston, G.E., 1995. Local advection of momentum, heat and moisture during the melt of patchy snow covers. Journal of Applied Meteorology 34, 1705–1715.
Lloyd, C.D., 2005. Assessing the effect of integrating elevation data into the estimation of monthly precipitation in Great Britain. Journal of Hydrology 308 (1–4), 128–150.
Lloyd, C.D., Atkinson, P.M., 1998. Scale and the spatial structure of landform: optimizing sampling strategies with geostatistics. In: Proceedings of the 3rd International Conference on GeoComputation. University of Bristol, United Kingdom, 17–19 September 1998. University of Bristol, Bristol, UK, 16 pp.
Lobeck, A.K., 1939. Geomorphology: An Introduction to the Study of Landscapes. McGraw–Hill, New York, 731 pp.
Lopez, C., 2000. Improving the elevation accuracy of digital elevation models: a comparison of some error detection procedures. Transactions in GIS 4 (1), 43–64.
Lopez-Moreno, J.I., Nogués-Bravo, D., Chueca-Cía, J., Julián-Andrés, A., 2006. Glacier development and topographic context. Earth Surface Processes and Landforms 31 (12), 1585–1594.
Lucieer, A., Fisher, P., Stein, A., 2003. Texture-based segmentation of high-resolution remotely sensed imagery for identification of fuzzy objects. In: Proceedings of the Seventh International Conference on Geocomputation. University of Southampton, Southampton, UK, 9 pp.
Ludwig, R., Schneider, P., 2006. Validation of digital elevation models from SRTM X-SAR for applications in hydrologic modeling. ISPRS Journal of Photogrammetry and Remote Sensing 60, 339–358.
Luebke, D., Reddy, M., Cohen, J., Varshney, A., Watson, B., Huebner, R., 2002. Level of Detail for 3D Graphics. Morgan Kaufmann Publishers, San Francisco, 432 pp.
Lundblad, E., Wright, D.J., Miller, J., Larkin, E.M., Rinehart, R., Battista, T., Anderson, S.M., Naar, D.F., Donahue, B.T., 2006. A benthic terrain classification scheme for American Samoa. Marine Geodesy 29 (2), 89–111.
Luo, W., 2000. Quantifying groundwater-sapping landforms with a hypsometric technique. Journal of Geophysical Research 105 (E1), 1685–1694.
Lydolph, P.E., 1977. Climate of the Soviet Union. World Survey of Climatology, vol. 7. Elsevier, Amsterdam, 443 pp.

Lynn, B.H., Tao, W.-K., Wetzel, P.J., 1998. A study of landscape-generated deep moist convection. Monthly Weather Review 126, 928–942.
Maathuis, B.H.P., Wang, L., 2006. Digital elevation model based hydro-processing. Geocarto International 21 (1), 21–26.
Machguth, H., Paul, F., Hoelzle, M., Haeberli, W., 2006. Distributed glacier mass balance modelling as an important component of modern multi-level glacier monitoring. Annals of Glaciology 43 (1), 335–343.
Maclure, M., Willett, W.C., 1987. Misinterpretation and misuse of the kappa statistic. American Journal of Epidemiology 126 (2), 161–169.
MacMillan, R.A., Moon, D.E., Coupé, R.A., 2007. Automated predictive ecological mapping in a Forest Region of B.C., Canada, 2001–2005. Geoderma 140 (4), 353–373.
MacMillan, R.A., 2004. A comparison of 1:20,000 DEM 25 m data and 5 m DEM data. Technical Report. LandMapper Environmental Solutions Inc., Edmonton, Canada, 17 pp.
MacMillan, R.A., Jones, R.K., McNabb, D.H., 2004. Defining a hierarchy of spatial entities for environmental analysis and modeling using digital elevation models (DEMs). Computers, Environment and Urban Systems 28 (3), 175–200.
MacMillan, R.A., Martin, T.C., Earle, T.J., McNabb, D.H., 2003. Automated analysis and classification of landforms using high-resolution digital elevation data: applications and issues. Canadian Journal of Remote Sensing 29 (5), 592–606.
MacMillan, R.A., Pettapiece, W.W., Nolan, S.C., Goddard, T.W., 2000. A generic procedure for automatically segmenting landforms into landform elements using DEMs, heuristic rules and fuzzy logic. Fuzzy Sets and Systems 113, 81–109.
MacMillan, R.A., Pettapiece, W.W., 1997. Soil Landscape Models: automated landscape characterization and generation of soil–landscape models. Research Report No. 1. Agriculture and Agri-Food Canada, Research Branch, Lethbridge, Canada, 75 pp.
Malamud, B.D., Turcotte, D.L., Guzzetti, F., Reichenbach, P., 2004. Landslide inventories and their statistical properties. Earth Surface Processes and Landforms 29 (6), 687–711.
Malberg, H., 1994. Meteorologie und Klimatologie. Eine Einführung. Springer, Berlin, 350 pp.
Maling, D.H., 1989. Measurements from Maps: Principles and Methods of Cartometry. Pergamon, Oxford, 577 pp.
Mandelbrot, B., 1977. Fractals: Form, Chance and Dimension. W.H. Freeman, San Francisco, 365 pp.
Mandelbrot, B., 1967. How long is the coast of Britain? Statistical self-similarity and fractional dimension. Science 156, 636–638.
Manis, G., Lowry, J., Ramsey, R.D., 2001. Preclassification: an ecologically predictive landform model. GAP Analysis Bulletin 10, 1–3.
Manning, G., Fuller, L.G., Flaten, D.N., Eilers, R.G., 2001. Wheat yield and grain protein variation within an undulating soil landscape. Canadian Journal of Soil Science 81 (3), 459–467.
Mark, D.M., Smith, B., 2004. A science of topography: from qualitative ontology to digital representations. In: Bishop, M.P., Shroder, J.F. (Eds.), Geographic Information Science and Mountain Geomorphology. Springer–Praxis, Chichester, England, pp. 75–97.
Mark, D.M., Aronson, P.B., 1984. Scale-dependent fractal dimensions of topographic surfaces: an empirical investigation, with applications in geomorphology and computer mapping. Mathematical Geology 16 (7), 671–683.
Mark, D.M., 1983. Relations between field-surveyed channel networks and map-based geomorphometric measures, Inez, Kentucky. Annals of the Association of American Geographers 73 (3), 358–372.
Mark, D.M., 1979. Topology of ridge patterns: randomness and constraints. Geological Society of America Bulletin Part I 90, 164–172.
Mark, D.M., 1975a. Computer analysis of topography: a comparison of terrain storage methods. Geografiska Annaler 57A, 179–188.
Mark, D.M., 1975b. Geomorphometric parameters: a review and evaluation. Geografiska Annaler 57A (3–4), 165–177.
Marks, D., Dozier, J., 1992. Climate and energy exchange at the snow surface in the alpine region of Sierra Nevada. 2. Snow cover energy balance. Water Resources Research 28, 3043–3054.
Marks, D., Dozier, J., 1979. A clear-sky longwave radiation model for remote Alpine areas. Archiv für Meteorologie, Geophysik und Bioklimatologie B 27, 159–187.

Marshall, T.R., Lee, P.F., 1994. Mapping aquatic macrophytes through digital image analysis of aerial photographs: an assessment. Journal of Aquatic Plant Management 32, 61–66.

Martinoni, D., 2002. Models and experiments for quality handling in digital terrain modelling. Ph.D. Thesis. University of Zürich.

Martz, L., Garbrecht, J., 1999. An outlet breaching algorithm for the treatment of closed depressions in a raster DEM. Computers and Geosciences 25 (7), 835–844.

Martz, L.W., de Jong, E., 1988. CATCH: a Fortran program for measuring catchment area from digital elevation models. Computers and Geosciences 14 (5), 627–640.

Mather, P.M., 1972. Areal classification in geomorphology. In: Chorley, R.J. (Ed.), Spatial Analysis in Geomorphology. Harper & Row, New York, pp. 305–322.

Maune, D.F. (Ed.), 2001. Digital Elevation Model Technologies and Applications: The DEM Users Manual. American Society for Photogrammetry and Remote Sensing, Bethesda, MD, 539 pp.

Maune, D.F., Lloyd, C.H., Guenther, G.C., 2001. DEM user applications. In: Maune, D.F. (Ed.), Digital Elevation Model Technologies and Applications: The DEM Users Manual. American Society for Photogrammetry and Remote Sensing, Bethesda, MD, pp. 367–394.

Maxwell, J.C., 1870. On hills and dales. The London, Edinburgh and Dublin Philosophical Magazine and Journal of Science Series 4 269, 421–427.

Maybury, K.P. (Ed.), 1999. Seeing the Forest and the Trees: Ecological Classification for Conservation. The Nature Conservancy, Arlington, VA, 40 pp.

McArdell, B.W., Graf, C., Swartz, M., 2004. Debris flow monitoring and modelling in Switzerland. In: Tropeano, D., Arattano, M., Maraga, F., Pelissero, C. (Eds.), Progress in Surface and Subsurface Water Studies at the Plot and Small Basin Scale. International Conference ERB2004 — Euromediterranean Conference. National Research Council of Italy, Turin, Italy, pp. 249–250.

McBratney, A.B., Whelan, B., Ancev, T., Bouma, J., 2005. Future directions of precision agriculture. Precision Agriculture 6 (1), 7–23.

McBratney, A.B., Mendonça Santos, M.L., Minasny, B., 2003. On digital soil mapping. Geoderma 117 (1–2), 3–52.

McBratney, A.B., 1998. Some considerations on methods for spatially aggregating and disaggregating soil information. Nutrient Cycling in Agroecosystems 50, 51–62.

McBratney, A.B., de Gruijter, J.J., 1992. A continuum approach to soil classification by modified fuzzy-k means with extragrades. Journal of Soil Science 43, 159–176.

McKean, J., Roering, J., 2004. Objective landslide detection and surface morphology mapping using high-resolution airborne laser altimetry. Geomorphology 57, 331–351.

McKenzie, N.J., Gessler, P.E., Ryan, P.J., O'Connell, D.A., 2000. The role of terrain analysis in soil mapping. In: Wilson, J.P., Gallant, J.C. (Eds.), Terrain Analysis: Principles and Applications. Wiley, pp. 245–265.

McKenzie, N.J., Ryan, P.J., 1999. Spatial prediction of soil properties using environmental correlation. Geoderma 89 (1–2), 67–94.

McQueen, J.T., Draxler, R.R., Rolph, G.D., 1995. Influence of grid size and terrain resolution on wind field predictions from an operational mesoscale model. Journal of Applied Meteorology 34 (10), 2166–2181.

Medor, W.E., Weaver, W.R., 1980. Two-stream approximations to radiative transfer in planetary atmospheres: a unified description of existing methods and a new improvement. Journal for Atmospheric Sciences 36, 630–643.

Meijerink, A.M.J., 1988. Data acquisition and data capture through terrain mapping units. ITC Journal 1, 23–44.

Melton, M.A., 1965. The geomorphic and paleoclimatic significance of alluvial deposits in southern Arizona. Journal of Geology 73, 1–38.

Melton, M.A., 1958. Use of punched cards to speed statistical analysis of geomorphic data. Geological Society of America Bulletin 69 (3), 355–358.

Menz, G., Richters, J., in press. Quantitative classification of landscape units in Northern Namibia using an ASTER digital elevation model. In: Bollig, M., Gruntkowski, N. (Eds.), Landscape in Interdisciplinary Research. Kluwer.

Miles, S.B., Keefer, D.K., 2001. Seismic landslide hazard for the cities of Oakland and Piedmont, California. Misc. Field Studies Map MF-2379. U.S. Geological Survey, 6 pp.

Miliaresis, G.C., Paraschou, C.V.E., 2005. Vertical accuracy of the SRTM DTED level 1 of Crete. International Journal of Applied Earth Observation and Geoinformation 7 (1), 49–59.

Miliaresis, G.C., 2001. Geomorphometric mapping of Zagros Ranges at regional scale. Computers and Geosciences 27 (7), 775–786.

Miliaresis, G.C., Argialas, D.P., 1999. Segmentation of physiographic features from the global digital elevation model/GTOPO30. Computers & Geosciences 25 (7), 715–728.

Miller, C.L., Laflamme, R.A., 1958. The digital terrain model — theory and application. Photogrammetric Engineering 24 (3), 433–442.

Miller, J., Franklin, J., 2002. Modeling the distribution of four vegetation alliances using generalized linear models and classification trees with spatial dependence. Ecological Modelling 157 (2–3), 227–247.

Milne, G., 1935. Some suggested units of classification and mapping particularly for East African soils. Soil Research 4, 183–198.

Milton, S.F., Wilson, C.A., 1996. The impact of parameterized subgrid-scale orographic forcing on systematic errors in a global NWP model. Monthly Weather Review 124, 2023–2045.

Minár, J., Evans, I.S., 2008. Elementary forms for land surface segmentation: the theoretical basis of terrain analysis and geomorphological mapping. Geomorphology 95 (3–4), 236–259.

Minasny, B., McBratney, A.B., 2006. A conditioned Latin hypercube method for sampling in the presence of ancillary information. Computers & Geosciences 32 (9), 1378–1388.

Minasny, B., McBratney, A.B., 2001. A rudimentary mechanistic model for soil formation and landscape development II. A two-dimensional model incorporating chemical weathering. Geoderma 103, 161–179.

Ministry of Environment, Lands and Parks, Surveys and Resource Mapping Branch, 1992. Digital Baseline Mapping at 1:20,000. British Columbia Baseline Specifications and Guidelines for Geomatics, Content Series, vol. 3. Geographic Data BC, Victoria, BC, Canada.

Mitas, L., Mitášová, H., 1999. Spatial interpolation. In: Longley, P., Goodchild, M.F., Maguire, D.J., Rhind, D.W. (Eds.), Geographical Information Systems: Principles, Techniques, Management and Applications, vol. 1. Wiley, pp. 481–492.

Mitas, L., Mitášová, H., 1998. Distributed soil erosion simulation for effective erosion prevention. Water Resources Research 34 (3), 505–516.

Mitášová, H., Mitas, L., Harmon, R., 2005. Simultaneous spline approximation and topographic analysis for lidar elevation data in open-source GIS. IEEE Geoscience and Remote Sensing Letters 2, 375–379.

Mitášová, H., Mitas, L., 2001. Multiscale soil erosion simulations for land use management. In: Harmon, R., Doe, W. (Eds.), Landscape Erosion and Landscape Evolution Modeling. Kluwer Academic/Plenum Publishers, pp. 321–347.

Mitášová, H., Brown, W.M., Mitas, L., Warren, S., 1997. Multi-dimensional GIS environment for simulation and analysis of landscape processes. In: ASAE Annual International Meeting. 10–14 August, 1997, Minneapolis, MN, USA, 19 pp.

Mitášová, H., Hofierka, J., Zlocha, M., Iverson, R.L., 1996. Modelling topographic potential for erosion and deposition using GIS. International Journal of Geographical Information Systems 10 (5), 629–641.

Mitášová, H., Mitas, L., Brown, W.M., Gerdes, D.P., Kosinovsky, I., Baker, T., 1995. Modeling spatially and temporally distributed phenomena: new methods and tools for GRASS GIS. International Journal of Geographical Information Systems 9 (4), 433–446.

Mitášová, H., Hofierka, J., 1993. Interpolation by regularized spline with tension, II Application to terrain modelling and surface geometry analysis. Mathematical Geology 25, 657–669.

Mitchell, C.W., 1991. Terrain Evaluation: An Introductory Handbook to the History, Principles, and Methods of Practical Terrain Assessment, 2nd edition. Longman, Harlow, 441 pp.

Mitchell, S.G., Montgomery, D.R., 2006a. Influence of a glacial buzzsaw on the height and morphology of the Cascade Range in central Washington State, USA. Quaternary Research 65, 96–107.

Mitchell, S.G., Montgomery, D.R., 2006b. Polygenetic topography of the Cascade Range, Washington State, USA. American Journal of Science 306 (11), 736–768.

Molander, C., Merritt, S., Corrubia, A., 2006. Marrying photogrammetry and LiDAR. Earth Observation Magazine 11 (6).

Montgomery, D.R., Brandon, M.T., 2002. Topographic controls on erosion rates in tectonically active mountain ranges. Earth and Planetary Science Letters 201 (3–4), 481–489.
Montgomery, D.R., 2001. Slope distributions, threshold hillslopes and steady-state topography. American Journal of Science 301, 432–454.
Montgomery, D.R., Balco, G., Willett, S.D., 2001. Climate, tectonics and the morphology of the Andes. Geology 29 (7), 579–582.
Montgomery, D.R., Sullivan, K., Greenberg, H.M., 1998. Regional test of a model for shallow landsliding. Hydrological Processes 12, 943–955.
Montgomery, D.R., Dietrich, W.E., 1994. A physically-based model for the topographic control on shallow landsliding. Water Resources Research 30 (4), 1153–1171.
Montgomery, D.R., Dietrich, W.E., 1992. Channel initiation and the problem of landscape scale. Science 255, 826–830.
Montgomery, D.R., Dietrich, W.E., 1989. Source areas, drainage density, and channel initiation. Water Resources Research 25 (8), 1907–1918.
Moore, I.D., Gessler, P.E., Nielsen, G.A., Peterson, G.A., 1993a. Soil attribute prediction using terrain analysis. Soil Science Society of America Journal 57 (2), 443–452.
Moore, I.D., Norton, T.W., Williams, J.E., 1993b. Modelling environmental heterogeneity in forested landscapes. Journal of Hydrology 150, 717–747.
Moore, I.D., Wilson, J.P., 1992. Length-slope factors for the Revised Universal Soil Loss Equation: simplified method of estimation. Journal of Soil and Water Conservation 47, 423–428.
Moore, I.D., Grayson, R.B., 1991. Terrain-based catchment partitioning and runoff prediction using vector elevation data. Water Resources Research 27, 1177–1191.
Moore, I.D., Grayson, R.B., Ladson, A.R., 1991a. Digital terrain modelling: a review of hydrological, geomorphological, and biological applications. Hydrological Processes 5 (1), 3–30.
Moore, I.D., Lees, B.G., Davey, S.M., 1991b. A new method for predicting vegetation distributions using decision tree analysis in a geographic information system. Environmental Management 15 (1), 59–71.
Moore, I.D., Burch, G.J., Mackenzie, D.H., 1988. Topographic effects on the distribution of surface soil water and the location of ephemeral gullies. Transactions of the ASAE 31 (4), 1098–1107.
Moore, I.D., Burch, G.J., 1986. Physical basis of the length-slope factor in the Universal Soil Loss Equation. Soil Science Society of America Journal 50 (5), 1294–1298.
Moran, C.J., Bui, E.N., 2002. Spatial data mining for enhanced soil modelling. International Journal of Geographical Information Systems 16 (6), 533–549.
Morisawa, M., 1985. The Geological Society of America Bulletin and the development of quantitative geomorphology. Geological Society of America Bulletin 100 (7), 1016–1022.
Morris, D., Heerdegen, R., 1988. Automatically derived catchment boundaries and channel networks and their hydrological applications. Geomorphology 1, 131–141.
Mortensen, N.G., Bowen, A.J., Antoniou, I., 2006. Improving WAsP predictions in (too) complex terrain. In: Proceedings of the European Wind Energy Conference. Athens (GR), 27 February – 2 March 2006. European Wind Energy Association, Brussels, 9 pp.
Mouginis-Mark, P.J., Garbeil, H., Boyce, J.M., Ui, C.S.E., Baloga, S.M., 2004. Geometry of Martian impact craters — first results from an interactive software package. Journal of Geophysical Research 109E, 1–9.
Mulla, D.J., 1988. Using geostatistics and spectral analysis to study spatial patterns in the topography of southeastern Washington state, USA. Earth Surface Processes and Landforms 13, 389–405.
Mun, E.Y., Von Eye, A., 2004. Analyzing Rater Agreement: Manifest Variable Methods. Lawrence Erlbaum Associates, Mahwah, NJ, 190 pp.
Murray, J., 1888. On the height of the lands and the depth of the Ocean. Scottish Geographical Magazine 4, 418–428.
Nachlinger, J., Sochi, K., Comer, P., Kittel, G., Dorfman, D., 2001. Great Basin: An Ecoregion-Based Conservation Blueprint. The Nature Conservancy, Reno, NV.
Nagihara, S., Mulligan, K.R., Xiong, W., 2004. Use of a three-dimensional laser scanner to digitally capture the topography of sand dunes in high spatial resolution. Earth Surface Processes and Landforms 29, 391–398.
Nelson, E.J., Jones, N.L., Miller, A.W., 1994. Algorithm for precise drainage-basin delineation. Journal of Hydraulic Engineering 120, 298–312.

Neteler, M., Mitášová, H., 2008. Open Source GIS: A GRASS GIS Approach, 3nd edition. The International Series in Engineering and Computer Science, vol. 773. Springer, New York, 406 pp.
Neter, J., Kutner, M.H., Nachtsheim, C.J., Wasserman, W. (Eds.), 1996. Applied Linear Statistical Models. 4th edition. McGraw–Hill, 1391 pp.
Neuenschwander, G., 1944. Morphometrische Begriffe; ein kritische Übersicht auf Grund der Literatur. Ph.D. Thesis. Universität Zürich.
Neustruev, S.S., 1930. Elements of soil geography. In: Genesis and Geography of Soils. Nauka, Moscow, Russia, pp. 149–314 (in Russian).
Newell, W., Clark, I., 2008. Geomorphic map of Worcester County, Maryland, interpreted from a LiDAR-based, digital elevation model. Open-File Report 2008-1005. U.S. Geological Survey, 34 pp. http://pubs.usgs.gov/of/2008/1005/.
Nikolov, N.T., Zeller, K.F., 1992. A solar radiation algorithm for ecosystem dynamic models. Ecological Modelling 61, 149–168.
Nilsen, T.H., 1975. Preliminary photo-interpretation maps of landslide and other surficial deposits of 56 7.5-minute quadrangles, Alameda, Contra Costa, and Santa Clara Counties, California. Open-File Report No. 75-277, scale 1:24,000. USGS.
Noetzli, J., Huggel, C., Hoelzle, M., Haeberli, W., 2006. GIS-based modelling of rock-ice avalanches from alpine permafrost areas. Computational Geosciences 10, 161–178.
Nogami, M., 1995. Geomorphometric measures for digital elevation models. Zeitschrift für Geomorphologie, Supplementband 101, 53–67.
Noma, A.A., Misulia, M.G., 1959. Programming topographic maps for automatic terrain model construction. Surveying and Mapping 19 (3), 355–366.
Norheim, R.A., Queija, V.R., Haugerud, R.A., 2002. Comparison of LiDAR and INSAR DEMs with dense ground control. In: Proceedings of the ESRI 2002 User Conference. ESRI, San Diego, 9 pp.
Northcote, K.H., 1984. Soil–landscapes, taxonomic units and soil profiles: a personal perspective on some unresolved problems of soil survey. Soil Survey and Land Evaluation 4, 1–7.
O'Brien, J.S., Julien, P.Y., Fullerton, W.T., 1993. Two-dimensional water flood and mudflow simulation. Journal of Hydraulic Engineering 119 (2), 244–261.
O'Callaghan, J.F., Mark, D.M., 1984. The extraction of drainage networks from digital elevation data. Computer Vision, Graphics, and Image Processing 28, 323–344.
Odeh, I.O.A., McBratney, A.B., Chittleborough, D.J., 1994. Spatial prediction of soil properties from landform attributes derived from a digital elevation model. Geoderma 63, 197–214.
Ogden, F.L., Julien, P.Y., 2002. CASC2D: a two-dimensional, physically-based, Hortonian hydrologic model. In: Singh, V.P., Frevert, D. (Eds.), Mathematical Models of Small Watershed Hydrology and Applications. Water Resources Publications, Littleton, CO, pp. 69–112.
Okabe, A., Boots, B., Sugihara, K., 2001. Spatial Tessellations: Concepts and Applications of Voronoi Diagrams, 2nd edition. Wiley Series in Probability and Mathematical Statistics. Wiley, New York, 696 pp.
Oke, T.R., 1988. Boundary Layer Climates, 2nd edition. Taylor & Francis, London, New York, 435 pp.
Oksanen, J., 2006a. Digital elevation model error in terrain analysis. Ph.D. Thesis. Faculty of Science, University of Helsinki.
Oksanen, J., 2006b. Uncovering the statistical and spatial characteristics of fine toposcale DEM error. International Journal of Geographical Information Science 20 (4), 345–369.
Oksanen, J., Minchin, P.R., 2002. Continuum theory revisited: what shape are species responses along ecological gradients? Ecological Modelling 157, 119–129.
Olaya, V., 2004. A gentle introduction to SAGA GIS. The SAGA User Group e.V., Gottingen, Germany, 208 pp.
O'Loughlin, E.M., 1986. Prediction of surface saturation zones in natural catchments by topographic analysis. Water Resources Research 22 (5), 794–804.
Omernick, J.M., 2004. Perspectives on the nature and definition of ecological regions. Environmental Management 34, 39–60.
O'Neill, M.P., Mark, D.M., 1987. On the frequency distribution of land slope. Earth Surface Processes and Landforms 12, 127–136.
Ordnance Survey, 1992. 1:50,000 Scale Height Data User Manual. Ordnance Survey, Southhampton.
Oreskes, N., Shrader-Frechette, K., Belitz, K., 1994. Verification, validation and confirmation of numerical models in the earth sciences. Science 263, 641–646.

Orlanski, I., 1975. A rational subdivision of scales for atmospheric processes. Bulletin of the American Meteorological Society 56, 527–530.

Oskin, M., Burbank, D.W., 2005. Alpine landscape evolution dominated by cirque retreat. Geology 33 (12), 933–936.

Outcalt, S.I., Hinkel, K.M., Nelson, F.E., 1994. Fractal physiography? Geomorphology 11 (2), 91–106.

Over, T.M., Gupta, V.K., 1996. A space–time theory of mesoscale rainfall using random cascades. Journal of Geophysical Research 101 (D21), 26319–26331.

Pack, R.T., Tarboton, D.G., Goodwin, C.N., 2001. Assessing terrain stability in a GIS using SINMAP. In: 15th Annual GIS Conference. International Association of Engineering Geology, Vancouver, British Columbia, 9 pp.

Pain, C.F., 2005. Size does matter: relationships between image pixel size and landscape process scales. In: Zerger, A., Argent, R.M. (Eds.), MODSIM 2005 International Congress on Modelling and Simulation. Modelling and Simulation Society of Australia and New Zealand, Melbourne, pp. 1430–1436.

Palacios-Velez, O.L., Cuevas-Renaud, B., 1986. Automated river-course, ridge and basin delineation from digital elevation data. Journal of Hydrology 86, 299–314.

Park, S.J., Vlek, P.L.G., 2002. Environmental correlation of three-dimensional soil spatial variability: a comparison of three adaptive techniques. Geoderma 109 (1–2), 117–140.

Park, S.J., McSweeney, K., Lowery, B., 2001. Identification of the spatial distribution of soils using a process-based terrain characterisation. Geoderma 103, 249–272.

Parks, J.M., 1966. Cluster analysis applied to multivariate geologic problems. Journal of Geology 74 (5), 703–715.

Parsons, A.J., 1988. Hillslope Form. Routledge, London, 212 pp.

Partsch, J., 1911. Schlesien: eine Landeskunde für das deutsche Volk, vol. 2. F. Hirt, Breslau, 690 pp.

Pasuto, A., Schrott, L. (Eds.), 1999. Recent Developments of Landslide Research in Europe. Geomorphology 30 (1–2), 211 pp. (special issue).

Paul, F., Huggel, C., Kääb, A., 2004. Combining satellite multispectral image data and a digital elevation model for mapping debris-covered glaciers. Remote Sensing of Environment 89 (4), 510–518.

Paz, J.O., Batchelor, W.D., Tylka, G.L., Hartler, R.G., 2001. A modeling approach to quantify the effects of spatial soybean limiting factors. Transaction of the ASAE 44 (5), 1329–1334.

Pebesma, E.J., 2004. Multivariable geostatistics in S: the gstat package. Computers & Geosciences 30 (7), 683–691.

Peckham, S.D., Gupta, V.K., 1999. A reformulation of Horton's laws for large river networks in terms of statistical self-similarity. Water Resources Research 35 (9), 2763–2777.

Peckham, S.D., 1998. Efficient extraction of river networks and hydrologic measurements from digital elevation data. In: Barndorff-Nielsen, O.E., et al. (Eds.), Stochastic Methods in Hydrology: Rain, Landforms and Floods. World Scientific, Singapore, pp. 173–203.

Peckham, S.D., 1995a. New results for self-similar trees with applications to river networks. Water Resources Research 31 (4), 1023–1029.

Peckham, S.D., 1995b. Self-similarity in the three-dimensional geometry and dynamics of large river basins. Ph.D. Thesis. University of Colorado at Boulder.

Péguy, C.P., 1948. Introduction a l'emploi des méthodes statistiques en géographie physique. Revue de Géographie Alpine 36, 5–101.

Péguy, C.P., 1942. Principes de morphométrie alpine. Revue de Géographie Alpine 30, 453–486.

Penck, A., 1894. Morphographie und Morphometrie. In: Morphologie der Erdoberfläche, vol. 1. Engelhorn, Stuttgart, pp. 33–95.

Pennock, D.J., 2003. Terrain attributes, landform segmentation, and soil redistribution. Soil and Tillage Research 69 (1–2), 15–26.

Pennock, D.J., Walley, F., Solohub, M., Si, B., Hnatowich, G., 2001. Topographically controlled yield response of Canola to nitrogen fertilizer. Soil Science Society of America Journal 65 (6), 1838–1845.

Pennock, D.J., Corré, M.D., 2000. Development and application of landform segmentation procedures. Soil and Tillage Research 58, 151–162.

Pennock, D.J., Anderson, D.W., de Jong, E., 1994. Landscape-scale changes in indicators of soil quality due to cultivation in Saskatchewan, Canada. Geoderma 64, 1–19.

Pennock, D.J., Zebarth, B.J., de Jong, E., 1987. Landform classification and soil distribution in hummocky terrain, Sasketchewan, Canada. Geoderma 40, 297–315.

Perez, R., Seals, R., Ineichen, P., Stewart, R., Menicucci, D., 1987. A new simplified version of the Perez diffuse irradiance model for tilted surfaces. Solar Energy 39, 221–231.
Petrie, G., Kennie, T.J.M., 1987. Terrain modelling in surveying and civil engineering. Computer-Aided Design 19 (4), 171–187.
Petzold, B., Reiss, P., Stüssel, W., 1999. Laser scanning — surveying and mapping agencies are using a new technique for the derivation of digital terrain models. ISPRS Journal of Photogrammetry and Remote Sensing 54, 95–104.
Peucker, T.K., Douglas, D.H., 1975. Detection of surface-specific points by local parallel processing of discrete terrain elevation data. Computer Graphics and Image Processing 4, 375–387.
Peuquet, D.J., 1984. A conceptual framework and comparison of spatial data models. Cartographica 21 (4), 66–113.
Pfeffer, K., Pebesma, E.J., Burrough, P.A., 2003. Mapping alpine vegetation using vegetation observations and topographic attributes. Landscape Ecology 18, 759–776.
Phillips, J.D., 1994. Deterministic uncertainty in landscapes. Earth Surface Processes and Landforms 19, 389–401.
Pichler, H., Steinacker, R., Lanzinger, A., 1990. Cyclogenesis induced by the Alps. Meteorology and Atmospheric Physics 43, 21–29.
Pielke, R.A., 1984. Mesoscale Meteorological Modeling. Academic Press, 612 pp.
Pike, R.J., Sobieszczyk, S., 2008. Soil slip/debris flow localized by site attributes and wind-driven rain in the San Francisco Bay region storm of January 1982. Geomorphology 94 (3–4), 290–313.
Pike, R.J., Howell, D.G., Graymer, R.W., 2003. Landslides and cities — an unwanted partnership. In: Heiken, G., Fakundiny, R., Sutter, J.F. (Eds.), Earth Science in the City — A Reader. In: Special Publications, vol. 56. AGU, pp. 187–254.
Pike, R.J., 2002. A bibliography of terrain modeling (geomorphometry), the quantitative representation of topography — supplement 4.0. Open-File Report 02-465. U.S. Geological Survey, Denver, 116 pp.
Pike, R.J., 2001a. Geometric signatures — experimental design, first results. In: Ohmori, H. (Ed.), DEMs and Geomorphology (Proceedings of the Symposia on New Concepts and Models in Geomorphology, and Geomorphometry, DEMS and GIS), vol. 1. Special Publications of the Geographic Information Systems Association (Japan) and Chuo University of Tokyo, Tokyo, Japan, pp. 50–51.
Pike, R.J., 2001b. Scenes into numbers — facing the subjective in landform quantification. In: Hoffman, R.R., Markman, A.B. (Eds.), Interpreting remote sensing imagery-human factors. Lewis Publishers (CRC), Boca Raton, FL, pp. 83–114.
Pike, R.J., 2001c. Topographic fragments of geomorphometry, GIS, and DEMs. In: Ohmori, H. (Ed.), DEMs and Geomorphology (Proceedings of the Symposia on New Concepts and Models in Geomorphology, and Geomorphometry, DEMS and GIS), vol. 1. Special Publications of the Geographic Information Systems Association (Japan) and Chuo University of Tokyo, Tokyo, Japan, pp. 34–35.
Pike, R.J., Graymer, R.W., Roberts, S., Kalman, N.B., Sobieszczyk, S., 2001. Map and map database of susceptibility to slope failure by sliding and earth flow in the Oakland area, California. Misc. Field Studies Map, MF-2385, scale 1:50,000. U.S. Geological Survey, 37 pp.
Pike, R.J., 2000a. Geomorphometry — diversity in quantitative surface analysis. Progress in Physical Geography 24 (1), 1–20.
Pike, R.J., 2000b. Nano-metrology and terrain modelling — convergent practice in surface characterisation. Tribology International 33 (9), 593–600.
Pike, R.J., 1995. Geomorphometry — progress, practice, and prospect. Zeitschrift für Geomorphologie, Supplementband 101, 221–238.
Pike, R.J., Acevedo, W., Card, D.H., 1989. Topographic grain automated from digital elevation models. In: Proceedings of the Ninth International Symposium on Computer Assisted Cartogtraphy. ASPRS/ASCM, Baltimore, MD, pp. 128–137.
Pike, R.J., 1988. The geometric signature: quantifying landslide-terrain types from Digital Elevation Models. Mathematical Geology 20, 491–511.
Pike, R.J., Wilson, S.E., 1971. Elevation–relief ratio, hypsometric integral, and geomorphic area-altitude analysis. Geological Society of America Bulletin 82, 1079–1084.
Pilouk, M., Tempfli, K., 1992. A digital image processing approach to creating DTMs from digitized contours. International Archives of Photogrammetry and Remote Sensing 29 (B4), 956–961.
Piotrowski, J.A., 1989. Relationship between drumlin length and width as a manifestation of the subglacial processes. Zeitschrift für Geomorphologie 33 (4), 429–441.

Pitty, A.F., 1969. A Scheme for Hillslope Analysis. Occasional Papers in Geography, vol. 9. University of Hull Publications, Hull, Yorkshire, 76 pp.

Planchon, O., Darboux, F., 2001. A fast, simple and versatile algorithm to fill the depressions of digital elevation models. Catena 46, 159–176.

Pleim, J.E., Xiu, A., Finkelstein, P.L., Otte, T.L., 2001. A coupled land-surface and dry deposition model and comparison to field measurements of surface heat, moisture, and ozone fluxes. Water, Air & Soil Pollution: Focus 1 (5–6), 243–252.

Pojar, J., Klinka, K., Meidinger, D.V., 1987. Biogeoclimatic ecosystem classification in British Columbia. Forest Ecology and Management 22, 119–154.

Preusser, A., 1984. Computing contours by successive solution of quintic polynomial equations. ACM Transactions on Mathematical Software 10 (4), 463–472.

Priestley, C.H.B., Taylor, T.J., 1972. On the assessment of surface heat flux and evaporation using large-scale parameters. Monthly Weather Review 100, 81–92.

Qi, F., Zhu, A.-X., Harrower, M., Burt, J.E., 2006. Fuzzy soil mapping based on prototype category theory. Geoderma 136, 774–787.

Queney, P., 1948. The problem of the airflow over mountains: a summary of theoretical studies. Bulletin of the American Meteorological Society 29, 16–26.

Quinn, P.F., Beven, K.J., Lamb, R., 1995. The ln(a/tan b) index: how to calcute it and how to use it within in the TOPMODEL framework. Hydrological Processes 9, 161–182.

Quinn, P., Beven, K., Chevallier, P., Planchon, O., 1991. The prediction of hillslope paths for distributed hydrological modeling using digital terrain models. Hydrological Processes 5, 59–79.

Raaflaub, L.D., Collins, M.J., 2006. The effect of error in gridded digital elevation models on the estimation of topographic parameters. Environmental Modelling & Software 21, 710–732.

Rabus, B., Eineder, M., Roth, A., Bamler, R., 2003. The shuttle radar topography mission — a new class of digital elevation models acquired by spaceborne radar. Photogrammetric Engineering and Remote Sensing 57 (4), 241–262.

Rana, S. (Ed.), 2004. Topological Data Structures for Surfaces: An Introduction for Geographical Information Science. Wiley, New York, 214 pp.

Rasemann, S., Schmidt, J., Schrott, L., Dikau, R., 2004. Geomorphometry in mountain terrain. In: Bishop, M.P., Shroder, J.F. (Eds.), GIS & Mountain Geomorphology. Springer, Berlin, pp. 101–145.

Reichenbach, P., Carrara, A., Guzzetti, F. (Eds.), 2002. Assessing and Mapping Landslide Hazards and Risk. Natural Hazards and Earth System Sciences 2. Copernicus Publications, 117 pp. (special issue).

Resource Inventory Committee (RIC), 2000. Standards for Predictive Ecosystem Mapping (PEM) — Digital Data Capture. Predictive Ecosystem Technical Standards and Database Manual, Version 1. Prepared by: PEM Data Committee for the TEM Alternatives Task Force. Government of British Columbia, Vancouver, Canada, 31 pp.

Resource Inventory Committee (RIC), 1999. Standards for Predictive Ecosystem Mapping: Inventory Standard, Version 1. Prepared by: Terrestrial Ecosystem Mapping Alternatives. Government of British Columbia, Vancouver, Canada, 43 pp.

Reuter, H.I., Wendroth, O., Kersebaum, K.-C., 2006. Optimisation of relief classification for different levels of generalisation. Geomorphology 77 (1–2), 79–89.

Reuter, H.I., Kersebaum, K.-C., Wendroth, O., 2005. Modelling of solar radiation influenced by topographic shading — evaluation and application for precision farming. Physics and Chemistry of the Earth 30 (1–3), 143–149.

Reuter, H.I., 2004. Spatial crop and soil landscape processes under special consideration of relief information in a loess landscape. Ph.D. Thesis. Universität Hannover.

Rhoads, B.L., Thorn, C.E. (Eds.), 1996. The Scientific Nature of Geomorphology, 27th Binghamton Symposium in Geomorphology, Proceedings. 27–29 September. Wiley, Chichester, UK, 481 pp.

Riaño, D., Chuvieco, E., Salas, J., Aguado, I., 2003. Assessment of different topographic corrections in Landsat-TM data for mapping vegetation types. IEEE Transactions on Geoscience and Remote Sensing 41 (5), 1056–1061.

Rich, P.M., Hetrick, W.A., Saving, S.C., 2002. Modeling topographic influences on solar radiation: a manual for the SOLLRALUX model. LA-12989-M. Los Alamos National Laboratory, Los Alamos, NM, 33 pp.

Rich, P.M., Dubayah, R., Hetrick, W.A., Saving, S.C., 1994. Using viewshed models to calculate intercepted solar radiation: applications in ecology. American Society for Photogrammetry and Remote Sensing Technical Papers, 524–529.
Richards, P.W., 1981. The Tropical Rain Forest. Cambridge University Press, Cambridge, 450 pp.
Richardson, L.F., 1961. The problem of contiguity: an appendix to Statistics of Deadly Quarrels. In: Yearbook of the Society for General Systems Research, vol. 6. Ann Arbor, MI, pp. 140–187.
Richter, H., 1962. Eine neue Methode der großmaßstäbigen Kartierung des Reliefs. Petermanns Geographische Mitteilungen 104, 309–312.
Rickenmann, D., 1999. Empirical relationships for debris flows. Natural Hazards 19, 47–77.
Rieger, W., 1998. A phenomenon-based approach to upslope contributing area and depressions in DEMs. Hydrological Processes 12, 857–872.
Rieger, W., 1992. Automated river line and catchment area extraction for DEM data. In: Proceedings of the 17th ISPRS Congress, vol. 28(B4). ISPRS, Washington, DC, pp. 642–649.
Rigollier, C., Bauer, O., Wald, L., 2000. On the clear sky model of the ESRA (European Solar Radiation Atlas) with respect to the Heliosat method. Solar Energy 68 (1), 33–48.
Rigon, R., Ghesla, E., Tiso, A., Cozzini, A., 2006. The HORTON Machine: A System for DEM Analysis. Quaderni del Dipartimento. Università degli Studi di Trento, Trento, Italy, 144 pp.
Rigon, R., Rinaldo, A., Rodriguez-Iturbe, I., Bras, R.L., Ijjasz-Vasquez, E., 1993. Optimal channel networks: a framework for the study of river basin morphology. Water Resources Research 29 (6), 1635–1646.
Robert, P.C., 1993. Characterisation of soil conditions at the field level for soil specific management. Geoderma 60, 57–72.
Robinson, A.H., 1982. Early Thematic Mapping in the History of Cartography. University of Chicago Press, Chicago, 266 pp.
Rodriguez, E., Morris, C.S., Belz, J.E., Chapin, E.C., Martin, J.M., Daffer, W., Hensley, S., 2005. An Assessment of the SRTM Topographic Products. Jet Propulsion Laboratory, Pasadena, CA, 143 pp.
Rodríguez-Iturbe, I., Rinaldo, A., 1997. Fractal River Basins — Chance and Self-Organization. Cambridge University Press, Cambridge, 564 pp.
Roe, G.H., 2005. Orographic precipitation. Annual Review of Earth and Planetary Science 33, 645–671.
Roeckner, E., Arpe, K., Bengtsson, L., Christoph, M., Claussen, M., Dümenil, L., Esch, M., Giogetta, M., Schlese, U., Schultz-Weida, U., 1996. The atmospheric general circulation model ECHAM4. Model description and simulation of the present-day climate. Report 218. MPI für Meteorologie, Hamburg, Germany, 90 pp.
Rosenberg, N., Crosson, P., Frederick, K., Easterling, W., McKenney, M., Bowes, M., Sedjo, R., Darmstadler, J., Katz, L., Lemon, K., 1993. The MINK methodology: background and baseline. Climatic Change 24, 7–22.
Rosenbloom, N.A., Doney, S.C., Schimel, D.S., 2001. Geomorphic evolution of soil texture and organic matter in eroding landscapes. Global Biogeochemical Cycles 15 (2), 365–381.
Rossiter, D.G., Hengl, T., 2002. Technical note. Creating geometrically-correct photo-interpretations, photomosaics, and base maps for a project GIS. ITC, Department of Earth System Analysis, Enschede, NL, 32 pp.
Roth, R., Kühn, J., Werner, A., 2001. Decision support system to derive site specific sowing rates for managing winter wheat within precision agriculture. In: Grenier, G., Blackmore, S. (Eds.), ECPA 2001. Third European Conference on Precision Agriculture, vol. 2. Ecole Nationale Supérieure Agronomique, AGRO Montpellier, Montpellier (France), pp. 701–706.
Rouse, W.R., Mills, P.F., Stewart, R.B., 1977. Evaporation in high latitudes. Water Resources Research 13 (6), 909–914.
Rouse, W.R., Stewart, R.B., 1972. A simple model for determining the evaporation from high latitude upland sites. Journal of Applied Meteorology 11, 1063–1070.
Rowbotham, D.N., Dudycha, D., 1998. GIS modelling of slope stability in Phewa Tal watershed, Nepal. Geomorphology 26, 151–170.
Rowe, J.S., 1996. Land classification and ecosystem classification. In: Sims, R.A., Corns, I.G.W., Klinka, K. (Eds.), Global to Local: Ecological Land Classification. Kluwer Academic Publishers, Dordrecht, The Netherlands, pp. 11–20.
Rowe, J.S., Barnes, B.V., 1994. Geo-ecosystems and bio-ecosystems. Bulletin of the Ecological Society of America 75 (1), 40–41.

Rowley, R., Kostelnick, J., Braaten, D., Li, X., Meisel, J., 2007. Risk of rising sea level to population and land area. EOS, Transactions, American Geophysical Union 88 (9), 105–107.
Ruhe, R.V., Walker, P.H., 1968. Hillslope models and soil formation II: open systems. In: Proceedings of 9th Congress of the International Soil Science Society, vol. 4. International Soil Science Society, Adelaide, Australia, pp. 551–560.
Ruhe, R.V., 1960. Elements of the soil landscape. In: Transactions of the 9th Congress of the International Society of Soil Science, vol. 4. International Soil Science Society, Madison, WI, pp. 165–170.
Russell, G.D., Hawkins, C.P., O'Neill, M.P., 1997. The role of GIS in selecting sites for riparian restoration based on watershed hydrology and land use. Restoration Ecology 54, 56–58.
Russell, W.E., Jordan, J.K., 1992. Ecological classification systems for classifying land capability in midwestern and northeastern US National Forests. In: Proceedings: Ecological Land Classification — Applications to Identify the Productive Potential of Southern Forests. January 7–9, 1992, Charlotte, NC. US Department of Agriculture, Asheville, NC, pp. 18–24.
Ryan, P.J., McKenzie, N.J., O'Connell, D., Loughhead, A.N., Leppert, P.M., Jacquier, D., Ashton, L., 2000. Integrating forest soils information across scales: spatial prediction of soil properties under Australian forests. Forest Ecology and Management 138, 139–157.
Rykiel, E.J., 1996. Testing ecological models: the meaning of validation. Ecological Modelling 90, 229–244.
Sætersdal, M., Gjerde, I., Blom, H.H., Ihlen, P.G., Myrseth, E.W., Pommeresche, R., Skartveit, J., Solhøy, T., Aas, O., 2003. Vascular plants as a surrogate species group in complementary site selection for bryophytes, macrolichens, spiders, carabids, staphylinids, snails, and wood living polypore fungi in a northern forest. Biological Conservation 115, 21–31.
Sakude, M.T., Schiavone, G.A., Morelos-Borja, H., Martin, G., Cortes, A., 1998. Recent advances on terrain database correlation testing. Proceedings of SPIE 3369, 364–376.
Salzmann, N., Kääb, A., Huggel, C., Allgöwer, B., Haeberli, W., 2004. Assessment of the hazard potential of ice avalanches using remote sensing and GIS-modelling. Norsk Geografisk Tidskrift 58, 74–84.
Sampl, P., Zwinger, T., 2004. Avalanche simulation with SAMOS. Annals of Glaciology 38 (1), 393–398.
Saunders, W.K., 1999. Preparation of DEMs for use in environmental modeling analysis. In: Proceedings of the Nineteenth Annual ESRI International User Conference. ESRI, San Diego, pp. 1–8.
Saunders, W.K., Maidment, D.R., 1996. A GIS assessment of nonpoint source pollution in the San Antonio-Nueces Coastal Basin. Online Report 96-1. Center for Research in Water Resources, University of Texas, Austin, TX, 152 pp.
Saupe, D., 1988. Algorithms for random fractals. In: Peitgen, H., Saupe, D. (Eds.), The Science of Fractal Images. Springer-Verlag, pp. 71–133.
Saye, S.E., van der Wal, D., Pye, K., Blott, S.J., 2005. Beach–dune morphological relationships and erosion/accretion: an investigation at five sites in England and Wales using LiDAR data. Geomorphology 72, 128–155.
Schaab, G., 2000. Modellierung und Visualisierung der räumlichen und zeitlichen Variabilität der Einstrahlungsstärke mittels eines Geo-Informationssystems. Ph.D. Thesis. Technische Universität Dresden.
Schaber, G.G., Pike, R.J., Berlin, G.L., 1979. Terrain-analysis procedures for modeling radar back-scatter. Open-File Report 79-1088. U.S. Geological Survey, 61 pp.
Scharmer, K., Greif, J., 2000. The European Solar Radiation Atlas: Volume 2, Database and Exploitation Software. Les Presses de l'Ècole des Mines, Paris, 110 pp.
Scheidegger, A.E., 1991. Theoretical Geomorphology, 3rd edition. Springer-Verlag, Berlin, 434 pp.
Schilling, S.P., 1998. LAHARZ: GIS programs for automated mapping of lahar-inundation hazard zones. Open-File Report 98-638. U.S. Geological Survey, 80 pp.
Schlicting, H., 1960. Boundary Layer Theory, 4th edition. McGraw–Hill, New York, 647 pp.
Schmidt, J., Andrew, R., 2005. Multi-scale landform characterization. Area 37, 341–350.
Schmidt, J., Hewitt, A., 2004. Fuzzy land element classification from DTMs based on geometry and terrain position. Geoderma 121 (3–4), 243–256.
Schmidt, J., Evans, I.S., Brinkmann, J., 2003. Comparison of polynomial models for land surface curvature calculation. International Journal of Geographical Information Science 17, 797–814.

Schmidt, J., Dikau, R., 1999. Extracting geomorphometric attributes and objects from digital elevation models — semantics, methods, future needs. In: Dikau, R., Saurer, H. (Eds.), GIS for Earth Surface Systems — Analysis and Modelling of the Natural Environment. Schweizbart'sche Verlagsbuchhandlung, pp. 153–173.

Schmidt, J., Hennrich, K., Dikau, R., 1998. Scales and similarities in runoff processes with respect to geomorphometry. In: Geocomputation 1998: Proceedings of the 3rd International Conference on GeoComputation. University of Bristol, United Kingdom, 17–19 September 1998. University of Bristol, Bristol, UK, 20 pp.

Schmidtlein, S., 2005. Imaging spectroscopy as a tool for mapping Ellenberg indicator values. Journal of Applied Ecology 42, 966–974.

Schneider, B., 2001. Phenomenon-based specification of the digital representation of terrain surfaces. Transactions in GIS 5 (1), 39–52.

Schneider, B., 1998. Geomorphologically plausible reconstruction of the digital representation of terrain surfaces from contour data. Ph.D. Thesis. Universtiy of Zürich (in German).

Schoorl, J.M., Veldkamp, A., 2005. Multiscale soil–landscape process modelling. In: Grunwald, S. (Ed.), Environmental Soil–Landscape Modeling: Geographic Information Technologies and Pedometrics. CRC Press, Boca Raton, FL, pp. 417–435.

Schoorl, J.M., Veldkamp, T., Bouma, J., 2002. Modeling soil and water redistribution in a dynamic landscape context. Soil Science Society of America Journal 66, 1610–1619.

Schoorl, J.M., Sonneveld, M.P.W., Veldkamp, A., 2000. Three dimensional landscape process modelling: the effect of DEM resolution. Earth Surface Processes and Landforms 25 (9), 1025–1034.

Schumm, S.A., 1956. Evolution of drainage systems and slopes in badlands at Perth Amboy. Geological Society of America Bulletin 67, 597–646.

Schweinfurth, H., 1956. Über klimatische Trockentäler im Himalaya. Erdkunde 10, 297–302.

Scott, P.J., 2004. An application of surface networks in surface texture. In: Rana, S. (Ed.), Topological Data Structures for Surfaces — An Introduction for Geographical Information Science. Wiley, Chichester, UK, pp. 157–166.

Scull, P., Franklin, J., Chadwick, O.A., 2005. The application of classification tree analysis to soil type prediction in a desert landscape. Ecological Modelling 181 (1), 1–15.

Scull, P., Franklin, J., Chadwick, O.A., McArthur, D., 2003. Predictive soil mapping: a review. Progress in Physical Geography 27 (2), 171–197.

Segurado, P., Araújo, M.B., 2005. An evaluation of methods for modeling species distributions. Journal of Biogeography 31, 1555–1568.

Selige, T., Böhner, J., Ringeler, A., 2006. Processing of SRTM X-SAR data to correct interferometric elevation models for land surface applications. In: Böhner, J., McCloy, K.R., Strobl, J. (Eds.), SAGA — Analyses and Modelling Applications, vol. 115. Verlag Erich Goltze GmbH, pp. 97–104.

Sellers, P.J., Dickinson, R.E., Randall, D.A., Betts, A.K., Hall, F., Berry, J., Collatz, G., Denning, A.S., Mooney, H.A., Nobre, C.A., Sato, N., Field, C.B., Henderson-Sellers, A., 1997. Modeling the exchanges of energy, water, and carbon between continents and the atmosphere. Science 275, 502–509.

Sequeira, V., Ng, K., Wolfart, E., Goncalves, J.G.M., Hogg, D., 1999. Automated reconstruction of 3D models from real environments. ISPRS Journal of Photogrammetry and Remote Sensing 54 (1), 1–22.

Seul, M., O'Gorman, L., Sammon, M.J., 2000. Practical Algorithms for Image Analysis: Descriptions, Examples, and Code. Cambridge University Press, Cambridge, 295 pp.

Shary, P.A., Sharaya, L.S., Mitusov, A.V., 2005. The problem of scale-specific and scale-free approaches in geomorphometry. Geografia Fisica e Dinamica Quaternaria 28 (1), 81–101.

Shary, P.A., Sharaya, L.S., Mitusov, A.V., 2002. Fundamental quantitative methods of land surface analysis. Geoderma 107 (1–2), 1–32.

Shary, P.A., 1995. Land surface in gravity points classification by a complete system of curvatures. Mathematical Geology 27 (3), 373–390.

Shary, P.A., 1991. The second derivative topographic method. In: Stepanov, I.N. (Ed.), The Geometry of Earth Surface Structures. Pushchino Research Centre Press, Pushchino, pp. 28–58 (in Russian).

Shary, P.A., Stepanov, I.N., 1991. On the second derivative method in geology. Doklady AN SSSR 319 (2), 456–460 (in Russian).

Shepherd, J.D., Dymond, J.R., 2003. Correcting satellite imagery for the variance of reflectance and illumination with topography. International Journal of Remote Sensing 24 (17), 3503–3514.

Sherman, L.K., 1932. Streamflow from rainfall by the unit-graph method. Engineering News Record 108, 501–505.
Shi, A.-X., Zhu, A.-X., Burt, J.E., Qi, F., Simonson, D., 2004. A case-based reasoning approach to fuzzy soil mapping. Soil Science Society of America Journal 68 (3), 885–894.
Shi, X., Zhu, A.-X., Wang, R.-X., 2005. Deriving fuzzy representations of some special terrain features based on their typical locations. In: Cobb, M., Petry, F., Robinson, V. (Eds.), Fuzzy Modeling with Spatial Information for Geographic Problems. Springer-Verlag, New York, pp. 233–252.
Shreve, R.L., 1974. Variation of main stream length with basin area in river networks. Water Resources Research 10 (6), 1167–1177.
Sims, R.A., Corns, I.G.W., Klinka, K., 1996. Global to Local: Ecological Land Classification. Kluwer Academic Publishers, Dordrecht, The Netherlands, 610 pp.
Sims, R.A. (Ed.), 1992. Forest Site Classification in Canada: A Current Perspective. Science and Sustainable Development Directorate, Forestry Canada, Ottawa, Ontario, 120 pp.
Sinowski, W., Auerswald, K., 1999. Using relief parameters in a discriminant analysis to stratify geological areas with different spatial variability of soil properties. Geoderma 89, 113–128.
Skidmore, A.K., Watford, F., Luckananurug, P., Ryan, P.J., 1996. An operational GIS expert system for mapping forest soils. Photogrammetric Engineering and Remote Sensing 62 (5), 501–511.
Skidmore, A.K., Ryan, P.J., Dawes, W., Short, D., O'Loughlin, E., 1991. Use of an expert system to map forest soils from a geographical information system. International Journal of Geographical Information Systems 5, 431–445.
Skidmore, A.K., 1990. Terrain position as mapped from a gridded digital elevation model. International Journal of Geographical Information Systems 4, 33–49.
Skidmore, A.K., 1989a. A comparison of techniques for calculating gradient and aspect from a gridded digital elevation model. International Journal of Geographical Information Systems 3, 323–334.
Skidmore, A.K., 1989b. An expert system classifies eucalypt forest types using Landsat Thematic Mapper and a digital terrain model. Photogrammetric Engineering and Remote Sensing 55, 1449–1464.
Slater, J.A., Garvey, G., Johnston, C., Haase, J., Heady, B., Kroenung, G., Little, J., 2006. The SRTM data finishing process and products. Photogrammetric Engineering and Remote Sensing 72 (3), 237–248.
Smirnova, T.G., Brown, J.M., Benjamin, S.G., 1997. Performance of different soil model configurations in simulating ground surface temperature and surface fluxes. Monthly Weather Review 125, 1870–1884.
Smith, B., Mark, D.M., 2006. Do mountains exist? Towards an ontology of landforms. Environment and Planning B: Planning and Design 30 (3), 411–427.
Smith, D., et al., 1999. The global topography of Mars and implications for surface evolution. Science 284 (5419), 1495–1503.
Smith, D.E., Zuber, M.T., Neumann, G.A., Lemoine, F.G., 1997. Topography of the Moon from the Clementine LiDAR. Journal of Geophysical Research 102 (E1), 1591–1611.
Smith, G.H., 1953. The relative relief of Ohio. The Geographical Review 25 (2), 272–284.
Smith, J.K., Chacón-Moreno, E.J., Jongman, R.H.G., Wenting, P., Loedeman, J.H., 2006. Effect of dyke construction on water dynamics in the flooding savannahs of Venezuela. Earth Surface Processes and Landforms 31, 81–96.
Smith, M.P., Zhu, A.X., Burt, J.E., Stiles, C., 2006. The effects of DEM resolution and neighborhood size on digital soil survey. Geoderma 137 (1–2), 58–68.
Smith, R.B., 1987. Aerial observations of the Yugoslavian bora. Journal of the Atmospheric Sciences 44, 269–297.
Smith, R.B., 1979. The influence of mountains on the atmosphere. Advances in Geophysics 21, 87–230.
Smith, R.E., 2002. Infiltration Theory for Hydrologic Applications, 1st edition. Water Resources Monograph, vol. 15. American Geophysical Union, Washington, DC, 212 pp.
Smith, R.E., 1990. Analysis of infiltration through a two-layer soil profile. Soil Science Society of America Journal 54 (5), 1219–1227.
Smith, S.E., 2005. Topographic mapping. In: Grunwald, S. (Ed.), Environmental Soil–Landscape Modeling: Geographic Information Technologies and Pedometrics, vol. 1. CRC Press, New York, pp. 155–182.
Smith, S.L., Holland, D.A., Longley, P., 2005. Quantifying interpolation errors in airborne laser scanning models. Geographical Analysis 37 (2), 200–224.

Smith, W.H.F., Sandwell, D.T., 1997. Global sea floor topography from satellite altimetry and ship depth soundings. Science Magazine 277 (5334), 1956–1962.
Sodano, E.M., 1965. General non-iterative solution of the inverse and direct geodetic problems. Journal of Geodesy 39 (1), 69–89.
Soil Survey Division Staff, 1993. Soil Survey Manual. United States Department of Agriculture, Washington, DC, 411 pp.
Soille, P., 2006. Morphological image compositing. IEEE Transactions on Pattern Analysis and Machine Intelligence 28 (5), 673–683.
Soille, P., 2004a. Morphological carving. Pattern Recognition Letters 25 (5), 543–550.
Soille, P., 2004b. Optimal removal of spurious pits in grid digital elevation models. Water Resources Research 40 (12), W12509.
Soille, P., Vogt, J., Colombo, R., 2003. Carving and adaptive drainage enforcement of grid digital elevation models. Water Resources Research 39 (12), 1366–1378.
Soille, P., Gratin, C., 1994. An efficient algorithm for drainage networks extraction on DEMs. Journal of Visual Communication and Image Representation 5 (2), 181–189, doi:10.1006/jvci.1994.1017.
Sokal, R.R., Sneath, P.H.A., 1963. Principles of Numerical Taxonomy. W.H. Freeman and Company, San Francisco, 359 pp.
Sood, C., Bhagat, R.M., 2005. Interfacing geographical information systems and pesticide models. Current Science 89, 1362–1370.
Speight, J.G., 1990. Landform. In: McDonald, R.C., Isbell, R.F., Speight, J.G., Walker, J., Hop, M.S. (Eds.), Australian Soil and Land Survey Field Handbook. Inkata Press, Melbourne, pp. 9–57.
Speight, J.G., 1974. A parametric approach to landform regions. In: Progress in Geomorphology, Special Publication, vol. 7. Institute of British Geographers, Alden & Mowbray Ltd at the Alden Press, Oxford, pp. 213–230.
Speight, J.G., 1971. Log-normality of slope distributions. Zeitschrift für Geomorphologie 15, 290–311.
Speight, J.G., 1968. Parametric description of land form. In: Stewart, G.A. (Ed.), Land Evaluation: Papers of a CSIRO Symposium. Macmillan, Melbourne, pp. 239–250.
Sperduto, M.B., Congalton, R.G., 1996. Predicting rare orchid (small whorled Pogonia) habitat using GIS. Photogrammetric Engineering and Remote Sensing 62 (11), 1269–1279.
Spronken-Smith, R.A., Sturman, A.P., Owens, I.F., 2003. Spatial variability of surface radiation fluxes in mountainous terrain. Journal of Applied Meteorology 42, 113–128.
Srinivasan, R., Engel, B.A., 1991. Effect of slope prediction methods on slope and erosion estimates. Journal of Applied Engineering in Agriculture 7 (6).
Staley, D., Wasklewicz, T., Blaszczynski, J., 2006. Surficial patterns of debris flow deposition on alluvial fans in Death Valley, CA using airborne laser swath mapping data. Geomorphology 74 (1–4), 152–163.
Steen, O., Coupé, R., 1997. A Field Guide to Forest Site Identification and Interpretation for the Cariboo Forest Region. Land Management Handbook, vol. 39. British Columbia Ministry of Forestry, Vancouver, Canada, 41 pp.
Stein, A., Riley, J., Halberg, N., 2001. Issues of scale for environmental indicators. Agriculture, Ecosystems & Environment 87 (2), 215–232.
Stevens, D., Dragicevic, S., Rothley, K., 2006. City: a GIS-CA modelling tool for urban planning and decision making. Environmental Modelling & Software 22 (6), 761–773.
Stevenson, F.C., 1996. A landscape-scale and small-plot assessment of the nitrogen and non-nitrogen rotation benefits of pea. Ph.D. Thesis. University of Saskatchewan, Department of Soil Science.
Stockwell, W.R., Middleton, P., Chang, J.S., Tang, X., 1990. The second generation regional acid deposition model chemical mechanism for regional air quality modeling. Journal of Geophysical Research 95, 16343–16367.
Stohlgren, T.J., Chase, T.N., Pielke Sr., R.A., Kittel, T.G.F., Baron, J.S., 1998. Evidence that local land use practices influence regional climate, vegetation, and stream flow patterns in adjacent natural areas. Global Change Biology 4, 495–504.
Strahler, A.N., 1957. Quantitative analysis of watershed geomorphology. American Geophysical Union Transactions 38 (6), 912–920.
Strahler, A.N., 1952. Hypsometric (area-altitude) analysis of erosional topography. Geological Society of America Bulletin 63, 1117–1142.

Strahler, A.N., 1950. Equilibrium theory of erosional slopes approached by frequency distribution analysis. American Journal of Science 248, 673–696 and 800–814.
Streit, U., 1981. Kriging — eine geostatistische Methode zur räumlichen Interpolation hydrologischer Informationen. Wasserwirtschaft 71, 219–223.
Stull, R.B., 1988. An Introduction to Boundary Layer Meteorology. Kluwer Academic Publishers, 666 pp.
Sulebak, J.R., Hjelle, Ø., 2003. Multiresolution spline models and their applications in geomorphology. In: Evans, I.S., Dikau, R., Tokunaga, R., Ohmori, H., Hirano, M. (Eds.), Concepts and Modelling in Geomorphology — International Perspectives. TERRAPUB, Tokyo, pp. 221–237.
Summerell, G.K., Vaze, J., Tuteja, N.K., Grayson, R.B., Beale, G., Dowling, T.I., 2005. Delineating the major landforms of catchments using an objective hydrological terrain analysis method. Water Resources Research 41, 1–12 (W12416).
Sumner, D.M., Jacobs, J.M., 2005. Utility of Penman–Monteith, Priestley–Taylor, reference evapotranspiration, and pan evaporation methods to estimate pasture evapotranspiration. Journal of Hydrology 308, 81–104.
Šúri, M., Hofierka, J., 2004. A new GIS-based solar radiation model and its application to photovoltaic assessments. Transactions in GIS 8 (2), 175–190.
Suryana, N., de Hoop, S., 1994. Hierarchical structuring of terrain mapping units. In: Proceedings of the Fifth European Conference and Exhibition on Geographic Information Systems, EGIS '94, vol. 1. EGIS Foundation, Utrecht, The Netherlands, pp. 869–877.
Svoray, T., Carmel, Y., 2005. Empirical method for topographic correction in aerial photographs. IEEE Geoscience and Remote Sensing Letters 2 (2), 211–214.
Swanson, D.K., 1990a. Landscape classes: higher-level map units for soil survey. Soil Survey Horizons 31, 52–54.
Swanson, D.K., 1990b. Soil landform units for soil survey. Soil Survey Horizons 31, 17–21.
Tachikawa, Y., Shiiba, M., Takasao, T., 1994. Development of a basin geomorphic information system using a TIN–DEM data structure. Water Resources Bulletin 30 (1), 9–17.
Tang, G., Shi, W., Zhao, M., 2002. Evaluation on the accuracy of hydrologic data derived from DEMs of different spatial resolution. In: Hunter, G.J., Lowell, K. (Eds.), Proceedings of the 5th International Symposium on Spatial Accuracy Assessment in Natural Resources and Environmental Sciences (Accuracy 2002). RMIT University, Melbourne, Australia, pp. 204–213.
Tappeiner, U., Tappeiner, G., Aschenwald, J., Tasser, E., Ostendorf, B., 2001. GIS-based modelling of spatial pattern of snow cover duration in an alpine area. Ecological Modelling 138, 265–275.
Tarboton, D.G., 1997. A new method for the determination of flow directions and upslope areas in grid digital elevation models. Water Resources Research 33 (2), 309–319.
Tarboton, D.G., Bras, R.L., Rodriguez-Iturbe, I., 1992. A physical basis for drainage density. Geomorphology 5 (1–2), 59–76.
Tarboton, D.G., Bras, R.L., Rodriguez-Iturbe, I., 1991. On the extraction of channel networks from digital elevation data. Hydrological Processes 5 (1), 81–100.
Taylor, P.A., 1987. Comments and further analysis on effective roughness lengths for use in numerical three-dimensional models. Boundary-Layer Meteorology 39, 403–418.
Temme, A.J.A.M., Schoorl, J.M., Veldkamp, A., 2006. An algorithm for dealing with depressions in dynamic landscape evolution models. Computers & Geosciences 32, 452–461.
Thelin, G.P., Pike, R.J., 1991. Landforms of the conterminous United States — a digital shaded-relief portrayal. Miscellaneous Investigations Map I-2206. U.S. Geological Survey, 16 pp.
Thomas, A.L., King, D., Dambrine, E., Couturier, A., Roque, J., 1999. Predicting soil classes with parameters derived from relief and geologic materials in a sandstone region of the Vosges mountains (Northeastern France). Geoderma 90 (3–4), 291–305.
Thomas, T.R., 1999. Rough Surfaces, 2nd edition. Imperial College Press/World Scientific Publishers, London, 294 pp.
Thomas Jr., W.L. (Ed.), 1956. Man's Role in Changing the Face of the Earth. University of Chicago Press, Chicago, IL, 1193 pp.
Thompson, J.A., Bell, J.C., Butler, C.A., 2001. Digital elevation model resolution: effects on terrain attribute calculation and quantitative soil–landscape modeling. Geoderma 100, 67–89.
Thompson, J.A., Bell, J.C., Butler, C.A., 1997. Quantitative soil–landscape modeling for estimating the areal extent of hydromorphic soils. Soil Science Society of America Journal 61 (3), 971–980.

Thomson, R.S., Shafer, S.L., Anderson, K.H., Stickland, L.E., Pelltier, R.T., Bartlein, P.J., Kerwin, M.W., 2004. Topographic, bioclimatic, and vegetation characteristics of three ecoregion classification systems in North America: comparisons along continent-wide transects. Environmental Management 34, 125–148.

Thorn, C.E., 1988. An Introduction to Theoretical Geomorphology. Unwin Hyman, Boston, 247 pp.

Thornbury, W.D., 1954. Principles of Geomorphology. Wiley, New York, 618 pp.

Thrun, S., et al., 2006a. Stanley: the robot that won the DARPA Grand Challenge. Journal of Field Robotics 23, 661–692.

Thrun, S., Montemerlo, M., Aron, A., 2006b. Probabilistic terrain analysis for high-speed desert driving. In: Sukhatme, G., Schaal, S., Burgard, W., Fox, D. (Eds.), Proceedings of the Robotics Science and Systems Conference. University of Pennsylvania, Philadelphia, PA, pp. 21–28.

Tim, U.S., Jolly, R., 1994. Evaluation of agricultural non-point source pollution using GIS and hydrologic/water quality modeling. Journal Environmental Quality 23, 25–35.

Tobler, W.R., 2000. The development of analytical cartography — a personal note. Cartography and Geographic Information Science 27 (3), 189–194.

Tobler, W.R., 1976. Analytical cartography. The American Cartographer 3 (1), 21–31.

Tobler, W.R., 1970. A computer movie simulating urban growth in the Detroit region. Economic Geography (Supplement) 46 (2), 234–240.

Tobler, W.R., 1969. An analysis of a digitalized surface. In: Davis, C.M. (Ed.), A Study of the Land Type. Report on Contract DA-31-124-ARO-D-456. University of Michigan, Department of Geography, MI, pp. 59–76, 86.

Tobler, W.R., 1968. A digital terrain library. Contract DA-31-124-ARO-D-456, Technical Report 08055-1-T. U.S. Army Research Office, University of Michigan Department of Geography, Ann Arbor, 23 pp.

Tomer, M.D., Anderson, J.L., 1995. Variation in soil water storage across a sand plain hillslope. Soil Science of America Proceedings 54 (9), 1091–1100.

Tomlin, C.D., 1990. Geographic Information Systems and Cartographic Modelling. Prentice–Hall, Englewood Cliffs, NJ, 572 pp.

Townshend, J.R.G. (Ed.), 1981. Terrain Analysis and Remote Sensing. Allen & Unwin, London, 232 pp.

Travis, M.R., Elsner, G.H., Iverson, W.D., Johnson, C.G., 1975. VIEWIT: Computation of Seen Areas, Slope, and Aspect for Land-Use Planning. USDA, Berkeley, CA, 67 pp.

Tribe, A., 1992a. Automated recognition of valley lines and drainage networks from grid digital elevation models — a review and a new method. Journal of Hydrology 139 (1–4), 263–293.

Tribe, A., 1992b. Problems in automated recognition of valley features from digital elevation models and a new method toward their resolution. Earth Surface Processes and Landforms 17, 437–454.

Tribe, A., 1991. Automated recognition of valley heads from digital elevation models. Earth Surface Processes and Landforms 16, 33–49.

Tricart, J., 1965. Principes et Méthodes de la Géomorphologie. Masson, Paris, 496 pp.

Tricart, J., Muslin, J., 1951. L'étude statistique des versants. Revue de Géomorphologie Dynamique 2, 173–182.

Tricart, J., 1947. Sur quelques indices géomorphométriques. Comptes Rendus Hebdomadaires des Séances de l'Académie des Sciences 225, 747–749.

Tripathi, S.C., Sayre, K.D., Kaul, J.N., Narang, R.S., 2003. Growth and morphology of spring wheat (Triticum aestivum L.) culms and their association with lodging: effects of genotypes, N levels and ethephon. Field Crops Research 84 (3), 271–290.

Troeh, F.R., 1965. Landform equations fitted to contour maps. American Journal of Science 263, 616–627.

Troeh, F.R., 1964. Landform parameters correlated to soil drainage. Soil Science of America Proceedings 59 (4), 808–812.

Troll, K., 1952. Die Lokalwinde der Tropengebirge und ihr Einfluss auf Niederschlag und Vegetation. Bonner Geographische Abhandlungen 9, 124–182.

Troutman, B.M., Karlinger, M.R., 1984. On the expected width function for topologically random channel networks. Journal of Applied Probability 21, 836–849.

Tucker, G.E., Lancaster, S.T., Gasparini, N.M., Bras, R.L., Rybarczyk, S.M., 2001. An object-oriented framework for distributed hydrologic and geomorphic modeling using triangulated irregular networks. Computers & Geosciences 27 (8), 959–973.

Tucker, L., MacCallum, R., 1997. Exploratory Factor Analysis. Ohio State University, Columbus, 500 pp.
Tufte, E.R., 1990. Envisioning Information. Graphics Press, Cheshire, CT, 126 pp.
Turner, A.K., Schuster, R.L. (Eds.), 1996. Landslides: Investigation and Mitigation. Transportation Research Board, Special Report 247. National Research Council, Washington, DC, 685 pp.
Tveito, O.E., Forland, E.J., Alexandersson, H., Drebs, A., Jónsson, T., Tuomenvirta, H., Vaarby Laursen, E., 2001. Nordic climate maps. DNMI Report 06/01. Klima, 29 pp.
Twery, M.J., Elmes, G.A., Yuill, C.B., 1991. Scientific exploration with intelligent GIS: predicting species composition from topography. Artificial Intelligence Applications in Natural Resource Management 5 (2), 45–53.
Tylor, A., 1875. Action of denuding agencies. Geological Magazine 22, 433–473.
U.S. Army Map Service, 1963. Computer production of terrain models. Communications of the Association for Computing Machinery 6 (4), 190–191.
Unit Geo Software Development, 2001. ILWIS 3.0 Academic User's Guide. International Institute for Geo-Information Science and Earth Observation (ITC), Enschede, 530 pp.
Vachaud, G., Chen, T., 2002. Sensitivity of computed values of water balance and nitrate leaching to within soil class variability of transport parameters. Journal of Hydrology 264, 87–100.
Van Alphen, B.J., Stoorvogel, J.J., 2000. A functional approach to soil characterization in support of precision agriculture. Soil Science Society of America Journal 64 (5), 1706–1713.
Van Den Eeckhaut, M., Vanwalleghem, T., Poesen, J., Govers, G., Verstraeten, G., Vandekerckhove, L., 2006. Prediction of landslide susceptibility using rare events logistic regression: a case study in the Flemish Ardennes (Belgium). Geomorphology 76 (3–4), 392–410.
Van der Beek, P.A., Braun, J., 1998. Numerical modelling of landscape evolution on geological time-scales — a parameter analysis and comparison with the south-eastern highlands of Australia. Basin Research 10 (1), 49–68.
van Engelen, V.W.P., Ting-tiang, W. (Eds.), 1995. Global and national soils and terrain digital databases (SOTER), procedures manual. World Soil Resources Report 74 Rev 1. Land and Water Division, Food and Agriculture Organization of the United Nations, Wageningen, 125 pp.
van Genuchten, M.T., 1980. A closed-form equation for predicting the hydraulic conductivity of unsaturated soils. Soil Science Society of America Journal 44 (5), 892–898.
van Meirvenne, M., van Cleemput, I., 2005. Pedometrical techniques for soil texture mapping at different scales. In: Grunwald, S. (Ed.), Environmental Soil–Landscape Modeling: Geographic Information Technologies and Pedometrics. CRC Press, Boca Raton, FL, pp. 323–341.
Van Niel, K., Laffan, S.W., Lees, B.G., 2004. Effect of error in the DEM on environment variables for predictive vegetation modelling. Journal of Vegetation Science 15, 747–756.
Van Niel, K., Laffan, S.W., 2003. Gambling with randomness: the use of pseudo-random number generators in GIS. International Journal of Geographical Information Science 17, 49–68.
van Zuidam, R.A., 1986. Aerial Photo-Interpretation in Terrain Analysis and Geomorphologic Mapping. Smits, Hague, 442 pp.
Vardavas, I.M., 1987. A simple model for rapidly computing terrestrial flux, solar flux and global mean surface temperature. Ecological Modelling 35, 189–221.
Veatch, J.O., 1935. Graphic and quantitative comparisons of land types. Journal of the American Society of Agronomy 27 (7), 505–510.
Ventura, S.J., Irvin, B.J., 2000. Automated Landform Classification Methods for Soil–Landscape Studies. In: Wilson, J.P., Gallant, J.C. (Eds.), Terrain Analysis: Principles and Applications. Wiley, pp. 267–294.
Vincenty, T., 1975. Direct and inverse solutions of geodesics on the ellipsoid with application of nested equations. Survey Review 23 (176), 88–93.
Vivoni, E.R., Ivanov, V.Y., Bras, R.L., Entekhabi, D., 2005. On the effects of triangulated terrain resolution on distributed hydrologic model response. Hydrological Processes 19 (11), 2101–2122.
Vivoni, E.R., Ivanov, V.Y., Bras, R.L., Entekhabi, D., 2004. Generation of triangulated irregular networks based on hydrological similarity. Journal of Hydrologic Engineering 9 (4), 288–302.
Vogt, J.V., Colombo, R., Paracchini, M.L., de Jager, A., Soille, P., 2003. CCM River and Catchment Database for Europe, Version 1.0, EUR 20756 EN. EC-JRC, Ispra, Italy, 32 pp.
Volkov, N.M., 1950. Printsipy i Metody Kartometrii. Izdateltsvo Akademiya Nauk SSSR, Moskva, 327 pp.

von Sonklar, C.E.I., 1873. Allgemeine Orographie: Die Lehre von den Relief-Formen der Erdoberfläche (General Orography: Lessons on the Relief Forms of the Earth's Surface). Wilhelm Braümuller, Wien, 254 pp.
Von Storch, H., 1995. Inconsistencies at the interface of climate impact studies and global climate research. Meteorologische Zeitschrift 4, 72–80.
Voronoi, G., 1907. Nouvelles applications des paramètres continus à la théorie des formes quadratiques. Journal für die Reine und Angewandte Mathematik 133, 97–178.
Voss, R.F., 1988. Fractals in nature: from characterisation to simulation. In: Peitgen, H., Saupe, D. (Eds.), The Science of Fractal Images. Springer-Verlag, pp. 21–70.
Vosshenrich, H.-H., Maquering, J., Gattermann, B., Täger-Fany, W., Sommer, C., 2001. Managment system for DGPS-supported primary soil tillage. In: Grenier, G., Blackmore, S. (Eds.), ECPA 2001 Third European Conference on Precision Agriculture. AGRO Montpellier, Montpellier, pp. 731–736.
Wahba, G., 1990. Spline Models for Observational Data. CBMS–NSF Regional Conference Series in Applied Mathematics. Society for Industrial and Applied Mathematics, Philadelphia, 180 pp.
Wahr, J.M., Oort, A.H., 1984. Friction- and mountain-torque estimates from global atmospheric data. Journal of the Atmospheric Sciences 41, 190–204.
Walcott, R., Summerfield, M., 2008. Scale dependence of hypsometric integrals: an analysis of southeast African basins. Geomorphology 96 (1–2), 174–186.
Walls, J.H., Houbolt, J.C., Press, H., 1954. Some measurements and power spectra of runway roughness. Technical Note 3305. NACA, Washington, DC.
Walsh, S.J., Butler, D.R., Allen, T.R., Malanson, G.P., 1994. Influence of snow patterns and snow avalanches on the alpine treeline ecotone. Journal of Vegetation Science 5 (5), 657–672.
Walter, C., McBratney, A.B., Donuaoui, A., Minasny, B., 2001. Spatial prediction of topsoil salinity in the Chelif Valley, Algeria, using local ordinary kriging with local variograms versus whole-area variogram. Australian Journal of Soil Research 39, 259–272.
Ware, C., 2004. Information Visualization: Perception for Design, 2nd edition. Academic Press, London, 486 pp.
Waren, S.D., Hohmann, M.G., Auerswald, K., Mitášová, H., 2004. An evaluation of methods to determine slope using digital elevation data. Catena 58, 215–233.
Warntz, W., 1975. Stream ordering and contour mapping. Journal of Hydrology 25, 209–227.
Warntz, W., 1966. The topology of a socio-economic terrain and spatial flows. Papers and Proceedings of the Regional Science Association 17 (1), 47–61.
Way, D.S., 1973. Terrain Analysis: A Guide to Site Selection Using Aerial Photographic Interpretation. Dowden, Hutchinson & Ross, Stroudsburg, PA, 392 pp.
Weaver, G.D., 1965. What is a landform? The Professional Geographer 17 (1), 11–13.
Webster, R., Oliver, M.A., 2001. Geostatistics for Environmental Scientists. Statistics in Practice. Wiley, Chichester, 265 pp.
Weibel, R., Brandli, M., 1995. Adaptive methods for the refinement of digital terrain models for geomorphometric applications. Zeitschrift für Geomorphologie, Supplementband 101, 13–30.
Weibel, R., Heller, M., 1991. Digital terrain modelling. In: Maguire, D.J., Goodchild, M.F., Rhind, D.W. (Eds.), Geographical Information Systems, vol. 1. Longman, London, pp. 269–297.
Weibel, R., Heller, M., 1990. A framework for digital terrain modeling. In: Proceedings of the Fourth International Symposium on Spatial Data Handling. University of Zürich, Zürich, Switzerland, pp. 219–229.
Weibel, R., DeLotto, J.S., 1988. Automated terrain classification for GIS modeling. In: Proceedings of GIS/LIS. San Antonio, NM, pp. 618–627.
Weischet, W., 1995. Einführung in die Allgemeine Klimatologie. Gebrüder Bornträger Verlagsbuchhandlung, Stuttgart.
Welty, J.R., Wicks, C.E., Wilson, R.E., 1984. Fundamentals of Momentum, Heat and Mass Transfer, 3rd edition. Wiley, New York, 212 pp.
Wendroth, O., Reuter, H.I., Kersebaum, K.-C., 2003. Predicting yield of barley across a landscape: a state-space modelling approach. Journal of Hydrology 272, 250–263.
Wenkel, K.-O., Brozio, S., Gebbers, R., 2002. Düngung (TP III-4). In: Werner, A., Jarfe, A. (Eds.), Precision Agriculture: Herausforderung an integrative Forschung. KTBL, pp. 251–274.
Wentworth, C.K., 1930. A simplified method of determining the average slope of land surfaces. American Journal of Science Series 5 (20), 184–194.

Wieczorek, G.F., Naeser, N.D. (Eds.), 2000. Debris-Flow Hazards Mitigation: Mechanics, Prediction, and Assessment. Balkema, Rotterdam, 1392 pp.
Wigmosta, M.S., Vail, L., Lettenmaier, D.P., 1994. A distributed hydrology-vegetation model for complex terrain. Water Resources Research 30, 1665–1679.
Wiken, E.B., Ironside, G., 1977. The development of ecological (biophysical) land classification in Canada. Landscape Planning 4, 273–275.
Willgoose, G., Hancock, G., Kuczera, G., 2003. A framework for the quantitative testing of landform evolution models. In: Wilcock, P.R., Iverson, R.M. (Eds.), Prediction in Geomorphology. In: Geophysical Monograph, vol. 135. American Geophysical Union, Washington, DC, pp. 195–216.
Williams, K.K., Zuber, M.T., 1998. Measurement and analysis of lunar basin depths from Clementine altimetry. Icarus 131 (1), 107–122.
Wilson, J.P., Gallant, J.C. (Eds.), 2000. Terrain Analysis: Principles and Applications. Wiley, New York, 303 pp.
Wilson, J.P., Repetto, P.L., Snyder, R.D., 2000. Effect of data source, grid resolution, and flow-routing method on computed topographic attributes. In: Wilson, J.P., Gallant, J.C. (Eds.), Terrain Analysis: Principles and Applications. Wiley, pp. 133–161.
Wilson, J.P., Gallant, J.C., 1996. EROS: a grid-based program for estimating spatially distributed erosion indices. Computers and Geosciences 22, 707–712.
Wilson, J.P., 1986. Estimating the topographic factor in the universal soil loss equation for watersheds. Journal of Soil and Water Conservation 41, 179–184.
Wischmeier, W.H., Smith, D.D., 1978. Predicting Rainfall Erosion Losses. Agriculture Handbook, vol. 537. United States Department of Agriculture. Science and Education Administration, 58 pp.
Wischmeier, W.H., Smith, D.D., 1958. Rainfall energy and its relationship to soil loss. Transactions American Geophysical Union 39, 285–291.
Wise, S.M., 2000. Assessing the quality for hydrological applications of digital elevation models derived from contours. Hydrological Processes 14 (11–12), 1909–1929.
Wise, S.M., 1998. The effect of GIS interpolation errors on the use of digital elevation models in geomorphology. In: Lane, S.N., Richards, K.S., Chandler, J.H. (Eds.), Landform Monitoring, Modelling and Analysis. Wiley, pp. 139–164.
Wolf, G., 2004. Topographic surfaces and surface networks. In: Rana, S. (Ed.), Topological Data Structures for Surfaces: An Introduction for Geographical Information Science, vol. 1. Wiley, pp. 15–30.
Wolf, P.R., Dewitt, B.A., 2000. Elements of Photogrammetry: With Applications in GIS, 3rd edition. McGraw–Hill, Boston, 624 pp.
Wolinsky, M.A., Pratson, L.F., 2005. Constraints on landscape evolution from slope histograms. Geology 33 (6), 477–480.
Wolock, D.M., Winter, T.C., McMahon, G., 2004. Delineation and evaluation of hydrologic-landscape regions in the United States using geographic information system tools and multivariate statistical analyses. Environmental Management 34, 71–88.
Wolock, D.M., Mccabe, G.J., 1995. Comparison of single and multiple flow direction algorithms for computing topographic parameters in Topmodel. Water Resources Research 31 (5), 1315–1324.
Wood, E.F., Lakshmi, V., 1993. Scaling water and energy fluxes in climate systems: three land-atmospheric modeling experiments. Journal of Climate 6 (5), 839–857.
Wood, J., Kirschenbauer, S., Doellner, J., Lopes, A., Bodum, L., 2004. Using 3D in visualization. In: Dykes, J., MacEachren, A., Kraak, M.-J. (Eds.), Exploring Geovisualization, vol. 1. Elsevier, pp. 283–301.
Wood, J., 1996. The geomorphological characterisation of digital elevation models. Ph.D. Thesis. Department of Geography, University of Leicester, Leicester, UK, 185 pp.
Wood, J., Fisher, P.F., 1993. Assessing interpolation accuracy in elevation models. IEEE Computer Graphics and Applications 13 (2), 48–56.
Wood, W.F., Snell, J.B., 1960. A quantitative system for classifying landforms. Technical Report EP-124. U.S. Army Quartermaster Research and Engineering Center, Natick, MA, 20 pp.
Wood, W.F., Snell, J.B., 1957. The dispersion of geomorphic data around measures of central tendency and its application. Research Study Report EA-8. U.S. Army Quartermaster Research and Development Center, Natick, MA, 5 pp.
Woodcook, C.E., Strahler, A.H., Jupp, D.L.B., 1988a. The use of variograms in remote sensing: real digital images. Remote Sensing of Environment 25, 349–379.

Woodcook, C.E., Strahler, A.H., Jupp, D.L.B., 1988b. The use of variograms in remote sensing: scene models and simulated images. Remote Sensing of Environment 25, 323–348.
Woodcock, N.H., 1977. Specification of fabric shapes using an eigenvalue method. Geological Society of America Bulletin 88, 1231–1236.
Wooldridge, S.W., 1958. The trend of geomorphology. Transactions, Institute of British Geographers 25, 29–35.
Woolhiser, D.A., Smith, R.E., Giraldez, J.V., 1996. Effects of spatial variability of saturated hydraulic conductivity on Hortonian overland flow. Water Resources Research 32 (3), 671–678.
Wysocki, D.A., Schoeneberger, P.J., LaGarry, H.E., 2000. Geomorphology of soil landscapes. In: Sumner, M.E. (Ed.), Handbook of Soil Science. CRC Press, Boca Raton, FL, pp. E–5–40.
Xiu, A., Pleim, J.E., 2001. Development of a land surface model. Part I: Application in a mesoscale meteorological model. Journal of Applied Meteorology 40, 192–209.
Yang, C., Peterson, C.L., Shropshire, G.J., Otawa, T., 1998. Spatial variability of field topography and wheat yield in the Palouse region of the Pacific Northwest. Transactions of the ASAE 41 (1), 17–27.
Yang, Z., Teller, J.T., 2005. Modeling the history of Lake of the Woods since 11,000 cal yr B.P. using GIS. Journal of Paleolimnology 33, 483–498.
Yoeli, P., 1967. The mechanisation of analytical hill shading. Cartographic Journal (Glasgow) 4 (2), 82–88.
Yokoyama, R., Shirasawa, M., Pike, R.J., 2002. Visualizing topography by openness: a new application of image processing to digital elevation models. Photogrammetric Engineering and Remote Sensing 68 (3), 251–266.
Young, A., 1972. Slopes. Oliver and Boyd, Edinburgh, 288 pp.
Young, A., 1964. Slope profile analysis. Zeitschrift für Geomorphologie 5, 17–27.
Young, M., 1978. Terrain analysis: program documentation. Report 6 on Grant DA-ERO-591-73-G0040. Statistical characterization of altitude matrices by computer. Department of Geography, University of Durham, Durham, UK, 27 pp.
Young, T., 1805. An essay on the cohesion of fluids. Philosophical Transactions of the Royal Society of London 95, 65–87.
Yu, Q., Tian, J., Liu, J., 2004. A NOVEL contour-based 3D terrain matching algorithm using wavelet transform. Pattern Recognition Letters 25 (1), 87–99.
Zaitchik, B.F., van Es, H.M., Sullivan, P.J., 2003. Modeling slope stability in Honduras: parameter sensitivity and scale of aggregation. Soil Science Society of America Journal 67, 268–278.
Zakrzewska, B., 1963. Trends and methods in landform geography. Annals of the Association of American Geographers 57, 128–165.
Zambon, M., Lawrence, R., Bunn, A., Powell, S., 2006. Effect of alternative splitting rules on image processing using classification tree analysis. Photogrammetric Engineering and Remote Sensing 72 (1), 25–30.
Zavoianu, I., 1985. Morphometry of Drainage Basins. Developments in Water Science, vol. 20. Elsevier, Amsterdam, 238 pp.
Zevenbergen, L.W., Thorne, C.R., 1987. Quantitative analysis of land surface topography. Earth Surface Processes and Landforms 12, 47–56.
Zhang, W.H., Montgomery, D.R., 1994. Digital elevation model grid size, landscape representation and hydrological simulations. Water Resources Research 30 (4), 1019–1028.
Zhang, Y., 1999. Optimisation of building detection in satellite images by combining multispectral classification and texture filtering. ISPRS Journal of Photogrammetry and Remote Sensing 54 (1), 50–60.
Zhang, Z., Kane, D.L., Hinzman, L.D., 2000. Development and application of a spatially-distributed Arctic hydrological and thermal process model (ARHYTHM). Hydrological Processes 14, 1017–1044.
Zhou, Q., Liu, X., 2004. Analysis of errors of derived slope and aspect related to DEM data properties. Computers & Geosciences 30 (4), 369–378.
Zhu, A.X., Hudson, B., Burt, J., Lubich, K., Simonson, D., 2001. Soil mapping using GIS, expert knowledge and fuzzy logic. Soil Science Society of America Journal 65, 1463–1472.
Zhu, A.X., 1997. A similarity model for representing soil spatial information. Geoderma 77, 217–242.
Zhu, A.X., Band, L.E., Vertessy, R., Dutton, B., 1997. Derivation of soil properties using a soil land inference model (SoLIM). Soil Science Society of America Journal 61, 523–533.

Zhu, A.X., Band, L.E., 1994. A knowledge-based approach to data integration for soil mapping. Canadian Journal of Remote Sensing 20, 408–418.

Zhu, J., Mohanty, B.P., 2004. Soil hydraulic parameter upscaling for steady-state flow with root water uptake. Vadose Zone Journal 3, 1464–1470.

Zinck, J.A., Valenzuela, C.R., 1990. Soil geographic database: structure and application examples. ITC Journal 1990 (3), 270–294.

Zukowskyj, P.M., Teeuw, R.M., 2000. Interpolated digital elevation models, differential global positioning system surveys and digital photogrammetry: a quantitative comparison of accuracy from a geomorphological perspective. In: Abrahart, R.J., Carlisle, B.H. (Eds.), Proceedings of the 5th International Conference on GeoComputation. University of Greenwich, Greenwich, UK, 15 pp.

Zwally, H.J., Schutz, B., Abdalati, W., Abshire, J., Bentley, C., Brenner, A., Bufton, J., Dezio, J., Hancock, D., Harding, D., Herring, T., Minster, B., Quinn, K., Palm, S., Spinhirne, J., Thomas, R., 2002. ICESat's laser measurements of polar ice, atmosphere, ocean, and land. Journal of Geodynamics 34 (3–4), 405–445.

INDEX

B

C

D

Z

Colour Plate Section

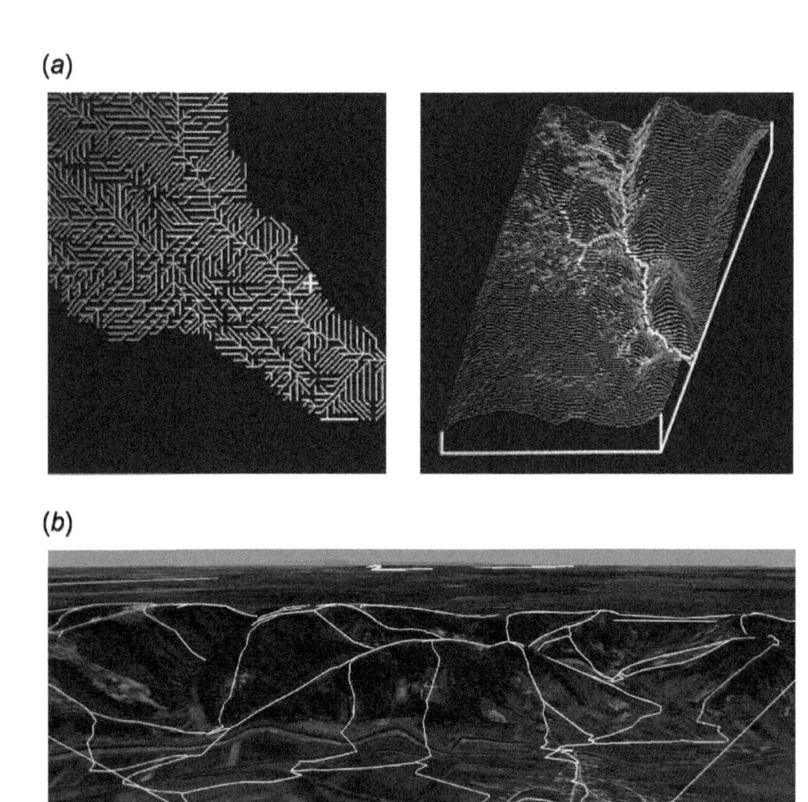

FIGURE 9 (CHAPTER 1) Geomorphometry then and now: (a) output from late-1980s DOS programme written to display land-surface properties: (left) map of local drainage direction, (right) cumulative upstream drainage elements draped over a DEM rendered in 3-D by parallel profiles. Courtesy of P.A. Burrough; (b) watershed boundaries for the Baranja Hill study area overlaid in Google Earth, an online geographical browser accessible to everyone. (See page 23 of this volume.)

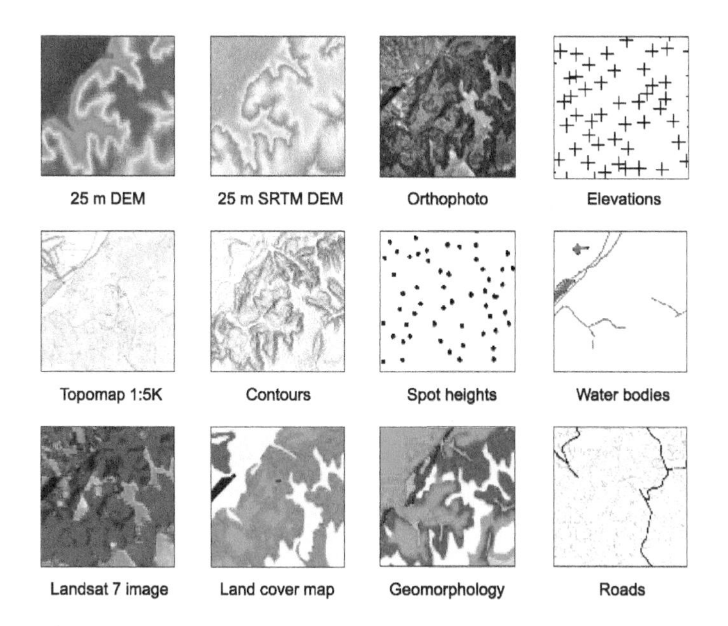

FIGURE 10 (CHAPTER 1) The "Baranja Hill" datasets. Courtesy of the Croatian State Geodetic Department (http://www.dgu.hr). (See page 27 of this volume.)

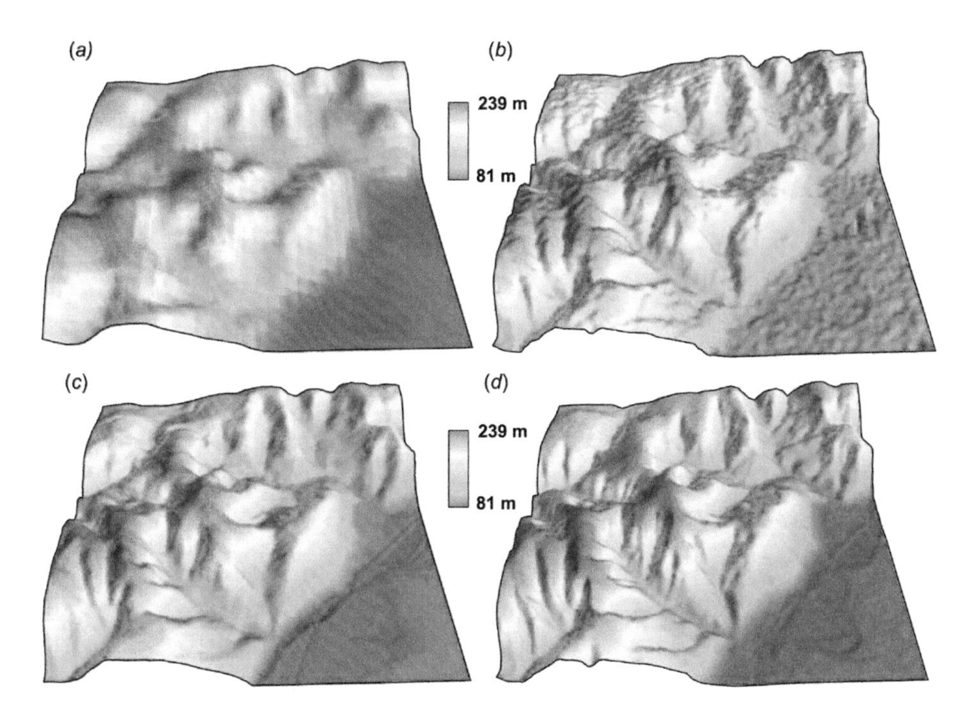

FIGURE 5 (CHAPTER 3) Comparison of DEMs from main sources for Baranja Hill: (a) 90 m resolution SRTM DEM, (b) 30 m resolution SRTM DEM, (c) DEM from 1:50,000 topo-map, and (d) DEM from 1:5000 topo-map. (See page 72 of this volume.)

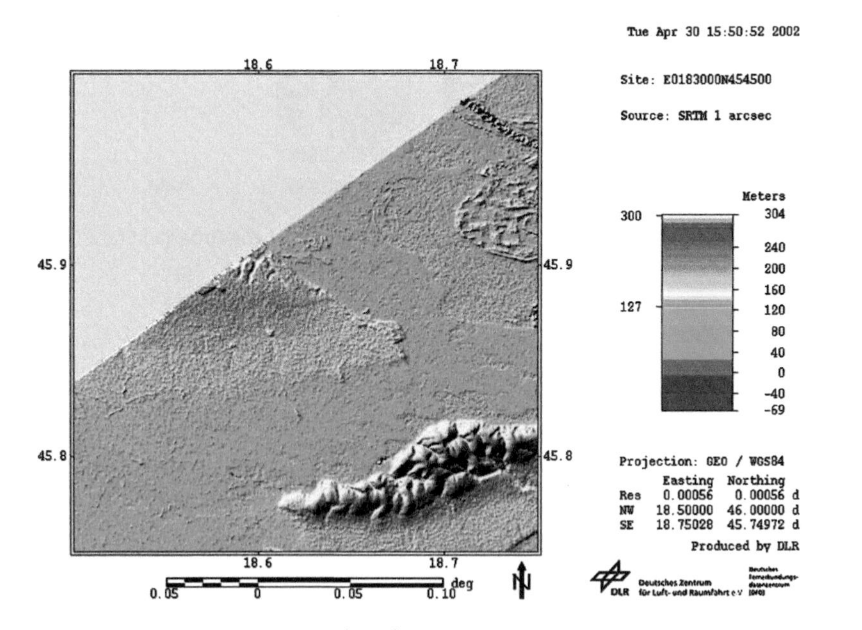

FIGURE 6 (CHAPTER 3) Example of a $15' \times 15'$ block of 1 arcsec SRTM DEM ordered for Baranja Hill. Courtesy of German Space Agency (DLR). (See page 77 of this volume.)

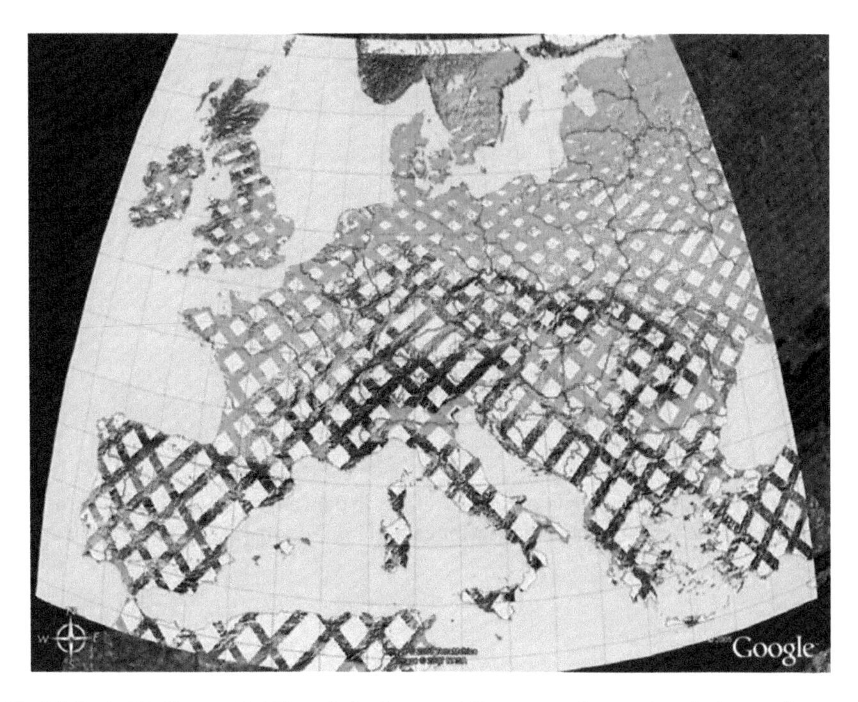

FIGURE 7 (CHAPTER 3) Availability of the 1 arcsec SRTM DEMs (C-Band Radar) over the European continent. Missing areas are were not acquired due to an energy shortage at the end of the mission (Rabus et al., 2003). To load the Google Earth map, visit geomorphometry.org. (See page 78 of this volume.)

FIGURE 8 (CHAPTER 3) Availability of the 30 m ASTER DEMs over the European continent (before January 2006). (See page 78 of this volume.)

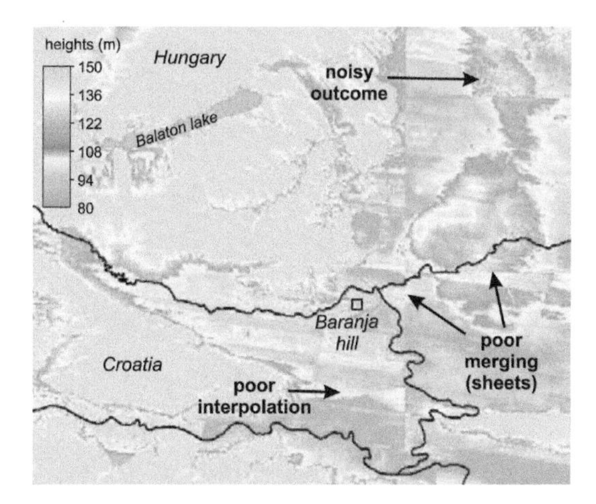

FIGURE 3 (CHAPTER 4) An example of local artefacts in part of the GTOPO DEM (1 km resolution). Such artefacts are only visible after careful inspection. (See page 91 of this volume.)

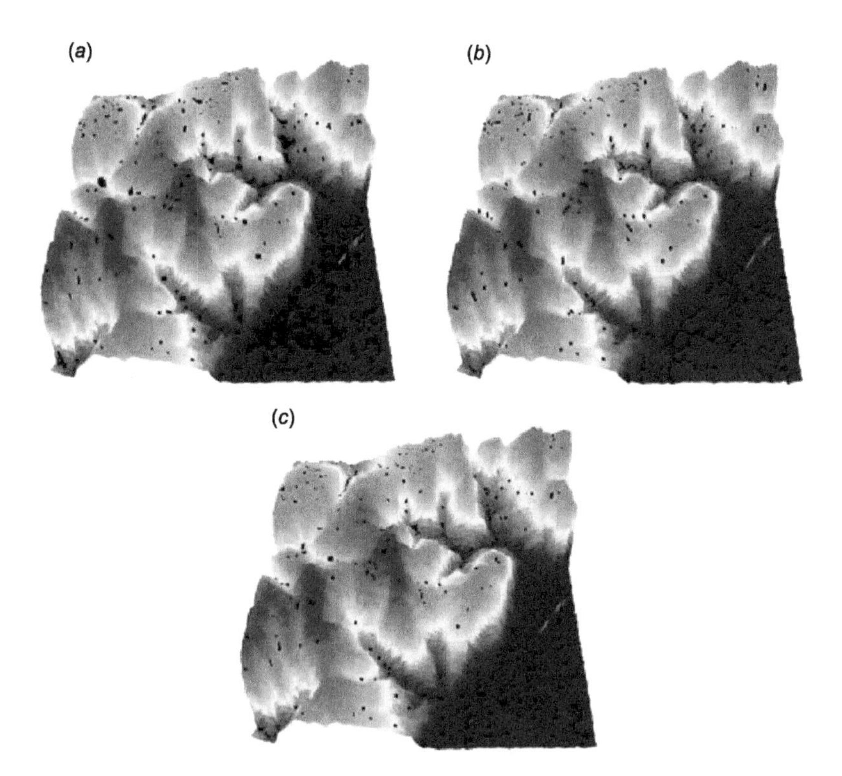

FIGURE 13 (CHAPTER 4) Three approaches to removing spurious sinks: (a) sink filling, (b) carving, and (c) the optimal combination of filling and carving. The detected sinks are indicated black. (See page 107 of this volume.)

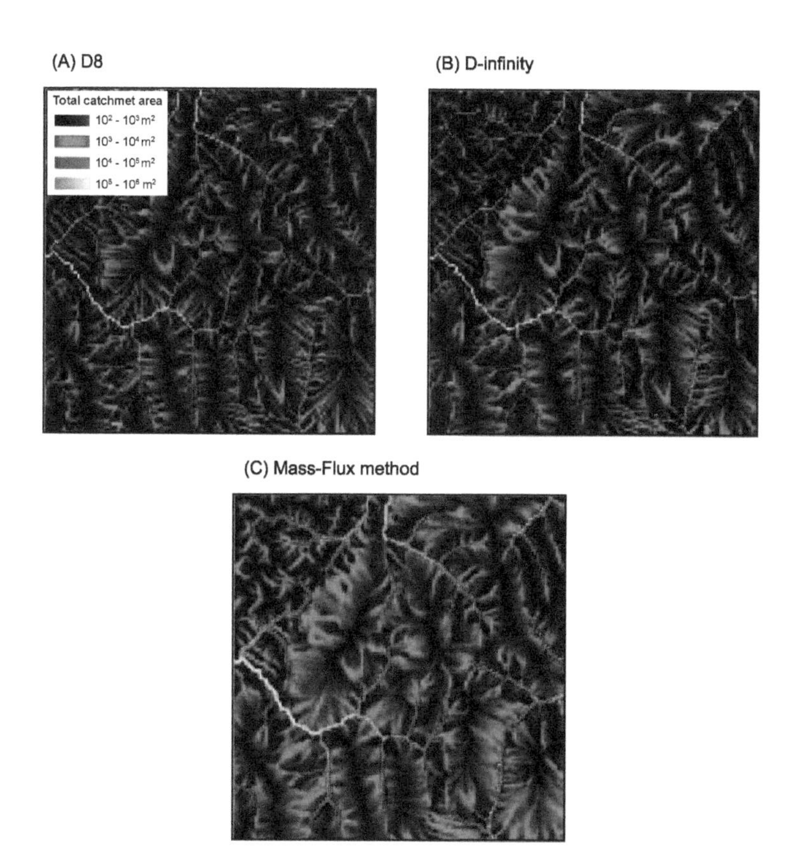

FIGURE 7 (CHAPTER 7) Total catchment area calculated for the Baranja Hill area using three different methods. (See page 183 of this volume.)

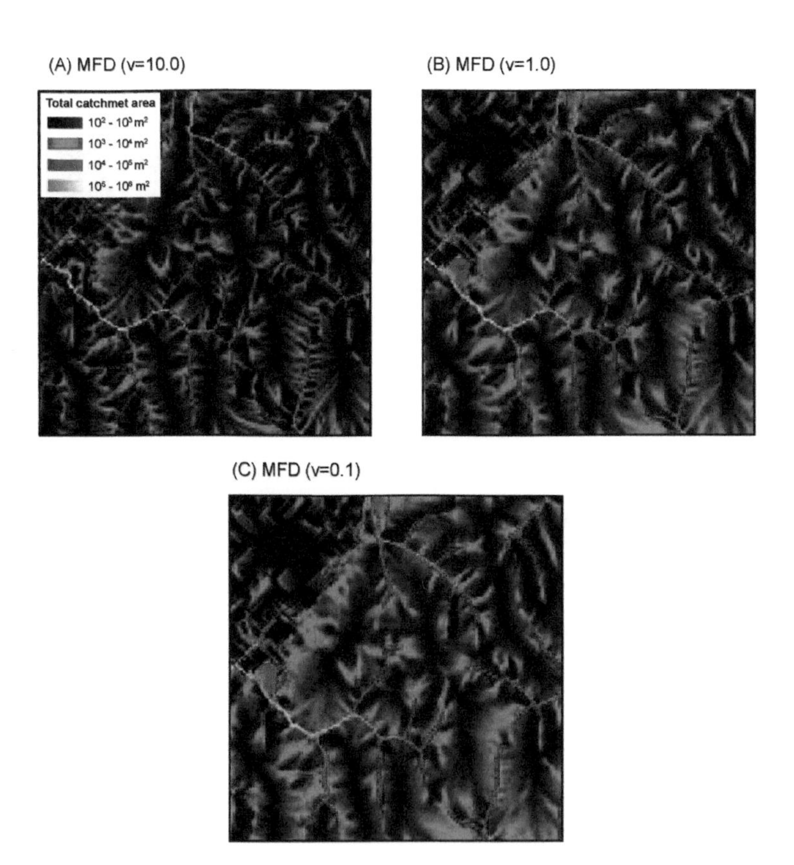

FIGURE 8 (CHAPTER 7) Total catchment area calculated for the Baranja Hill area using MFD and three different dispersion exponents. (See page 184 of this volume.)

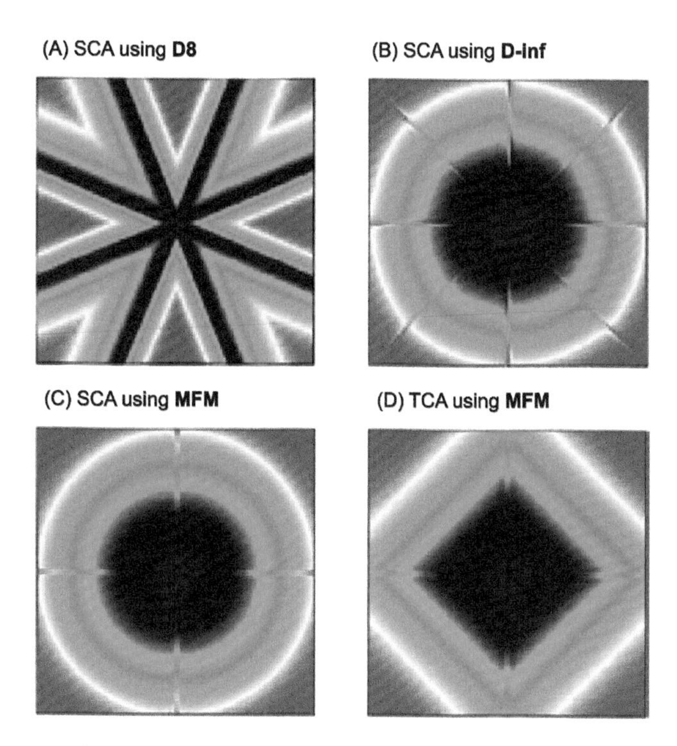

FIGURE 9 (CHAPTER 7) Parts (A)–(C) show the specific contributing area (SCA) calculated for the DEM of a cone sing D8, D-Infinity and MFM. The strong grid bias inherent in D8 is readily visible from the star pattern (A). Part (D) of this figure shows the total contributing area (TCA) calculated using MFM. This counter-intuitive result is correct because of the different flow widths of pixels (see Figure 6). When divided by the flow width, the SCA (C) shows the right circular pattern. (See page 185 of this volume.)

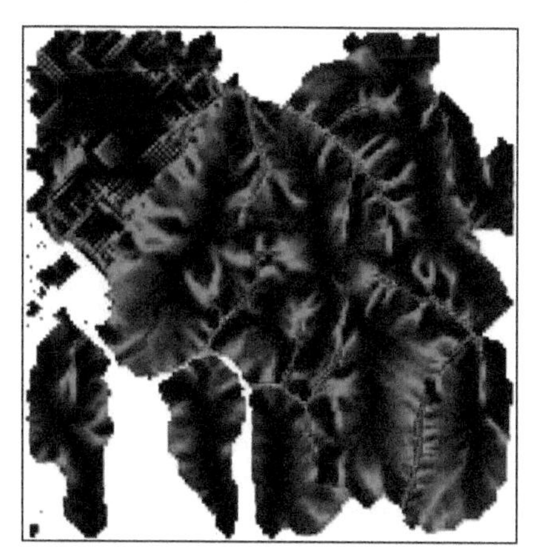

FIGURE 11 (CHAPTER 7) Edge-contaminated areas (white) have been removed from the calculated total contributing area. Both, the flow accumulation as well as the edge-contamination were computed using MFD. Other, less dispersive methods result in a smaller area of edge contamination. (See page 186 of this volume.)

FIGURE 12 (CHAPTER 7) Wetness index calculated for the Baranja Hill. Values range from 3 (dark) to 20 (yellow); the data is linearly stretched. (See page 187 of this volume.)

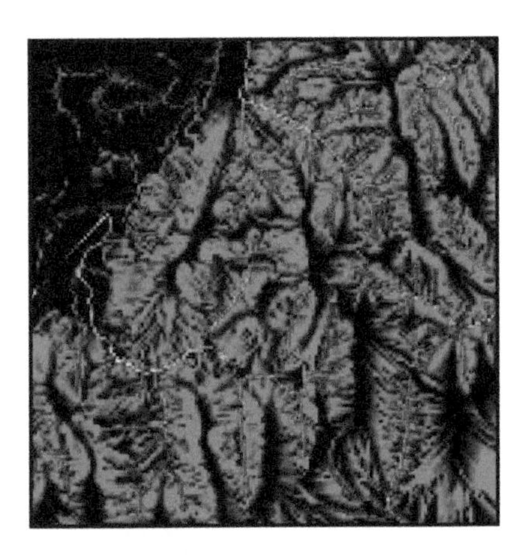

FIGURE 13 (CHAPTER 7) Stream power index calculated for the Baranja Hill. Values range from 1 (dark) to 12,000 (yellow); the data is stretched using logarithmic display. (See page 187 of this volume.)

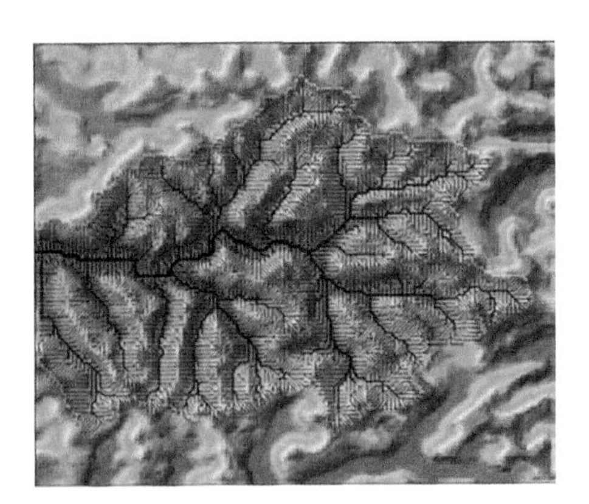

FIGURE 14 (CHAPTER 7) Complete drainage lines for one catchment. In the background, elevation is represented by colour. (See page 188 of this volume.) © 2004 Rivix LLC, used with permission.

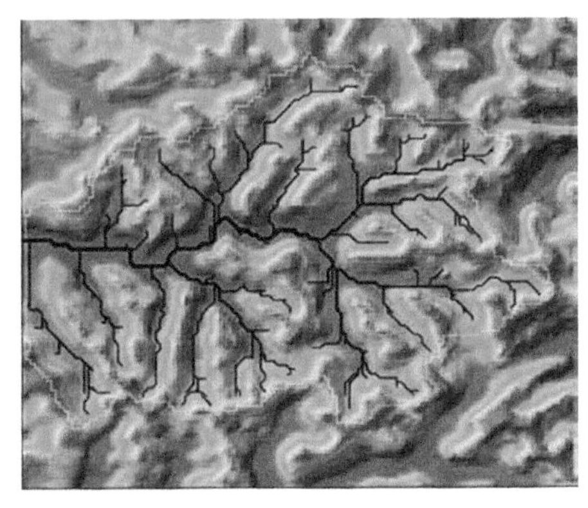

FIGURE 15 (CHAPTER 7) Drainage lines pruned by Horton–Strahler order. (See page 189 of this volume.) © 2004 Rivix LLC, used with permission.

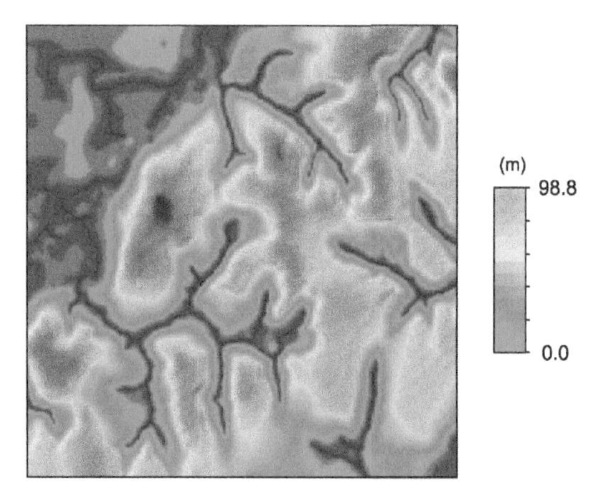

FIGURE 12 (CHAPTER 8) Altitude above channel lines. (See page 217 of this volume.)

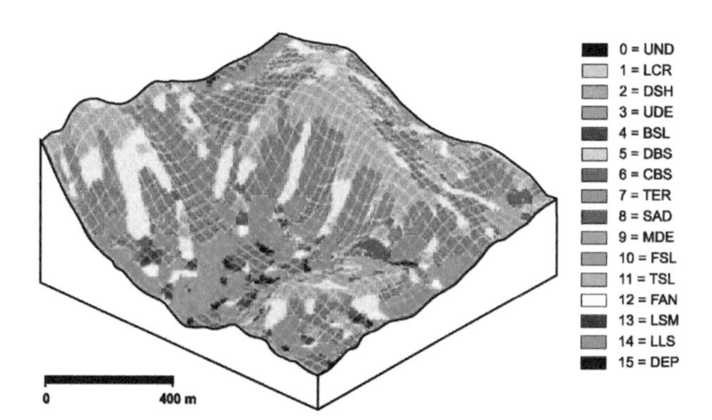

FIGURE 10 (CHAPTER 9) Illustration of landform elements extracted from land-surface parameters: 64 ha site in Alberta, Canada. See further Section 2 in Chapter 24. (See page 243 of this volume.)

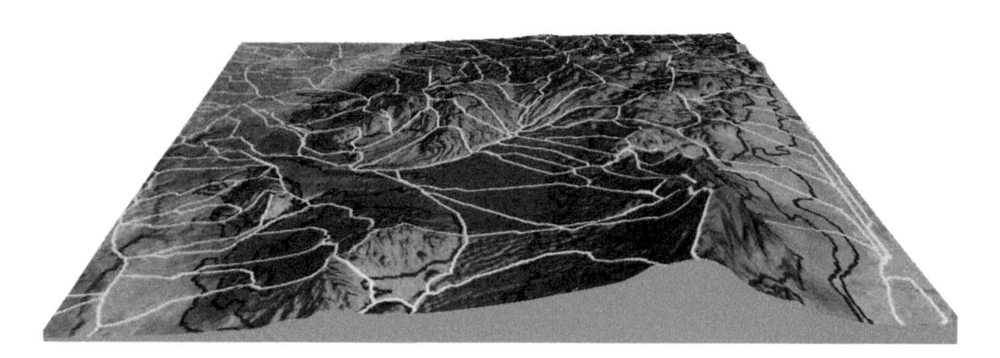

FIGURE 12 (CHAPTER 9) Illustration of possibilities and problems with using hillslopes as basic spatial entities for classifying repeating landform types. See text for detailed discussion. (See page 250 of this volume.)

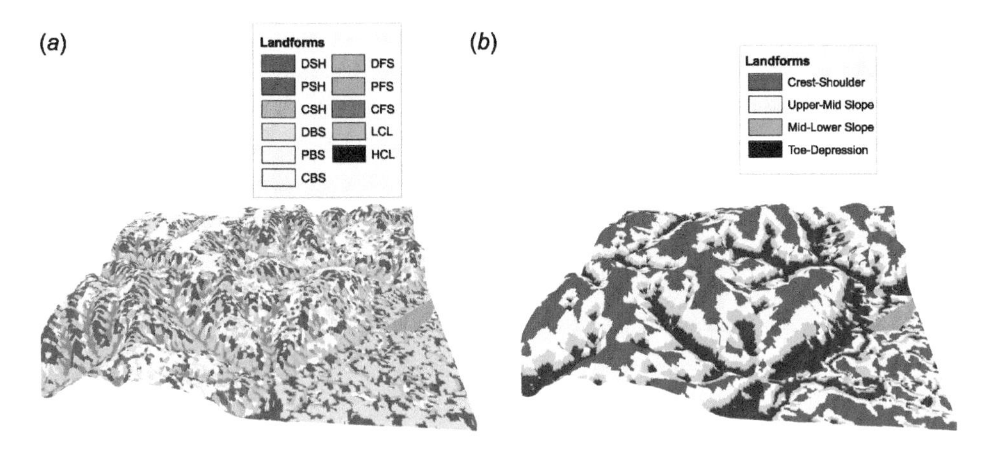

FIGURE 8 (CHAPTER 11) Landform classification as shown above using (a) `pennock97.aml` and (b) `simplelfabc.aml` scripts for the Baranja Hill Case study with a resolution of 10 m. (See page 286 of this volume.)

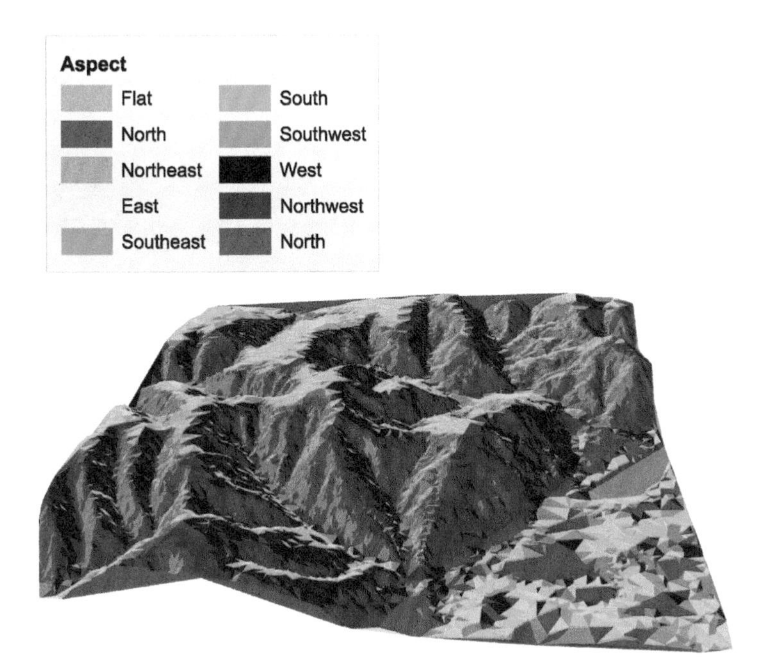

FIGURE 9 (CHAPTER 11) Aspect classes calculated for the Baranja Hill DEM TIN. (See page 288 of this volume.)

FIGURE 6 (CHAPTER 12) Convergence Index. (See page 301 of this volume.)

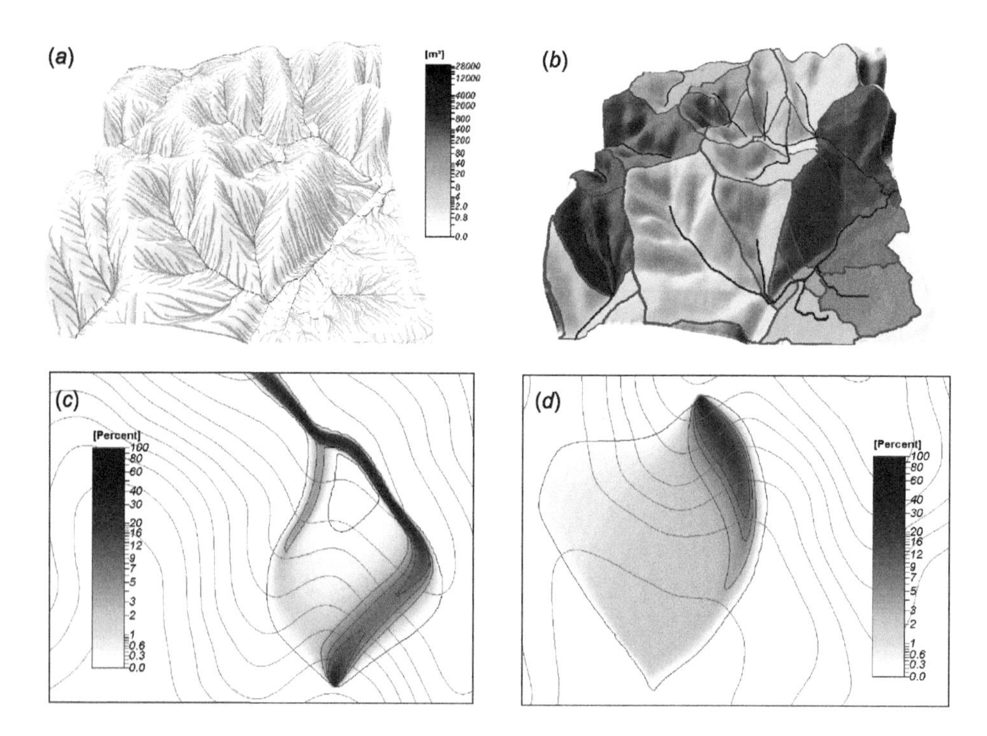

FIGURE 8 (CHAPTER 12) Hydrological analysis in SAGA: (a) catchment areas (DEMON, each 100th cell), (b) watershed basins, (c) downslope area (FD8) and (d) upslope area (FD8). (See page 303 of this volume.)

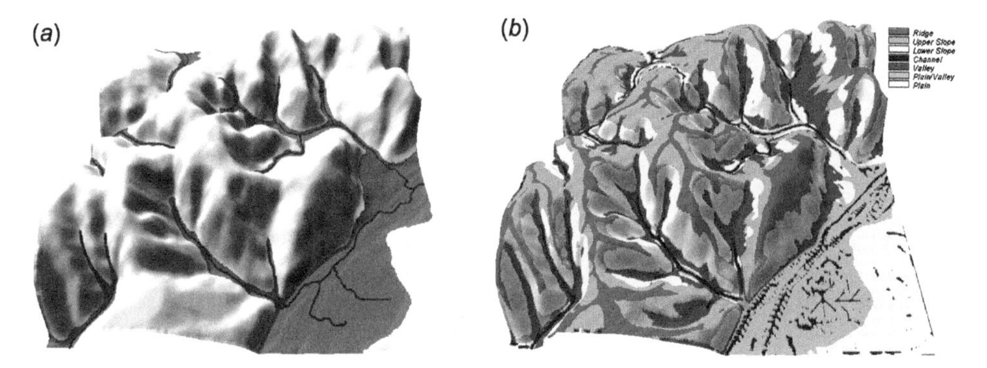

FIGURE 12 (CHAPTER 12) (a) Flood plain map calculated using a threshold buffer, (b) terrain classification using Cluster Analysis. (See page 306 of this volume.)

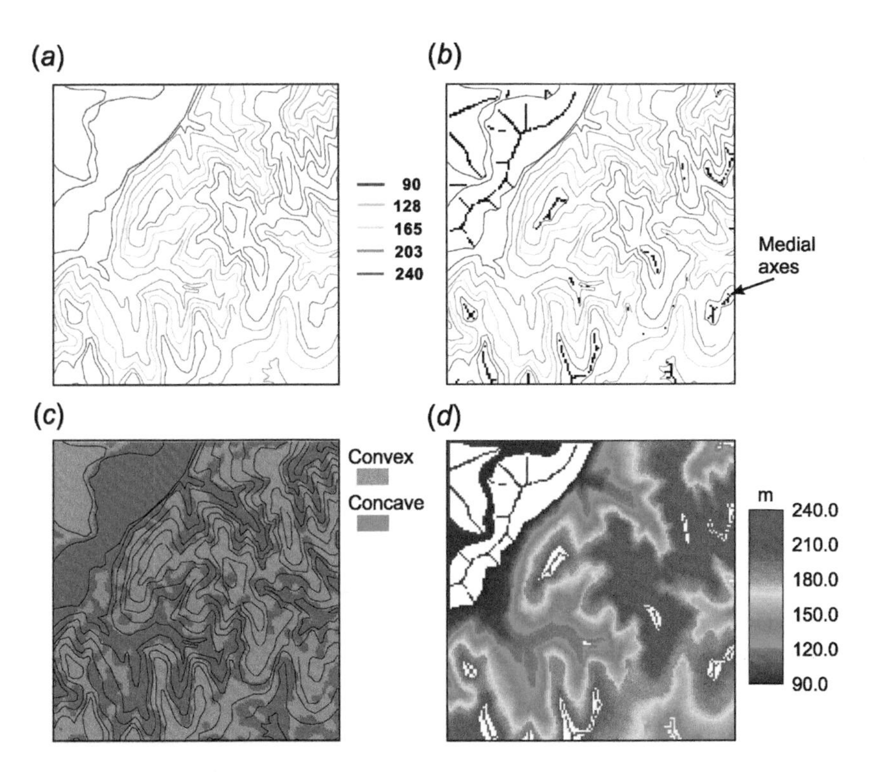

FIGURE 3 (CHAPTER 13) Addition of medial axes: (a) original (bulk) contour data; (b) detected medial axes in problematic areas (*padi*-terraces); (c) extrapolated shape of the land surface; and (d) temporary terrace-free map prior to interpolation of the remaining undefined pixels. (See page 316 of this volume.)

FIGURE 4 (CHAPTER 13) Visualization of the DEMs using the multi illuminated angles in ILWIS. (See page 319 of this volume.)

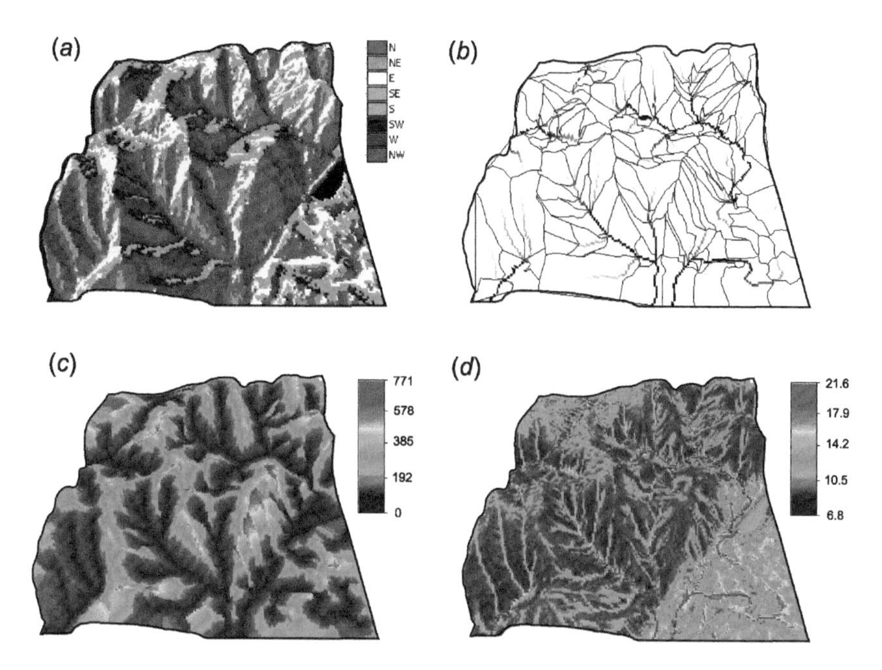

FIGURE 8 (CHAPTER 13) Extraction of hydrological parameters and objects using the built-in operations: (a) flow direction, (b) flow accumulation with catchment lines, (c) overland flow length and (d) wetness index. All calculated using the Deterministic-8 algorithm. (See page 323 of this volume.)

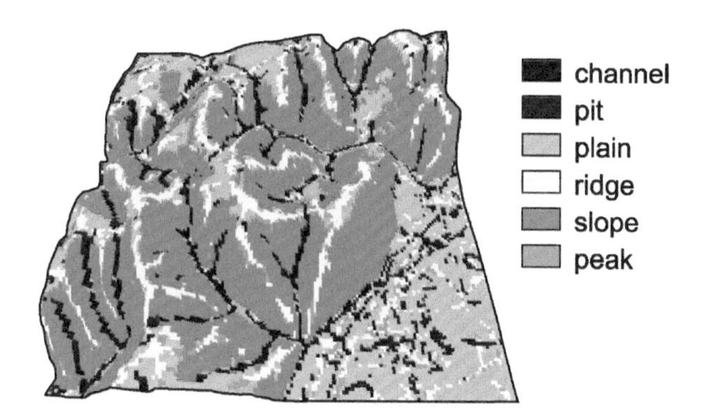

FIGURE 12 (CHAPTER 13) Study area classified into the generic landforms. (See page 330 of this volume.)

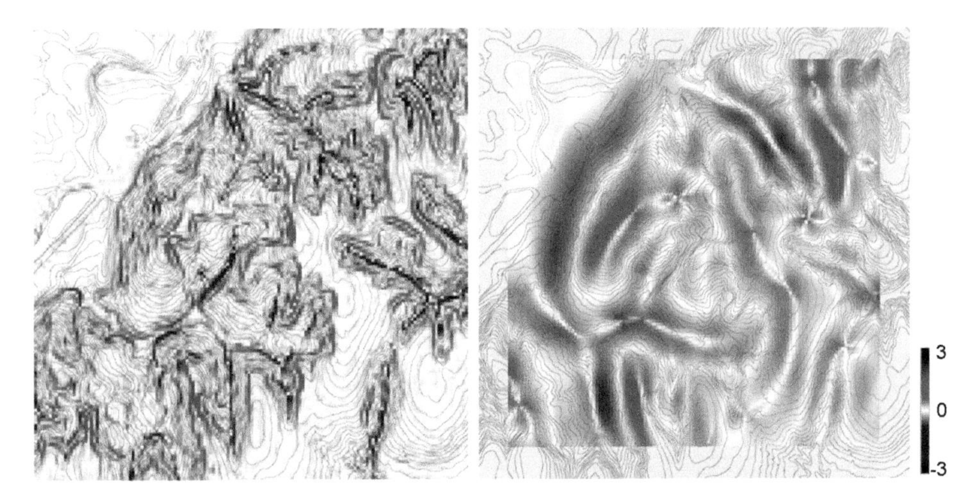

FIGURE 4 (CHAPTER 14) Profile curvature (per 100 m) measured over 75 and 625 m spatial extents. (See page 337 of this volume.)

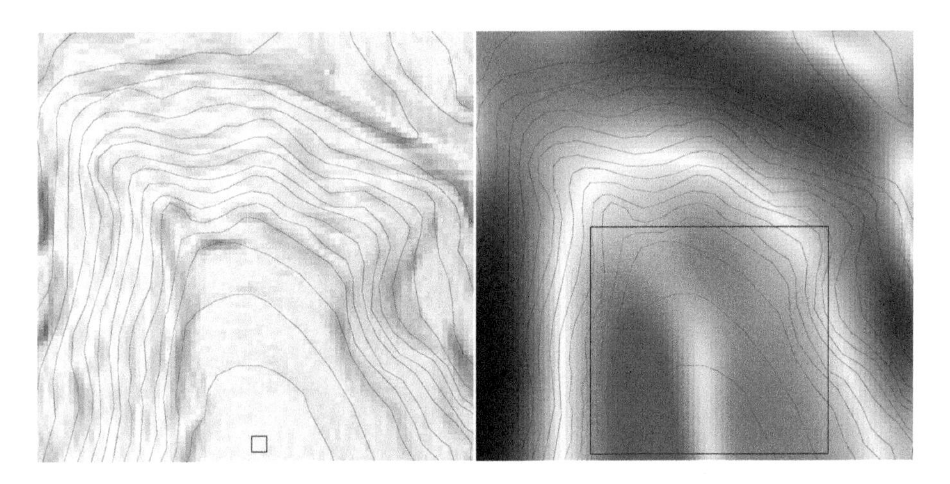

FIGURE 5 (CHAPTER 14) Profile curvature (per 100 m) measured from the Baranja Hill 5 m DEM at contrasting spatial scales. The square in the bottom centre of each image represents the size of the window used for processing (15 and 275 m respectively). (See page 338 of this volume.)

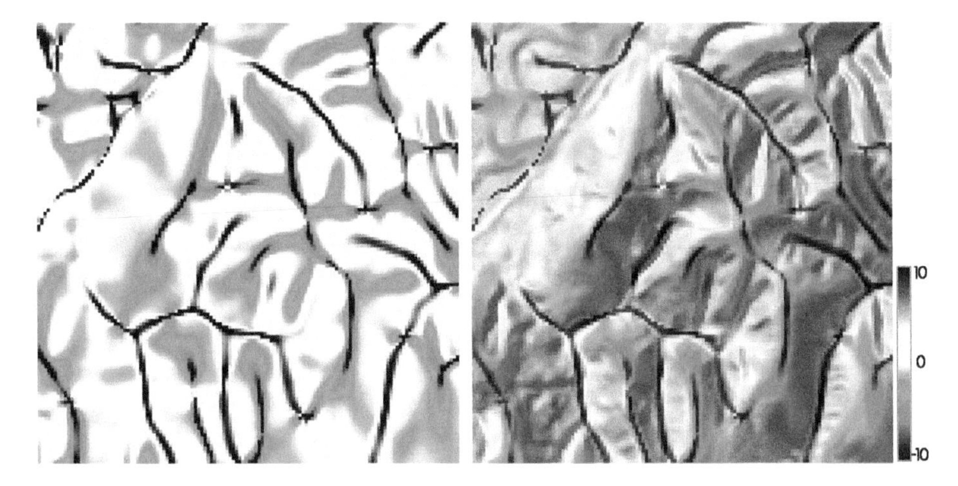

FIGURE 6 (CHAPTER 14) Plan curvature (per 100 m) of the Baranja Hill 25 m DEM measured at the 275 m window scale. The image on the left shows only plan curvature. The image on the right shows the same measure but with colour intensity, representing local shaded relief of the underlying surface. (See page 339 of this volume.)

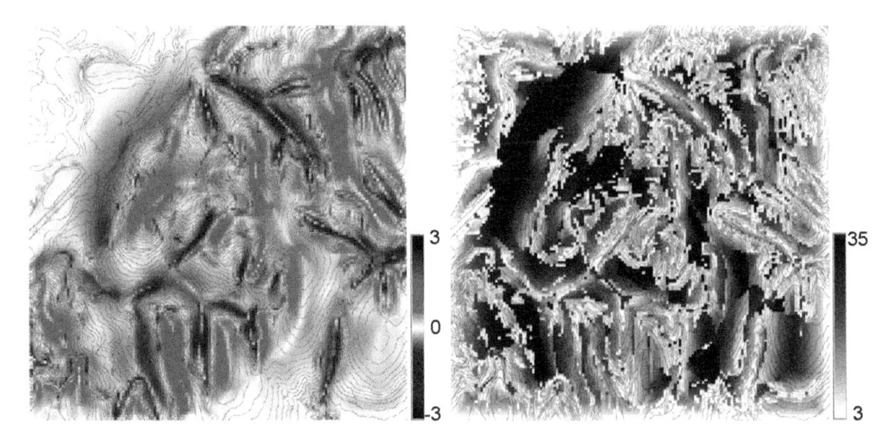

FIGURE 11 (CHAPTER 14) Maximum absolute profile curvature (per 100 m) measured over all scales between 75 m and 1.7 km (window sizes 3 to 35). The image to the right shows the window scale (in pixels) at which the most extreme value of profile curvature occurs. (See page 348 of this volume.)

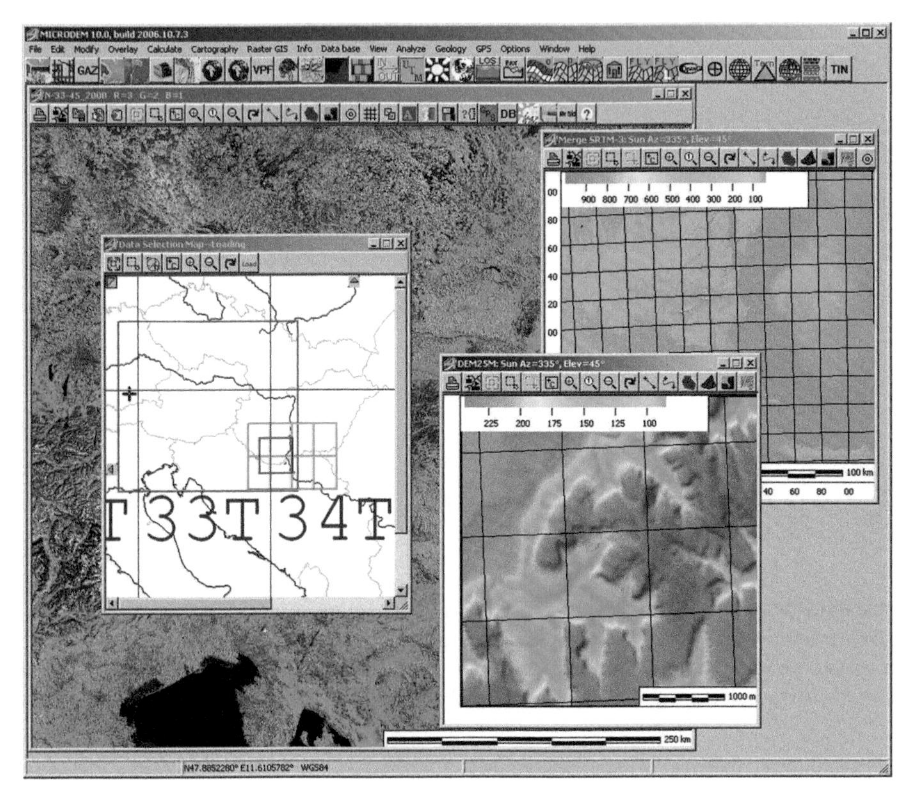

FIGURE 1 (CHAPTER 15) The main window of MicroDEM, with standard Windows controls and four active child windows. The centre left window is an index map showing eastern Europe with available Landsat imagery outlined by the large red rectangle, SRTM data shown in green, and the Baranja Hill DEM barely visible at this scale. Selecting the small box in red opened two DEMs, one a merge of 4 SRTM cells, and the satellite image visible in the background. (See page 353 of this volume.)

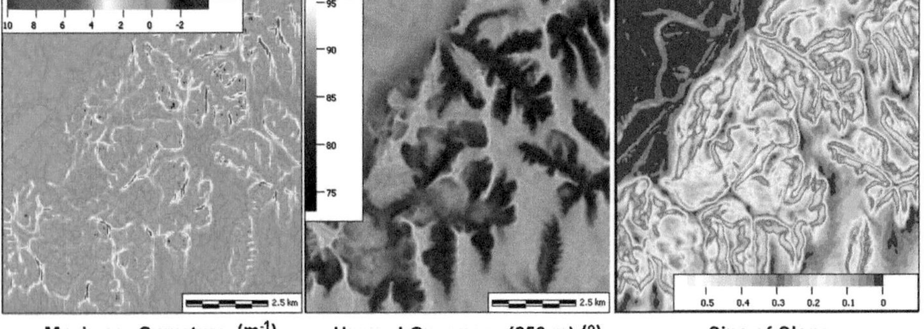

FIGURE 6 (CHAPTER 15) Sample maps of land-surface parameters created with MicroDEM. From left to right these show three options for colour coding: a continuous colour scale, a greyscale, and a discrete colour scale. These maps also show the options for placement and orientation of legend and scale bar. (See page 358 of this volume.)

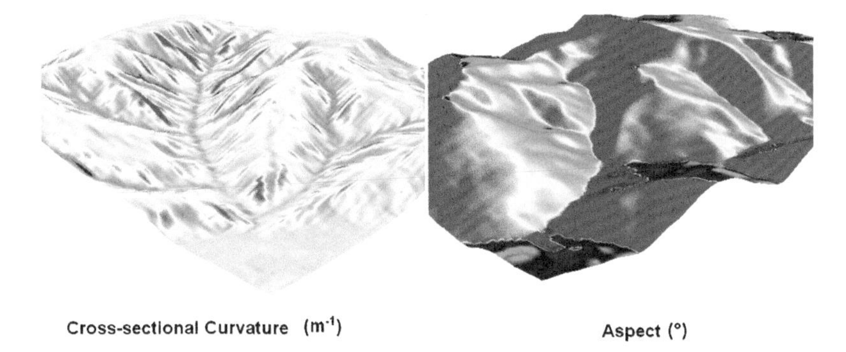

FIGURE 7 (CHAPTER 15) Sample land-surface parameters draped on the Baranja Hill DEM. (See page 358 of this volume.)

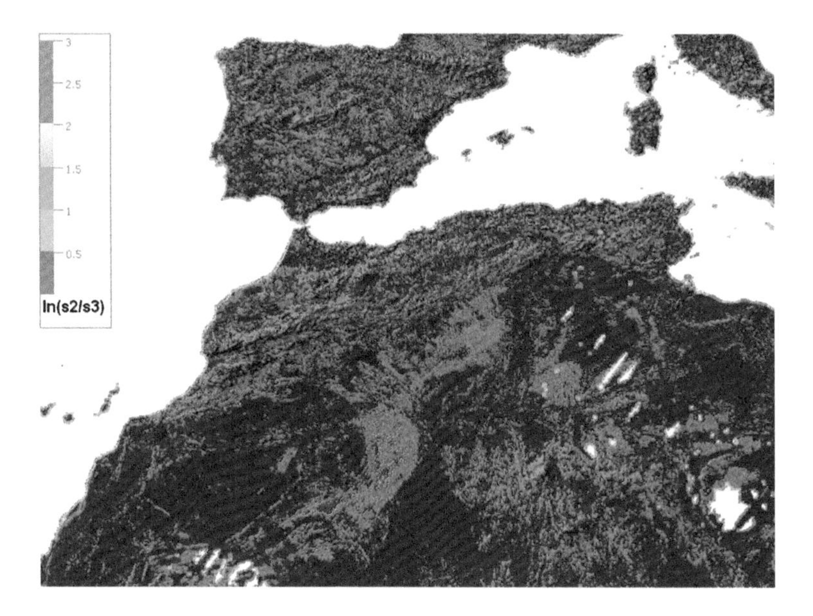

FIGURE 11 (CHAPTER 15) Organisation map of North Africa, with colour displaying the degree of organisation (red highly, to blue poorly organized), draped on shaded topography. Note the large void regions where dry sand led to no radar returns. (See page 362 of this volume.)

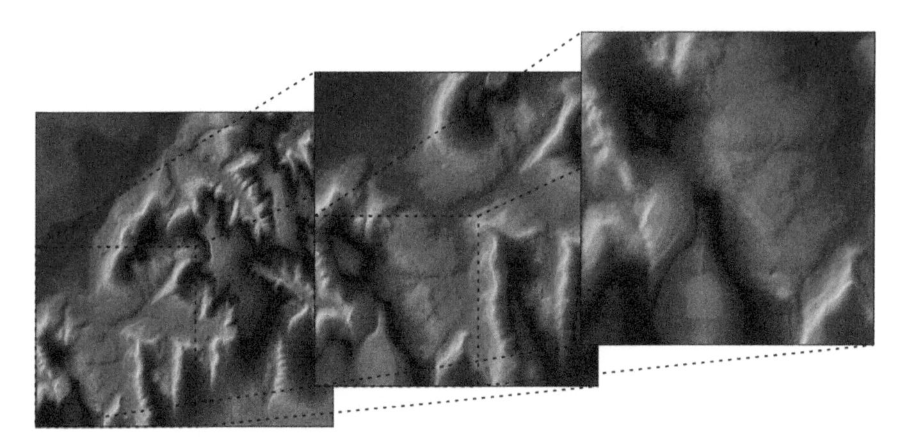

FIGURE 2 (CHAPTER 16) TAS can apply a histogram equalisation stretch dynamically as an image is zoomed into. (See page 371 of this volume.)

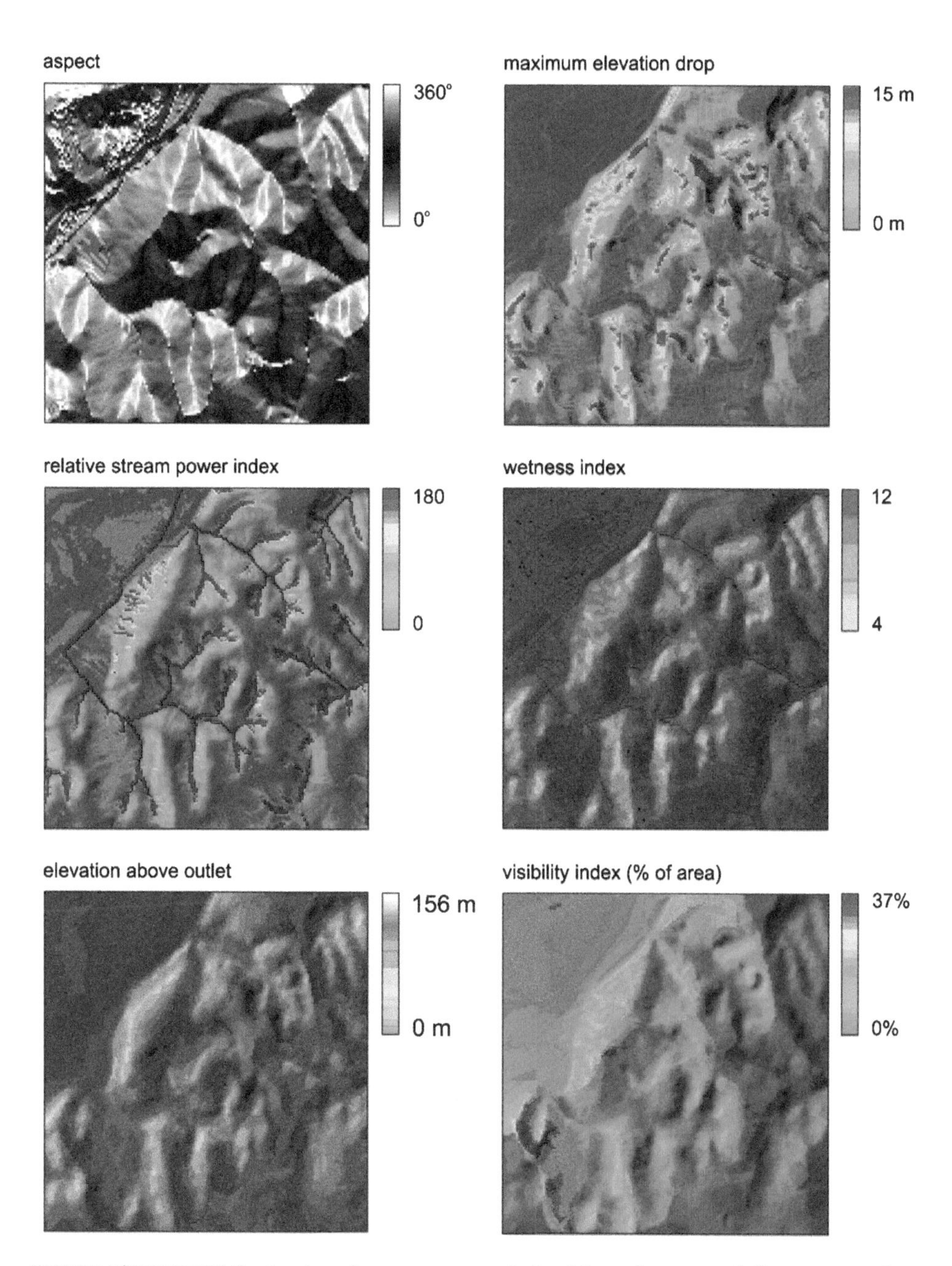

FIGURE 4 (CHAPTER 16) Land-surface parameters derived from the Baranja hill SRTM DEM. (See page 374 of this volume.)

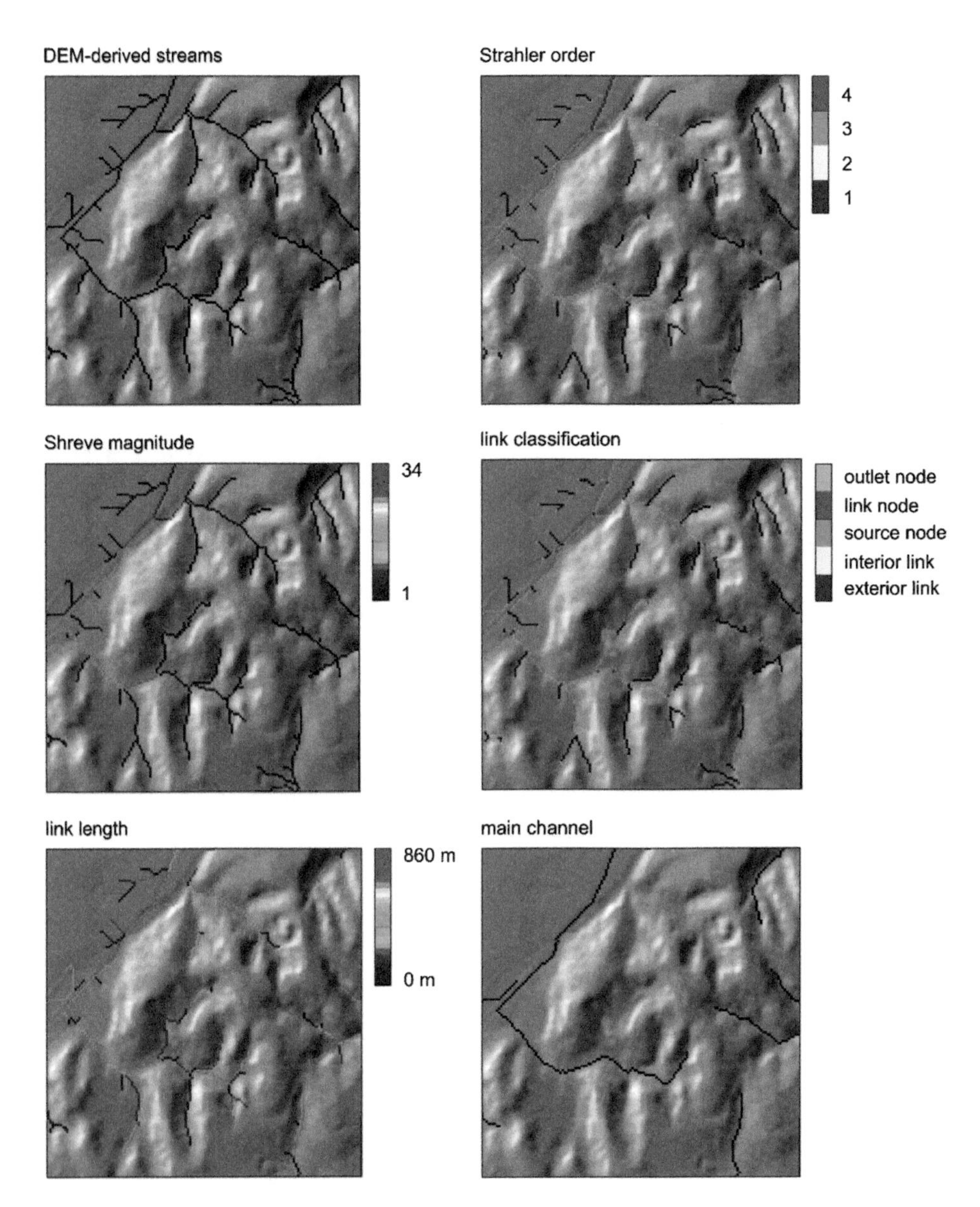

FIGURE 5 (CHAPTER 16) Stream morphometrics calculated for a stream network derived from the Baranja Hill DEM. (See page 376 of this volume.)

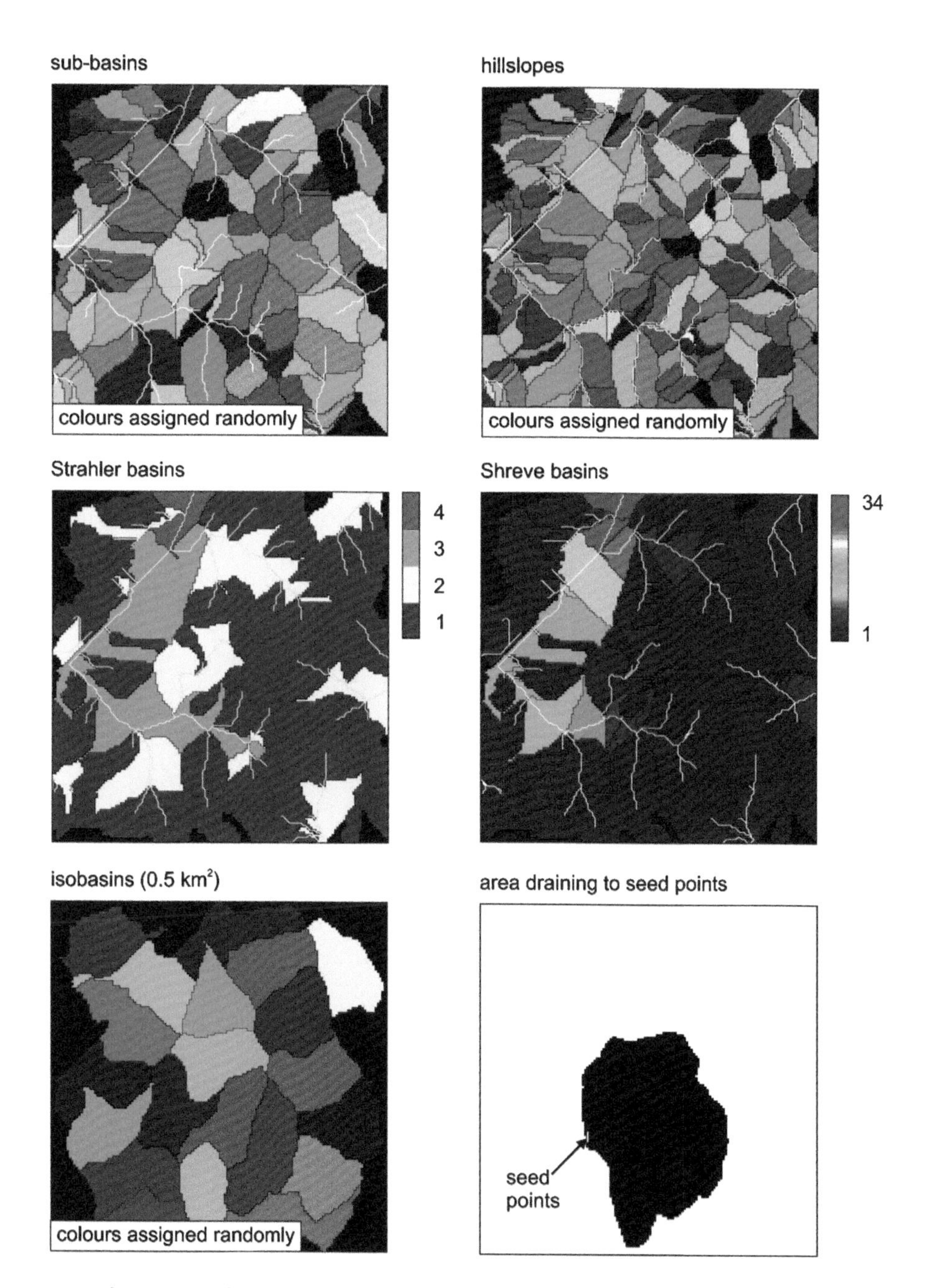

FIGURE 6 (CHAPTER 16) Various means of extracting watersheds for the Baranja Hill DEM. (See page 378 of this volume.)

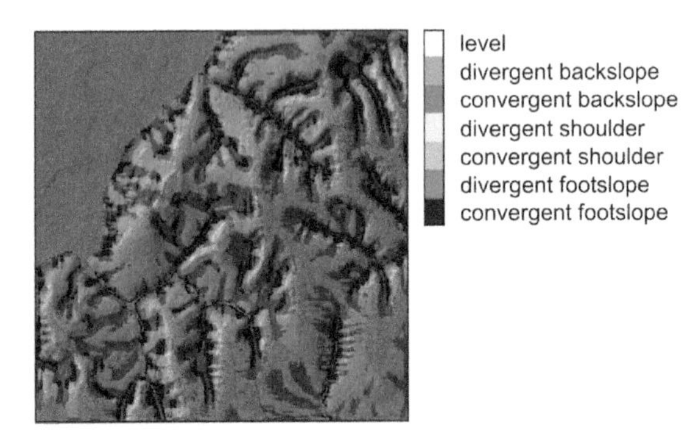

FIGURE 7 (CHAPTER 16) Automated landform classification of the Baranja Hill 25 m SRTM DEM, based on the crisp classification scheme of Pennock et al. (1987). The DEM was pre-processed by running a 21×21 mean filter to remove fine-scale topographic variation. (See page 379 of this volume.)

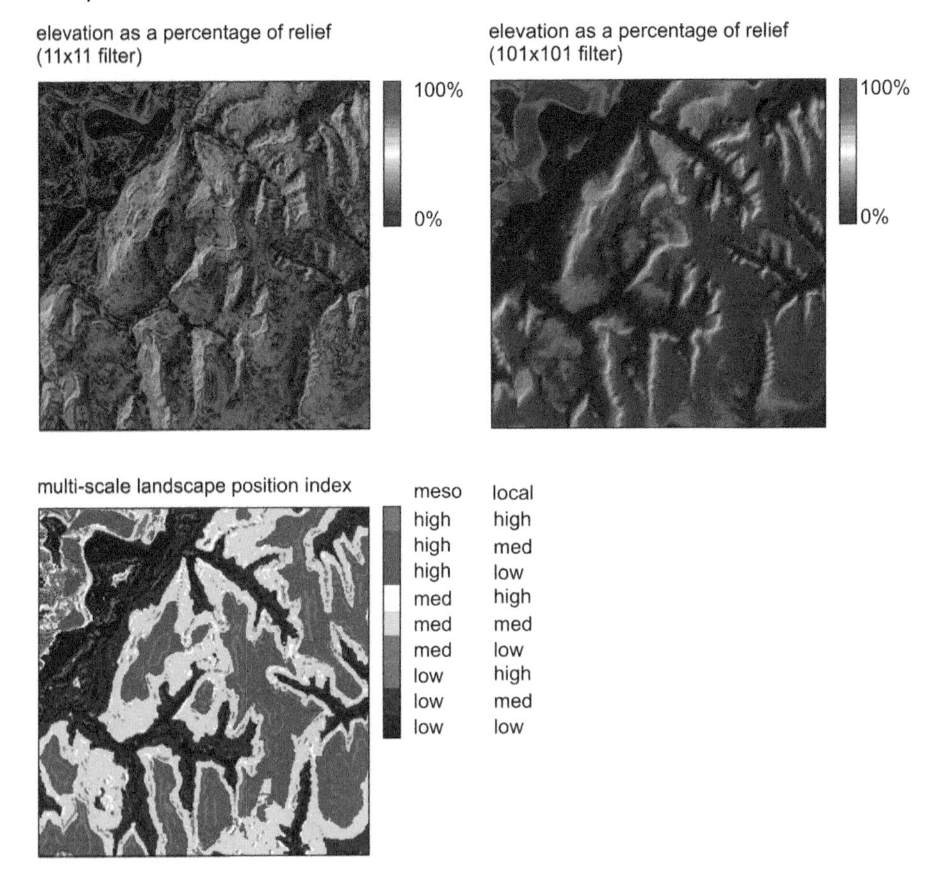

FIGURE 9 (CHAPTER 16) Elevation as a percentage of local relief (EPR) calculated using an 11×11 (a) and a 101×101 (b) filter and a multi-scale landscape position index (c). Images have been derived from the sample script applied to the Baranja Hill 25 m SRTM DEM. (See page 382 of this volume.)

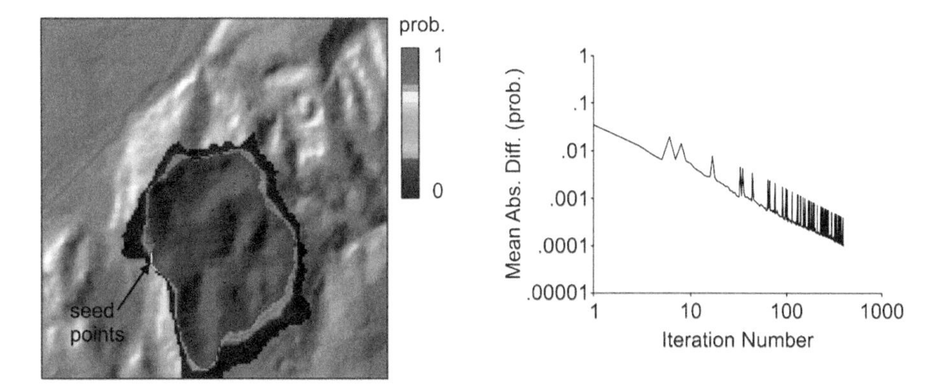

FIGURE 10 (CHAPTER 16) Results of a Monte-Carlo uncertainty analysis of the watershed area of a group of seed points in the Baranja Hill 25 m SRTM DEM. (See page 383 of this volume.)

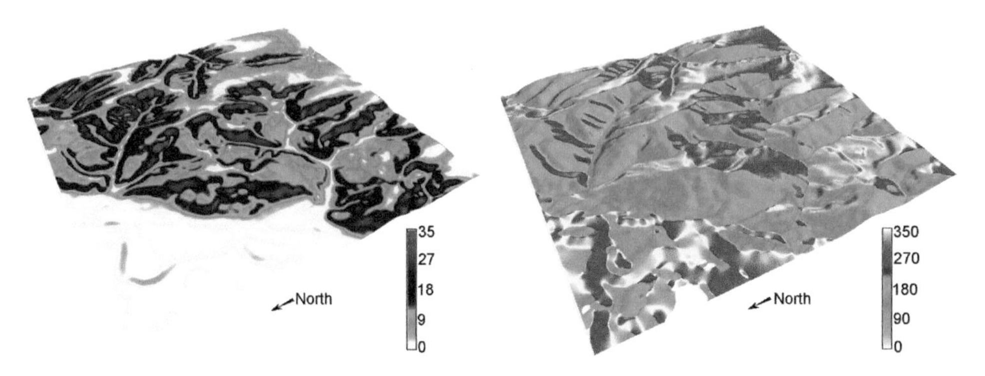

FIGURE 4 (CHAPTER 17) Slope steepness [°]. (See page 395 of this volume.)

FIGURE 5 (CHAPTER 17) Aspect [°]. (See page 395 of this volume.)

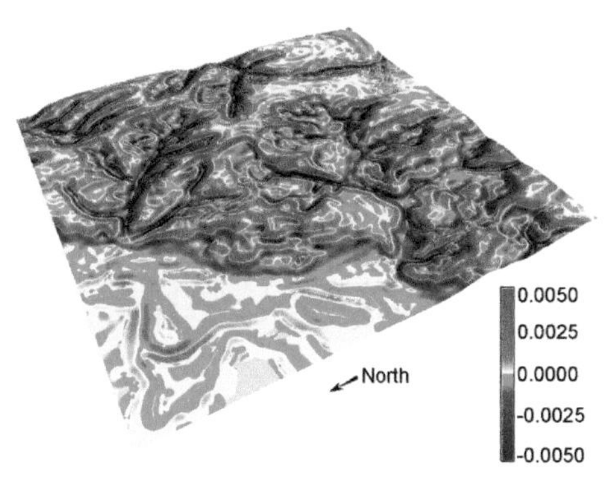

FIGURE 6 (CHAPTER 17) Profile curvature [m^{-1}]. (See page 395 of this volume.)

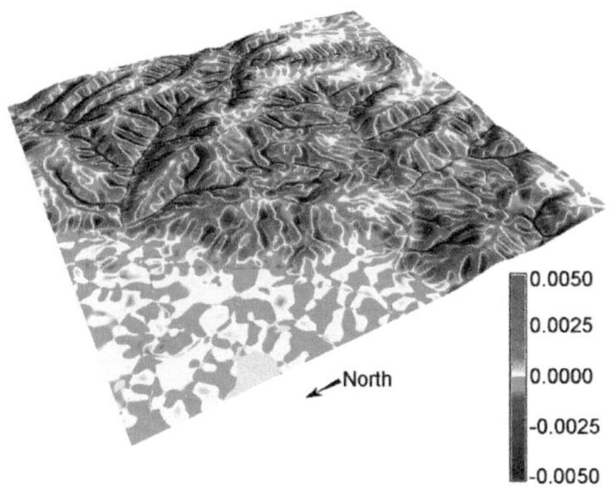

FIGURE 7 (CHAPTER 17) Tangential curvature [m^{-1}]. (See page 395 of this volume.)

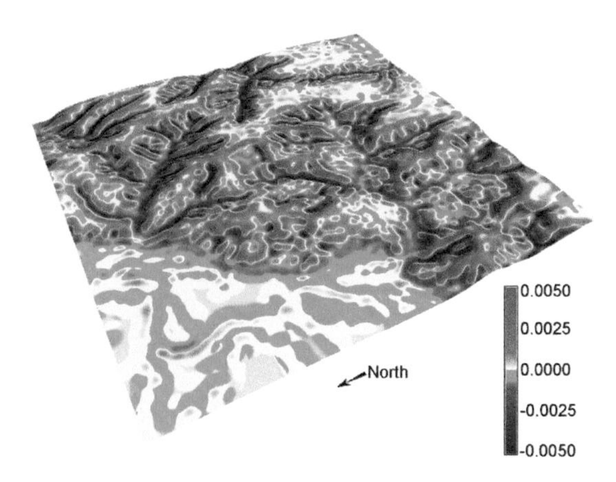

FIGURE 8 (CHAPTER 17) Mean curvature [m^{-1}]. (See page 395 of this volume.)

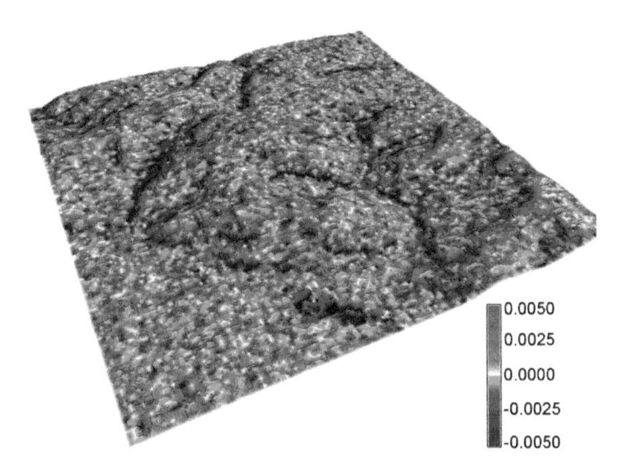

FIGURE 9 (CHAPTER 17) Profile curvature [m^{-1}] computed directly from SRTM data using `r.slope.aspect`. (See page 396 of this volume.)

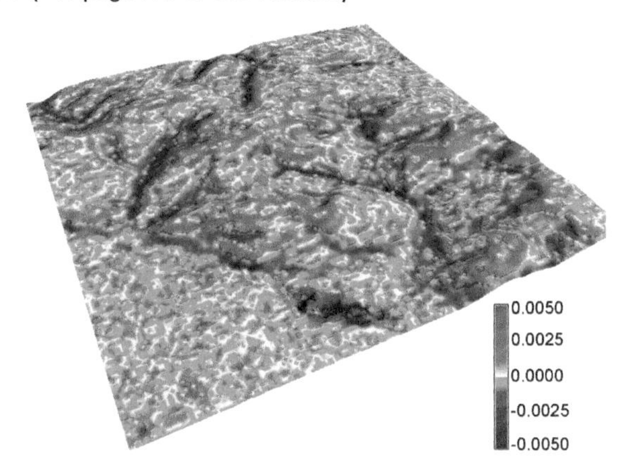

FIGURE 10 (CHAPTER 17) Profile curvature [m^{-1}] from smoothed SRTM data using `r.resamp.rst`. (See page 396 of this volume.)

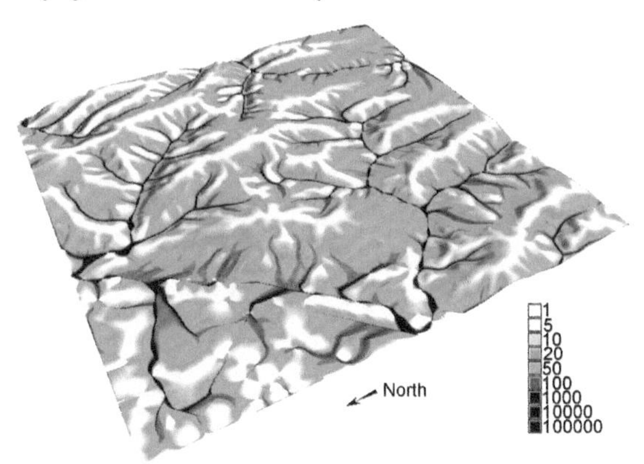

FIGURE 11 (CHAPTER 17) Flow accumulation [-] generated by `r.terraflow`. (See page 398 of this volume.)

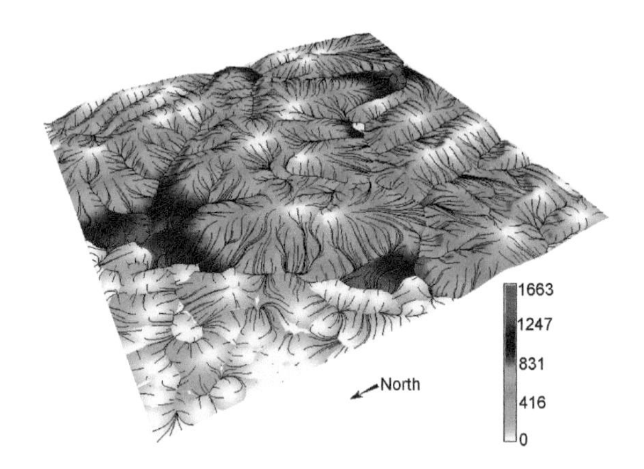

FIGURE 12 (CHAPTER 17) Flowpath lengths [m] and flowlines generated by `r.flow`. (See page 399 of this volume.)

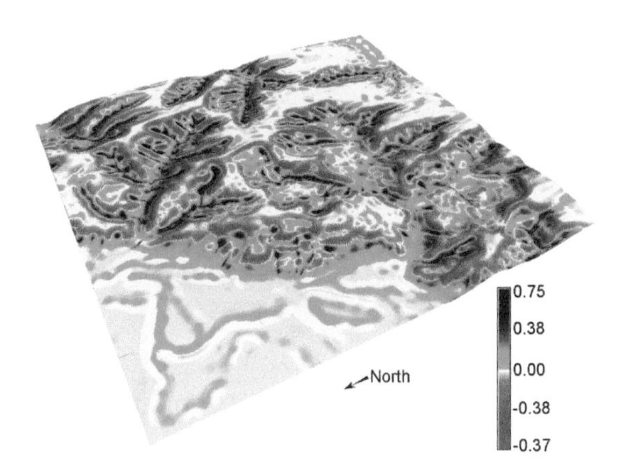

FIGURE 14 (CHAPTER 17) Topographic soil erosion index [-]. (See page 401 of this volume.)

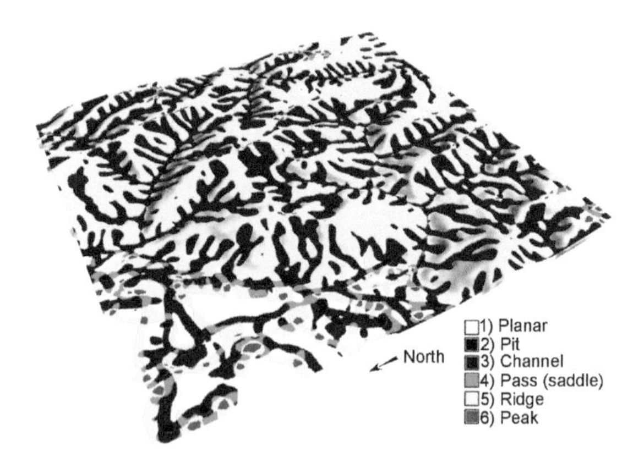

FIGURE 16 (CHAPTER 17) Basic land-surface features extracted using `r.param.scale`. (See page 404 of this volume.)

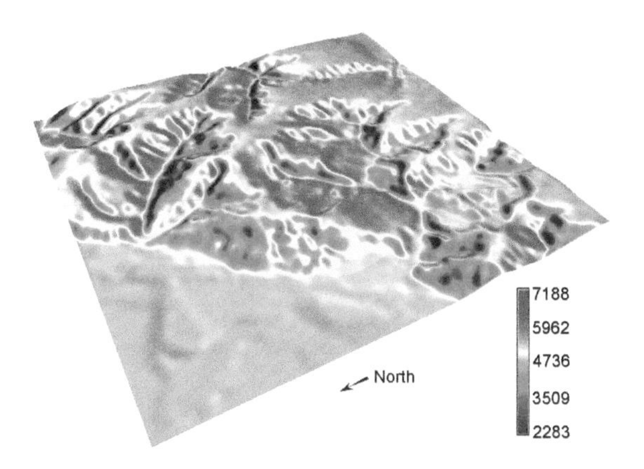

FIGURE 17 (CHAPTER 17) Global solar radiation for spring equinox [Wh/m^2]. (See page 405 of this volume.)

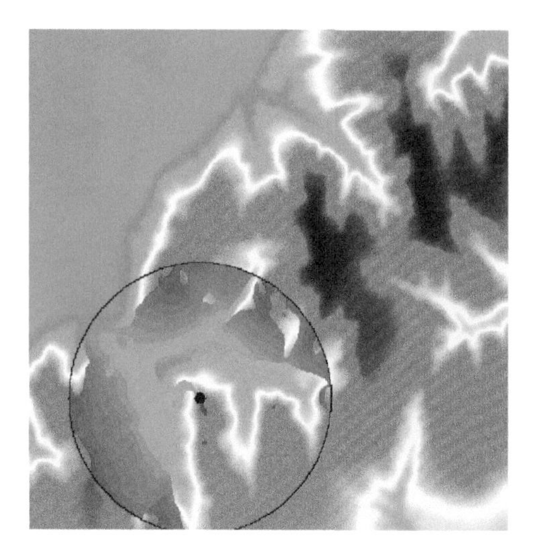

FIGURE 18 (CHAPTER 17) Visibility analysis using `r.los`. (See page 405 of this volume.)

FIGURE 19 (CHAPTER 17) Random fractal surface generated by `r.surf.fractal`. (See page 406 of this volume.)

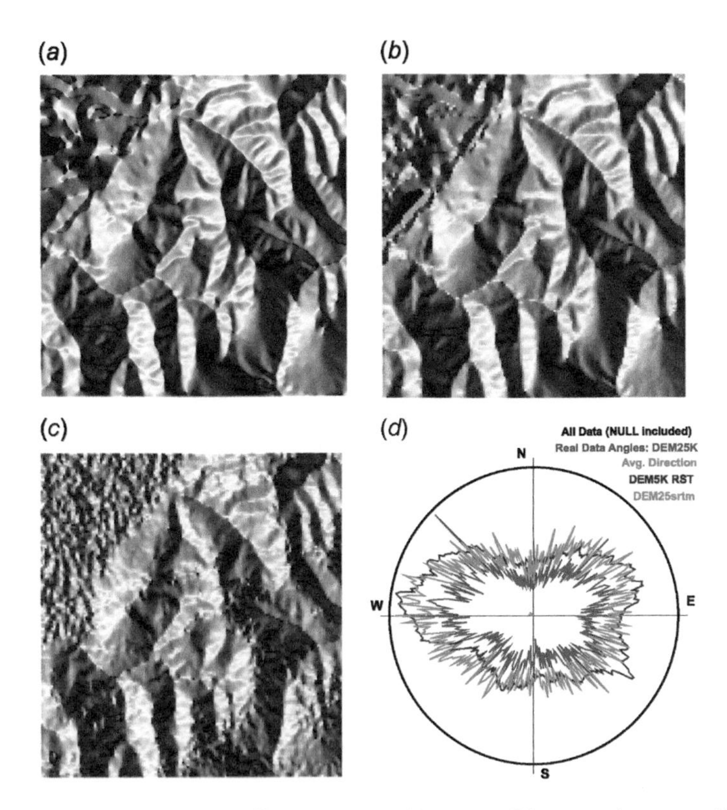

FIGURE 20 (CHAPTER 17) Baranja Hill aspect maps: (a) DEM25, (b) DEM5K (generated by `v.surf.rst`), (c) DEM25-SRTM, and (d) a combined polar diagram of all aspect maps from `d.polar`. (See page 408 of this volume.)

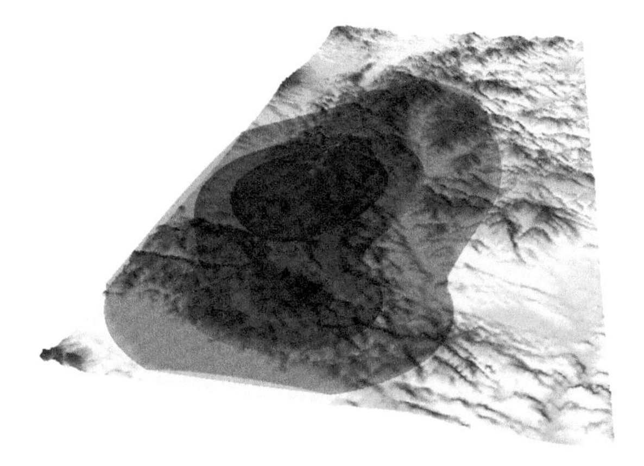

FIGURE 21 (CHAPTER 17) Volume interpolation and isosurface visualisation of precipitation (isosurfaces of 1100, 1200, 1250 mm/year are shown) using `v.vol.rst`. (See page 409 of this volume.)

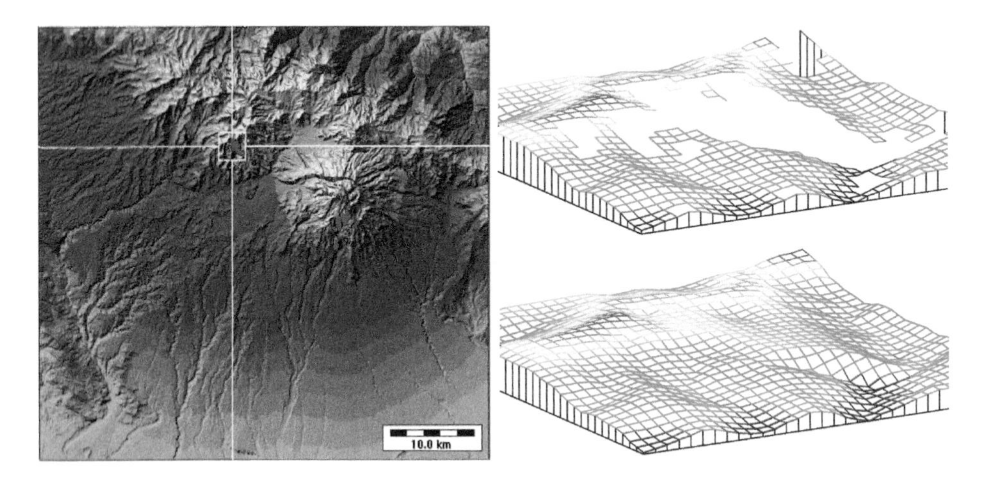

FIGURE 2 (CHAPTER 18) A yellow box and crosshairs on a shaded relief image shows the location of a hole (red) in an SRTM DEM for Volcan Baru, Panama. The two images on the right show wire mesh surface plots of the area near the hole, before and after using the Repair Bad Values tool. (See page 417 of this volume.)

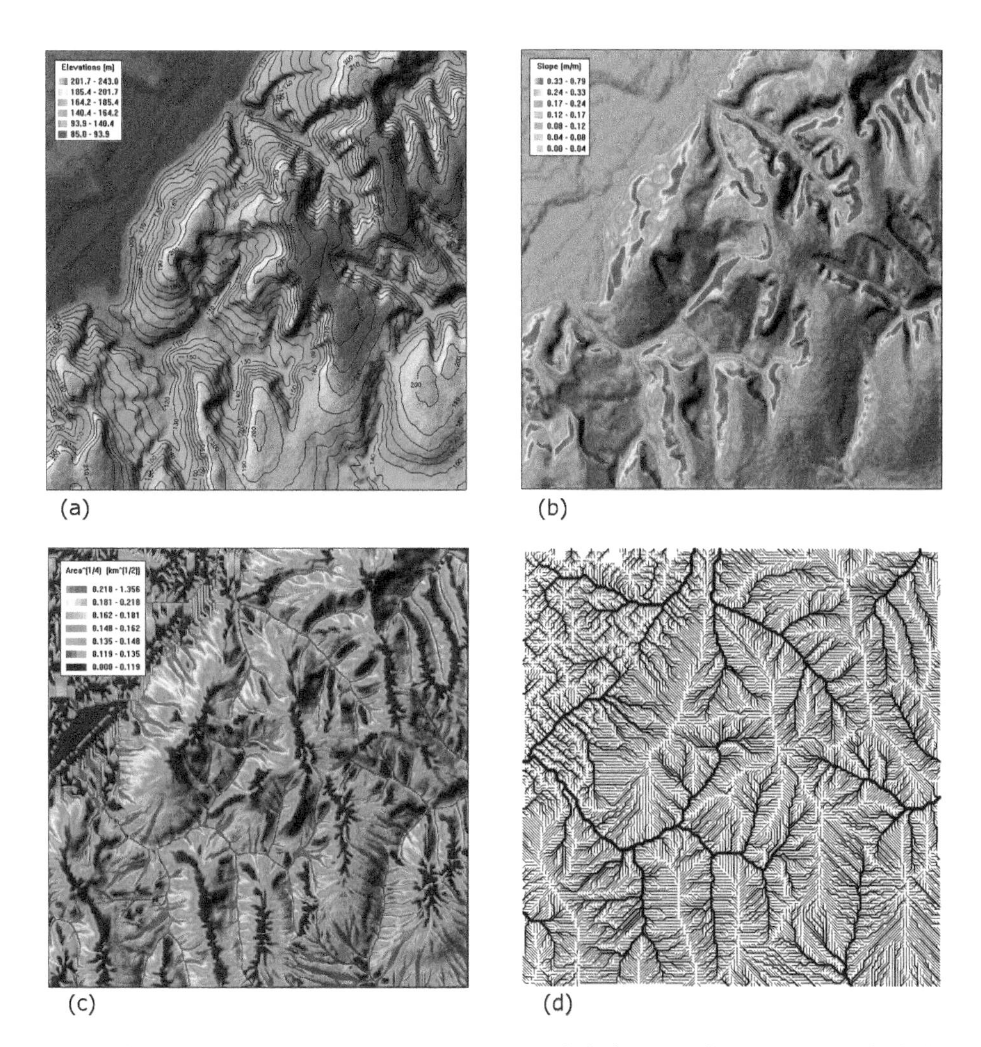

FIGURE 3 (CHAPTER 18) (a) Shaded relief image with labeled contour line overlay; (b) Shaded image of a D8 slope grid; (c) Shaded image of a total contributing area grid, extracted using the mass flux method; (d) Drainage pattern obtained by plotting all D8 flow vectors; (e) Watershed subunits with overlaid contours and channels (blue), using a D8 area threshold of 0.025 km^2; (f) Shaded image of plan curvature, extracted using the method of Zevenbergen–Thorne. (See page 418 of this volume.)

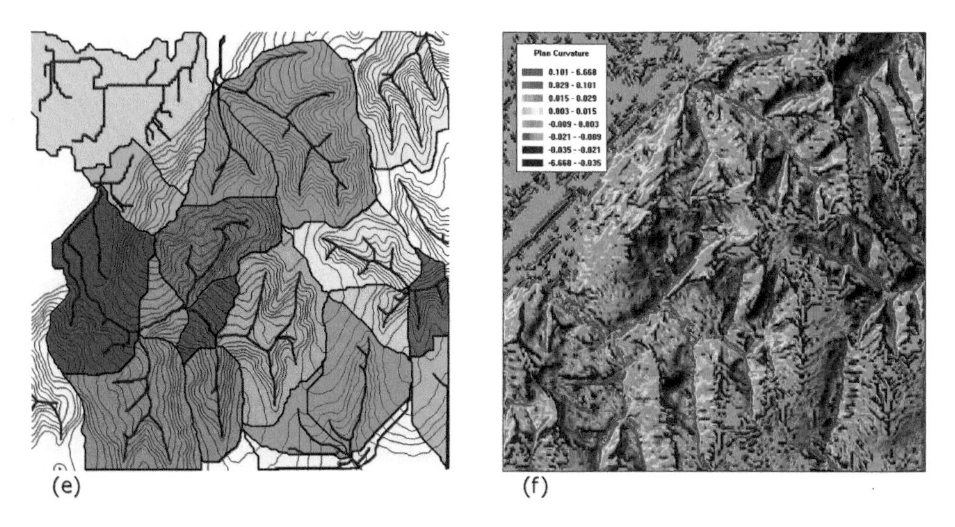

FIGURE 3 (CHAPTER 18) *(continued)*

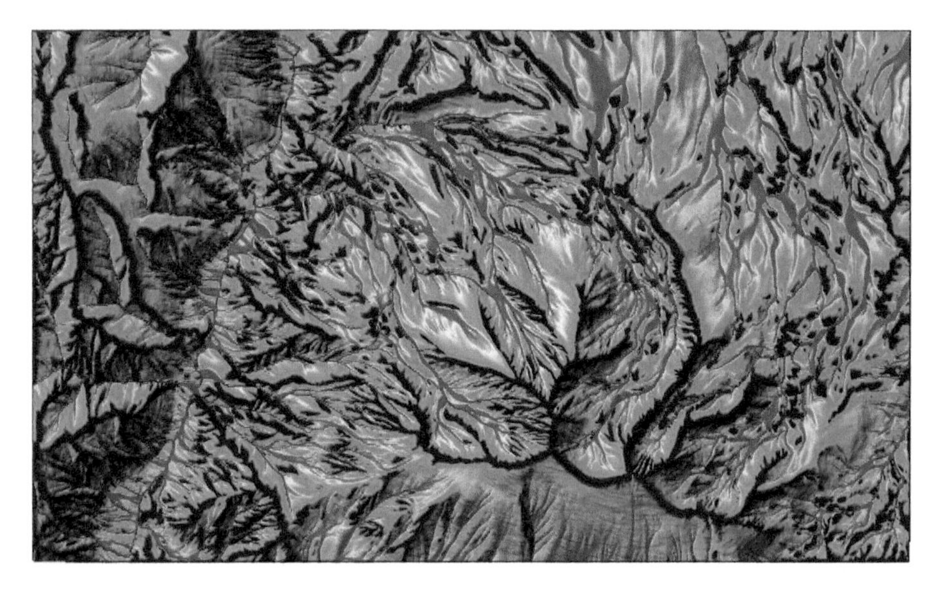

FIGURE 5 (CHAPTER 18) A relief-shaded image of a TCA grid for Mt. Sopris, Colorado, that was created using the Mass Flux method. Areas with a large TCA are shown in red while areas with a small TCA value (e.g. ridgelines) are shown in blue and purple. Complex flow paths are clearly visible and results are superior to both the D8 and D-infinity methods. (See page 422 of this volume.)

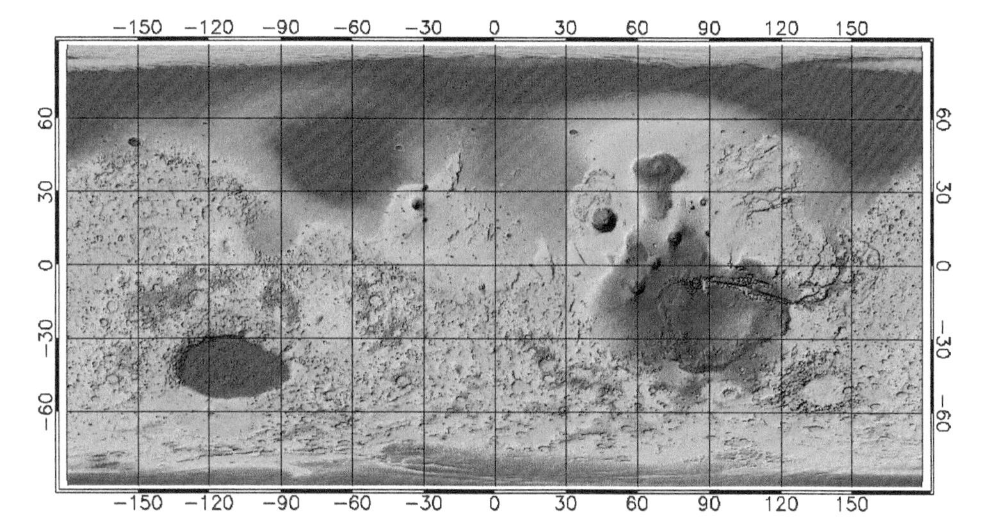

FIGURE 8 (CHAPTER 18) High-resolution MOLA (Mars Orbiter Laser Altimeter) DEM displayed in RiverTools: colour shaded relief image for planet Mars shown by the cylindrical equidistant map projection. (See page 426 of this volume.)

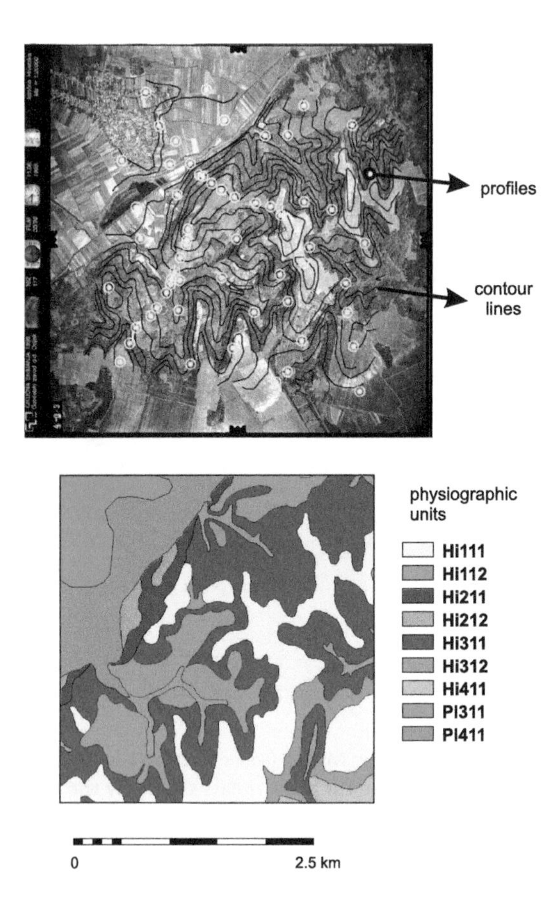

FIGURE 2 (CHAPTER 20) A traditional soil delineation drawn on an aerial photo overlain by contour lines (above) and the derived soil map with soil mapping units (below) for Baranja Hill region (Croatia). The lines are delineated manually and points show the location of soil profile observations. (See page 464 of this volume.)

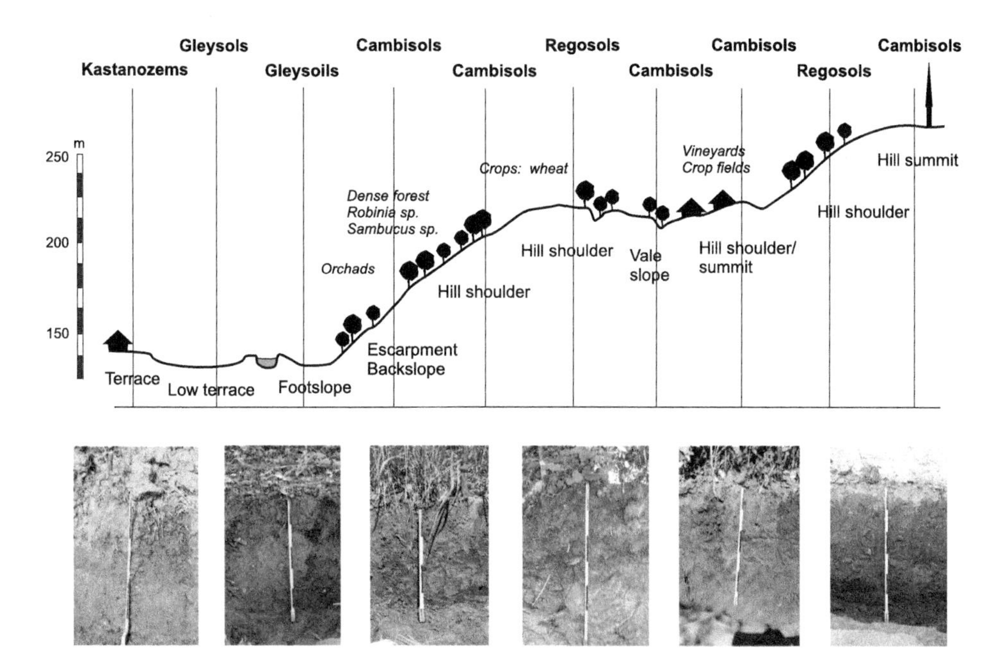

FIGURE 3 (CHAPTER 20) Vertical zonation of soils in the Baranja Hill: from deep, drained soils (Kastanozems), to saturated (Gleysoils) and shallow eroded soils (Regosols). (See page 467 of this volume.)

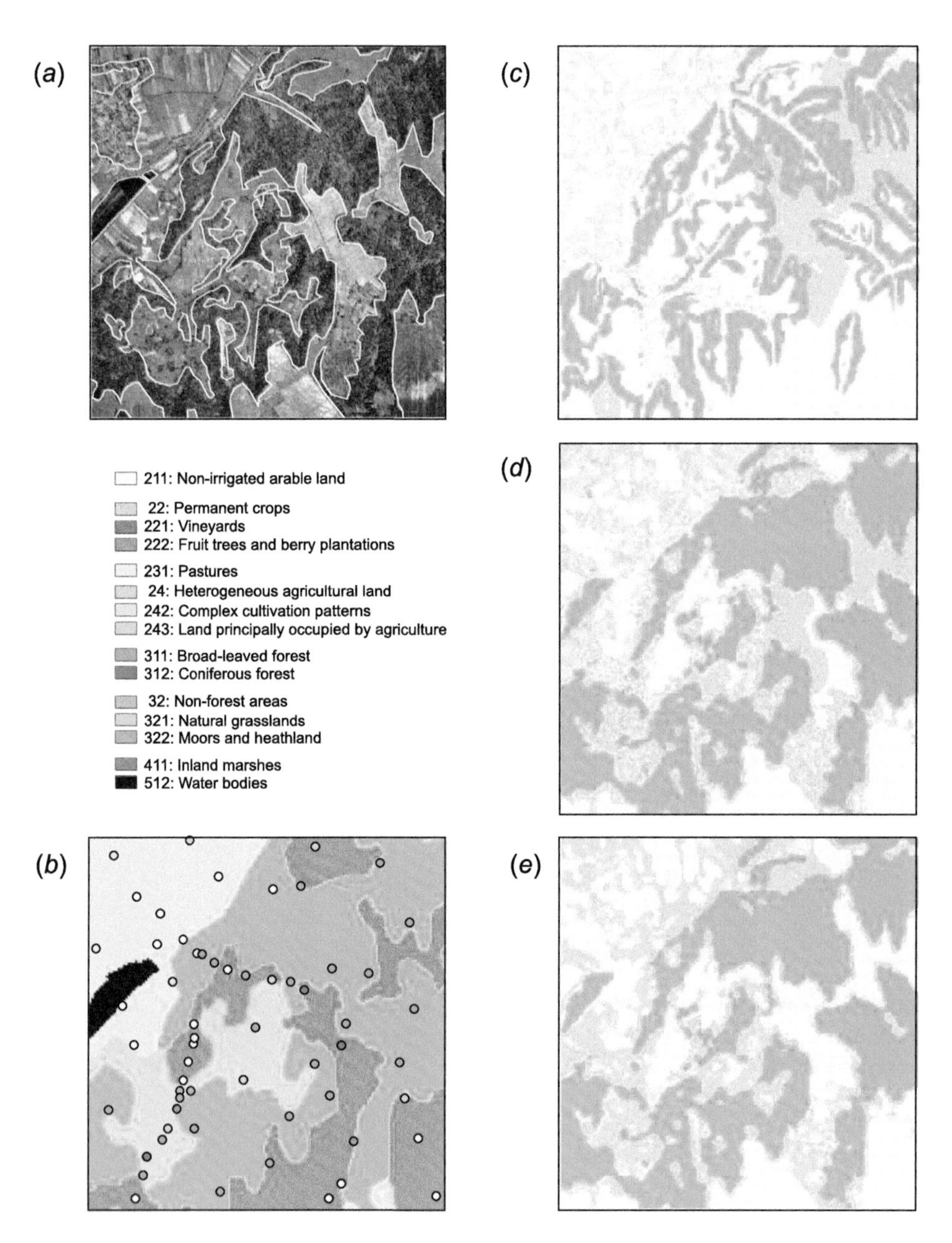

FIGURE 5 (CHAPTER 21) An automated extraction of land-cover classes: (a) an orthophoto of the Baranja Hill area, overlaid with manually digitised land-cover areas; (b) land-cover classes from the CLC 2000 Croatia (www.azo.hr) and field observations; (c) the land-cover of the study area, predicted using land-surface parameters only; (d) the land-cover of the study area predicted using land-surface parameters plus RS data; (e) the land-cover of the study area, predicted using RS data only. (See page 493 of this volume.)

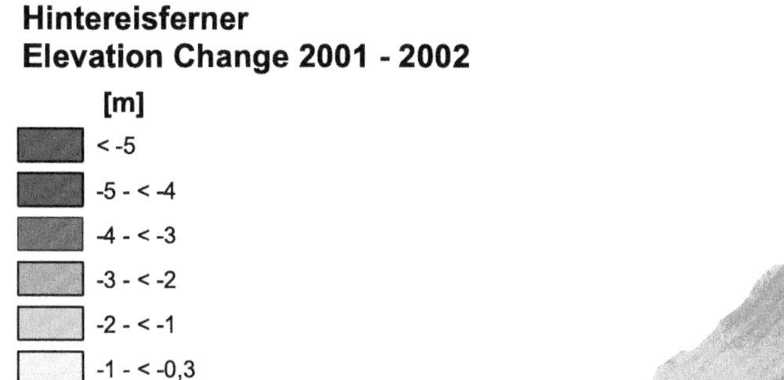

FIGURE 2 (CHAPTER 22) Net elevation change on Hintereisferner in the 2001–2002 budget year. Reprinted from Geist and Stötter (2007). Used with permission (http://www.borntraeger-cramer.de). (See page 505 of this volume.)

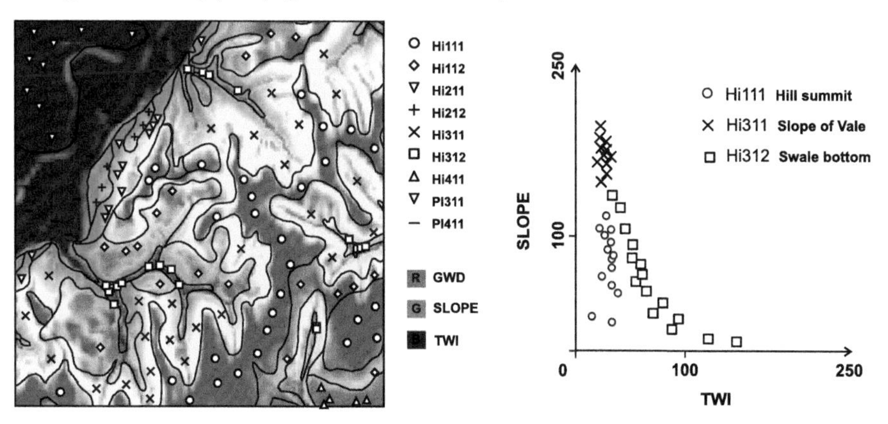

FIGURE 6 (CHAPTER 22) Training points displayed in geographical (left) and feature (right) space. The false colour composite (DEM, SLOPE, TWI) can be used to interactively select the most typical locations for each landform class (in this case manually delineated units). The values for TWI and SLOPE in the right plot have been stretched to the 0–255 scale. (See page 514 of this volume.)

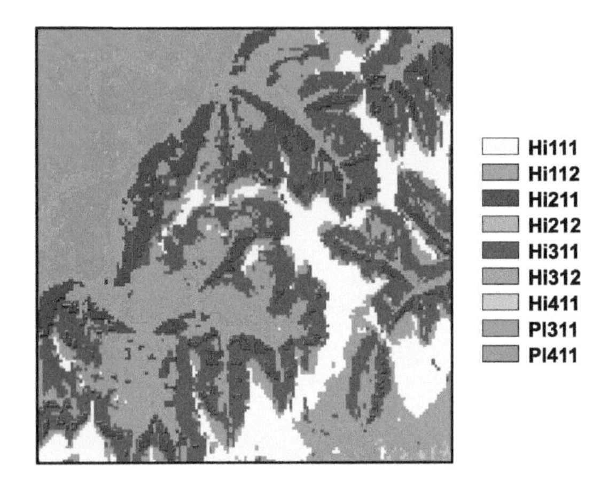

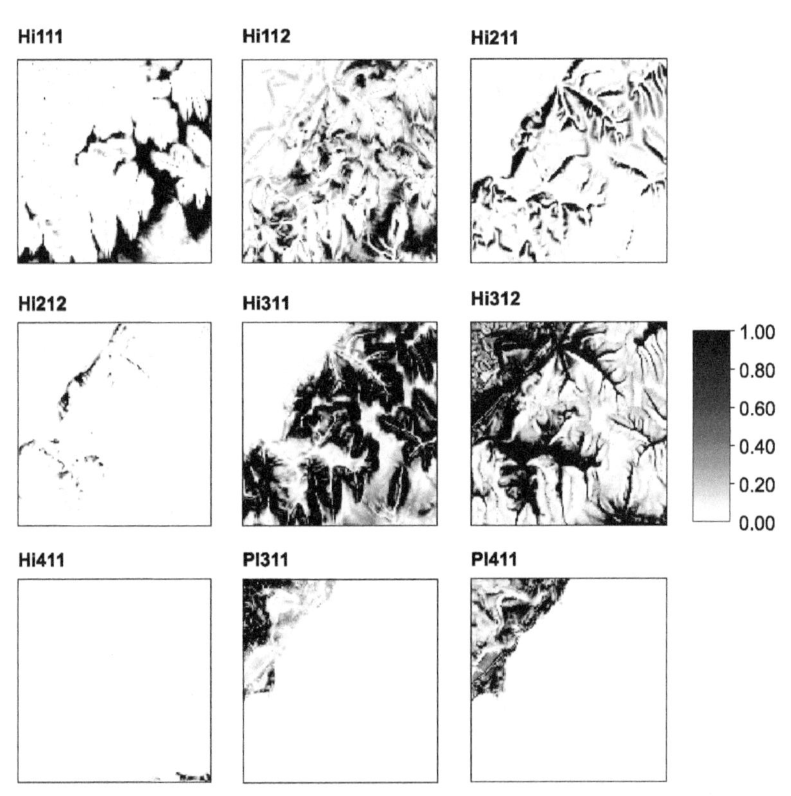

FIGURE 7 (CHAPTER 22) Results of supervised classification using maximum likelihood classifier (above) and memberships derived using fuzzy k-means classification (below). Hi111 (Hill summit), Hi112 (Hill shoulder), Hi211 (Escarpment scarp), Hi212 (Escarpment colluvium), Hi311 (Valley slope), Hi312 (Valley bottom), Hi411 (Glacis slope), Pl311 (High terrace) and P411 (Low terrace). Compare with Figure 6. (See page 517 of this volume.)

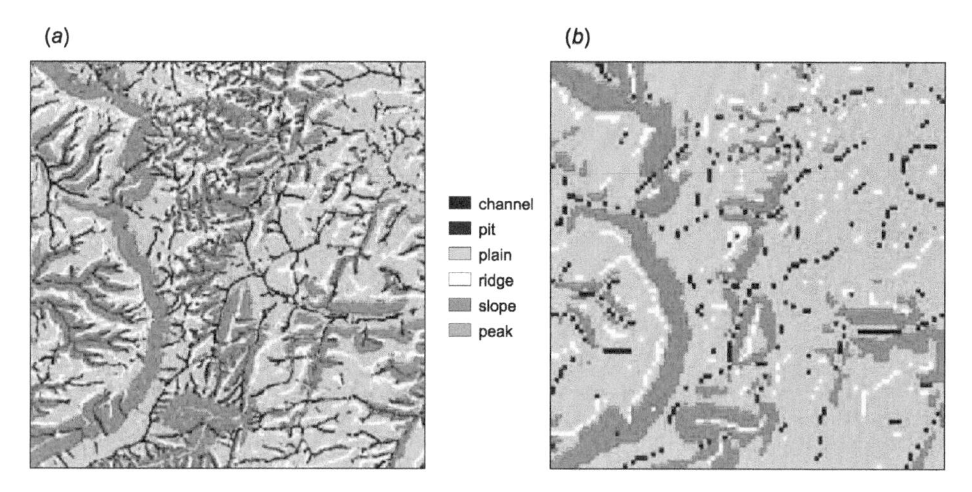

FIGURE 10 (CHAPTER 22) Extraction of landform elements for the 10×10 km Ebergötzen study area, Germany using the 25 m DEM (a) and the 90 m SRTM DEM (b). (See page 520 of this volume.)

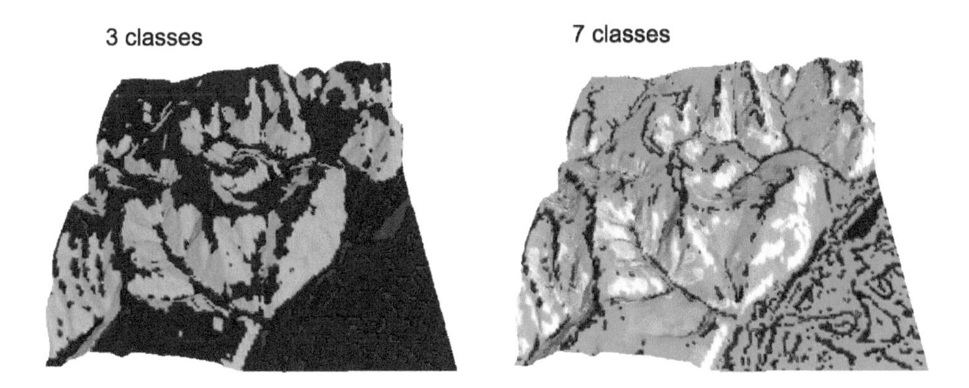

FIGURE 12 (CHAPTER 22) Landforms extracted using unsupervised fuzzy k-means classification with 3 and 7 classes in the FuZME package. Because the classification is unsupervised, the legend can be constructed only *a posteriori*. (See page 522 of this volume.)

FIGURE 1 (CHAPTER 23) The track and deposits left by the June 2001 flow of debris that overwhelmed the Swiss village of Täsch. Reproduced by permission of SWISSTOPO (BA081244). (See page 535 of this volume.)

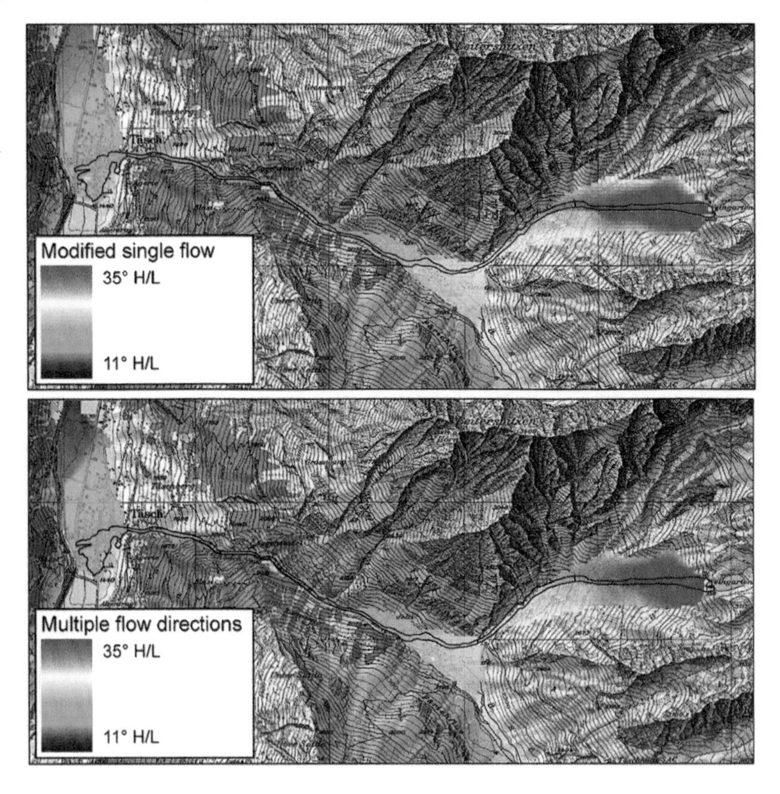

FIGURE 3 (CHAPTER 23) Modelling H/L angles using the MSF (top) and the MFD (bottom) models. Map and DEM reproduced by permission of SWISSTOPO (BA081244). (See page 537 of this volume.)

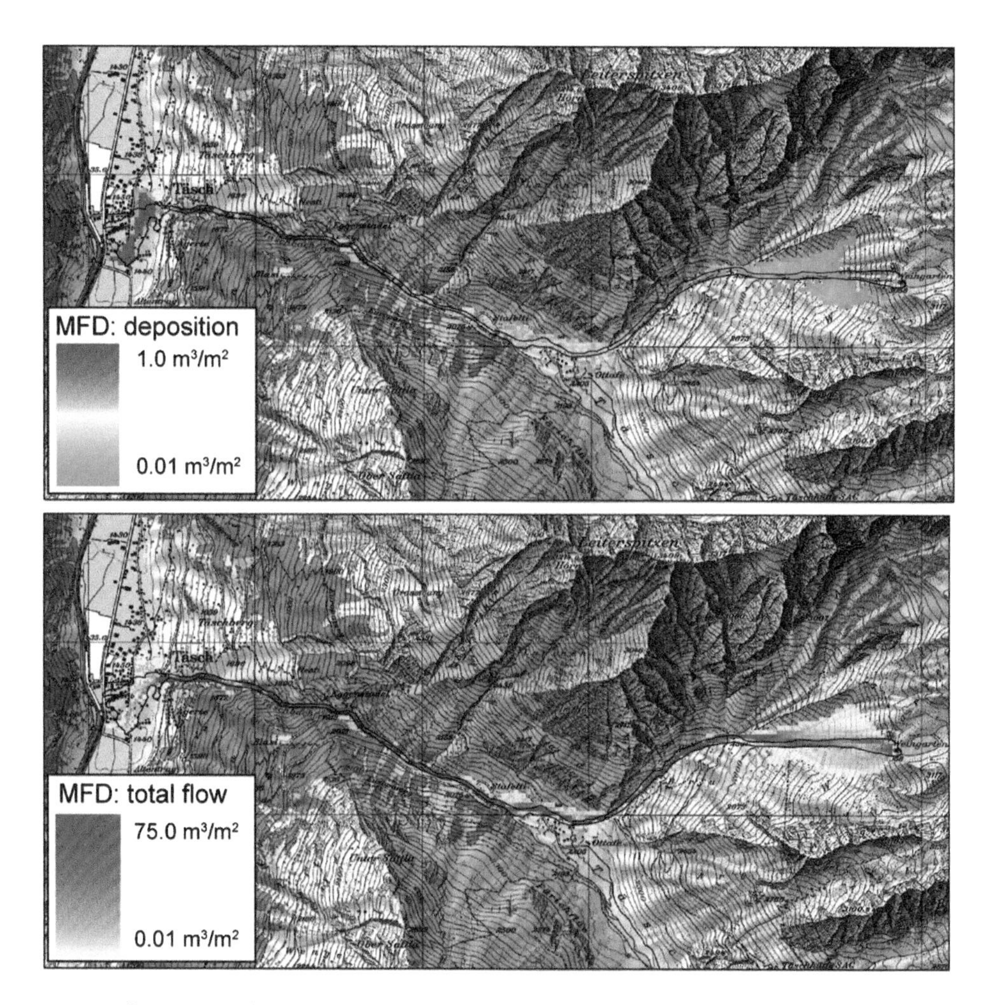

FIGURE 4 (CHAPTER 23) Deposition and the total volume of flow as modelled by the MFD deposition approach (map and DEM data reproduced by permission of SWISSTOPO). (See page 538 of this volume.)

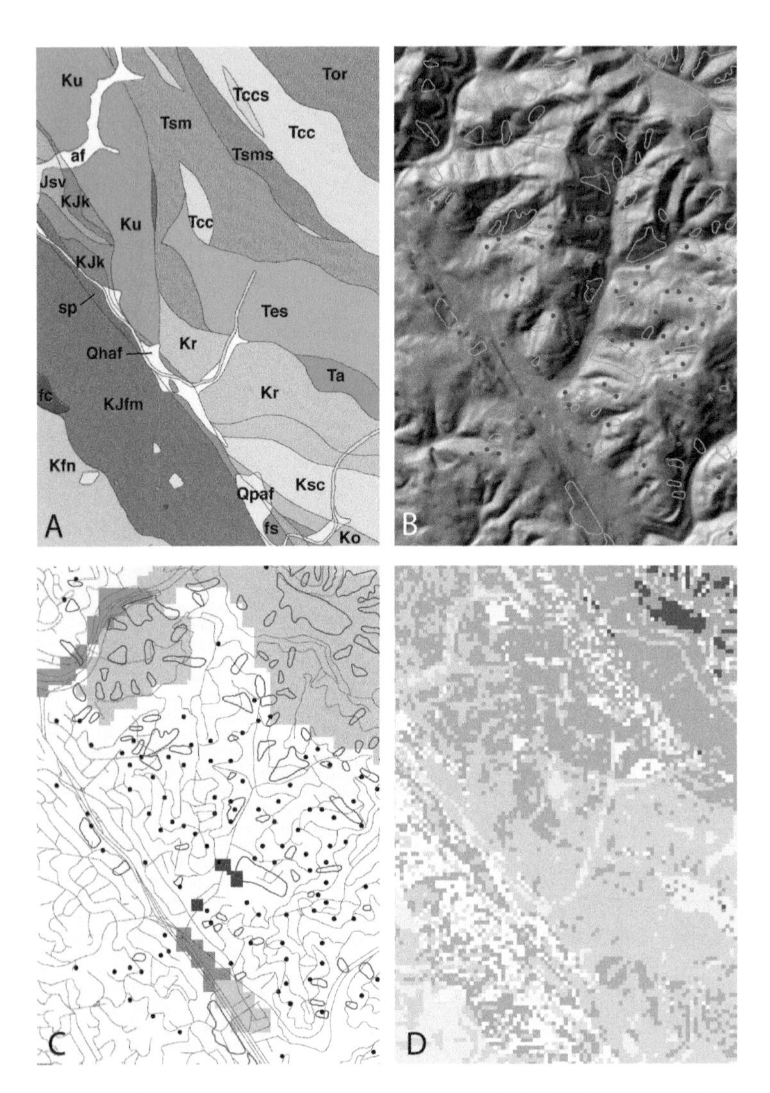

FIGURE 7 (CHAPTER 23) Features illustrating preparation of a landslide-susceptibility map for a part of the city of Oakland, California (Pike et al., 2001); the area shown in the four maps is about 2 km across. (A) Geology, showing 21 of the 25 map units in Table 1; the NNW-striking Hayward Fault Zone lies along the eastern edge of unit KJfm. (B) Inventory of old landslide deposits (orange polygons) and locations of post-1967 landslides (red dots) on uplands east of the fault and on gentler terrain to the west; shaded relief is from a 10 m DEM. (C) Old landslide deposits and recent landslides overlain on 1995 land use (100 m resolution): yellow, residential land; green, forest; tan, scrub vegetation; blue, major highway; pink, school; orange, commercial land; brown, public institution; white, vacant and mixed-use land; road net in grey. (D) Values of relative susceptibility at 30-m resolution mapped in eight intervals from low to high as grey, 0.00; purple, 0.01–0.04; blue, 0.05–0.09; green, 0.10–0.19; yellow, 0.20–0.29; light-orange, 0.30–0.39; orange, 0.40–0.54; red, 0.55. Low to moderate values 0.05–0.20 predominate in this 9 km^2 sample of the study area. (See page 543 of this volume.)

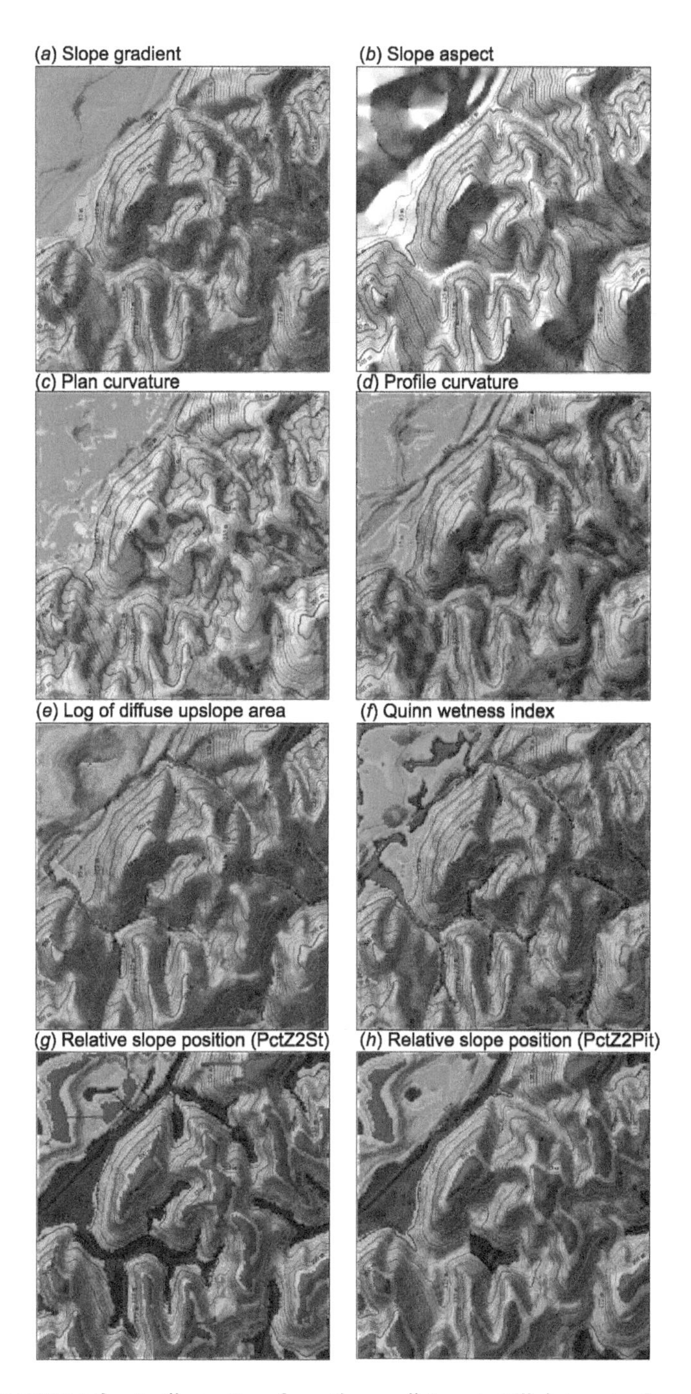

FIGURE 3 (CHAPTER 24) An illustration, from the small Baranja Hill data-set, of several of the more frequently used land-surface parameters in the PEM process. (See page 564 of this volume.)

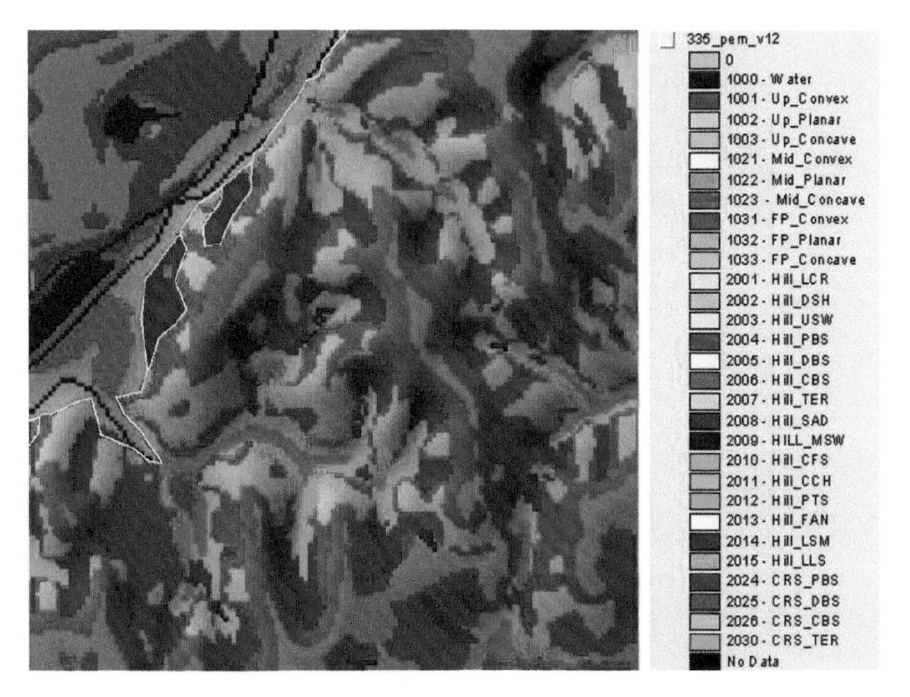

FIGURE 4 (CHAPTER 24) An illustration of the results of applying a hypothetical set of ecological–landform classification rules to the small data set from Baranja Hill. See Table 3 for an explanation of legend classes. (See page 571 of this volume.)

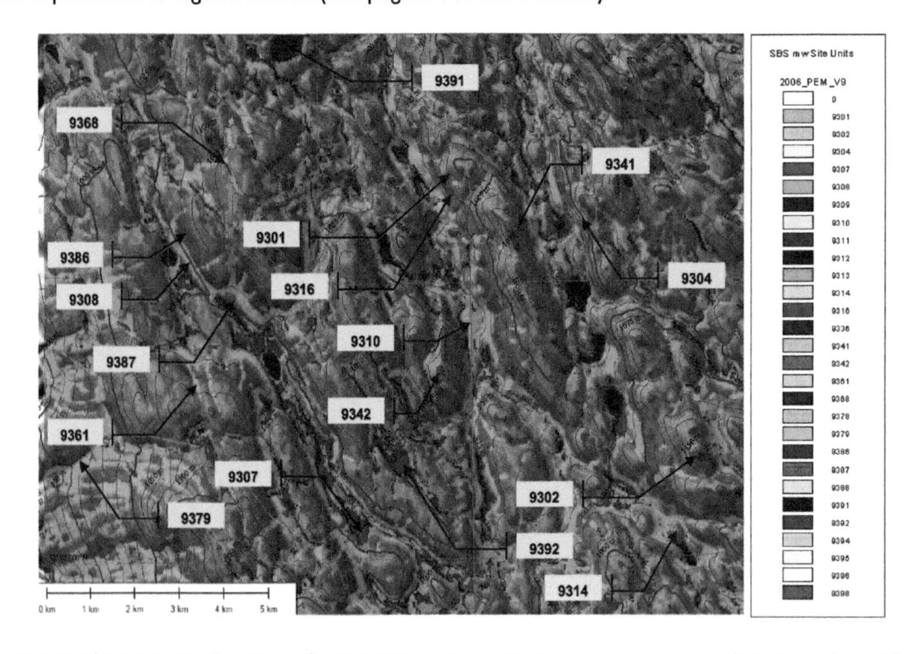

FIGURE 5 (CHAPTER 24) Part of a 1:20,000 scale predictive ecosystem map (PEM) produced for an area in the former Cariboo Forest Region of BC, Canada. (See page 575 of this volume.)

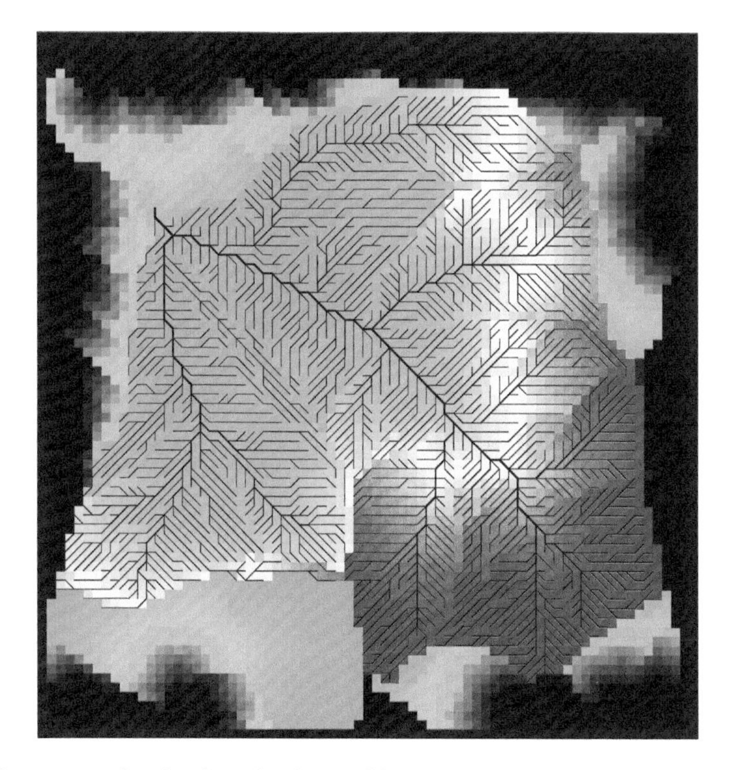

FIGURE 4 (CHAPTER 25) Flow lines for the small basin near the north edge of the Baranja DEM, as extracted from a DEM by the D8 method. The flow lines are overlaid on a colour image that shows flow distance to the basin outlet. (See page 596 of this volume.)

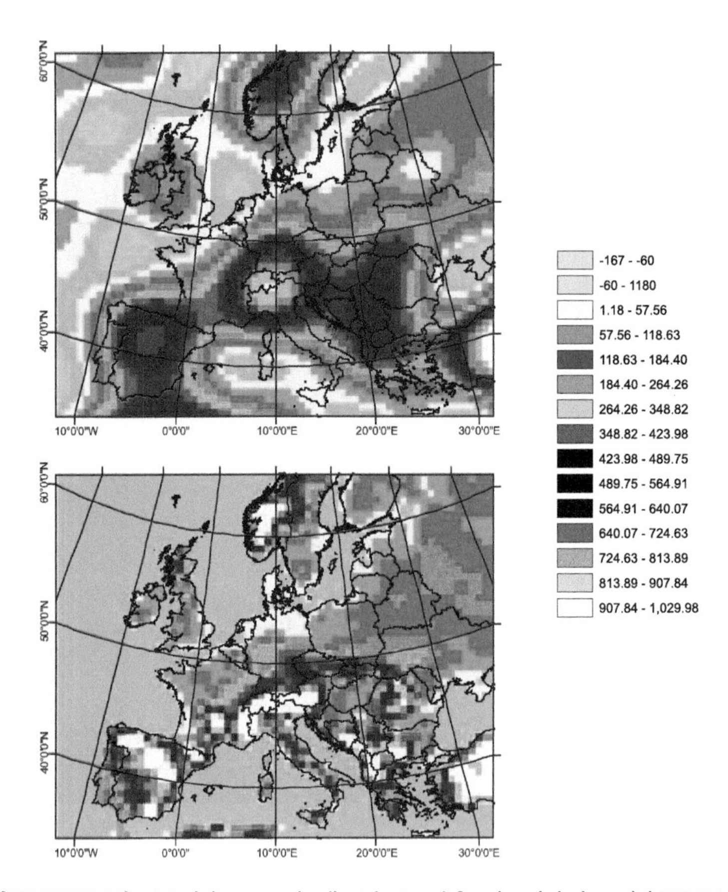

FIGURE 2 (CHAPTER 26) Model orography (height in m) for the global model ECHAM4, resolution about 250 km (top), and for the present regionalisation simulation with MM5, resolution 60 km (bottom). (See page 613 of this volume.)

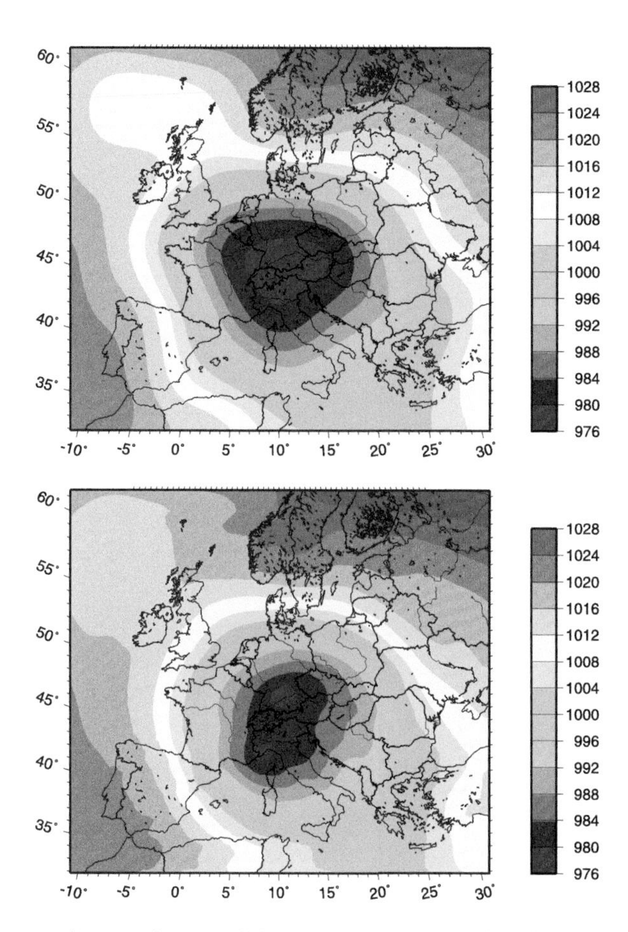

FIGURE 3 (CHAPTER 26) Distribution of the sea level pressure in hPa for April 5th 1991, 00 UTC, in the global simulation with ECHAM4 (top) and in the regional simulation with MM5 (bottom). (See page 614 of this volume.)

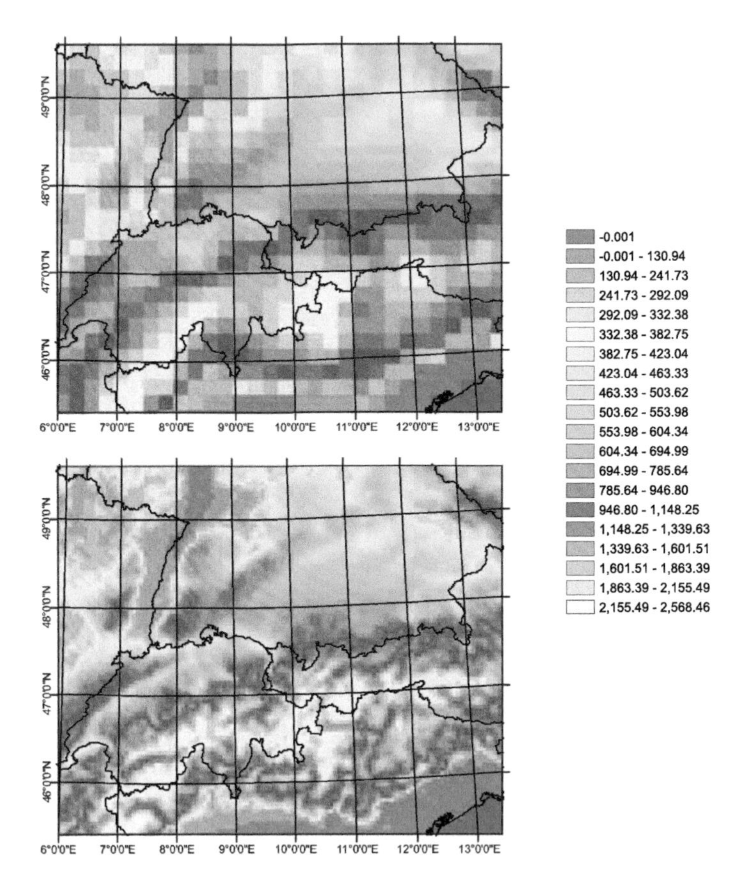

FIGURE 4 (CHAPTER 26) Model orographies in the nested regional MM5-simulations: (top) simulation S_1 with a horizontal resolution of 19.2 km and (bottom) simulation S_2 with a resolution of 4.8 km. (See page 615 of this volume.)

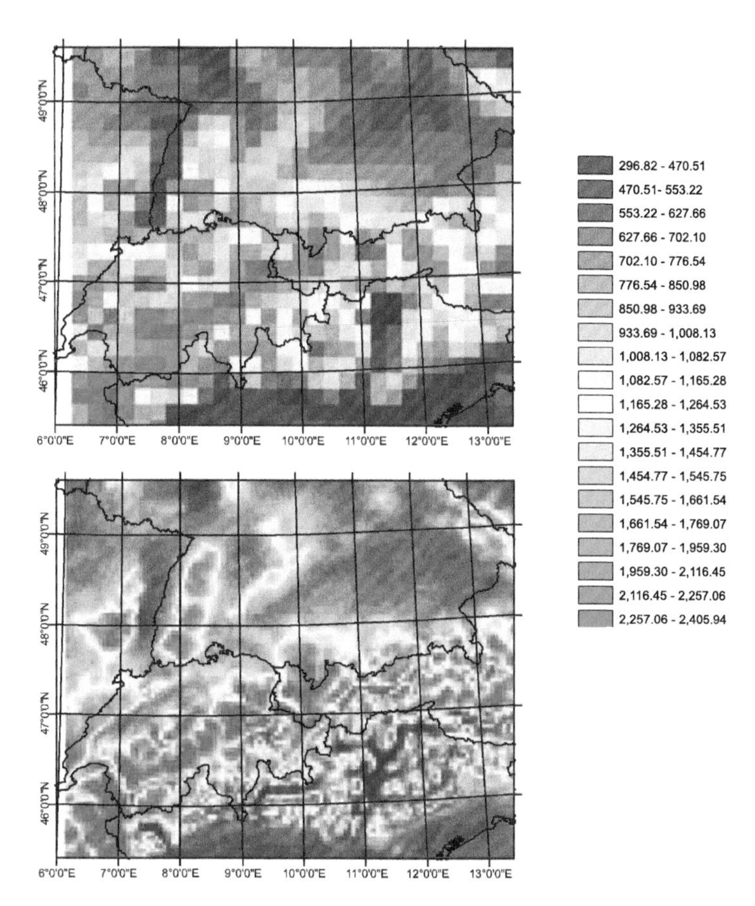

FIGURE 5 (CHAPTER 26) Annual precipitation sum (in mm) for the year 1990 as calculated by the MM5-simulations S_1 (top) and S_2 (bottom). (See page 616 of this volume.)

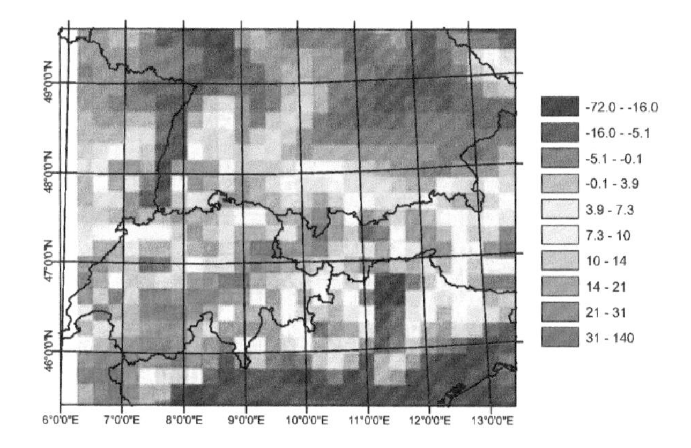

FIGURE 6 (CHAPTER 26) Scale reduced difference ($S_2 - S_1$) of the annual precipitation sum (in mm) for the year 1990. (See page 617 of this volume.)

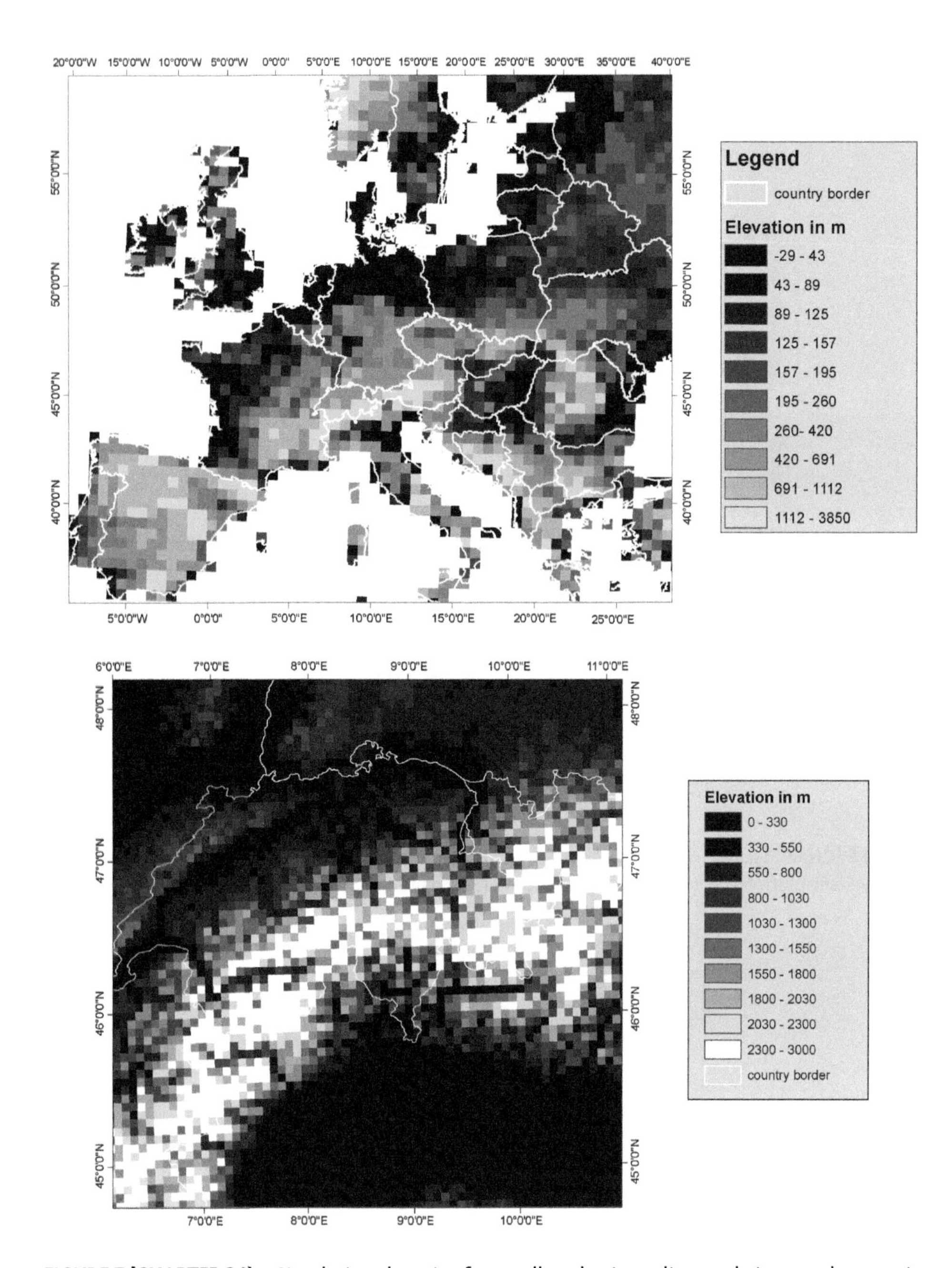

FIGURE 7 (CHAPTER 26) Simulation domains for small-scale air quality study in complex terrain following a one-way nesting strategy. The horizontal spatial resolution is 54 km (domain D_1, top frame), 6 km (domain D_2, middle frame), and 1 km (domain D_3, bottom frame). (See page 618 of this volume.)

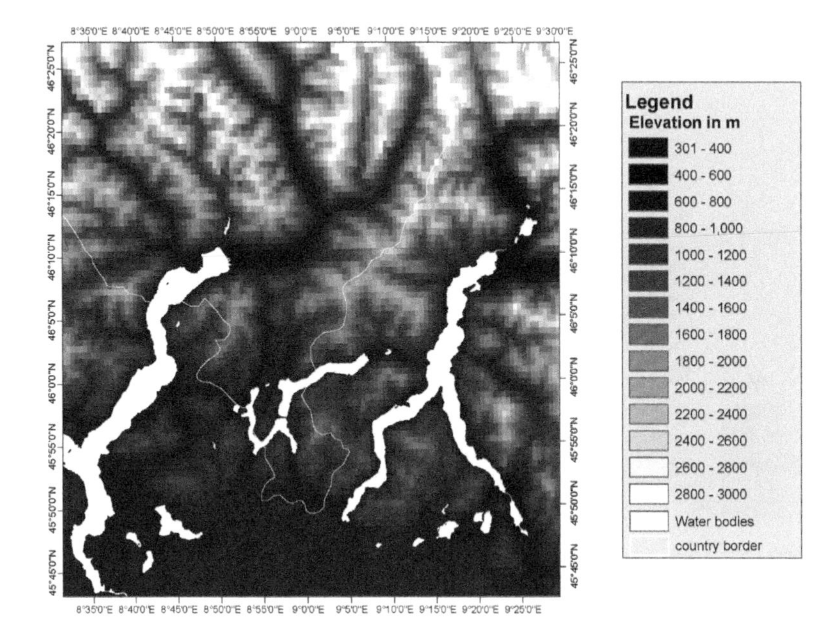

FIGURE 7 (CHAPTER 26) *(continued)*

FIGURE 8 (CHAPTER 26) The figure shows for August 19, 1996, 1900 GMT, for the whole D_3 (see also Figure 7) the horizontal winds at 1000 m above sea level (arrows, length indicating strength and arrowheads indicating the direction of the flow) and the ozone concentration at the same height (colours, red: high, green: medium, blue: low, white: terrain height higher than 1000 m above sea level). Black lines in white areas give terrain height in 400 m interval (first line: 1200 m above sea level). (See page 620 of this volume.)

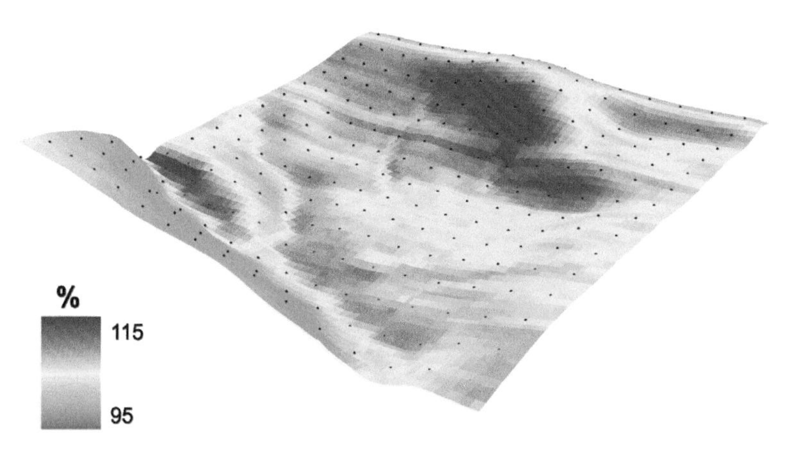

FIGURE 2 (CHAPTER 27) Simulated spatial distribution of solar radiation relative to a neighbouring automatic weather station for the field site Sportkomplex. (See page 628 of this volume.)

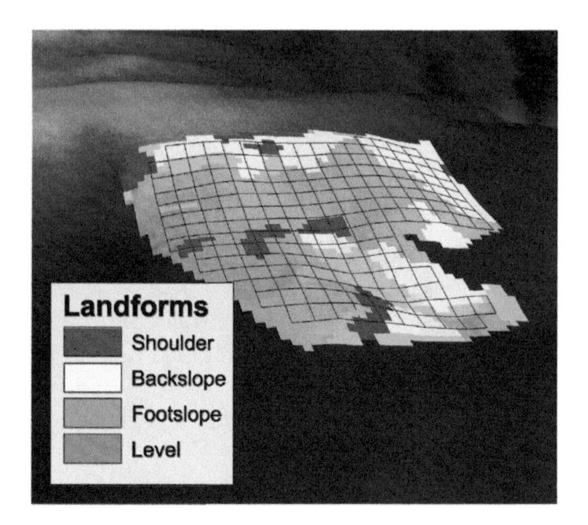

FIGURE 6 (CHAPTER 27) Distribution of landforms at the field site Bei Lotte. (See page 632 of this volume.)

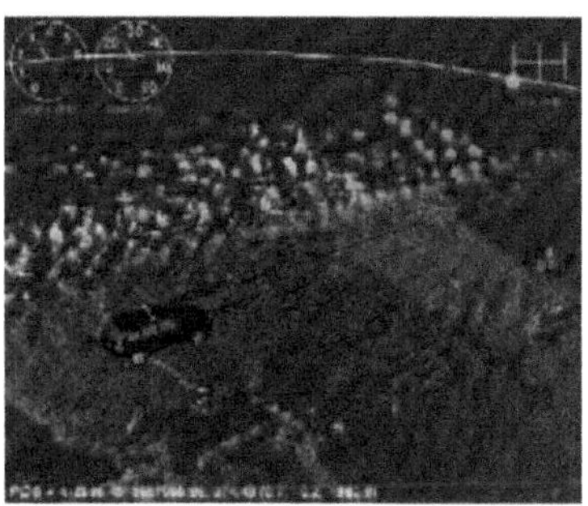

FIGURE 1 (CHAPTER 28) Geomorphometry on-the-fly. Laser sensors (top) scan terrain in front of a robotic vehicle as it moves, recording x, y, z location, distance, and time. The resulting data are integrated into 3-D point clouds (bottom) and continuously sorted by the changing relation among the three variables to separate drivable terrain from potential obstacles. Reprinted from Thrun et al. (2006a). (See page 639 of this volume.)

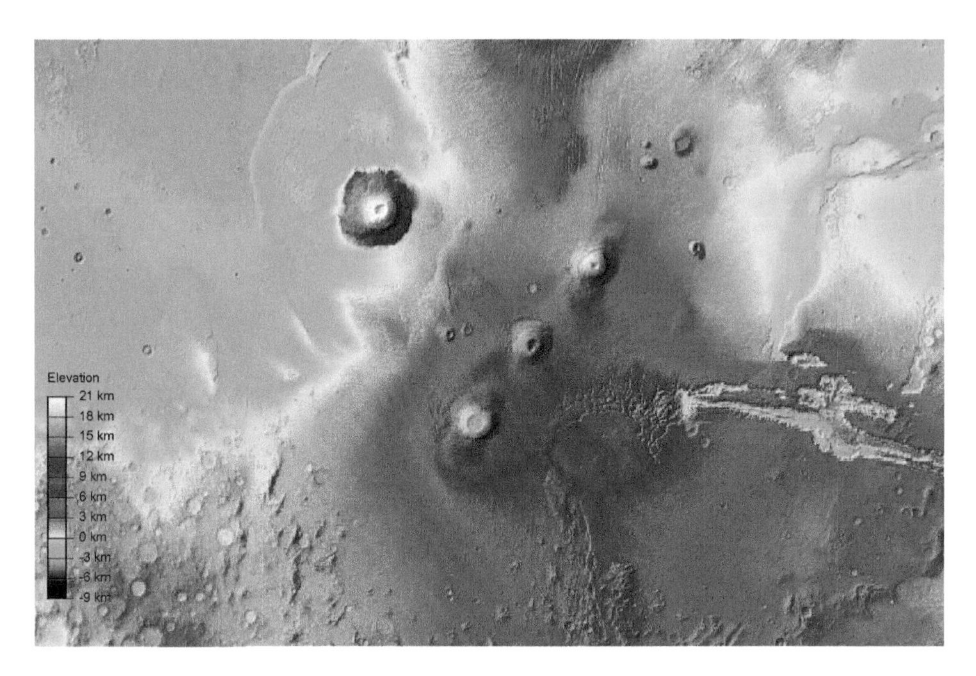

FIGURE 3 (CHAPTER 28) Out-of-this-world geomorphometry of volcanoes, canyons and, craters. Shaded relief image of Mars from 40°N to 40°S and 60°W to 180°W colour-coded by elevation, from the MOLA 1/128° DEM (available via http://mars.google.com). Image credit: NASA/JPL/GSFC/Arizona State University. (See page 642 of this volume.)

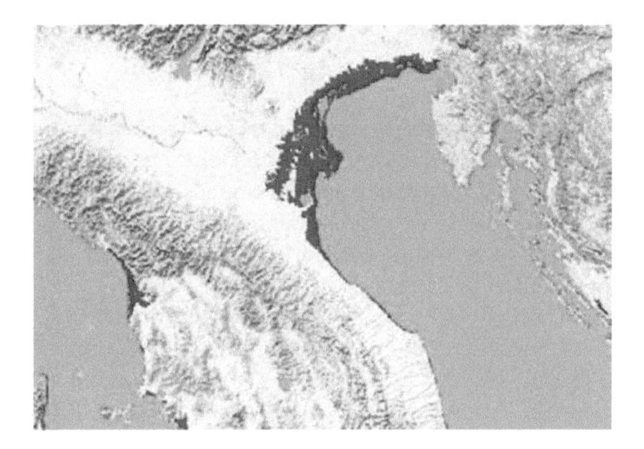

FIGURE 5 (CHAPTER 28) Mapping flood risk by geomorphometry. Inundation of southern European coast (in red), given a potential 3-m rise in sea level, estimated from the 1-km GLOBE DEM (http://www.cresis.ku.edu/). Image credit: Center for Remote Sensing of Ice Sheets. (See page 648 of this volume.)

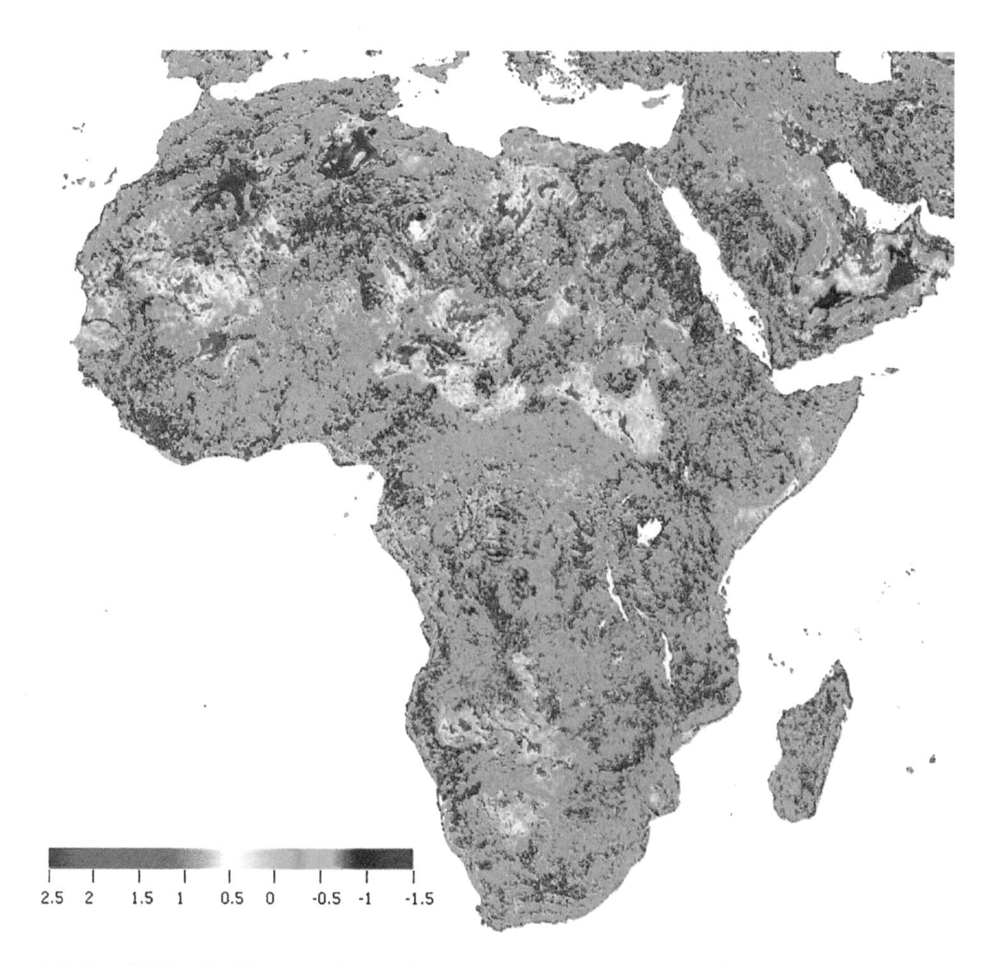

FIGURE 6 (CHAPTER 28) Page from a future geomorphometric atlas? Elevation kurtosis computed for Africa from the SRTM 3" data set in regions measuring 2.5'×2.5'. Zero kurtosis indicates a more or less statistically normal (bell-shaped) distribution of elevations; positive kurtosis denotes fewer-than-normal very high and low points, whereas negative kurtosis indicates the opposite — more high and low elevations than normal but fewer points in between. The highest values of kurtosis (red) occur in river deltas like the Nile, Mesopotamia, and isolated dune fields in the Sahara and the Empty Quarter of the Arabian peninsula. The lowest values (blue/purple) occur in fields of uniform linear dunes where most topography is either crest or trough. The SRTM data allow creation of global maps (like this), but their detail can only be appreciated when visualised at continental or finer scales. Courtesy of P.L. Guth. (See page 649 of this volume.)

Printed and bound by CPI Group (UK) Ltd, Croydon, CR0 4YY
08/05/2025
01864825-0001